NEUTRINOS IN PHYSICS AND ASTROPHYSICS

FROM 10^{-33} TO 10^{28} CM

TASI 98

NEUTRINOS IN PHYSICS AND ASTROPHYSICS

FROM 10^{-33} TO 10^{28} CM

TASI 98

Boulder, Colorado, USA   1 – 26 June 1998

Editor

Paul Langacker

University of Pennsylvania

World Scientific
Singapore • New Jersey • London • Hong Kong

Published by

World Scientific Publishing Co. Pte. Ltd.

P O Box 128, Farrer Road, Singapore 912805

USA office: Suite 1B, 1060 Main Street, River Edge, NJ 07661

UK office: 57 Shelton Street, Covent Garden, London WC2H 9HE

British Library Cataloguing-in-Publication Data
A catalogue record for this book is available from the British Library.

NEUTRINOS IN PHYSICS AND ASTROPHYSICS

ISBN 981-02-3887-8

Printed in Singapore by World Scientific Printers

Preface and Introduction

The Theoretical Advanced Study Institute (TASI) in elementary particle physics has been held each summer since 1984, and since 1989 has been permanently hosted by the University of Colorado at Boulder. Typically some fifty of the best advanced theory students in the United States have attended, and in recent years nearly all of the outstanding students in the country have had an opportunity to participate. The topic of the school varies from year to year, but there have typically been courses of lectures in phenomenology, mathematical physics, and often astrophysics and experimental topics. It has been highly successful, expecially in introducting the students to a broader range of ideas than they normally experience in their home institutions while working on their (often narrow) dissertation topics.

The topic of the 1998 TASI was *Neutrinos in Physics and Astrophysics: from* 10^{-33} *to* 10^{+28} *cm*, with the title emphasizing neutrinos as the central thread in the study of many aspects of particle physics and astrophysics. Neutrino interactions test the standard electroweak theory and its TeV scale extensions, and probe the structure of the nucleon and of the CKM matrix. Searches for neutrino mass and other intrinsic properties probe new physics at very short distance scales. The weak interactions of neutrinos imply for them a unique role in studying the early universe, the core of the Sun, type-II supernovae, and active galactic nuclei, and suggest the possibility of small neutrino masses contributing to the missing matter in the Universe, especially on very large distance scales.

The timing was especially appropriate, because during the period of the Institute the Superkamiokande Collaboration announced compelling evidence for non-zero neutrino mass from the zenith distribution of atmospheric neutrinos at the *Neutrino 98* conference in Takayama, Japan.

There were courses of lectures on current topics in particle physics, as well as on all aspects of neutrino physics and astrophysics. These were

- Overview of Neutrino Physics and Astrophysics; P. Langacker, Pennsylvania

- Particle Physics

  - Introduction to the Standard Model (5); G. Altarelli, CERN
  - Supersymmetry and Grand Unification (8); N. Polonsky, Rutgers
  - Superstring Theory and its Implications (8); K. Dienes, CERN
  - Collider Physics (4); D. Zeppenfeld, Wisconsin

v

- Extended Gauge Structures in String Theories (2); M. Cvetic, Pennsylvania

- Neutrino Properties and Theories of Neutrino Mass (6); P. Langacker, Pennsylvania

- Laboratory Searches for Neutrino Mass (3), and Solar and Atmospheric Neutrino Experiments (2); T. Bowles, LANL

- Astrophysics

  - Introduction to Neutrino Astrophysics (3); W. Haxton, INT, Seattle

  - Stellar Structure and Evolution (4), and The Standard Solar Model (2); M. Pinsonneault, Ohio State

  - Introduction to Cosmology (5), and Big Bang Nucleosynthesis (1); G. Steigman, Ohio State

  - Helioseismology (1); S. Basu, IAS, Princeton

  - Explanations of the Solar Neutrino Experiments (3); N. Hata, IAS, Princeton

  - High Energy Astrophysical Neutrinos (3); F. Halzen, Wisconsin

  - Supernova Neutrinos (2); A. Burrows, Arizona

  - Supernova 1987A (1); A. Mann, Pennsylvania

  - Neutrinos and Dark Matter (2); C.-P. Ma, Pennsylvania

  - Gravity Waves (2); D. Sigg, MIT

Most of these were written up for these proceedings, which we hope will be a useful reference and text for students and researchers in neutrino physics. There was also a very successful student seminar series, which is detailed at the end of this volume.

The efforts of many people were necessary to ensure the success of the school. In particular, I would like to thank the following for their help:

- The general director, K. T. Mahanthappa, and the members of the local organizing committee, Shanta deAlwis, Tom DeGrand, Anna Hasenfratz, Tamas Kovacs, and Sechul Oh.

- Mary Dang and Kathy Oliver for secretarial help.

- Thomas DeGrand for making computers accessible to participants and leading the hikes.

- Mu-Chun Chen for help in daily operation of TASI.

- Rellen Hardtke and Manoj Kaplinghat for organizing the student seminars.

- NSF, DoE and the University of Colorado for financial support and providing facilities.

Paul Langacker
Program Director
April 23 , 2000

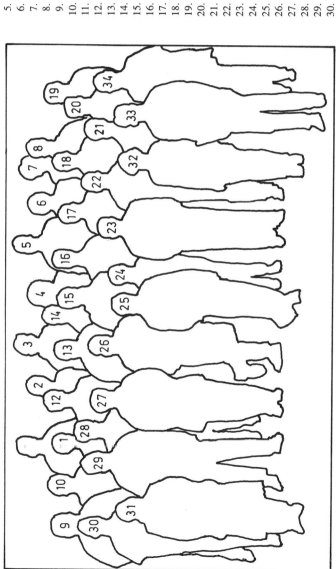

1. Kirk Schneider
2. Lara Pasquali
3. Stuart Wick
4. Rellen Hardtke
5. Stefan Recksiegel
6. Eun Kyung Paik
7. Manoj Kaplinghat
8. Laith Haddad
9. Keith Dienes
10. Jing Wang
11. Deirdre Black
12. Michael Plümacher
13. José Herman Muñoz
14. Tyce DeYoung
15. Sven Bergmann
16. Steen Hansen
17. Sadek Mansour
18. Stephen C. Gibbons
19. Jae Park
20. Sechul Oh
21. James N. Hormuzdiar
22. Parvez Anandam
23. K.T. Mahanthappa
24. Irina Mocioiu
25. Anja Werthenbach
26. Máximo Ave Pernas
27. Norma Quiroz
28. Hyun-Min Lee
29. Paul Langacker
30. Nir Polonsky
31. Sharada Iyer
32. Mu-Chun Chen
33. Mitesh Patel
34. Nurur Rahman

CONTENTS

OVERVIEW OF NEUTRINO PHYSICS AND ASTROPHYSICS

PAUL LANGACKER

Department of Physics and Astronomy
University of Pennsylvania, Philadelphia PA 19104-6396, USA
E-mail: pgl@langacker.hep.upenn.edu

The importance of non-zero neutrino mass as a probe of particle physics, astrophysics, and cosmology is emphasized. The present status and future prospects for the solar and atmospheric neutrinos are reviewed, and the implications for neutrino mass and mixing in 2, 3, and 4-neutrino schemes are discussed. The possibilities for significant mixing between ordinary and light sterile neutrinos are described.

1   Neutrino Mass

Neutrino mass [1] and properties are superb simultaneous probes of particle and astrophysics:

- Decays and scattering processes involving neutrinos have been powerful probes of the existence and properties of quarks, tests of QCD, of the standard electroweak model and its parameters, and of possible TeV-scale physics.

- Fermion masses in general are one of the major mysteries/problems of the standard model. Observation or nonobservation of the neutrino masses introduces a useful new perspective on the subject.

- Nonzero ν masses are predicted in most extensions of the standard model. They therefore constitute a powerful window on new physics at the TeV scale, intermediate scales (e.g., 10^{12} GeV), or the Planck scale.

- There may be a hot dark matter component to the universe. If so, neutrinos would be (one of) the most important things in the universe.

- The neutrino masses must be understood to fully exploit neutrinos as a probe of the Solar core, of supernova dynamics, and of nucleosynthesis in the big bang, in stars, and in supernovae.

2   Theory of Neutrino Mass

There are a confusing variety of models of neutrino mass. Here, I give a brief survey of the principle classes and of some of the terminology. For more detail, see [2].

A Weyl two-component spinor is a left (L)-handed[a] particle state, ψ_L, which is necessarily associated by CPT with a right (R)-handed antiparticle state[b] ψ_R^c. One refers to active (or ordinary) neutrinos as left-handed neutrinos which transform as $SU(2)$ doublets with a charged lepton partner. They therefore have normal weak interactions, as do their right-handed anti-lepton partners,

$$\begin{pmatrix} \nu_e \\ e^- \end{pmatrix}_L \overset{\text{CPT}}{\longleftrightarrow} \begin{pmatrix} e^+ \\ \nu_e^c \end{pmatrix}_R. \tag{1}$$

Sterile[c] neutrinos are $SU(2)$-singlet neutrinos, which can be added to the standard model and are predicted in most extensions. They have no ordinary weak interactions except those induced by mixing with active neutrinos. It is usually convenient to define the R state as the particle and the related L anti-state as the antiparticle.

$$N_R \overset{\text{CPT}}{\longleftrightarrow} N_L^c. \tag{2}$$

(Sterile neutrinos will sometimes also be denoted ν_s.)

Mass terms describe transitions between right (R) and left (L)-handed states. A Dirac mass term, which conserves lepton number, involves transitions between two distinct Weyl neutrinos ν_L and N_R:

$$-L_{\text{Dirac}} = m_D(\bar{\nu}_L N_R + \bar{N}_R \nu_L) = m_D \bar{\nu}\nu, \tag{3}$$

where the Dirac field is defined as $\nu \equiv \nu_L + N_R$. Thus a Dirac neutrino has four components ν_L, ν_R^c, N_R, N_L^c, and the mass term allows a conserved lepton number $L = L_\nu + L_N$. This and other types of mass terms can easily be generalized to three or more families, in which case the masses become matrices. The charged current transitions then involve a leptonic mixing matrix (analogous to the Cabibbo-Kobayashi-Maskawa (CKM) quark mixing matrix), which can lead to neutrino oscillations between the light neutrinos.

For an ordinary Dirac neutrino the ν_L is active and the N_R is sterile. The transition is $\Delta I = \frac{1}{2}$, where I is the weak isospin. The mass requires $SU(2)$ breaking and is generated by a Yukawa coupling

$$L_{\text{Yukawa}} - h_\nu(\bar{\nu}_e \bar{e})_L \begin{pmatrix} \varphi^0 \\ \varphi^- \end{pmatrix} N_R + H.C. \tag{4}$$

---

[a]The subscripts L and R really refer to the left and right chiral projections. In the limit of zero mass these correspond to left and right helicity states.
[b]Which is referred to as the particle or the antiparticle is a matter of convenience.
[c]Sterile neutrinos are often referred to as "right-handed" neutrinos, but that terminology is confusing and inappropriate when Majorana masses are present.

One has $m_D = h_\nu v/\sqrt{2}$, where the vacuum expectation value (VEV) of the Higgs doublet is $v = \sqrt{2}\langle\varphi^o\rangle = (\sqrt{2}G_F)^{-1/2} = 246$ GeV, and h_ν is the Yukawa coupling. A Dirac mass is just like the quark and charged lepton masses, but that leads to the question of why it is so small: one requires $h_{\nu_e} < 10^{-10}$ to have $m_{\nu_e} < 10$ eV.

A Majorana mass, which violates lepton number by two units ($\Delta L = \pm 2$), makes use of the right-handed antineutrino, ν_R^c, rather than a separate Weyl neutrino. It is a transition from an antineutrino into a neutrino. Equivalently, it can be viewed as the creation or annihilation of two neutrinos, and if present it can therefore lead to neutrinoless double beta decay. The form of a Majorana mass term is

$$-L_{\text{Majorana}} = \frac{1}{2}m_T(\bar{\nu}_L\nu_R^c + \bar{\nu}_R^c\nu_L) = \frac{1}{2}m_T\bar{\nu}\nu$$
$$= \frac{1}{2}m_T(\bar{\nu}_L C\bar{\nu}_L^T + H.C.), \tag{5}$$

where $\nu = \nu_L + \nu_R^c$ is a self-conjugate two-component state satisfying $\nu = \nu^c = C\bar{\nu}^T$, where C is the charge conjugation matrix. If ν_L is active then $\Delta I = 1$ and m_T must be generated by either an elementary Higgs triplet or by an effective operator involving two Higgs doublets arranged to transform as a triplet.

One can also have a Majorana mass term

$$-L_{\text{Majorana}} = \frac{1}{2}m_N(\bar{N}_L^c N_R + \bar{N}_R N_L^c) \tag{6}$$

for a sterile neutrino. This has $\Delta I = 0$ and thus can be generated by the VEV of a Higgs singlet[d].

Some of the principle classes of models for neutrino mass are:

- A triplet majorana mass m_T can be generated by the VEV v_T of a Higgs triplet field. Then, $m_T = h_T v_T$, where h_T is the relevant Yukawa coupling. Small values of m_T could be due to a small scale v_T, although that introduces a new hierarchy problem. The simplest implementation is the Gelmini-Roncadelli (GR) model[3], in which lepton number is spontaneously broken by v_T. The original GR model is now excluded by the LEP data on the Z width.

- A very different class of models are those in which the neutrino masses are zero at the tree level (typically because no sterile neutrino or elementary

---

[d]In principle this could also be generated by a bare mass, but this is usually forbidden by higher symmetries in extensions of the standard model.

Higgs triplets are introduced), but only generated by loops [4], *i.e.*, radiative generation. Such models generally require the *ad hoc* introduction of new scalar particles at the TeV scale with nonstandard electroweak quantum numbers and lepton number-violating couplings. They have also been introduced in an attempt to generate large electric or magnetic dipole moments. They also occur in some supersymmetric models with cubic R parity violating terms in the superpotential [5].

- In the seesaw models [6], a small Majorana mass is induced by mixing between an active neutrino and a very heavy Majorana sterile neutrino M_N. The light (essentially active) state has a naturally small mass

$$m_\nu \sim \frac{m_D^2}{M_N} \ll m_D. \tag{7}$$

There are literally hundreds of seesaw models, which differ in the scale M_N for the heavy neutrino (ranging from the TeV scale to grand unification scale), the Dirac mass m_D which connects the ordinary and sterile states and induces the mixing (e.g., $m_D \sim m_u$ in most grand unified theory (GUT) models, or $\sim m_e$ in left-right symmetric models), the patterns of m_D and M_N in three family generalizations, etc. One can also have mixings with heavy neutralinos in supersymmetric models with R parity breaking [5], induced either by bilinears connecting Higgs and lepton doublets in the superpotential or by the expectation values of scalar neutrinos.

- Superstring models often predict the existence of higher-dimensional (non-renormalizable) operators (NRO) such as

$$-L_{\text{eff}} = \bar{\psi}_L H \left(\frac{S}{M_{\text{str}}} \right)^P \psi_R + H.C., \tag{8}$$

where H is the ordinary Higgs doublet, S is a new scalar field which is a singlet under the standard model gauge group, and $M_{\text{str}} \sim 10^{18}$ GeV is the string scale. In many cases S will acquire an intermediate scale VEV (e.g., 10^{12} GeV), leading to an effective Yukawa coupling

$$h_{\text{eff}} \sim v \left(\frac{\langle S \rangle}{M_{\text{str}}} \right)^P \ll v. \tag{9}$$

Depending on the dimensions P of the various operators and on the scale $\langle S \rangle$, it may be possible to generate an interesting hierarchy for the quark and charged lepton masses and to obtain naturally small Dirac neutrino masses [7].

Similarly, one may obtain triplet and singlet Majorana neutrino masses, m_T and m_N by analogous higher-dimensional operators. The former are small. Depending on the operators[7] the latter may be either small, leading to the possibility of significant mixing between ordinary and sterile neutrinos[8], or large, allowing a conventional seesaw.

- Mixed models, in which both Majorana and Dirac mass terms are present, will be further discussed in the section on sterile neutrinos.

3 Solar Neutrinos

Tremendous progress has been made recently in solar neutrinos[9]. For many years there was only one experiment, while now there are a number that are running or finished, and more are coming on line soon. The original goal of using the solar neutrinos to study the properties of the solar core underwent a 30 year digression on the study of the properties of the neutrino itself. The quality of the experiments themselves and of related efforts on helioseismology, nuclear cross sections, and solar modeling is such that the revised goal of simultaneously studying the properties of the Sun and of the neutrinos is feasible.

3.1 Experiments

The experimental situation is very promising. We now have available the results of five experiments, Homestake (chlorine)[10], Kamiokande[11], GALLEX[12], SAGE[13], and Superkamiokande[11]. Especially impressive are the successful ^{51}Cr source experiments for SAGE and GALLEX (which probe a combination of the extraction efficiencies and the neutrino absorption cross section, yielding $0.95 \pm 0.07^{+0.04}_{-0.03}$ of the expected rate), and the successful ^{71}As spiking experiment completed at the end of the GALLEX run to test the extraction efficiency (yielding $R = 1.00 \pm 0.01$ for the ratio of actual to expected extractions).

Coming soon, there should be results from SNO, Borexino, The Gallium Neutrino Observatory (GNO), and the next phase of SAGE, which will yield much more detailed, precise, or model independent information on the ^8B (SNO[14]), ^7Be (Borexino[15]), and pp (GNO, SAGE) neutrinos. Future generations of even more precise experiments should especially be sensitive to the ^7Be and pp neutrinos[16]. The overall goal of the program should be very ambitious, i.e., to measure the arriving flux of $\nu_e, \nu_{\mu+\tau}$, and ν_s (sterile neutrinos), and even possible antineutrinos, for each of the initial flux components, as well as to measure or constrain possible spectral distortions, day-night (earth) effects, seasonal and solar cycle variations, and mixed (*e.g.,* simultaneous spectral and day-night) effects.

3.2 Interpretation

The observed fluxes are in strong disagreement with the predictions of the standard solar model (SSM). The overall rates are compared with the predictions of the new Bahcall-Pinsonneault 1998 (BP 98) model[17] in Table 1, where it is seen that all of the fluxes are much lower than the expectations. BP 98 contains a number of refinements compared to earlier theoretical calculations, but the most important changes are a 20% (1.3 σ) lower ^8B flux, as described below, and 1.1 σ decreases in the ^{37}Cl and ^{71}Ga capture rates.

Table 1: Results of Solar neutrino experiments, compared with the predictions of BP 98. The chlorine and gallium results are in units of SNU ($10^{-36}s^{-1}$ captures per target atom), and the water Cerenkov results are in units of $10^6/cm^2s$.

	experiment	BP-98
Homestake (chlorine)	2.56 ± 0.23	$7.7^{+1.2}_{-1.0}$
GALLEX, SAGE (gallium)	72.2 ± 5.6	129^{+8}_{-6}
Kamiokande, SuperK ($\nu e \rightarrow \nu e$)	2.44 ± 0.10	$5.15 \times (1^{+0.19}_{-0.14})$

Recent results in helioseismology [17,18,19,20,21] leave little room for deviations from the standard solar model. The eigen-frequencies effectively measure the sound speed T/μ, where T and μ are respectively the temperature and density, as a function of radial position, down to 5% of the solar radius. The results agree with the predictions of BP 98 to $\sim 10^{-3}$, even though T and μ individually vary by large values over the radius of the Sun. This leaves very little room for non-standard solar models (NSSM), which would typically have to deviate by several percent to have much impact on the neutrino flux predictions. The only aspect of the SSM relevant to the neutrino fluxes that is not severely constrained are nuclear cross sections, especially S_{17} and S_{34}, which are respectively proportional to the cross sections for ^8B and ^7Be production, and to the absorption cross sections for the radiochemical experiments.

The experimental and theoretical status of the nuclear cross sections were critically examined at a workshop at the Institute for Nuclear Theory in 1997 (INT 97) [21,22]. The participants recommended a lower S_{17}, by relying on the best documented individual measurements rather than an average, and also a larger uncertainty in S_{34}, both of which were incorporated in BP 98. Haxton has recently argued [23] that there are still considerable uncertainties in the Ga absorption cross sections, but this possibility is strongly disfavored by the ^{51}Cr source and ^{71}As spiking experiments.

Even the relatively large shift in S_{17} advocated by INT 97 and used by BP 98 does little to change the basic disagreement between the observations and the standard solar model. Even if a particular NSSM could be consistent with helioseimology, it would be difficult to account for the observations. The Kamiokande and Superkamiokande results can be regarded (in the absence of neutrino oscillations) as a measurement of the ^8B flux. Subtracting this "experimental" ^8B flux from either the gallium or chlorine predictions, the observed fluxes are still inconsistent with the observed solar luminosity.

This line of reasoning is developed in the "model-independent" analyses of the neutrino flux components [24,25,26], which can be viewed as a measurement of "global" spectral distortions. The idea is that all plausible astrophysical or nuclear physics modifications of the standard solar model do not significantly distort the spectral shape of the pp or ^8B neutrinos: all that they can do is modify the overall magnitude of the pp, ^7Be, ^8B, and minor flux components. Furthermore, the observed solar luminosity places a linear constraint on the pp, ^7Be, and CNO fluxes (provided that the time scale for changes in the solar core is long compared to the 10^4 yr required for a photon to diffuse to the surface).

By combining the different experiments, each class of which has a different spectral sensitivity, one concludes that

$$\frac{\phi(^7Be)}{\phi(^7Be)_{\text{SSM}}} \ll \frac{\phi(^8B)}{\phi(^8B)_{\text{SSM}}}, \tag{10}$$

where SSM refers to the standard solar model predictions. The same result holds even if one discards any one of the three types of experiment (chlorine, gallium, water), or ignores the luminosity constraint. No plausible astrophysical model has succeeded in suppressing ^7Be neutrinos significantly more than ^8B neutrinos, mainly because ^8B is made from ^7Be. Models with a lower core temperature or with a lower S_{17} do not come anywhere near the data. The Cumming-Haxton model [27] with large ^3He diffusion comes closest, but even that is far from the data. That model is probably also excluded by helioseismology, but Haxton has argued [23] that final judgment should wait until a self-consistent model with ^3He diffusion is constructed to be compared with the helioseismology data.

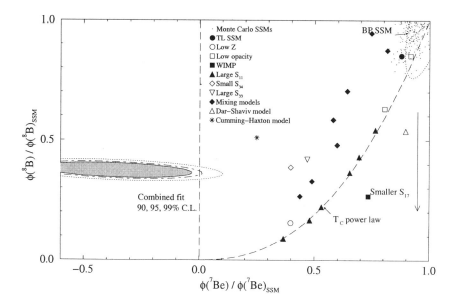

Figure 1: Allowed regions for the ^7Be and ^8B fluxes (normalized by BP 98), compared with the predictions and uncertainties in the SSM and various non-standard solar models. Courtesy of N. Hata.

3.3 Possible Solutions

As discussed in the previous section, an astrophysical/nuclear explanation of the solar neutrinos experiments is unlikely. The most likely particle physics explanations include:

- A matter enhanced (MSW) transition of ν_e into ν_μ or ν_τ. There are the familiar small (SMA) and large (LMA) mixing angle solutions[24] with $\Delta m^2 \sim 10^{-5}$ eV^2, as well as the low mass (LOW) solution with $\Delta m^2 \sim 10^{-7}$ eV^2 and near maximal mixing. The latter is a very poor fit, but sometimes shows up in fits at the 99% cl.

- There is also a small mixing angle MSW solution for ν_e into a sterile neutrino ν_s. The major difference between $\nu_{\mu,\tau}$ and ν_s, and the reason there is no LMA solution, is that in the first case the $\nu_{\mu,\tau}$ can scattering elastically from electrons in the water Cerenkov experiments, with about $1/6^{th}$ the ν_e cross section, leading to a lower survival probability for ν_e than for astrophysical

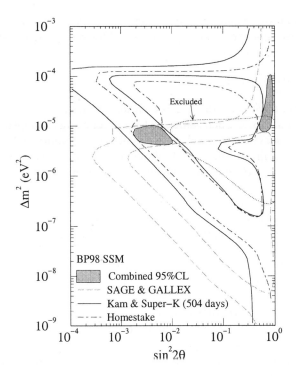

Figure 2: Allowed MSW solutions, not including Superkamiokande spectral data. Courtesy of N. Hata.

or sterile neutrino solutions. There is also a small difference for the MSW conversion rate for sterile neutrinos in the Sun, but that is proportional to the neutron density, and is much less important.

- The vacuum ("just so"[28]) oscillation solutions[24], with near maximal mixing and $\Delta m^2 \sim 10^{-10}$ eV^2 are another possibility. These are somewhat fine-tuned, with Δm^2 such that the Earth-Sun distance is at roughly half an oscillation length, L_{osc}, or an odd multiple. Since $L_{\text{osc}} = 4\pi E/\Delta m^2$, one expects a significant variation of the ν_e survival probability with neutrino energy.

- The above solutions are such that only two neutrinos are important for the Solar neutrinos. However, it is possible that transitions between all three neutrinos are important. There could be generalized MSW solutions involving more than one value of Δm^2, or mixed MSW and vacuum solutions [29]. In both cases, there could be considerably different spectral distortions than in the two-neutrino case.

Other possibilities include:

- Maximal mixing [30] (i.e., vacuum oscillations with $\Delta m^2 \gg 10^{-10}$ eV^2), combined with a low S_{17}. Such solutions lead to an energy independent suppression of the ν_e survival probability. Even allowing a suppressed ^8B production rate, this possibility is viable only if one ignores (or greatly expands the uncertainties in) the Homestake Chlorine experiment.

- RSFP [31] (resonant spin flavor precession), involving rotations of left handed neutrinos into sterile right handed neutrinos, combined with MSW flavor transitions. These were motivated by possible hints (not confirmed by other experiments) of time dependence correlated with the Sunspot activity in the chlorine experiment. This could only occur if there are extremely large neutrino electric or magnetic dipole moments or transition moments, which would present a considerable challenge to the model builder. Although such effects have not been reported by other groups, there is still a somewhat surprising difference in rates observed by the GALLEX collaboration in their third and fourth data taking intervals. However, this could also be a statistical fluctuation. In any case, such RSFP effects could be probed experimentally by studying the $\bar{\nu}_e$ and $\bar{\nu}_\mu$ spectra [32].

- Flavor changing neutral current effects [33], possibly generated by R-parity violating terms in supersymmetry, could be an alternative means of generating enhanced neutrino flavor conversions in the Sun.

- The possible violation of Lorentz invariance [34] could affect not only the Solar neutrinos, but could also be relevant to the observed ultra high energy cosmic rays.

- There could be a lepton flavor dependent violation of the the equivalence principle [35].

Perhaps the most important possibility or complication is that more than one thing could be going on simultaneously. There could be any of the above effects in conjunction with non-standard properties of the Sun or nuclear cross sections. Many but not all such NSSM possibilities are excluded by helioseismology and neutrino source experiments. While it is very unlikely than such effects could by themselves account for the data, their combination with new neutrino properties could considerably confuse the interpretation of future experimental results. This is one or the reasons that it is important to have as many independent precise experimental results as possible.

3.4 Needs

To distinguish the many possibilities we need as much precise data as possible. Especially useful are observables that are independent of or insensitive to the initial ν_e fluxes, and therefore to the astrophysical and nuclear cross section uncertainties. Such observables include:

- The neutral to charged current interaction ratio (NC/CC), which will be measured by SNO for deuteron dissociation. Since the NC cross section is the same for all active neutrinos, the NC rate measures the sum of the ν_e, ν_μ, and ν_τ fluxes, while the CC only measures ν_e. An anomalous NC/CC ratio would provide definitive evidence for transitions of ν_e into ν_μ or ν_τ, either by MSW or vacuum oscillations. Although the NC measurement is difficult, SNO should have the requisite sensitivity. A confirmation could be obtained by comparing the SNO CC rate with the fluxes determined in $\nu e \to \nu e$ measurements, since $\nu_{\mu,\tau}$ also contribute to the latter, with about $1/6^{\text{th}}$ the ν_e cross section. (The Borexino experiment will similarly allow an indirect determination of the transitions of ^7Be neutrinos into $\nu_{\mu,\tau}$ by comparing with the ν_e flux inferred from radiochemical experiments.)

  Transitions of ν_e into a sterile neutrino ν_s would not lead to an anomalous NC/CC ratio. This would make it much harder to verify ν_s transitions, but would serve as evidence for sterile neutrinos if MSW or vacuum oscillations are established by other means.

- There is no known astrophysical mechanism that can significantly distort the ^8B neutrino spectrum from the expected β decay shape. Not only would a spectral distortion establish a non-astrophysical solution to the solar neutrinos, but it would be a powerful probe of the mechanism. Study of the ^8B spectrum can be viewed as a cleaner extension (by individual experiments) of the "global" spectral distortion inferred from the combined experiments.

  One expects significant spectral distortions for the MSW SMA solution, for vacuum oscillations, and for hybrid solutions, but not for the LMA solution. The ratio of observed to expected spectrum can be conveniently parametrized by the first two moments [36], i.e., a linear approximation, for the SMA case, while the other cases can exhibit more complicated shapes. Measurement of the spectral distortion is very difficult, and requires excellent energy calibrations and extending the measurement to as low an energy as possible.

Both SuperKamiokande and SNO have the capability to measure a spectral distortion. SuperK has the advantage of higher statistics. However, the ν energy is shared between the final electron and neutrino, so any spectral distortion is partially washed out in the observed e^- spectrum. SNO, on the other hand, has the advantage that the electron in the CC reaction carries all of the neutrino energy (plus the known binding energy), leading to a harder electron spectrum and an essentially direct measurement of the ν spectrum.

One of the highlights of this conference was the preliminary new statistics-limited SuperKamiokande spectrum, from around 6.7 to 14.5 MeV, obtained after a series of careful calibrations of their detector using an electron Linac[11]. The lower energy data are consistent with no distortions, but there is evidence for a significant excess of events in the three energy bins above 13 MeV. These data, for the first time, give a statistically significant indication of a spectral distortion: the no oscillation hypothesis (and also LMA solution) is disfavored at the 95-99% CL level. The SMA MSW solution is also a very poor fit, although it is allowed at 95% CL. The best fit favors vacuum oscillations. The favored $\Delta m^2 \sim 4 \times 10^{-10}$ eV^2 gives a much better fit to the data than for the lower range Δm^2 around 10^{-10} eV^2 found in recent global analyses of the total event rates. However, new studies based on BP 98 with its larger S_{17} allow a larger Δm^2, consistent with the spectral distortions.

An alternate interpretation of the high energy excess is that the flux of hep neutrinos ($^3He + p \to {}^4He + e^+ + \nu_e$) has been seriously underestimated. Their flux would have to be larger by a factor of twenty or so from the usual estimates for them to contribute significantly to the excess, but it has been emphasized that there is no direct experimental measure of or rigorous theoretical bound on the cross section [37]. The issue can be resolved by a careful study of the energy range 14-18.8 MeV, above the endpoint of the ^8B spectrum. (The highest energy SuperK bin is centered above this endpoint, but there is a significant energy uncertainty.)

The SuperK spectrum has important implications, but it is still preliminary. In additional to finalizing the analysis, additional lower energy points are expected that should help clarify the situation.

- For some regions of MSW parameters, one expects an asymmetry between day and night event rates due to regeneration of ν_e at night as the converted neutrinos travel through the Earth[38]. Superkamionde has binned their data for daytime and for a number of different nighttime zenith angles (i.e., different paths through the earth). They see no evidence for a zenith angle dependence, and their overall day-night asymmetry is

$$\frac{D - N}{D + N} = -0.023 \pm 0.020 \pm 0.014, \tag{11}$$

where D (N) refers to day (night) rates and the first (second) error is statistical (systematic). The absence of an effect excludes a significant region of MSW parameter space independent of the details of the solar model (and with only a small uncertainly from the Earth's density profile). This excludes the lower Δm^2 part of the LMA solution, but has little impact on the SMA solution. (The part of the SMA solution with the largest $\sin^2 2\theta$ was expected to have a barely observable day-night asymmetry, but the effect is predicted to be smaller with the new BP 98 fluxes, which shift the SMA region to slightly smaller $\sin^2 2\theta$.)

Several authors have emphasized recently that the Earth effect is signficantly enhanced for neutrinos passing through the core of the Earth[39]. (There is an analogous effect for atmospheric neutrinos.) This parametric (or oscillation length) resonance, in which the oscillation length is comparable to the diameter of the core, was included automatically in previous numerical studies, but not explicitly commented on. It is larger for transitions into $\nu_{\mu,\tau}$ than for ν_s. Since relatively few of the solar neutrinos pass through the core for the existing high latitude detectors, it has been suggested that there should be a dedicated experiment at low latitude[40].

- For vacuum oscillations[24], the Earth-Sun distance is typically at a node of the oscillations. This is somewhat fine-tuned, leading to the name "just so". Since the oscillation length is $4\pi E/\Delta m^2$ there is a strong energy dependence to the survival probability. One also expects a strong seasonal variation, due to the eccentricity of the Earth's orbit. However, the seasonal variation can be partially washed out as one averages over energies, so one should ideally measure the spectral shape binned with respect to the time of the year[41].

- RSFP could lead to long term variations in the neutrino flux, e.g., correlated with Sunspots or Solar magnetic fields. Other changing magnetic effects could conceivably alter the solar neutrinos in other ways, e.g., by changing the local density. Only the Homestake experiment has seen any significant hint of a time variation, and that hint has been considerably weakened by more recent Homestake data. Nevertheless, it is conceivable that time dependent effects are energy dependent, and therefore different experiments have different sensitivity. They could also have been somewhat hidden in the water Cerenkov experiments because of the neutral current. It would be useful to run all of the experiments simultaneously through a solar cycle.

14

- RSFP [31] could also lead to the production of $\bar{\nu}_e$ which can be observed in the SNO detector by the delayed coincidence of the γ ray emitted by the capture of the neutron from $\bar{\nu}_e p \rightarrow e^+ n$.

3.5 Outlook

The model independent observables that can be measured by SuperK and SNO for the ^8B neutrinos should go far towards distinguishing the different possibilities. However, it will be especially difficult to establish transitions into sterile neutrinos. It will also be very difficult to sort out what is happening in a three-flavor or hybrid scenario, such as MSW transitions combined with non-standard solar physics. For these reasons, we would like to have accurate information on spectral distortions, day-night effects (especially for neutrinos passing through the Earth core), NC/CC ratios, and absolute fluxes arriving at the Earth for the ^7Be and pp neutrinos as well. There is a strong need for the next generations of experiments.

A challenging but realistic goal is to simultaneously establish the neutrino mechanism(s) (e.g., MSW SMA solution), determine the neutrino parameters, *and* study the Sun [42]. Even with existing data, if one assumed two-flavor MSW but allowed an arbitrary solar core temperature T_C, it was possible to simultaneously determine the MSW parameters (with larger uncertainties than when the SSM is assumed) and T_C, with the result that $T_C = 0.99^{+0.02}_{-0.03}$ with respect to the SSM prediction of 1.57×10^7 K [25]. In the future, it should be possible to determine the neutrino parameters and simultanously the ^8B and ^7Be fluxes, for comparison with the SSM predictions. It will also be possible to constrain density fluctuations in the Sun [43], which can smear out the MSW affects. However, recent estimates suggest that such effects are negligible [44].

To fully exploit the future data, it will be important to carry out global analyses of all of the observables in all of the experiments (possibly incorporating helioseismology data as well). Global analyses are difficult because of difficulties with systematic errors. However, they often contain more information than the individual experiments, and allow uniform treatment of theoretical uncertainties. For this purpose, it is important that each experiment publish all of their data, such as double binning the data with respect to energy and zenith angle, including full systematics and correlations.

4 Atmospheric Neutrinos

Although the prediction for the absolute number of μ or e produced by the interactions of neutrinos produced in cosmic ray interactions in the atmosphere has a theoretical uncertainty of around 20%, it is believe that the ratio

$N(\mu)/N(e)$ can be predicted to within 5% [45]. To zeroth approximation, the ratio is just two, independent of the details of the cosmic ray flux or interactions, because each produced pion decays into two ν_μ and one ν_e (I am not distinguishing ν from $\bar{\nu}$), and for energies large compared to m_μ the interaction cross sections are the same. Of course, the actual ratio depends on the neutrino energies, and therefore on the details of the hadronic energies, polarization of the intermediate muons from π decay, etc.

For years the ratio R of observed $N(\mu)/N(e)$, normalized by the predicted value, found in the water Cerenkov experiments (Kamiokande, IMB, SuperKamiokande) has been around 0.6 [46]. This has recently been confirmed by the higher SuperK [47] statistics ($R = 0.63(3)(5)$ for sub-GeV events and 0.65(5)(8) for multi-GeV), and independently by the iron calorimeter experiment at Soudan [48] (0.58(11)(5)) and by Macro [49] (0.53(15) for upward events and 0.71(21) for stopping or downgoing events). This depletion of μ events suggests the possibility of ν_μ oscillations into ν_e, ν_τ, or ν_s, with near-maximal mixing ($\sin^2 2\theta > 0.8$) and $\Delta m^2 \sim 10^{-3} - 10^{-2}$ eV^2.

To confirm oscillations, more detailed information is needed. Already, the CHOOZ [50] (France) reactor $\bar{\nu}_e$ disappearance experiment excludes the $\nu_\mu \to \nu_e$ interpretation of the atmospheric neutrino anomaly for $\Delta m^2 > 10^{-3}$ eV^2. This should be extended by the coming Palo Verde experiment [51], and the planned KamLand [52] experiment at Kamiokande (sensitive to many nearby reactors) should extend the sensitivity down to the MSW LA solar neutrino range.

In the future [53], there will also be accelerator long baseline experiments for $\nu_\mu \to \nu_{e,\tau}$ appearance, or ν_μ disappearance (into $\nu_{e,\tau,s}$). The KEK to Kamiokande (K2K) experiment will be sensitive to ν_μ disappearance down to $\Delta m^2 \sim 5 \times 10^{-3}$ eV^2, while the Fermilab to Soudan (MINOS) experiment will probe both appearance and disappearance down to 10^{-3} eV^2. There are also proposals for a CERN to Gran Sasso experiment (ICARUS, OPERA), which would be sensitive to most of the parameter range suggested by Superkamiokande. These experiments should be able to confirm or refute the atmospheric neutrino oscillations, except[e] possibly for the smallest $\Delta m^2 \sim 10^{-3}$ eV^2.

Much more detailed information can be derived from the atmospheric neutrino data itself, by searching for indications of the $\sin^2(1.27\Delta m^2 L/E)$ dependence of the transition probability characteristic of neutrino oscillations. (L is the distance traveled and E is the neutrino energy.) This can be studied by considering the zenith angle distribution for fixed neutrino energy (in practice, the data is divided into sub-GeV and mutli-GeV bins), or by up-down asymmetries $(U - D)/(U + D)$, where U and D are respectively the number

---

[e]The long baseline experiments were proposed when the earlier Kamiokande results suggested a somewhat larger $\Delta m^2 \sim 10^{-2}$ eV^2.

of up and downgoing muons or electrons [54]. The data can also be plotted as a function of L/E, but that is less direct since the full neutrino energy is not measured on an event by event basis in the water Cerenkov experiments.

The Kamiokande collaboration observed an indication of oscillations in their zenith angle distribution for contained events [55]. However, the new Superkamiokande zenith angle distributions for contained events have much better statistics. They strongly indicate a zenith angle distribution in muon events consistent with oscillations, with an enhanced effect in the multi-GeV sample, consistent with expectations. There is no anomaly or excess in the electron events. This implies that ν_μ is oscillating into ν_τ or possibly a sterile neutrino ν_s, and not into ν_e. The latter result confirms the conclusions of CHOOZ. (Subdominant oscillations into ν_e in three-neutrino schemes are still possible.) The SuperK events virtually establish neutrino oscillations. Independent evidence is obtained by the zenith angle distributions for upward through-going muon events from SuperK, MACRO[f], and very preliminary results from SOUDAN.

Future atmospheric neutrino observations could possibly shed further light on the question of whether ν_μ is oscillating into ν_τ or into ν_s, although they are all very difficult. These include (a) subtle (e.g., parametric resonance) effects on neutrinos propagating through the Earth's core [56], which would affect $\nu_\mu \to \nu_s$, but not $\nu_\mu \to \nu_\tau$ (because ν_μ and ν_τ have the same neutral current interactions). In either scenario, secondary $\nu_\mu \to \nu_e$ oscillations would also be modified by Earth core effects. (b) The NC/CC ratio, including its zenith angle distribution and up-down asymmetry [57]. The NC rate could in principle be measured in $\nu N \to \nu \pi^0 X$, although this is a very difficult measurement. The preliminary SuperK result [11] $R(\pi^0/e) = 0.93(7)(19)$ on the ratio or π^0 to e events compared to expectations slightly favors $\nu_\mu \to \nu_\tau$ but does not exclude $\nu_\mu \to \nu_s$. (c) Direct observation of events in which ν_τ produces a τ would establish $\nu_\mu \to \nu_\tau$ oscillations [58]. However, this is extremely difficult.

There may also be significant three neutrino effects. For example, even if the dominant transition for the atmospheric neutrinos involves $\nu_\mu \to \nu_\tau$, there could be important subdominant ν_e effects.

There have been several careful phenomenological analyses of the atmospheric neutrino data in two neutrino and three neutrino mixing schemes [59]. One important theoretical issue posed by the atmospheric neutrinos is why is there nearly maximal mixing (i.e., $\sin^2 2\theta \sim 1$), when most theoretical schemes involving hierarchies of neutrino masses, as well as the analogs in the quark mixing sector, yield small mixings.

---

[f] The MACRO results [49] are not in very good agreement with oscillations, but the oscillation hypothesis nevertheless fits much better than the no-oscillation case.

5 Implications for Neutrino Mixing

5.1 The Global Picture

Various scenarios for the neutrino spectrum are possible, depending on which of the experimental indications one accepts. The simplest scheme, which accounts for the Solar (S) and Atmospheric (A) neutrino results, is that there are just three light neutrinos, all active, and that the mass eigenstates ν_i have masses in a hierarchy, analogous to the quarks and charged leptons. In that case, the atmospheric and solar neutrino mass-squared differences are measures of the mass-squares of the two heavier states, so that $m_3 \sim (\Delta m_{\text{atm}}^2)^{1/2} \sim 0.03 - 0.1$ eV; $m_2 \sim (\Delta m_{\text{solar}}^2)^{1/2} \sim 0.003$ eV (for MSW) or $\sim 10^{-5}$ eV (vacuum oscillations), and $m_1 \ll m_2$. The weak eigenstate neutrinos $\nu_a = (\nu_e, \nu_\mu, \nu_\tau)$ are related to the mass eigenstates ν_i by a unitary transformation $\nu_a = U_{ai}\nu_i$. If one makes the simplest assumption (from the Superkamiokande and CHOOZ data), that the ν_e decouples entirely from the atmospheric neutrino oscillations, $U_{e3} = 0$, (of course, one can relax this assumption somewhat) and ignores possible CP-violating phases [60], then

$$
\begin{pmatrix} \nu_e \\ \nu_\mu \\ \nu_\tau \end{pmatrix} = \begin{pmatrix} 1 & 0 & 0 \\ 0 & c_\alpha & -s_\alpha \\ 0 & s_\alpha & c_\alpha \end{pmatrix} \times \begin{pmatrix} c_\theta & -s_\theta & 0 \\ s_\theta & c_\theta & 0 \\ 0 & 0 & 1 \end{pmatrix} \begin{pmatrix} \nu_1 \\ \nu_2 \\ \nu_3 \end{pmatrix}, \tag{12}
$$

where α and θ are mixing angles associated with the atmospheric and solar neutrino oscillations, respectively, and where $c_\alpha \equiv \cos\alpha$, $s_\alpha \equiv \sin\alpha$, and similarly for c_θ, s_θ.

For maximal atmospheric neutrino mixing, $\sin^2 2\alpha \sim 1$, this implies $c_\alpha = s_\alpha = 1/\sqrt{2}$, so that

$$
U = \begin{pmatrix} c_\theta & -s_\theta & 0 \\ \frac{s_\theta}{\sqrt{2}} & \frac{c_\theta}{\sqrt{2}} & -\frac{1}{\sqrt{2}} \\ \frac{s_\theta}{\sqrt{2}} & \frac{c_\theta}{\sqrt{2}} & \frac{1}{\sqrt{2}} \end{pmatrix}. \tag{13}
$$

For small θ, this implies that $\nu_{3,2} \sim \nu_{+,-} \equiv (\nu_\tau \pm \nu_\mu)/\sqrt{2}$ participate in atmospheric oscillations, while the solar neutrinos are associated with a small additional mixing between ν_e and ν_-. Another limit, suggested by the possibility of vacuum oscillations for the solar neutrinos, is $\sin^2 2\theta \sim 1$, or $c_\theta = s_\theta = 1/\sqrt{2}$, yielding

$$
U = \begin{pmatrix} \frac{1}{\sqrt{2}} & -\frac{1}{\sqrt{2}} & 0 \\ \frac{1}{2} & \frac{1}{2} & -\frac{1}{\sqrt{2}} \\ \frac{1}{2} & \frac{1}{2} & \frac{1}{\sqrt{2}} \end{pmatrix}, \tag{14}
$$

which is referred to as bi-maximal mixing [61]. A number of authors have discussed this pattern and how it might be obtained from models, as well as how much freedom there is to relax the assumptions of maximal atmospheric and solar mixing (the data actually allow $\sin^2 2\alpha \gtrsim 0.8$ and $\sin^2 2\theta \gtrsim 0.6$) or the complete decoupling of ν_e from the atmospheric neutrinos. Another popular pattern,

$$U = \begin{pmatrix} \frac{1}{\sqrt{2}} & -\frac{1}{\sqrt{2}} & 0 \\ \frac{1}{\sqrt{6}} & \frac{1}{\sqrt{6}} & -\frac{2}{\sqrt{6}} \\ \frac{1}{\sqrt{3}} & \frac{1}{\sqrt{3}} & \frac{1}{\sqrt{3}} \end{pmatrix}, \tag{15}$$

known as democratic mixing [62], yields maximal solar oscillations and near-maximal (8/9) atmospheric oscillations.

In this hierarchical pattern, the masses are all too small to be relevant to mixed dark matter (in which one of the components of the dark matter is hot, i.e., massive neutrinos) or to neutrinoless double beta decay ($\beta\beta_{0\nu}$). However, the solar and atmospheric oscillations only determine the differences in mass squares, so a variant on this scenario is that the three mass eigenstates are nearly degenerate rather than hierarchical [63], with small splittings associated with Δm_{atm}^2 and $\Delta m_{\text{solar}}^2$. For the common mass m_{av} in the 1-several eV range, the hot dark matter could account for the dark matter on large scales (with another, larger, component of cold dark matter accounting for smaller structures) [64]. If the neutrinos are Majorana they could also lead to $\beta\beta_{0\nu}$ [65]. Current limits imply an upper limit of

$$\langle m_{\nu_e} \rangle = \sum_i \eta_i U_{ei}^2 |m_i| < 0.46 - 1 \text{ eV}, \tag{16}$$

on the effective mass for a mixture of light Majorana mass eigenstates, where η_i is the CP-parity of ν_i and the uncertainty on the right is due to the nuclear matrix elements. (There is no constraint on Dirac neutrinos.) The combination of small $\langle m_{\nu_e} \rangle \ll m_{\text{av}}$, maximal atmospheric mixing, and $U_{e3} = 0$ would imply cancellations, so that $\eta_1\eta_2 = -1$ and $c_\theta = s_\theta = 1/\sqrt{2}$, i.e., maximal solar mixing. Even the more stringent limit in (16) is large enough that there is room to relax all of these assumptions considerably. Nevertheless, there is strong motivation to try to improve the $\beta\beta_{0\nu}$ limits.

The LSND experiment [66] has reported evidence for $\nu_\mu \to \nu_e$ and $\bar{\nu}_\mu \to \bar{\nu}_e$ oscillations with $\Delta m_{\text{LSND}}^2 \sim 1 \text{ eV}^2$ and small mixing $\sim 10^{-3} - 10^{-2}$, while the KARMEN experiment sees no candidates. KARMEN [67] is sensitive to most of the same parameter range as LSND, although there is a small window of oscillation parameters for which both experiments are consistent. A resolution of the situation may have to wait for the mini-BOONE experiment at Fermilab. However, it is interesting to consider the implications if the LSND result is confirmed. In that case, there are three distinct mass-squared differences,

$\Delta m^2_{\text{LSND}} \sim 1$ eV^2, $\Delta m^2_{\text{atm}} \sim 10^{-3} - 10^{-2}$ eV^2, and $\Delta m^2_{\text{solar}} \sim 10^{-5}$ eV^2 (MSW) or 10^{-10} eV^2 (vacuum), implying the need for a fourth neutrino[g]. Since the Z lineshape measurements at LEP only allow 2.992 ± 0.011 light, active neutrinos[71], any light fourth neutrino would have to be sterile, ν_s.

Several mass patterns for the four neutrinos have been suggested[72]. (There course also be more than four light neutrinos[73].) To be consistent with both LSND and CHOOZ, states containing ν_μ and ν_e must be separated by about 1 eV. Assuming the atmospheric neutrinos involve $\nu_\mu \to \nu_\tau$, one could have nearly degenerate $\nu_{+,-} \equiv (\nu_\tau \pm \nu_\mu)/\sqrt{2}$ at around 1 eV, with the solar neutrinos described by a dominantly ν_s state at ~ 0.003 eV or $\sim 10^{-5}$ eV and a much lighter (dominantly) ν_e. (Solar neutrinos can be accounted for by a SMA MSW solution or possibly by vacuum oscillations, but not by a LMA MSW.) Alternatively, one could reverse the pairing, with a nearly degenerate ν_s and ν_e at ~ 1 eV, and $\nu_{+,-}$ around 0.03-0.1 eV. The other models involve $\nu_\mu \to \nu_s$ with near-maximal mixing for the atmospheric neutrinos, and $\nu_e \to \nu_\tau$ for the solar neutrinos. Again, there are two possibilities, with the nearly degenerate $\nu_s - \nu_\tau$ pair around 1 eV and a lighter $\nu_\tau - \nu_e$, or the other way around.

All of these patterns involve two neutrinos in the eV range, and therefore the possibility of a significant hot dark matter component. The two which have the (dominantly) ν_e state around 1 eV could contribute to $\beta\beta_{0\nu}$ if the neutrinos are Majorana. A very small $\langle m_{\nu_e} \rangle$ due to cancellations would suggest near maximal mixing for the solar neutrinos, but this could again be relaxed significantly given all of the uncertainties.

6 Particle Physics Implications: From the Top Down

Almost all extensions of the standard model predict non-zero neutrino mass at some level, often in the observable $10^{-5} - 10$ eV range. It is therefore difficult to infer the underlying physics from the observed neutrino masses. However, the neutrino mass spectrum should be extremely useful for top-down physics; i.e., the predicted neutrino masses and mixings should provide an important test, complementary to, e.g., the sparticle, Higgs, and ordinary fermion spectrum, of any concrete fundamental theory with serious predictive power.

---

[g]There have been several attempts to get by with only three neutrinos. However, attempts to take $\Delta m^2_{\text{solar}} = \Delta m^2_{\text{atm}}$[68] fail because they lead to an unacceptable energy-independent suppression of the solar neutrinos. Similarly, $\Delta m^2_{\text{atm}} = \Delta m^2_{\text{LSND}}$ were marginally compatible with the earlier Kamiokande atmospheric data[69], but do not describe the zenith angle distortions (and lower Δm^2_{atm}) observed by Superkamiokande. There is still a possibility of combining a three neutrino scheme with anomalous interactions[70], which could, e.g., affect the zenith distribution and allow a larger Δm^2_{atm}, or affect the LSND results and allow a lower Δm^2_{LSND}.

Prior to the precision Z-pole measurements at LEP and SLC there were two promising paths for physics beyond the standard model: compositeness at the TeV scale (e.g., dynamical symmetry breaking, composite Higgs, or composite fermions), or unification, which most likely would have led to deviations from the standard model prediction at the few % or few tenths of a % level, respectively. The absence of large deviations [74] strongly supports the unification route, which is the domain of supersymmetry, grand unification, and superstring theory. The implication is that non-zero neutrino masses are most likely not the result of unexpected new physics at the TeV scale, such as by loop effects associated with new ad hoc scalar fields. (They could, however, be due to neutrino-neutralino mixing or loop effects in supersymmetric models with R parity breaking.) Alternatively, they could be associated with new physics at very high energy scales, most likely either seesaw models or higher dimensional operators.

7 Ordinary-Sterile Neutrino Mixing

As discussed in Section 5.1 the combination of solar neutrinos, atmospheric neutrino oscillations, and the LSND results, if confirmed, would most likely imply the mixing of ordinary active neutrinos with one (or more) light sterile neutrinos. One difficulty is that the sterile neutrinos could have been produced in the early universe by the mixings. For the range of mass differences and mixings relevant to LSND and the atmospheric neutrinos, the sterile neutrino would have been produced prior to nucleosynthesis, changing the freezeout temperature for $\nu_e n \leftrightarrow e^- p$ and leading to too much ^4He [75]. However, Foot and Volkas have recently [76] argued that MSW effects involving sterile neutrinos could amplify a small lepton asymmetry, leading to an excess of ν_e compared to $\bar{\nu}_e$, reducing the ^4He. It has also been argued that ordinary-sterile neutrino mixing could facilitate heavy element synthesis by r-processes in the ejecta of neutrino-heated supernova explosions [77,75].

Most extensions of the standard model predict the existence of sterile neutrinos. For example, simple $SO(10)$ and E_6 grand unified theories predict one or two sterile neutrinos per family, respectively. The only real questions are whether the ordinary and sterile neutrinos of the same chirality mix significantly with each other, and whether the mass eigenstate neutrinos are sufficiently light. When there are only Dirac masses, the ordinary and sterile states do not mix because of the conserved lepton number. Pure Majorana masses do not mix the ordinary and sterile sectors either. In the seesaw model the mixing is negligibly small, and the (mainly) sterile eigenstates are too heavy to be relevant to oscillations. The only way to have significant mixing and small mass eigenstates is for the Dirac and Majorana neutrino mass terms to be extremely small and to also be comparable to each other. This appears to require two miracles in conventional models of neutrino mass.

One promising possibility involves the generation of neutrino masses from higher-dimensional operators in theories involving an intermediate scale [7], as described in Section 2. Depending on the intermediate scale and the dimensions of the operators naturally small Dirac and Majorana masses are possible, and in some cases they are automatically of the same order of magnitude [8]. Another interesting possibility [78] involves sterile neutrinos associated with a parallel hidden sector of nature as suggested in some superstring and supergravity theories. Other mechanisms in which one can obtain ordinary-sterile neutrino mixing are described in [79].

8 Conclusions

- Neutrino mass is an important probe of particle physics, astrophysics, and cosmology.

- There are several experimental indications or suggestions: (a) The Superkamiokan and other results on atmospheric neutrinos provide strong evidence for ν_μ oscillations. (b) The combination of solar neutrino experiments implies a global spectral distortion, strongly supporting neutrino transitions or oscillations. The preliminary SuperK results on the ^8B spectrum suggests a spectral distortion, most consistent with vacuum oscillations but possibly with small angle MSW. (c) LSND has candidate events in both decay at rest and decay in flight. The non-observation of candidates by KARMEN is close to being an experimental contradiction, also there is still a small parameter space consistent with both. (d) Mixed dark matter is an interesting hint for eV scale masses, but is not established.

- In the future many solar neutrino experiments and (model independent) observables will be needed to identify the mechanism, determine the neutrino parameters, and simultaneously study the Sun. This program is complicated by possible three neutrino effects, possible sterile neutrinos, and the possibility that there are both neutrino mass effects and nonstandard solar physics (although the latter is constrained by helioseismology). Experiments that are sensitive to the pp and ^7Be neutrinos are needed. Important observables include neutral to charged current ratios, spectral distortions, day-night effects (possibly involving parametric core enhancement), and seasonal variations (especially for vacuum oscillations).

- For the atmospheric neutrinos, we need more detailed spectral and zenith angle information, and the neutral to charged current ratio as a function of the zenith angle. Independent information, including possible ν_τ appearance, for the same parameter range should be forthcoming from long baseline experiments.

- The planned Mini-BOONE experiment at Fermilab should clarify the LSND-KARMEN situation.

- Future cosmic microwave anisotropy experiments and large scale sky surveys should be able to determine whether neutrinos contribute significantly to the dark matter.

- Significant improvements in $\beta\beta_{0\nu}$ would be very powerful probes of the Majorana nature of neutrinos in the mass ranges suggested by the LSND and atmospheric neutrino results.

- Most extensions of the standard model predict nonzero neutrino masses, so it is difficult to determine their origin in a "bottom-up" matter. However, the neutrino spectrum will be a powerful constraint on "top-down" calculations of fundamental models.

- The possibility of mixing between ordinary and light sterile neutrinos should be taken seriously.

ACKNOWLEDGMENTS

This work was supported by U.S. Department of Energy Grant No. DOE-EY-76-02-3071. I am grateful to Naoya Hata for collaborations on the implications of Solar neutrinos.

1. The present article is updated from P. Langacker, proceedings of *Neutrino 98*, Nucl. Phys. Proc. Suppl. **77**, 241 (1999).
2. For detailed reviews, see G. Gelmini and E. Roulet, Rept. Prog. Phys. **58**,1207 (1995); P. Langacker in *Testing The Standard Model*, ed. M. Cvetic and P. Langacker (World, Singapore, 1991) p. 863; B. Kayser, F. Gibrat Debu, and F. Perrier, *The Physics of Massive Neutrinos*, (World Scientific, Singapore, 1989); Report of the *DPF Long Range Study: Neutrino Mass Working Group*, P. Langacker, R. Rameika, and H. Robertson, conveners, Feb. 1995.
3. G. B. Gelmini and M. Roncadelli, *Phys. Lett.* **99B**, 411 (1981); H. Georgi *et al.*, *Nucl. Phys.* **B193**, 297 (1983).

23

4. A. Zee, *Phys. Lett.* **93B**, 389 (1980). For later references, see [2].

5. L. J. Hall and M. Suzuki, Nucl. Phys. **B231**, 419 (1984); T. Banks et al., Phys. Rev. **D52**, 5319 (1995); R. Hempfling, Nucl. Phys. **B478**, 3 (1996); B. de Carlos and P. L. White, Phys. Rev. **D54**, 3427 (1996); A. Yu Smirnov and F. Vissani, Nucl. Phys. **B460**, 37 (1996); R, M. Borzumati et al., Phys. Lett. **B384**, 123 (1996); H. P. Nilles and N. Polonsky, Nucl. Phys. **B499**, 33 (1997); E. Nardi, Phys. Rev. **D55**, 5772 (1997).

6. M. Gell-Mann, P. Ramond, and R. Slansky, in *Supergravity*, ed. F. van Nieuwenhuizen and D. Freedman, (North Holland, Amsterdam, 1979) p. 315; T. Yanagida, *Proc. of the Workshop on Unified Theory and the Baryon Number of the Universe*, KEK, Japan, 1979; S. Weinberg, *Phys. Rev. Lett.* **43**, 1566 (1979).

7. G. Cleaver, M. Cvetic, J. R. Espinosa, L. Everett, P. Langacker, *Phys. Rev.* **D57**, 2701 (1998).

8. P. Langacker, Phys. Rev. **D58**, 093017 (1998).

9. For a general background, see J. N. Bahcall, *Neutrino Astrophysics*, (Cambridge Univ. Press, Cambridge, 1989).

10. K. Lande, proceedings of *Neutrino 98*, Takayama, Japan, June 1998.

11. Y. Suzuki, *ibid.*

12. T. Kirsten, *ibid.*

13. V. N. Gavrin, *ibid.*

14. A. McDonald, *ibid.*

15. L. Oberauer, *ibid.*

16. R. E. Lanou, *ibid.*; R. Raghavan, *ibid.*

17. J. N. Bahcall, S. Basu, M. H. Pinsonneault, *Phys. Lett.* **B433**, 1 (1998).

18. V. Castellani et al., Nucl. Phys. Proc. Suppl. **70**, 301 (1999).

19. S. Degl'Innocenti and B. Ricci, Astropart. Phys. **8**, 293 (1998).

20. D. Gough, *Neutrino 98*.

21. J. N. Bahcall, *Neutrino 98*.

22. E. Adelberger et al., astro-ph/9805121.

23. W. Haxton, *Neutrino 98*.

24. See [25,26] and references theirin.

25. N. Hata and P. Langacker, *Phys. Rev.* **D56**, 6107 (1997).

26. J. N. Bahcall, PI. Krastev, and A. Yu. Smirnov, Phys. Rev. **D58** (1998), hep-ph/9807216.

27. A. Cumming and W. C. Haxton, *Phys. Rev. Lett.* **77**, 4286 (1996).

28. S. L. Glashow and L. M. Krauss, *Phys. Lett.* **B190**, 199 (1987).

29. Q. Y. Liu and S. T. Petcov, *Phys. Rev.* **D56**, 7392 (1997); K. S. Babu, Q. Y. Liu, and A. Yu. Smirnov, *Phys. Rev.* **D57**, 5825 (1998).

30. A. J. Baltz, A. S. Goldhaber, and M. Goldhaber, hep-ph/9806540.

31. For a recent review, see E. Kh. Akhmedov, hep-ph/9705451.
32. S. Pastor, V. B. Semikoz, and J. W. F. Valle, *Phys. Lett.* **B423**, 118 (1998)
33. S. Bergmann and A. Kagan, hep-ph/9803305; G. Brooijmans, hep-ph/9808498.
34. S. L. Glashow et al., *Phys. Rev.* **D56**, 2433 (1997).
35. A. Halprin and C. N. Leung, *Phys. Lett.* **B416**, 361 (1998).
36. J. N. Bahcall, P. I. Krastev, and E. Lisi, *Phys. Rev.* **C55**, 494 (1997); G. L. Fogli, E. Lisi, and D. Montanino, *Phys. Lett.* **B434**, 333 (1998); B. Faid et al., hep-ph/9805293.
37. J. N. Bahcall and P. I. Krastev, *Phys. Lett.* **B436**, 243 (1998); G. Fiorentini et al., astro-ph/9810083.
38. For recent discussions, see J. N. Bahcall and P. I. Krastev, *Phys. Rev.* **C56**, 2839 (1997) and [24].
39. E. Akhmedov, P. Lipari, and M. Lusignoli, *Phys. Lett.* **B300**, 128 (1993); S. T. Petcov, *Phys. Lett.* **B434**, 321 (1998); M. Maris and S. T. Petcov, hep-ph/9803244, Phys. Rev. **D58** (1998); M. Chizhov, M. Maris, and S. T. Petcov, hep-ph/9810501; E. K. Akhmedov, hep-ph/9805272; A. Smirnov, *Neutrino 98*; S. T. Petcov, *ibid.*
40. J. M. Gelb, W.-k. Kwong, and S. P. Rosen, *Phys. Rev. Lett.* **78**, 2296 (1997).
41. S. L. Glashow, P. J. Kernan, and L. M. Krauss, hep-ph/9808470.
42. N. Hata and P. Langacker, *Phys. Rev.* **D52**, 420 (1995).
43. A. B. Balantekin, J. M. Fetter, and F. N. Loreti, *Phys. Rev.* **D54**, 3941 (1996); H. Nunokawa et al., *Nucl. Phys.* **B472**, 495 (1996); C. P. Burgess and D. Michaud, Ann. Phys. (NY) **256**, 1 (1997).
44. P. Bamert, C. P. Burgess, and D. Michaud, *Nucl. Phys.* **B513**, 319 (1998).
45. P. Lipari, T. Stanev, and T. K. Gaisser, Phys. Rev. **D58** (1998); T. K. Gaisser, *Neutrino 98*.
46. References to earlier work may be found in [47].
47. T. Kajita, hep-ex/9810001, *Neutrino 98*; Y. Fukuda et al., *Phys. Rev. Lett.* **81**, 1562 (1998).
48. E. Peterson, *Neutrino 98*.
49. F. Ronga, hep-ex/9810008, *Neutrino 98*. M. Ambrosio et al., *Phys. Lett.* **B434**, 451 (1998).
50. C. Bomporad, *Neutrino 98*; M. Apollonio et al., *Phys. Lett.* **B420**, 397 (1998).
51. A. Piepke, *Neutrino 98*.
52. A. Suzuki, *Neutrino 98*.
53. Articles by S. Wojcicki, F. Pietropaolo, and K. Nishikawa in *Neutrino 98*.

4. J. W. Flanagan, J. G. Learned, and S. Pakvasa, *Phys. Rev.* **D57**, 2649 (1998).
5. Y. Fukuda et al., *Phys. Lett.* **B335**, 237 (1994).
6. E. K. Akhmedov et al., hep-ph/9808270; P. Lipari and M. Lusignoli, Phys. Rev. **D58** (1998), hep-ph/9803440; Q. Y. Liu, S. P. Mikheyev, and A. Yu. Smirnov, hep-ph/9803415; Q. Y. Liu and A. Yu. Smirnov, *Nucl. Phys.* **B524**, 505 (1998); and [39].
7. F. Vissani and A. Yu. Smirnov, *Phys. Lett.* **B432**, 376 (1998); J. G. Learned, S. Pakvasa, and J. L. Stone, *Phys. Lett.* **B435**, 131 (1998); M. Goldhaber and M. Diwan, unpublished.
8. L. J. Hall and H. Murayama, hep-ph/ 9810468.
9. M. C. Gonzalez-Garcia, H. Nunokawa, O. L. G. Peres, and J. W. F. Valle, hep-ph/9807305; O. Yasuda, *Neutrino 98*, and hep-ph/9809205; V. Barger, T. J. Weiler, and K. Whisnant, hep-ph/9807319; G. L. Fogli, E. Lisi, A. Marrone, and G. Scioscia, hep-ph/9808205.
50. See G. C. Branco, M. N. Rebelo, and J. I. Silva-Marcos, hep-ph/9810328, and references theirin.
51. V. Barger, S. Pakvasa, T. J. Weiler, and K. Whisnant, *Phys. Lett.* **B437**, 107 (1998); M. Jezabek and Y. Sumino, hep-ph/9807310; R. N. Mohapatra and S. Nussinov, hep-ph/9809415; and [30,60].
52. H. Fritzsch and Z.-z. Xing, hep-ph/9808272 and 9807234.
53. D. O. Caldwell and R. N. Mohapatra, Phys. Rev. D **48**, 3259 (1993), **50**, 3477 (1994); J. T. Peltoniemi and J. W. F. Valle, Nucl. Phys. B **406**, 409 (1993). For recent discussions, see D. Caldwell, hep-ph/9804367; H. Georgi and S. L. Glashow, hep-ph/9808293.
54. For recent discussions, see J. Primack, Science **280**, 1398 (1998); E. Gawiser and J. Silk, Science **280**, 1405 (1998).
55. H. V. Klapdor-Kleingrothaus, *Neutrino 98*; R. N. Mohapatra, *Neutrino 98*.
56. D. H. White, *Neutrino 98*; C. Athanassopoulos et al., *Phys. Rev. Lett.* **81**, 1774 (1998).
57. K. Eitel et al., hep-ex/9809007, *Neutrino 98*.
58. A. Acker and S. Pakvasa, *Phys. Lett.* **B397**, 209 (1997).
59. C. Cardall and G. Fuller, Phys. Rev. D **53**, 4421 (1996); G. L. Fogli et al., Phys. Rev. D **54**, 3667 (1996), **56** 4365 (1997); K.S. Babu, J. C. Pati, and F. Wilczek, Phys. Lett. B **359**, 351 (1995), erratum: **364**, 251 (1995).
70. E. Ma and P. Roy, *Phys. Rev. Lett.* **80**, 4637 (1998); L. M. Johnson and D. W. McKay, *Phys. Lett.* **B433**, 355 (1998).
71. C. Caso et al., Eur. Phys. J. **C3**,1 (1998).
72. D. O. Caldwell and R. N. Mohapatra[63]; J. T. Peltoniemi and J. W. F. Valle[63]; S. C. Gibbons et al., *Phys. Lett.* **B430**, 296 (1998); Q. Y. Liu and

A. Yu. Smirnov, *Nucl. Phys.* **B524**, 505 (1998); V. Barger, T. J. Weiler and K. Whisnant, *Phys. Lett.* **B427**, 97 (1998); S. M. Bilenkii, C. Giunti and W. Grimus, *Neutrino 98*; E. J. Chun, C. W. Kim, and U. W. Lee, *Phys. Rev.* **D58** (1998), hep-ph/9802209; N. Okada and O. Yasuda, Int. J. Mod. Phys. **A12**,3669 (1997).

73. A. Geiser, CERN-EP-98-056; W. Krolikowski, hep-ph/9808307 and 9808207 D. Suematsu, hep-ph/9808409.

74. For a recent study, see J. Erler and P. Langacker, Eur. Phys. J. **C3**, 9((1998).

75. For a complete set of references, see [8].

76. R. Foot and R. R. Volkas, *Phys. Rev.* **D55**, 5147 (1997), **D56**, 665: (1997); N. F. Bell, R. Foot, and R. R. Volkas, *Phys. Rev.* **D58** (1998) hep-ph/9805259.

77. D. O. Caldwell, G. M. Fuller, and Y.-Z. Qian, in preparation; H. Nunokawa et al., *Phys. Rev.* **D56**, 1704 (1997); J. Peltoniemi, hep-ph/9511323.

78. B. Brahmachari and R. N. Mohapatra, *Phys. Lett.* **B437**, 100 (1998); R. N. Mohapatra, hep-ph/9808236.

79. U. Sarkar, hep-ph/9808277 and 9807466; N. Arkani-Hamed and Y. Grossman, hep-ph/9806223; K. Benakli and A. Yu Smirnov, *Phys. Rev. Lett.* **79**, 4314 (1997); E. Ma, *Phys. Lett.* **B380**, 286 (1996); R. Foot and R. R. Volkas, *Phys. Rev.* **D52**, 6595 (1995); J. T. Peltoniemi and J. W. F. Valle[63]; J. T. Peltoniemi, D. Tommasini, and J. W. F. Valle, *Phys. Lett.* **B298**, 383 (1993).

THE STANDARD ELECTROWEAK THEORY AND BEYOND

G. ALTARELLI

Theoretical Physics Division, CERN
CH-1211 Geneva 23
and
Università di Roma Tre, Rome, Italy
E-mail: gual@mail.cern.ch

CONTENT

1 Introduction

These lectures on electroweak (EW) interactions start with a short summary of the Glashow–Weinberg–Salam theory and then cover in detail some main subjects of present interest in phenomenology.

The modern EW theory inherits the phenomenological successes of the

$(V - A) \otimes (V - A)$ four-fermion low-energy description of weak interactions, and provides a well-defined and consistent theoretical framework including weak interactions and quantum electrodynamics in a unified picture.

As an introduction, we recall some salient physical features of the weak interactions. The weak interactions derive their name from their intensity. At low energy the strength of the effective four-fermion interaction of charged currents is determined by the Fermi coupling constant G_F. For example, the effective interaction for muon decay is given by

$$\mathcal{L}_{\text{eff}} = (G_F/\sqrt{2}) \left[\bar{\nu}_\mu \gamma_\alpha (1 - \gamma_5)\mu\right] \left[\bar{e}\gamma^\alpha (1 - \gamma_5)\nu_e\right] , \tag{1}$$

with [1]

$$G_F = 1.16639(1) \times 10^{-5} \text{ GeV}^{-2} . \tag{2}$$

In natural units $\hbar = c = 1$, G_F has dimensions of $(\text{mass})^{-2}$. As a result, the intensity of weak interactions at low energy is characterized by $G_F E^2$, where E is the energy scale for a given process ($E \approx m_\mu$ for muon decay). Since

$$G_F E^2 = G_F m_p^2 (E/m_p)^2 \simeq 10^{-5} (E/m_p)^2 , \tag{3}$$

where m_p is the proton mass, the weak interactions are indeed weak at low energies (energies of order m_p). Effective four fermion couplings for neutral current interactions have comparable intensity and energy behaviour. The quadratic increase with energy cannot continue for ever, because it would lead to a violation of unitarity. In fact, at large energies the propagator effects can no longer be neglected, and the current–current interaction is resolved into current–W gauge boson vertices connected by a W propagator. The strength of the weak interactions at high energies is then measured by g_W, the $W - \mu - \nu_\mu$ coupling, or, even better, by $\alpha_W = g_W^2/4\pi$ analogous to the fine-structure constant α of QED. In the standard EW theory, we have

$$\alpha_W = \sqrt{2}\, G_F\, \frac{m_W^2}{\pi} = \frac{\alpha}{\sin^2 \theta_W} \cong 1/30 . \tag{4}$$

That is, at high energies the weak interactions are no longer so weak.

The range r_W of weak interactions is very short: it is only with the experimental discovery of the W and Z gauge bosons that it could be demonstrated that r_W is non-vanishing. Now we know that

$$r_W = \frac{\hbar}{m_W c} \simeq 2.5 \times 10^{-16} \text{ cm} , \tag{5}$$

corresponding to $m_W \simeq 80$ GeV. This very large value for the W (or the Z) mass makes a drastic difference, compared with the massless photon and the infinite range of the QED force. The direct experimental limit on the photon mass is[1] $m_\gamma < 2 \ 10^{-16}$ eV. Thus, on the one hand, there is very good evidence that the photon is massless. On the other hand, the weak bosons are very heavy. A unified theory of EW interactions has to face this striking difference.

Another apparent obstacle in the way of EW unification is the chiral structure of weak interactions: in the massless limit for fermions, only left-handed quarks and leptons (and right-handed antiquarks and antileptons) are coupled to W's. This clearly implies parity and charge-conjugation violation in weak interactions.

The universality of weak interactions and the algebraic properties of the electromagnetic and weak currents [the conservation of vector currents (CVC), the partial conservation of axial currents (PCAC), the algebra of currents, etc.] have been crucial in pointing to a symmetric role of electromagnetism and weak interactions at a more fundamental level. The old Cabibbo universality for the weak charged current:

$$J_\alpha^{\text{weak}} = \bar{\nu}_\mu \gamma_\alpha (1 - \gamma_5)\mu + \bar{\nu}_e \gamma_\alpha (1 - \gamma_5)e + \cos\theta_c \ \bar{u}\gamma_\alpha(1 - \gamma_5)d +$$
$$\sin\theta_c \ \bar{u}\gamma_\alpha(1 - \gamma_5)s + \cdots , \tag{6}$$

suitably extended, is naturally implied by the standard EW theory. In this theory the weak gauge bosons couple to all particles with couplings that are proportional to their weak charges, in the same way as the photon couples to all particles in proportion to their electric charges [in Eq. (6), $d' = \cos\theta_c \ d + \sin\theta_c \ s$ is the weak-isospin partner of u in a doublet. The (u, d') doublet has the same couplings as the (ν_e, ℓ) and (ν_μ, μ) doublets].

Another crucial feature is that the charged weak interactions are the only known interactions that can change flavour: charged leptons into neutrinos or up-type quarks into down-type quarks. On the contrary, there are no flavour-changing neutral currents at tree level. This is a remarkable property of the weak neutral current, which is explained by the introduction of the Glashow-Iliopoulos-Maiani mechanism and has led to the successful prediction of charm.

The natural suppression of flavour-changing neutral currents, the separate conservation of e, μ and τ leptonic flavours, the mechanism of CP violation through the phase in the quark-mixing matrix, are all crucial features of the Standard Model. Many examples of new physics tend to break the selection rules of the standard theory. Thus the experimental study of rare flavour-changing transitions is an important window on possible new physics.

In the following sections we shall see how these properties of weak interactions fit into the standard EW theory.

2 Gauge Theories

In this section we summarize the definition and the structure of a gauge Yang–Mills theory. We will list here the general rules for constructing such a theory. Then in the next section these results will be applied to the EW theory.

Consider a Lagrangian density $\mathcal{L}[\phi, \partial_\mu \phi]$ which is invariant under a D dimensional continuous group of transformations:

$$\phi' = U(\theta^A)\phi \qquad (A = 1, 2, ..., D) . \tag{7}$$

For θ^A infinitesimal, $U(\theta^A) = 1 + ig \sum_A \theta^A T^A$, where T^A are the generators of the group Γ of transformations in the (in general reducible) representation of the fields ϕ. Here we restrict ourselves to the case of internal symmetries, so that T^A are matrices that are independent of the space–time coordinates. The generators T^A are normalized in such a way that for the lowest dimensional non-trivial representation of the group Γ (we use t^A to denote the generators in this particular representation) we have

$$\text{tr}(t^A t^B) = \frac{1}{2}\delta^{AB} . \tag{8}$$

The generators satisfy the commutation relations

$$[T^A, T^B] = iC_{ABC}T^C . \tag{9}$$

In the following, for each quantity V^A we define

$$\mathbf{V} = \sum_A T^A V^A . \tag{10}$$

If we now make the parameters θ^A depend on the space–time coordinates $\theta^A = \theta^A(x_\mu)$, $\mathcal{L}[\phi, \partial_\mu \phi]$ is in general no longer invariant under the gauge transformations $U[\theta^A(x_\mu)]$, because of the derivative terms. Gauge invariance is recovered if the ordinary derivative is replaced by the covariant derivative:

$$D_\mu = \partial_\mu + ig\mathbf{V}_\mu , \tag{11}$$

where V_μ^A are a set of D gauge fields (in one-to-one correspondence with the group generators) with the transformation law

$$\mathbf{V}'_\mu = U\mathbf{V}_\mu U^{-1} - (1/ig)(\partial_\mu U)U^{-1} . \tag{12}$$

For constant θ^A, $\mathbf{V}$ reduces to a tensor of the adjoint (or regular) representation of the group:

$$\mathbf{V}'_\mu = U\mathbf{V}_\mu U^{-1} \simeq \mathbf{V}_\mu + ig[\theta, \mathbf{V}_\mu] , \tag{13}$$

which implies that

$$V'^C_\mu = V^C_\mu - gC_{ABC}\theta^A V^B_\mu , \tag{14}$$

where repeated indices are summed up.

As a consequence of Eqs. (11) and (12), $D_\mu\phi$ has the same transformation properties as ϕ:

$$(D_\mu\phi)' = U(D_\mu\phi) . \tag{15}$$

Thus $\mathcal{L}[\phi, D_\mu\phi]$ is indeed invariant under gauge transformations. In order to construct a gauge-invariant kinetic energy term for the gauge fields V^A, we consider

$$[D_\mu, D_\nu]\phi = ig\{\partial_\mu\mathbf{V}_\nu - \partial_\nu\mathbf{V}_\mu + ig[\mathbf{V}_\mu, \mathbf{V}_\nu]\}\phi \equiv ig\mathbf{F}_{\mu\nu}\phi , \tag{16}$$

which is equivalent to

$$F^A_{\mu\nu} = \partial_\mu V^A_\nu - \partial_\nu V^A_\mu - gC_{ABC}V^B_\mu V^C_\nu . \tag{17}$$

From Eqs. (10), (15) and (16) it follows that the transformation properties of $F^A_{\mu\nu}$ are those of a tensor of the adjoint representation

$$\mathbf{F}'_{\mu\nu} = U\mathbf{F}_{\mu\nu}U^{-1} . \tag{18}$$

The complete Yang–Mills Lagrangian, which is invariant under gauge transformations, can be written in the form

$$\mathcal{L}_{\mathrm{YM}} = -\frac{1}{4}\sum_A F^A_{\mu\nu}F^{A\mu\nu} + \mathcal{L}[\phi, D_\mu\phi] . \tag{19}$$

For an Abelian theory, as for example QED, the gauge transformation reduces to $U[\theta(x)] = \exp[ieQ\theta(x)]$, where Q is the charge generator. The associated gauge field (the photon), according to Eq. (12), transforms as

$$V'_\mu = V_\mu - \partial_\mu\theta(x) . \tag{20}$$

In this case, the $F_{\mu\nu}$ tensor is linear in the gauge field V_μ so that in the absence of matter fields the theory is free. On the other hand, in the non-Abelian case the $F_{\mu\nu}^A$ tensor contains both linear and quadratic terms in V_μ^A, so that the theory is non-trivial even in the absence of matter fields.

3 The Standard Model of Electroweak Interactions

In this section, we summarize the structure of the standard EW Lagrangian and specify the couplings of $W^\pm$ and Z, the intermediate vector bosons.

For this discussion we split the Lagrangian into two parts by separating the Higgs boson couplings:

$$\mathcal{L} = \mathcal{L}_{\text{symm}} + \mathcal{L}_{\text{Higgs}} . \tag{21}$$

We start by specifying $\mathcal{L}_{\text{symm}}$, which involves only gauge bosons and fermions:

$$\mathcal{L}_{\text{symm}} = -\frac{1}{4} \sum_{A=1}^{3} F_{\mu\nu}^A F^{A\mu\nu} - \frac{1}{4} B_{\mu\nu} B^{\mu\nu} + \bar{\psi}_L i\gamma^\mu D_\mu \psi_L$$
$$+ \bar{\psi}_R i\gamma^\mu D_\mu \psi_R . \tag{22}$$

This is the Yang–Mills Lagrangian for the gauge group $SU(2) \otimes U(1)$ with fermion matter fields. Here

$$B_{\mu\nu} = \partial_\mu B_\nu - \partial_\nu B_\mu \quad \text{and} \quad F_{\mu\nu}^A = \partial_\mu W_\nu^A - \partial_\nu W_\mu^A - g\epsilon_{ABC} W_\mu^B W_\nu^C \tag{23}$$

are the gauge antisymmetric tensors constructed out of the gauge field B_μ associated with $U(1)$, and W_μ^A corresponding to the three $SU(2)$ generators; ϵ_{ABC} are the group structure constants [see Eqs. (9)] which, for $SU(2)$, coincide with the totally antisymmetric Levi-Civita tensor (recall the familiar angular momentum commutators). The normalization of the $SU(2)$ gauge coupling g is therefore specified by Eq. (23).

The fermion fields are described through their left-hand and right-hand components:

$$\psi_{L,R} = [(1 \mp \gamma_5)/2]\psi, \quad \bar{\psi}_{L,R} = \bar{\psi}[(1 \pm \gamma_5)/2] , \tag{24}$$

with γ_5 and other Dirac matrices defined as in the book by Bjorken–Drell. In particular, $\gamma_5^2 = 1, \gamma_5^\dagger = \gamma_5$. Note that, as given in Eq. (24),

$$\bar{\psi}_L = \psi_L^\dagger \gamma_0 = \psi^\dagger[(1 - \gamma_5)/2]\gamma_0 = \bar{\psi}[\gamma_0(1 - \gamma_5)/2]\gamma_0 = \bar{\psi}[(1 + \gamma_5)/2] .$$

The matrices $P_\pm = (1 \pm \gamma_5)/2$ are projectors. They satisfy the relations $P_\pm P_\pm = P_\pm, P_\pm P_\mp = 0, P_+ + P_- = 1$.

The sixteen linearly independent Dirac matrices can be divided into γ_5-even and γ_5-odd according to whether they commute or anticommute with γ_5. For the γ_5-even, we have

$$\bar{\psi}\Gamma_E\psi = \bar{\psi}_L\Gamma_E\psi_R + \bar{\psi}_R\Gamma_E\psi_L \qquad (\Gamma_E \equiv 1, i\gamma_5, \sigma_{\mu\nu}) , \qquad (25)$$

whilst for the γ_5-odd,

$$\bar{\psi}\Gamma_O\psi = \bar{\psi}_L\Gamma_O\psi_L + \bar{\psi}_R\Gamma_O\psi_R \qquad (\Gamma_O \equiv \gamma_\mu, \gamma_\mu\gamma_5) . \qquad (26)$$

In the Standard Model (SM) the left and right fermions have different transformation properties under the gauge group. Thus, mass terms for fermions (of the form $\bar{\psi}_L\psi_R +$ h.c.) are forbidden in the symmetric limit. In particular, all ψ_R are singlets in the Minimal Standard Model (MSM). But for the moment, by ψ_R we mean a column vector, including all fermions in the theory that span a generic reducible representation of $SU(2) \otimes U(1)$. The standard EW theory is a chiral theory, in the sense that ψ_L and ψ_R behave differently under the gauge group. In the absence of mass terms, there are only vector and axial vector interactions in the Lagrangian that have the property of not mixing ψ_L and ψ_R. Fermion masses will be introduced, together with $W^\pm$ and Z masses, by the mechanism of symmetry breaking. The covariant derivatives $D_\mu\psi_{L,R}$ are explicitly given by

$$D_\mu\psi_{L,R} = \left[\partial_\mu + ig\sum_{A=1}^{3} t^A_{L,R}W^A_\mu + ig'\frac{1}{2}Y_{L,R}B_\mu\right]\psi_{L,R} , \qquad (27)$$

where $t^A_{L,R}$ and $1/2Y_{L,R}$ are the $SU(2)$ and $U(1)$ generators, respectively, in the reducible representations $\psi_{L,R}$. The commutation relations of the $SU(2)$ generators are given by

$$[t^A_L, t^B_L] = i\,\epsilon_{ABC}t^C_L \quad \text{and} \quad [t^A_R, t^B_R] = i\epsilon_{ABC}t^C_R . \qquad (28)$$

We use the normalization (8) [in the fundamental representation of $SU(2)$]. The electric charge generator Q (in units of e, the positron charge) is given by

$$Q = t^3_L + 1/2\,Y_L = t^3_R + 1/2\,Y_R . \qquad (29)$$

Note that the normalization of the $U(1)$ gauge coupling g' in (27) is now specified as a consequence of (29).

All fermion couplings to the gauge bosons can be derived directly from Eqs. (22) and (27). The charged-current (CC) couplings are the simplest. From

$$g(t^1 W^1_\mu + t^2 W^2_\mu) = g\left\{[(t^1 + it^2)/\sqrt{2}](W^1_\mu - iW^2_\mu)/\sqrt{2}] + \text{h.c.}\right\}$$

$$= g\left\{[(t^+ W^-_\mu)/\sqrt{2}] + \text{h.c.}\right\} , \qquad (30)$$

where $t^\pm = t^1 \pm it^2$ and $W^\pm = (W^1 \pm iW^2)/\sqrt{2}$, we obtain the vertex

$$V_{\bar{\psi}\psi W} = g\bar{\psi}\gamma_\mu\left[(t^+_L/\sqrt{2})(1 - \gamma_5)/2 + (t^+_R/\sqrt{2})(1 + \gamma_5)/2\right]\psi W^-_\mu + \text{h.c.} \quad (31)$$

In the neutral-current (NC) sector, the photon A_μ and the mediator Z_μ of the weak NC are orthogonal and normalized linear combinations of B_μ and W^3_μ:

$$A_\mu = \cos\theta_W B_\mu + \sin\theta_W W^3_\mu ,$$
$$Z_\mu = -\sin\theta_W B_\mu + \cos\theta_W W^3_\mu . \qquad (32)$$

Equations (32) define the weak mixing angle θ_W. The photon is characterized by equal couplings to left and right fermions with a strength equal to the electric charge. Recalling Eq. (29) for the charge matrix Q, we immediately obtain

$$g \sin\theta_W = g' \cos\theta_W = e , \qquad (33)$$

or equivalently,

$$\text{tg } \theta_W = g'/g \qquad (34)$$

Once θ_W has been fixed by the photon couplings, it is a simple matter of algebra to derive the Z couplings, with the result

$$\Gamma_{\bar{\psi}\psi Z} = g/(2 \cos\theta_W)\bar{\psi}\gamma_\mu[t^3_L(1 - \gamma_5) + t^3_R(1 + \gamma_5) - 2Q \sin^2\theta_W]\psi Z^\mu , \quad (35)$$

where $\Gamma_{\bar{\psi}\psi Z}$ is a notation for the vertex. In the MSM, $t^3_R = 0$ and $t^3_L = \pm 1/2$.

In order to derive the effective four-fermion interactions that are equivalent, at low energies, to the CC and NC couplings given in Eqs. (31) and (35), we anticipate that large masses, as experimentally observed, are provided for $W^\pm$ and Z by $\mathcal{L}_{\text{Higgs}}$. For left–left CC couplings, when the momentum transfer squared can be neglected with respect to m^2_W in the propagator of Born diagrams with single W exchange, from Eq. (31) we can write

$$\mathcal{L}^{\text{CC}}_{\text{eff}} \simeq (g^2/8m^2_W)[\bar{\psi}\gamma_\mu(1 - \gamma_5)t^+_L\psi][\bar{\psi}\gamma^\mu(1 - \gamma_5)t^-_L\psi] . \qquad (36)$$

By specializing further in the case of doublet fields such as $\nu_e - e^-$ or $\nu_\mu - \mu^-$, we obtain the tree-level relation of g with the Fermi coupling constant G_F measured from μ decay [see Eq. (2)]:

$$\frac{G_F}{\sqrt{2}} = \frac{g^2}{8m_W^2} \ . \tag{37}$$

By recalling that $g \sin\theta_W = e$, we can also cast this relation in the form

$$m_W = \frac{\mu_{\text{Born}}}{\sin\theta_W} \ , \tag{38}$$

with

$$\mu_{\text{Born}} = \left(\frac{\pi\alpha}{\sqrt{2}G_F}\right)^{1/2} \simeq 37.2802 \text{ GeV} \ , \tag{39}$$

where α is the fine-structure constant of QED ($\alpha \equiv e^2/4\pi = 1/137.036$).

In the same way, for neutral currents we obtain in Born approximation from Eq. (35) the effective four-fermion interaction given by

$$\mathcal{L}_{\text{eff}}^{\text{NC}} \simeq \sqrt{2} \ G_F \rho_0 \bar{\psi}\gamma_\mu[...]\psi\bar{\psi}\gamma^\mu[...]\psi \ , \tag{40}$$

where

$$[...] \equiv t_L^3(1 - \gamma_5) + t_R^3(1 + \gamma_5) - 2Q\sin^2\theta_W \tag{41}$$

and

$$\rho_0 = m_W^2/m_Z^2 \ \cos^2\theta_W \ . \tag{42}$$

All couplings given in this section are obtained at tree level and are modified in higher orders of perturbation theory. In particular, the relations between m_W and $\sin\theta_W$ [Eqs. (38) and (39)] and the observed values of ρ ($\rho = \rho_0$ at tree level) in different NC processes, are altered by computable EW radiative corrections, as discussed in Section 6.

The gauge-boson self-interactions can be derived from the $F_{\mu\nu}$ term in $\mathcal{L}_{\text{symm}}$, by using Eq. (32) and $W^\pm = (W^1 \pm iW^2)/\sqrt{2}$. Defining the three-gauge-boson vertex as in Fig. 1, we obtain ($V \equiv \gamma, Z$)

$$\Gamma_{W^-W^+V} = ig_{W^-W^+V}[g_{\mu\nu}(q - p)_\lambda + g_{\mu\lambda}(p - r)_\nu + g_{\nu\lambda}(r - q)_\mu] \ , \tag{43}$$

with

$$g_{W^-W^+\gamma} = g \ \sin\theta_W = e \quad \text{and} \quad g_{W^-W^+Z} = g \ \cos\theta_W \ . \tag{44}$$

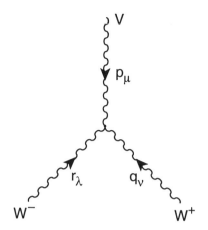

Figure 1: The three-gauge boson vertex: $V = \gamma, Z$

This form of the triple gauge vertex is very special: in general, there could be departures from the above SM expression, even restricting us to $SU(2) \otimes U(1)$ gauge symmetric and C and P invariant couplings. In fact some small corrections are already induced by the radiative corrections. But, in principle, more important could be the modifications induced by some new physics effect. The experimental testing of the triple gauge vertices is presently underway at LEP2 and limits on departures from the SM couplings have also been obtained at the Tevatron and elsewhere (see Section 12).

We now turn to the Higgs sector of the EW Lagrangian. Here we simply review the formalism of the Higgs mechanism applied to the EW theory. In the next section we shall make a more general and detailed discussion of the physics of the EW symmetry breaking. The Higgs Lagrangian is specified by the gauge principle and the requirement of renormalizability to be

$$\mathcal{L}_{\text{Higgs}} = (D_\mu \phi)^\dagger (D^\mu \phi) - V(\phi^\dagger \phi) - \bar{\psi}_L \Gamma \psi_R \phi - \bar{\psi}_R \Gamma^\dagger \psi_L \phi^\dagger , \qquad (45)$$

where ϕ is a column vector including all Higgs fields; it transforms as a reducible representation of the gauge group. The quantities Γ (which include all coupling constants) are matrices that make the Yukawa couplings invariant under the Lorentz and gauge groups. The potential $V(\phi^\dagger \phi)$, symmetric under $SU(2) \otimes U(1)$, contains, at most, quartic terms in ϕ so that the theory is renormalizable:

$$V(\phi^\dagger \phi) = -\frac{1}{2}\mu^2 \phi^\dagger \phi + \frac{1}{4}\lambda (\phi^\dagger \phi)^2 \qquad (46)$$

As discussed in the next section, spontaneous symmetry breaking is induced if the minimum of V which is the classical analogue of the quantum mechanical vacuum state (both are the states of minimum energy) is obtained for non-vanishing ϕ values. Precisely, we denote the vacuum expectation value (VEV) of ϕ, i.e. the position of the minimum, by v:

$$\langle 0|\phi(x)|0\rangle = v \neq 0 . \tag{47}$$

The fermion mass matrix is obtained from the Yukawa couplings by replacing $\phi(x)$ by v:

$$M = \bar{\psi}_L \, \mathcal{M}\psi_R + \bar{\psi}_R \mathcal{M}^\dagger \psi_L , \tag{48}$$

with

$$\mathcal{M} = \Gamma \cdot v . \tag{49}$$

In the MSM, where all left fermions ψ_L are doublets and all right fermions ψ_R are singlets, only Higgs doublets can contribute to fermion masses. There are enough free couplings in Γ, so that one single complex Higgs doublet is indeed sufficient to generate the most general fermion mass matrix. It is important to observe that by a suitable change of basis we can always make the matrix $\mathcal{M}$ Hermitian, γ_5-free, and diagonal. In fact, we can make separate unitary transformations on ψ_L and ψ_R according to

$$\psi'_L = U\psi_L, \quad \psi'_R = V\psi_R \tag{50}$$

and consequently

$$\mathcal{M} \to \mathcal{M}' = U^\dagger \mathcal{M} V . \tag{51}$$

This transformation does not alter the general structure of the fermion couplings in $\mathcal{L}_{\text{symm}}$.

If only one Higgs doublet is present, the change of basis that makes $\mathcal{M}$ diagonal will at the same time diagonalize also the fermion–Higgs Yukawa couplings. Thus, in this case, no flavour-changing neutral Higgs exchanges are present. This is not true, in general, when there are several Higgs doublets. But one Higgs doublet for each electric charge sector i.e. one doublet coupled only to u-type quarks, one doublet to d-type quarks, one doublet to charged leptons would also be all right, because the mass matrices of fermions with different charges are diagonalized separately. For several Higgs doublets in a given charge sector it is also possible to generate CP violation by complex phases in the Higgs couplings. In the presence of six quark flavours, this CP-violation mechanism is not necessary. In fact, at the moment, the simplest model with only one Higgs doublet seems adequate for describing all observed phenomena.

We now consider the gauge-boson masses and their couplings to the Higgs. These effects are induced by the $(D_\mu\phi)^\dagger(D^\mu\phi)$ term in $\mathcal{L}_{\text{Higgs}}$ [Eq. (45)], where

$$D_\mu\phi = \left[\partial_\mu + ig\sum_{A=1}^{3} t^A W_\mu^A + ig'(Y/2)B_\mu\right]\phi .\tag{52}$$

Here t^A and $1/2Y$ are the $SU(2)\otimes U(1)$ generators in the reducible representation spanned by ϕ. Not only doublets but all non-singlet Higgs representations can contribute to gauge-boson masses. The condition that the photon remains massless is equivalent to the condition that the vacuum is electrically neutral:

$$Q|v\rangle = (t^3 + \frac{1}{2}Y)|v\rangle = 0 .\tag{53}$$

The charged W mass is given by the quadratic terms in the W field arising from $\mathcal{L}_{\text{Higgs}}$, when $\phi(x)$ is replaced by v. We obtain

$$m_W^2 W_\mu^+ W^{-\mu} = g^2|(t^+ v/\sqrt{2})|^2 W_\mu^+ W^{-\mu} ,\tag{54}$$

whilst for the Z mass we get [recalling Eq. (32)]

$$\frac{1}{2}m_Z^2 Z_\mu Z^\mu = |[g\cos\theta_W t^3 - g'\sin\theta_W(Y/2)]v|^2 Z_\mu Z^\mu ,\tag{55}$$

where the factor of $1/2$ on the left-hand side is the correct normalization for the definition of the mass of a neutral field. By using Eq. (53), relating the action of t^3 and $1/2Y$ on the vacuum v, and Eqs. (34), we obtain

$$\frac{1}{2}m_Z^2 = (g\cos\theta_W + g'\sin\theta_W)^2|t^3 v|^2 = \left(\frac{g^2}{\cos^2\theta_W}\right)|t^3 v|^2 .\tag{56}$$

For Higgs doublets

$$\phi = \begin{pmatrix} \phi^+ \\ \phi^0 \end{pmatrix} , \quad v = \begin{pmatrix} 0 \\ v \end{pmatrix} ,\tag{57}$$

we have

$$|t^+ v|^2 = v^2, \quad |t^3 v|^2 = 1/4v^2 ,\tag{58}$$

so that

$$m_W^2 = \frac{1}{2}g^2 v^2, \quad m_Z^2 = \frac{1}{2}\frac{g^2 v^2}{\cos^2\theta_W} .\tag{59}$$

Note that by using Eq. (37) we obtain

$$v = 2^{-3/4} G_F^{-1/2} = 174.1 \text{ GeV} .$$ (60)

It is also evident that for Higgs doublets

$$\rho_0 = m_W^2 / m_Z^2 \cos^2 \theta_W = 1 .$$ (61)

This relation is typical of one or more Higgs doublets and would be spoiled by the existence of Higgs triplets etc. In general,

$$\rho_0 = \frac{\sum_i ((t_i)^2 - (t_i^3)^2 + t_i) v_i^2}{\sum_i 2(t_i^3)^2 v_i^2} ,$$ (62)

for several Higgses with VEVs v_i, weak isospin t_i, and z-component t_i^3. These results are valid at the tree level and are modified by calculable EW radiative corrections, as discussed in Section 6.

In the minimal version of the SM only one Higgs doublet is present. Then the fermion–Higgs couplings are in proportion to the fermion masses. In fact, from the Yukawa couplings $g_{\phi \bar{f} f}(\bar{f}_L \phi f_R + h.c.)$, the mass m_f is obtained by replacing ϕ by v, so that $m_f = g_{\phi \bar{f} f} v$. In the minimal SM three out of the four Hermitian fields are removed from the physical spectrum by the Higgs mechanism and become the longitudinal modes of W^+, W^-, and Z. The fourth neutral Higgs is physical and should be found. If more doublets are present, two more charged and two more neutral Higgs scalars should be around for each additional doublet.

The couplings of the physical Higgs H to the gauge bosons can be simply obtained from $\mathcal{L}_{\text{Higgs}}$, by the replacement

$$\phi(x) = \begin{pmatrix} \phi^+(x) \\ \phi^0(x) \end{pmatrix} \rightarrow \begin{pmatrix} 0 \\ v + (H/\sqrt{2}) \end{pmatrix} ,$$ (63)

[so that $(D_\mu \phi)^\dagger (D^\mu \phi) = 1/2(\partial_\mu H)^2 + ...$], with the result

$$\mathcal{L}[H, W, Z] = g^2 \left(\frac{v}{\sqrt{2}} \right) W_\mu^+ W^{-\mu} H + \left(\frac{g^2}{4} \right) W_\mu^+ W^{-\mu} H^2$$
$$+ \left[\frac{g^2 v Z_\mu Z^\mu}{2\sqrt{2} \cos^2 \theta_W} \right] H$$
$$+ \left[\frac{g^2}{8 \cos^2 \theta_W} \right] Z_\mu Z^\mu H^2 .$$ (64)

In the minimal SM the Higgs mass $m_H^2 \sim \lambda v^2$ is of order of the weak scale v. We will discuss in sect. 8 the direct experimental limit on m_H from LEP, which is $m_H \gtrsim m_Z$. We shall also see in sect.12 , that, if there is no physics beyond the SM up to a large scale Λ, then, on theoretical grounds, m_H can only be within a narrow range between 135 and 180 GeV. But the interval is enlarged if there is new physics nearby. Also the lower limit depends critically on the assumption of only one doublet. The dominant decay mode of the Higgs is in the $b\bar{b}$ channel below the WW threshold, while the W^+W^- channel is dominant for sufficiently large m_H. The width is small below the WW threshold, not exceeding a few MeV, but increases steeply beyond the threshold, reaching the asymptotic value of $\Gamma \sim 1/2m_H^3$ at large m_H, where all energies are in TeV.

4   The Higgs Mechanism

The gauge symmetry of the Standard Model was difficult to discover because it is well hidden in nature. The only observed gauge boson that is massless is the photon. The gluons are presumed massless but are unobservable because of confinement, and the W and Z weak bosons carry a heavy mass. Actually a major difficulty in unifying weak and electromagnetic interactions was the fact that e.m. interactions have infinite range ($m_\gamma = 0$), whilst the weak forces have a very short range, owing to $m_{W,Z} \neq 0$.

The solution of this problem is in the concept of spontaneous symmetry breaking, which was borrowed from statistical mechanics.

Consider a ferromagnet at zero magnetic field in the Landau–Ginzburg approximation. The free energy in terms of the temperature T and the magnetization $\mathbf{M}$ can be written as

$$F(\mathbf{M}, T) \simeq F_0(T) + 1/2 \; \mu^2(T)\mathbf{M}^2 + 1/4 \; \lambda(T)(\mathbf{M}^2)^2 + \dots . \tag{65}$$

This is an expansion which is valid at small magnetization. The neglect of terms of higher order in $\vec{M}^2$ is the analogue in this context of the renormalizability criterion. Also, $\lambda(T) > 0$ is assumed for stability; F is invariant under rotations, i.e. all directions of $\mathbf{M}$ in space are equivalent. The minimum condition for F reads

$$\partial F/\partial M = 0, \quad [\mu^2(T) + \lambda(T)\mathbf{M}^2]\mathbf{M} = 0 . \tag{66}$$

There are two cases. If $\mu^2 > 0$, then the only solution is $\mathbf{M} = 0$, there is no magnetization, and the rotation symmetry is respected. If $\mu^2 < 0$, then another solution appears, which is

$$|\mathbf{M}_0|^2 = -\mu^2/\lambda .$$ (67)

The direction chosen by the vector $\mathbf{M}_0$ is a breaking of the rotation symmetry. The critical temperature T_{crit} is where $\mu^2(T)$ changes sign:

$$\mu^2(T_{\text{crit}}) = 0 .$$ (68)

It is simple to realize that the Goldstone theorem holds. It states that when spontaneous symmetry breaking takes place, there is always a zero-mass mode in the spectrum. In a classical context this can be proven as follows. Consider a Lagrangian

$$\mathcal{L} = |\partial_\mu \phi|^2 - V(\phi)$$ (69)

symmetric under the infinitesimal transformations

$$\phi \to \phi' = \phi + \delta\phi, \quad \delta\phi_i = i\delta\theta t_{ij}\phi_j .$$ (70)

The minimum condition on V that identifies the equilibrium position (or the ground state in quantum language) is

$$(\partial V/\partial\phi_i)(\phi_i = \phi_i^0) = 0 .$$ (71)

The symmetry of V implies that

$$\delta V = (\partial V/\partial\phi_i)\delta\phi_i = i\delta\theta(\partial V/\partial\phi_i)t_{ij}\phi_j = 0 .$$ (72)

By taking a second derivative at the minimum $\phi_i = \phi_i^0$ of the previous equation, we obtain

$$\partial^2 V/\partial\phi_k\partial\phi_i(\phi_i = \phi_i^0)t_{ij}\phi_j^0 + \frac{\partial V}{\partial\phi_i}(\phi_i = \phi_i^0)t_{ik} = 0 .$$ (73)

The second term vanishes owing to the minimum condition, Eq. (71). We then find

$$\partial^2 V/\partial\phi_k\partial\phi_i \, (\phi_i = \phi_i^0)t_{ij}\phi_j^0 = 0 .$$ (74)

The second derivatives $M_{ki}^2 = (\partial^2 V/\partial\phi_k\partial\phi_i)(\phi_i = \phi_i^0)$ define the squared mass matrix. Thus the above equation in matrix notation can be read as

$$M^2 t\phi^0 = 0 ,$$ (75)

which shows that if the vector $(t\phi^0)$ is non-vanishing, i.e. there is some generator that shifts the ground state into some other state with the same energy, then $t\phi^0$ is an eigenstate of the squared mass matrix with zero eigenvalue. Therefore, a massless mode is associated with each broken generator.

When spontaneous symmetry breaking takes place in a gauge theory, the massless Goldstone mode exists, but it is unphysical and disappears from the spectrum. It becomes, in fact, the third helicity state of a gauge boson that takes mass. This is the Higgs mechanism. Consider, for example, the simplest Higgs model described by the Lagrangian

$$\mathcal{L} = -\frac{1}{4}\,F_{\mu\nu}^2 + |(\partial_\mu - ieA_\mu)\phi|^2 + \frac{1}{2}\mu^2\phi^*\phi - (\lambda/4)(\phi^*\phi)^2 \ . \tag{76}$$

Note the 'wrong' sign in front of the mass term for the scalar field ϕ, which is necessary for the spontaneous symmetry breaking to take place. The above Lagrangian is invariant under the $U(1)$ gauge symmetry

$$A_\mu \to A_\mu' = A_\mu - (1/e)\partial_\mu\theta(x), \quad \phi \to \phi' = \phi\,\exp[i\theta(x)] \ . \tag{77}$$

Let $\phi^0 = v \neq 0$, with v real, be the ground state that minimizes the potential and induces the spontaneous symmetry breaking. Making use of gauge invariance, we can make the change of variables

$$\phi(x) \to (1/\sqrt{2})[\rho(x) + v]\,\exp[i\zeta(x)/v] \ ,$$
$$A_\mu(x) \to A_\mu - (1/ev)\partial_\mu\zeta(x). \tag{78}$$

Then $\rho = 0$ is the position of the minimum, and the Lagrangian becomes

$$\mathcal{L} = -\frac{1}{4}F_{\mu\nu}^2 + \frac{1}{2}e^2v^2A_\mu^2 + \frac{1}{2}e^2\rho^2A_\mu^2 + e^2\rho vA_\mu^2 + \mathcal{L}(\rho) \ . \tag{79}$$

The field $\zeta(x)$, which corresponds to the would-be Goldstone boson, disappears, whilst the mass term $\frac{1}{2}e^2v^2A_\mu^2$ for A_μ is now present; ρ is the massive Higgs particle.

The Higgs mechanism is realized in well-known physical situations. For a superconductor in the Landau–Ginzburg approximation the free energy can be written as

$$F = F_0 + \frac{1}{2}\mathbf{B}^2 + |(\nabla - 2ie\mathbf{A})\phi|^2/4m - \alpha|\phi|^2 + \beta|\phi|^4 \ . \tag{80}$$

Here $\mathbf{B}$ is the magnetic field, $|\phi|^2$ is the Cooper pair (e^-e^-) density, $2e$ and $2m$ are the charge and mass of the Cooper pair. The 'wrong' sign of α

leads to $\phi \neq 0$ at the minimum. This is precisely the non-relativistic analogue of the Higgs model of the previous example. The Higgs mechanism implies the absence of propagation of massless phonons (states with dispersion relation $\omega = kv$ with constant v). Also the mass term for $\mathbf{A}$ is manifested by the exponential decrease of $\mathbf{B}$ inside the superconductor (Meissner effect).

5   The CKM Matrix

Weak charged currents are the only tree level interactions in the SM that change flavour: by emission of a W an up-type quark is turned into a down-type quark, or a ν_l neutrino is turned into a l^- charged lepton (all fermions are letf-handed). If we start from an up quark that is a mass eigenstate, emission of a W turns it into a down-type quark state d' (the weak isospin partner of u) that in general is not a mass eigenstate. In general, the mass eigenstates and the weak eigenstates do not coincide and a unitary transformation connects the two sets:

$$\begin{pmatrix} d' \\ s' \\ b' \end{pmatrix} = V \begin{pmatrix} d \\ s \\ b \end{pmatrix} \tag{81}$$

V is the Cabibbo-Kobayashi-Maskawa matrix. Thus in terms of mass eigenstates the charged weak current of quarks is of the form:

$$J_\mu^+ \propto \bar{u}\gamma_\mu(1 - \gamma_5)t^+Vd \tag{82}$$

Since V is unitary (i.e. $VV^\dagger = V^\dagger V = 1$) and commutes with T^2, T_3 and Q (because all d-type quarks have the same isospin and charge) the neutral current couplings are diagonal both in the primed and unprimed basis (if the Z down-type quark current is abbreviated as $\bar{d}'\Gamma d'$ then by changing basis we get $\bar{d}V^\dagger \Gamma V d$ and V and Γ commute because, as seen from eq.(41), Γ is made of Dirac matrices and T_3 and Q generator matrices). This is the GIM mechanism that ensures natural flavour conservation of the neutral current couplings at the tree level.

For N generations of quarks, V is a NxN unitary matrix that depends on N^2 real numbers (N^2 complex entries with N^2 unitarity constraints). However, the $2N$ phases of up- and down-type quarks are not observable. Note that an overall phase drops away from the expression of the current in eq.(82), so that only $2N-1$ phases can affect V. In total, V depends on $N^2-2N+1 = (N-1)^2$ real physical parameters. A similar counting gives $N(N-1)/2$ as the number of independent parameters in an orthogonal NxN matrix. This implies that in

V we have $N(N-1)/2$ mixing angles and $(N-1)^2 - N(N-1)/2$ phases: for $N = 2$ one mixing angle (the Cabibbo angle) and no phase, for $N = 3$ three angles and one phase etc.

Given the experimental near diagonal structure of V a convenient parametrisation is the one proposed by Maiani. One starts from the definition:

$$|d'\rangle = c_{13}|d_C\rangle + s_{13}e^{-i\phi}|b\rangle \qquad (83)$$

where $c_{13} \equiv cos\theta_{13}$, $s_{13} \equiv sin\theta_{13}$ (analogous shorthand notations will be used in the following), d_C is the Cabibbo down quark and $\theta_{12} \equiv \theta_C$ is the Cabibbo angle (experimentally $s_{12} \equiv \lambda \sim 0.22$).

$$|d_C\rangle = c_{12}|d\rangle + s_{12}|s\rangle \qquad (84)$$

Note that in a four quark model the Cabibbo angle fixes both the ratio of couplings $(u \to d)/(\nu_e \to e)$ and the ratio of $(u \to d)/(u \to s)$. In a six quark model one has to choose which to keep as a definition of the Cabibbo angle. Here the second definition is taken and, in fact the $u \to d$ coupling is given by $V_{ud} = c_{13}c_{12}$ so that it is no longer specified by θ_{12} only. Also note that we can certainly fix the phases of u, d, s so that a real coefficient appears in front of d_C in eq.(83). We now choose a basis of two orthonormal vectors, both orthogonal to $|d'\rangle$:

$$|s_C\rangle = -s_{12}|d\rangle + c_{12}|s\rangle, \qquad |v\rangle = -s_{13}e^{i\phi}|d_C\rangle + c_{13}|b\rangle \qquad (85)$$

Here $|s_C\rangle$ is the Cabibbo s quark. Clearly s' and b' must be othonormal superpositions of the above base vectors defined in terms of an angle θ_{23}:

$$|s'\rangle = c_{23}|s_C\rangle + s_{23}|v\rangle, \qquad |b'\rangle = -s_{23}|s_C\rangle + c_{23}|v\rangle \qquad (86)$$

The general expression of V_{ij} can be obtained from the above equations. But a considerable notational simplification is gained if one takes into account that from experiment we know that $s_{12} \equiv \lambda$, $s_{23} \sim o(\lambda^2)$ and $s_{13} \sim o(\lambda^3)$ are increasingly small and of the indicated orders of magnitude. Thus, following Wolfenstein one can set:

$$s_{12} \equiv \lambda, \qquad s_{23} = A\lambda^2, \qquad s_{13}e^{-i\phi} = A\lambda^3(\rho - i\eta) \qquad (87)$$

As a result, by neglecting terms of higher order in λ one can write down:

$$V = \begin{bmatrix} V_{ud} & V_{us} & V_{ub} \\ V_{cd} & V_{cs} & V_{cb} \\ V_{td} & V_{ts} & V_{tb} \end{bmatrix} \sim \begin{bmatrix} 1 - \frac{\lambda^2}{2} & \lambda & A\lambda^3(\rho - i\eta) \\ -\lambda & 1 - \frac{\lambda^2}{2} & A\lambda^2 \\ A\lambda^3(1 - \rho - i\eta) & -A\lambda^2 & 1 \end{bmatrix}.$$

$$(88)$$

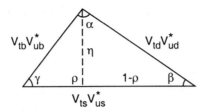

Figure 2: The Bjorken triangle corresponding to eq.(90)

Indicative values of the CKM parameters as obtained from experiment are (a survey of the current status of the CKM parameters can be found in ref.[1]):

$$\lambda = 0.2196 \pm 0.0023$$
$$A = 0.82 \pm 0.04$$
$$\sqrt{\rho^2 + \eta^2} = 0.4 \pm 0.1; \qquad \eta \sim 0.3 \pm 0.2 \qquad (89)$$

In the SM the non vanishing of the η parameter is the only source of CP violation. The most direct and solid evidence for η non vanishing is obtained from the measurement of ϵ in K decay. Unitarity of the CKM matrix V implies relations of the form $\sum_a V_{ba} V_{ca}^* = \delta_{bc}$. In most cases these relations do not imply particularly instructive constraints on the Wolfenstein parameters. But when the three terms in the sum are of comparable magnitude we get interesting information. The three numbers which must add to zero form a closed triangle in the complex plane, with sides of comparable length. This is the case for the t-u triangle (Bjorken triangle) shown in fig.2:

$$V_{td} V_{ud}^* + V_{ts} V_{us}^* + V_{tb} V_{ub}^* = 0 \qquad (90)$$

All terms are of order λ^3. For η=0 the triangle would flatten down to vanishing area. In fact the area of the triangle, J of order $J \sim \eta A^2 \lambda^6$, is the Jarlskog invariant (its value is independent of the parametrization). In the SM all CP violating observables must be proportional to J.

We have only discussed flavour mixing for quarks. But, clearly, if neutrino masses exist, as indicated by neutrino oscillations, then a similar mixing matrix must also be introduced in the leptonic sector (see section 10.2).

6 Renormalisation and Higher Order Corrections

The Higgs mechanism gives masses to the Z, the $W^\pm$ and to fermions while the Lagrangian density is still symmetric. In particular the gauge Ward identities and the conservation of the gauge currents are preserved. The validity of these relations is an essential ingredient for renormalisability. For example the massive gauge boson propagator would have a bad ultraviolet behaviour:

$$W_{\mu\nu} = \frac{-g_{\mu\nu} + \frac{q_\mu q_\nu}{m_W^2}}{q^2 - m_W^2} \qquad (91)$$

But if the propagator is sandwiched between conserved currents J_μ the bad terms in $q_\mu q_\nu$ give a vanishing contribution because $q_\mu J^\mu = 0$ and the high energy behaviour is like for a scalar particle and compatible with renormalisation.

The fondamental theorem that in general a gauge theory with spontaneous symmetry breaking and the Higgs mechanism is renormalisable was proven by 't Hooft. For a chiral theory like the SM an additional complication arises from the existence of chiral anomalies. But this problem is avoided in the SM because the quantum numbers of the quarks and leptons in each generation imply a remarkable (and apparently miracoulous) cancellation of the anomaly, as originally observed by Bouchiat, Iliopoulos and Meyer. In quantum field theory one encounters an anomaly when a symmetry of the classical lagrangian is broken by the process of quantisation, regularisation and renormalisation of the theory. For example, in massless QCD there is no mass scale in the classical lagrangian. Thus one would predict that dimensionless quantities in processes with only one large energy scale Q cannot depend on Q and must be constants. As well known this naive statement is false. The process of regularisation and renormalisation necessarily introduces an energy scale which is essentially the scale where renormalised quantities are defined. For example the renormalised coupling must be defined from the vertices at some scale. This scale μ cannot be zero because of infrared divergences. The scale μ destroys scale invariance because dimensionless quantities can now depend on Q/μ. The famous Λ_{QCD} parameter is a tradeoff of μ and leads to scale invariance breaking. Of direct relevance for the EW theory is the Adler-Bell-Jackiw chiral anomaly. The classical lagrangian of a theory with massless fermions is invariant under a U(1) chiral transformations $\psi' = e^{i\gamma_5\theta}\psi$. The associated axial Noether current is conserved at the classical level. But, at the quantum level, chiral symmetry is broken due to the ABJ anomaly and the current is not conserved. The chiral breaking is introduced by a clash between chiral symmetry, gauge invariance

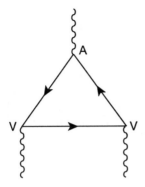

Figure 3: Triangle diagram that generates the ABJ anomaly

and the regularisation procedure. The anomaly is generated by triangular fermion loops with one axial and two vector vertices (fig.3). For neutral currents (Z and γ) the axial coupling is proportional to the 3rd component of weak isospin t_3, while vector couplings are proportional to a linear combination of t_3 and the electric charge Q. Thus in order for the chiral anomaly to vanish all traces of the form $tr\{t_3QQ\}$, $tr\{t_3t_3Q\}$, $tr\{t_3t_3t_3\}$ (and also $tr\{t_+t_-t_3\}$etc., when charged currents are included) must vanish, where the trace is extended over all fermions in the theory that can circulate in the loop. Now all these traces happen to vanish for each fermion family separately. For example take $tr\{t_3QQ\}$. In one family there are, with $t_3 = +1/2$, three colours of up quarks with charge $Q = +2/3$ and one neutrino with $Q = 0$ and, with $t_3 = -1/2$, three colours of down quarks with charge $Q = -1/3$ and one l^- with $Q = -1$. Thus we obtain $tr\{t_3QQ\} = 1/2\ 3\ 4/9 - 1/2\ 3\ 1/9 - 1/2\ 1 = 0$. This impressive cancellation suggests an interplay among weak isospin, charge and colour quantum numbers which appears as a miracle from the point of view of the low energy theory but is more understandable from the point of view of the high energy theory. For example in GUTs there are similar relations where charge quantisation and colour are related: in the 5 of SU(5) we have the content $(d, d, d, e^+, \bar{\nu})$ and the charge generator has a vanishing trace in each SU(5) representation (the condition of unit determinant, represented by the letter S in the SU(5) group name, translates into zero trace for the generators). Thus the charge of d quarks is -1/3 of the positron charge because there are three colours.

Since the SM theory is renormalisable higher order perturbative corrections can be reliably computed. Radiative corrections are very important for precision EW tests. The SM inherits all successes of the old V-A theory of charged currents and of QED. Modern tests focus on neutral current processes, the W mass and the measurement of triple gauge vertices. For Z physics and the W mass the state of the art computation of radiative corrections include the complete one loop diagrams and selected dominant two loop corrections. In addition some resummation techniques are also implemented, like Dyson resummation of vacuum polarisation functions and important renormalisation group improvements for large QED and QCD logarithms. We now discuss in more detail sets of large radiative corrections which are particularly significant [2].

A set of important quantitative contributions to the radiative corrections arise from large logarithms [e.g. terms of the form $(\alpha/\pi \ln (m_Z/m_{f_\ell}))^n$ where f_ℓ is a light fermion]. The sequences of leading and close-to-leading logarithms are fixed by well-known and consolidated techniques (β functions, anomalous dimensions, penguin-like diagrams, etc.). For example, large logarithms dominate the running of α from m_e, the electron mass, up to m_Z. Similarly large logarithms of the form $[\alpha/\pi \ln (m_Z/\mu)]^n$ also enter, for example, in the relation between $\sin^2 \theta_W$ at the scales m_Z (LEP, SLC) and μ (e.g. the scale of low-energy neutral-current experiments). Also, large logs from initial state radiation dramatically distort the line shape of the Z resonance as observed at LEP and SLC and must be accurately taken into account in the measure of the Z mass and total width.

For example, a considerable amount of work has deservedly been devoted to the theoretical study of the Z line-shape. The present experimental accuracy on m_Z obtained at LEP is $\delta m_Z = \pm 2.1$ MeV (see table 1 , sect.7). This small error was obtained by a precise calibration of the LEP energy scale achieved by taking advantage of the transverse polarization of the beams and implementing a sophisticated resonant spin depolarization method. Similarly, a measurement of the total width to an accuracy $\delta\Gamma = \pm 2.4$ MeV has by now been achieved. The prediction of the Z line-shape in the SM to such an accuracy has posed a formidable challenge to theory, which has been successfully met. For the inclusive process $e^+e^- \rightarrow f\bar{f}X$, with $f \neq e$ (for simplicity, we leave Bhabha scattering aside) and X including γ's and gluons, the physical cross-section can be written in the form of a convolution [2]:

$$\sigma(s) = \int_{z_0}^1 dz \, \hat{\sigma}(zs)G(z,s) , \qquad (92)$$

where $\hat{\sigma}$ is the reduced cross-section, and $G(z,s)$ is the radiator function that describes the effect of initial-state radiation; $\hat{\sigma}$ includes the purely weak corrections, the effect of final-state radiation (of both γ's and gluons), and also non-factorizable terms (initial- and final-state radiation interferences, boxes, etc.) which, being small, can be treated in lowest order and effectively absorbed in a modified $\hat{\sigma}$. The radiator $G(z, s)$ has an expansion of the form

$$G(z, s) = \delta(1 - z) + \alpha/\pi(a_{11}L + a_{10}) + (\alpha/\pi)^2(a_{22}L^2 + a_{11}L + a_{20})$$

$$+ \ldots + (\alpha/\pi)^n \sum_{i=0}^{n} a_{ni}L^i , \tag{93}$$

where $L = \ln s/m_e^2 \simeq 24.2$ for $\sqrt{s} \simeq m_Z$. All first- and second-order terms are known exactly. The sequence of leading and next-to-leading logs can be exponentiated (closely following the formalism of structure functions in QCD). For $m_Z \approx 91$ GeV, the convolution displaces the peak by $+110$ MeV, and reduces it by a factor of about 0.74. The exponentiation is important in that it amounts to a shift of about 14 MeV in the peak position.

A very remarkable class of contributions among the one loop EW radiative corrections are those terms that increase quadratically with the top mass. The sensitivity of radiative corrections to m_t arises from the existence of these terms. The quadratic dependence on m_t (and on other possible widely broken isospin multiplets from new physics) arises because, in spontaneously broken gauge theories, heavy loops do not decouple. On the contrary, in QED or QCD, the running of α and α_s at a scale Q is not affected by heavy quarks with mass $M \gg Q$. According to an intuitive decoupling theorem[3] , diagrams with heavy virtual particles of mass M can be ignored at $Q \ll M$ provided that the couplings do not grow with M and that the theory with no heavy particles is still renormalizable. In the spontaneously broken EW gauge theories both requirements are violated. First, one important difference with respect to unbroken gauge theories is in the longitudinal modes of weak gauge bosons. These modes are generated by the Higgs mechanism, and their couplings grow with masses (as is also the case for the physical Higgs couplings). Second the theory without the top quark is no more renormalisable because the gauge symmetry is broken since the doublet (t,b) would not be complete (also the chiral anomaly would not be completely cancelled). With the observed value of m_t the quantitative importance of the terms of order $G_F m_t^2/4\pi^2\sqrt{2}$ is substancial but not dominant (they are enhanced by a factor $m_t^2/m_W^2 \sim 5$ with respect to ordinary terms). Both the large logarithms and the $G_F m_t^2$ terms have a simple structure and are to a large extent universal, i.e. common to a wide class of processes. In particular the $G_F m_t^2$ terms appear in vacuum polarisation diagrams which are universal and in the $Z \to b\bar{b}$ vertex which is not (this vertex is connected with the top quark which runs in the loop, while other types of heavy particles could in principle also contribute to vacuum polarisation diagrams). Their study is important for an understanding of the pattern of radiative corrections. One can also derive approximate formulae (e.g. improved Born approximations), which can be useful in cases where a limited precision may be adequate. More in general, another very important consequence of non decoupling is that precision tests of the electroweak theory may be sensitive to new physics even if the new particles are too heavy for their direct production.

While radiative corrections are quite sensitive to the top mass, they are unfortunately much less dependent on the Higgs mass. If they were sufficiently sensitive by now we would precisely know the mass of the Higgs. But the dependence of one loop diagrams on m_H is only logarithmic: $\sim G_F m_W^2 \log(m_H^2/m_W^2)$. Quadratic terms $\sim G_F^2 m_H^2$ only appear at two loops and are too small to be important. The difference with the top case is that the difference $m_t^2 - m_b^2$ is a direct breaking of the gauge symmetry that already affects the one loop corrections, while the Higgs couplings are "custodial" $SU(2)$ symmetric in lowest order.

The basic tree level relations:

$$\frac{g^2}{8m_W^2} = \frac{G_F}{\sqrt{2}}, \qquad g^2 \sin^2 \theta_W = e^2 = 4\pi\alpha \tag{94}$$

can be combined into

$$\sin^2 \theta_W = \frac{\pi\alpha}{\sqrt{2}G_F m_W^2} \tag{95}$$

A different definition of $\sin^2 \theta_W$ is from the gauge boson masses:

$$\frac{m_W^2}{m_Z^2 \cos^2 \theta_W} = \rho_0 = 1 \implies \sin^2 \theta_W = 1 - \frac{m_W^2}{m_Z^2} \tag{96}$$

where $\rho_0 = 1$ assuming that there are only Higgs doublets. The last two relations can be put into the convenient form

$$(1 - \frac{m_W^2}{m_Z^2})\frac{m_W^2}{m_Z^2} = \frac{\pi\alpha}{\sqrt{2}G_F m_Z^2} \tag{97}$$

These relations are modified by radiative corrections:

$$(1 - \frac{m_W^2}{m_Z^2})\frac{m_W^2}{m_Z^2} = \frac{\pi\alpha(m_Z)}{\sqrt{2}G_F m_Z^2} \frac{1}{1 - \Delta r_W}$$

$$\frac{m_W^2}{m_Z^2 \cos^2 \theta_W} = 1 + \rho_m \tag{98}$$

In the first relation the replacement of α with the running coupling at the Z mass $\alpha(m_Z)$ makes Δr_W completely determined by the purely weak corrections. This relation defines Δr_W unambigously, once the meaning of $\alpha(m_Z)$ is specified. On the contrary, in the second relation $\Delta\rho_m$ depends on the definition of $\sin^2\theta_W$ beyond the tree level. For LEP physics $\sin^2\theta_W$ is usually defined from the $Z \to \mu^+\mu^-$ effective vertex. At the tree level we have:

$$Z \to f^+f^- = \frac{g}{2\cos\theta_W}\bar{f}\gamma_\mu(g_V^{f^*} - g_A^f\gamma_5)f \tag{99}$$

with $g_A^{f2} = 1/4$ and $g_V^f/g_A^f = 1 - 4|Q_f|\sin^2\theta_W$. Beyond the tree level a corrected vertex can be written down in the same form of eq.(99) in terms of modified effective couplings. Then $\sin^2\theta_W \equiv \sin^2\theta_{eff}$ is in general defined through the muon vertex:

$$g_V^\mu/g_A^\mu = 1 - 4\sin^2\theta_{eff}$$

$$\sin^2\theta_{eff} = (1 + \Delta k)s_0^2, \qquad s_0^2c_0^2 = \frac{\pi\alpha(m_Z)}{\sqrt{2}G_Fm_Z^2}$$

$$g_A^{\mu2} = \frac{1}{4}(1 + \Delta\rho) \tag{100}$$

Actually, since in the SM lepton universality is only broken by masses and is in agreement with experiment within the present accuracy, in practice the muon channel is replaced with the average over charged leptons.

We end this discussion by writing a symbolic equation that summarises the status of what has been computed up to now for the radiative corrections Δr_W, $\Delta\rho$ and Δk:

$$\Delta r_W, \Delta\rho, \Delta k = g^2\frac{m_t^2}{m_W^2}(1+\alpha_s+\alpha_s^2)+g^2(1+\alpha_s+\sim\alpha_s^2)+g^4\frac{m_t^4}{m_W^4}+g^4\frac{m_t^2}{m_W^2}+\cdots \tag{101}$$

The meaning of this relation is that the one loop terms of order g^2 are completely known, together with their first order QCD corrections (the second order QCD corrections are only estimated for the g^2 terms not enhanced by m_t^2/m_W^2), and the terms of order g^4 enhanced by the ratios m_t^4/m_W^4 or m_t^2/m_W^2 are also known.

In recent years new powerful tests of the SM have been performed mainly at LEP but also at SLC and at the Tevatron. The running of LEP1 was terminated in 1995 and close-to-final results of the data analysis are now available. The SLC is still running. The experiments at the Z resonance have enormously improved the accuracy in the electroweak neutral current sector. The top quark has been at last found at the Tevatron and the errors on m_Z and $\sin^2 \theta_{eff}$ went down by two and one orders of magnitude respectively since the start of LEP in 1989. The LEP2 programme is in progress. The validity of the SM has been confirmed to a level that we can say was unexpected at the beginning. In the present data there is no significant evidence for departures from the SM, no convincing hint of new physics (also including the first results from LEP2). The impressive success of the SM poses strong limitations on the possible forms of new physics. Favoured are models of the Higgs sector and of new physics that preserve the SM structure and only very delicately improve it, as is the case for fundamental Higgs(es) and Supersymmetry. Disfavoured are models with a nearby strong non perturbative regime that almost inevitably would affect the radiative corrections, as for composite Higgs(es) or technicolor and its variants.

7   Status of the Data

The relevant electro-weak data together with their SM values are presented in table 1 [4,5,6]. The SM predictions correspond to a fit of all the available data (including the directly measured values of m_t and m_W) in terms of m_t, m_H and $\alpha_s(m_Z)$, described later in sect.8, table 4.

Other important derived quantities are, for example, N_ν the number of light neutrinos, obtained from the invisible width: $N_\nu = 2.994(11)$, which shows that only three fermion generations exist with $m_\nu < 45 \; GeV$. This is one of the most important results of LEP. Other important quantities are the leptonic width Γ_l, averaged over e, μ and τ: $\Gamma_l = 83.90(10) MeV$ and the hadronic width $\Gamma_h = 1742.3(2.3) MeV$.

For indicative purposes, in table the "pulls" are also shown, defined as: pull = (data point- fit value)/(error on data point). At a glance we see that the agreement with the SM is quite good. The distribution of the pulls is statistically normal. The presence of a few $\sim 2\sigma$ deviations is what is to be expected. However it is maybe worthwhile to give a closer look at these small discrepancies

One persistent feature of the data is the difference between the values of $\sin^2 \theta_{eff}$ measured at LEP and at SLC (although the discrepancy is going down in the most recent data). The value of $\sin^2 \theta_{eff}$ is obtained from a set of combined asymmetries. From asymmetries one derives the ratio $x = g_V^l/g_A^l$ of the vector and axial vector couplings of the Z, averaged over the charged leptons. In turn $\sin^2 \theta_{eff}$ is defined by $x = 1 - 4\sin^2 \theta_{eff}$. SLD obtains x from the single measurement of A_{LR}, the left-right asymmetry, which requires longitudinally polarized beams. The distribution of the present measurements of $\sin^2 \theta_{eff}$ is shown in fig.4. The LEP average, $\sin^2 \theta_{eff} = 0.23187(24)$, differs by 2.2σ from the SLD value $\sin^2 \theta_{eff} = 0.23101(31)$. The most precise individual measurement at LEP is from A_b^{FB}: the combined LEP error on this quantity is comparable to the SLD error, but the two values are 2.5σ's away. One might attribute this to the fact that the b measurement is more delicate and affected by a complicated systematics. In fact one notices from fig.4 that the value obtained at LEP from A_l^{FB}, the average for l=e, μ and τ, is somewhat low (indeed quite in agreement with the SLD value). However the statement that LEP and SLD agree on leptons while they only disagree when the b quark is considered is not quite right. First, the low value of $\sin^2 \theta_{eff}$ found at LEP from A_l^{FB} turns out to be entirely due to the τ lepton channel which leads to a central value different than that of e and μ. The e and μ asymmetries, which are experimentally simpler, are perfectly on top of the SM fit. Second, if we take only e and μ asymmetries at LEP and disregard the b and τ measurements the LEP average becomes $\sin^2 \theta_{eff} = 0.23168(36)$, which is still 1.4σ away from the SLD value. Thus it is difficult to find a simple explanation for the SLD-LEP discrepancy on $\sin^2 \theta_{eff}$. In the following we will tentatively use the official average

$$\sin^2 \theta_{eff} = 0.23155 \pm 0.00019 \qquad (102)$$

obtained by a simple combination of the LEP-SLC data. One could be more conservative and enlarge the error because of the larger dispersion, but the difference would not be too large. Also, this dispersion has decreased in the most recent data. The data-taking by the SLD experiment is still in progress and also at LEP seizable improvements on A_τ and A_b^{FB} are foreseen as soon as the corresponding analyses will be completed. We hope to see the difference to be further reduced in the end.

From the above discussion one may wonder if there is evidence for something special in the τ channel, or equivalently if lepton universality is really supported by the data. Indeed this is the case: the hint of a difference in A_τ^{FB} with respect to the corresponding e and μ asymmetries is not confirmed by the measurements of A_τ and Γ_τ which appear normal. In principle the fact that an anomaly shows up in A_τ^{FB} and not in A_τ and Γ_τ is not unconceivable because the FB lepton asymmetries are very small and very precisely measured. For example, the extraction of A_τ^{FB} from the data on the angular distribution of τ's could be biased if the imaginary part of the continuum was altered by some

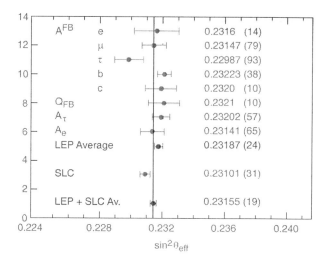

Figure 4: A summary of $\sin^2 \theta_{eff}$ measurements)

non universal new physics effect. But a more trivial experimental problem is at the moment the most plausible option.

A similar question can be asked for the b couplings. We have seen that the measured value of A_b^{FB} is 1.8σ's below the SM fit. At the same time R_b which used to show a major discrepancy is now only about 1σ's away from the SM fit (as a result of the more sophisticated second generation experimental techniques). There is a -2.5σ deviation on the measured value of A_b vs the SM expectation. That somewhat depends on how the data are combined. Let us discuss this point in detail. A_b can be measured directly at SLC by taking advantage of the beam longitudinal polarization. The SLC value (see Table 1 is 2.2σ's below the SM value. At LEP one measures $A_b^{FB} = 3/4\ A_e A_b$. One can then derive A_b by inserting a value for A_e. The question is what to use for A_e: the LEP value obtained, using lepton universality, from the measurements of A_l^{FB}, A_τ, A_e: $A_e = 0.1470(27)$, or the combination of LEP and SLD etc Since we are here concerned with the b couplings it is perhaps wiser to obtain A_b from LEP by using the SM value for A_e (that is the pull-zero value of table 1): $A_e^{SM} = 0.1467(15)$. With the value of A_b derived in this way from LEP (which is 1.7σ's below the SM value) we finally obtain

$$A_b = 0.890 \pm 0.018 \quad (\text{LEP} + \text{SLD}, A_e = A_e^{SM} : -2.5\sigma) \quad (103)$$

Table 1: Data on precision electroweak tests

Quantity	Data (August'98)	Pull
m_Z (GeV)	91.1867(21)	0.1
Γ_Z (GeV)	2.4939(24)	-0.8
σ_h (nb)	41.491(58)	0.3
R_h	20.765(26)	0.7
R_b	0.21656(74)	0.9
R_c	0.1735(44)	0.3
A_{FB}^l	0.01683(96)	0.7
A_τ	0.1431(45)	-0.8
A_e	0.1479(51)	0.25
A_{FB}^b	0.0990(21)	-1.8
A_{FB}^c	0.0709(44)	-0.6
A_b (SLD direct)	0.867(35)	-1.9
A_c (SLD direct)	0.647(40)	-0.5
$\sin^2 \theta_{eff}$ (LEP-combined)	0.23187(24)	1.3
$A_{LR} \rightarrow \sin^2 \theta_{eff}$	0.23101(31)	-1.8
m_W (GeV) (LEP2+$p\bar{p}$)	80.39(60)	-0.4
$1 - \frac{m_W^2}{m_Z^2}$ (νN)	0.2253(21)	1.1
Q_W (Atomic PV in Cs)	-72.11(93)	1.2
m_t (GeV)	173.8(5.0)	0.5

In the SM A_b is so close to 1 because the b quark is almost purely left-handed. A_b only depends on the ratio $r = (g_R/g_L)^2$ which in the SM is small: $r \sim 0.033$. To adequately decrease A_b from its SM value one must increase r by a factor of about 1.6, which appears large for a new physics effect. Also such a large change in r must be compensated by decreasing g_L^2 by a small but fine-tuned amount in order to counterbalance the corresponding large positive shift in R_b. In view of this the most likely way out is that A_b^{FB} and A_b have been a bit underestimated at LEP and actually there is no anomaly in the b couplings. Then the LEP value of $\sin^2 \theta_{eff}$ would slightly move down, in the direction of decreasing the SLD-LEP discrepancy.

8 Precision Electroweak Data and the Standard Model

For the analysis of electroweak data in the SM one starts from the input parameters: some of them, α, G_F and m_Z, are very well measured, some other ones, $m_{f_{light}}$, m_t and $\alpha_s(m_Z)$ are only approximately determined while m_H is largely unknown. With respect to m_t the situation has much improved since the CDF/D0 direct measurement of the top quark mass. From the input parameters one computes the radiative corrections to a sufficient precision to match the experimental capabilities. Then one compares the theoretical predictions and the data for the numerous observables which have been measured, checks the consistency of the theory and derives constraints on m_t, $\alpha_s(m_Z)$ and hopefully also on m_H.

Some comments on the least known of the input parameters are now in order. The only practically relevant terms where precise values of the light quark masses, $m_{f_{light}}$, are needed are those related to the hadronic contribution to the photon vacuum polarisation diagrams, that determine $\alpha(m_Z)$. This correction is of order 6%, much larger than the accuracy of a few per mille of the precision tests. Fortunately, one can use the actual data to in principle solve the related ambiguity. But we shall see that the left over uncertainty is still one of the main sources of theoretical error. As is well known [2], the QED running coupling is given by:

$$\alpha(s) = \frac{\alpha}{1 - \Delta\alpha(s)} \tag{104}$$

$$\Delta\alpha(s) = \Pi(s) = \Pi_\gamma(0) - \mathrm{Re}\Pi_\gamma(s) \tag{105}$$

where $\Pi(s)$ is proportional to the sum of all 1-particle irreducible vacuum polarization diagrams. In perturbation theory $\Delta\alpha(s)$ is given by:

$$\Delta\alpha(s) = \frac{\alpha}{3\pi} \sum_f Q_f^2 N_{Cf} \left(\log \frac{2}{m_f^2} - \frac{5}{3} \right) \tag{106}$$

where $N_{Cf} = 3$ for quarks and 1 for leptons. However, the perturbative formula is only reliable for leptons, not for quarks (because of the unknown values of the effective quark masses). Separating the leptonic, the light quark and the top quark contributions to $\Delta\alpha(s)$ we have:

$$\Delta\alpha(s) = \Delta\alpha(s)_\ell + \Delta\alpha(s)_h + \Delta\alpha(s)_t \tag{107}$$

with:

$$\Delta\alpha(s)_\ell = 0.0331421 \; ; \; \Delta\alpha(s)_t = \frac{\alpha}{3\pi} \frac{4}{15} \frac{m_Z^2}{m_t^2} = -0.000061 \tag{108}$$

Note that in QED there is decoupling so that the top quark contribution approaches zero in the large m_t limit. For $\Delta\alpha(s)_h$ one can use eq.(105) and the Cauchy theorem to obtain the representation:

$$\Delta\alpha(m_Z^2)_h = -\frac{\alpha m_Z^2}{3\pi} \mathrm{Re} \int_{4m_\pi^2}^\infty \frac{ds}{s} \frac{R(s)}{s - m_Z^2 - i\epsilon} \tag{109}$$

where $R(s)$ is the familiar ratio of the hadronic to the pointlike $\ell^+\ell^-$ cross-section from photon exchange in e^+e^- annihilation. At s large one can use the perturbative expansion for $R(s)$ while at small s one can use the actual data. In recent years there has been a lot of activity on this subject and a number of independent new estimates of $\alpha(m_Z)$ have appeared in the literature [7]. A consensus has been established and the value used at present is

$$\alpha(m_Z)^{-1} = 128.90 \pm 0.09 \qquad (110)$$

As I said, for the derivation of this result th QCD theoretical prediction is actually used for large values of s where the data do not exist. But the sensitivity of the dispersive integral to this region is strongly suppressed, so that no important model dependence is introduced. More recently some analyses have appeared where one studied by how much the error on $\alpha_s(m_Z)$ is reduced by using the QCD prediction down to $\sqrt{s} = m_\tau$, with the possible exception of the regions around the charm and beauty thresholds [8]. These attempts were motivated by the apparent success of QCD predictions in τ decays, despite the low τ mass (note however that the relevant currents are V-A in τ decay but V in the present case). One finds that the central value is not much changed while the error in eq.(110) is reduced from 0.09 down to something like 0.03-0.04, but, of course, at the price of more model dependence. For this reason, in the following, we shall use the more conservative value in eq.(110).

As for the strong coupling $\alpha_s(m_Z)$ the world average central value is by now quite stable. The error is going down because the dispersion among the different measurements is much smaller in the most recent set of data. The most important determinations of $\alpha_s(m_Z)$ are summarised in table 2 [9]. For all entries, the main sources of error are the theoretical ambiguities which are larger than the experimental errors. The only exception is the measurement from the electroweak precision tests, but only if one assumes that the SM electroweak sector is correct. My personal views on the theoretical errors are reflected in the table 2. The error on the final average is taken by all authors between ± 0.003 and ± 0.005 depending on how conservative one is. Thus in the following our reference value will be

$$\alpha_s(m_Z) = 0.119 \pm 0.004 \qquad (111)$$

Finally a few words on the current status of the direct measurement of m_t. The present combined CDF/D0 result is

$$m_t = 173.8 \pm 5.0 \ GeV \qquad (112)$$

Table 2: Measurements of $\alpha_s(m_Z)$. In parenthesis we indicate if the dominant source of errors is theoretical or experimental. For theoretical ambiguities our personal figure of merit is given.

Measurements	$\alpha_s(m_Z)$	
R_τ	0.122 ± 0.006	(Th)
Deep Inelastic Scattering	0.116 ± 0.005	(Th)
Y_{decay}	0.112 ± 0.010	(Th)
Lattice QCD	0.117 ± 0.007	(Th)
$Re^+e^-(\sqrt{s} < 62 \text{ GeV})$	0.124 ± 0.021	(Exp)
Fragmentation functions in e^+e^-	0.124 ± 0.012	(Th)
Jets in e^+e^- at and below the Z	0.121 ± 0.008	(Th)
Z line shape (Assuming SM)	0.120 ± 0.004	(Exp)

The error is so small by now that one is approaching a level where a more careful investigation of the effects of colour rearrangement on the determination of m_t is needed. One wants to determine the top quark mass, defined as the invariant mass of its decay products (i.e. b+W+ gluons + γ's). However, due to the need of colour rearrangement, the top quark and its decay products cannot be really isolated from the rest of the event. Some smearing of the mass distribution is induced by this colour crosstalk which involves the decay products of the top, those of the antitop and also the fragments of the incoming (anti)protons. A reliable quantitative computation of the smearing effect on the m_t determination is difficult because of the importance of non perturbative effects. An induced error of the order of 1 GeV on m_t could reasonably be expected. Thus further progress on the m_t determination demands tackling this problem in more depth.

In order to appreciate the relative importance of the different sources of theoretical errors for precision tests of the SM, we report in table 3 a comparison for the most relevant observables. What is important to stress is that the ambiguity from m_t, once by far the largest one, is by now smaller than the error from m_H. We also see from table 3 that the error from $\Delta\alpha(m_Z)$ is expecially important for $\sin^2\theta_{eff}$ and, to a lesser extent, is also sizeable for Γ_Z and ϵ_3.

The most important recent advance in the theory of radiative corrections is the calculation of the $o(g^4 m_t^2/m_W^2)$ terms in $\sin^2\theta_{eff}$, m_W and, more recently in $\delta\rho$[10]. The result implies a small but visible correction to the predicted values but expecially a seizable decrease of the ambiguity from scheme dependence (a typical effect of truncation). These callculations are now implemented in the fitting codes used in the analysis of LEP data. The fitted value of the Higgs

Table 3: Errors from different sources: Δ_{now}^{exp} is the present experimental error; $\Delta\alpha^{-1}$ is the impact of $\Delta\alpha^{-1} = \pm0.09$; Δ_{th} is the estimated theoretical error from higher orders; Δm_t is from $\Delta m_t = \pm6\,\mathrm{GeV}$; Δm_H is from $\Delta m_H = 60\text{--}1000$ GeV; $\Delta\alpha_s$ corresponds to $\Delta\alpha_s = \pm0.003$. The epsilon parameters are defined in sect.9.1.

Parameter	Δ_{now}^{exp}	$\Delta\alpha^{-1}$	Δ_{th}	Δm_t	Δm_H	$\Delta\alpha_s$
Γ_Z (MeV)	±2.4	±0.7	±0.8	±1.4	±4.6	±1.7
σ_h (pb)	58	1	4.3	3.3	4	17
$R_h \cdot 10^3$	26	4.3	5	2	13.5	20
Γ_l (keV)	100	11	15	55	120	3.5
$A_{FB}^l \cdot 10^4$	9.6	4.2	1.3	3.3	13	0.18
$\sin^2\theta \cdot 10^4$	19	2.3	0.8	1.9	7.5	0.1
m_W (MeV)	60	12	9	37	100	2.2
$R_b \cdot 10^4$	7.4	0.1	1	2.1	0.25	0
$\epsilon_1 \cdot 10^3$	1.2		~0.1			0.2
$\epsilon_3 \cdot 10^3$	1.2	0.5	~0.1			0.12
$\epsilon_b \cdot 10^3$	2.1		~0.1			1

mass is lowered by about 30 GeV due to this effect.

We now discuss fitting the data in the SM. As the mass of the top quark is now rather precisely known from CDF and D0 one must distinguish two different types of fits. In one type one wants to answer the question: is m_t from radiative corrections in agreement with the direct measurement at the Tevatron? Similarly how does m_W inferred from radiative corrections compare with the direct measurements at the Tevatron and LEP2? For answering these interesting but somewhat limited questions, one must clearly exclude the measurements of m_t and m_W from the input set of data. Fitting all other data in terms of m_t, m_H and $\alpha_s(m_Z)$ one finds the results shown in the second column of table 4[5]. The extracted value of m_t is typically a bit too low. For example, as shown in the table 4, from all the electroweak data except the direct production results on m_t and m_W, one finds $m_t = 158 \pm 9\,\mathrm{GeV}$. There is a strong correlation between m_t and m_H. $\sin^2\theta_{eff}$ and m_W drive the fit to small values of m_H. Then, at small m_H the widths, in particular the leptonic width (whose prediction is nearly independent of α_s) drive the fit to small m_t. In a more general type of fit, e.g. for determining the overall consistency of the SM or the best present estimate for some quantity, say m_W, one should of course not ignore the existing direct determinations of m_t and m_W. Then, from all the available data, by fitting m_t, m_H and $\alpha_s(m_Z)$ one finds the values shown in the last column of table 4.

This is the fit also referred to in table 1. The corresponding fitted values

Table 4: Standard Model fits of electroweak data.

Parameter	LEP(incl.m_W)	All but m_W, m_t	All Data
m_t (GeV)	160+13 − 10	158+9 − 8	171.3 ± 4.9
m_H (GeV)	66+142 − 38	34+45 − 16	84+91 − 51
$log[m_H(GeV)]$	1.82+0.50 − 0.37	1.53+0.37 − 0.28	1.92+0.32 − 0.41
$\alpha_s(m_Z)$	0.121 ± 0.003	0.120 ± 0.003	0.119 ± 0.003
χ^2/dof	4.2/9	14/12	16.4/15

of $\sin^2 \theta_{eff}$ and m_W are:

$$\sin^2 \theta_{eff} = 0.23156 \pm 0.00019; \quad m_W = 80.370 \pm 0.027 \; GeV \qquad (113)$$

The fitted value of $\sin^2 \theta_{eff}$ is practically identical to the LEP+SLD average. The error of 27 MeV on m_W clearly sets up a goal for the direct measurement of m_W at LEP2 and the Tevatron.

As a final comment we want to recall that the radiative corrections are functions of $log(m_H)$. It is truly remarkable that the fitted value of $log(m_H)$ is found to fall right into the very narrow allowed window around the value 2 specified by the lower limit from direct searches, $m_H >\sim 90 \; GeV$, and the theoretical upper limit in the SM $m_H < 600 - 800 \; GeV$ (see later). Note that if the Higgs is removed from the theory, $\log m_H \to \log \Lambda$ + constant, where Λ is a cutoff or the scale of the new physics that replaces the Higgs. The control of the finite terms is lost. Thus the fact that from experiment, one finds $\log m_H \sim 2$ is a strong argument in favour of the precise form of the Higgs mechanism as in the SM. The fulfilment of this very stringent consistency check is a beautiful argument in favour of a fundamental Higgs (or one with a compositeness scale much above the weak scale).

9 A More General Analysis of Electroweak Data

We now discuss an update of the epsilon analysis [6] which is a method to look at the data in a more general context than the SM. The starting point is to isolate from the data that part which is due to the purely weak radiative corrections. In fact the epsilon variables are defined in such a way that they are zero in the approximation when only effects from the SM Λat the tree level plus pure QED and pure QCD corrections are taken into account. This very simple version of improved Born approximation is a good first approximation according to the data and is independent of m_t and m_H. In fact the whole m_t

and m_H dependence arises from weak loop corrections and therefore is only contained in the epsilon variables. Thus the epsilons are extracted from the data without need of specifying m_t and m_H. But their predicted value in the SM or in any extension of it depend on m_t and m_H. This is to be compared with the competitor method based on the S, T, U variables. The latter cannot be obtained from the data without specifying m_t and m_H because they are defined as deviations from the complete SM prediction for specified m_t and m_H. Of course there are very many variables that vanish if pure weak loop corrections are neglected, at least one for each relevant observable. Thus for a useful definition we choose a set of representative observables that are used to parametrize those hot spots of the radiative corrections where new physics effects are most likely to show up. These sensitive weak correction terms include vacuum polarization diagrams which being potentially quadratically divergent are likely to contain all possible non decoupling effects (like the quadratic top quark mass dependence in the SM). There are three independent vacuum polarization contributions. In the same spirit, one must add the $Z \to b\bar{b}$ vertex which also includes a large top mass dependence. Thus altogether we consider four defining observables: one asymmetry, for example A_{FB}^l, (as representative of the set of measurements that lead to the determination of $\sin^2 \theta_{eff}$), one width (the leptonic width Γ_l is particularly suitable because it is practically independent of α_s), m_W and R_b. Here lepton universality has been taken for granted, because the data show that it is verified within the present accuracy. The four variables, ϵ_1, ϵ_2, ϵ_3 and ϵ_b are defined in one to one correspondence with the set of observables A_l^{FB}, Γ_l, m_W, and R_b. The definition is so chosen that the quadratic top mass dependence is only present in ϵ_1 and ϵ_b, while the m_t dependence of ϵ_2 and ϵ_3 is logarithmic. The definition of ϵ_1 and ϵ_3 is specified in terms of A_l^{FB} and Γ_l only. Then adding m_W or R_b one obtains ϵ_2 or ϵ_b. We now specify the relevant definitions in detail.

9.1 Basic Definitions and Results

We start from the basic observables m_W/m_Z, Γ_l and A_l^{FB} and Γ_b. From these four quantities one can isolate the corresponding dynamically significant corrections Δr_W, $\Delta\rho$, Δk and ϵ_b, which contain the small effects one is trying to disentangle and are defined in the following. First we introduce Δr_W as obtained from m_W/m_Z by the relation:

$$(1 - \frac{m_W^2}{m_Z^2})\frac{m_W^2}{m_Z^2} = \frac{\pi\alpha(m_Z)}{\sqrt{2}G_F m_Z^2(1 - \Delta r_W)} \tag{114}$$

Here $\alpha(m_Z) = \alpha/(1 - \Delta\alpha)$ is fixed to the central value $1/128.90$ so that the effect of the running of α due to known physics is extracted from $1 - \Delta r = (1 - \Delta\alpha)(1 - \Delta r_W)$. In fact, the error on $1/\alpha(m_Z)$, as given in eq.(110) would then affect Δr_W. In order to define $\Delta\rho$ and Δk we consider the effective vector and axial-vector couplings g_V and g_A of the on-shell Z to charged leptons, given by the formulae:

$$\Gamma_l = \frac{G_F m_Z^3}{6\pi\sqrt{2}}(g_V^2 + g_A^2)(1 + \frac{3\alpha}{4\pi}),$$

$$A_l^{FB}(\sqrt{s} = m_Z) = \frac{3g_V^2 g_A^2}{(g_V^2 + g_A^2)^2} = \frac{3x^2}{(1 + x^2)^2}. \tag{115}$$

Note that Γ_l stands for the inclusive partial width $\Gamma(Z \to l\bar{l} + \text{photons})$. We stress the following points. First, we have extracted from $(g_V^2 + g_A^2)$ the factor $(1 + 3\alpha/4\pi)$ which is induced in Γ_l from final state radiation. Second, by the asymmetry at the peak in eq.(115) we mean the quantity which is commonly referred to by the LEP experiments (denoted as A_{FB}^0 in ref.[5]), which is corrected for all QED effects, including initial and final state radiation and also for the effect of the imaginary part of the γ vacuum polarization diagram. In terms of g_A and $x = g_V/g_A$, the quantities $\Delta\rho$ and Δk are given by:

$$g_A = -\frac{\sqrt{\rho}}{2} \sim -\frac{1}{2}(1 + \frac{\Delta\rho}{2}),$$

$$x = \frac{g_V}{g_A} = 1 - 4\sin^2\theta_{eff} = 1 - 4(1 + \Delta k)s_0^2. \tag{116}$$

Here s_0^2 is $\sin^2\theta_{eff}$ before non pure-QED corrections, given by:

$$s_0^2 c_0^2 = \frac{\pi\alpha(m_Z)}{\sqrt{2}G_F m_Z^2} \tag{117}$$

with $c_0^2 = 1 - s_0^2$ ($s_0^2 = 0.231095$ for $m_Z = 91.188\ GeV$).

We now define ϵ_b from Γ_b, the inclusive partial width for $Z \to b\bar{b}$ according to the relation

$$\Gamma_b = \frac{G_F m_Z^3}{6\pi\sqrt{2}}\beta(\frac{3 - \beta^2}{2}q_{bV}^2 + \beta^2 g_{bA}^2)N_C R_{QCD}(1 + \frac{\alpha}{12\pi}) \tag{118}$$

where $N_C = 3$ is the number of colours, $\beta = \sqrt{1 - 4m_b^2/m_Z^2}$, with $m_b = 4.7$ GeV, R_{QCD} is the QCD correction factor given by

$$R_{QCD} = 1 + 1.2a - 1.1a^2 - 13a^3\ ; \quad a = \frac{\alpha_s(m_Z)}{\pi} \tag{119}$$

and g_{bV} and g_{bA} are specified as follows

$$g_{bA} = -\frac{1}{2}(1 + \frac{\Delta\rho}{2})(1 + \epsilon_b),$$

$$\frac{g_{bV}}{g_{bA}} = \frac{1 - 4/3\sin^2\theta_{eff} + \epsilon_b}{1 + \epsilon_b}. \tag{120}$$

This is clearly not the most general deviation from the SM in the $Z \to b\bar{b}$ but ϵ_b is closely related to the quantity $-Re(\delta_{b-vertex})$ where the large m_t corrections are located in the SM.

As is well known, in the SM the quantities Δr_W, $\Delta\rho$, Δk and ϵ_b, for sufficiently large m_t, are all dominated by quadratic terms in m_t of order $G_F m_t^2$. As new physics can more easily be disentangled if not masked by large conventional m_t effects, it is convenient to keep $\Delta\rho$ and ϵ_b while trading Δr_W and Δk for two quantities with no contributions of order $G_F m_t^2$. We thus introduce the following linear combinations:

$$\epsilon_1 = \Delta\rho,$$

$$\epsilon_2 = c_0^2\Delta\rho + \frac{s_0^2\Delta r_W}{c_0^2 - s_0^2} - 2s_0^2\Delta k,$$

$$\epsilon_3 = c_0^2\Delta\rho + (c_0^2 - s_0^2)\Delta k. \tag{121}$$

The quantities ϵ_2 and ϵ_3 no longer contain terms of order $G_F m_t^2$ but only logarithmic terms in m_t. The leading terms for large Higgs mass, which are logarithmic, are contained in ϵ_1 and ϵ_3. In the Standard Model one has the following "large" asymptotic contributions:

$$\epsilon_1 = \frac{3G_F m_t^2}{8\pi^2\sqrt{2}} - \frac{3G_F m_W^2}{4\pi^2\sqrt{2}}\tan^2\theta_W \ln\frac{m_H}{m_Z} + ...,$$

$$\epsilon_2 = -\frac{G_F m_W^2}{2\pi^2\sqrt{2}}\ln\frac{m_t}{m_Z} + ...,$$

$$\epsilon_3 = \frac{G_F m_W^2}{12\pi^2\sqrt{2}}\ln\frac{m_H}{m_Z} - \frac{G_F m_W^2}{6\pi^2\sqrt{2}}\ln\frac{m_t}{m_Z} ....,$$

$$\epsilon_b = -\frac{G_F m_t^2}{4\pi^2\sqrt{2}} + ... \tag{122}$$

The relations between the basic observables and the epsilons can be linearised, leading to the approximate formulae

$$\frac{m_W^2}{m_Z^2} = \frac{m_W^2}{m_Z^2}|_B(1 + 1.43\epsilon_1 - 1.00\epsilon_2 - 0.86\epsilon_3),$$

Table 5: Values of the epsilons in the SM as functions of m_t and m_H as obtained from recent versions of ZFITTER and TOPAZ0. These values (in 10^{-3} units) are obtained for $\alpha_s(m_Z)$ = 0.119, $\alpha(m_Z)$ = 1/128.90, but the theoretical predictions are essentially independent of $\alpha_s(m_Z)$ and $\alpha(m_Z)$.

m_t (GeV)	ϵ_1 m_H (GeV) =			ϵ_2 m_H (GeV) =			ϵ_3 m_H (GeV) =			ϵ_b All m_H
	70	300	1000	70	300	1000	70	300	1000	
150	3.55	2.86	1.72	−6.85	−6.46	−5.95	4.98	6.22	6.81	−4.50
160	4.37	3.66	2.50	−7.12	−6.72	−6.20	4.96	6.18	6.75	−5.31
170	5.26	4.52	3.32	−7.43	−7.01	−6.49	4.94	6.14	6.69	−6.17
180	6.19	5.42	4.18	−7.77	−7.35	−6.82	4.91	6.09	6.61	−7.08
190	7.18	6.35	5.09	−8.15	−7.75	−7.20	4.89	6.03	6.52	−8.03
200	8.22	7.34	6.04	−8.59	−8.18	−7.63	4.87	5.97	6.43	−9.01

$$
\begin{aligned}
\Gamma_l &= \Gamma_l|_B(1 + 1.20\epsilon_1 - 0.26\epsilon_3), \\
A_l^{FB} &= A_l^{FB}|_B(1 + 34.72\epsilon_1 - 45.15\epsilon_3), \\
\Gamma_b &= \Gamma_b|_B(1 + 1.42\epsilon_1 - 0.54\epsilon_3 + 2.29\epsilon_b).
\end{aligned}
\tag{123}
$$

The Born approximations, as defined above, depend on $\alpha_s(m_Z)$ and also on $\alpha(m_Z)$. Defining

$$
\delta\alpha_s = \frac{\alpha_s(m_Z) - 0.119}{\pi}; \quad \delta\alpha = \frac{\alpha(m_Z) - \frac{1}{128.90}}{\alpha},
\tag{124}
$$

we have

$$
\begin{aligned}
\frac{m_W^2}{m_Z^2}\Big|_B &= 0.768905(1 - 0.40\delta\alpha), \\
\Gamma_l|_B &= 83.563(1 - 0.19\delta\alpha)\text{MeV}, \\
A_l^{FB}|_B &= 0.01696(1 - 34\delta\alpha), \\
\Gamma_b|_B &= 379.8(1 + 1.0\delta\alpha_s - 0.42\delta\alpha).
\end{aligned}
\tag{125}
$$

Note that the dependence on $\delta\alpha_s$ for $\Gamma_b|_B$, shown in eq.(125), is not simply the one loop result for $m_b = 0$ but a combined effective shift which takes into account both finite mass effects and the contribution of the known higher order terms.

The important property of the epsilons is that, in the Standard Model, for all observables at the Z pole, the whole dependence on m_t (and m_H) arising from one-loop diagrams only enters through the epsilons. The same is actually true, at the relevant level of precision, for all higher order m_t-dependent corrections. Actually, the only residual m_t dependence of the various observables not included in the epsilons is in the terms of order $\alpha_s^2(m_Z)$ in the pure QCD correction factors to the hadronic widths. But this one is quantitatively irrelevant, especially in view of the errors connected to the uncertainty on the value of $\alpha_s(m_Z)$. The theoretical values of the epsilons in the SM from state of the art radiative corrections, also including the recent development of ref.[10], are given in table 5. It is important to remark that the theoretical values of the epsilons in the SM, as given in table 5, are not affected, at the percent level or so, by reasonable variations of $\alpha_s(m_Z)$ and/or $\alpha(m_Z)$ around their central values. By our definitions, in fact, no terms of order $\alpha_s^n(m_Z)$ or $\alpha \ln m_Z/m$ contribute to the epsilons. In terms of the epsilons, the following expressions hold, within the SM, for the various precision observables

$$
\begin{aligned}
\Gamma_T &= \Gamma_{T0}(1 + 1.35\epsilon_1 - 0.46\epsilon_3 + 0.35\epsilon_b), \\
R &= R_0(1 + 0.28\epsilon_1 - 0.36\epsilon_3 + 0.50\epsilon_b), \\
\sigma_h &= \sigma_{h0}(1 - 0.03\epsilon_1 + 0.04\epsilon_3 - 0.20\epsilon_b), \\
x &= x_0(1 + 17.6\epsilon_1 - 22.9\epsilon_3), \\
R_b &= R_{b0}(1 - 0.06\epsilon_1 + 0.07\epsilon_3 + 1.79\epsilon_b).
\end{aligned}
\tag{126}
$$

where $x = g_V/g_A$ as obtained from A_l^{FB}. The quantities in eqs.(123–126) are clearly not independent and the redundant information is reported for convenience. By comparison with the computed radiative corrections we obtain

$$
\begin{aligned}
\Gamma_{T0} &= 2489.46(1 + 0.73\delta\alpha_s - 0.35\delta\alpha) \ MeV, \\
R_0 &= 20.8228(1 + 1.05\delta\alpha_s - 0.28\delta\alpha), \\
\sigma_{h0} &= 41.420(1 - 0.41\delta\alpha_s + 0.03\delta\alpha) \ nb, \\
x_0 &= 0.075619 - 1.32\delta\alpha, \\
R_{b0} &= 0.2182355.
\end{aligned}
\tag{127}
$$

Note that the quantities in eqs.(127) should not be confused, at least in principle, with the corresponding Born approximations, due to small "non universal" electroweak corrections. In practice, at the relevant level of approximation, the difference between the two corresponding quantities is in any case significantly smaller than the present experimental error.

In principle, any four observables could have been picked up as defining variables. In practice we choose those that have a more clear physical significance and are more effective in the determination of the epsilons. In fact, since Γ_b is actually measured by R_b (which is nearly insensitive to α_s), it is preferable

Table 6: Experimental values of the epsilons in the SM from different sets of data. These values (in 10^{-3} units) are obtained for $\alpha_s(m_Z) = 0.119 \pm 0.003$, $\alpha(m_Z) = 1/128.90 \pm 0.09$, the corresponding uncertainties being included in the quoted errors.

ϵ 10^3	Only def. quantities	All asymmetries	All High Energy	All Data
ϵ_1 10^3	4.1 ± 1.2	4.3 ± 1.2	4.0 ± 1.1	3.7 ± 1.1
ϵ_2 10^3	-8.2 ± 2.1	-8.8 ± 1.9	-9.0 ± 2.0	-9.3 ± 2.0
ϵ_3 10^3	3.3 ± 1.9	4.4 ± 1.2	4.2 ± 1.1	3.9 ± 1.1
ϵ_b 10^3	-4.3 ± 1.9	-4.4 ± 1.9	-4.8 ± 1.9	-4.6 ± 1.9

to use directly R_b itself as defining variable, as we shall do hereafter. In practice, since the value in eq.(127) is practically indistinguishable from the Born approximation of R_b, this determines no change in any of the equations given above but simply requires the corresponding replacement among the defining relations of the epsilons.

9.2 Experimental Determination of the Epsilon Variables

The values of the epsilons as obtained, following the specifications in the previous sect.9.1, from the defining variables m_W, Γ_l, A_l^{FB} and R_b are shown in the first column of table 6.

To proceed further and include other measured observables in the analysis we need to make some dynamical assumptions. The minimum amount of model dependence is introduced by including other purely leptonic quantities at the Z pole such as A_τ, A_e (measured from the angular dependence of the τ polarization) and A_{LR} (measured by SLD). For this step, one is simply assuming that the different leptonic asymmetries are equivalent measurements of $\sin^2 \theta_{eff}$. We add, as usual, the measure of A_b^{FB} because this observable is dominantly sensitive to the leptonic vertex. We then use the combined value of $\sin^2 \theta_{eff}$ obtained from the whole set of asymmetries measured at LEP and SLC given in eq.(102). At this stage the best values of the epsilons are shown in the second column of table 6. In figs. 5-8 we report the 1σ ellipses in the indicated ϵ_i ϵ_j planes that correspond to this set of input data.

All observables measured on the Z peak at LEP can be included in the analysis provided that we assume that all deviations from the SM are only contained in vacuum polarization diagrams (without demanding a truncation of the q^2 dependence of the corresponding functions) and/or the $Z \to b\bar{b}$ vertex.

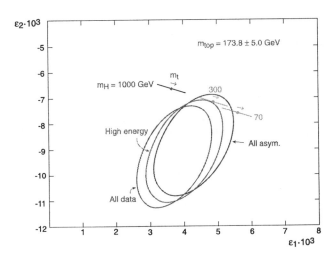

Figure 5: Data vs theory in the ϵ_2-ϵ_1 plane. The origin point corresponds to the "Born" approximation obtained from the SM at tree level plus pure QED and pure QCD corrections. The predictions of the full SM (also including the improvements of ref.[10]) are shown for m_H = 70, 300 and 1000 GeV and $m_t = 175.6 \pm 5.5$ GeV (a segment for each m_H with the arrow showing the direction of m_t increasing from -1σ to $+1\sigma$). The three $1 - \sigma$ ellipses (38% probability contours) are obtained from a) "All Asymm." :Γ_l, m_W and $\sin^2 \theta_{eff}$ as obtained from the combined asymmetries (the value in eq. (102)); b) "All High En.": the same as in a) plus all the hadronic variables at the Z; c) "All Data": the same as in b) plus the low energy data.

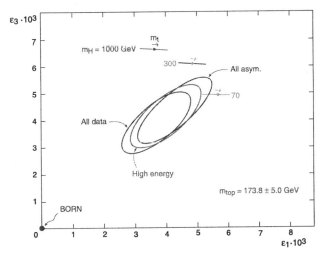

Figure 6: Data vs theory in the ϵ_3-ϵ_1 plane (notations as in fig.5)

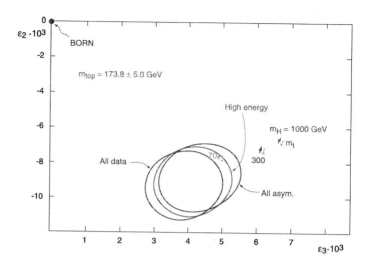

Figure 7: Data vs theory in the ϵ_2-ϵ_3 plane (notations as in fig.5)

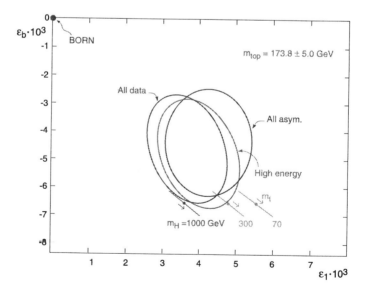

Figure 8: Data vs theory in the ϵ_b-ϵ_1 plane (notations as in fig.5)

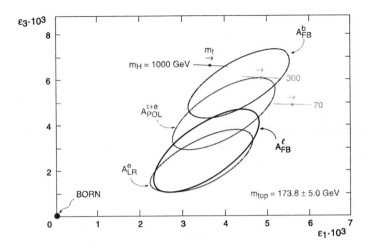

Figure 9: Data vs theory in the ϵ_3-ϵ_1 plane (notations as in fig.5). The ellipse indicated with "Average" corresponds to the case "All high en" of fig.6 and is obtained from the combined value of $sin^2\theta_{eff}$. The other ellipses are obtained by replacing the combined $sin^2\theta_{eff}$ with the values obtained in turn from each individual asymmetry as shown by the labels.

From a global fit of the data on m_W, Γ_T, R_h, σ_h, R_b and $sin^2\theta_{eff}$ (for LEP data, we have taken the correlation matrix for Γ_T, R_h and σ_h given by the LEP experiments [5], while we have considered the additional information on R_b and $sin^2\theta_{eff}$ as independent) we obtain the values shown in the third column of table 6. The comparison of theory and experiment at this stage is also shown in figs. 5-8. More detailed information is shown in fig. 9, which refers to the level when also hadronic data are taken into account. But in fig.9 we compare the results obtained if $sin^2\theta_{eff}$ is extracted in turn from different asymmetries among those listed in fig.4. The ellipse marked "average" is the same as the one labelled "All high en." in fig.6 and corresponds to the value of $sin^2\theta_{eff}$ which is shown on the figure (and in eq.(102)). We confirm that the value from A_{LR} is far away from the SM given the experimental value of m_t and the bounds on m_H and would correspond to very small values of ϵ_3 and of ϵ_1. We see also that while the τ FB asymmetry is also on the low side, the combined e and μ FB asymmetry are right on top of the average. Finally the b FB asymmetry is on the high side.

To include in our analysis lower energy observables as well, a stronger hypothesis needs to be made: vacuum polarization diagrams are allowed to vary from the SM only in their constant and first derivative terms in a q^2 expansion. In such a case, one can, for example, add to the analysis the ratio R_ν of neutral to charged current processes in deep inelastic neutrino scattering on nuclei[11,12] the "weak charge" Q_W measured in atomic parity violation experiments on Cs[13] and the measurement of g_V/g_A from $\nu_\mu e$ scattering[14]. In this way one obtains the global fit given in the fourth column of table 6 and shown in figs. 5-8. In fig. 10 we see the ellipse in the ϵ_1-ϵ_3 plane that is obtained from the low energy data by themselves. It is interesting that the tendency towards low values of ϵ_1 and ϵ_3 is present in the low energy data as in the high energy ones. Note that the low energy data by themselves are actually compatible with the "Born" approximation. With the progress of LEP the low energy data, while important as a check that no deviations from the expected q^2 dependence arise, play a lesser role in the global fit. This does not mean that they are not important. For example, the measured parity violation in atomic physics provides the best limits on possible new physics in the electron-quark sector. When HERA suggested the presence of leptoquarks, the limits from atomic parity violation practically excluded all possible parity violating four fermiom electron-quark contact terms. So low energy data are no more powerful enough to improve the determination of the parameters if the SM is assumed, but they are a very powerful constraint on new physics models. The best values of the ϵ's from all the data are at present:

$$\epsilon_1 \ 10^3 = 3.7 \pm 1.1$$
$$\epsilon_2 \ 10^3 = -9.3 \pm 2.0$$
$$\epsilon_3 \ 10^3 = 3.9 \pm 1.1$$
$$\epsilon_b \ 10^3 = -4.6 \pm 1.9. \tag{128}$$

Note that the present ambiguity on the value of $\delta\alpha^{-1}(m_Z) = \pm 0.09$ corresponds to an uncertainty on ϵ_3 (the other epsilons are not much affected) given by $\Delta\epsilon_3 \ 10^3 = \pm 0.6$. Thus the theoretical error is still confortably less than the experimental error. In fig.11 we present a summary of the experimental values of the epsilons as compared to the SM predictions as functions of m_t and m_H, which shows agreement within 1σ, but the central value of ϵ_1, ϵ_2 and ϵ_3 are all low, while the central value of ϵ_b is shifted upward with respect to the SM as a consequence of the still imperfect matching of R_b. A number of interesting features are clearly visible from figs.5-11. First, the good agreement with the SM and the evidence for weak corrections, measured by the distance of the data from the improved Born approximation point (based on tree level SM plus pure QED or QCD corrections). There is by now a solid evidence for departures from the improved Born approximation where all the epsilons vanish. In other words a clear evidence for the pure weak radiative corrections

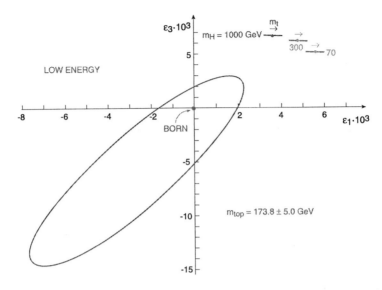

Figure 10: Data vs theory in the ϵ_3-ϵ_1 plane (notations as in fig.5). Here the ellipse from the low energy data by themselves is plotted (deep inelastic neutrino scattering, atomic parity violation and $\nu_\mu - e$ scattering.

has been obtained and LEP/SLC are now measuring the various components of these radiative corrections. For example, some authors [15] have studied the sensitivity of the data to a particularly interesting subset of the weak radiative corrections, i.e. the purely bosonic part. These terms arise from virtual exchange of gauge bosons and Higgses. The result is that indeed the measurements are sufficiently precise to require the presence of these contributions in order to fit the data. Second, the general results of the SM fits are reobtained from a different perspective. We see the preference for light Higgs manifested by the tendency for ϵ_3 to be rather on the low side. Since ϵ_3 is practically independent of m_t, its low value demands m_H small. If the Higgs is light then the preferred value of m_t is somewhat lower than the Tevatron result (which in the epsilon analysis is not included among the input data). This is because also the value of $\epsilon_1 \equiv \delta\rho$, which is determined by the widths, in particular by the leptonic width, is somewhat low. In fact ϵ_1 increases with m_t and, at fixed m_t, decreases with m_H, so that for small m_H the low central value of ϵ_1 pushes m_t down. Note that also the central value of ϵ_2 is on the low side, because the experimental value of m_W is a little bit too large. Finally, we see that adding the hadronic quantities or the low energy observables hardly makes a difference in the ϵ_i-ϵ_j plots with respect to the case with only the leptonic variables

72

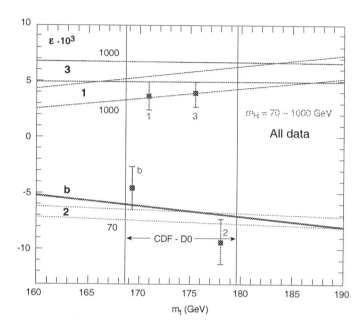

Figure 11: The bands (labeled by the ϵ index) are the predicted values of the epsilons in the SM as functions of m_t for $m_H = 70 - 1000$ GeV (the m_H value corresponding to one edge of the band is indicated). The CDF/D0 experimental 1-σ range of m_t is shown. The esperimental results for the epsilons from all data are displayed (from the last column of table 6). The position of the data on the m_t axis has been arbitrarily chosen and has no particular meaning.

being included (the ellipse denoted by "All Asymm."). But, for example for the ϵ_1-ϵ_3 plot, while the leptonic ellipse contains the same information as one could obtain from a $\sin^2 \theta_{eff}$ vs Γ_l plot, the content of the other two ellipses is much larger because it shows that the hadronic as well as the low energy quantities match the leptonic variables without need of any new physics. Note that the experimental values of ϵ_1 and ϵ_3 when the hadronic quantities are included also depend on the input value of α_s given in eq.(111).

The good agreement of the fitted epsilon values with the SM impose strong constraints on possible forms of new physics. Consider, for example, new quarks or leptons. Mass splitted multiplets contribute to $\Delta\epsilon_1$, in analogy to the t-b quark doublet. Recall that $\Delta\epsilon_1 \sim +9.5 \ 10^{-3}$ for the t-b doublet, which is about eight σ's in terms of the present error [16]. Even mass degenerate multiplets are strongly constrained. They contribute to $\Delta\epsilon_3$ according to [17]

$$\Delta\epsilon_3 \sim N_C \frac{G_F m_W^2}{8\pi^2\sqrt{2}} \frac{4}{3}(T_{3L} - T_{3R})^2 \tag{129}$$

For example a new left-handed quark doublet, degenerate in mass, would contribute $\Delta\epsilon_3 \sim +1.3 \ 10^{-3}$, that is about one σ, but in the wrong direction, in the sense that the experimental value of ϵ_3 favours a displacement, if any, with negative sign. Only vector fermions ($T_{3L} = T_{3R}$) are not constrained. In particular, naive technicolour models , that introduce several new technifermions, are strongly disfavoured because they tend to produce large corrections with the wrong sign to ϵ_1, ϵ_3 and also to ϵ_b [18].

10 Why Beyond the Standard Model?

Given the striking success of the SM why are we not satisfied with that theory? Why not just find the Higgs particle, for completeness, and declare that particle physics is closed? The main reason is that there are strong conceptual indications for physics beyond the SM. There are also some phenomenological hints.

10.1 Conceptual Problems with the Standard Model

It is considered highly unplausible that the origin of the electro-weak symmetry breaking can be explained by the standard Higgs mechanism, without accompanying new phenomena. New physics should be manifest at energies in the TeV domain. This conclusion follows fron an extrapolation of the SM at very high energies. The computed behaviour of the $SU(3) \otimes SU(2) \otimes U(1)$ couplings with energy clearly points towards the unification of the electroweak and strong forces (Grand Unified Theories: GUTs) at scales of energy $M_{GUT} \sim 10^{14} - 10^{16}$ GeV which are close to the scale of quantum gravity, $M_{Pl} \sim 10^{19}$ GeV [19]. One can also imagine a unified theory of all interactions also including gravity (at present superstrings provide the best attempt at such a theory). Thus GUTs and the realm of quantum gravity set a very distant energy horizon that modern particle theory cannot anymore ignore. Can the SM without new physics be valid up to such large energies? This appears unlikely because the structure of the SM could not naturally explain the relative smallness of the weak scale of mass, set by the Higgs mechanism at $m \sim 1/\sqrt{G_F} \sim 250$ GeV with G_F being the Fermi coupling constant. The weak scale m is $\sim 10^{17}$ times smaller than M_{Pl}. Even if the weak scale is set near 250 GeV at the classical level, quantum fluctuations would naturally shift it up to where new physics starts to apply, in particular up to M_{Pl} if there was no new physics up to gravity. This so-called hierarchy problem [20] is related to the presence of fundamental scalar fields in the theory with quadratic mass divergences and no protective extra symmetry at m=0. For fermions, first, the divergences are logaritmic and, second, at m=0 an additional symmetry, i.e. chiral symmetry, is restored. Here, when talking of divergences we are not worried of actual infinities. The theory is renormalisable and finite once the dependence on the cut off is absorbed in a redefinition of masses and couplings. Rather the hierarchy problem is one of naturalness. If we consider the cut off as a manifestation of new physics that will modify the theory at large energy scales, then it is relevant to look at the dependence of physical quantities on the cut off and to demand that no unexplained enormously accurate cancellation arise.

According to the above argument the observed value of $m \sim 250$ GeV is indicative of the existence of new physics nearby. There are two main possibilities. Either there exist fundamental scalar Higgses but the theory is stabilised by supersymmetry, the boson-fermion symmetry that would downgrade the degree of divergence from quadratic to logarithmic. For approximate supersymmetry the cut off is replaced by the splitting between the normal particles and their supersymmetric partners. Then naturalness demands that this splitting (times the size of the weak gauge coupling) is of the order of the weak scale of mass, i.e. the separation within supermultiplets should be of the order of no more than a few TeV. In this case the masses of most supersymmetric partners of the known particles, a very large managerie of states, would fall, at least in part, in the discovery reach of the LHC. There are consistent, fully formulated field theories constructed on the basis of this idea, the simplest one being the MSSM [21]. Note that all normal observed states are those whose masses are forbidden in the limit of exact $SU(2) \otimes U(1)$. Instead for all SUSY partners the masses are allowed in that limit. Thus when supersymmetry is broken in the TeV range but $SU(2) \otimes U(1)$ is intact only s-partners take mass while all normal particles remain massless. Only at the lower weak scale the masses of ordinary particles are generated. Thus a simple criterium exists to understand the difference between particles and s-particles.

The other main avenue is compositeness of some sort. The Higgs boson is not elementary but either a bound state of fermions or a condensate, due to a new strong force, much stronger than the usual strong interactions, responsible for the attraction. A plethora of new "hadrons", bound by the new strong force would exist in the LHC range. A serious problem for this idea is that nobody sofar has been able to build up a realistic model along these lines, but that could eventually be explained by a lack of ingenuity on the theorists side. The most appealing examples are technicolor theories [18]. These models where inspired by the breaking of chiral symmetry in massless QCD induced by quark condensates. In the case of the electroweak breaking new heavy techniquarks must be introduced and the scale analogous to Λ_{QCD} must be about three orders of magnitude larger. The presence of such a large force relatively nearby has a strong tendency to clash with the results of the electroweak precision tests. Another interesting idea is to replace the Higgs by a $t\bar{t}$ condensate [22]. The Yukawa coupling of the Higgs to the $t\bar{t}$ pair becomes a four fermion $\bar{t}t\bar{t}t$ coupling with the corresponding strenght. The strong force is in this case provided by the large top mass. At first sight this idea looks great: no fundamental scalars, no new states. But, looking closely, the advantages are largely illusory. First, in the SM the required value of m_t is too large $m_t \geq 220 \ GeV$ or so. Also a tremendous fine tuning is required, because m_t would naturally be of the order of M_{GUT} or M_{Pl} if no new physics is present (the hierarchy problem in a different form!). Supersymmetry could come to the rescue in this case also. In a minimal SUSY version the required value of the top mass is lowered [23], $m_t \sim 205 \sin \beta \ GeV$. But the resulting theory is physically indistinguishable from the MSSM with small $\tan \beta$, at least at low energies [24]. This is because a strongly coupled Higgs looks the same as a $t\bar{t}$ pair.

The hierarchy problem is certainly not the only conceptual problem of the SM. There are many more: the proliferation of parameters, the mysterious pattern of fermion masses and so on. But while most of these problems can be postponed to the final theory that will take over at very large energies, of order M_{GUT} or M_{Pl}, the hierarchy problem arises from the unstability of the low energy theory and requires a solution at relatively low energies. A supersymmetric extension of the SM provides a way out which is well defined, computable and that preserves all virtues of the SM. The necessary SUSY breaking can be introduced through soft terms that do not spoil the stability of scalar masses. Precisely those terms arise from supergravity when it is spontaneously broken in a hidden sector [25]. But alternative mechanisms of SUSY breaking are also being considered [26]. In the most familiar approach SUSY is broken in a hidden sector and the scale of SUSY breaking is very large of order $\Lambda \sim \sqrt{G_F^{-1/2} M_P}$ where M_P is the Planck mass. But since the hidden sector only communicates with the visible sector through gravitational interactions the splitting of the SUSY multiplets is much smaller, in the TeV energy domain, and the Goldstino is practically decoupled. In an alternative scenario the (not so much) hidden sector is connected to the visible one by ordinary gauge interactions. As these are much stronger than the gravitational interactions, Λ can be much smaller, as low as 10-100 TeV. It follows that the Goldstino is very light in these models (with mass of order or below 1 eV typically) and is the lightest, stable SUSY particle, but its couplings are observably large. The radiative decay of the lightest neutralino into the Goldstino leads to detectable photons. The signature of photons comes out naturally in this SUSY breaking pattern: with respect to the MSSM, in the gauge mediated model there are typically more photons and less missing energy. Gravitational and gauge mediation are extreme alternatives: a spectrum of intermediate cases is conceivable. The main appeal of gauge mediated models is a better protection against flavour changing neutral currents. In the gravitational version even if we accept that gravity leads to degenerate scalar masses at a scale near M_{Pl} the running of the masses down to the weak scale can generate mixing induced by the large masses of the third generation fermions [27].

10.2 Hints from Experiment

Unification of Couplings

At present the most direct phenomenological evidence in favour of supersymmetry is obtained from the unification of couplings in GUTs. Precise LEP data on $\alpha_s(m_Z)$ and $\sin^2 \theta_W$ confirm what was already known with less accuracy: standard one-scale GUTs fail in predicting $\sin^2 \theta_W$ given $\alpha_s(m_Z)$ (and $\alpha(m_Z)$) while SUSY GUTs [28] are in agreement with the present, very precise, experimental results. According to the recent analysis of ref. [29], if one starts from the known values of $\sin^2 \theta_W$ and $\alpha(m_Z)$, one finds for $\alpha_s(m_Z)$ the results:

$$\alpha_s(m_Z) = 0.073 \pm 0.002 \quad \text{(Standard GUTS)}$$
$$\alpha_s(m_Z) = 0.129 \pm 0.010 \quad \text{(SUSY GUTS)} \tag{130}$$

to be compared with the world average experimental value $\alpha_s(m_Z)$ =0.119(4).

Dark Matter

There is solid astrophysical and cosmological evidence [30], [31] that most of the matter in the universe does not emit electromagnetic radiation, hence is "dark". Some of the dark matter must be baryonic but most of it must be non baryonic. Non baryonic dark matter can be cold or hot. Cold means non relativistic at freeze out, while hot is relativistic. There is general consensus that most of the non baryonic dark matter must be cold dark matter. A couple of years ago the most likely composition was quoted to be around 80% cold and 20% hot. At present it appears to me that the need of a sizeable hot dark matter component is more uncertain. In fact, recent experiments have indicated the presence of a previously disfavoured cosmological constant component in $\Omega = \Omega_m + \Omega_\Lambda$ [30]. Here Ω is the total matter-energy density in units of the critical density, Ω_m is the matter component (dominated by cold dark matter) and Ω_Λ is the cosmological component. Inflationary theories almost inevitably predict $\Omega = 1$ which is consistent with present data. At present, still within large uncertainties, the approximate composition is indicated to be $\Omega_m \sim 0.4$ and $\Omega_\Lambda \sim 0.6$ (baryonic dark matter gives $\Omega_b \sim 0.05$).

The implications for particle physics is that certainly there must exist a source of cold dark matter. By far the most appealing candidate is the neutralino, the lowest supersymmetric particle, in general a superposition of photino, Z-ino and higgsinos. This is stable in supersymmetric models with R parity conservation, which are the most standard variety for this class of models (including the Minimal Supersymmetric Standard Model:MSSM). A neutralino with mass of order 100 GeV would fit perfectly as a cold dark matter candidate. Another common candidate for cold dark matter is the axion, the elusive particle associated to a possible solution of the strong CP problem along the line of a spontaneously broken Peccei-Quinn symmetry. To my knowledge and taste this option is less plausible than the neutralino. One favours supersymmetry for very diverse conceptual and phenomenological reasons, as described in the previous sections, so that neutralinos are sort of standard by now. For hot dark matter, the self imposing candidates are neutrinos. If we demand a density fraction $\Omega_\nu \sim 0.1$ from neutrinos, then it turns out that the sum of stable neutrino masses should be around 5 eV.

Baryogenesis

Baryogenesis is interesting because it could occur at the weak scale[32] but not in the SM. For baryogenesis one needs the three famous Sakharov conditions[33]: B violation, CP violation and no termal equilibrium. In principle these conditions could be verified in the SM. B is violated by instantons when kT is of the order of the weak scale (but B-L is conserved). CP is violated by the CKM phase and out of equilibrium conditions could be verified during the electroweak phase transition. So the conditions for baryogenesis appear superficially to be present for it to occur at the weak scale in the SM. However, a more quantitative analysis[34],[35] shows that baryogenesis is not possible in the SM because there is not enough CP violation and the phase transition is not sufficiently strong first order, unless $m_H < 80\ GeV$, which is by now excluded by LEP. Certainly baryogenesis could also occur below the GUT scale, after inflation. But only that part with $|B - L| > 0$ would survive and not be erased at the weak scale by instanton effects. Thus baryogenesis at $kT \sim 10^{12} - 10^{15}\ GeV$ needs B-L violation at some stage like for m_ν. The two effects could be related if baryogenesis arises from leptogenesis[36] then converted into baryogenesis by instantons. While baryogenesis at a large energy scale is thus not excluded it is interesting that recent studies have shown that baryogenesis at the weak scale could be possible in the MSSM[35]. In fact, in this model there are additional sources of CP violations and the bound on m_h is modified by a sufficient amount by the presence of scalars with large couplings to the Higgs sector, typically the s-top. What is required is that $m_h \sim 80 - 110\ GeV$ (in the LEP2 range!), a s-top not heavier than the top quark and, preferentially, a small $\tan\beta$.

Neutrino Masses

Recent data from Superkamiokande[37] (and also MACRO[38]) have provided a more solid experimental basis for neutrino oscillations as an explanation of the atmospheric neutrino anomaly. In addition the solar neutrino deficit is also probably an indication of a different sort of neutrino oscillations. Results from the laboratory experiment by the LNSD collaboration[39] can also be considered as a possible indication of yet another type of neutrino oscillation. But the preliminary data from Karmen [40] have failed to reproduce this evidence. The case of LNSD oscillations is far from closed but one can tentatively assume, pending the results of continuing experiments, that the signal will not persist. Then solar and atmospheric neutrino oscillations can possibly be explained in terms of the three known flavours of neutrinos without invoking extra sterile species. Neutrino oscillations for atmospheric neutrinos require $\nu_\mu \rightarrow \nu_\tau$ with $\Delta m_{atm}^2 \sim 2 \ 10^{-3} \ eV^2$ and a nearly maximal mixing angle $\sin^2 2\theta_{atm} \geq 0.8$. In most of the Superkamiokande allowed region the bound by Chooz [41] essentially excludes $\nu_e \rightarrow \nu_\mu$ oscillations for atmospheric neutrino oscillations. Furthermore the last results from Superkamiokande allow a solution of the solar neutrino deficit in terms of ν_e disappearance vacuum oscillations (as opposed to MSW [42] oscillations within the sun) with $\Delta m_{sol}^2 \sim 10^{-10} \ eV^2$ and again nearly maximal mixing angles. Among the large and small angle MSW solutions the small angle one is perhaps more likely at the moment (with [43] $\Delta m_{sol}^2 \sim 0.5 \ 10^{-5} \ eV^2$ and $\sin^2 2\theta_{sol} \sim 5.5 \ 10^{-3}$) than the large angle MSW solution. Of course experimental uncertainties are still large and the numbers given here are presumably only indicative. But by now it is very unlikely that all this evidence for neutrino oscillations will disappear or be explained away by astrophysics or other solutions. The consequence is that we have a substantial evidence that neutrinos are massive.

In a strict minimal standard model point of view neutrino masses could vanish if no right handed neutrinos existed (no Dirac mass) and lepton number was conserved (no Majorana mass). In Grand Unified theories both these assumptions are violated. The right handed neutrino is required in all unifying groups larger than SU(5). In SO(10) the 16 fermion fields in each family, including the right handed neutrino, exactly fit into the 16 dimensional representation of this group. This is really telling us that there is something in SO(10)! The SU(5) alternative in terms of $\bar{5} + 10$, without a right handed neutrino, is certainly less elegant. The breaking of $|B - L|$, B and L is also a generic feature of Grand Unification. In fact, the see-saw mechanism [44] explains the smallness of neutrino masses in terms of the large mass scale where $|B - L|$ and L are violated. Thus, neutrino masses, as would be proton decay, are important as a probe into the physics at the GUT scale.

Oscillations only determine squared mass differences and not masses. The case of three nearly degenerate neutrinos is the only one that could in principle accomodate neutrinos as hot dark matter together with solar and atmospheric neutrino oscillations. According to our previous discussion, the common mass should be around 1-3 eV. The solar frequency could be given by a small 1-2 splitting, while the atmospheric frequency could be given by a still small but much larger 1,2-3 splitting. A strong constraint arises in the degenerate case from neutrinoless double beta decay which requires that the ee entry of m_ν must obey $|(m_\nu)_{11}| \leq 0.46$ eV. As observed in ref. [45], this bound can only be satisfied if double maximal mixing is realized, i.e. if also solar neutrino oscillations occur with nearly maximal mixing. We have mentioned that it is not at all clear at the moment that a hot dark matter component is really needed[30]. However the only reason to consider the fully degenerate solution is that it is compatible with hot dark matter. Note that for degenerate masses with $m \sim 1 - 3$ eV we need a relative splitting $\Delta m/m \sim \Delta m_{atm}^2/2m^2 \sim 10^{-3}-10^{-4}$ and an even smaller one for solar neutrinos. It is difficult to imagine a natural mechanism compatible with unification and the see-saw mechanism to arrange such a precise near symmetry.

If neutrino masses are smaller than for cosmological relevance, we can have the hierarchies $|m_3| >> |m_{2,1}|$ or $|m_1| \sim |m_2| >> |m_3|$. Note that we are assuming only two frequencies, given by $\Delta_{sun} \propto m_2^2 - m_1^2$ and $\Delta_{atm} \propto m_3^2 - m_{1,2}^2$. We prefer the first case, because for quarks and leptons one mass eigenvalue, the third generation one, is largely dominant. Thus the dominance of m_3 for neutrinos corresponds to what we observe for the other fermions. In this case, m_3 is determined by the atmospheric neutrino oscillation frequency to be around $m_3 \sim 0.05$ eV. By the see-saw mechanism m_3 is related to some large mass M, by $m_3 \sim m^2/M$. If we identify m with either the Higgs vacuum expectation value or the top mass (which are of the same order), as suggested for third generation neutrinos by Grand Unification in simple SO(10) models, then M turns out to be around $M \sim 10^{15}$ GeV, which is consistent with the connection with GUT's. If solar neutrino oscillations are determined by vacuum oscillations, then $m_2 \sim 10^{-5}$ eV and we have that the ratio m_2/m_3 is well consistent with $(m_c/m_t)^2$.

A lot of attention is being devoted to the problem of a natural explanation of the observed nearly maximal mixing angle for atmospheric neutrino oscillations and possibly also for solar neutrino oscillations, if explained by vacuum oscillations[46]. Large mixing angles are somewhat unexpected because the observed quark mixings are small and the quark, charged lepton and neutrino mass matrices are to some extent related in GUT's. There must be some special interplay between the neutrino Dirac and Majorana matrices in the see-saw mechanism in order to generate maximal mixing. It is hoped that looking for a natural explanation of large neutrino mixings can lead us to decripting some interesting message on the physics at the GUT scale.

11   Comparing the Data with the Minimal Supersymmetric Standard Model

The MSSM [21] is a completely specified, consistent and computable theory. There are too many parameters to attempt a direct fit of the data to the most general framework. So we consider two significant limiting cases: the "heavy" and the "light" MSSM.

The "heavy" limit corresponds to all s-particles being sufficiently massive, still within the limits of a natural explanation of the weak scale of mass. In this limit a very important result holds [47]: for what concerns the precision electroweak tests, the MSSM predictions tend to reproduce the results of the SM with a light Higgs, say $m_H \sim 100$ GeV. So if the masses of SUSY partners are pushed at sufficiently large values the same quality of fit as for the SM is guaranteed. Note that for $m_t \sim 175 \; GeV$ and $m_H \sim 70 \; GeV$ the values of the four epsilons computed in the SM lead to a fit of the corresponding experimental values with $\chi^2 \sim 4$, which is reasonable for $d.o.f = 4$. This value corresponds to the fact that the central values of ϵ_1, ϵ_2, ϵ_3 and $-\epsilon_b$ are all below the SM value by about 1σ, as can be seen from fig.11.

In the "light" MSSM option some of the superpartners have a relatively small mass, close to their experimental lower bounds. In this case the pattern of radiative corrections may sizeably deviate from that of the SM [51]. The potentially largest effects occur in vacuum polarisation amplitudes and/or the $Z \to b\bar{b}$ vertex. In particular we recall the following contributions :

i) a threshold effect in the Z wave function renormalisation [47] mostly due to the vector coupling of charginos and (off-diagonal) neutralinos to the Z itself. Defining the vacuum polarisation functions by $\Pi_{\mu\nu}(q^2) = -ig_{\mu\nu}[A(0) + q^2 F(q^2)] + q_\mu q_\nu$ terms, this is a positive contribution to $\epsilon_5 = m_Z^2 F'_{ZZ}(m_Z^2)$,the prime denoting a derivative with respect to q^2 (i.e. a contribution to a higher derivative term not included in the usual epsilon formalism). The ϵ_5 correction shifts ϵ_1, ϵ_2 and ϵ_3 by $-\epsilon_5$, $-c^2\epsilon_5$ and $-c^2\epsilon_5$ respectively, where $c^2 = \cos^2\theta_W$, so that all of them are reduced by a comparable amount. Correspondingly all the Z widths are reduced without affecting the asymmetries. This effect falls down particularly fast when the lightest chargino mass increases from a value close to $m_Z/2$. Now that we know, from the LEP2 runs, that the chargino mass is not smaller than m_Z its possible impact is drastically reduced.

ii) a positive contribution to ϵ_1 from the virtual exchange of split multiplets of SUSY partners, for example of the scalar top and bottom superpartners [48], analogous to the contribution of the top-bottom left-handed quark doublet. From the experimental value of m_t not much space is left for this possibility, and the experimental value of ϵ_1 is an important constraint on the spectrum. This is especially true now that the rather large lower limits on the chargino mass reduce the size of a possible compensation from ϵ_5 .For example, if the stop is light then it must be mainly a right-handed stop. Also large values of $\tan\beta$ are disfavoured because they tend to enhance the splittings among SUSY partner multiplets. In general it is simpler to decrease the predicted values of ϵ_2 and ϵ_3 by taking advantage of ϵ_5 than to decrease ϵ_1, because the negative shift from ϵ_5 is most often counterbalanced by the increase from the effect of split SUSY multiplets.

iii) a negative contribution to ϵ_b due to the virtual exchange of a charged Higgs [49]. If one defines, as customary, $\tan\beta = v_2/v_1$ (v_1 and v_2 being the vacuum expectation values of the Higgs doublets giving masses to the down and up quarks, respectively), then, for negligible bottom Yukawa coupling or $\tan\beta << m_t/m_b$, this contribution is proportional to $m_t^2/\tan^2\beta$.

iv) a positive contribution to ϵ_b due to virtual chargino–s-top exchange [50] which in this case is proportional to $m_t^2/\sin^2\beta$ and prefers small $\tan\beta$. This effect again requires the chargino and the s-top to be light in order to be sizeable.

With the recent limits set by LEP2 on the masses of SUSY partners the above effects are small enough that other contributions from vertex diagrams could be comparable. Thus in the following we will only consider the experimental values of the epsilons obtained at the level denoted by "All Asymmetries" which only assumes lepton universality.

We have analysed the problem of what configurations of masses in the "light" MSSM are favoured or disfavoured by the present data ([6],updating ref.[52]). We find that no lower limits on the masses of SUSY partners are obtained which are better than the direct limits. One exception is the case of s-top and s-bottom masses, which are severely constrained by the ϵ_1 value and also, at small $\tan\beta$, by the increase at LEP2 of the direct limit on the Higgs mass. Charged higgs masses are also rather severely constrained. Since the central values of ϵ_1, ϵ_2 and ϵ_3 are all below the SM it is convenient to make ϵ_5 as large as possible. For this purpose light gaugino and s-lepton masses are favoured. We find that for $m_{\chi_1^+} \sim 90 - 120 \ GeV$ the effect is still sizeable. Also favoured are small values of $\tan\beta$ that allow to put s-lepton masses relatively low, say, in the range 100-500 GeV, without making the split in the isospin doublets too large for ϵ_1. Charged Higgses must be heavy because they contribute to ϵ_b with the wrong sign. A light right-handed s-top could help on R_b for a higgsino-like chargino. But one needs small mixing (the right-handed s-top must be close to the mass eigenstate) and beware of the higgs mass constraint at small $\tan\beta$ (a higgs mass above 83 GeV, the range of LEP2 at $\sqrt{s} = 183$ GeV, starts being a strong constraint at small $\tan\beta$). So we prefer in the following to keep the s-top mass large. The limits on $b \to s\gamma$ also prefer heavy charged higgs and s-top[53].

12 The LEP2 Programme and the Search for the Higgs and New Physics

The LEP2 programme has started in the second part of 1996. At first the energy has been fixed at 161 GeV, which is the most favourable energy for the measurement of m_W from the cross-section for $e^+e^- \to W^+W^-$ at threshold. Then gradually the energy was brought up to 172, 183, 189 GeV. It will be increased up to a maximum of about 200 GeV to be reached in mid '99. An integrated luminosity of about 150 pb^{-1} per year is now achievable (in fact more was achieved in 1998). LEP2 has been approved to run until the end of 2000, before the shutdown for the installation of the LHC. The main goals of LEP2 are the search for the Higgs and for new particles, the measurement of m_W and the investigation of the triple gauge vertices WWZ and $WW\gamma$. A complete updated survey of the LEP2 physics is collected in the two volumes of ref.[51].

An important competitor of LEP2 is the Tevatron collider. In mid 2000 the Tevatron will start RunII with the purpose of collecting a few fb^{-1} of integrated luminosity at 2 TeV. The competition is especially on the search of new particles, but also on m_W and the triple gauge vertices. For example, for supersymmetry while the Tevatron is superior for gluinos and squarks, LEP2 is strong on Higgses, charginos, neutralinos and sleptons. There are plans for RunIII to start in 2002 or so with the purpose of collecting of the order 5 fb^{-1} of integrated luminosity per year. Then the Tevatron could also hope to find the Higgs before the LHC starts if the Higgs mass is close to the LEP2 range.

Concerning the Higgs it is interesting to recall that the large value of m_t has important implications on m_H both in the minimal SM[55-57] and in its minimal supersymmetric extension[58,59]. I will now discuss the restrictions on m_H that follow from the observed value of m_t.

It is well known[55]–[57] that in the SM with only one Higgs doublet a lower limit on m_H can be derived from the requirement of vacuum stability. The limit is a function of m_t and of the energy scale Λ where the model breaks down and new physics appears. Similarly an upper bound on m_H (with mild dependence on m_t) is obtained[61] from the requirement that up to the scale Λ no Landau pole appears. The lower limit on m_H is particularly important in view of the search for the Higgs at LEP2. Indeed the issue is whether one can reach the conclusion that if a Higgs is found at LEP2, i.e. with $m_H \leq m_Z$, then the SM must break down at some scale $\Lambda > 1$ TeV.

The possible instability of the Higgs potential $V[\phi]$ is generated by the quantum loop corrections to the classical expression of $V[\phi]$. At large ϕ the derivative $V'[\phi]$ could become negative and the potential would become unbound from below. The one-loop corrections to $V[\phi]$ in the SM are well known and change the dominant term at large ϕ according to $\lambda\phi^4 \rightarrow (\lambda + \gamma \log \phi^2/\Lambda^2)\phi^4$. The one-loop approximation is not enough for our purposes, because it fails at large enough ϕ, when $\gamma \log \phi^2/\Lambda^2$ becomes of order 1. The renormalization group improved version of the corrected potential leads to the replacement $\lambda\phi^4 \rightarrow \lambda(\Lambda)\phi'^4(\Lambda)$ where $\lambda(\Lambda)$ is the running coupling and $\phi'(\mu) = \exp \int^t \gamma(t')dt'\phi$, with $\gamma(t)$ being an anomalous dimension function and $t = \log\Lambda/v$ (v is the vacuum expectation value $v = (2\sqrt{2}G_F)^{-1/2}$). As a result, the positivity condition for the potential amounts to the requirement that the running coupling $\lambda(\Lambda)$ never becomes negative. A more precise calculation, which also takes into account the quadratic term in the potential, confirms that the requirements of positive $\lambda(\Lambda)$ leads to the correct bound down to scales Λ as low as ~ 1 TeV. The running of $\lambda(\Lambda)$ at one loop is given by:

$$\frac{d\lambda}{dt} = \frac{3}{4\pi^2}[\lambda^2 + 3\lambda h_t^2 - 9h_t^4 + \text{gauge terms}] , \qquad (131)$$

with the normalization such that at $t = 0, \lambda = \lambda_0 = m_H^2/2v^2$ and the top Yukawa coupling $h_t^0 = m_t/v$. We see that, for m_H small and m_t large, λ decreases with t and can become negative. If one requires that λ remains positive up to $\Lambda = 10^{15}$–10^{19} GeV, then the resulting bound on m_H in the SM with only one Higgs doublet is given by[56]:

$$m_H > 134 + 2.1\,[m_t - 173.8] - 4.5\,\frac{\alpha_s(m_Z) - 0.119}{0.006} . \qquad (132)$$

Summarizing, we see that the discovery of a Higgs particle at LEP2, or $m_H \lesssim 100\,GeV$, would imply that the SM breaks down at a scale Λ of the order of a few TeV. It can be shown [55] that the lower limit is not much relaxed even if strict vacuum stability is replaced by some sufficiently long metastability.

The upper limit on the Higgs mass in the SM is important for assessing the chances of success of the LHC as an accelerator designed to solve the Higgs problem. The upper limit [61] arises from the requirement that the Landau pole associated with the non asymptotically free behaviour of the $\lambda \phi^4$ theory does not occur below the scale Λ. The initial value of λ at the weak scale increases with m_H and the derivative is positive at large m_H. Thus if m_H is too large the Landau pole occurs at too low an energy. The upper limit on m_H has been recently reevaluated [60]. For $m_t \sim 175\ GeV$ one finds $m_H \lesssim 180\ GeV$ for $\Lambda \sim M_{GUT} - M_{Pl}$ and $m_H \lesssim 0.5 - 0.8\ TeV$ for $\Lambda \sim 1\ TeV$. Actually, for $m_t \sim$ 174 GeV, only a small range of values for m_H is allowed, $130 < m_H < \sim$ 200 GeV, if the SM holds up to $\Lambda \sim M_{GUT}$ or M_{Pl}.

A particularly important example of theory where the above bounds do not apply and in particular the lower bound is violated is the MSSM, which we now discuss. As is well known [21], in the MSSM there are two Higgs doublets, which implies three neutral physical Higgs particles and a pair of charged Higgses. The lightest neutral Higgs, called h, should be lighter than m_Z at tree-level approximation. However, radiative corrections [62] increase the h mass by a term proportional to m_t^4 and logarithmically dependent on the stop mass . Once the radiative corrections are taken into account the h mass still remains rather small: for $m_t = 174$ GeV one finds the limit (for all values of tg β) $m_h <$ 130 GeV [59]. Actually there are reasons to expect that m_h is well below the bound. In fact, if h_t is large at the GUT scale, which is suggested by the large observed value ot m_t and by a natural onsetting of the electroweak symmetry breaking induced by m_t, then at low energy a fixed point is reached in the evolution of m_t. The fixed point corresponds to $m_t \sim 205 \sin \beta$ GeV (a good approximate relation for tg $\beta = v_{up}/v_{down} < 10$). If the fixed point situation is realized, then m_h is considerably below the bound, as shown in ref.[59].

In conclusion, for $m_t \sim 174$ GeV, we have seen that, on the one hand, if a Higgs is found at LEP the SM cannot be valid up to M_{Pl}. On the other hand, if a Higgs is found at LEP, then the MSSM has good chances, because this model would be excluded for $m_h > 130$ GeV.

For the SM Higgs, which plays the role of a benchmark also important for a more general context, the LEP2 reach has been studied in detail [54]. At 200 GeV with about 150 pb^{-1} per experiment one can discover or exclude a SM Higgs up to about 105 GeV of mass. In the MSSM a more complicated discussion is needed because there are several Higgses and the parameter space is multidimensional. Only the lightest MSSM Higgs h is accessible at LEP2. The dominant production processes are $e^+e^- -> hZ$ and $e^+e^- -> hA$, where A is the CP odd MSSM Higgs particle. They are nicely complementary. At given m_h within the range of interest, at large $\tan\beta$ the first process is the relevant one, while the second determines the bound at small $\tan\beta$. The absolute lower limit on m_h for a given beam energy and integrated luminosity is always below the limit on the SM Higgs, because the crossections are smaller. For example the present limit is around 90 GeV for the SM Higgs and around 80 GeV for the MSSM Higgs. It is interesting that by the end of LEP2 one will have completely explored the region at small $\tan\beta$ (below a value of about 5), which is a particularly likely region.

A main goal of LEP2 is the search for supersymmetry. For charginos the discovery range at LEP2 is only limited by the beam energy for practically all values of the parameters. In fact the typical limit is at present about 90 GeV. Thus every increase of the beam energy is directly translated into the upper limit in chargino mass for discovery or exclusion. For the Tevatron the discovery range is much more dependent on the position in parameter space. For some limited regions of this space, with 1 fb^{-1} of integrated luminosity, the discovery range for charginos at the Tevatron goes well beyond $m_\chi = 100\ GeV$, i.e. the boundary of LEP2, but in much of the parameter space LEP2 at the maximum energy would be sensitive to larger chargino masses..

The stop is probably the lightest squark. For a light stop the most likely decay modes are $\tilde{t} \to b\chi^+$ if kinematically allowed, otherwise $\tilde{t} \to c\chi$. At LEP2 the discovery range is up to about $(E_{beam} - 10)$ GeV. At the Tevatron there is some difference between the two possible decay modes and some dependence on the position in the $\tilde{t} - \chi$ or the $\tilde{t} - \chi^+$ planes, but in general the Tevatron is very powerful for s-quarks and gluinos and much of the LEP2 range is already excluded by the Tevatron.

By now most of the discovery potential of LEP2 for supersymmetry has been already deployed. For example, the limit on the chargino mass was about 45 GeV after LEP1 and is now about 90 GeV and can only improve up to 100 GeV. For the Higgs the experimental task is more demanding and so one is only a bit more than half way through: the lower limit on the SM Higgs was around 67 GeV after LEP1, is now about 90 GeV and could go up to 105 GeV or so. So there are still fair chances for LEP2 to find the Higgs, especially because the attainable range of masses is particularly likely in the MSSM.

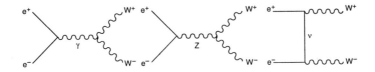

Figure 12: Lowest order Feynman diagrams for $e^+e^- \to W^+W^-$.

The study of the W^+W^- crosssection is a very important chapter of LEP2 physics [54]. In the Born approximation three Feynman diagrams contribute to the crosssection, as shown in fig.12. In the two s-channel exchange diagrams the triple gauge vertices $WW\gamma$ and WWZ appear, while the third is the t-channel neutrino exchange that only involves well established charged current couplings. One loop radiative corrections have also been computed . It is interesting that if we take neutrino exchange alone or neutrino plus γ exchanges alone the crosssection increases much faster with energy than the complete result. This corresponds to the good convergence properties of the SM which in fact is renormalisable. Indeed, the WW crosssection is related to the imaginary part of the WW loop contribution to the amplitude for $e^+e^- \to e^+e^-$. The good large energy behavior of the former crosssection is related to the convergence ot the latter loop correction. The data neatly confirm the SM prediction as shown in fig.13. Thus the WW crosssection supports the specific form of the triple gauge vertices as predicted by the SM. More detailed studies with large statistics are useful to set bounds on possible departures from the exact SM predicted couplings. In fact the study of triple gauge vertices is another major task of LEP2. The capabilities of LEP2 in this domain are comparable to those of the LHC. LEP2 can push down the existing direct limits considerably down. For given anomalous couplings the departures from the SM are expected to increase with energy. For the energy and the luminosity available at LEP2, given the accuracy of the SM established at LEP1, it is however not very likely, to find signals of new physics in the triple gauge vertices. The measurement of m_W is been done at LEP2 from the W^+W^- cross-section at threshold and from direct reconstruction of the W mass from the final state after W decay. At present m_W is known with an error of ± 60 MeV from the combined LEP2 and Tevatron direct measurements (see table 1), with the same error of ± 90 MeV at LEP2 and at the Tevatron. From the fit to all electroweak data one finds $m_W = 80370 \pm 27$ MeV (see eq.(113)), in agreement with the direct measurement. As a consequence the goal for LEP2 is to measure m_W with an accuracy $\delta m_W \leq \pm(30 - 40)$ MeV, in order to provide an additional significant check of the theory.

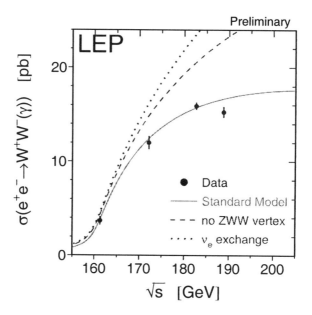

Figure 13: Data vs theory for the WW cross-section measured at LEP2. The solid line is the SM prediction. The dashed and dotted lines refer to only a subset diagrams as indicated.

For the threshold method[54] the minimum of the statistical error is obtained for $\sqrt{s} = 2m_W + 0.5$ GeV $= 161$ GeV, which in fact was the initial operating energy of LEP2. At threshold the WW crossesection is dominated by the neutrino t-channel exchange which is quite model independent. The total error of this method is dominated by the statistics. With the collected luminosity at 161 GeV of ~ 10 pb^{-1} per experiment, the present combined result is $m_W = (80.4\pm0.2\pm0.03)$ GeV[5]. Thus with the available data at threshold this method is not sufficient by itself.

In principle the direct reconstruction method can use the totally hadronic or the semileptonic final states $e^+e^- \to W^+W^- \to jjjj$ or $jjl\nu$. The total branching ratio of the hadronic modes is 49%, while that of the $\ell = e, \mu$ semileptonic channels is 28%. The hadronic channel has more statistics but could be severely affected by non-perturbative strong interaction effects: colour recombination among the jets from different W's and Bose correlations among mesons in the final state from WW overlap. Colour recombination is perturbatively small. But gluons with $E < \Gamma_W$ are important and non-perturbative effects could be relatively large, of the order of 10–100 MeV. Similarly for Bose correlations. One is not in a position to really quantify the associated uncertainties. Fortunately the direct reconstruction from the semi-leptonic channels can, by itself, lead to a total error $\delta m_W = \pm44$ MeV, for the combined four experiments, each with 500 pb^{-1} of luminosity collected at $\sqrt{s} \geq 175$ GeV. Thus the goal of measuring m_W with an accuracy below $\delta m_W = \pm50$ MeV can be fulfilled, and it is possible to do better by learning from the data how to limit the error from colour recombination and Bose correlations.

13 Conclusion

Today in particle physics we follow a double approach: from above and from below. From above there are, on the theory side, quantum gravity (that is superstrings), GUT theories and cosmological scenarios. On the experimental side there are underground experiments (e.g. searches for neutrino oscillations and proton decay), cosmic ray observations, satellite experiments (like COBE, IRAS etc) and so on. From below, the main objectives of theory and experiment are the search of the Higgs and of signals of particles beyond the Standard Model (typically supersymmetric particles). Another important direction of research is aimed at the exploration of the flavour problem: study of CP violation and rare decays. The general expectation is that new physics is close by and that should be found very soon if not for the complexity of the necessary experimental technology that makes the involved time scale painfully long.

References

1. Particle Data Group, *The Europ. Phys. J.* **C3**, 1 (1998).
2. For reviews of radiative corrections for LEP1 physics, see, for example:
   G. Altarelli, R. Kleiss and C. Verzegnassi (eds.), Z Physics at LEP1 (CERN 89-08, Geneva, 1989), Vols. 1–3;
   Precision Calculations for the Z Resonance, ed. by D. Bardin, W. Hollik and G. Passarino, CERN Rep. 95-03 (1995);
   M.I. Vysotskii, V.A. Novikov, L.B. Okun and A.N. Rozanov, hep-ph/9606253 or *Phys. Usp.* **39**, 503 (1996).
3. T. Appelquist and J. Carazzone, *Phys. Rev.* **D11** (1975) 2856.
4. See, for example:
   D. Karlen, Proceedings of ICHEP,98, Vancouver, 1998.
5. The LEP Elelectroweak Working Group, LEPEWWG/98-01, (1998).
6. For a complete list of references see (which sect.7-9 are an update of):
   G. Altarelli, R. Barbieri and F. Caravaglios, *Int. J. Mod. Phys.* **A13**, 1031 (1998).
7. F. Jegerlehner, *Z.Phys.* **C32**, 195 (1986);
   B.W. Lynn, G. Penso and C. Verzegnassi, *Phys. Rev.* **D35**, 42 (1987);
   H. Burkhardt et al., *Z.Phys.* **C43**, 497 (1989);
   F. Jegerlehner, *Progr. Part. Nucl. Phys.* **27**, 32 (1991);
   M.L. Swartz, Preprint SLAC-PUB-6710 (1994);
   M.L. Swartz, *Phys. Rev.* **D53**, 5268 (1996);
   A.D. Martin and D. Zeppenfeld, *Phys.Lett.* **B345**, 558 (1995);
   R.B. Nevzorov, A.V. Novikov and M.I. Vysotskii, hep-ph/9405390;
   H. Burkhardt and B. Pietrzyk, *Phys. Lett.* **B356**, 398 (1995);
   S. Eidelman and F. Jegerlehner, *Z.Phys.* **C67**, 585 (1995);
   R. Alemany, M. Davier and A. Hocker, *Europ. Phys. J.* **C2**, 123 (1998).
8. S. Groote et al., hep-ph/9802374;
   M. Davier and A. Hocker, *Phys. Lett.* **B419**, 419 (1998);
   J.H. Kuhn and M. Steinhauser, hep-ph/9802241;
   J. Erler, hep-ph/9803453;
   M. Davier and A. Hocker, hep-ph/9805470.
9. See, for example:
   Y. Dokshitser, Proceedings of ICHEP,98, Vancouver, 1998.
10. G. Degrassi, P. Gambino and A. Vicini, *Phys. Lett.* **B383**, 219 (1996);
    G. Degrassi, P. Gambino and A. Sirlin, *Phys. Lett.* **B394**, 188 (1997);

G. Degrassi, P. Gambino, M. Passera and A. Sirlin, hep-ph/9708311.

11. CHARM Collaboration, J.V. Allaby et al., *Phys. Lett.* **B177**, 446 (1986);
    Z. *Phys.* **C36**, 611 (1987);
    CDHS Collaboration, H. Abramowicz et al., *Phys. Rev. Lett.* **57**, 298 (1986);
    A. Blondel et al., Z. *Phys.* **C45**, 361 (1990;
    CCFR Collaboration, K. McFarland, hep-ex/9701010.

12. NuTeV Collaboration, K.S.McFarland et al., hep-ex/9806013.

13. C.S. Wood et al., *Science* **275** 1759 (1997).

14. CHARM II Collaboration, P. Vilain et al., *Phys. Lett.* **B335**, 246 (1997).

15. P. Gambino and A. Sirlin, *Phys. Rev. Lett.* **73**, 621 (1994);
    S.Dittmaier et al., *Nucl. Phys.* **B426**, 249 (1994);
    S. Dittmaier, D. Schildknecht and G. Weiglein, *Nucl. Phys.* **B465**, 3 (1996).

16. L. Alvarez-Gaumé, J. Polchinski and M. Wise, *Nucl. Phys.* **B221**, 495 (1983);
    R. Barbieri and L. Maiani, *Nucl. Phys.* **B224**, 32 (1983).

17. M. Veltman, *Nucl. Phys.* **B123**, 589 (1977);
    S. Bertolini and A. Sirlin, *Nucl. Phys.* **B248**, 589 (1984).

18. For a recent review, see R.S. Chivukula, hep-ph/9803219.

19. G.G.Ross,"Grand Unified Theories", Benjamin, 1985;
    R.N. Mohapatra,"Unification and Supersymmetry" Springer-Verlag, 1986.

20. E. Gildener, *Phys. Rev.* **D14**, 1667 (1976);
    E. Gildener and S. Weinberg, *Phys. Rev.* **D15**, 3333 (1976).

21. H.P. Nilles, *Phys. Rep.* **C110**, 1 (1984);
    H.E. Haber and G.L. Kane, *Phys. Rep.* **C117**, 75 (1985);
    R. Barbieri, *Riv. Nuovo Cim.* **11**, 1 (1988).

22. For a review, see, for example, C.T. Hill in Perspectives on Higgs Physics, ed. G. Kane, World Scientific, Singapore, 1993, and references therein.

23. W.A. Bardeen, T.E. Clark and S.T. Love, *Phys. Lett.* **B237**, 235 (1990);
    M. Carena et al., *Nucl. Phys.* **B369**, 33 (1992).

24. A. Hasenfratz et al., UCSD/PTH 91-06(1991).

25. A. Chamseddine, R. Arnowitt and P. Nath, *Phys. Rev. Lett.* **49**, 970 (1982);
    R. Barbieri, S. Ferrara and C. Savoy, *Phys. Lett.* **110B**, 343 (1982);
    E. Cremmer et al., *Phys. Lett.* **116B**, 215 (1982).

26. For a review, see G.F.Giudice and R.Rattazzi, hep-ph/9801271 and references therein.

27. A.Nelson, Proceedings of ICHEP'98, Vancouver, 1998.

28. S. Dimopoulos, S. Raby and F. Wilczek, *Phys.Rev.* **D24**, 1681 (1981); S. Dimopoulos and H. Georgi, *Nucl. Phys.*B193, 150 (1981) 1; L.E. Ibáñez and G.G. Ross, *Phys. Lett.* **105B**, 439 (1981).

29. P. Langacker and N.Polonsky, hep-ph/9503214.

30. R. Kolb, Proceedings of ICHEP'98, Vancouver, 1998.

31. M. Spiro, Proceedings of ICHEP'98, Vancouver, 1998.

32. V.A. Kuzmin, V.A. Rubakov and M.E. Shaposhnikov,*Phys. Lett.* **155B**, 36 (1985); M.E.Shaposhnikov,*Nucl. Phys.*B287, 757 (1987); *Nucl. Phys.*B299, 797 (1988).

33. A.D. Sakharov, *JETP Lett.* **91B**, 24 (1967).

34. A.G. Cohen, D.B. Kaplan and A.E. Nelson, *Annu. Rev. Part. Sci.* **43**, 27 (1993); M. Quirós, *Helv. Phys. Acta* **67**, 451 (1994); V.A. Rubakov and M.E. Shaposhnikov, hep-ph/9603208.

35. M. Carena, M. Quirós, and C.E.M. Wagner, *Nucl. Phys.* **B524**, 3 (1998).

36. See, for example, M. Fukugita and T.Yanagida, *Phys. Lett.* **B174**, 45 (1986); G. Lazarides and Q.Shafi, *Phys. Lett.* **B258**, 305(1991).

37. Y. Fukuda et al., hep-ex/9805006, hep-ex/9805021 and hep-ph/9807003.

38. M. Ambrosio et al., hep-ex/9807005.

39. C. Athanassopoulos et al., *Phys. Rev. Lett.* **77**, 3082 (1996), nucl-ex/9706006 and nucl-ex/9709006.

40. B. Armbruster et al., *Phys. Rev.* **C57**, 3414 (1998) and G. Drexlin, talk at Wein'98.

41. M. Apollonio et al., *Phys. Lett.* **B420**, 397 (1998).

42. L. Wolfenstein, *Phys. Rev.* **D17**, 2369 (1978); S.P. Mikheyev and A. Yu Smirnov, *Sov. J. Nucl. Phys.* **42**, 913 (1986).

43. G.L. Fogli, E. Lisi and D. Montanino, hep-ph/9709473; J.N. Bahcall, P.I. Krastev and A. Yu Smirnov, hep-ph/9807216; R. Barbieri, L.J. Hall, D. Smith, A. Strumia and N. Weiner, hep/ph 9807235.

44. M. Gell-Mann, P. Ramond and R. Slansky, in Supergravity, ed. by D.Freedman et al, North Holland, 1979; T Yanagida, *Pruy. Thuu. Phyu.* **B135**, 66 (1978).

45. F. Vissani, hep-ph/9708483; H. Georgi and S.L. Glashow, hep-ph/9808293.

46. For an introduction to this problem and list of references, see for example, G. Altarelli and F. Feruglio, hep-ph/9807353, hep-ph/9809596.

47. R. Barbieri, F. Caravaglios and M. Frigeni, *Phys. Lett.* **B279**, 169

(1992).

48. R. Barbieri and L. Maiani, *Nucl. Phys.* **B224**, 32 (1983); L. Alvarez-Gaumé, J. Polchinski and M. Wise, *Nucl. Phys.* **B221**, 495 (1983).

49. W. Hollik, *Mod. Phys. Lett.* **A5**, 1909 (1990).

50. A. Djouadi et al., *Nucl. Phys.* **B349**, 48 (1991); M. Boulware and D. Finell, *Phys. Rev.* **D44**, 2054 (1991). The sign discrepancy between these two papers appears now to be solved in favour of the second one.

51. See, e.g., S.Pokorski,Proceedings of ICHEP'96, Warsaw, 1996. See also, P.H. Chankowski, J. Ellis, S. Pokorski, *Phys. Lett.* **B423**, 327 (1998).

52. G. Altarelli, R. Barbieri and F. Caravaglios, *Phys. Lett.* **B314**, 357 (1993).

53. A. Brignole, F. Feruglio and F. Zwirner, *Z. Phys.* **C71**, 679 (1996).

54. G. Altarelli, T. Sjöstrand and F. Zwirner (eds.),"Physics at LEP2", CERN Report 95-03.

55. M. Sher, *Phys. Rep.* **179**, 273 (1989); *Phys. Lett.* **B317**, 159 (1993).

56. G. Altarelli and G. Isidori, *Phys. Lett.* **B337**, 141 (1994).

57. J.A. Casas, J.R. Espinosa and M. Quirós, *Phys. Lett.* **B342**, 171 (1995).

58. J.A. Casas et al., *Nucl. Phys.* **B436**, 3 (1995); E**B439**, 466 (1995).

59. M. Carena and C.E.M. Wagner, *Nucl. Phys.* **B452**, 45 (1995).

60. T. Hambye and K. Riesselmann, *Phys. Rev.* **D55**, 7255 (1997).

61. See, for example, M. Lindner, *Z. Phys.* **31**, 295 (1986).

62. H. Haber and R. Hempfling, *Phys. Rev. Lett.* **66**, 1815 (1991); J. Ellis, G. Ridolfi and F. Zwirner, *Phys. Lett.* **B257**, 83 (1991); Y. Okado, M. Yamaguchi and T. Yanagida, *Progr. Theor. Phys. Lett.* **85**, 1 (1991); R. Barbieri, F. Caravaglios and M. Frigeni, *Phys. Lett.* **B258**, 167 (1991). For a 2-loop improvement, see also: R. Memplfling and A.H. Hoang, *Phys. Lett.* **B331**, 99 (1994); J. Heinmeyer, W. Hollik and G. Weiglin, hep-ph/9807423.

ESSENTIAL SUPERSYMMETRY

NIR POLONSKY[a]

Department of Physics and Astronomy, Rutgers University
Piscataway, NJ 08854-8016, USA

Theoretical and experimental motivations, status and prospects of low-energy supersymmetry are discussed. It is shown by explicit construction that the stabilization of any perturbative theory which contains fundamental scalar bosons naturally leads to the notion of supersymmetry. The supersymmetric algebra is briefly described and the resulting constraints on the low-energy theory are emphasized. The minimal supersymmetric extension of the standard model is then pedagogically defined and discussed, and its experimental status is summarized. Dimensionless couplings are extrapolated to high energies and the unification scale is found and interpreted. Renormalization of the models is discussed and the linkage between high and low energies is demonstrated and argued to provide a potential probe of Planck scale physics. Electroweak symmetry breaking is shown to occur dynamically and to successfully predict a heavy t-quark. The low-energy Higgs potential is discussed more generally and special attention is given to the lightest Higgs boson. The upper bound on its mass in minimal and extended models is reviewed and explained. The Higgs mass is shown to provide a crucial test of nearly all supersymmetric theories. Generic low-energy flavor violations (the so-called flavor problem) are described and are used to constrain and classify consistent high-energy models for the s-particle spectrum parameters which softly break supersymmetry at low-energies. Generation of these parameters via tree-level local-supersymmetry interactions (supergravity mediation) or gauge quantum corrections (gauge mediation) is outlined. Possible lepton-number violation and scenarios for neutrino mass generation are also described.

[a]Address after September 1, 1999: Center for Theoretical Physics, Massachusetts Institute of Technology, 77 Massachusetts Avenue, Cambridge, MA 02139 USA.

1 Stepping Beyond The Standard Model

These lecture notes aim to provide an introductory yet comprehensive discussion of the phenomenology associated with supersymmetric extensions of the Standard Model of electroweak and strong interactions (SM). The choice of topics and of their presentation meant to draw readers with various backgrounds and research interests, as represented in this school, and to provide a pedagogical tour of supersymmetry, structure and phenomena. Given the ongoing intensive efforts to understand and discover such phenomena or to alternatively falsify the paradigm predicting it, and given the central role of supersymmetry in particle physics studies, taking such a journey is indeed a well motivated endeavor – if not a necessity.

Before embarking on our mission, however, it is useful to recall and establish our starting point: The Standard Model. In this lecture, the main ingredients of the SM will be summarized and briefly discussed; the hints for an additional structure within experimental reach will be emphasized, and the known theoretical possibilities for such a structure, particularly supersymmetry, will be introduced. We conclude this lecture with a general overview of the reminder of these lectures.

1.1 Basic Ingredients: Fermions and Bosons

The Standard Model of Strong and Electroweak Interactions (SM) was extensively outlined and discussed in Altarelli's lectures [1], where the reader can find an in-depth discussion and references. The SM provides the cornerstone of elementary particle physics for more than two decades, particularly so after it was soundly established in recent years by the many experiments at the various high-energy collider facilities. Even though the reader's familiarity with the SM is assumed throughout these lectures, it is useful to recall some of the its main ingredients, particularly those that force one to look for its extensions.

Indeed, the most direct evidence for an extended structure was provided most recently by experimentally establishing neutrino flavor oscillations. (See Langacker's lectures for details and references [2].) However, let us refrain from discussing neutrino mass and mixing until the very end of these lectures, and adopt until then the SM postulation of massless neutrinos. More intrinsic and fundamental indications for additional structure (that will ultimately have to also explain the neutrino spectrum) stem from the SM (so-far untested) postulation of a scalar field, the Higgs boson, whose vacuum expectation value (vev) spontaneously breaks the SM gauge symmetry.

The SM is a theory of fermions and of gauge bosons mediating the $SU(3)_c \times SU(2)_L \times U(1)_Y$ color and electroweak gauge interactions of the fermions. However, the $SU(2)_L \times U(1)_Y$ weak isospin × hypercharge electroweak symmetry is spontaneously broken and is not respected by the vacuum, as is manifested in the massive W and Z physical gauge bosons. The Higgs boson parameterizes the spontaneous electroweak symmetry breaking, and its inclusion in the weak-scale spectrum restores unitarity and perturbative consistency of WW scattering. Its gauge-invariant Yukawa interactions allow one to simultaneously explain, at least technically so, the $SU(2)_L \times U(1)_{\frac{Y}{2}}$ violating fermion mass spectrum. The SM employs the by-now standard quantum-field theory tools, and it commutes with, rather than incorporates, gravity. Table 1 lists the matter, Higgs and gauge fields which constitute the SM particle (or field) content (the graviton is listed for completeness). The table also serves to establish the relevant notation. The fermion flavor index $a = 1, 2, 3$ indicates the three identical generations (or families) of fermions which are distinguishd only by their mass spectrum. It is important to note that the $U(1)$ hypercharge is *de facto* assigned such that each family constitutes an anomaly-free set of fermions. All fields (or their linear combinations corresponding to the physical eigenstates : B, $W \to \gamma$, Z, $W^{\pm}$; see below) aside from the Higgs boson have been essentially observed. (Indirect evidence for the existence of the Higgs from precision measurement of electroweak observables, however, has been recently acquiring statistical significance.) It is not surprising therefore that it is the symmetry breaking sector which is the least understood. It will require most of our attention in the remaining of this section.

Given the charge assignments, it is straightforward to write down the gauge invariant scalar potential and Yukawa interactions. The scalar potential is given by

$$V(H) = -m^2 HH^{\dagger} + \lambda(HH^{\dagger})^2, \tag{1}$$

where $\lambda > 0$ is an arbitrary quartic coupling. If $m^2 < 0$ then the global minimum of the potential is at the origin and all symmetries are preserved. However, for $m^2 > 0$, as we shall assume, the Higgs boson acquires a non-vanishing *vev* $\langle H \rangle = (\nu, 0)/\sqrt{2}$ where $\nu \equiv \sqrt{(m^2/\lambda)}$. The electroweak $SU(2)_L \times U(1)_Y$ symmetry is now spontaneously broken into $U(1)_Q$ of QED with $Q = T_3 + Y/2$. (A $SU(2)$ rotation was used to fix the Higgs *vev* in its conventionally chosen direction). The Higgs boson in this case is a Goldsone boson and its charged and neutral CP-odd components are absorbed by the electroweak gauge bosons which, as given by experiment, are massive with $M_{W^{\pm}}^2 = g^2 \nu^2/4$ and $M_Z^2 = (g^2 + g'^2)\nu^2/4$. Here, g and g' are the conventional notation for the $SU(2)$ and hypercharge gauge couplings, respectively. One defines the weak angle $\tan\theta_W \equiv g'/g$ so that $Z = \cos\theta_W W_3 - \sin\theta_W B$ and the massless photon of the unbroken QED is given by the orthogonal combination. (The charged mass eigenstates are $W^{\pm} = (W_1 \mp iW_2)/\sqrt{2}$.) The QED cou-

Table 1: The SM field content. Our notation $(Q_c, Q_L)_{Q_{Y/2}}$ lists color, weak isospin and hypercharge assignments of a given field, respectively, and $Q_c = 1$, $Q_L = 1$ or $Q_{Y/2} = 0$ indicate a singlet under the respective group transformations. The T_3 isospin operator are $+1/2$ $(-1/2)$ when acting on the upper (lower) component of an isospin doublet (and zero otherwise).

Sector	Spin	Field
$SU(3)$ gauge bosons (gluons)	1	$g \equiv (8, 1)_0$
$SU(2)$ gauge bosons		$W \equiv (1, 3)_0$
$U(1)$ gauge boson		$B \equiv (1, 1)_0$
Chiral Matter (Three Families: $a = 1, 2, 3$)	$\frac{1}{2}$	$Q_a \equiv \begin{pmatrix} u \\ d \end{pmatrix}_{L_a} \equiv (3, 2)_{\frac{1}{6}}$
		$U_a \equiv u^c_{L_a} \equiv (\bar{3}, 1)_{-\frac{2}{3}}$
		$D_a \equiv d^c_{L_a} \equiv (\bar{3}, 1)_{\frac{1}{3}}$
		$L_a \equiv \begin{pmatrix} \nu \\ e^- \end{pmatrix}_{L_a} \equiv (1, 2)_{-\frac{1}{2}}$
		$E_a \equiv (e^-)^c_{L_a} \equiv (1, 1)_1$
Symmetry Breaking	0	$H \equiv \begin{pmatrix} H^0 \\ H^- \end{pmatrix} \equiv (1, 2)_{-\frac{1}{2}}$
Gravity (The Graviton)	2	$G \equiv (1, 1)_0$

pling is $e = g \sin\theta_W$. Note the (tree level) mass relation $M_W = M_Z \cos\theta_W$. (We will mention below the quantum corrections to this relation.) From the Fermi constant, $G_F = g^2/4\sqrt{2}M_W^2$, measured in muon decay one can extract $\nu = (\sqrt{2}G_F)^{-1/2} = 246$ GeV. (Our normalization is the conventional one for the one-Higgs doublet SM and will be modified when discussing the two-Higgs doublet supersymmetric extensions.)

The gauge invariant Yukawa interactions are of the from $\psi_L H \psi_R$ or of the form $\psi_L i\tau_2 H^* \psi_R$, where we labeled the (left-handed) chiral matter transforming under $SU(2)_L$ and the (right-handed) chiral matter which is a weak-isopspin singlet with the L and R subscripts, respectively. One has

$$\mathcal{L}_{\text{Yukawa}} = y_{l_{ab}} L_a H E_b + y_{d_{ab}} Q_a H D_b + y_{u_{ab}} Q_a (i\tau_2 H) U_b + h.c., \qquad (2)$$

where $SU(2)$ indices are implicit and are contracted with the antisymmetric tensor ϵ_{ij}: $\epsilon_{12} = +1 = -\epsilon_{21}$. Note that the choice of the hypercharge sign for H is somewhat arbitrary since one can always define $\bar{H} = i\tau_2 H^*$ which carries the opposite hypercharge (τ_2 is the Pauli matrix and $\bar{H} = (H^+, -H^{0*})^T$ given our choice). This will not be the case in the supersymmetric extension. One can rewrite the Lagrangian (2) in terms of the physical CP-even component η (with mass $\sqrt{2}m$) and the *vev* of $H(x) \rightarrow ((\nu + \eta(x), 0)/\sqrt{2})^T$ to find the Physical Higgs Yukawa interactions and the fermion mass terms $m_{f_{ab}} = y_{f_{ab}}\nu/\sqrt{2}$. Had the spectrum contained a SM singlet fermion $N \equiv (1, 1)_0$ then a neutrino Yukawa/mass term $y_{n_{ab}} L_a (i\tau_2 H) N_b + h.c.$ could also be written. N would then be the right-handed neutrino. However, in the SM neutrinos are assumed massless and lepton L and baryon B numbers are automatically consreved by its interactions (e.g., see eqs. (2) and (1)) so that the neutrinos and the proton are stable. (Particularly, there are no $\Delta L = 2$ Majorana mass terms $\sim N_a N_b$.) Since L and B are accidental but exact symmetries of the SM, this holds to all orders. We note in passing that the introduction of a right-handed neutrino N also allows for a new anomaly-free gauged $U(1)$ symmetry $U(1)_{Q'}$ with $Q' = B - L$.

Both Yukawa couplings and fermion masses are written above as a 3×3 matrices in the corresponding (up, down, lepton) flavor space. In order to obtain the physical mass eigenstates one has to perform independent unitary transformations on all vectors in field-space, e.g., $(u, c, t)_L$ and $(d, s, b)_L$ (employing standard flavor symbols) are now rotated by different transformation U and V and weak and mass eigenstates are not identical. It results in the flavor changing charged currents (interacting with $W^\pm$) $J^{CC} \propto U^\dagger V \equiv V_{CKM}$. The 3×3 Cabibbo-Kobayashi-Maskawa V_{CKM} (unitary) matrix contains three independent angles and one phase parameter. On the other hand, unitarity guarantees that no flavor changing neutral currents (FCNC) arise, $J^{NC} \propto U^\dagger U$, $V^\dagger V \equiv I$.

In addition, the absence of right-handed weak-singlet neutrinos and the corresponding Yukawa terms implies that the flavor rotation of the neutrinos are not physical and can always be choosen so that the left-handed neutrinos align with the charged leptons. Hence no leptonic flavor-violating (LFV) currents arise in the SM at any order in perturbation theory. Both predictions provide a frutile ground for testing extensions of the SM, which often predict new sources of FCNC and/or LFV.

Even though the SM is extremely successful in explaining all observed phenomena to date, it does not provide any guidelines in choosing its various *a-priori* free parameters: its rank, the gauge couplings, the Higgs parameters, and ultimately, the weak scale $\sim g\nu$. The fermion mass spectrum is indeed successfully related to the spontaneous gauge symmetry breaking. Nevertheless, the chiral symmetries of the SM, which would have forbidden massive fermions, are explicitly broken by hand in (2) and all flavor parameters - fermion masses and CKM angles and phase – have to be fixed by hand: The SM does not guide us as to the origins of flavor (nor as to the family duplication). Charge quantization and the quark-lepton distinction also remain mysterious (even though they can be argued to provide an anomaly-free set of fermions, given the gauge symmetries). One may argue that these (and the exclusion of gravity) are sufficient reasons to view the SM as only a "low-energy" limit of an extended theory, which may provide some of the answers. However, a stronger motivation to adopt the effective theory view emerges from the discussion of quantum corrections to the Higgs potential (1), and in fact, it will restrict the possible frameworks in which one can address all other fundamental questions. We turn to this subject next.

1.2 The Hierarchy Problem

Using standard field theory tools, one renormalizes the SM Lagrangian in order to account for quantum (*i.e.*, loop) corrections. Indeed, this procedure is successfully carried out in the case of the extrapolation of $SU(3)_c \times U(1)_Q$ theory to the weak scale and its embedding in the $SU(3)_c \times SU(2)_L \times U(1)_Y$ SM, as was described by Altarelli [1]. While most corrections are at most logarithmically divergent and nence correspond to finite shifts of parameters or to their well-understood logarithmic renormalization, there is an important exception. The model now contains the Higgs boson whose lightness is not protected by either gauge or chiral symmetries. The leading corrections to the Higgs two-point function (and hence, to its mass parameter in (1)) depend quadratically on the ultra-violet cut-off. Adopting the view that the SM itself is only an effective low-energy theory, the renormalization procedure is again understood as extrapolating the model towards the more fundamental ultra-violet scale. The quadratic dependence on this scale, however, is forcing the theory to reside at the ultra violet scale, hence leading to an apparent paradox with crucial implications. It amounts to a failure to consistently accommodate fundamental scalar fields within the framework of the SM, undermining the whole notion of an ultra-violet independent low-energy theory.

Figure 1: The gauge interaction contribution to the quadratic divergence.

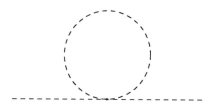

Figure 2: The scalar self interaction contribution to the quadratic divergence.

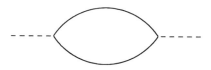

Figure 3: The Yukawa interaction contribution to the quadratic divergence.

More specifically, one needs to consider, at one-loop order, the three classes of quadratically divergent one-loop contributions shown in Figs. 1 – 3. The first divergence results from the Higgs $SU(2) \times U(1)$ gauge interactions. (Since we are interested in the ultra-violet dependence we will only consider the full $SU(2) \times U(1)$ theory above the weak scale.) The gauge bosons and the Higgs fields themselves are propagating in the loops. A naive power counting gives for the correction to the Higgs mass

$$\delta m^2_{\text{gauge}} \sim C_W \frac{g^2}{16\pi^2} \Lambda^2_{\text{UV}}, \tag{3}$$

where C_W is a numerical coefficient whose value is immaterial here, we included only the leading contributions in the ultra-violet cut-off Λ_{UV} (neglecting logarithmic dependences), and similarly for the hypercharge contribution which we omit. A second contribution results from the Higgs Yukawa interactions. In this case fermions are propagating in the loops and one has

$$\delta m^2_{\text{Yukawa}} \sim - \sum_f C_f \frac{y^2_f}{16\pi^2} \Lambda^2_{\text{UV}}, \tag{4}$$

where the summation is over all fermion flavors. (It is important to note that even small Yukawa couplings are not negligible once $\Lambda_{\text{UV}} \to \infty$.) Note the negative sign due to the fermion loop. Lastly, The Higgs self interaction leads to another contribution which is proportional to its quartic coupling,

$$\delta m^2_{\text{Higgs}} \sim C_H \frac{\lambda}{16\pi^2} \Lambda^2_{\text{UV}}. \tag{5}$$

Summing all independent contributions one finds

$$-m^2_{\text{tree}} + \frac{1}{16\pi^2} \left\{ C_W g^2 + C_H \lambda - \sum_f C_f h^2_f \right\} \Lambda^2_{\text{UV}} = -\lambda (246 \text{ GeV})^2. \tag{6}$$

The "one-loop improved" relation (6) essentially dependens on all of SM free parameters. (The $SU(3)$ coupling g_s enters at higher loop orders which are not shown here.) Hence, even though cancellations are technically possible (by adjusting m^2_{tree} or λ) and despite the fact that in the presence of a tree-level mass one can simply subtract the infinity by introducing an appropriate counter-term, such procedures do not seem reasonable unless $\Lambda_{\text{UV}} \sim \mathcal{O}(v)$. For example, for a cut-off of the order of magnitude of the Planck mass $\Lambda_{\text{UV}} \sim M_{\text{Planck}} \sim 10^{19}$ GeV one needs to fine-tune the value of v by 10^{17} orders of magnitude! This naturalness problem is often referred to as the hierarchy problem: Naively one expects that electroweak symmetry breaking would have occured near the Planck scale and not at the precisely measured electroweak scale $M_{\text{weak}} \sim g\nu \sim 100$ GeV. Since the SM does not contain any means which one could use in order to address the issue, it follows that the resolution of the hierarchy problem and any understanding of the specific choice of the electroweak scale must lie outside the SM! It is intriguing, however, that the instability of the infra-red SM Higgs potential implies that the model cannot be fundamental and at the same time severely constrains its possible extrapolations and embeddings, which we discuss in the following section.

1.3 Possible Avenues

Once it has been argued that the SM cannot be a fundamental theory, one may ask what possible avenues could be pursued in order to extend it? Our discussion of the hierarchy problem clearly indicates two possibilities: An electroweak-scale cut-off on quantum corrections in the SM or a theory which miraculously does not contain corrections of the form of eqs. (3) – (5). Both options imply the prediction of a new field content at or near the electroaweak scale, which is associated with either the nearby more fundamental "cut-off" scale or with the cancellation of the quadratically divergent corrections. This realization is a main driving force, side by side with the search for the Higgs boson itself, in current studies of particle physics at electroweak energies and beyond, and provides the basis for these lectures.

Various proposals exist that postulate $\Lambda_{\text{UV}} \sim \nu$. They can be roughly divided into two groups, frameworks that propose a fundamental Higgs boson and frameworks in which the Higgs boson is a meson, a composite of fermions. In the latter case the notion of a cut-off scale and of a condensation scale are entangled and, in fact, the solution to the hierarchy problem is the elimination of scalar bosons from the fundamental theory. It duplicates (but at much higher energies) the QCD picture in which the low-energy degrees of freedom are mesons and baryons while at high energy these are the quarks and gluons. Under this category fall all versions of technicolor models, top-color models, and their various offsprings. (For a recent review, see Refs. [3,4].) The models predict not only additional matter, but also strong and rich dynamics "nearby" (not far from current experimental energies). Strong dynamics, in general, does not completely decouple from its nearby low-energy limit, leading, in most cases, to many predictions at observable energies (the weak scale and below in this case). For example, the models typically predict new FCNC as well as (oblique) corrections to the W and Z masses and interactions (which lead to calculable shifts from SM tree-level relations such as $M_W = M_Z \cos\theta_W$ mentioned above). While a detailed critique can be found in Altarelli's lectures, no corresponding deviations from SM predictions (when including its own quantum corrections to such observables) were found, seriously under-mining the case for such an embedding of the SM. One should bare in mind, however, that by its nature non-perturbative strong dynamics does not always allow reliable calculations.

The most recent proposal for a framework with a low-energy cut-off and a fundamental Higgs boson corresponds to theories with extra low-energy com-pacitified dimensions[5,6]. In this framework, our four-dimensional universe re-sides on a four-dimensional submanifold of an extended space-time volume. Such theories must contain towers of $M \sim \mathcal{O}(\text{TeV})$ or heavier Kaluza-Klein states which parameterize the excitations of those states propagating in the extra dimensions. Such non-local states may include the some of the graviton (and other fields describing quantum gravity); the usual and extended gauge fields; and/or the usual and extended matter and Higgs fields, leading to a large variety of possibilities and constraints on the size and number compact-ified dimensions $R \gg M_{\text{Planck}}^{-1}$. The relation between the four-dimensional to $4 + n$-dimensional Planck constants reads

$$M_{\text{Planck}}^{4d} = (M_{\text{Planck}}^{(4+n)d})^{\frac{n+2}{2}} R^{\frac{n}{2}}, \tag{7}$$

relating the size and number of compactified dimensions and explaining the weakness of gravity as we know it in terms of extra large dimensions rather then a large Planck mass. (The limit $\sim (M_{\text{Planck}}^{(4+n)d}) \lesssim (M_{\text{Planck}}^{4d})$ occurs in what is often called M-theory[1,7] which will be mentioned when discussing unifica-tion.) The hierarchy problem is not eliminated but only rephrased in a differ-ent language: How to stabilize the large extra dimensions. The stabilization question is the hierarchy problem of models with extra large (or equivalently, low-energy) dimensions. It now cannot be solved without the introduction of quantum gravity; on the other hand, the only candidate to describe quantum gravity, namely string theory, does not provide the answer. Furthermore, if the $d + n$ Planck mass is not sufficiently close to the weak-scale the usual hierarchy problem re-appears. Many issues have been recently discussed, including the unification of couplings, proton decay, $etc.$ (which, however, are difficult to discuss due to the lack of a concrete framework and, in the case of couplings, the strong dependence on the details of the full theory, $i.e.$, boundary condi-tions) and on the other hand, collider signatures and low-energy constraints. It is intriguing that a possibility of low-energy extra dimensions may still be con-sistent with all data, though significant constraints were derived most recently in the literature. (Note in particular that the issue of possible new contribu-tions to FCNC is difficult to address in the absence of concrete realizations of these ideas.) Nevertheless, the data do not indicate or suggest such a scheme. The generic prediction in this scheme of large effective modifications of the

SM gauge interactions, particularly, the QCD $SU(3)_c$ interactions (due, for example, to graviton emission into the bulk of the extended volume), can be tested in the next generation of hadron and lepton colliders if the Planck mass is not too large.

On a different yet somewhat similar note, it was also argued (but no demonstrated) that one could construct theories which are conformally invariant above the weak scale and hence contain no dimensionful parameters and no hierarchy problem[8]. Again, this is only a rephrasing of the problem since conformal invariance must be broken and at the right amount.

Unlike in all of the above schemes, one could maintain perturbativity until very high energies if dangerous quantum corrections are cancelled to all orders in perturbation theory. Pursuing this alternative, one arrives at the notion of weak-scale supersymmetry. Supersymmetry ensures the desired cancellations by relating bosonic and fermionic degrees of freedom (e.g., for any chiral fermion the spectrum contains a complex scalar field, the s-fermion) and couplings (e.g., quartic couplings depend on gauge and Yukawa couplings). Henceforth, it is able to "capitalize" on the sign difference between bosonic and fermionic loops, which is insignificant as it stands in the SM result (6). This is the avenue that will be pursued in these lectures, and the elimination of the dangerous corrections will be shown below explicitly in a simple example. A detailed discussion will follow in the succeeding chapters. Next, however, we elaborate on the status of (weak-scale) supersymmetry, which further motivates our choice.

We note in passing that it is possible to link together various avenues. Theories with extra dimensions may very well contain weak-scale supersymmetry, depending on the relation between the (string) compactification scale and the supersymmetry-breaking scale[9]. In fact, it is difficult to believe that the stabilization of the large extra dimensions does not involve supersymmetry in some form or another. It was also suggested that supersymmetric QCD-like theories may play some role in defining the low-energy theory (for example, that some heavy fermions and their corresponding s-fermions are composite [10,11].) (Of course, strong dynamics may reside in extra dimensions and all three avenues may be realized simultaneously.) These possibilities, though interesting, will not be explored in any detail in these lectures.

1.4 More on Supersymmetry

Aside from its elegant and nearly effortless solution to the hierarchy problem (which will occupy us in the next chapter), supersymmetric models suggest that electroweak symmetry is broken in only a slightly stronger fashion (*i.e.*, two Higgs doubles acquire a *vev*) than predicted in the SM (where only one Higgs doublet is present). Furthermore, the more complicated Higgs sector imitates in many cases the SM situation with $m_\eta \lesssim 130 - 160$ GeV. This is, in fact, a crucial element in testing this framework and we will dedicate attention to this issue during these lectures. Typically, one finds that the theory is effectively a direct sum of the SM sector and of a superparticle (s-particle) sector, where the latter decouples to large extent from electroweak symmetry breaking (*i.e.*, its spectrum is to a very good approximation $SU(2)_L \times U(1)_Y$ conserving and is nearly independent of electroweak symmetry breaking *vevs*). The s-particle mass scale provides in a sense the desired cut-off scale above which the presence of the s-particles and the restoration of supersymmetric relations among couplings ensure cancellation of quadratically divergent quantum corrections. Hence, the super-particle mass scale is predicted to be $M_{\rm SUSY} \sim \mathcal{O}(100\,{\rm GeV} - 1\,{\rm TeV})$ scale with details varying among models. The mass-scale $M_{\rm SUSY}$ is also a measure in a sense of supersymmetry breaking in nature (e.g., the observation that a fermion and the corresponding s-fermion boson cannot be degenerate in mass) and it implies that supersymmetry survives (and describes the effective theory) down to the weak scale. (More correctly, it is $M_{\rm SUSY}$ that defines in this picture the weak scale $\nu \sim M_{\rm SUSY}$.) These issues will be discussed in some detail in the following chapters.

The electroweak symmetry conserving structure of the heavy s-particle spectrum sharply distinguishes the models from, e.g., technicolor models. It implies that the heavy s-particles do not contribute significantly even at the quantum level to electroweak symmetry (or more correctly, the related custodial $SU(2)$) breaking effects (such as the relation $M_W = M_Z \cos\theta_W$). This is consistent with the data, which is in nearly perfect agreement with SM predictions. (This is generically true if all relevant superparticle masses are above the 200 GeV mark.) Recent analysis of various electroweak observables (*i.e.*, various cross-sections, partial widths, left-right and forward-backward asymmetries, etc.), including their SM quantum corrections, suggests that the SM (or the SM-like Higgs boson in supersymmetry) is below $200 - 300$ GeV[12,1], which is consistent with the predicted range in supersymmetry, and more generally, with the notion of a fundamental Higgs particle (rather that a condensate with a mass $m_\eta \sim \Lambda_{\rm UV}$). (In models with strong dynamics the Higgs boson is typically heavier by a factor of $2 - 4$, while the situation is unclear in models with extra dimensions.)

The situation with the Higgs mass was described in great detail by Altartelli, and it was stressed that one is able to reach such conclusions regarding the Higgs mass largely because the t-quark (top) mass $m_t = 174 \pm 5$ GeV is measured and well known, hence, reducing the number of unknown parameters that appear in the expression to the quantum mechanically corrected SM predictions of the different observables. Indeed, it was m_t which appears quadratically in such expressions (rather than only logarithmically as the Higgs mass) that was initially determined by a similar analysis long before it was measured directly after the t-quark discovery [13] in 1995. (For example, Langacker and Luo found [14] $m_t = 124 \pm 37$ GeV in 1991.) The most important lesson learned was that the top is heavy, $m_t \sim \nu$ (in analogy to the current lesson that the Higgs boson is light, with light and heavy assuming different meaning in each case). Once we discuss the renormalization of supersymmetric models and the dynamical realization of electroweak symmetry breaking that follows, we will find that a crucial element in this realization is a sufficiently large t-Yukawa coupling, i.e., a sufficiently large m_t. Indeed, this observation corresponds to the successful prediction of a heavy top (though there are small number of cases in which this requirement can be evaded).

A somewhat similar but more speculative issue has to do with the extrapolation to high energy of the SM gauge couplings. Once the s-particle loops are included in the extrapolation at their predicted mass scale of $\mathcal{O}(100\,\text{GeV} - 1\,\text{TeV})$, then one finds that the coupling unify two orders of magnitude below the Planck scale with a perturbative value. This enables the consistent discussion of grand-unified and string theories, which claim such a unification, in this framework. (Supersymmetry is a natural consequence of (super)string theory, though its survival to low energies is an independent conjecture.) This result was known qualitatively for a long time, but it was shown more recently to hold once the increasing precision in the measurements of the gauge coupling at the weak scale is taken into account, and it re-focused the attention of many on low-energy supersymmetry.

On the other hand, like any other low-energy extension, new potentially large contributions to FCNC (and in some cases to LNV) arise. Their suppression to acceptable levels is definitely possible, but strongly restricts the "model-space". The absence of large low-energy FCNC is the most important information extracted from the data so far with regard to supersymmetry, and as we shall see, one's choice of how to satisfy the FCNC constraints (the "flavor problem") defines to great extent the model.

In conclusion, once large contributions to FCNC are absent the whole framework is consistent with all data, and furthermore, it is successful with regard to certain indirect probes. Its ultimate test, however, lies in direct searches in existing and future collider experiments. (See Zeppenfeld's lectures [15] for a discussion of collider searches and Halzen's lectures [16] for the discussion of possible non-collider searches.) In the course of these lectures all of the above statements and claims regarding supersymmetry will be discussed in some detail and presented using the tools and notions that we are about to define and develop.

1.5 Overview

Even though we narrowed down the focus of these lectures to (weak-scale) supersymmetry, it would still be impossible to cover in depth all that it entails. In the following, we chose to stress the more elementary details over the technicalities, the physical picture over over its formal derivation. While some issues are covered in depth, many others are only sketched or referred to, leaving the details to either optional exercises or further readings. Our mission is to gain acquaintance with the fundamental picture (regardless of the reader's background), and we attempt to do so in an intuitive fashion while providing a comprehensive and honest review of the related issues. Many reviews, lecture notes, and text books are available and references will be given to some of those as well as some early and recent research papers on the subjects discussed. (We do not attempt to give in these notes a complete reference list!) Some areas are particularly well covered by reviews, e.g., the algebra on the one hand, and experimental searches on the other hand, and hence will not be explored intensively here.

After providing a phenomenological "bottom-up" construction of supersymmetry, we turn to a brief discussion of the formalism – providing intuitive understanding as well as the necessary terminology. We then supersymmetrize the Standard Model, paying attention to some of the details, while only briefly discussing other aspects (such as searches). We then turn to an exploration of a small number of issues such as renormalization, unification, the Higgs boson, the flavor problem and the model space, attempting to leave the reader with a clear view of current research topics and avenues. Upon concluding, we return and comment on neutrinos and supersymmetry. An extended version of these lectures will appear at a later time.

2 "Bottom-Up" Construction

Though supersymmetry corresponds to a self-consistent field-theoretical framework, one can arrive at its crucial elements simply by requiring a theory which is at most logarithmically divergent: Logarithmic divergences imply only weak dependence on the ultra-violate cut-off and can be consistently absorbed in tree-level quantaties and thus understood as simply scaling (or renormalization) of the theory. Hence, unlike quadratic divergences – which imply strong dependence on the ultra-violate cut-off and need to be eliminated, they do not destabilize the infra-red theory and can be treated consistently.

The elimination of the quadratic divergences from the theory renders it "supersymmetric", hence the importance of supersymmetry to the discussion of any consistent high-energy extension of the SM. This bottom-up construction of "supersymmetry" is an instructive exercise. The most minimal example is of a (conserved) U(1) gauge theory, which will suffice here and will be considered below. This exercise could repeated for the case of a (conserved) non-Abelian gauge theory and in the case of SM (with a spontaneously broken non-Abelian gauge symmetry) which are technically more evolved. Suggested readings for this section include lecture notes Refs. [17,18,19].

We set to build a theory which contains at least one complex scalar (Higgs) field ϕ^+ (with a U(1) charge of $+1$) and which is at most logarithmically divergent. In particular, we would like to extend the field content and fix the couplings such that all quadratic divergences are eliminated. For simplicity, assume a massless theory and zero external momenta $p^2 = 0$, but all terms which are consistent with the symmetry are allowed. (The Feynamn gauge, with the gauge boson propagator given by $D_{\mu\nu} = -ig_{\mu\nu}/q^2$, is employed below.)

As argued above, the most obvious "trouble spot" is the scalar two-point function (Figs. 1 – 3). A U(1) theory with only scalar (Higgs) field(s) receives at one-loop positive contributions $\propto \Lambda_{UV}^2$ from the propagation of the gauge and Higgs bosons in the loops (Figs. 1 and 2). They are readily evaluated in the U(1) case to read

$$\delta m_\phi^2|_{\text{boson}} = + \left\{ (4 - 1)g^2 + \lambda \frac{N}{2} \right\} \int \frac{d^4q}{(2\pi)^4} \frac{1}{q^2}, \tag{8}$$

where g and λ are the gauge and quartic couplings, respectively, $N = 8$ is a combinatorial factor ($N = 2$ for non-identical external and internal fields), and keeping only leading terms the integral is replaced by $\Lambda_{UV}^2/16\pi^2$.

Obviously, the theory must contain fermions which could generate a negative contribution to δm_ϕ^2, if cancellation is to be achieved. A fermion loop equivalent to Fig. 3 exists and could cancel (8) if

1. There is a chiral fermion ψ_V with the quantum numbers (aside from spin) of the gauge boson (*i.e.*, neutral). (It follows ψ_V is a real Majorana fermion.)

2. There is one chiral fermion ψ_ϕ^+ with the quantum numbers (aside from spin) of the Higgs boson (i.e., with U(1) charge of $+1$).

3. There is a Yukawa interaction $a\sqrt{[(3g^2 + 4\lambda]}\phi\bar{\psi}_\phi\psi_V$ that leads to a negative contribution equivalent to Fig. 3, and where a is (at this point) a free proportionality coefficient.

Note the equal number (two) of bosonic and fermionic degrees of freedom (*d.o.f*) with the same quantum numbers which is implied by the first two conditions. The first and second requirements also guarantee that the above Yukawa interaction is allowed. In particular, the requirement of identical quantum number assignments guarantees its gauge invariance!

Let us redfine $\lambda \equiv bg^2$ and accordingly $ag\sqrt{(3 + 4b)} \equiv cg$. Thus, one arrives at

$$\delta m_\phi^2|_{\text{fermion}} = -2c^2 g^2 \int \frac{d^4q}{(2\pi)^4} \frac{1}{q^2}. \tag{9}$$

The coefficients b and c must be predetermined or otherwise a cancellation is always possible and would simply correspond to fine-tuning (of the coefficient this time). Even though the determination of the proportionality coefficient cannot be done unambiguously at the level of our current discussion, it follows from the consideration of the logarithmic divergences and the requirement that any proportionality relation is scale-invariant (i.e., that it is preserved under renormalization group evolution: $db/d\ln\Lambda = dc/d\ln\Lambda = 0$). One finds $c = \sqrt{2}$ and $b = 1/2$. However, after substitution in the above expressions the $\mathcal{O}(\Lambda_{\text{UV}}^2)$ terms still do not cancel! Indeed, this is a result of inconsistencies in our construction above.

Once fermions are introduced, one has to ensure that the (massless) fermion content is such that the theory is anomaly free, that is $\text{Tr}Q_{\psi_i} = 0$ (where the trace is taken over the U(1) charges of the fermions). The gauge fermion ψ_V carries no charge (it carries charge in non-Abelian theories, but being in the traceless adjoint representation it still does not contribute) so only the introduction of one additional chiral fermion ψ_ϕ^- with a U(1) charge of -1 is required in order to render the theory anomaly free. Even though by itself ψ_ϕ^- does not lead to new one-loop quadratically divergent contribution to the ϕ^+ two-point function, our newly adopted principle of equal number of bosonic and fermionic *d.o.f.* suggests that it should be accompanied by a complex scalar Higgs boson

ϕ^- which also carries a U(1) charge of -1. Hence, the most general gauge invariant quartic potential is that of a two Higgs model,

$$V = \lambda_1 |\phi^+|^2 |\phi^+|^2 + \lambda_2 |\phi^-|^2 |\phi^-|^2 + \lambda_3 |\phi^+|^2 |\phi^-|^2 + \lambda_4 |\phi^+ \phi^-|^2$$
$$+ \{ \lambda_5 (\phi^+ \phi^-)^2 + \lambda_6 |\phi^+|^2 \phi^+ \phi^- + \lambda_7 |\phi^-|^2 \phi^+ \phi^- + \text{h. c.} \}, \qquad (10)$$

where we adopted standard notation for the couplings. So far, we fixed λ_1. It is straightforward to show that the same arguments lead to $\lambda_2 = \lambda_1$. Redefining $\tilde{\lambda}_3 = \lambda_3 + \lambda_4$, similar wave-function renormalization and scale-invariance arguments fix $\tilde{\lambda}_3 = \tilde{b} g^2$ with $\tilde{b} = -1$. (Note that $\lambda_{6,7}$ (λ_5) may be fixed by consideration of one-loop (two-loop) wave-function mixing between ϕ^+ and ϕ^- and must vanish.)

Thus, we finally arrive at the desired result: A vanishing total contribution

$$\delta m_\phi^2 |_{\text{total}} = g^2 \left\{ 3 - 2c^2 + 4b^2 + \tilde{b}^2 \right\} \int \frac{d^4 q}{(2\pi)^4} \frac{1}{q^2} = 0, \qquad (11)$$

where the last term is from ϕ^- circulating in the loop (with $N = 2$). One concludes that in a consistent theory with (i) equal number of bosonic and fermionic $d.o.f.$, and (ii) scale-invariant relations among the respective couplings ("supersymmetry") the scalar two-point function is at most logarithmically divergent!

This is a remarkable result that suggests more generally the interaction terms, rewritten in a more compact form,

$$\mathcal{L}_{\text{int}} = \sqrt{2} g \sum_i Q_i \left(\phi_i \bar{\psi}_i \bar{\psi}_V + \text{h. c.} \right) - \frac{g^2}{2} \left\{ \sum_i Q_i |\phi_i|^2 \right\}^2, \qquad (12)$$

where we denoted explicitly the dependence on the U(1) charge Q_i. (This dependence is already apparent above in $\tilde{b} = Q_{\phi^+} Q_{\phi^-}$.) The most important lesson is that the theory contains only one free coupling: The gauge coupling. This could have been anticipated from our introductory discussion of divergences, and further suggests that the above Lagrangian should be derived in a supersymmtric theory from its gauge Lagrangian. (This will be illustrated in the following sections.)

Our discussion so far dealt with the elimination of the $\mathcal{O}(\Lambda_{\text{UV}})^2$ terms at one-loop order. They re-appear at higher loop orders, for example, two-loop $\mathcal{O}((g^2/16\pi^2)^2 \Lambda_{\text{UV}}^2)$ contributions to the two-point function, etc. It is straightforward to show, though it is quite evolved technically, that the condition found above are not only sufficient for the elimination of the undesired divergences at one-loop order, but rather suffice for their elimination order by order. Thus,

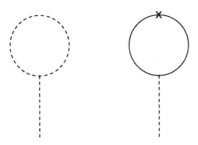

Figure 4: The divergent contributions to the one-point function of the singlet.

"supersymmetry", which postulates equal number of bosonic and fermionic $d.o.f.$ and specific scale-invariant relations among the various couplings, is indeed free of quadratic divergences at all orders in perturbation theory. This is a derivative of the non-renormalization theorems of supersymmetry.

The above prescription extends also to non-gauge interactions. For example, consider a theory with a gauge-singlet complex scalar and a gauge-singlet chiral fermion fields, ϕ_s and ψ_s, respectively (or in a more compact terminology, a superfield S with a scalar and a chiral fermion components ϕ_s and ψ_s, respectively): The theory contains equal number of fermionic and bosonic $d.o.f.$ The interaction Lagrangian

$$\mathcal{L}_{\text{int}} = y\phi_s\psi_s\bar{\psi}_s + A\phi_s^3 + A'\phi_s^*\phi_s^2 + \text{h. c.} - \lambda(|\phi_s|^2)^2 \tag{13}$$

leads to quadratically divergent contributions to the scalar two-point function (from the equivalents of diagrams 1c and 1d),

$$\delta m_{\phi_s}^2|_{\text{total}} = 4(\lambda - y^2) \int \frac{d^4q}{(2\pi)^4} \frac{1}{q^2}. \tag{14}$$

(Note the additional factor of 2 in the fermion loop from contraction of Weyl indices.) The cancellation of (14) leads to the relation $\lambda = y^2$ among the dimensionless couplings, which can be shown to also be scale-invariant. Again we find the same lesson of constrained quartic couplings!

The singlet theory contains another potential quadratic divergence, in the scalar one-point function (the tadpole). The contribution now depends on the dimensionful trilinear couplings A'. Their dimensionality prevented them from contributing to (14), but if non-vanishing they will contribute to the quadratically divergent tadpole. Their contribution may be off-set by a fermion loop $\propto \slashed{q} + \mu$, but only if the fermion is massive (Dirac) fermion with (a Dirac) mass $\mu \neq 0$. Both divergent loop contributions to the singlet one-loop function are illustrated in Fig. 4. Evaluation of the loop integral gives for the leading contributions to the tadpole

$$\delta T_{\phi_s} = \left\{ y \mathrm{Tr} I \mu - 4A' \right\} \int \frac{d^4 q}{(2\pi)^4} \frac{1}{q^2}. \tag{15}$$

The trace over the identity is four, leading to the (scale-invariant) condition $A' = y\mu$ so that the tadpole is canceled (at leading order).

Again, the relevant relations, e.g., $\lambda = y^2$, can be shown to extend to non-singlet Yukawa theories such as the SM.

Note that the trilinear interaction $A\phi_s^3$ as well as a mass term $m^2|\phi_s|^2$ do not lead to any quadratic divergences nor upset their cancellations as described above. Yet, one defines the supersymmetric limit as $m^2 = \mu^2$ and $A = 0$, corresponding to a fermion-boson mass degeneracy.

The singlet theory described above is essentially the Wess-Zumino model [20], which is widely regarded as the discovery of supersymmetry and from which stems the field-theoretical formulation of supersymmetry.

Exercises

1. Using the Feynamn rules of a $U(1)$ (QED) theory, confirm the numerical factors in eqs. (8) and (9).

2. Repeat our exercise for ϕ^- and show $\lambda_2 = \lambda_1$.

3. Repeat the $U(1)$ example in the Landau gauge, $D_{\mu\nu} = -\frac{i}{q^2}\left\{g_{\mu\nu} - \frac{q_\mu q_\nu}{q^2}\right\}$.

4. Show that $\lambda_{6,7} = 0$ by considering the loop corrections to the $m_{+-}^2 \phi^+ \phi^-$ term. Repeat the exercise for λ_5 which is constrained by two-loop corrections.

5. Show that a term $A\phi_s^3$ does not lead to any new quadratic divergences.

6. Show that a term $m^2 \neq \mu^2$ (boson fermion non-degeneracy) does not lead to any new quadratic divergences (but only to finite corrections $\propto m^2 - \mu^2$).

3 Supersymmetry

In the previous section it was shown that in trying to eliminate quadratic divergences in two- and one-point functions of a scalar field one is led to postulate an equal number of bosonic and fermionic $d.o.f.$ as well as specific relation among the couplings in the theory. (Uniqueness was achieved once the relations were required to be scale invariant.) From the consideration of the scalar-vector interaction one is led to postulate new gauge-Yukawa and gauge-quartic interactions. From the consideration of a (non-gauge) Yukawa interaction one is able to fix the scalar tri-linear and quartic couplings. These relations suggest small number of fundamental couplings, and hence a small number of fundamental (super)fields: Arranging fermions and bosons in spin multiplets, the superfields, which interact with each other. Hence, in this supersymmetric language the Lagrangian will contain a small number of terms even though when rewriting the interactions and kinetic terms in terms of the individual boson and fermion fields the Lagrangian will contain many terms with presumably mysterious correlations among their couplings.

The field theoretical formalism that realizes this elegant idea is supersymmetry, with the spin multiplets as its fundamental building blocks which are subject to the supersymmetric transformations; notions which bring back to mind low-energy flavor symmetries and their fundamental building blocks, the isospin multiplets. Isospin transformations act on the proton-neutron doublet. supersymmetric "spin" transformations act on the boson-fermion multiplet, the superfield. In order to define supersymmetry as a consistent (at this point, global) symmetry one needs to extend the usual space-time to superspace by the introduction of (not surprisingly) equal number of "fermionic" (or spinorial) coordinates

$$\{x_\mu\}_{\mu=1,...,4} \to \left\{ x_\mu, \begin{pmatrix} \theta_\alpha \\ \bar{\theta}_{\dot{\beta}} \end{pmatrix} \right\} \begin{array}{c} \mu = 1,...,4 \\ \alpha, \dot{\beta} = 1,2. \end{array}$$

The new coordinate θ is spin $1/2$ object, i.e., a Grassmann variable which carries two $d.o.f.$ (and which, as shown below, carries mass dimensions).

The formalism of supersymmetric field theories can be found in the literature: Refs. [21,22,23,24,25,18,26] provide a useful sample of such reviews, while Wess and Bagger [27] provide a complete though somewhat condensed discussion. It is not our purpose here to introduce it in an elaborate way, but rather to outline it for completeness and to illustrate some of its more intuitive aspects, which are the more relevant ones for the discussion in the rest of these lectures. We begin with a brief introduction to Grassmann algera. We will then turn define a chiral superfield and to illustrate the supersymmetry algebra. We then will fill some of the details such as kinetic and interaction terms. We conclude with a brief discussions of a gauge theory and, finally, of gauging supersymmetry itself.

3.1 Grassmann Algebra

The Grassmann variables are anticommuting variables,

$$\{\theta_{i_\alpha}, \theta_{j_\beta}\} = 0 \tag{16}$$

with the antisymmetric tensor

$$\epsilon_{\alpha\beta} = \begin{pmatrix} 0 & 1 \\ -1 & 0 \end{pmatrix}$$

as metric. One has, for example, $\theta_{i\,\alpha} = \epsilon_{\alpha\beta}\theta_i^\beta$, where Greek indices correspond to spinorial indices while Latin indices (which are sometimes omitted) label the variable. It is customary to reserve a special notation to the hermitian conjugate (as above) $(\theta_i^\alpha)^\dagger = \bar{\theta}_i^{\dot\alpha}$. The Pauli matrices, which act on Grassmann variables, are then written as $\sigma^\mu_{\alpha\dot\alpha}$.

One defines derivatives and integration with respect to the Grassmann variables as

$$\frac{d\theta_{i_\alpha}}{d\theta_{j_\beta}} = \delta_i^j \delta_\alpha^\beta, \tag{17}$$

$$\int d\theta_i = 0; \quad \int d\theta_{i_\alpha}\theta_{j_\beta} = \delta_{ij}\delta_{\alpha\beta}. \tag{18}$$

In Eq. (18) the first relation follows from the invariance requirement $\int d\theta f(\theta) = \int d\theta f(\theta + \alpha)$ where f is an arbitrary function, e.g., $f(\theta) = \theta$ and θ and α are two distinct Grassmann variables. The second relation is simply a choice of normalization. The derivative operator is often denoted with the short-hand notation ∂_{θ_j}.

The function $f(\theta)$ can be expanded as a polynomial in θ. However, the anticommuting nature of θ,

$$\theta^\alpha \theta_\alpha \theta_\alpha = \tfrac{1}{2}\theta^\alpha \{\theta_\alpha, \theta_\alpha\} = 0,$$

implies that at most it can be a quadratic polynomial in θ. In particular, any expansion is finite! Using its polynomial form and the above definitions it follows that

$$\int d\theta_i f(\theta_i) = \frac{df(\theta_i)}{d\theta_i}. \tag{19}$$

Another useful relation that will be implicitly used in some the our expansions is

$$\theta^\alpha \theta^\beta = -\frac{1}{2} \epsilon^{\alpha\beta} \theta^\alpha \theta_\alpha. \tag{20}$$

3.2 The chiral superfield: The L-Representation

Equipped with the algebraic tools, our obvious task is to embed the usual bosons and fermions into spin multiplets, the superfields. Begin by considering a Higgs boson, i.e., a (complex) scalar field $\phi(x_\mu)$. Requiring the corresponding superfield $\Phi(x_\mu, \theta_\alpha, \bar{\theta}^{\dot\beta})$ (which is a function defined in superspace rather than only space-time) to carry the same quantum numbers and dimensions as ϕ and expanding it, for example, in powers of θ_α (with x_μ-dependent coefficients and redefining $x_\mu \to x_\mu + i\bar\theta \sigma_\mu \theta$), one arrives at the following embedding:

$$\Phi(x_\mu, \theta_\alpha) = \phi(x_\mu) + \theta^\alpha \psi_\alpha(x_\mu) + \theta^\alpha \theta_\alpha F(x_\mu). \tag{21}$$

(We will return to the issue of an expansion in $\bar\theta^{\dot\beta}$ below.) From the spinorial nature of θ_α one has that ψ_α is also a spin 1/2 object (so that their product is a scalar), which can be identified with the usual chiral fermion. Now we can deduce the dimensionality of θ: $[\theta] = [\Phi] - [\psi] = [\phi] - [\psi] = -1/2$. Lastly, the coefficient of the θ^2 term must also be a (complex) scalar. (Note that if it was a vector field, which could also allow for the total spin of this term to be zero, it would have carried an additional space-time index unlike Φ.) Its mass dimension is also uniquely determined $[F] = [\Phi] - 2[\theta] = 2$.

The coefficient $F(x)$ cannot be identified with any known physical $d.o.f$ (there are no fundamental scalars whose mass dimension equals two!) and its presence implies four bosonic vs. only two fermionic $d.o.f.$, seemingly violating the supersymmetric principle found in the previous section. In order to understand the role of F and the resolution of this issue, let us first find the (supersymmetry) transformation law.

Consider the transformation $\hat{S}(\alpha, \bar\alpha)$ acting on the superfield $\Phi(x, \theta)$ of eq. (21) (where here α is also a Grassmann variable). Listing all objects at our disposal with dimensionality greater than $[\phi] - [\alpha] = 3/2$ but less than $[F] - [\alpha] = 5/2$ we have a spin 1/2 mass-dimension 3/2 object ψ; spin zero mass-dimension 2 objects F and $\partial_\mu \phi$; and a spin 1/2 mass-dimension 5/2 object $\partial_\mu \psi$. Using the Pauli matrices to contract space-time and spinorial indices one immediately arrives at the transformation law (up to the bracketed coefficients)

$$\delta\phi = \alpha^\alpha \psi_\alpha, \tag{22}$$

$$\delta\psi_\alpha = (2)\alpha_\alpha F + (2i)\sigma^\mu_{\dot\alpha\alpha}\bar\alpha^{\dot\alpha}\partial_\mu\phi, \tag{23}$$

$$\delta F = (-i)(\partial_\mu\psi^\alpha)\sigma^\mu_{\dot\alpha\alpha}\bar\alpha^{\dot\alpha}. \tag{24}$$

The "component fields" transforms into each other, which is possible since the superspace coordinate carries mass dimensions and spin. Hence, the superfield Φ transforms onto itself.

The (linearized) transformation law can be written in a more compact fashion:

$$S(\alpha,\ \bar\alpha)\Phi(x,\ \theta) = \left\{\alpha\frac{\partial}{\partial\theta} + \bar\alpha\frac{\partial}{\partial\bar\theta} + 2i\theta\sigma^\mu\bar\alpha\partial_\mu\right\}\Phi(x,\ \theta), \tag{25}$$

where the second term on the left-hand side is zero and the idendtity (20) was used to derive (24) from the last term. Rewriting the differential transformation in a standard form $S(\alpha,\ \bar\alpha)\Phi(x,\ \theta) = \{\alpha Q + \bar Q\bar\alpha\}\Phi(x,\ \theta)$, where Q is the supercharge, i.e., the generator of the supersymmetric algebra, on can identify the generators

$$Q_\alpha = \frac{\partial}{\partial\theta_\alpha}; \quad \bar Q_{\dot\alpha} = -\frac{\partial}{\partial\bar\theta_{\dot\alpha}} + 2i\theta^\alpha\sigma^\mu_{\dot\alpha\alpha}\partial_\mu, \tag{26}$$

and $P_\mu = i\partial/\partial x^\mu = i\partial_\mu$.

The generators are fermions rather than bosons! Their anticommutation relations can be found by performing successive transformations of Φ. After carefully re-arranging the result, it reads

$$\left\{Q_\alpha,\ \bar Q_{\dot\beta}\right\} = 2\sigma^\mu_{\alpha\dot\beta}P_\mu, \tag{27}$$

$$\{Q_\alpha, Q_\beta\} = \left\{\bar Q_{\dot\alpha}, \bar Q_{\dot\beta}\right\} = 0. \tag{28}$$

Relation (27) implies that in order to close the algebra, one need to include space-time. (This already hints towards a relation to gravity as is indeed one finds in the local case.) Thus, the algebra includes the usual commutation relation

$$[P_\mu,\ P_\nu] = 0 \tag{29}$$

as well as

$$[Q_\alpha,\ P_\nu] = [\bar Q_{\dot\alpha},\ P_\nu] = 0 \tag{30}$$

which is only the statement that the supersymmetry is a global symmetry. (More generally, supersymmetry provides a unique consistent extension of the

Poincare group.) Such an algebra which includes both commutation and anti-commutation relations among its generators is referred to as a graded Lie algebra.

The algebra described here has one supersymmetric charge, Q. Hence, this is a $N = 1$ (global) supersymmetry. In general, theories with more supersymmetries can be constructed. However, already at $N = 2$ theories do not contain chiral matter and cannot describe nature at electroweak energies. $N = 1$ is unique in this sense and $N \geq 2$ theories are not considered in these lectures (though one may consider explicit breaking of their vectorial symmetries and an effective $N = 1$ low-energy theory).

Though it is not our intention to discuss the algebra in great detail, some general lessons can be drawn by observation. the commutation between the charge and the Hamiltonian $[Q, H] = [Q, P_0] = 0$ implies that supersymmetric transformations which transform bosons to fermions and vice versa do not change the "energy levels", i.e., there is a boson-fermion mass degeneracy. The Hamiltonian can be written as $H = P_0 = (1/2)\{Q, \bar{Q}\} = QQ^\dagger \geq 0$ and hence is semi-positive definite. If (global) supersymmetry is conserved then the vacuum is invariant under the transformation $Q|0\rangle = 0$ and hence the vacuum energy $\langle 0|H|0\rangle = 0$ is zero, regardless whether, for example, the gauge symmetry is conserved or not. If (global) supersymmetry is broken spontanuously then the vacuum energy $\sim \|Q|0\rangle\|^2 > 0$ is positive. The semi-positive vacuum energy is an order parameter of (global) supersymmetry breaking. The Goldstone particle of supesymmetry is given by the operation of its (super) current $\sim Q_\alpha$ on the vacuum, and hence it is a fermion ψ_α, the Goldstino.

The superfield $\Phi(x_\mu, \theta_\alpha)$ is called a chiral superfield since it contains a chiral fermion. It constitutes a representation of the algebra as is evident from the closed relations that we were able to deduce – mostly by simple arguments. (The arbitrary coefficients in (22) – (24) were fixed by requiring consistency of the algebra.) It is called the left- (or L-)representation. The more general chiral representation, which contains both chiral (left-handed) and anti-chiral (right-handed) fermions, is related to it by $\Phi(x_\mu, \theta_\alpha, \bar{\theta}_{\dot{\alpha}}) = \Phi_L(x_\mu + i\theta_\alpha \sigma_\mu^{\alpha\dot{\alpha}} \bar{\theta}_{\dot{\alpha}}, \theta_\alpha)$ where the shift in x_μ is to be treated as a translation, and the object previously was denoted by Φ is now denoted by Φ_L. There is also a right (or R-)representation given by $\Phi_R(x_\mu, \bar{\theta}_{\dot{\alpha}}) = \Phi(x_\mu + i\theta_\alpha \sigma_\mu^{\alpha\dot{\alpha}} \bar{\theta}_{\dot{\alpha}}, \theta_\alpha, \bar{\theta}_{\dot{\alpha}})$. In these lectures the L-representation will be mostly used.

It is instructive to verify that the L-representation is closed. For example, we saw that $S(\alpha, \bar{\alpha})\Phi_L(x_\mu, \theta) = \Phi'_L(x_\mu, \theta)$. Similarly, it is straightforward to show that $\Phi^2 = \phi^2 + 2\theta\phi\psi + \theta^2(2\phi F - (1/2)\psi^2)$ is also a left-handed field, where $\theta^2 = \theta^\alpha \theta_\alpha$ ($\bar{\theta}^2 = \bar{\theta}_{\dot{\alpha}} \bar{\theta}^{\dot{\alpha}}$). Also, multiplying $\Phi^2\Phi$ one has $\Phi^3 = \cdots + \theta^2(\phi^2 F - (1/2)\phi\psi^2)$, a result which will be useful below.

Next the vector representation will be outlined before returning to discuss the dimension two scalar fields that seems to have appeared in the spectrum.

3.3 The vector superfield

A vector field $V_\mu(x_\mu)$ is to be embedded in a real quantity, a vector superfield $V(x_\mu, \theta_\alpha, \bar{\theta}_{\dot{\alpha}}) = V^\dagger(x_\mu, \theta_\alpha, \bar{\theta}_{\dot{\alpha}})$. Hence, aside from vector bosons, it can contain only real scalar and Majorana fermion fields. If indeed it contains spin zero and 1/2 fields, it cannot carry itself a space-time index. Also, it must depend on real combinations of θ and $\bar{\theta}$. In particular, its expansion contains terms up to a $\theta^2\bar{\theta}^2$ term. One then arrives at the following form (coefficients are again fixed by the consistency of the algebra)

$$V(x_\mu, \theta_\alpha, \bar{\theta}_{\dot{\alpha}}) = -\theta_\alpha \sigma_\mu^{\alpha\dot{\alpha}} \bar{\theta}_{\dot{\alpha}} V^\mu(x_\mu) + i\theta^2 \bar{\theta}^{\dot{\alpha}} \bar{\lambda}_{\dot{\alpha}} - i\bar{\theta}^2 \theta^\alpha \lambda_\alpha + \frac{1}{2}\theta^2\bar{\theta}^2 D. \quad (31)$$

The vector superfield V is a dimensionless spin zero object! Its component fields include, aside from the vector boson, also a Majorana fermion λ (the gaugino) and a real scalar field D, again with mass dimension $[D] = 2$. It again contains four bosonic vs. only two fermionic d.o.f., a problem we already encountered in the case of a chiral superfiled, and again we postpone its resolution.

In fact, the expansion (31) is not the most general possible expansion. The following terms: C, $i\theta\chi - i\bar{\theta}\bar{\chi}$, and $i\theta^2(A + iB) - i\bar{\theta}^2(A - iB)$, where A, B, C are real scalar fields and χ is again a Majorana fermion, are dimensionless, spin zero, real terms that could appear in V (in addition to other derivative terms). We implicitly eliminated the additional d.o.f. by fixing the gauge. While in ordinary gauge theory the gauge parameter is a real scalar field, in supersymmetry it has to be a chiral superfiled, hence it contains four bosonic and two fermionic d.o.f.. Subtracting one bosonic d.o.f. to be identified with the usual gauge parameter, three bosonic and two fermionic d.o.f. are available for fixing the "supersymmetric gauge". These can then be used to eliminate A, B, C, χ from the spectrum. This corresponds to the convenient Wess-Zumino gauge which is used in these lectures.

Similar considerations to the ones used above in order to write the chiral field transformation law (22) – (23) can be used to write the transformation law $S(\alpha, \bar{\alpha})V(x_\mu, \theta, \bar{\theta})$ of the component field of the vector superfiled,

$$\delta V_\mu = (i)\left[\alpha\sigma_\mu\bar{\lambda} + \bar{\alpha}\sigma_\mu\lambda\right], \quad (32)$$

$$\delta\lambda = \bar{\alpha}D + \alpha\sigma^{\mu\nu}\left[\partial_\mu V_\nu - \partial_\nu V_\mu\right], \quad (33)$$

$$\delta D = \sigma^\mu\partial_\mu\left[-\alpha\bar{\lambda} + \bar{\alpha}\lambda\right], \quad (34)$$

where spinorial indices are suppressed and $\sigma^{\mu\nu} = (i/2)[\sigma^\mu, \sigma^\nu]$, as usual.

An important object that transforms in the vectorial representation is $\Phi_L \Phi_L^\dagger$. In order to convince oneself, simply note that it is a real field. In more detail, note that $\Phi_L^\dagger$ must depend on $\bar\theta$ only, and is therefore in the right-handed representation, $\Phi_L^\dagger = \Phi_L^*(x_\mu + 2i\theta\sigma_\mu\bar\theta, \theta, \bar\theta)$. Hence, $\Phi_L \Phi_L^\dagger$ expansion contains terms up to $\theta^2\bar\theta^2$. Of particular interest to our discussion below is its coefficient, the D-term, $\Phi_L \Phi_L^\dagger = \cdots + \theta^2\bar\theta^2[FF^* - \phi\partial^\mu\partial_\mu\phi^* + (i/2)\psi\sigma_\mu\partial^\mu\bar\psi]$.

In turn, it is also possible to construct a chiral object from vectorial fields. A useful object is (for a U(1) theory) $W_\alpha = (\partial_{\bar\theta})^2(\partial_\theta + 2i\sigma_\mu\bar\theta\partial^\mu V)$. For our purposes it is enough show that W_α is a chiral field. This is obvious, since from the anti-commutation we have $\partial_{\bar\theta} W_\alpha = 0$ so it is independent $\bar\theta$ and hence is a left-handed chiral superfield.

3.4 Interaction and kinetic terms

After discussing chiral and vector superfields, understanding their expansion in terms of component fields and learning some simple manipulations of these super-objects, we are in position to do the final count of the physical $d.o.f.$; resolve the tantalizing access of bosonic $d.o.f.$ which are manifested in the form of dimension-two scalar fields; and derive the usual interaction and kinetic terms for the component fields. We begin with a non-gauge theory.

In fact, the two issues are strongly related. Observe that the F-term of the bilinear Φ^2 given by $\int d^2\theta\Phi^2$ includes terms which resemble mass terms, while $\int d^2\theta\Phi^3$ includes Yukawa-like terms. Also, δF given in (24) is a total derivative, for any chiral superfield. That is, the mysterious F-field contains interaction and mass terms on the one hand, and, on the other hand, $\delta \int d^4xF = \int d^4x(\text{total} - \text{derivative}) = 0$ so that $\int d^4xF$ is invariant under supersymmetric transformations and hence, is a good candidate to describe the potential. Indeed, if it is a non-propagating (auxiliary) $d.o.f.$ which encodes the potential then there are no physical spin-two bosonic $d.o.f.$

Clearly, one needs to identify the kinetic terms in order to verify this conjecture. A reasonable guess would be that they are given by the self-conjugate combination $\Phi\Phi^\dagger$. Indeed, the corresponding D-term given above contains kinetic terms for ϕ and ψ, but not for F. Below it is shown that the D-field itself also transforms (for any vector superfield) as a total derivative, so again it is not a coincidence that a term that could be consistently used to write the Lagrangian contains the appropriate kinetic terms. The F-fields are therefore auxiliary non-propagating fields that could be eliminated from the theory. That is, they do not correspond to any physical $d.o.f.$ (But what about the D-fields? The kinetic terms of the gauge-fields are given by the F-term of

W^2, which can be shown to not contain a kinetic term for the D-field: The D-term is again an auxiliary field which does not represent a physical $d.o.f.$)

More specifically, consider a theory described by a Lagrangian

$$\int d^2\theta \left[\mu\Phi^2 + y\Phi^3 + h.c.\right] + \int d^2\theta d^2\bar\theta \Phi\Phi^\dagger, \qquad (35)$$

We can then describe the component (ordinary) field theory by performing the integration over the super-space coordinates and extracting the corresponding F and D terms:

$$F_{\Phi^2} + F_{\Phi^2}^* : 2\mu\phi F - \frac{1}{2}\mu\psi\psi + h.c.; \qquad (36)$$

$$F_{\Phi^3} + F_{\Phi^3}^* : 3y\phi^2 F - \frac{3}{2}y\phi\psi\psi + h.c.; \qquad (37)$$

$$D : |\partial_\mu\phi|^2 - \frac{i}{2}\bar\psi\,\partial\!\!\!/\psi - FF^*; \qquad (38)$$

The Lagrangian $\mathcal{L} = T - V$ is given by their sum.

Indeed, no kinetic terms appear for F. Hence, one can eliminate the auxiliary $d.o.f.$ by solving $\partial V/\partial F = 0$ and substituting the solution

$$\frac{\partial V}{\partial F} = 0 \Rightarrow F^* = -(2\mu\phi + 3y\phi^2), \qquad (39)$$

back into the potential V. One finds for the scalar potential

$$V(\phi) = FF^* = 4|\mu|^2\phi\phi^* + 6(y\mu^*\phi^*\phi^2 + h.c.) + 9|y|^2|\phi\phi^*|^2 \qquad (40)$$

where all other interaction, mass and kinetic terms are already given explicitly in $(36) - (38)$.

It is now instructive to compare our result (40) to our conjectured "finite" theory from the previous lecture. In order to do so it is useful to use standard normalizations, i.e., $\mu \to \mu/2$; $y \to y/3$; and most importantly $\psi \to \psi/\sqrt{2}$. Not surprisingly, our Lagrangian includes all of (and only) the $d.o.f.$ previously conjectured and it exhibits the anticipated structure of interactions. In particular trilinear and quartic couplings are not arbitrary and are given by the masses and Yukawa couplings, as required. Quartic couplings are (semi-)positive definite.

3.5 The superpotential and the Kahler potential

Let us define a function
$$W = \mu\Phi^2 + y\Phi^3, \qquad (41)$$

then $F^* = -\partial W/\partial \Phi$ where the replacement $\Phi \to \phi$ is understood at the last step. The scalar potential is then simply

$$V = |\partial W/\partial \Phi|^2.$$

Also note that the potential terms involving the fermions is simply given by

$$-(1/2)\sum_{IJ}(\partial^2 W/\partial \Phi_I \partial \Phi_J)\psi_I \psi_J + h.c.$$

(These general "recipes" will not be justified here.) The function W is the superpotential which describes our theory. In general, the superpotential could be any analytic (holomorphic) function. In particular, it can depend on any number of chiral superfields Φ_I, but not on their complex conjugates $\Phi_I^\dagger$ (the holomorphicity property). For example, the most general renormalizable superpotential describing a theory with one singlet field is $W = C + l\Phi + \mu\Phi^2 + y\Phi^3$. That is, W has mass dimension $[W] = 3$ and any terms which involve higher powers of Φ are necessarily non-renormalizable terms. Indeed, $W = \kappa\Phi^4 \Rightarrow V = |\partial W/\partial \Phi|^2 = 8\kappa^2\phi^6$ implying that $[\kappa] = -1$ is a non-renormailzable coupling. The holomorphicity property of W is at the core on the non-renormalization theorems that protect its parameters from divergences (and of which some aspects were shown in the previous lecture but from a different point of view).

The non-gauge theory is then characterized by its superpotential, but also by its kinetic terms. Above, it was shown that the kinetic terms are given in the one singlet model simply by $\Phi\Phi^\dagger$. More generally, however, one could write a functional form

$$\Phi\Phi^\dagger \to K(\Phi, \Phi^\dagger) = K_J^I \Phi_I \Phi^J + \left(H^{IJ}\Phi_I \Phi_J + h.c.\right)$$

$$+\text{Non-renormalizable terms} \tag{42}$$

(with $\Phi^J = \Phi_J^\dagger$) which is a dimension $[K] = 2$ real function: The Kahler potential. The canonical kinetic terms $K_J^I \partial_\mu \phi_I \partial^\mu \phi^J = \sum_I |\partial_\mu \phi_I|^2$ are given by the "minimal" Kahler potential, which is a renormalizable function with $K_J^I = \delta_J^I$, as we implicitly assumed above. Note that holomorphic terms (i.e., terms which do not depend on $\Phi^\dagger$'s) such as $H^{IJ}\Phi_I\Phi_J + h.c.$ do not alter the normalization of the kinetic terms since $\int d^2\theta d^2\bar\theta \left\{H^{IJ}\Phi_I\Phi_J + h.c.\right\} = 0$. (These terms, however, could play an interesting role in the case of local supersymmetry, i.e., supergravity.) The Kahler potential which determines the normalization of the kinetic terms cannot share the holomorphicity of the superpotential. Hence, W and K are very different functions which together define the (non-gauge) theory.

3.6 The case of a gauge theory

The more relevant case of a gauge theory is obviously more technically evolved. Here, we will only outline the derivation of the component Lagrangian of an interacting gauge theory, stressing those element which will play a role in the following lectures. The pure gauge theory is described by

$$\int d^2\theta \frac{1}{2g^2} W^\alpha W_\alpha, \tag{43}$$

leading to the usual gauge kinetic terms, as well as kinetic terms for the gaugino λ, (setting for simplicity the gauge coupling $g = 1$)

$$F_{W^2} : -\frac{1}{4}|F^{\mu\nu}|^2 + \frac{1}{2}D^2 - \frac{i}{2}\bar\lambda \,/\!\!\!D\lambda. \tag{44}$$

There are kinetic term for the physical gauge bosons and fermions, but not for the D fields. The D fields are indeed auxiliary fields.

More generally, $(1/2)W^\alpha W_\alpha \to f_{\alpha\beta} W^\alpha W^\beta$ and $f_{\alpha\beta}$ is a holomorphic (analytic) function which determines the normalization of the gauge kinetic terms, hence, it is called the gauge-kinetic function. It could be a constant pre-factor or a function of a dynamical (super)field $f_{\alpha\beta}(\Phi)$. In either case it is essentially equivalent to the gauge coupling, $f_{\alpha\beta} = 1/2g^2$ (but here $g = 1$), only that in the latter case the gauge couplings is determined by the vev of the field Φ – the dilaton.

As before, the usual kinetic terms are described by the D-terms, however, they are now written as $\int d^2\theta d^2\bar\theta \Phi e^{2gQV} \Phi^\dagger$ where for simplicity the gauge coupling will be set again $g = 1$; $Q = 1$ is the charge, and $\Phi e^{2V} \Phi^\dagger = \Phi\Phi^\dagger + \Phi 2V\Phi^\dagger + \cdots$ contains the gauge-covariant kinetic terms $(\partial_\mu \to D_\mu \equiv \partial_\mu + iV_\mu)$. (More generally, $\Phi e^{2QV} \Phi^\dagger \to K(\Phi_I, e^{2Q_J V}\Phi^J)$.) That is,

$$D : |D_\mu\phi|^2 - \frac{i}{2}\bar\psi \,/\!\!\!D\psi + \phi^*D\phi + i\phi^*\lambda\psi - i\bar\lambda\bar\psi\phi - FF^*; \tag{45}$$

Note that when recovering the explicit dependence on the gauge coupling, the strength of the gauge Yukawa interaction $\phi^*\lambda\psi - \bar\lambda\bar\psi\phi$ is g. In the standard normalization $\psi \to \psi/\sqrt{2}$ it is $\sqrt{2}g$.

Consider now a theory

$$\int d^2\theta \left[(\mu\Phi^+\Phi^- + h.c.) + \frac{1}{2}W^\alpha W_\alpha \right] + \int d^2\theta d^2\bar\theta \left\{ \Phi^+ e^{2V}\Phi^{+\dagger} + \Phi^- e^{-2V}\Phi^{-\dagger} \right\}, \tag{46}$$

where the matter fields carry $+1$ and -1 charges. One has

$$\frac{\partial V}{\partial F^+} = 0 \Rightarrow F^{+*} = -\mu\phi^- \tag{47}$$

$$\frac{\partial V}{\partial F^-} = 0 \Rightarrow F^{-\,*} = -\mu\phi^+ \tag{48}$$

$$\frac{\partial V}{\partial D} = 0 \Rightarrow -D = \phi^{+\,*}\phi^+ - \phi^{-\,*}\phi^-, \tag{49}$$

where in deriving (49) we used eqs. (44) and (45). The scalar potential is now readily derived,

$$V(\phi) = FF^* + \frac{1}{2}D^2 = |\mu|^2\phi^+\phi^{+\,*} + |\mu|^2\phi^-\phi^{-\,*} + \frac{1}{2}|\phi^+\phi^{+\,*} - \phi^-\phi^{-\,*}|^2. \tag{50}$$

The quartic potential is now dictated by the gauge coupling and again is positive definite. Note that the quartic D-potential (the last term in (50)) has a flat direction $\langle\phi^+\rangle = \langle\phi^-\rangle$, potentially destabilizing the theory: This is a common phenomenon is supersymmetric field theories.

The derivation of the interactions in the case of a gauge theory essentially completes our sketch of the formal derivation of the ingredients that were required in the previous section in order to guarantee that the theory is at most logarithmically divergent. Specifically, quartic and gaugino couplings were found to be proportional to the gauge coupling (or its square) and with the appropriate coefficients. One is led to conclude that supersymmetry is indeed a natural habitat of weakly coupled theories. (Consequently, supersymmetry also provides powerful tool for the study of strongly coupled theories, an issue that will not be addressed in these lectures.) The next step is then the supersymmetrization of the Standard Model. Beforehand, however, we comment on gauging supersymmetry itself (though for the most part these lectures will be concerned with global supersymmetry).

3.7 Gauging supersymmetry

The same building blocks are used in the case of local supersymmetry, which is referred to as supergravity, only that the gravity multiplet (which includes the graviton and the spin 3/2 gravitino) is now a gauge multiplet; with the gravitino absorbing the Goldstino once supersymmetry is spontaneously broken and becoming massive with a mass

$$m_{3/2} = \frac{\langle W \rangle}{M_P^2} e^{\frac{\langle K \rangle}{2M_P^2}}. \tag{51}$$

The reduced Planck mass $M_P = M_{\text{Planck}}/\sqrt{8\pi} \simeq 2.4 \times 10^{18}$ GeV is the inverse of the supergravity expansion parameter (rather than the Planck mass itself) and it provides the fundamental mass scale in the theory. Hence, local supersymmetry corresponds to gauging gravity (and hence, supergravity). The appearance of momentum in the anticommutation relations of global supersymmetry already hinted in this direction. Gravity is now included automatically, opening the door for gauge-gravity unification. However, supergravity is a non-renormalizable theory and is still expected to provide only a "low-energy" limit of the theory of quantum gravity (for example, of a (super)string theory).

The gravitino mass could be given by an expectation value of any superpotential term, even a term that does not describe interactions of known light fields. Because supergravity is a theory of gravity, all fields in all sectors of the theory still "feel" the massive gravitino (since (super)gravity is modified). For example, one often envisions a scenario in which supersymmetry is broken spontaneously in a hidden sector – hidden in the sense that it interacts only gravitationally with the SM sector – and the SM sector appears as globally supersymmetric but with explicit supersymmetry breaking terms that are functions of the gravitino mass.

The F-terms are still good order parameters for supersymmetry breaking though they take a more complicated form (since they are now given by a covariant derivative of the superpotential):

$$-F_\Phi^* = \frac{\partial W}{\partial \Phi} + \frac{\partial K}{\partial \Phi}\frac{W}{M_P^2}, \tag{52}$$

where in the canonical normalization one has $\partial K/\partial \Phi = \phi^*$. Indeed, $|F| > 0$ if supersymmetry is broken. The scalar potential (or vacuum energy) now reads

$$V = e^{K/M_P^2}\left\{F_I K_J^I F^J - \frac{3}{M_P^2}|W|^2\right\}, \tag{53}$$

and is not a good order parameter as it could take either sign, or preferably vanish, even if $|F| > 0$. This is in fact a blessing in disguise as it allows one to cancel the cosmological constant (that was fixed in the global case to $\langle V\rangle = \langle|F|^2\rangle$) by tuning

$$\langle W\rangle = \frac{1}{\sqrt{3}}\langle F\rangle M_P. \tag{54}$$

One can define the scale of spontaneous supersymmetry breaking $M_{SUSY}^2 = \langle F\rangle \exp[\langle K\rangle/2M_P^2]$. The cancellation of the cosmological constant then gives $M_{SUSY}^2 = \sqrt{3}m_{3/2}M_P$ as a geometric mean of the gravitino and supergravity scales. Note that regardless of the size of M_{SUSY}, supersymmetry breaking is communicated to the whole theory (to the full superpotential) at the supergravity scale M_P, as is evident from the form of the potential (53). (The symbol M_{SUSY} itself is also used to denote the mass scale of the superpartners of the SM particles, as we discuss in the following lectures. In that case

M_{SUSY} is the scale of explicit breaking in the SM sector. The two meaning should not be confused.)

Of course, the supergravity scalar potential has also the usual D-term contribution $V_D = (1/4f_{\alpha\beta})D_\alpha D_\beta$, and the D-terms are also order parameters of supersymmetry breaking. (However, it seems more natural to fine-tune the supepotential to cancel its first derivative (the F-term) in order to eliminate the cosmological constant, rather than to cancel the square of the D-terms.)

We will briefly return to supergravity when discussing the origins of the explicit supersymmetry breaking in the SM sector, but otherwise the basic tools of global supersymmetry will suffice to our purposes.

Exercises

1. Since Φ_L by itself constitutes a representation of the algebra, its component fields cannot transform for example to a vectorial object V_μ. Show this explicitly from the consideration of dimensions, space-time and spinorial indices.

2. Use the above definition of Φ_R to show that it is independent of θ_α.

3. Confirm the expansion of Φ_L^2 and Φ_L^3.

4. Examine the expansion of the vector superfield (31), show that $V = V^\dagger$, and confirm all the assertion regarding spin and mass dimension of the component fields.

5. Derive the transformation law (32) –(34) (aside from coefficients) from dimensional and spin arguments.

6. Derive the expression for $\Phi_L^\dagger = \Phi_L^*(x_\mu + 2i\theta\sigma_\mu\bar\theta, \theta, \bar\theta) = [1 - \delta x_\mu P^\mu + (1/2)(\delta x_\mu P^\mu)^2]\Phi_L^*(x_\mu, \theta, \bar\theta)$ with $\delta x_\mu = 2i\theta\sigma_\mu\bar\theta$. Use your result to derive the D-term for $\Phi_L\Phi_L^\dagger$ by performing the integration $\int d^2\theta d^2\bar\theta \Phi_L\Phi_L^\dagger$.

7. Verify that our derivation of the non-gauge potential agrees with the theory conjectured in the previous section.

8. Find the correctly renormalized Yukawa couplings y, $W = (y/3)\Phi_I^3$, if $K_J^I = k_I\delta_J^I$ (The fields have to be rescaled so that the kinetic terms have their canonical form.); what about $W = y_{IJK}\Phi_I\Phi_J\Phi_K$? Show that this normalization is also the correct one for the quartic potential.

9. Show that $\int d^2\theta d^2\bar\theta \{H^{IJ}\Phi_I\Phi_J + h.c.\} = 0$.

10. Recover the explicit dependence on the gauge coupling in the expressions in Sec. 3.6. Show that the relevant quartic coupling in is proportional to $g^2/2$ and is indeed positive definite.

11. Introduce a Yukawa term $y\Phi^0\Phi^+\Phi^-$ to the model of eq. (46) and work out the scalar potential and the component Yukawa interactions. What are the quartic couplings?

12. Show that the Lagrangians (35), (43) and (46) are total derivatives.

13. Show that all the supergravity expressions reduce to the global supersymmetry case in the limit $M_P \to \infty$ in which supergravity effects decouple. In this limit, for example, supersymmetry breaking is not communicated from the hidden to the SM sector.

4 Supersymmetrizing The Standard Model

The discussion in the previous sections suggests that the SM could still be a sensible theory at the ultra violet and at the same time insensitive to the ultra-violet cut-off scale if supersymmetry is realized near the infra red regime (specifically, near the Fermi scale). Let us then consider this possibility and construct a (minimal) supersymmetric extension of the SM.

Nearly all of the necessary ingredients are already present within the SM: scalar bosons, (chiral) fermions and gauge bosons. Each of the SM gauge-boson multiplets requires the presence of a real (Majorana) fermion with the same quantum numbers, its *gaugino* superpartner; each SM fermion requires the presence of a complex scalar boson with the same quantum numbers, its *sfermion* superpartner (which also inherits its chirality label); the Higgs boson requires the introduction of the *Higgsino*, its fermion superpartner. It is straightforward to convince oneself that the theory will not be anomaly free unless two Higgsino doublets (and hence, by supersymmetry, two Higgs doublets) with opposite hypercharge are introduced so that the trace over the Higgsino hypercharge vanishes in analogy to the vanishing hypercharge trace of each SM generation. (Recall our construction in Section 2.) The minimal supersymmetric extension of the Standard Model (MSSM) as was outlined above is the extension with minimal new matter content, and it must contain a two Higgs doublet model (2HDM) (see eq. (10)). Note that the MSSM contains also Majorana fermions. Discovery of supersymmetry will not only establish the existence of fundamental scalar fields but also of Majorana fermions!

We list, following our previous notation, the MSSM field content in Table 2. For completeness we also include the gravity multiplet with the spin 2 graviton G and its spin 3/2 gravitino superpartner $\widetilde{G}$, as well as a generic supersymmetry breaking scalar superfield X (with an auxiliary component $\langle F_X \rangle \neq 0$) whose fermion partner $\widetilde{X}$ is the Goldstone particle of supersymmetry breaking, the Goldstino. (X parameterizes, for example, the hidden sector mentioned in Sec. 3.7.) It is customary to name a sfermion with an "s" suffix attached to the name of the corresponding fermion, e.g., *top*-quark $\rightarrow$ *stop*-squark and τ-lepton $\rightarrow$ *stau*-slepton. The gaugino partner of a gauge boson V is typically named V-ino, for example, W-boson $\rightarrow$ wino. The naming scheme for the mass eigenstates will be discussed after the diagonalization of the mass matrices. All superpartners of the SM particles are denoted as s-particles, or simply *sparticles*.

Given the above matter content and the constrained relations between the couplings that follow from supersymmetry, the MSSM is guaranteed not to contain quadratic divergences. In order to achieve that and have a sensible

Table 2: The Minimal Supersymmetric SM (MSSM) field content. Our notation is explained in Table 1. Each multiplet $(Q_c, Q_L)_{Q_{Y/2}}$ is listed according to its color, weak isospin and hypercharge assignments. Note that chirality is now associated, by supersymmetry, also with the scalar bosons. In particular, the superpartner of a fermion f_L^c is a sfermion $\tilde{f}_R^*$ and that of a fermion f_L is a sfermion $\tilde{f}_L$. The model contains two Higgs doublets. In the case of the matter (Higgs) fields, the same symbol will be used for the superfield as for its fermion (scalar) component.

Multiplet	Boson	Fermion
Gauge Fields		
$(8, 1)_0$	g	$\tilde{g}$
$(1, 3)_0$	W	$\tilde{W}$
$(1, 1)_0$	B	$\tilde{B}$
Matter Fields		
$(3, 2)_{\frac{1}{6}}$	$\tilde{Q}_a$	Q_a
$(\bar{3}, 1)_{-\frac{2}{3}}$	$\tilde{U}_a$	U_a
$(\bar{3}, 1)_{\frac{1}{3}}$	$\tilde{D}_a$	D_a
$(1, 2)_{-\frac{1}{2}}$	$\tilde{L}_a$	L_a
$(1, 1)_1$	$\tilde{E}_a$	E_a
Symmetry Breaking		
$(1, 2)_{-\frac{1}{2}}$	H_1	$\tilde{H}_1$
$(1, 2)_{\frac{1}{2}}$	H_2	$\tilde{H}_2$
Gravity and Supersymmetry Breaking		
$(1, 1)_0$	G	$\tilde{G}$
$(1, 1)_0$	X	$\tilde{X}$

formalism to treat a fundamental scalar we were forced to introduce a complete scalar replica of the SM matter which, however, supersymmetry enables us to understand in terms of a boson-fermion symmetry. By doing so, one gives up in a sense spin as a good quantum number. For example, the Higgs doublet H_1 and the lepton doublets L_a are indistinguishable at this level: In the SM the former is a scalar field while the latter are fermions. The anomaly cancellation considerations expalined above do not allow us to identify H_1 with L_a, i.e., the Higgs boson with a slepton (or equivalently, a lepton with a Higgsino). The possibility of lepton-Higgs mixing arises naturally if no discrete symmetries are imposed to preserve lepton number, a subject that we will return to below. Spin and the SM spectrum correspond in this framework to the low-energy limit. That is, supersymmetry must be broken at a scale below the typical cut-off scale that regulates the quadratic divergences $\Lambda \sim (4\pi/\alpha)M_W$. If the stop-squark, for example, is heavier, then divergences must be again fine-tuned away – undermining our original motivation. As we shall see, consistency with experiment does not allow the stop to be much lighter.

As explaine in the previous section, the boson-fermion symmetry and the duplication of the spectrum follow naturally once the building blocks used to describe nature (at electroweak energies and above) are the (chiral and vector) superfields. However, from the low-energy point of view it is the *component field* iteractions which are relevant. Let us then elaborate on the structure of the *component field* iteractions, which is dictated by the supersymmetry invariant interactions of the parent superfields.

Before doing so, it is useful fix the normalization of the Higgs fields, as thier expectation values appear in essentaily all mass matrices of the MSSM fields. It is costumary to define in 2HDM in general, and in the MSSM in particular

$$\sqrt{\langle H_1^0 \rangle^2 + \langle H_2^0 \rangle^2} = \tfrac{\sqrt{2}}{g}M_W \simeq 174 \text{ GeV} \equiv \nu.$$

Note that a $1/\sqrt{2}$ factor is now absorbed in the definiton of H_i (and ν) in comparison to the Higgs doublet definition in the SM. The *vev* ν is not a free parameter but is fixed by the Fermi scale (or equivalently by the W or Z mass). However, the ratio of the two *vev's*

$$\tan \beta = \tfrac{\langle H_1^0 \rangle}{\langle H_2^0 \rangle} \geq 1$$

is a free parameter. Its positive sign corresponds a conventioanl phase choice that we adopt, and its lower bound stems from pertubativity of Yukawa couplings. Correspondingly, one has $\langle H_1^0 \rangle = \nu \cos \beta$ and $\langle H_2^0 \rangle = \nu \sin \beta$.

The supersymmatrization of the standard model is also discussed in many of the review and notes listed in the previous lectures (particualrly, Refs.[21,18,26]), as well as in Refs.[28] and [29].

4.1 Yukawa, gauge-Yukawa, and quartic interactions

Identifying the gauge kinetic function $f_{\alpha\beta} = \delta_{\alpha\beta}/2g_a^2$ for a gauge group a, our previous exercise in Section 3.6 showed that the fermion f coupling to a vector boson V_a with a coupling g_a implies, by supersymmetry, the following interaction terms:

$$g_a V_a \bar{f} T_a^i f \to \begin{cases} \sqrt{2} g_a \lambda_a \tilde{f}^* T_a^i f + \text{h.c.} \\ \\ \frac{g_a^2}{2} \sum_I \left| \tilde{f}_I^* T_a^i \tilde{f}_I \right|^2 \end{cases},$$

in addition to usual gauge interactions. Here, we explicitly denote the gauge group generators T_a^i which are taken in the appropriate representation (of the fermion f) and which are reduced to the fermion charge in the case of a U(1) considered above. The gaugino is denoted by λ_a and, converting to standard conventions, it defines to absorb a factor of i which appeared previously in the gaugino-fermion-sfermion vertex. Also, the fermion component of the chiral superfield absorbs a factor of $1/\sqrt{2}$ as explained in the previous section. The gaugino matter interaction is often referred to as gauge-Yukawa interaction. The quartic interaction is given by the square of the auxiliary D field and hence includes a summation over all the scalar fields which transform under the gauge group, weighted by their respective charge. This leads to "mixed" quartic interactions, for example, between two squarks and two Higgs bosons.

The matter (Yukawa) interactions, and as we learned – the corresponding tri-linear and quartic terms in the scalar potential, are described by the superpotential. Recall that the superpotential W is a holomorphic function so it cannot contain complex conjugate fields. Particularly, no terms involving H_1^* can be written. Indeed, we were already forced to introduce $H_2 \sim H_1^*$, so a Yukawa/mass term for the up quarks involving H_2 can be written. The holomorphicity property offers, however, an independent reasoning for introducing two Higgs doublets with opposite hypercharge. Keeping in mind the generalization $V \sim y\phi_i\psi_j\psi_k \to W \sim y\Phi_I\Phi_J\Phi_K$, the MSSM Yukawa superpotential is readily written

$$W_{\text{Yukawa}} = y_{l_{ab}} \epsilon_{ij} H_1^i L_a^j E_b + y_{d_{ab}} \epsilon_{ij} H_1^i Q_a^j D_b - y_{u_{ab}} \epsilon_{ij} H_2^i Q_a^j U_b, \tag{55}$$

were SU(2) indices are explicitly displayed (while SU(3) indices are suppressed) and our phase convention is such that all "mass" terms and Yukawa couplings are positive.

It is instructive to explicitly write all the component interactions encoded in the superpotential (55). Decoding the SU(2) structure and omitting flavor indices, one has

$$W_{\text{Yukawa}} = y_l(H_1^0 E_L^- E - H_1^- N_L E) + y_d(H_1^0 D_L D - H_1^- U_L D) \\ + y_u(H_2^0 U_L U - H_2^+ D_L U), \tag{56}$$

with self explanatory definitions of the various superfield symbols. We note in passing that explicit decoding of the SU(2) indices need to be carried out cautiously in certain cases, for example, when studying the vacuum structure which usually requires the application of SU(2) rotations. However, it is acceptable here. Considering next, for example, the first term $y_l H_1^0 E_L E$, one can derive the component interaction $\sim (\partial^2 W/\partial \Phi_I \Phi_J)\psi_I \psi_J + \text{h.c.} + |\partial W/\partial \Phi_I|^2$ (where the substitution $\Phi \to \phi$ in $\partial^n W$ is understood),

$$\mathcal{L}_{\text{Yukawa}} = y_l \left\{ H_1^0 E_L^- E + \widetilde{H}_1^0 \widetilde{E}_L^- E + \widetilde{H}_1^0 E_L^- \widetilde{E} \right\} + \text{h.c.}, \tag{57}$$

$$V_{\text{scalar(quartic)}} = y_l^2 \left\{ |\, \widetilde{E}_L^- \widetilde{E} \,|^2 + |\, H_1^0 \widetilde{E}_L^- \,|^2 + |\, H_1^0 \widetilde{E} \,|^2 \right\}. \tag{58}$$

A single Yukawa interaction in the SM $H_1^0 E_L^- E$ leads, by supersymmetry, to two additional Yukawa terms and three quartic terms in the MSSM (or any other supersymmetric extension).

It is important to note that the holomorphicity of the superpotential which forbids terms $\sim H_1^* QU$, $H_2^* QD$, $H_2^* LE$ automatically implies that the MSSM contains a 2HDM of type II, that is a model in which each fermion flavor couples only to one of the Higgs doublets. This is a crucial requirement to suppressing dangerous tree-level contributions to FCNC from operators such as $QQUD$ which result from virtual Higgs exchange in a general 2HDM, but which do not appear in a type II model. This requirement is automatically satisfies by the virtue of the superpotential properties.

4.8 The Higgs mixing parameter

While no "supersymmetric" (i.e., a holomorphic superpotential) mass involving the (SM) matter fields can be written, a Higgs mass term $\mu\epsilon_{ij}H_1^i H_2^j$ is gauge invariant and is allowed. The dimension $[\mu] = 1$ parameter mixes the two doublets and acts as a Dirac mass for the Higgsinos. The Dirac fermion and the Higgs bosons are then degenerate, as implied by supersymmetry, with mass

$|\mu|$. The μ-parameter can carry an arbitrary phase. Hence, choosing $\mu > 0$, one has for the MSSM superpotential

$$
\begin{aligned}
W_{\text{MSSM}} &= W_{\text{Yukawa}} \pm \mu\epsilon_{ij} H_1^i H_2^j \\
&= y_{l_{ab}} \epsilon_{ij} H_1^i L_a^j E_b + y_{d_{ab}} \epsilon_{ij} H_1^i Q_a^j D_b \\
&\quad - y_{u_{ab}} \epsilon_{ij} H_2^i Q_a^j U_b \pm \mu\epsilon_{ij} H_1^i H_2^j.
\end{aligned} \tag{59}
$$

Following the standard procedure it is straightforward to derive the new interaction and mass terms, which complement the usual Yukawa and quartic potential terms,

$$
\mathcal{L}_{\text{Dirac}} = \mp\mu\widetilde{H}_1^- \widetilde{H}_2^+ \pm \mu\widetilde{H}_1^0 \widetilde{H}_2^0 + \text{h.c.}, \tag{60}
$$

$$
V_{\text{scalar(mass)}} = \mu^2 \left(H_1 H_1^\dagger + H_2 H_2^\dagger \right), \tag{61}
$$

$$
V_{\text{scalar(tri-linear)}} = \pm\mu y_l H_2 \widetilde{L}^\dagger \widetilde{E}^\dagger \pm \mu y_d H_2 \widetilde{Q}^\dagger \widetilde{D}^\dagger \mp \mu y_u H_1 \widetilde{Q}^\dagger \widetilde{U}^\dagger + \text{h.c.}, \tag{62}
$$

where SU(2) indices were also suppressed and the fermion mass terms were written in SU(2) components, as it will be useful later when constructing the fermion mass matrices. Note that the tri-linear terms arise from the cross terms in $|\partial W/\partial H_2|^2 = |\pm\mu H_1 - y_u QU|^2$ and $|\partial W/\partial H_1|^2 = |\pm\mu H_2 + y_l LE + y_d QD|^2$.

Could it be that $\mu \simeq 0$? Naively, one may expect that in this case the Higgsinos are massless and therefore Z-boson decays to a Higgsino pair $Z \rightarrow \widetilde{H}\widetilde{H}$ should have been observed at the Z resonance (from total or invisible width measurements, for example). In fact, once electroweak symmetry breaking effects are taken into account (see below) the two charged Higgsinos as well as one neutral Higgsino are massive with M_W and M_Z masses, respectively. Their mass follows from the gauge-Yukawa interaction terms $g\langle H_i^0 \rangle \widetilde{W}\widetilde{H}$ and $g\langle H_i^0 \rangle \widetilde{Z}\widetilde{H}$, (setting $\tan\beta = 1$). While the Z decays are kinematicly forbidden in this case, charged Higgsinos should have been produced in pairs at the WW production threshold at the Large Electron Positron (LEP) collider at CERN. The absence of anomalous events at the WW threshold allows one to exclude this possibility and to bound $|\mu|$ from below. (This observation will become clearer after defining the fermion mass eigenstates; For a discussion see Ref.[30].)

Obviously, $|\mu|$ cannot be too large or the Higgs doublets will decouple, re-introducing the hierarchy problem. As we proceed we will see that μ must encode information on the ultra-violet in order for it to be a small parameter of the order of the Fermi scale. Hence, it may contain one of the keys to unveiling the high-energy theory.

4.3 The supersymmetric limit of the MSSM

Having established the MSSM matter-matter and matter-gaugino interactions (in the supersymmetric limit in which only D and F terms are considered), the new particle spectrum can be written down. In the limit of conserved electroweak $SU(2) \times U(1)$ symmetry all particles – fermions and bosons, new and old – are massless. It is electroweak symmetry breaking (EWSB) that is responsible for all mass generation, as is required by the boson - fermion degeneracy. In fact, supersymmetry is not conserved but is spontaneously broken once electroweak symmetry is broken since the non-vanishing Higgs *vevs* generate non-vanishing F and D *vevs*, the order parameters of supersymmetry breaking, as will be shown below. Therefore, some non-degeneracy is expected, and our limit is to be understood as the MSSM with no explicit supersymmetry breaking. Let us then postulate (as will see, it does not occur in this limit) that the Higgs fields acquire *vevs* which break spontaneously $SU(2) \times U(1)$ and consider the impact on supersymmetry breaking and on the sparticle spectrum.

Consider, for example, the sfermion spectrum. The spectrum can be organized by flavor (breaking $SU(2)$ doublets to their components), each flavor sector constitute a left-handed $\widetilde{f}_L$ and right-handed $\widetilde{f}_R$ sfermions (and in principle all sfermions with the same QED and QCD quantum numbers could further mix in a 6×6 subspaces). One has in each sector three possible mass terms $m_{LL}^2 \widetilde{f}_L \widetilde{f}_L^\dagger$, $m_{RR}^2 \widetilde{f}_R \widetilde{f}_R^\dagger$, $m_{LR}^2 \widetilde{f}_L \widetilde{f}_R$ + h.c. with $m_{LL,\,RR}^2 \sim \langle D \rangle \sim \nu^2$ generated by substituting the Higgs *vev* in the quartic potential and $m_{LR}^2 \sim \langle F \rangle \sim \mu\nu$ generated by substituting the Higgs *vev* in the tri-linear potential. The mass matrix for the sfermions $(\widetilde{f}_L, \widetilde{f}_R)$

$$
m_{\widetilde{f}}^2 = \left(\begin{array}{cc} m_{LL}^2 & m_{LR}^2 \\ m_{LR}^{2\,*} & m_{RR}^2 \end{array} \right)
$$

then reads in this limit

$$
\left(\begin{array}{cc} m_f^2 + M_Z^2 \cos 2\beta \left[T_3^f - Q_{f_L} \sin^2 \theta_W \right] & m_f \mu^* \tan \beta \ (\text{or } 1/\tan \beta) \\ m_f \mu \tan \beta \ (\text{or } 1/\tan \beta) & m_f^2 - M_Z^2 \cos 2\beta \times Q_{f_R} \sin^2 \theta_W \end{array} \right).
$$

The sneutrino is an exception since there are no singlet (right-handed) neutrinos and sneutrinos, so its mass is simply $m_{LL}^2 \widetilde{\nu}_L \widetilde{\nu}_L^*$.

It is instructive to examine all contributions in some detail (see also exercises). The diagonal fermion mass term is the F-contribution $|y_f \langle H \rangle \widetilde{f}_{L,\,R}|^2$. Note that this F-term itself has no *vev* and hence does not break supersymmtry but only lead to a degenerate mass contribution to the fermion and to the re-spective sfermions. The second diagonal contribution is that of the D terms which has non-vanishing vev

$$\langle D \rangle^2 = \frac{g^2 + g'^2}{8} \left(\langle H_2 \rangle^2 - \langle H_1 \rangle^2 \right)^2 = \frac{1}{4} M_Z^2 \nu^2 |\cos 2\beta|^2 = \langle V_{\text{MSSM}} \rangle^2, \qquad (63)$$

where $\cos 2\beta < 0$ appears only as absolute value (so that $\langle V_{\text{MSSM}} \rangle \geq 0$ as re-quired by (global) supersymmetry), $T_3 = \pm 1/2$ for the Higgs weak isospin was used. The above relations hold in general and not only in the supersymmetric limit. Note that the limit $\tan \beta \to 1$ corresponds to $\langle D \rangle^2 \to 0$ and supersym-metry is recovered. Indeed in this limit the diagonal sfermion masses squared are given by the respective fermion mass squared. This limit also corresponds to a flat direction in field space (as alluded to in the previous lecture) along which the (tree-level) potential vanishes. A flat direction must correspond to a massless (real) scalar field in the spectrum (its zero mode) so there is a boson whose (tree-level) mass is proportional to the D-term and vanishes as $\cos 2\beta \to 0$. This is the (model-independent) light Higgs boson of supersymme-try. This is a crucial point in the phenomenology of the models which we will return to when discussing the Higgs sector in lecture 5. It is then straightfor-ward, after rewriting the hypercharge in term of electric charge Q_f, to arrive to the contribution to the sfermion mass squared One can readily calculate the numerical coefficients $T_3 - Q_f \sin^2 \theta_W$ using $\sin^2 \theta_W \simeq 0.23$ for the weak angle to find $T_3 - Q_f \sin^2 \theta_W \simeq 0.34, 0.14, -0.42, -0.08, -0.27, -0.23, 0.5$ for $f = u_L, u_L^*, d_L, d_L^*, e_L, e_L^*, \nu$ (and their generational replicas).

Lastly, the off-diagonal terms (often referred to as left-right mixing) are supersymmtry breaking terms which split the sfermion spectrum from the fermion spectrum even in the absence of D-term contributions. These terms arise from the cross-terms in F_{H_i} as do the supersymmetry breaking $vevs$ $\langle F_{H_{1,2}} \rangle^2 = y_f^2 \mu^2 \langle H_{2,1} \rangle^2$. The $\tan \beta$ dependence in mass squared matrix as-sumes $f = d, l$ ($f = u$), i.e., a superpotential coupling to H_1 (H_2). The pres-ence of the left-right terms implies that the sfermion mass eigenstates have no well-defined chirality association.

Since supesymmetry is spontaneously broken by the Higgs $vevs$, the Hig-gsinos provide the Goldstino. This is the Achilles heal of this scheme: Let us now define the supertrace function

$$\text{STR}_I \mathcal{F}(O_I) \equiv \sum_I N_{C_I} (-1)^{2S_I} (2S_I + 1) \mathcal{F}(O_I)$$

which sums over any function $\mathcal{F}$ of an object O_I which carries spin S_I. (The summation is also over color and isospin factor which here we pre-summed in the color factor N_C.) In the $\langle D \rangle \to 0$ limit, the supertrace summation over the mass eigenvalues of the fermions and sfermions in each flavor sector is zero! It implies that after spontaneous supersymmetry breaking (in this limit) one has $m^2_{\tilde{f}_{1,2}} = m^2_f \pm \Delta$ where $\Delta = \langle F \rangle$ is given by the F-vev that spontaneously break supersymmetry, in our case the left-right mixing term. Hence, unless $|\langle F \rangle| < m_e \sim 0$, which it cannot be given the lower bound on $|\mu|$ discussed previously, a sfermion must acquire a negative mass squared so that the SM does not correspond to a minimum of the potential (and furthermore, QED and QCD could be broken in the vacuum). This situation is, of course, intolerable. It arises because the matter fields couple directly to the Goldstino supermultiplet. (It is a general (supertrace) theorem for any spontaneous braking of global supersymmetry.) Another general implication is that the fermions do not feel the spontaneous supersymmetry breaking (at tree level). Turning on the D-contributions does not improve the situation. Eventhough their sum over a single sector does not vanish, gauge invariance guarantees that their sum over each family vanishes, i.e., negative contributions to some eigenvalues of the mass squared mass, as indeed we saw above.

This comes as no surprise, as the Higgs squared mass matrix in this limit simply reads

$$m^2_H = \begin{pmatrix} \mu^2 & 0 \\ 0 & \mu^2 \end{pmatrix}$$

and has no negative eigenvalues. Thus, it cannot produce the SM Higgs potential and the Higgs fields do no acquire a vev, so our $ad\ hoc$ assumption of electroweak symmetry breaking cannot be justified. Note that a negative eigenvalue could arise, however, from an appropriate off-diagonal term. Had we extended the model by including a singlet superfield S, replacing $W = \mu H_1 H_2$ with $W = y_S S H_1 H_2 - \mu_s^2 S$, then electroweak symmetry could be broken while supersymmetry is preserved. One has for the Higgs potential $V = |F_S|^2 = |y_S H_1 H_2 - \mu_s^2|^2$. Its minimization gives $\langle H_1^0 \rangle = \langle H_2^0 \rangle = \mu_s/\sqrt{y_S}$ and $\langle S \rangle = 0$. Electroweak symmetry is broken along the flat direction $V = \langle F_S \rangle^2 = 0$ so that supersymmetry is conserved with $m^2_{\tilde{f}} = m^2_f$ (and with no left-right mixing!). This is a counter example to our observation that in general electroweak breaking implies supersymmetry breaking. Models in which μ is replaced by a dynamical singlet field are often called the next to minimal extension (NMSSM). The specific NMSSM model described here is, however, still far from leading to a realistic model.

Of course, one need not go through the exercise of electroweak symmetry breaking in order to convince oneself that the models as they are manifested in this limit are inconsistent. The gluino partner of the gluon, for example, cannot receive its mass from the colorless Higgs bosons and remains massless at tree level, even if electroweak symmetry was successfully broken. It would receive a $\mathcal{O}(\text{GeV})$ or smaller mass from quantum corrections if the winos and bino are massive. Such a light gluino would alter SM QCD predictions at the level of current experimental sensitivity [31] while there is no indication for its existence. (It probably cannot be ruled out if it is the lightest supersymmetric particle [32], in which case it must hadronize.)

We must conclude that a realistic model must contain some other source of supersymmetry breaking. The most straightforward approach is then to parameterize this source in terms of explicit breaking in the low-energy effective Lagrangian.

4.4 Soft supersymmetry breaking

Our previous exercises in electroweak and supersymmetry breaking lead to a very concrete "shopping list" of what a realistic model should contain:

- Positive squared masses for the sfermions;

- Gaugino masses, particularly for the gluino;

- Possibly an off-diagonal Higgs mass term $m^2 H_1 H_2$ as it could lead to the desired negative eigenvalue.

We can contrast our "shopping list" with what we learned that we cannot have (in order to preserve the cancellation of quadratic divergences):

- Arbitrary quartic couplings;

- Arbitrary tri-linear singlet couplings in the scalar potential $m s \phi \phi^*$;

- Arbitrary fermion masses.

The two last items are in fact related (and fermion masses are constrained only in models with singlets). Hence, both items do not apply to the MSSM (but apply in the NMSSM). We conclude that acceptable spectrum can be accommodated without altering cancellation of divergences! Such explicit breaking of supersymmetry is soft breaking in the sense that no quadratic divergences are re-introduced above the (explicit) supersymmetry breaking scale.

Let us elaborate and show that sfermion mass parameters indeed are soft. For example, consider the effect of the soft operator $m^2 \phi \phi^*$ and its modification of our previous calculation of the quadratic divergence due to the quartic scalar interaction $y^2|\phi_1|^2|\phi_2|^2$:

$$y^2 \int \frac{d^4q}{(2\pi^4)} \frac{1}{q^2} \to y^2 \int \frac{d^4q}{(2\pi^4)} \frac{1}{q^2 - m^2}$$

$$\sim y^2 \int \frac{d^4q}{(2\pi^4)} \left\{ \frac{1}{q^2} + \frac{m^2}{q^4} \right\} \sim y^2 \int \frac{d^4q}{(2\pi^4)} \frac{1}{q^2} + i\frac{y^2}{16\pi^2} m^2 \ln \frac{\Lambda_{\rm UV}^2}{m^2}, \quad (64)$$

a result which translates to a logarithmically divergent mass correction

$$\Delta m_i^2 = -y^2 \frac{m_j^2}{16\pi^2} \ln \frac{\Lambda_{\rm UV}^2}{m^2}, \quad (65)$$

and similarly for the g^2 quartic interaction. Note the negative over-all sign of the correction! The logarithmic correction is nothing but the one-loop renormalization of the mass squared parameters due to its Yukawa interactions, and it is negative. The implications are clear - a negative squared mass, for example in the weak-scale Higgs potential, may be a a result of quantum corrections - a subject which deserves a dedicated discussion, and which we will return to in the next lecture. (Of course, integrating (64) properly one finds also finite corrections to the mass parameters which will not be discussed here.)

An important implication of the above discussion is that light particles are protected from corrections due to the decoupling of heavy particles (as heavy as the ultra-violet cut-off scale), as long as the decoupling is within a supersymmetric regime, i.e., $(m_{\rm heavy-boson}^2 - m_{\rm heavy-fermion}^2)/(m_{\rm heavy-boson}^2 + m_{\rm heavy-fermion}^2) \to 0$. This ensures the sensibility of the discussion of grand-unified theories or any other theory with heavy matter and where there is tree-level mixing between light and heavy matter (leading to one-loop corrections). Corrections due to decoupling of heavy particles are still proportional only to the soft parameters which therefore must appear with the same order of magnitude in both the light and heavy spectrum. This persists also at higher loop orders. (The only caveat being mixing among heavy and light singlets.)

Our potential $V = V_F + V_D + V_{\rm SSB}$ now contains superpotential contributions $V_F = |F|^2$, gauge contributions $V_D = |D|^2+$ gaugino-Yukawa interactions, and a contribution that explicitly but softly break supersymmetry, the soft supersymmetry breaking (SSB) terms which most generally consist of

$$V_{\rm SSB} = m_{j*}^{i\,2}\phi_i\phi^{j*} + \left\{ B^{ij}\phi_i\phi_j + A^{ijk}\phi_i\phi_j\phi_k + C_{i*}^{jk}\phi^{i*}\phi_j\phi_k + \right.$$

$$\left. + \frac{1}{2}M_\alpha\lambda_\alpha\lambda_\alpha + {\rm h.c.} \right\}. \quad (66)$$

The soft parameters were originally classified by Girardello and Grisaru[33], while more recent and general discussions include Refs[34,35,36]. Note that the tri-linear couplings A and C and the gaugino mass M carry one mass dimension, and also carry phases. The parameters C are not soft if the model contain singlets (e.g., in the NMSSM) but are soft otherwise, in particular in the MSSM. In fact it can be shown that they are equivalent to explicit supersymmetry breaking matter fermion masses. Nevertheless, they appear naturally only in special classes of models and will be omitted hereafter. The scale of the soft supersymmetry breaking parameters $m \sim \sqrt{|B|} \sim |A| \sim |M| \sim m_{\text{SSB}}$ is dictated by the quadratic divergence that is cut off by the mass scale they set (see eq. (65)) and hence is given by

$$m^2_{SSB} \lesssim \frac{16\pi^2}{y_t^2} m^2_{\text{Weak}} \simeq (1\,\text{TeV})^2. \tag{67}$$

Ultimately, the explicit breaking is to be understood as the imprints of spontaneous supersymmetry breaking at higher energies rather than be put by hand. For the purpose of defining the MSSM, however, a general parameterization is sufficient.

Once substituting all flavor indices in the potential (66), e.g., $B^{ij}\phi_i\phi_j \rightarrow BH_1H_2$ and (suppressing family indices) $A^{ijk}\phi_i\phi_j\phi_k \rightarrow A_U H_2 \widetilde{Q}\widetilde{U} + A_D H_1 \widetilde{Q}\widetilde{D} + A_E H_1 \widetilde{L}\widetilde{E}$, one finds that the MSSM contains many more parameters in addition to the 17 free parameters of the SM. The Higgs sector can be shown to still be described by only two free parameters. However, the gauge sector contains three new gaugino mass parameters, which could carry three independent phases; the scalar spectrum is described by five 3×3 hermitian matrices with six independent real parameters and three independent phases each (where we constructed the most general matrix in family space for the Q, U, D, L, and E sfermions); and tri-linear interactions are described (in family space) by three 3×3 arbitrary matrices A_U, A_D, A_E with nine real parameters and nine phases (and equivalently for the C matrices). More careful examinations reveals that four phases can be eliminated by field redefinitions and hence are not physical. Hence, the model, setting all $C = 0$ (allowing arbitrary C coefficients) is described by 77 (104) parameters if all phases are zero and by 122 (176) parameters if phases (and therefore CP violation aside from that encoded in the CKM matrix) are admitted.

4.5 R-parity

While supersymmetrizing the SM we followed a simple guideline of writing the minimal superpotential that consistently reproduces the SM Lagrangian.

Once we realized that supersymmetry must be broken explicitly at the weak scale, we introduced SSB parameters which conserved all of the SM local and global symmetries. While it is the most straightforward procedure, it does not lead to the most general result. In the SM lepton L and baryon B number are accidental symmetries and are preserved to all orders in perturbation theory. (For example, models for baryogenesis at the electroweak phase transition rely on non-perturbative baryon number violation.) Once all fields are elevated to superfields, one can interchange a lepton and Higgs doublets $L \leftrightarrow H_1$ in the superpotential and in the scalar potential, leading to violation of lepton number by one unit by renormalizable operators. The $\Delta L = 1$ superpotential operators, for example, are

$$W_{\Delta L=1} = \pm \mu_{L_i} L_i H_2 + \frac{1}{2} \lambda_{ijk} [L_i, L_j] E_k + \lambda'_{ijk} L_i Q_j D_k, \qquad (68)$$

where we noted explicitly the antisymmetric nature of the λ operators. Similarly, one can write SSB $\Delta L = 1$ operators.

Lepton number violation is indeed constrained by experiment, but is allowed at a reasonable level, and in particular, electroweak-scale $\Delta L = 1$ operators could lead to a $\Delta L = 2$ Majorana neutrino spectrum, hence, extending the SM and the MSSM in a desired direction. However, baryon number is also not an accidental symmetry in the MSSM and the

$$W_{\Delta B=1} = \frac{1}{2} \lambda''_{ijk} U_i [D_j, D_k] \qquad (69)$$

operators are also allowed by gauge invariance. The combination of lepton and baryon number violation allows for tree-level decay of the proton from $\widetilde{s}$ or $\widetilde{b}$ exchange $(U)UD \to (U)\widetilde{D} \to (U)QL$, i.e., $p \to \pi l, Kl$. (The squark plays the role of a lepto-quark field in this case and the model is a special case of a scalar lepto-quark theory. See the scalar exchange diagram in Fig. 7 below.) The proton life time is given by $\tau_{p \to e\pi} \sim 10^{-16 \pm 1} \text{yr}(m_{\widetilde{q}}/1 \text{ TeV})^4 (\lambda' \lambda'')^{-2}$. It is constrained by the observed proton stability $\tau_{p \to e\pi} \gtrsim 10^{33}$ yr, leading to the constraint $\lambda' \lambda'' \lesssim 10^{-25}$. The constraint is automatically satisfied if either $\lambda' = 0$ (only B violation) or $\lambda'' = 0$ (only L violation), leaving room for many possible and interesting extensions of the MSSM. (For a review, see Ref. [37].)

The L and B conserving MSSM, however, is again the most general allowed extension (with minimal matter) if one imposes L nad B by hand. It is sufficient for that purpose to postulate a discrete Z_2 (mirror) symmetry, *matter parity*,

$$P_M(\Phi_I) = (-)^{3(B_I - L_I)}. \qquad (70)$$

Matter parity is a discrete subgroup of the anomaly free $U(1)_{B-L}$ Abelian symmetry, and hence can be an exact symmetry if $U(1)_{B-L}$ is gauged at some high energy. Equivalently, one often impose a discrete Z_2 R-parity[38],

$$R_P(x) = P_M(x) \times (-)^{2J_x} = (-)^{2J_x+3B_x+L_x}, \qquad (71)$$

where J denotes spin. R-parity is a discrete subgroup of a $U(1)_R$ Abelian symmetry under which the supersymmetric coordinate transforms with a charge $R(\theta) = -1$ (which is a conventional normalization), i.e., $\theta \to e^{i\alpha}\theta$. R-symmetry does not commute with supersymmetry (and extends its algebra). In particular, it distinguishes the superfield components. Hence, R-parity efficiently separates all SM (or more correctly, its 2HDM type II extension) particles with charge $R_P(x_{SM}) = (+)$ from their superpartners, the sparticles, with charge $R_P(\text{sparticle}) = (-)$, and hence, is more often used. Matter or R-parity correspond to a maximal choice that guarantees stability of the proton. Other choices that do not conserve $B-L$, but only B or L exist and correspond to Z_3 or higher symmetries (for example, baryon and lepton parities[39,40] and the theta-parity[41]).

Though the minimal supersymmetric extension is defined by its minimal matter content, it is often defined by also its minimal interaction content, as is the case for R_P invariant models. We will assume, as is customary, that R_P is an exact symmetry of the low-energy theory (unless otherwise is stated).

Imposing R-parity on the model dictate to large extent the phenomenology of the model. In particular, the lightest superpartner (LSP) (or equivalently, the lightest $R_P = (-)$ particle) must be stable: It cannot decay to only $R_P = (+)$ ordinary particles, since such a vertex is not invariant under the symmetry and hence cannot exist. (This is not the case for any of the other symmetries that can stabilize the proton and which admit additional superpotential or potential terms.) This is a most important observation with strong implications:

1. The LSP is stable and hence has to be neutral (e.g., a bino, wino, Higgsino, sneutrino, or gravitino) or at least bound to a stable neutral state (as would be the case if the gluino is the LSP and it then hadronizes).

2. Sparticles are produced in the collider in pairs.

3. Once a sparticle is produced in the collider, it decays to an odd number of sparticles. In particular, its decay chain must conclude with the LSP. The stable LSP escapes the detector, leading in many cases to a distinctive large missing energy signature. (If a light gravitino is the LSP, for instance, there may not be a large missing energy signatures but other distinctive signatures such as hard photons exist.)

4. The neutral LSP (in most cases) is a weakly interacting massive particle (WIMP) and it could constitute cold dark matter (CDM) with a sufficient density. Roughly speaking, its relic density is proportional to the inverse of its annihilation rate which in turn is given by a cross section which is proportional to the inverse of the exchanged sparticle mass squared. Thus, the relic density is proportional to the square of sparticle mass scale, a relation which leads to model dependent upper bounds of the order of $\mathcal{O}(\text{TeV})$. (In practice, the density is a complicated function of various sparticle mass parameters.) It should be stressed that the proximity of this upper bound and the one derived from the fine-tuning of the Higgs potential is tantalizing and suggestive. More generally, it relates the CDM to the sparticle mass scale, but other particles in other sectors which may be linked to the SM sector only gravitationally could also have similar masses and provide CDM candidates. An interesting proposal raised recently is that of an axino dark matter [42], which would weaken any upper bound on the mass of the ordinary superpartners. (The axino is the fermion partner of the axion of an anomalous Peccei-Quinn or R-symmetry, on which we do no elaborate in these lectures.)

The collider phenomenology of the models was addressed in Zeppenfeld's lectures [15] and will not be studied here in detail. See also Tata's lecture notes[43,18]. Models of CDM are reviewed, for example, in Refs. [44,45], and search for energetic neutrinos from LSP annihilation in the sun was discussed in Halzen's lectures [16].

4.6 Mass eigenstates and experimental status

We conclude this lecture with a transformation from current to mass eigenstates, which is not a trivial transformation given electroweak symmetry breaking (EWSB): As in our "warm-up" case of a supersymmetric limit (Sec. 4.3), interaction eigenstates with different electroweak charges (and hence, chirality) mix once electroweak symmetry is broken.

The Higgs mass matrix is now

$$m_H^2 = \begin{pmatrix} m_{H_1}^2 + \mu^2 & m_{12}^2 \\ m_{12}^{*\,2} & m_{H_2}^2 + \mu^2 \end{pmatrix}, \tag{72}$$

and it has a negative eigenvalue for $m_1^2 m_2^2 < |m_{12}^2|^2$, where $m_i^2 = m_{H_i}^2 + \mu^2$, so that electroweak symmetry can be broken. This condition is automatically satisfied for $m_{H_2}^2 \lesssim -\mu^2$ which often occurs due to negative quantum corrections proportional to the t-quark Yukawa couplings. (This is the radiative symmetry breaking mechanism, which we will return to in Sec. 5.2.) One neutral CP odd and two charged $d.o.f.$ are absorbed in the Z and $W^\pm$ gauge bosons, respectively, and two CP even, one CP odd (A^0) and one complex charged Higgs remain. Folding in EWSB constraints, the spectrum is described by only two parameters which are often taken to be $\tan\beta$ and $m_{A^0}^2 = m_1^2 + m_2^2$. One of the CP even states is the (model-independent) light Higgs boson of supersymmetry which parmeterizes the $\tan\beta = 1$ flat direction mentioned above. This is readily seen in the limit in which all other Higgs $d.o.f.$ form (approximately) a degenerate $SU(2)$ doublet which is heavy with a mass $\sim |\mu| \gg M_W$. In this case, EWSB is SM-like, with the remaining physical CP-even state receiving mass which is proportional to its quartic coupling λ, now given by the gauge couplings $\lambda = (g_2^2 + g'^2)/8$. In general, its mass $m_{h^0} \leq M_Z |\cos 2\beta|$ at tree-level. Large $\mathcal{O}(100\%)$ loop corrections, again proportional to the large t-quark Yukawa coupling, lift the (flat direction and the) bound to $m_{h^0} \lesssim \sqrt{2} M_Z \sim 130$ GeV. Though the quartic coupling, and hence the mass, can be somewhat larger in extended models, e.g., in some versions of the the NMSSM, as long as perturbativity is maintained one has $m_{h^0} \lesssim 160 - 200$ GeV where the upper range is achieved only in a small class of (somewhat ad hoc) models. We discuss the Higgs sector in more detail in Sec. 5.3.

Sparticles and not Higgs bosons, however, would provide the evidence for supersymmetry (though the absence of a light Higgs boson can rule perturbative low-energy supersymmetry out). The Dirac-like neutral Higgsinos mix with the other two neutral fermions, the Majorana bino and neutral wino (the latter can be rewritten as a photino and zino, i.e., as linear combinations aligned with the photon and the Z). The physical eigenstates are the neutralinos. Their mass and mixing is given by the diagonalization of the neutralino (tree-level) mass matrix

$$m_{\tilde{\chi}^0} = \begin{pmatrix} M_1 & 0 & -M_Z c_\beta s_W & M_Z s_\beta s_W \\ 0 & M_2 & M_Z c_\beta c_W & M_Z s_\beta c_W \\ -M_Z c_\beta s_W & M_Z c_\beta c_W & 0 & \mu \\ -M_Z s_\beta s_W & -M_Z s_\beta c_W & \mu & 0 \end{pmatrix}, \quad (73)$$

where M_1 and M_2 are the bino and wino SSB mass parameters, respectively, and $s_\beta = \sin\beta$, $c_\beta = \cos\beta$ and similarly for the weak angle denoted by a subscript W. The neutralino mass matrix is written in the basis $(-i\tilde{B}, -i\tilde{W}^0, \tilde{h}_1^0, \tilde{h}_2^0)$. The off-diagonal terms correspond the the gauge-Yukawa interaction terms with the Higgs repalced by its vev. Note that in the limit $\mu \to 0$ the wino and bino do not mix at tree-level. Similarly, the charged Higgs and charged gaugino states mix to form the physical mass eigenstates, the charginos. The chargino mass and mixing is determined by the chargino mass matrix

$$m_{\tilde{\chi}^{\pm}} = \begin{pmatrix} M_2 & \sqrt{2}M_W s_\beta \\ \sqrt{2}M_W c_\beta & -\mu \end{pmatrix}, \tag{74}$$

The gluino, of course, cannot mix and has a SSB Majorana mass M_3.

Finally, we can rewrite the sfermion mass matrix for the sfermions $(\tilde{f}_L, \tilde{f}_R)$. The mass-squared matrix

$$m_{\tilde{f}}^2 = \begin{pmatrix} m_{LL}^2 & m_{LR}^2 \\ m_{LR}^{2\,*} & m_{RR}^2 \end{pmatrix} \tag{75}$$

was previously given in the supersymmetric limit. Including the SSB interactions one has

$$m_{LL}^2 = m_{\tilde{f}_L}^2 + m_f^2 + M_Z^2 \cos 2\beta \left[T_3^f - Q_{f_L} \sin^2 \theta_W \right], \tag{76}$$

$$m_{RR}^2 = m_{\tilde{f}_R}^2 + m_f^2 + M_Z^2 \cos 2\beta \times Q_{f_R} \sin^2 \theta_W, \tag{77}$$

$$m_{LR}^2 = m_f \left(A_f - \mu^* \tan \beta \right) \; [\text{or } m_f \left(A_f - \mu^*/ \tan \beta \right)], \tag{78}$$

where the first term in (76) and in (77) is the SSB squared-mass parameter, the tri-liner SSB parameter is implicitly assumed to be proportional to the Yukawa coupling $\hat{A}_f = y_f A_f$, which is then factored out (the assumption is trivial in the case of one generation but it constitutes a strong constraint in the case of inter-generational mixing), and $\mu \tan \beta$ ($\mu/ \tan \beta$) terms appear in the down-squark and slepton mass matrices (up-squark mass matrix).

The A-terms are not invariant under $SU(2) \times U(1)$ and the sfermion doublet masses are now split. Neglecting fermion masses, one has the sum rule $m_{\tilde{e}_L}^2 - m_{\tilde{\nu}_L}^2 = m_{\tilde{d}_L}^2 - m_{\tilde{u}_L}^2 = -\cos^2 \theta_W M_Z^2 \cos 2\beta = -M_W^2 \cos 2\beta > 0$. This sum rule is modified if there is an extended gauge structure with more than just the hypercharge D-term. The mass matrices can further be combined to three 6×6 matrices and one 3×3 matrix (for the sneutrinos) just as in the supersymmetric limit discussed in Sec. 4.3. Note that the SSB A-terms $AH\tilde{f}_L\tilde{f}_R$ are holomorphic (and do not involve complex conjugate fields) unlike the supersymmetric tri-linear interactions $y\mu H^*\tilde{f}_L\tilde{f}_R$ (and SSB C-terms), a property which particularly relevant for the stability of the vacuum as well as to the dependence of the relevant couplings of the physical eigenstates on $\tan \beta$.

The sfermion left-right mixing angle is conveniently given by

$$\tan 2\theta_{\tilde{f}} = \frac{2m_{LR}^2}{m_{LL}^2 - m_{RR}^2}. \qquad (79)$$

Obviously, significant left-right mixing is possible if either the corresponding fermion is heavy (as in the case of the stop squarks) or if $\tan\beta$ is large and the corresponding fermion is not very light (as could be the case for the sbottom squarks). Observe, however, that in the limit $m_{\tilde{f}}^2 \gg \langle H \rangle^2$ the sfermion mixing is suppressed and mass eigenstates align with the current eigenstates.

Model-independent limits (*i.e.*, independent of the decay mode) on the sparticles were given by the total width measuremnt at the Z pole at LEP. These limits constrain the sparticles to be heavier than $40 - 45$ GeV, with the exception of the lightest neutralino, whose couplings to the Z could be substantially suppressed and could still be as light as ~ 20 GeV. Also, no limit on the (heavy) gluino, which does not couple to the Z at tree level, is derived. LEP runs at higher energy further constrain many of the sparticles to be heavier than $70 - 80$ GeV. However, now the constraints are model-dependent since off-resonance production involves also a t-cahnnel exchange (which introduce strong model dependence) and in the absebce of universal tool such as the Z width, searches must assume in advance the decay chain and its final products. The FNAL Tevatron can constrain efficently the strongly interacting squarks and gluino with lower bounds of $200 - 250$ GeV, but again, these bounds contain model-dependent assumptions. Many specialized searches assuming unconventional decay chains were (+and are) conducted in recent years both at the Tevatron and LEP. They are summarized and updated updated limits are given periodicly by the Particle Data Group [46]. Obviously, a significant gap remains between current limits (and particularly so, the model-independent limits) and the theoretically suggested range of $\mathcal{O}(1\,\text{TeV})$. Though some sparticles may still be discovered at future LEP and Tevatron runs, it is the Large Hadron Collider, currently under consttruction at CERN, which will explore the $\mathcal{O}(1\,\text{TeV})$ regime.

Exercises

1. Draw all the Feynamn diagrams that stem from that of the vector-fermion-fermion interaction and which describe the gauge-matter interactions. What are the Feynman rules for the case of a U(1)?

2. Confirm that $\text{Tr}Y^3 = 0$ for hypercharge in the MSSM, as well as the mixed anomaly traces $\text{Tr}SU(N)^2U(1)$ where $N = 2, 3$.

3. Derive the complete MSSM component-field Yukawa Lagrangian.

4. Derive the complete MSSM quartic potential, including gauge (D) and Yukawa (F) contributions. Use the relation $\sigma_{ij}^a\sigma_{kl}^a = 2\delta_{il}\delta_{jk} - \delta_{ij}\delta_{kl}$ among the $SU(2)$ generators to write the Higgs quartic potential explicitly. Map it onto the general form of a 2HDM potential Eq. (10) and show that (at tree level) $\lambda_1 = \lambda_2 = (1/8)(g'^2 + g_2^2)$; $\lambda_3 = -(1/4)(g'^2 - g_2^2)$; $\lambda_4 = -(1/2)g_2^2$; and $\lambda_5 = \lambda_6 = \lambda_7 = 0$ (where g' and g_2 are the hypercharge and $SU(2)$ couplings, respectively).

5. Show that at the quantum level there could be non-diagonal corrections to $f_{\alpha\beta}$, for example, in a theory involving U(1)×U(1)'. (Consider a loop correction that mixes the gauge boson propagators.)

6. Derive the D-term expectation value eq. (63). Rewrite the hypercharg in terms of the electric charge, weak isospin, and the weak angle to derive the D term contribution to the sfermion mass squared. Confirm its flavor-dependent numerical coefficient.

7. Derive the Higgs squared-mass matrix in the supersymmetric limit.

8. Derive the Higgs squared-mass matrix in the model with a singlet S. Show that it has an off-diagonal element and that it has a negative eigenvalue. Show, by considering F_{H_i} contributions to the scalar potential that indeed $\langle S \rangle = 0$ and no left-right sfermion mixing arises.

9. Show that a supersymmetry breaking fermion mass, if allowed by the gauge symmetries (for example, consider a toy model with two singlets $W = S_1 S_2^2 + m S_1^2$ giving a supersymmetry breaking mass to the fermion component of S_2) can be recast, after appropriate redefinitions, as a $ms\phi\phi^*$ term in the scalar potential.

10. Show that all the soft prameters are indeed soft by naive counting, where possible (e.g., in the case of tri-linear scalar interactions) or otherwise by integration.

11. Show that in a model with a singlet (consider, for example, our toy model above) an interaction $Cs\phi_1\phi_2$ is not soft but leads to a quadtratically divergent linear term in the scalar potential.

12. Substitute the SM flavor indices in the soft potential (66) and confirm our counting of free parameters.

13. Write the most general gauge invariant (R-parity violating) SSB potential. How many paramters describe the MSSM if R-parity is not imposed?

14. Using the R-charge assignemnt of the coordinate θ, what are the R-charges of the various component fields of the chiral and vector superfields? and of the superpotential? Show that requiring that the potential is R-invariant forbids gaugino masses and tri-linear scalar terms. Therefore, these parameters must carry R-charge and their presence breaks the $U(1)_R$ symmetry. Note that all of the above terms are invariant under the R_P subgroup!

15. Diagonalize the neutralino and chargino mass matrices in the limits $\mu \rightarrow \infty$ (gaugino region), $M_1 \sim M_2 \rightarrow \infty$ (Higgsino region), and $\mu \sim M_1 \sim M_2 \rightarrow 0$.

16. Extend the neutralino and chargino mass matrices to include mixing with neutrinos and charged leptons, respectiveley, for $\mu_{L_{ij}} \neq 0$. The neutrino mass is given by the ratio of the determinant of this matrix and that of the usual neutralino mass matrix.

5 Origins, Renormalization, And Some Phenomenology

In previous lectures supersymmetry was motivated, constructed, and applied to the Standard model of particle physics. The minimal model was defined according to its particle content and superpotential, but it was shown to contain many arbitrary parameters that break supersymmetry near the Fermi scale. Indeed, many questions still remain unanswered:

- Does the model remain perturbative and if so, up to what scale?

- What is the high-energy scale to which we keep referring to as the ultra-violet cut-off scale?

- What is the origin of the soft supersymmetry breaking parameters and can they be organized and their number reduced?

- Are there indirect signatures of supersymmetry that could be tested at past and current low-energy experiments?

- Could one distinguish an MSSM Higgs boson from a SM one?

- Can the model be extended to incorporate neutrino mass and mixing?

These and many other question have been extensively studied in recent years. While some possible answers and proposals were put forward, no standard *high-energy* supersymmetric model exists. On the contrary, the challenge ahead is the deciphering of the high-energy theory from the low-energy data if and when supersymmetry is discovered and established. Here, we will confine ourselves to a discussion of only a few of these questions. Yet, we will attempt to provide a comprehensive pictures of the model space, the related phenomenology, and of the status of the models.

Though no evidence for supersymmetry has been detected as of the writing of this manuscript, indications in support of the framework have been accumulating and include the unification of gauge couplings; the heavy t-quark; and electroweak data strong preference of (i) a light Higgs and (ii) no "new-physics" quantum modifications to various observables. While individually each argument carries little weight, the combination of all indications is definitely intriguing. The first three topics will be explored next, and in a way that provides the reader with a critical view on the one hand and a road map to many of the other issued not discussed here in detail. (Electroweak quantum corrections were discussed by Altarelli [1], and were investigated in detail in Ref. [47].) We will then turn to the question of lack of low-energy evidence.

In particular, the absence of observable contribution to flavor changing neutral currents will be shown to provide an organizing principle for high-energy frameworks, which themselves attempt to organize the model many parameters. The latter is referred to as the flavor problem and understanding its solution (which is straightforward in the SM given is special structure but not in any of its extensions) goes to the heart of the question of origins of the low-energy theory, as is the case in any other extension of the SM. Useful reviews not mentioned so far include Refs.[48,49,50]

5.1 Unification

Shortly after the SM was established and asymptotic freedom realized, it was suggested that the SM semi-simple product group of $SU(3) \times SU(2) \times U(1)$ is embedded at some high energies in a unique simple group, for example $SU(5)$ (which like the SM group has rank 5 and hence is the smallest simple group that can contain the SM gauge structure). Aside from implications for quark-lepton unification, Higgs fields, and the proton stability (to which we will return) it predicts first and foremost that the seemingly independent SM gauge couplings originate from a single coupling of the unified simple group and that their infrared splitting is therefore due to only the scaling (or renormalization) of the corresponding quantum field theory from high to low energies. Independently, it was also realized in the context of string theory that the theory just below the string scale often has a unique (or unified) value for all gauge couplings [6,7,51], regardless of whether the SM group itself truly unifies (in the sense of embedding all SM fields in representation of some simple group).

The question is then obvious: Do the measured low-energy couplings unify at some ultra-violet energies? Clearly, if this is the case then their unification automatically defines the ultra-violet scale, and supergravity may further enable gauge-gravity unification. This question is readily addressed using the renormalization group formalism which is beautifully confirmed by the data for energies up to the LEP center-of-mass energy of ~ 190 GeV. Once the particle content of the model is specified, the β function coefficients b can be calculated and the couplings can be extrapolated by integrating the renormaization group equation $dg/d\ln\Lambda = (b/16\pi^2)g^3$ (given here at one-loop). Such integration is nothing but the scaling of the theory, e.g., between the weak scale, where the couplings are precisely known, and some high-energy scale, assuming a specific particle content. Its conclusions are meaningful if and only if the low-energy couplings are measured to such precision so that their experimental errors allow only a sufficiently small range of values for each of the extrapolated high-energy couplings as to determine whether all

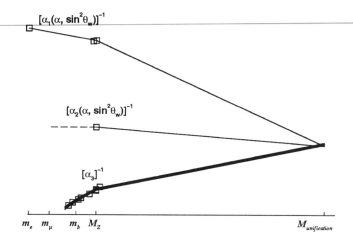

Figure 5: The SM gauge couplings evolve from their measured low-energy value (indicated by the square symbols) to high energies using SM three-loop (for scales $\Lambda < M_Z$) and MSSM two-loop (for scales $\Lambda > M_Z$) β-functions. Their renormalization within the SM is confirmed by the data, while extrapolation to higher energies, assuming the MSSM, leads to their unification at nearly a point.

three coupling intersects to a satisfactory precision at a point, and if perturbativity (which is used in deriving the equations) is maintained. This decade brought about the most precise determination of the gauge couplings at the Z pole (see Altarelli's lectures[1]), therefore enabling one to address the question. Assuming only the SM matter content at all energies, the three couplings fail to unify – their integration curves $g(\ln \Lambda)$ intersects in pairs only. (Of course, one may question the validity of such a framework to begin with in the absence of any understanding as to how is the hierarchy problem is resolved in this case.) Extending the SM to the MSSM and using the MSSM particle content begining at some point between the weak scale and a TeV scale, the three SM gauge coupling unify at a point at a scale $M_U \simeq 3 \times 10^{16}$ GeV and with a value $g_U \simeq 0.7$ or $\alpha_U \simeq 0.04$, defining the unification scale (to be used as the ultra-violet cut off?) as well as confirming perturbativity, and hence consistency, of the framework. This is illustrated in Fig. 5.

Let us repeat this exercise in some detail. The renormalization of the couplings is given at two-loop order by

$$\frac{dg_i}{\ln \Lambda} = \frac{b_i}{16\pi^2} g_i^3 + \frac{b_{ij}}{(16\pi^2)^2} g_i^3 g_j^2 - \frac{a_{i\alpha}}{(16\pi^2)^2} g_i^3 y_\alpha^2. \tag{80}$$

The first term is the one-loop term given above, while the second and third terms sum gauge and Yukawa two-loop corrections, respectively, and we did not include any higher loop terms which are negligible in the MSSM. (The two-loop equation is free of any scheme dependences, though negligible corrections appear when matching the scheme used to describe the data to a scheme which can be used in supersymmetry.) The Yukawa terms, which are negative ($a_{i\alpha} > 0$), correspond to only small $\mathcal{O}(1\%)$ but model ($\tan\beta$-) dependent corrections, which we again neglect here, while the gauge two-loop terms correspond to $\mathcal{O}(10\%)$ corrections, which we will include in the results. The one loop coefficients are the most important. In supersymmetry they are conveniently written as a function of the chiral superfields quantum numbers,

$$b_i = \sum_a T_i(\Phi_a) - 3C_i, \qquad (81)$$

where the Dynkin index $T_i(\Phi_a) = 1/2\,(Q^2)$ for a superfield Φ_a in the fundamental representation of $SU(N)$ (with a $U(1)$ charge Q), and the Casimir coefficient for the adjoint representation $C_i = N\,(0)$ for $SU(N)\,(U(1))$. It is now straightforward to find in the MSSM

$$\begin{pmatrix} b_1 \\ b_2 \\ b_3 \end{pmatrix} = N_{\text{family}} \begin{pmatrix} 2 \\ 2 \\ 2 \end{pmatrix} + N_{\text{Higgs}} \begin{pmatrix} \frac{6}{10} \\ 1 \\ 0 \end{pmatrix} - \begin{pmatrix} 0 \\ 6 \\ 9 \end{pmatrix} = \begin{pmatrix} +\frac{66}{10} \\ +1 \\ -3 \end{pmatrix}, \qquad (82)$$

where N_{family} and N_{Higgs} correspond to the number of chiral families and Higgs doublet pairs, respectively and the index $i = 1, 2, 3$ correspond to the $U(1)$, $SU(2)$, and $SU(3)$, coefficients. Here, we chose the "unification" normalization of the hypercharge $U(1)$ factor, $Y \to \sqrt{3/5}Y$ and $g' \to \sqrt{5/3} \equiv g_1$. This is the correct normalization if each chiral family is to be embedded in a representation(s) of the unified group. In this case, the trace over a family over the generators T^α for each of the SM groups i, $\text{Tr}(T_i^\alpha(\Phi_a)T_i^\beta(\Phi_a)) = N\delta^{\alpha\beta}$, has to be equal to the same trace but when taken over the unified group generators, and hence, has to be equal to same N for all subgroups. In the MSSM, both non-Abelian factors have $N = 2$, so the $U(1)$ normalization is chosen so that $\text{Tr}(Q_a^2) = 2$ for each family. On the same footing, adding any complete multiplet of a unified group modify b_1, b_2, and b_3 exactly by the same amount. (Note, however, that asymptotic freedom of QCD may be lost if adding such multiplets. $SU(2)$ is already not asymptotically free in the MSSM.) This relation also explains, in the context of unification, the quantization of hypercharge which is dictated by the embedding of the $U(1)$ in a non-Abelian group. (Such an embedding of the SM group in a single non-Abelian group automatically implies an anomaly free theory, and hence, this relation is equivalent to the low-energy quantization based on the anomaly constraints.)

Neglecting the Yukawa term, the coupled two-loop equations can be solved in iterations,

$$\frac{1}{\alpha_i(M_Z)} = \frac{1}{\alpha_U} + b_i t + \frac{1}{4\pi} \sum_{j=1}^{3} \frac{b_{ij}}{b_j}(1 + b_j \alpha_U t) - \Delta_i, \qquad (83)$$

where the integration time is conveniently defined $t \equiv (1/2\pi)\ln(M_U/M_Z)$, and we assume, for simplicity, that the MSSM β-functions can be used from the Z scale and on. The threshold function Δ_i compensates for this naive assumption and takes into account the actual sparticle spectrum. It can also contain contributions from additional super-heavy particles and from operators that may appear near the unification scale.

These equation can be recast in terms of the fine-structure constant $\alpha(M_Z)$ and the weak angle $\sin^2 \theta_W \equiv s_W^2(M_Z)$. Using their precise experimental values one can then calculate (up to threshold corrections) the unification scale, the value of the unified coupling, and the value of the low-energy strong coupling. (Since there are only two high-energy parameters, one equation can be used to predict one weak scale coupling!) The first two are predictions that only test the perturbativity and consistency of the extrapolation. (Note that the unification scale is sufficiently below the Planck scale so that gravitational correction may only constitute a small perturbation which can be summed, in principle, in Δ_i.) One finds, given current values,

$$\alpha_3(M_Z) = \frac{5(b_1 - b_2)\alpha(M_Z)}{(5b_1 + 3b_2 - 8b_3)s_W^2(M_Z) - 3(b_2 - b_3)} + \cdots \qquad (84)$$

$$= \frac{7\alpha(M_Z)}{15s_W^2(M_Z) - 3} + \text{two-loop and threshold corrections}$$

$$= 0.116 + 0.014 - (0.000 - 0.003) \pm \delta$$

$$= 0.127 - 0.130 \pm \delta, \qquad (85)$$

where in the third line one-loop, two-loop gauge, two-loop Yukawa, and threshold corrections (denoted by δ) are listed. Note that any additional complete multiplets of the unified group in the low-energy spectrum would factor out from the (one-loop) prediction. This is true only at one loop and only for the $\alpha_3(M_Z)$ and t predictions.

Comparing the predicted value to the experimental one (whose precision increased dramatically in the recent couple of years) $\alpha_3(M_Z) = 0.118 \pm 0.003$, one finds a $\sim 8\%$ discrepancy, which determines the role that structure near the weak scale, the unification/gravity scale, or intermediate scales, can play. In fact, only $\sim 3\%$ corrections are allowed at the unification scale, since the QCD renormalization amplifies the corrections to the required 8%. This is the most likely conclusion from the discrepancy, but we will not discuss here in detail the many possible sources of such a small perturbation which could appear in the form of super-heavy non-degenerate spectrum, non-universal corrections to the gauge kinetic function once the grand-unified theory is integrated out, string corrections (in the case of a string interpretation), etc. Let us instead comment on the possible threshold structure at the weak scale (ignoring the possibility of additional complete multiplet at intermediate energies which will modify the two-loop correction). Unless some particles are within tens of GeV from the weak scale, only the (leading-)logarithm corrections need to be considered (either by direct calculations or using the renormalization group formalism). This is only the statement that the sparticle spectrum is typically expected to preserve the custodial $SU(2)$ symmetry of the SM and hence not to contribute to the oblique and other (non-universal) parameters that measure its breaking and which would contribute non-logarithmic corrections to δ. The leading logarithm threshold can be summed in a straightforward but far from intuitive way in the threshold parameter $\hat{M}_{SUSY}$,

$$\delta = -\frac{19}{28}\frac{\alpha_3^2(M_Z)}{\pi}\ln\frac{\hat{M}_{SUSY}}{M_Z} \simeq -0.003\ln\frac{\hat{M}_{SUSY}}{M_Z}. \tag{86}$$

If all sparticles are degenerate and heavy with a mass $\hat{M}_{SUSY} \simeq 3$ TeV then the predicted and experimental values of the strong coupling are the same. However, more careful examination shows that $\hat{M}_{SUSY}$ is more closely related to the gaugino and Higgsino spectrum (since the sfermion families correspond complete multiplets of $SU(5)$) and furthermore, typical models for the spectrum give $10 \lesssim \hat{M}_{SUSY} \lesssim 300$ GeV even though the sparticles themselves are in the hundreds of GeV range. The threshold correction may then be even positive! (This is particularly true when non-logarithmic corrections are included.) While high-energy contributions to δ are likely to resolve the discrepancy, they also render the unification result insensitive to the exact value of $\hat{M}_{SUSY}$, since, e.g., additional deviation of $\sim 3\%$, from $\hat{M}_{SUSY} \ll M_Z$, correspond to only $\sim 1\%$ corrections at the unification scale. Hence, even though the sparticle threshold corrections may not resolve the discrepancy, they are very unlikely destroy the successful unification. This level of ultra-violet sensitivity more then allows one to trust the result. If the factor of $2\pi t \simeq 30$ corresponds to a power-law (rather than logarithmic) renormalizations, as in some model in which the gauge theory is embedded in a theory with intermediate-energy extra dimensions, the sensitivity is amplified by more than an order of magnitude, undermining any predictive power in that case. The predictive power in the minimal framework is unique and one of the pillars in its support.

One could then take the point of view that the SM is to be embedded at a grand-unification scale $M_U \simeq 3 \times 10^{16}$ GeV in a simple grand-unified (GUT) group, $SU(3)_c \times SU(2)_L \times U(1)_{Y/2} \subseteq SU(5); SO(10); SU(6); E_6 \cdots$ where the rank five and six options were specified. This group may or may not be further embedded in a string theory, for example. Alternatively, the unification scale may be interpreted as a direct measurement of the string scale. The latter interpretation is one of the forces that brought about a revolution in string model building, which generically had the string scale an order of magnitude higher, and motivated M-theory model building and other non-traditional approaches[7], which are reviewed in this volume by Dienes[6]. Embedding of a GUT group with three chiral generations; and the appropriate Higgs representations that can break it to the SM group; as well as provide the SM Higgs doublets; in a string theory a decade or so above the unification scale is also possible, in principle, but encounters many difficulties and was not yet demonstrated to a satisfactory level (see Dienes's lectures [6]). In the reminder of our discussion of unification we will outline (leaving out many of the details) the embedding in a GUT group, leaving open the question of string theory embedding.

Grand-unification, in contrast to only coupling unification, requires one to embed all matter and gauge fields in representations of the large GUT group, for example, all the SM gauge (super)fields are embedded in the case of $SU(5)$ $(SO(10))$ in the **24** (**45**) dimensional adjoint representation while each family can be embedded in $\bar{\mathbf{5}} + \mathbf{10}$ (**16**) anti-fundamental and antisymmetric (spinor) representations. The Higgs fields of the SM can be embedded in $\bar{\mathbf{5}} + \mathbf{5}$ (one or more fundamental **10**'s). The Higgs sector must also contain a higher dimension representation(s) (which may be taken to be the adjoint) that can break the symmetry spontaneously to its SM subgroup, and whose interaction must be described by the appropriate superpotential. (Note that the GUT theory resides in the globally supersymmetric SM sector, and therefore the GUT $\rightarrow$ SM minimum is degenerate with the GUT conserving or any other minimum. It is the supergravity effects that typically can be parameterized by the (small) soft parameters, that must lift the degeneracy and pick the correct SM minimum.) All other issues stem from the choice of embedding. First and foremost, quarks and leptons unify in the sense that they are embedded in the same GUT representations (e.g., the down singlet and the lepton doublet are embedded in the $\bar{\mathbf{5}}$ of $SU(5)$). This translates to simple boundary conditions for ratios of their Yukawa couplings, e.g., $y_D/y_E = 1$ at the unification scale. This relation can then be renormalized down to the weak scale and tested, as is shown in Fig. 6. Indeed, it is found to be correct for the third family couplings ($b - \tau$ unification) for either $\tan \beta \simeq 1 - 2$ (large top Yukawa coupling

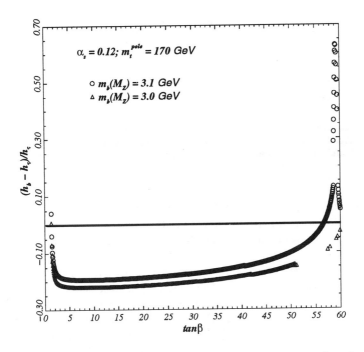

Figure 0. The unification-scale difference $y_b - y_\tau$ (denoted here $h_b - h_\tau$) is shown in y_τ (denoted here h_τ) units for $m_b(M_Z) = 3$ GeV, $\alpha_s(M_Z) = 0.12$, $m_t^{pole} = 170$ GeV and as a function of $\tan\beta$. The zero line corresponds to $b - \tau$ unification. For comparison, we also show the difference for $m_b(M_Z) = 3.1$ GeV (which for $\alpha_s(M_Z) = 0.12$ is inconsistent with $m_b(m_b) < 4.45$ GeV). Note the rapid change near the (naive) small and large $\tan\beta$ solutions, which is a measure of the required tuning in the absence of threshold corrections. Also note that in most of the parameter space the unification-scale leptonic coupling is the larger coupling.

$y_t(M_Z) \simeq 1/\sin\beta$) or $\tan\beta \gtrsim 50$ (large bottom Yukawa coupling $y_b(M_Z) \simeq 0.017\tan\beta$), and for a much larger parameter space once finite corrections to the quark masses from sparticle loops are included. These relations fail for the lighter families where fermion masses could be understood as setting the scale for the perturbations for these relations (either from higher dimensional Higgs representations or from Planck-mass suppressed operators, or both.) Secondly, additional non-SM matter and gauge (super)fields appear, and in most cases must be rendered heavy. For example, in $SO(10)$ the **16** contains also a singlet right-handed neutrino, which is desired in order to understand neutrino masses. The Higgs doublets are embedded together with color triplets which interacts with SM matter with the some Yukawa couplings as the Higgs doublets (e.g.,

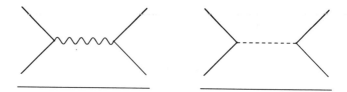

Figure 7: Tree-level proton decay via "lepto-quark" gauge boson and scalar exchange $qq' \to q''l$. The straight line represents the spectator quark which is contained in proton and after its decay hadronizes with q'' to form a meson.

$W \sim y_{U_{ij}} 10_i 10_j 5$ in $SU(5)$ and the **5** contains the H_2 Higgs doublet and a color triplet with an hypercharge $-1/3$). Also, the **24** contains not only the SM gauge content but also the $(3, 2)_{-5/6}$ $\bar{X}$ and $\bar{Y}$ vector superfields as well as the complex conjugate representation, which become massive after the spontaneous symmetry breaking. These are lepto-quark field which connect lepton and quark fields and therefore lead to proton decay, e.g., $p \to \pi^0 e^+$, at tree level. So do the color triplet Higgs fields, which therefore must also become massive with a mass near the unification scale. (Relevant diagrams are illustrated in Fig. 7.) The constraint from the proton life time measurement,

$$\tau_P \simeq 3 \times 10^{31 \pm 1} \text{ yr } \left(\frac{M_U}{4.6 \times 10^{14} \text{ GeV}} \right)^4 \gg 10^{33} \text{ yr}, \qquad (87)$$

is, however, easily satisfied for $M_U \simeq 3 \times 10^{16}$ GeV. Nevertheless, supersymmetry implies also heavy color triplet Higgsinos which can induce proton decay $p \to K^+ \bar{\nu}$ radiatively from quantum correction involving the colored Higgsino and MSSM gauginos and sfermions circulating in the loop. (The colored Higgsino exchange can be described by the effective dimension five operators in the superpotential $W \sim (QQQL + UDUE)/M_U$ whose F component leads to the vertices of the form $\tilde{f}\tilde{f}ff/M_U$.) Such diagrams $\propto (g^2 y^2/16\pi^2)(1/\tilde{m}M_U)$, where $\tilde{m} \sim M_{\text{Weak}}$ is a typical superpartner mass scale, proved to generically dominate proton decay in these models since the corresponding amplitude is suppressed by only two (rather than four) powers of the superheavy mass scale, and pose one of the most severe constraints on the models. Of course, on the other hand the large predicted (in relative terms) amplitudes for radiative proton decay provide an opportunity to test the unifcation framework and supersymmetry simultaneously. (It is interesting to note that these operators are forbidden if instead of R-parity a discrete Z_3 baryon parity is imposed which truly conserve baryon number but allows R-parity lepton number violating operators.) This leads us to perhaps the most difficult problem facing the grand-unification framework. Supersymmetry guarantees that once the Higgs doublets and color triplets are split so that the former are light and the latter are heavy, this hirerachy is preserved to all orders in perturbation theory. Nevertheless, it does not specify how such a split may occur. This is the doublet-triplet splitting problem which is philosophically, though (becuase of supersymmetry non-renormalization theorems) not technically, a manifistation of the hierarchy problem. More generally, it is an aspect of the problem of fixing the μ-parameter $\mu = \mathcal{O}(M_{\text{Weak}})$, which was mentioned breifly in the previous lecture. Like all other issues raised, extensive model-building efforts and many innovative solutions exist [50], but will not be reviewed here. They typically involve extending the models specific representations and symmetries.

Recommended reading include Ref. [52], which lists all earlier works. A sample of other recent research papers is given in Ref. [53]. Grand-unified theories were reviewed in the previous school by Mohapatra [50].

5.2 The heavy top and radiative symmetry breaking

Aside from gauge couplings, the top mass, and hence the top Yukawa (baring in mind the possibility of $\pm\mathcal{O}(5 - 10\%)$ finite corrections from sparticle loops), is also precisely measured $y_t(m_t) \simeq 0.95/\sin\beta$. Again, one can ask whether such a large weak-scale coupling remains perturbative (when properly renormalized) up to the unification scale. The answer is positive for the following reason: Upon examining the one-loop renormalization group equation for y_t,

$$\frac{dy_t}{d\ln\Lambda} \simeq \frac{y_t}{16\pi^2}\left\{-\frac{16}{3}g_3^2 + 6y_t^2\right\}, \tag{88}$$

where we neglected all other couplings, one finds a fixed point behavior [54]. For $y_{t_{\text{fixed}}}^2 = (16/18)g_3^2 \simeq (1.15)^2$ the right hand side equals zero (where we used, as an approximation, the weak-scale value of g_3) and y_t freezes at this value and is not renormalized any further. This is the top-Yukawa quasi-fixed point value, which it can be shown to always reach if $y_t(M_U) \gtrsim 0.5$. (The value of $y_{t_{\text{fixed}}}$ diminishes if there are any other large Yukawa couplings such as a large y_b, right-handed neutrino couplings, R_P-violating couplings, singlet couplings in the NMSSM, etc.) On the other hand, weak-scale values $y_t > y_{t_{\text{fixed}}}$ imply $y_t \gg 1$ at intermediate energies below the unification scale. This then gives a lower bound on $\tan\beta \gtrsim 1.8$. (An upper bound $\tan\beta \lesssim 60$ is derived by applying similar consideration to the bottom-Yukawa coupling.) Renormalization group study, however, shows that the fixed point is not reached to any (arbitrarily small) initial value of $y_t(M_U)$, and in fact, even when it is reached it is not an exact value but rather a small $\mathcal{O}(1\%)$ sensitivity for the boundary conditions remains. Hence, it is a quasi-fixed point (which serves more as an upper bound to insure consistency of perturbation theory up to the unification scale). Nevertheless, its existence can beautifully explain $y_t(m_t) \sim 1$ for a large range of initial values at the unification scale. It is this behavior that plays a crucial role in the successful prediction of $b - \tau$ unification discussed in Sec. 5.1.

However, whether or not the top-Yukawa does not saturate its quasi-fixed point, given the heavy top mass it is a large coupling. This is an essential and necessary ingredient in the renormalization of the SSB parameters from some (at this point, arbitrary) boundary conditions, for example, at the unification scale. (It is implicitly assumed here that these parameters appear in the effective theory and are "hard" already at high-energies, but we will return to this point in Sec. 5.5.) Here, we focus only on the renormalization of the terms relevant for electroweak symmetry breaking; the radiative symmetry breaking (RSB) mechanism. A large y_t coupling was postulated more than a decade ago as a mean to achieve RSB [55] and in that sense is a prediction of the MSSM framework which is successfully confirmed by the experimentally measured heavy top mass $m_t \simeq \nu$.

In order to reproduce the SM Lagrangian properly a negative mass squared in the Higgs potential is required. Indeed, the $m_{H_2}^2$ parameter is differentiated from all other squared mass parameters once we include the Yukawa interactions. Consider the coupled renormalization group equations, including, for simplicity, only gauge and top-Yukawa effects. (More generally, the b-quark, τ-lepton, right-handed neutrino, singlet and R_P-violating couplings may not be negligible.) Then, the one-loop evolution of $m_{H_2}^2$ (and of the coupled parameters $m_{U_3}^2$ and $m_{Q_3}^2$) with respect to the logarithm of the momentum-scale is given by

$$\frac{\partial m_{H_2}^2}{\partial \ln \Lambda} = \frac{1}{8\pi^2}(3y_t^2 \Sigma_{m^2} - 3g_2^2 M_2^2 - g_1^2 M_1^2), \tag{89}$$

and

$$\frac{\partial m_{U_3}^2}{\partial \ln \Lambda} = \frac{1}{8\pi^2}(2y_t^2 \Sigma_{m^2} - \frac{16}{3}g_3^2 M_3^2 - \frac{16}{9}g_1^2 M_1^2), \tag{90}$$

$$\frac{\partial m_{Q_3}^2}{\partial \ln \Lambda} = \frac{1}{8\pi^2}(y_t^2 \Sigma_{m^2} - \frac{16}{3}g_3^2 M_3^2 - 3g_2^2 M_2^2 - \frac{1}{9}g_1^2 M_1^2), \tag{91}$$

where $\Sigma_{m^2} = [m_{H_2}^2 + m_{Q_3}^2 + m_{U_3}^2 + A_t^2]$, and we denote the SM $SU(3)$, $SU(2)$ and $U(1)$ gaugino masses by $M_{3,2,1}$, as before.

The one-loop gaugino mass renormalization obeys

$$\frac{d}{d\ln\Lambda}\left(\frac{M_i^2}{g_i^2}\right) = 0, \tag{92}$$

and its solution simply reads $M_i(M_i) = (\alpha_i(M_i)/\alpha_i(\Lambda_{\rm UV}))M_i(\Lambda_{\rm UV})$ where a typical choice is $\Lambda_{\rm UV} = M_U$. Note that in unified theories the gaugino mass boundary conditions are given universally by the mass of the single gaugino of the GUT group so that $M_3 : M_2 : M_1 \simeq 3 : 1 : 1/2$ at the weak scale (where the numerical ratios are the ratios $\alpha_i(M_Z)/\alpha_U$). This is gaugino mass unification. (It also holds, but for different reasons, in many string models.)

Given the heavy t-quark, one has $y_t \sim 1 \sim g_3$. (In fact, for near quasi-fixed point values typically $y_t > g_3$ at high energies.) While QCD loops still dominate the evolution of the stop masses squared $m_{Q_3}^2$ and $m_{U_3}^2$, Yukawa loops dominate the evolution of $m_{H_2}^2$. On the one hand, the stop squared masses and Σ_{m^2} increase with the decreasing scale. On the other hand, the greater they increase the more the Higgs squared mass decreases with scale and, given the integration or evolution time $2\pi t \sim 30$, it is rendered negative at the weak scale. The m_3^2 Higgs doublet mixing ensures that both Higgs doublets have non-vanishing expectation values. This is a simplistic description of the mechanism of radiative electroweak symmetry breaking. In fact, the sizeable y_t typically renders the Higgs squared mass too negative and some (fine?) tuning (usually of μ) is required in order to extract correctly the precisely known electroweak scale. (The Higgs potential and its minimization are discussed in the next section.) An example of the renormalization group evolution of the SSB (and μ) parameter is illustrated in Fig. 8, taken from C. Kolda. For illustration, universal (see Sec. 5.5) boundary conditions are assumed at the unification scale M_U.

In the quasi-fixed point scenario, it is possible to solve analytically for the low energy values of the soft scalar masses in terms of the high scale boundary conditions. To conclude our brief discussion, we quote the solutions including, for completeness, also the solutions for sfermions and Higgs SSB masses which are not affected by the large y_t (For example, see Carena et al. [53]):

$$m_{H_2}^2 \simeq m_{H_2}^2(M_U) + 0.52M_{1/2}^2 - 3\Delta m^2$$
$$m_{H_1}^2 \simeq m_{H_1}^2(M_U) + 0.52M_{1/2}^2$$
$$m_{Q_i}^2 \simeq m_{Q_i}^2(M_U) + 7.2M_{1/2}^2 - \delta_i\Delta m^2$$
$$m_{U_i}^2 \simeq m_{U_i}^2(M_U) + 6.7M_{1/2}^2 - \delta_i 2\Delta m^2 \tag{93}$$
$$m_{D_i}^2 \simeq m_{D_i}^2(M_U) + 6.7M_{1/2}^2$$

160

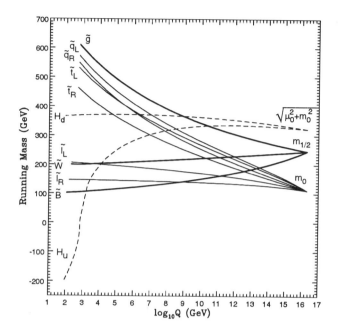

Figure 8: The renormalization group evolution of SSB masses and μ for a representative case with universal boundary conditions at the unification scale M_U.

$$m_{L_i}^2 \simeq m_{L_i}^2(M_U) + 0.52 M_{1/2}^2$$
$$m_{E_i}^2 \simeq m_{E_i}^2(M_U) + 0.15 M_{1/2}^2 \, ,$$

where

$$\Delta m^2 \simeq \frac{1}{6} \left[m_{H_2}^2(M_U) + m_{Q_3}^2(M_U) + m_{U_3}^2(M_U) \right] r$$
$$+ M_{1/2}^2 \left(\frac{7}{3} r - r^2 \right) + \frac{1}{3} A_0 \left(\frac{1}{2} A_0 - 2.3 M_{1/2} \right) r (1 - r) \, , \quad (94)$$

and, for simplicity, and as is costumary, we have assumed a common gaugino mass $M_{1/2}$ and tri-linear scalar coupling A_0 at the high scale, which is conveniently identified here with the unification scale M_U. (Numerical coefficients will change otherwise) We will not discuss in detail the renormalization of the A-parameters, which are assumed here to be proportinal to the Yukawa couplings. However, we note that their renormalization also exhibits fixed points.

The subscript i is a generational index; $\delta_1 = \delta_2 = 0$ and $\delta_3 = 1$. Finally, the parameter $r = [y_t/y_{t_{\text{fixed}}}]^2 \leq 1$ is a measure of the proximity of the top Yukawa coupling to its quasi-fixed point value at the weak scale. It is interesting to observe that for vanishing gaugino masses the system (89) – (91) has a zero fixed point (i.e., it is insensitive to its boundary conditions m_0^2) as long as $[m_{H_u}^2(M_U), m_{U_3}^2(M_U), m_{Q_3}^2(M_U)] = m_0^2 [3, 2, 1]$. Such fixed point, in the presence of Yukawa quasi-fixed points, are a more general phenomena with important consequences for the upper bound on m_0 [56].

The general (two-loop) renormalization group equations are given in Ref. [57]. Other recent related papers include Ref. [58].

5.3 The Higgs potential and the light Higgs boson

Now, that we understand how it is that a negative mass squared is generated radiatively for an arbitrary boundary conditions, we are in position to write and minimize the Higgs potential. In principle, it is far from clear that the Higgs bosons rather than some sfermion receive *vevs*. Aside from the sneutrino (whose *vev* may break lepton number, leading to the generation of neutrino masses) all other sfermions cannot have non-vanishing expectation values or QED and/or QCD would be spontaneously broken. Furthermore, there could be some direction in this many field space in which the complete scalar potential (which involves all Higgs and sfermion fields) is not bounded from below even though the subset corresponding to the pure Higgs potential (involving only the Higgs fields) is bounded. We refrain here from a discussion of these issues but only state that they lead to constrain on the parameter space, for example, on the ratios of the SSB parameters $|A_f/m_{\tilde{f}}| \lesssim 3-6$. For the purpose of our discussion, let us simply assume that any such constraints are satisfied (as is usually the case) and focus on the Higgs potential.

The Higgs part of the MSSM (weak-scale) scalar potential reads, assuming for simplicity that all the parameters are real,

$$V(H_1, H_2) = (m_{H_1}^2 + \mu^2)|H_1|^2 + (m_{H_2}^2 + \mu^2)|H_2|^2$$
$$-m_3^2(H_1 H_2 + h.c.) + \frac{\lambda^{\text{MSSM}}}{2}(|H_2|^2 - |H_1|^2)^2 + \Delta V, \qquad (95)$$

where $m_{H_1}^2$, $m_{H_2}^2$, and m_3^2 (μ) are the soft (supersymmetric) mass parameters renormalized down to the weak scale (*i.e.*, this is what is often called the one-loop improved tree-level potential), $m_3^2 > 0$, $\lambda^{\text{MSSM}} = \frac{g_2^2 + g'^2}{4}$ is given by the hypercharge and $SU(2)$ D-terms, and we suppress $SU(2)$ indices. The one-loop correction $\Delta V = \Delta V^{one-loop}$ (which, in fact, is a threshold correction to the one-loop improved tree level potential) can be absorbed to a good approximation in redefinitions of the tree-level parameters.

A broken $SU(2) \times U(1)$ (along with the constraint $m_{H_1}^2 + m_{H_2}^2 + 2\mu^2 \geq 2|m_3^2|$ from vacuum stability) requires

$$(m_{H_1}^2 + \mu^2)(m_{H_2}^2 + \mu^2) \leq |m_3^2|^2, \tag{96}$$

as is automatically the case if $m_{H_2}^2 < 0$ while $m_{H_1}^2 > 0$, which was the situation discussed in Sec. 5.2. The minimization conditions then give

$$\mu^2 = \frac{m_{H_1}^2 - m_{H_2}^2 \tan^2 \beta}{\tan^2 \beta - 1} - \frac{1}{2} M_Z^2, \tag{97}$$

$$m_3^2 = -\frac{1}{2} \sin 2\beta \left[m_{H_1}^2 + m_{H_2}^2 + 2\mu^2 \right]. \tag{98}$$

By writing eq. (97) we subscribed to the convenient notion that μ is determined by the precisely known value $M_Z = 91.18$ GeV. This is a mere convenience. Renormalization cannot mix the supersymmetric μ parameter and the SSB parameters, and hence μ can be treated as purely a low-energy parameter. Nevertheless it highlights the μ-problem, why is a supersymmetric mass parameter is exactly of the order of magnitude of the SSB parameters (rather than Λ_{UV}, for example) [59]. We touched upon it in the context of GUT's and doublet-triple splitting, but it is a much more general puzzle whose solution must encode some information on the ultra-violet theory which explains this relation. (Several answers were proposed in the literature, including Ref. [59,60,61] and various variants of the NMSSM) It also highlights the fine-tuning issue whose rough measure is the ratio $|\mu/M_Z|$. Typically $|m_{H_2}^2|$ is a large parameter controlled by the stop renormalization, which itself is controlled by QCD and gluino loops. One often finds that a phenomenologically acceptable value of μ is $|\mu| \simeq |M_3|$ and that M_Z is then determined by a cancellation between two $\mathcal{O}(\mathrm{TeV})$ parameters. Clearly, this is a product of our practical decision to fix M_Z rather than extract it. All it tells us is that M_Z (or ν) is a special rather than arbitrary value. The true tuning problem is instead in the typical relation $|\mu| \simeq |M_3|$ which is difficult to understand. Fine-tuning is difficult to quantify, and each of its definitions in the literature has its own merits and conceptual difficulties so caution is in place when applying such esthetic notion to actual calculations, an application which we will avoid.

Using the minimization equations the pseudo-scalar mass-squared matrix (72) (the corresponding CP-even and charged Higgs matrices receive also contribution from the D-terms, or equivalently, from the quartic terms) is now

$$M_{PS}^2 = m_3^2 \begin{pmatrix} \tan \beta & -1 \\ -1 & 1/\tan \beta \end{pmatrix}. \tag{99}$$

Its determinant vanishes due to the massless Goldstone boson. It has a positive mass-squared eigenvalue $m_{A^0}^2 = \mathrm{Tr} M_{PS}^2 = m_3^2/((1/2)\sin 2\beta) = m_1^2 + m_2^2$, where as before $m_i^2 \equiv m_{H_i}^2 + \mu^2$ for $i = 1, 2$. Electroweak symmetry breaking is then confirmed. The angle β is now seen to be the rotation angle between the current and mass eigenstates. the CP-even Higgs tree-level mass matrix reads

$$M_{H^0}^2 = m_{A^0}^2 \begin{pmatrix} s_\beta^2 & -s_\beta c_\beta \\ -s_\beta c_\beta & c_\beta^2 \end{pmatrix} + M_Z^2 \begin{pmatrix} c_\beta^2 & -s_\beta c_\beta \\ -s_\beta c_\beta & s_\beta^2 \end{pmatrix}, \qquad (100)$$

with eigenvalues

$$m_{h^0, H^0}^{2\,T} = \frac{1}{2}\left[m_{A^0}^2 + M_Z^2 \mp \sqrt{(m_{A^0}^2 + M_Z^2)^2 - 4 m_{A^0}^2 M_Z^2 \cos^2 2\beta} \right]. \qquad (101)$$

Note that at this level there is a sum rule for the neutral Higgs eigenvalues: $m_{H^0}^2 + m_{h^0}^2 = m_{A^0}^2 + M_Z^2$.

There are two particularly interesting limits to eq. (101) In the limit $\tan\beta \to 1$ one has $|\mu| \to \infty$ and the $SU(2) \times U(1)$ breaking is driven by the m_3^2 term. In practice, one avoids the divergent limit by taking $\tan\beta \gtrsim 1.1$, as is also required for the perturbativity of the top-Yukawa coupling. For $\tan\beta \to \infty$ one has $m_3 \to 0$ so that the symmetry breaking is driven by $m_{H_2}^2 < 0$.

The $\tan\beta \to 1$ case corresponds to an approximate $SU(2)_{L+R}$ custodial symmetry of the vacuum: Turning off hypercharge and flavor mixing, and if $y_t = y_b = y$, then one can rewrite the t and b Yukawa terms as $SU(2)_L \times SU(2)_R$ invariant [62],

$$y \begin{pmatrix} t_L \\ b_L \end{pmatrix}_a \epsilon_{ab} \begin{pmatrix} H_1^0 & H_2^+ \\ H_1^- & H_2^0 \end{pmatrix}_{bc} \begin{pmatrix} -b_L^c \\ t_L^c \end{pmatrix}_c \qquad (102)$$

(where in the SM $H_2 = i\tau_2 H_1^*$) and $SU(2)_L \times SU(2)_R \to SU(2)_{L+R}$ for $\nu \neq 0$. However, $y_t \neq y_b$ and the different hypercharges of t_L^c and b_L^c explicitly break the left-right symmetry, and therefore the residual custodial symmetry. In the MSSM, on the other hand, H_1 is distinct from H_2 and if $\nu_1 \neq \nu_2$ (where $\nu_i = \langle H_i^0 \rangle$) $SU(2)_L \times SU(2)_R \to U(1)_{T_{3L}+T_{3R}}$. The $SU(2)_{L+R}$ symmetry is preserved if $\beta = \frac{\pi}{4}$ and is maximally broken if $\beta = \frac{\pi}{2}$. (This is the same approximate custodial symmetry which was mentioned above in the context of the smallness of quantum corrections to electroweak observables and couplings, but in the Higgs sector.) The symmetry is broken at the loop level so that one expects in any case $\tan\beta$ slightly above unity. As a result of the symmetry,

$$M_{H^0}^2 \approx \mu^2 \times \begin{pmatrix} 1 & -1 \\ -1 & 1 \end{pmatrix}, \qquad (103)$$

and it has a massless tree-level eigenvalue, $m_{h^0}^T \approx 0$. This is, of course, a well known result of the tree-level formula when taking $\beta = \frac{\pi}{4}$. The mass is then determined by the loop corrections which are well known (to two-loop) $m_{h^0} \approx \Delta_{h^0} \propto h_t m_t$. The heavier CP-even Higgs boson mass eigenvalue equals approximately $\sqrt{2}|\mu|$. (The loop corrections are less relevant here as typically $m_{H^0}^2 \gg \Delta_{H^0}^2$). The custodial symmetry (or the large μ parameter) dictates this case a degeneracy $m_{A^0} \approx m_{H^0} \approx m_{H^+} \approx \sqrt{2}|\mu|$. (The tree-level corrections to that relation are of order $(M_{W,Z}/m_{A^0})^2$.) That is, at a scale $Q \approx \sqrt{2}|\mu|$ the heavy Higgs doublet H is decoupled, and the effective field theory below that scale has only one SM-like ($\nu_{h^0} = \nu$) Higgs doublet, h ($= \eta$ od Sec. 1) which contain a light physical state. This is a special case of the MSSM in which all other Higgs bosons (and sparticles) decouple.

The Higgs sector in large $\tan\beta$ case exhibits an approximate $O_4 \times O_4$ symmetry [63]. For $m_3 \to 0$ (which is the situation in case (2)) there is no mixing between H_1 and H_2 and the Higgs sector respects the $O_4 \times O_4$ symmetry (up to gauge-coupling corrections). The symmetry is broken to $O_3 \times O_3$ for $\nu_1 \neq \nu_2 \neq 0$ and the six Goldstone bosons are the three SM Goldstone bosons, A^0, and $H^\pm$. The symmetry is explicitly broken for $g_2 \neq 0$ (so that $m_{H^+} = M_W$) and is not exact even when neglecting gauge couplings (i.e., $m_3 \neq 0$). Thus, A^0 and $H^\pm$ are massive pseudo-Goldstone bosons, i.e., , $m_{H^+}^2 - M_W^2 \approx m_{A^0}^2 = C \times m_3^2$. However, $C = -\frac{2}{\sin 2\beta}$ and it can be large, which is a manifestation of the fact that $O_4 \times O_4 \to O_4 \times O_3$ for $\nu_1 = 0$. (The limit $m_3 \to 0$ corresponds also to a $U(1)$ Peccei-Quinn symmetry under which the combination $H_1 H_2$ is charged.) In the case $\beta \to \frac{\pi}{2}$ one has $m_{h^0}^T \approx M_Z$ (assuming $m_{A^0} \geq M_Z$). When adding the loop corrections $m_{h^0} \lesssim \sqrt{2} M_Z \approx 130$ GeV.

Before concluding the discussion of the Higgs sector, let us examine the lightness of the SM-like Higgs boson from a different perspective, as well as the one-loop corrections to its mass. Including one-loop corrections, the general upper bound is derived

$$m_{h^0}^2 \leq M_Z^2 \cos^2 2\beta + \frac{3\alpha m_t^4}{4\pi s^2(1-s^2)M_Z^2}\left\{ \ln\left(\frac{m_{\tilde{t}_1}^2 m_{\tilde{t}_2}^2}{m_t^4}\right) + \Delta_{\theta_t} \right\} \qquad (104)$$

where

$$\Delta_{\theta_t} = \left(m_{\tilde{t}_1}^2 - m_{\tilde{t}_2}^2\right) \frac{\sin^2 2\theta_t}{2m_t^2} \ln\left(\frac{m_{\tilde{t}_1}^2}{m_{\tilde{t}_2}^2}\right)$$

$$+ \left(m_{\tilde{t}_1}^2 - m_{\tilde{t}_2}^2\right)^2 \left(\frac{\sin^2 2\theta_t}{4m_t^2}\right)^2 \left[2 - \frac{m_{\tilde{t}_1}^2 + m_{\tilde{t}_2}^2}{m_{\tilde{t}_1}^2 - m_{\tilde{t}_2}^2}\ln\left(\frac{m_{\tilde{t}_1}^2}{m_{\tilde{t}_2}^2}\right)\right], \qquad (105)$$

and where $m_{\tilde{t}_i}^2$ are the eigenvalues of the t-scalar (stop) mass-squared matrix,

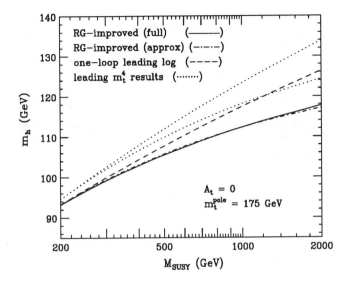

Figure 9: The upper bound to the mass of the light CP-even Higgs boson of the MSSM is plotted as a function of the common supersymmetric mass M_{SUSY} (in the absence of squark mixing). The one-loop leading logarithmic result [dashed line] is compared with the renormalization-group improved result, which was obtained by a numerical computation [solid line] and by the simple recipe [dot-dashed line]. Also shown are a leading m_t^4 result of [higher dotted line], and its renormalization-group improvement [lower dotted line]. For mor details, see Ref. [65].

θ_t is the left-right stop mixing angle, and we have neglected other loop contributions. The tree-level mass squared, $m_{h^0}^{T\,2}$, and the loop correction, $\Delta_{h^0}^2$, are bounded by the first and second terms on the r.h.s. of Eq. (104), respectively. In the absence of mixing $\Delta_{\theta_t} = 0$. For $\tan\beta \to 1$ one obtains $m_{h^0}^T \to 0$, and thus $m_{h^0} \approx \Delta_{h^0}$.

Clearly, and as we observed before, the tree-level mass vanishes as $\tan\beta \to 0$. In this limit, the D-term (expectation value) vanishes as well as the value of the tree-level potential which is now quadratic in the fields. It correspond to a flat direction of the potential and the massless real-scalar is its ground state. Once we have identified the flat direction it is clear that the upper bound must be proportional to $\cos 2\beta$ so that the mass vanishes once the flat direction is realized. Its proportionality to M_Z is only the manifestation that the quartic couplings are the gauge couplings. Hence, the lightness of the Higgs boson is a model independent statement. The above upper bound is modified if and only if the Higgs potential contains terms (aside from the loop corrections) that lift the flat direction, for example, this is the case in the NMSSM or if

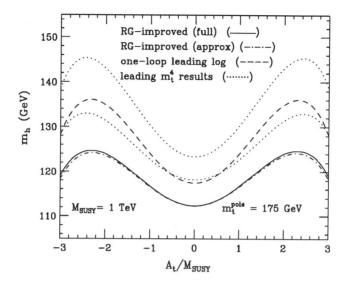

Figure 10: The upper bound to the mass of the light CP-even Higgs boson of the MSSM plotted as a function of A_t/M_{SUSY} where A_t is defined here as $A_t = m^2_{LR}/m_t$. Squark-mixing effects are incorporated. See the caption to Fig. 9.

the gauge structure is extended by an Abelian factor [64] SM $\to$ SM $\times U(1)'$ (an extension that could still be consistent with gauge coupling unification though this is not straightforward). However, as long as one requires all coupling to stay perturbative in the ultra-violet then the additional conributions to the Higgs mass are still modest leading to $m_{h^0} \lesssim 150 - 200$ GeV, where the upper bound of $190 - 200$ GeV was shown to be saturated only in somewhat contrived constructions.

Nevertheless, the flat direction is always lifted by quantum corrections, the most important of which is given in (104). These correction may be viewed as effective quartic couplings that have to be introduced to the effective theory once the stops, for example, are integrated out of the theory at a few hundred GeV or higher scale. These couplings are proportional to the large Yukawa couplings (for example from integrating out loops induced by $y_t^2 \tilde{t}^2 H^2$ quartic F-terms in the scalar potential, as in Fig. 2). Note that even though one finds in many cases $\mathcal{O}(100\%)$ corrections to the light Higgs mass (and hence $m_{h^0} \lesssim M_Z \to m_{h^0} \lesssim \sqrt{2}M_Z$) this does not signal the breakdown of perturbation theory. It is only that the tree-level mass (approximately) vanishes. Indeed, two-loop corrections are much smaller (and shift m_{h^0} by typically only a few GeV) and are often negative.

The existence of a model-independent light Higgs boson is therefore a prediction of the framework. It is encouraging to note that it seems to be consistent with current data. The W mass measurement and other electroweak observable strongly indicate that the SM-like Higgs is light $m_{h^0} \lesssim 200 - 300$ GeV where the best fitted values are near 100 GeV [12]. Serches at the LEP experiment bound the SM-like Higgs mass from below $m_{h^0} \gtrsim 90$ GeV [46]. Depending on final luminosity and center-of-mass energy, the model-independent light Higgs may be probed already in future LEP and Tevatron runs, but it may be that its discovery (or exclusion) must wait until the LHC is operative (where, however, detection of a Higgs boson in this mass range may be difficult).

The predicted mass range of the light Higgs boson in the MSSM is illustrated in Figs. 9 and 10 taken from Ref. [65]. Further Discussion of the Higgs sector and references can also be found, for example, in Refs. [66,67].

5.4 The flavor problem - An organizing principle

In spite of its success in producing the SM (and the Higgs potential) as its low-energy limit while allowing extrapolation and unification at high energies, the most general MSSM Lagrangian is still described by many arbitrary parameters, in particular, once flavor and CP conservation are not imposed on the soft parameters. Aside from the obvious loss of predictive power, many generic models are actually in an apparent conflict with the low-energy data: An arbitrary choice of parameters can result in unacceptably large flavor changing neutral currents (FCNC's) ("The Flavor Problem") and, if large phases are present, also large (either flavor conserving or violating) contributions to CP-violating amplitudes, e.g., the flavor conserving electron and neutron dipole moments ("The CP Problem"). Clearly, one has to identify those special cases which evade these constraints, and by doing so the whole framework regains predictive power. Let us first elaborate on the problems, focusing on the flavor problem. The flavor-conserving CP problem [68] is, in practice, less constraining; may be resolved independently if no new large physical phases are present in the weak-scale Lagrangian (i.e., they are absent in the high-energy theory or alternatively large relative phases could be diminished by renormalization effects could be "renormalized away"); and may even be resolved in some (very) special cases by cancellations [69] (though tuning cancellations is somewhat equivalent to assuming small phases).

The flavor problem stems from the very basic supersymmetrization of the SM. Recall that even though each fermion flavor is coupled in the SM with an arbitrary 3×3 Yukawa (or equivalently, mass) matrix, the unitary diagonalization matrix guarantees that the fermion-fermion coupling to a neutral gauge boson is basis independent and flavor blind at tree-level (the GIM mechanism). This is also true to a sfermion-sfermion coupling to a neutral gauge boson which is now multiplied by the sfermion unitary rotation matrices (the super-GIM mechanism). However, by supersymmetry, there exists also a fermion-sfermion-gaugino vertex. If both fermion and sfermion mass matrices are diagonalized, then the gaugino vertex is rotated by two independent unitary matrices and, in general, their product $U_f V_{\tilde{f}}^{\dagger} \neq I$ is not trivial. The rotations maintain in this case some non-trivial flavor structure and the gaugino-fermion-sfermion vertex is not flavor diagonal. That is, generically the theory contains, for example, a $g_s \tilde{g} d \tilde{s}^*$ vertex (in the fermion mass basis)!

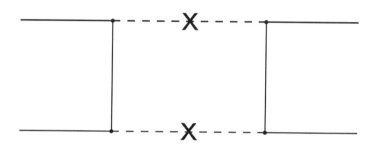

Figure 11: A generic "box" diagram contribution to meson mixing with gauginos and squarks circulating in the loop. The Flavor changing is approximated here as flavor changing squark-mass insertions, which are denoted explicitly.

Given the flavor violating gaugino vertex, it is straightforward to construct flavor violating quantum corrections for example, to $K - \bar{K}$ mixing, which violate flavor by two units: See Fig. 11 for illustration. (Adding phases, new contributions to the SM parameter ϵ_K, and, in some cases, also to the parameter ϵ_K'/ϵ_K, arise; the former provides a strong constraint while the latter, which corresponds to flavor violation by only one unit, has been recently measured with high accuracy and could actually accommodate new-physics contributions.) Such effects are more conveniently evaluated in the fermion mass basis in which the sfermion squared-mass matrix retain its inter-generational off-diagonal entries,

$$m_{\tilde{f}}^2 = \begin{pmatrix} m_{\tilde{f}_1}^2 & \Delta_{\tilde{f}_1 \tilde{f}_2} & \Delta_{\tilde{f}_1 \tilde{f}_3} \\ \Delta_{\tilde{f}_2 \tilde{f}_1} & m_{\tilde{f}_2}^2 & \Delta_{\tilde{f}_2 \tilde{f}_3} \\ \Delta_{\tilde{f}_3 \tilde{f}_1} & \Delta_{\tilde{f}_3 \tilde{f}_2} & m_{\tilde{f}_3}^2 \end{pmatrix}, \tag{106}$$

and the sfermions of a given sector f mix. The matrix (106) corresponds to the CKM rotated original sfermion mass-squared matrix. In the (realistic) limit of $\Delta \ll m_{\tilde{f}_i}^2$, the off-diagonal elements are given by $\Delta_{\tilde{f}_i \tilde{f}_j} \simeq m_{\tilde{f}_i}^2 - m_{\tilde{f}_j}^2$. Additional flavor-violating left-right mixing arises, in principle, from the A-matrices but here we ignore chirality labels. When this (insertion) approximation is applied to K meson mixing one has, for example, for the $K_L - K_S$ mass difference

$$\frac{\Delta m_K}{m_K f_K^2} \simeq \frac{2}{3} \frac{\alpha_3}{216} \frac{1}{\widetilde{m}^2} \left(\frac{\Delta_{\tilde{d}\tilde{s}}}{\widetilde{m}^2} \right)^2, \tag{107}$$

where we used standard notation for the kaon mass m_K and for the relevant form factor f_K, and for simplicity dimensionless functions are omitted. This leads to the constraint on the product of a typical sqaurk mass $\widetilde{m}$ and the inter-generational mixing

$$\left(\frac{500 \text{ GeV}}{\widetilde{m}^2} \right) \left(\frac{\Delta_{\tilde{d}\tilde{s}}}{\widetilde{m}^2} \right) \lesssim 10^{-(2-3)}, \tag{108}$$

where the complete amplitude, including higher order corrections, was evaluated [70] in deriving (108). Weaker constraints are derived from B- and from D-meson mixing.

Another set of observables which should be mentioned is given by magnetic moments with virtual sparticles circulating in the loop, shown in Fig. 12. In the supersymmetric limit this operator (i.e., the sum of SM and new-physics contributions) vanishes - the operator cannot be written in a supersymmetric fashion and hence must vanish. This implies that SM and new-physics contribution to any anomalous and transition magnetic moments are of the same order of magnitude and appear at the same loop order. To begin with consider flavor conserving magnetic moments, for example, the anomalous muon magnetic moment a_μ [71]. The flavor-conserving muon magnetic moment will be probed in the next few years at the Brookhaven E821 experiment far beyond current bounds, offering a rare opportunity for evidence of new physics. Supersymmetry generically contributes

$$a_\mu^{\text{SUSY}} \sim \pm \frac{\alpha_2}{4\pi} \frac{m_\mu^2 M_Z \tan \beta}{\widetilde{m}^3}, \tag{109}$$

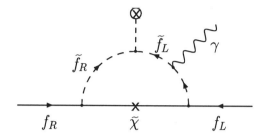

Figure 12: One-loop radiative magnetic moment operator (with the chiral violation provided by the sfermion left-right mixing).

where $\widetilde{m}$ is a typical superpartner (in this case, chargino) mass scale, and the sign is determined by the sign of the μ parameter. (In the special case of models in which the muon mass comes about radiatively rather from a tree-level Yukawa coupling one has [35] $a_\mu^{\text{SUSY}} \sim +m_\mu^2/3\widetilde{m}^2$, which is significantly enhanced.) This is a most interesting "flavor-conserving" $\tan\beta$-dependent constraint/prediciton in supersymmetry. Similarly, the same loop operator, but with an added phase, corresponds to a (flavor-conserving) dipole moment, for example, the electric dipole moment constraints slepton masses to be in the multi-TeV range if the corresponding "soft" phase ϕ (for example a gaugino or A-parameter phase, or their relative phase) is large,

$$\frac{\widetilde{m}}{\sqrt{(M_Z/\widetilde{m})\tan\beta\sin\phi}} \gtrsim 2 \text{ TeV}. \qquad (110)$$

Returning to flavor violation, the same magnetic operator could connect two different fermion flavors leading to magnetic transition operators which violate flavor by one unit and where the SM and supersymmetry contribution arise at the same level. The most publicized case is that of the $b \to s\gamma$ operator [72]; the first such operators whose amplitude was measured and where one hopes that the precision of the next generation of experiments would allow to disentangle SM from new-physics contributions,

$$BR(b \to s\gamma) \simeq BR(b \to ce\bar{\nu})\frac{|V_{ts}^* V_{tb}|^2}{|V_{cb}|^2}\frac{6\alpha}{\pi}\left[\text{SM} \pm \frac{m_t^2\tan\beta}{\widetilde{m}^2}\right]^2, \qquad (111)$$

where the first contribution is SM and the second arises in supersymmetry from from CKM rotations V (which, by supersymmetry, apply also to the chargino and neutralino vertices) and is always present (with a sign given by the relative sign of A_t and μ, and $\widetilde{m}$ of the order of the stop mass). There could also be a contribution which is intrinsically flavor violating supersymmetric contribution (see Gabbiani et al.[73]),

$$BR(b \to s\gamma) \simeq \frac{2\alpha_3^2\alpha}{81\pi^2}\mathcal{F}\frac{m_b^3\tau_B}{\widetilde{m}^2}\left(\frac{\Delta_{\widetilde{s}\widetilde{b}}}{\widetilde{m}^2}\right)^2 \qquad (112)$$

with $\mathcal{F} \simeq \mathcal{O}(1-10)$, $\tau_B \sim 10^{-12}$ sec the B-meson mean life time, and $\widetilde{m}$ of the order of the gluino and sbottom masses. (We did not write the charged Higgs contribution which in the limit in which sparticle contributions decouple – as is the case for the squarks in the gauge mediation framework which is discussed in the next section – can constrain its mass, but otherwise is typically a secondary contribution). While $b \to s\gamma$ has been observed and the branching ratio measured, it does not yet provide strong constraints or a signal of new physics, in particular, if both contributions (111) and (112) are considered. On the other hand, moderate to strong constraints arise in the most interesting case of individual-lepton number (lepton-flavor) violating (but total lepton-number conserving) magnetic transition operators such as $\mu \to e\gamma$. These vanish in the SM where lepton number is conserved. Thus, an observation of such a process will provide a clear indication for new physics! In supersymmetry slepton squared- mass matrices and tri-linear couplings may not respect these accidental flavor symmetries of the SM, allowing for such processes. (Lepton-flavor violations could further be related to specific models of total lepton number violation and neutrino masses and/or grand-unified models where quarks and leptons are predicted to have similar flavor structure at high energies.) The latter experiments are therefore of extreme importance, in particular, given that lepton-flavor violation in (atmospheric) neutrino oscillations was recently established experimentally [74]. However, the experiments are also extremely difficult given the small amplitudes expected for such processes (which are not enhanced by large QCD or Yukawa couplings). Typical amplitudes constrain, in the case of $\mu \to e\gamma$,

$$\widetilde{m} \gtrsim 600 \text{ GeV } \sqrt{\left(\frac{M_Z}{\widetilde{m}}\right)\left(\frac{\Delta_{\widetilde{e}\widetilde{\mu}}}{\widetilde{m}^2}\right)} \tan\beta, \qquad (113)$$

where $\widetilde{m}$ here is of the order of the slepton mass.

Flavor violations (and magnetic moments) in supersymmetry and the corresponding signals and constraints have been studied by many authors, and a sample of papers in listed in Ref.[73]. Supersymmetry is not the only extension of the SM which is constrained by flavor (and CP) conservation. In fact, such constraints are generic to any extension since by its definition new-physics extends the SM unique strucutre which eliminates tree-level FCNC's – a property which is difficult to maintain. Nevertheless, unlike strongly-coupled theories or theories with extra dimensions (whether supersymmetric or not), in the case of perturbative supersymmetry the theory is well-defined and calculable, and hence, the problem is well-defined and calculable (as illustrated above), leaving no place to hide.

The satisfaction of the above and similar constraints derived from low-energy observables leads to the consideration of a small number of families of special models, and hence, provides an organizing principle as well as predictive power. Clearly, realistic models require nearly flavor-conserving gaugino couplings. Models with the right balance of the universality limit $\Delta_{\tilde{f}_i \tilde{f}_j} \to 0$ (in the physical fermion mass basis) and the decoupling limit $\tilde{m} \gg M_Z$ can achieve that. The former limit is difficult to justify without a governing principle (though such a principle may be a derivative of the ultra-violet theory and only seem as arbitrary from the low-energy point of view); the latter limit may seem naively as contradictory to the notion of low-energy supersymmetry and its solution to the hierarchy problem. Nevertheless, these are the keys for the resolution of the flavor problem. We conclude this lecture with a brief review of the realization of these limits in various frameworks for the origins of the SSB parameters.

5.5 Models

We begin by postulating that supersymmetry is broken spontaneously in some "hidden" sector of the theory which contains the Goldstino and which interacts with the SM "observable" sector only via a specific agent. The agent is then the messenger of supersymmetry breaking. These messenger interactions cannot be renormalizable tree-level interactions or otherwise the SM could couple directly to the Goldstino multiplet and one would encounter many of the problems that plagued our attempt in Sec. 4.3 to break supersymmetry spontaneously in the SM sector. The messengers and their interactions then decouple at a scale $\Lambda_{\text{mediation}}$ which effectively serves the scale of the mediation of the SSB parameters in the low-energy theory, i.e., this is the scale below which these parameters can be considered as "hard" and treated with the renormalization-group formalism (as we did in Sec. 5.2 in our discussion of radiative symmetry breaking). This general "hidden-messenger-observable" phenomenological framework describes most of the models and will suffice for our purposes. It is the details of the messenger interactions and the mediation scale that impact the spectrum parameters and the flavor problem and its solution. (It is important to recall that gauge, Yukawa, and quartic interactions are still dictated by supersymmetry at all energies above the actual sparticle mass scale, and that only the dimensionful parameters interest us in this section.)

In order to review the different classes of models, let us re-write the soft parameters (and for completeness, also the μ-parameter) in a generalized operator form, using the tools of (global) supersymmetry described in previous lectures. We can express the most general set of operators in an expansion in powers of $\sqrt{F_{X,Z}}$, where X and Z are singlet and non-singlet superfields which spontaneously break (or parameterize such breaking) supersymmetry (and hence, contribute (if not provide) to the Goldstino component of the physical gravitino); their F-components are order parameters of supersymmetry breaking as seen in the SM sector; and $M \simeq \mathcal{O}(\Lambda_{\mathrm{mediation}})$. The fields X and Z which couple to the SM sector are the messengers of supersymmetry breaking, and the interaction which couple them to the SM sector is its agent. The gravitino mass can be shown (by the local-supersymmetry condition of a vanishing cosmological constant) to be $m_{3/2} = \sum_i F_i/\sqrt{3}M_P$ where M_P is as before the local-supersymmetry expansion parameter: the reduced Planck mass $M_P = M_{\mathrm{Planck}}/\sqrt{8\pi} \simeq 2.4 \times 10^{18}$ GeV, but otherwise no direct application of local supersymmetry is needed. The leading terms in this expansion which generate the μ parameter and soft terms have the following form:

$$\text{Scalar masses}: \ a \int d^2\theta d^2\bar\theta \, \Phi_i^\dagger \Phi_i \left[\frac{X^\dagger X}{M^2} + \frac{Z^\dagger Z}{M^2} + \cdots \right] \quad (114)$$

$$\mu \ \text{parameter}: \ b \int d^2\theta d^2\bar\theta \, H_1 H_2 \left[\frac{X^\dagger}{M} + \cdots \right]$$

$$\text{and} \ c \int d^2\theta \, H_1 H_2 \left[\frac{X^n}{M^{n-1}} + c' \frac{\hat{W}^\alpha \hat{W}_\alpha}{M^2} + \cdots \right] \quad (115)$$

$$\text{Higgs mixing } (m_3^2): \ d \int d^2\theta d^2\bar\theta \, H_1 H_2 \left[\frac{X^\dagger X}{M^2} + \frac{Z^\dagger Z}{M^2} + \cdots \right] \quad (116)$$

$$\text{Gaugino masses}: \ e \int d^2\theta \, W^\alpha W_\alpha \left[\frac{X}{M} + \cdots \right] \quad (117)$$

$$A-\text{terms}: \ f \int d^2\theta \, \Phi_i \Phi_j \Phi_k \left[\frac{X}{M} + \cdots \right]$$

$$\text{and} \ g \int d^2\theta d^2\bar\theta \, \Phi_i^\dagger \Phi_i \left[\frac{X}{M} + \cdots \right] , \quad (118)$$

where $a - g$ are dimensioless coefficients, Φ are observable (SM) sector fields, the W_α ($\hat{W}_\alpha$) are gauge vector supermultiplets containing the standard model gauginos (a hidden-sector gauge field which condenses), the Φ_i are standard model chiral superfields, and X and Z represent supersymmetry-breaking gauge singlet and non-singlet superfields, respectively. These terms give SSB parameters and the μ parameter when the X and Z fields get F-term vevs: $X \to F_X \theta^2$, $Z \to F_Z \theta^2$ (and, in the second source for A-terms, $\Phi_i^\dagger \to F_{\Phi_i}^* \bar{\theta}^2 \sim \Phi_j \Phi_k \bar{\theta}^2$). The μ-parameters could also arise from a supersymmetry conserving vev of x, but typically such a vev is a hidden superpotential $X^3 \leftrightarrow \langle W_{\text{hidden}} \rangle$ vev which (in local supersymmetry) also breaks supersymmetry. The case $n = 1$ is eq. (115) corresponds to the NMSSM and one has to arrange $\langle x \rangle \simeq M_{\text{Weak}}$. In this case $m_3^2 \sim A\mu$. The case $n = 2$ corresponds to $\langle x \rangle \simeq \sqrt{MM_{\text{Weak}}}$. The case $n = 3$ with $X^3 \equiv \langle W_{\text{hidden}} \rangle$ can be related to first operator in (115) via redefinitions (a Kahler transformation in local supersymmetry). Observe that if only one singlet field participates in supersymmetry breaking then the SSB parameters have only one common phase, and at most one additional phase arises in the μ parameter. Since it can be shown that two phases can always be rotated away and are not physical, there is no CP problem in this case.

It is widely thought that if supersymmetry is realized in nature then it is a local (super)symmetry, *i.e.*, a supergravity theory. (In particular, in this case the cosmological constant is not an order parameter and could be fine-tuned to zero.) Supergravity interactions provide in this case the theory of (quantum) gravity at some ultra-violet scale (though it is a non-renoemalizable theory and hence is probably not the ultimate theory of gravity). As a theory of gravity, supergravity interacts with all sectors of the theory and hence provides an ideal agent of supersymmetry breaking. In fact, even if other agents/messengers exist, supergravity mediation is always operative once supersymmetry is gauged and localized. It mediation scale is, however, always the reduced Planck mass. From the operator equations (114) – (118) and $m_{3/2} \sim F/M_P$ one observes that their contribution to the SSB parameters is always $\sim m_{3/2}$, so it could be the leading contribution for a sufficiently heavy gravitino $m_{3/2} \gtrsim M_Z$, but otherwise its contributions are suppressed. It may be argued that the hierarchy problem is then the problem of fixing the gravitino mass (which may be solved by dynamically breaking supersymmetry in the hidden sector).

Let us then consider the case of a "heavy" gravitino, $m_{3/2} \simeq \mathcal{O}(100-1000)$ GeV (and $\mathcal{O}(1)$ coefficients). Supersymmetry is spontaneously broken in this case at a scale $\sqrt{F} \sim m_{3/2}M_P \sim 10^{10-11}$ GeV, though the scale of the mediation is always fixed in supergravity to M_P. The hidden sector is truly hidden in this case in the sense that interacts only gravitationally with the observable sector (though supergravity interactions could take various forms). Supergravity mediation leads to some obvious observations. First, if $\mu = 0$ in the ultra-violet theory, it is induced once supersymmetry is broken and is of the order of the gravitino mass, resolving the general μ-problem[59,60,61]. (More solutions arise if the model is such that $\langle x \rangle \simeq \sqrt{F}$ or supersymmetry is broken by gaugino condensation in the hidden sector[30,75] $\langle \hat{W}^\alpha \hat{W}_\alpha \rangle \simeq \langle F \rangle^{3/2}$.) Secondly, since the gravitino mass determines the weak scale $m_{\tilde{f}} \simeq m_{3/2} \simeq M_Z$, the decoupling limit cannot be realized. Hence, in order for the SSB parameters to conserve flavor, supergravity has to conserve flavor so that the universality limit can be realized. Naively, gravity is flavor blind. In general, however, supergravity (and string theory) is not! Re-writing, for example, the quadratic operators with flavor indices, $a_{\alpha\beta ij} Z_\alpha Z_\beta^\dagger \Phi_i \Phi_j^\dagger$, in the universality limit $a_{\alpha\beta ij} = \hat{a}_{\alpha\beta} \delta_{ij}$ (where at least each flavor sector is described by a unique $\hat{a}_{\alpha\beta} \delta_{ij}$ combination, though different flavor sectors could be distinguished by an overall factor). In particular one has to forbid any other hidden-observable mixing (which, however, is suggested by the solution of the μ-problem) such as $(ZZ^\dagger)(\Phi\Phi^\dagger)^2/M^4$ which lead to quadratically divergent corrections (proportional to the gravitino mass) which, in turn, spoil the universality at the percentile level[76,75]. It was argued that the universality may be a result of a flavor (contentious or discrete) symmetry which is respected by the hidden sector and supergravity or a result of string theory. If the symmetry is exact (which most probably requires it to be gauged) then universality can hold to all orders. If universality is a result of a "string miracle" (such as supersymmetry breaking by a stabilized dilaton[77]) then it is not expected to be exact beyond the leading order. A minimal assumption that is often made as a "best first guess" is that of total universality, A-term proportionality (to Yukawa couplings), and gaugino unification, which lead to only four ultra-violet parameters (in addition to possible phases and the sign of the μ-parameter): $m_\phi^2(\Lambda_{\text{UV}}) = m_0^2$, $A_{ijk}(\Lambda_{\text{UV}}) = A_0 y_{ijk}$, $M_i(\Lambda_{\text{UV}}) = M_{1/2}$, $m_3^2(\Lambda_{\text{UV}}) = B_0\mu$, which are all of the order of the gravitino mass[78]. (μ is not a free parameter in this case but is fixed by the M_Z constraint, only its sign is a free parameter.) This framework is sometimes called minimal supergravity. It may be that only the operator (117) is present at the mediation scale ($m_0 = A_0 = 0$) and that all other SSB parameters are induced radiatively. In this case, the gauge interactions guarantee universality in 3×3 subspaces: GIM universality. (Such models are highly predictive and are not favored phenomenologically.) Universality (whether in each 3×3 subspace or for all fields) is the mystery of supergravity models, but since the mediation itself (and the μ-parameter) are trivially given in this framework, it cannot be discounted and the price may be worth paying.

It was also proposed recently that supergravity mediation is carried out only at the quantum level[79] (anomaly mediation) via supergravity quantum corrections which generically appear in the theory. The relevant coefficients are then given by loop factors with specific pre-factors (e.g., $e \sim b_i g_i^2$ is given by the one-loop β-function and there is no gaugino unifcation). These proposals, though economic and elegant, still face difficulties in deriving a consistent scalar spectrum and the μ-parameter. A universal lesson from these proposals, however, is that gaugino unification in supergravity may be a fiction. (For a general study of corrections to gaugino unification, see Ref. [80].)

Nevertheless, it may be that supergravity mediation is sufficiently suppressed by a light gravitino mass (and hence no assumption on its flavor structure is needed): One assumes the decoupling limit of supergravity. Then, a new mediation mechanism and messenger sector are required. An attractive option is that the messenger interactions are gauge interactions so that universality is an automatic consequence. This is the gauge-mediation framework. The hidden sector (which is not truly hidden now) communicates via, e.g., new (messenger) gauge interactions with a messenger sector, which, in turn, communicates via the ordinary gauge interactions with the observable sector. It is sufficient to postulate that the new gauge and messenger Yukawa interactions mediate the supersymmety breaking to a (SM) singlet messenger $X = x + \theta^2 F_X$, which parameterizes the supersymmetry breaking in the messenger sector. (Some other hidden fields, however, with $F > F_X$ may dominate the massive gravitino.) The singlet X interacts also with SM non-singlet messenger fields V and $\bar{V}$. The Yukawa interaction $\lambda X V \bar{V}$ communicates the supersymmetry breaking to the messengers V and $\bar{V}$ via the mass matrix,

$$M_{v\bar{v}}^2 \sim \begin{pmatrix} \lambda^2 x^2 & \lambda F_X \\ \lambda F_X^* & \lambda^2 x^2 \end{pmatrix}. \tag{119}$$

In turn, the vector-like pair V and $\bar{V}$, which transforms under the SM gauge group (for example, they transform as 5 and $\bar{5}$ of $SU(5)$, *i.e.*, as down singlets and lepton doublets and their complex conjugates), communicates the supersymmetry breaking to the ordinary MSSM fields via gauge loops. The gauge loops commute with flavor and, thus, the spectrum is charge dependent but flavor diagonal, if one ensures that all other possible contributions to the soft spectrum are absent or are strongly suppressed. This lead to (GIM-)universality. Such models are often referred to as "gauge mediation of supersymmetry breaking" or messenger models [81]. This scenario is conveniently described by the above operator equations with $M \sim \langle x \rangle$ and $e \sim \alpha/4\pi$ a generic one-loop factor and $a \sim (\alpha/4\pi)^2$ a two-loop factor. All other coefficients generically equal zero at the messenger scale. The sparticle spectrum, and hence, the weak scale, are given in this framework by $M_Z \sim m_{\text{sparticle}} \sim (\alpha_i/4\pi)(F_X/x)$, where α_i is the relevant gauge coupling at the scale $\Lambda_{\text{mediation}} \sim x$. One then has $F_x/x \sim (4\pi/\alpha)M_Z \sim 10^5$ GeV. In the minimal version one assumes $\Lambda_{\text{mediation}} \sim F_x/x \sim 10^5$ GeV; a similar scale for the spontaneous supersymmetry breaking in the hidden sector; as well as $F_X \sim x^2$; resulting in an one-scale model. The latter assumption could be relaxed as long as the phenomenologically determined ratio $F_x/x \sim (4\pi/\alpha)M_Z \sim 10^5$ GeV remains fixed. In this case, $\Lambda_{\text{mediation}} \sim x$ could be at a much higher scale[82]. Also, the scale of spontaneous supersymmetry breaking in the hidden sector could be one or two orders of magnitude higher than $\sqrt{F_X}$, for example, if F_X is induced radiatively by a much larger F-term of a hidden field (in which case X contributes negligibly to the massive gravitino).

The messengers induce gaugino masses at one-loop ($e \sim \alpha/4\pi$) and scalar masses at two-loops ($a \sim (\alpha/4\pi)^2$) (the messengers obviously cannot couple via gauge interactions to the other chiral fields at one-loop) creating the desired relation $m_{\tilde{f}}^2 \sim M_\lambda^2$ between scalar and gaugino masses. In addition, there is a mass hierarchy $\sim \alpha_3/\alpha_2/\alpha_1$ between the heavier strongly interacting sparticles and the only weakly interacting sparticles which are lighter. (In detail, it depends on the charges of the messengers). The fact that A parameters arise only via renormalization is of no concern. However, the Higgs mixing parameters do not arise from gauge interactions. One may introduce new Yukawa interactions, but then generically $b \simeq d \simeq (y^2/16\pi^2)$ for some generic Yukawa coupling y, leading to a hierarchy problem $|m_3^2| \sim |\mu|\Lambda_{\text{mediation}}$. This overshadows the otherwise success of gauge mediation, and the possible resolutions[83] are somewhat technically involved and will not be presented here. Note that this framework has a small number of "ultra-violet" parameters and it is highly predictive (less so in extended versions). The gravitino is very light and is the LSP (with signatures such as neutralino decays to energetic photons and gravitino missing energy or a charged slepton escaping the detector, decaying only outside the detector to a lepton and gravitino). Also, the SSB parameters are not "hard" at higher scales such as the unification scale. Radiative symmetry breaking relies in this case on mass hierarchy mentioned above $m_{\tilde{q}}^2(\Lambda_{\text{mediation}}) \sim (\alpha_3/4\pi)^2\Lambda_{\text{mediation}}^2 \gg m_{\tilde{H}}^2(\Lambda_{\text{mediation}}) \sim (\alpha_2/4\pi)^2\Lambda_{\text{mediation}}^2$ rather than on a large logarithm, and its solutions take a slightly different form than eqs. (93).

Can the gauge-mediation order parameter F_X/x be induced using only supergravity interactions? (Supergravity serves in such a case only as a trigger for the generation of the SSB parameters.) The answer is yes, since an operator of the form

$$\int d^2\theta d^2\bar{\theta}\, X \left[\frac{Z^\dagger Z \Phi^\dagger \Phi}{M_P^4} + h.c. \cdots\right] \tag{120}$$

leads to a quadratically divergent tadpole loop which is cut-off at M_P, and hence to a scalar potential of the form $V(x) = (|F_Z|^2/M_P)x + |\partial W/\partial X|^2 \sim m_{3/2}^2 M_P x + x^4$ (where in the last step we assumed $W(X) = X^3/3$). This is only a sketch of this hybrid frameork [76] in which a supergravity linear term in a singlet X can trigger a gauge mediation framework for $\sqrt{F_Z} \sim 10^{8\pm1}$ GeV (leading to $x \sim 10^5$ GeV and $F_X \sim x^2$). The clear benefit is that the hidden sector (Z in this case) remains hidden while X, which couples as usual $XV\bar{V}$, is a true observable sector singlet field. The triggering gravity mediation (leading to the linear term) is carried out, as in anomaly mediation, only at the quantum level (but with a very different source than in anomaly mediation).

A different hybrid approach is that of superheavy supersymmetry (or the $2 - 1$) framework, which as implied by its name, relies on decoupling in order to weaken the universality constraint. The conflict between naturalness and experimental constraints is resolved in this case by observing that, roughly speaking, naturalness restricts the masses of scalars with large Yukawa couplings, while experiment constrains the masses of scalars with small Yukawa couplings [84]. Naturalness affects particles which are strongly coupled to the Higgs sector, while experimental constraints are strongest in sectors with light fermions which are plentifully produced. This suggests that naturalness and experimental constraints may be simultaneously satisfied by an "inverted hierarchy" approach, in which light fermions have heavy superpartners, and heavy (third family) fermions have "light" $\mathcal{O}(M_Z)$ superpartners (hence, $2 - 1$ framework). Therefore, the third generation scalars (and Higgs) with masses $m_{\text{light}} \lesssim 1$ TeV satisfy naturalness constraints, while first and second generation scalars at some much higher scale m_{heavy} avoid many experimental difficulties. A number of possibilities have been proposed to dynamically generate scalar masses at two hierarchically separated scales. Usually one assumes that $m_{\text{light}} \simeq m_{3/2}$ is generated by the usual supergravity mediation, while m_{heavy} is generated by a different "more important" (in terms of the relative contribution) mechanism which, however, discriminates among the generations. (Note that more generally it is enough that only the first generation sfermions are heavy and their mixing with the second generation is "see-saw" suppressed $\sim (m_{\text{light}}/m_{\text{heavy}})^2$, or, alternatively naturalness allows the stau, in some cases, to be heavy.) Such a "more important" mechanism may arise from the D-terms of a ultra-violet gauged flavor but anomalous $U(1)$ (a case which we did not discuss in these lectures) with a non-vanishing D-term, $F_Z F_Z^\dagger \to \langle D \rangle^2$ in eq. (114) and $\langle D \rangle / M \gg m_{3/2}$ has to be arranged [85]. Alternatively, there could be a flavor (or horizontal) messenger mechanism at some intermediate energies [76,86] (and in this case the the gauged flavor symmetry is anomaly free and it is broken only at intermediate energies). The messenger model in this case follows our discussion above only that the messengers are SM singlets but charged under the flavor (horizontal) symmetry. In both of these examples a gauge horizontal (flavor) symmetry discriminates among the generations and does not affect the MSSM gaugino masses, for example, the horizontally neutral third generation sfermion and Higgs field do not couple the horizontal D-term or to the horizontal messengers. The coefficients a in eq. (114) are in these cases flavor dependent but do not mix the light and heavy sfermions, a mixing which is protected by the horizontal gauge symmetry. A fundamentally different approach[56] is that $F_Z \gg F_X$ (a limit which realizes an effective $U(1)_R$ symmetry in the low-energy theory) and all the boundary conditions for all the scalars are in the multi-TeV range (while gaugino and A-terms are much lighter). The light stop squarks and Higgs fields, for example, are then driven radiatively and asymptotically to m_{light}. Indeed, and not surprisingly, one find that in the presence of large Yukawa couplings there is such a zero fixed point which requires, however, that the respective sfermion and Higgs boundary conditions have specific ratios (i.e., it is realized along a specific direction in field space as pointed out in Sec. 5.2). Here, the hierarchy is indeed inverted as the light scale is reached by large Yukawa renormalization. In practice, all of these solutions are constrained by terms which couple the light and heavy fields and which are proportional to small Yukawa couplings or which arise only at two-loops. Such terms are generically suppressed, but are now enhanced by the heavy sfermions. The importance of such effects is highly model dependent and they constrain each of the possible realization in a different fashion, leaving more than sufficient room for model building. Even though such a realization of supersymmetry would leave some sfermions beyond the kinematic reach of the next generation of collider experiments, some sfermion and gauginos should be discovered. It was noted that by measuring the ratio of the gaugino-fermion-(light)sfermion coupling and the gauge coupling (the superoblique parameters[87]) one can confirm in many cases the presence of the heavy states via quantum corrections to these ratios, providing an handle on these and other models with multi-TeV fields.

The last class of models which we mention includes models in which the Yukawa and sfermion squared mass (and tri-linear A) matrices are aligned in field space, and hence, are diagonalized simultaneously[88]. Thus, $\Delta_{ij} = 0$ without universality. It was proposed that such an alignment may arise dynamically once the (Coleman-Weinberg) effective potential is minimized with respect to some low-energy moduli X and Z. However, such a mechanism tends to be unstable to higher-order corrections. Alternatively, it could be that such alignment is a result of some high-energy symmetry principle. Realization of this idea tend be cumbersome, in contrast to the simplicity of the assumption, and it is difficult to envision conclusive tests of specific symmetries (though the alignment idea itself may be tested (ruled out) in sfermion oscillations[89] which cannot arise for $\Delta_{ij} \equiv 0$.) Nevertheless, this is another possibility in which case F_X/M and its square are aligned with the Yukawa matrices and are diagonal (only) in the physical mass basis of fermions.

Though the origion of the SSB parameters remains elusive, discovry of supersymmetry will open the door to probing the mediation scale and mechanism, and hence, to a whole new (ultra-violet) arena.

Exercises

1. Calculate the β-function coefficients in the MSSM (eq. (82)). At what order the assumption of R_P conservation affect the calculation?

2. Confirm the hypercharge $U(1)$ normalization.

3. Solve in iterations the two-loop renormalization group equation for the gauge coupling, neglecting Yukawa couplings. Rewrite the solutions as predictions for t, M_U and $\alpha_3(M_Z)$.

4. Use the strong coupling to predict the weak angle. Show that at the unification scale one has the boundary condition $s_W^2(M_U) = g'^2/(g'^2 + g_2^2)|_{M_U} = 3/8$. Compare your prediction to the data.

5. Count degrees of freedom and show that the **16** of $SO(10)$ contains a singlet, the right-handed neutrino.

6. Introduce group theory (QCD and $SU(2)$) and generation indices to the proton decay operator $QQQL$ and show that one Q must be of the second generation or otherwise the group theory forces the operators to vanish once antisymmetric indices are properly summed. (Why not third generation?)

7. Confirm the one-loop gaugino-mass unification relation $M_3 : M_2 : M_1 \simeq 3 : 1 : 1/2$ at the weak scale.

8. Omitting all terms aside from QCD in the stop renormalization group equation, and using the renormalization group equations for the gauge couplings and for the gaugino mass, show that if $m_Q(\Lambda_{\rm UV}) = 0$ then $m_Q^2(m_Q) \simeq 7 M_3^2(\Lambda_{\rm UV}) \simeq M_3(M_3)^2$, where m_Q correspond to the stop doublet mass and M_3 to the gluino mass. Compare to the analytic solution with $\delta_i = 0$. Include all gauge terms and assume gaugino mass unification, can you derive the difference between the renormalized m_Q^2 and m_U^2 ?

9. Derive the minimization conditions of the Higgs potential given above by first organizing the minimization equations as

$$m_1^2 = m_3^2 \frac{\nu_2}{\nu_1} + \frac{1}{4}\left(g_2^2 + g'^2\right)\left(\nu_2^2 - \nu_1^2\right) \tag{121}$$

$$m_2^2 = m_3^2 \frac{\nu_1}{\nu_2} + \frac{1}{4}\left(g_2^2 + g'^2\right)\left(\nu_2^2 - \nu_1^2\right), \tag{122}$$

where $m_i^2 \equiv m_{H_i}^2 + \mu^2$ for $i = 1, 2$.

10. Derive the tree-level charged Higgs mass-squared matrix $M_{H^\pm}^2 - M_{PS}^2(1+ (1/2)M_W^2 \sin 2\beta)$, show that it has a massless eigenvalue (the Goldstone boson) and derive the sum rule $m_{H^\pm}^2 = m_{A^0}^2 + M_W^2$.

11. Show that the rotation angle of the CP-even Higgs states is given by

$$\sin 2\alpha = -\frac{m_{A^0}^2 + M_Z^2}{m_{H^0}^2 - m_{A^0}^2} \sin 2\beta. \tag{123}$$

One can define a decoupling limit $\alpha \to \beta - (\pi/2)$, which is reached for $m_{A^0} \gtrsim 200$ GeV. Show that in this limit the heavy Higgs eigenstates form a doublet that does not participate in electroweak symmetry breaking.

12. Derive the tree-level bound $m_{h^0} \le M_Z |\cos 2\beta|$ from eq. (101).

13. In a common version of the NMSSM, $W = \mu H_1 H_2$ is replaced with $W = \lambda S H_1 H_2 + (\kappa/3)S^3$. Derive the extended Higgs potential (for the doublets and singlet) in this case and its minimization conditions. Show that the potential exhibits a discrete Z_3 symmetry broken only spontaneously by the Higgs vev's, which is the (cosmological) downfall of the model due to the associated domain wall problems at a post inflationary epoch. Write down the CP-even Higgs and neutralino extended mass matrices in this model. Find the tree-level light Higgs mass in the limit $\tan\beta = 1$ and show that it does not vanish (and that indeed the potential does not have a flat direction in this case).

14. Convince yourself that the anomalous magnetic operator

$$(eQ_f/2m_f)a_f\bar{f}\sigma_{\mu\nu}fF^{\mu\nu}$$

has no supersymmetric analogue.

15. Integrate the operator equations (114) – (118) to find the SSB and μ-parameters.

16. Solve for the sfermion and Higgs mass-squared parameters in the case that supergravity mediation generates only the gaugino mass. Assume instead complete universality at the supergravity scale $m_{\tilde{f}}^2 = m_H^2 = m_0^2$ for all sfermion and Higgs fields. In what fashion gaugino masses and Yukawa couplings break universality? Why is the breakdown of univesality by corrections proportional to Yukawa couplings is not in conflict with low-energy data?

17. Supergravity mediation in the presence of a grand unified theory induces the SSB parameters for that theory, not for the MSSM. How many independent soft parameters are in this case at the unification scale (with and without universality) for $SU(5)$; $SO(10)$? Renormalization within a grand-unified epoch introduces radiative correction which are proportional to the large top-Yukawa coupling to the stau squared mass ($m_{\tilde{\tau}_L}^2$ or $m_{\tilde{\tau}_R}^2$?), which breaks universality in the slepton sector. The breakdown of slepton universality induces, in turn, contributions to low-energy lepton flavor violation such as the $\mu \to e\gamma$ process. For sufficiently light sleptons ($m_{\tilde{l}} \lesssim 300 - 400$ GeV) it may be observable in the next generation of experiments, if built.

18. Derive the messenger mass matrix (119).

19. Show that in the minimal gauge mediation model described above $e = (\alpha_i/4\pi)T_i$ for gaugino i, and normalizing Dynkin index T_i to unity (e.g., for 5 and $\bar{5}$ of $SU(5)$ messengers) that for each sfermion (and Higgs) $a = 2\sum_i(\alpha_i/4\pi)^2 C_i$ where $C_1 = (3/5)Y^2$, $C_2 = 3/4$ for an $SU(2)$ doublet and $C_3 = 4/3$ for an $SU(3)$ triplet are the Casimir operators.

20. Derive the relation $|m_3^2| \sim |\mu|\Lambda_{\text{mediation}}$ in gauge mediation.

21. Calculate F_X in the hybrid supergravity - gauge-mediation model. Calculate the gravitino mass in this and in the "traditional" gauge-mediation models and show that supergravity effects $\sim m_{3/2}^2$ are indeed negligible.

22. Show that the solution for the soft parameters has a zero stop-Higgs fixed point if $\left[m_{H_2}^2(\Lambda_{\text{UV}}), m_{U_3}^2(\Lambda_{\text{UV}}), m_{Q_3}^2(\Lambda_{\text{UV}})\right] = m_0^2\,[3, 2, 1]$.

184

6 Summary, Outlook, And Neutrinos

In conclusion, we have argued for and motivated supersymmetry as a reasonable, and even likely, perturbative extension of the standard model. The formalism was described, and it was shown that it could be arrived at based only on low-energy considerations. The standard model was supersymmetrized in some detail, and a review of some of the extrapolation, model building, and phenomenology aspects was given, stressing the successful gauge coupling unification, radiative (electroweak) symmetry breaking in the presence of a heavy top, the lightness of the Higgs boson, and the predictive power that is encompassed in the resolution of the flavor problem. It was not our aim to equip the reader with calculation tools or to review each and every aspect or possibility. Rather, we attempted to enable the interested reader to form an educated opinion and identify areas of interest for further study and exploration. Many of the issues discussed were presented from an angle different from what is customary, for the potential benefit of both the novice and expert readers.

The theme and focus of this school is neutrino physics. The confirmation of neutrino oscillations provides a concrete and first proof of physics beyond the SM. If such physics realizes supersymmetry, an option we explored in these lectures, then neutrino mass and mixing (and lepton flavor violation) has to be realized in a supersymmetric fashion. We therefore conclude these lecture with a brief review of some of the avenues for neutrino mass generation in supersymmetry.

Unless neutrinos have "boring" Dirac masses, which only imply super-light sterile right-handed neutrinos and an extension of the usual fermion mass hierarchy problem, there must be some source of lepton number violation. It is straightforward to "supersymmetrize" old ideas of a heavy right-handed neutrino with a large $\Delta L = 2$ Majorana mass M_R which mixes with the SM neutrinos via the usual $\mathcal{O}(M_W)$ Dirac mass term, leading to $m_\nu \sim M_W^2/M_R$; the see-saw mechanism[90]. This mechanism does not require supersymmetry, but is trivially embedded in supersymmetry. Unified model typically imply further that the Dirac τ-neutrino mass is of the order of the top mass (top - neutrino unification), $m_{\nu_\tau} \sim m_t^2/M_R$. (Also, $t - \nu$ unification excludes the $\tan\beta \sim 1$ solutions to $b-\tau$ unification (Fig. 6) and constrains $\tan\beta$ from below due to cancellation of large top-Yukawa effects by the large neutrino-Yukawa effects. Hence, it predicts that the light-Higgs mass saturates its upper bound of 130 GeV or so.) A natural choice for the scale M_R is the unification scale, and indeed simple fits to the data favor $M_R \sim 10^{13-15}$ GeV^2. This was a subject of intense activity in the recent year. For an example of a (unified) model along these line, see Ref.[91], where a double see-saw mechanism $M_R \sim M_U^2/M_P$ and

$m_\nu \sim M_W^2/M_R \sim M_W^2 M_P/M_U^2$ was proposed. Another possibility [61] is that the scale M_R is related to the scale of spontaneous supersymmetry breaking in the hidden sector.

If indeed the right-handed neutrino is present, in supersymmetry there is also a right-handed sneutrino. Since the heavy right-handed neutrino decouples in a global supersymmetric regime, the sneutrino is also heavy, but it still can have a small SSB mass (though not in the case of low-energy gauge mediation). If the right-handed neutrino superfield couples with a large Yukawa coupling, it can affect the renormalization of superpotential and SSB parameters, leaving its imprints in the weak-scale parameters and disturbing slepton (mass) universality. (Relevant references were included in Ref. [73].) This in turn offers an interesting complementarity between neutrino physics, lepton flavor violation experiments, and the slepton spectrum.

The relation between neutrino physics and supersymmetry is even more fundamental and extensive if the neutrino mass originates from R-partiy violation [41]. As argued before, supersymmetry essentially encodes lepton - Higgs duality, which is typically removed by hand when imposing R-parity. Generically, the theory contains explicit $\Delta L = 1$ breaking via superpotential operators, SSB operators, and as a consequence, L could also be broken spountaneously by sneutrino *vevs*. One needs to apply $\Delta L = 1$ operation twice in order to induce a Majorana neutrino mass. In general, such models can admit [41] a one-loop rndiative mass ($\propto \lambda'^2$) and a tree-level mass induced by neutrino-neutralino tree-level mixing ($\propto \langle \tilde{\nu} \rangle \mu_L$). The tree-level mass ($\propto \langle \tilde{\nu} \rangle \mu_L$) is intriguing: Only a supersymmetric neutrino mass $W_{\Delta L=1} = \mu_L L H_2$ is allowed in the superpotential (aside from the usual Higgs mass), while the neutrino is also the only SM fermion which is left massless by the superpotential Yukawa terms. Hence, it offers an interesting complementarity between Yukawa and mass terms in the superpotential, and an elegant and economic realization of R-partiy violation. It can also be shown to be related to the μ-problem. (For a recent discussion see Ref. [92] and references therein.) Here, however, consider the radiative Majorana neutrino mass illustrated in Fig. 13.

One obtains

$$\frac{m_\nu}{\text{MeV}} \sim \lambda'^2 \left(\frac{300\,\text{GeV}}{m_{\tilde{b}}} \right) \left(\frac{m_{LR}^2}{m_b\, m_{\tilde{b}}} \right), \tag{124}$$

where a b-quark and $\tilde{b}$-squark are assumed to circulate in the loop, and m_{LR}^2 is the $\tilde{b}$ left–right mixing squared mass. The neutrino mass may vanish in the limit of a continuous $U(1)_R$ symmetry, which suppresses A and μ terms and corresponds in our case to $m_{LR}^2 \ll m_b\, m_{\tilde{b}}$. (Other contributions to the neutrino mass may arise from tree-level neutrino–neutralino mixing and from an

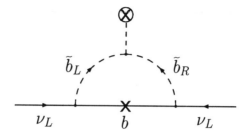

Figure 13: A one-loop contribution to the Majorana neutrino mass arising from the $\Delta L = 1$ λ'_{333} Yukawa operator.

independent see-saw mechanism, but it is assumed that no accidental cancellations among various contributions to the neutrino mass take place.) Imposing laboratory limits on the ν_e mass one can, for example, derive severe constraints on λ'_{133} [93], but λ'_{333} could still be $\mathcal{O}(1)$ [94]. One has to consider a λ' matrix in order to obtain neutrino mixing, but clearly, the complexity of the parameter space admits many possible scenarios for mixing. (Note that in the case that R-parity violation originates from a μ-term $W_{\Delta L=1} = \mu_L L H_2$ then the λ and λ' couplings appear by rotations of the usual Yukawa couplings once leptons and Higgs bosons are appropriately defined at low energies.)

R-partiy violation can then explain the neutrino spectrum as an electroweak (rather than GUT) effect. This comes at a price and with a reward. The price is that the sparticles are not stable and the LSP is not a dark matter candidate (but some other sector may still provide a dark matter candidate whose mass is of the order of the gravitino mass). The obvious reward is the many new channels are available at colliders to produce squarks and sleptons [37], and the information that the corresponding cross section and branching ratio measurements may provide for neutrino physics, an issue that was recently stressed in Ref. [95].

This last observation stresses the importance of confronting theory and experiment. Whether supersymmetry is realized in nature and if so, in what form, can and must be determined by experiment. Experimental efforts in this direction are ongoing and should result in an answer or a clear indication within the next decade. Discovery will only signal the start of a new era dedicated to deciphering the ultra-violet theory from the data: Supersymmetry, by its nature, relates the infra red to the ultra violet. By doing so it provides us with an opportunity to take a glimpse at scales we cannot reach, answer fundamental questions regarding unification, gravity, and much more. However, the more one is familiar with the ultra-violet possibilities and their infra-red implications, the more one can direct experiment towards a possible discovery. Clearly, a lot of work is yet to be done and there is room for new ideas to be put forward. It is my hope that these lectures convince and enable one to follow (if not participate in) the story of weak-scale supersymmetry, and new physics in general, as it unveils.

Acknowledgments

I thank the organizers of the school for their invitation and for giving me the privilege to return to the school as a lecturer, and I thank the participants for their interest, questions and the stimulating discussions (especially for those who gave up their free afternoon and came to learn about lepton number violation). I am grateful to Myck Schwetz for his comments on the manuscript, Francesca Borzumati for her help with axodraw, and Howard Haber and Chris Kolda for providing me with Figs. 9, 10 and Fig. 8, respectively. I would also like to thank my collaborators over the years for all they taught me, in particular, Paul Langacker for his support and advice, Jan Louis and Hans Peter Nilles for all they taught me about supersymmetry, and Francesca Borzumati and Jonathan Feng for sharing time and again their vast knowledge. Last, yet most importantly, I thank my wife and son, Janice and Nevo, for their patience, tolerance and support during the tedious preparation of the oral and written versions of these lectures.

References

1. G. Altarelli, "The standard electroweak theory and beyond," hep-ph/9811456, in this volume.
2. P. Langacker, in this volume.
3. R.S. Chivukula, "Models of electroweak symmetry breaking: Course," hep-ph/9803219.
4. C. Quigg, "Electroweak symmetry breaking and the Higgs sector," hep-ph/9905369.
5. These ideas were recently considered (though from quite different perspective) by:
   K.R. Dienes, E. Dudas and T. Gherghetta, "Extra space-time dimensions and unification," Phys. Lett. **B436**, 55 (1998);
   N. Arkani-Hamed, S. Dimopoulos and G. Dvali, "Phenomenology, astrophysics and cosmology of theories with submillimeter dimensions and TeV scale quantum gravity," Phys. Rev. **D59**, 086004 (1999).
6. K.R. Dienes, in this volume.
7. H.P. Nilles, "On the Low-energy limit of string and M theory," *Lectures given at Theoretical Advanced Study Institute in Elementary Particle Physics (TASI 97): Supersymmetry, Supergravity and Supercolliders, Boulder, CO, 1-7 Jun 1997.*
8. P.H. Frampton and C. Vafa, "Conformal approach to particle phenomenology," hep-th/9903226.
9. For example, see A. Pomarol and M. Quiros, "The Standard model from extra dimensions," Phys. Lett. **B438**, 255 (1998);
   I. Antoniadis, S. Dimopoulos, A. Pomarol and M. Quiros, "Soft masses in theories with supersymmetry breaking by TeV compactification," Nucl. Phys. **B544**, 503 (1999).
10. A.E. Nelson and M.J. Strassler, "A Realistic supersymmetric model with composite quarks," Phys. Rev. **D56**, 4226 (1997);
    N. Arkani-Hamed, M.A. Luty and J. Terning, "Composite quarks and leptons from dynamical supersymmetry breaking without messengers," Phys. Rev. **D58**, 015004 (1998).
11. Other proposals to strong electroweak dynamics within supersymmetry include
    G.F Giudice and A. Kusenko, "A Strongly interacting phase of the minimal supersymmetric model," Phys. Lett. **B439**, 55 (1998);
    K. Choi and H.D. Kim, "Dynamical solution to the μ-problem at TeV scale," hep-ph/9906363.
12. For example, see J. Erler and P. Langacker, "Electroweak model and

constraints on new physics: In Review of Particle Physics," Eur. Phys. J. **C3** (1998) 90.

13. F. Abe et al. [CDF Collaboration], "Study of t anti-t production p anti-p collisions using total transverse energy," Phys. Rev. Lett. **75**, 3997 (1995); S. Abachi et al. [D0 Collaboration], "Observation of the top quark," Phys. Rev. Lett. **74**, 2632 (1995).

14. P. Langacker and M. Luo, "Implications of precision electroweak experiments for m_t, $\rho(0)$, $\sin^2 \theta_W$ and grand unification," Phys. Rev. **D44**, 817 (1991).

15. D. Zeppenfeld, "Collider physics," hep-ph/9902307, in this volume.

16. F. Halzen, "Lectures on neutrino astronomy: Theory and experiment," astro-ph/9810368, in this volume.

17. M. Dine, "Supersymmetry phenomenology (with a broad brush)," *Lectures given at Theoretical Advanced Study Institute in Elementary Particle Physics (TASI 96): Fields, Strings, and Duality, Boulder, CO, 2-28 Jun 1996*, hep-ph/9612389.

18. X. Tata, "What is supersymmetry and how do we find it?," *Lectures given at 9th Jorge Andre Swieca Summer School: Particles and Fields, Sao Paulo, Brazil, 16-28 Feb 1997*, hep-ph/9706307.

19. M. Drees, "An Introduction to supersymmetry," *Lectures given at Inauguration Conference of the Asia Pacific Center for Theoretical Physics (APCTP), Seoul, Korea, 4-19 Jun 1996*, hep-ph/9611409.

20. J. Wess and B. Zumino, "A Lagrangian model invariant under supergauge transformations," Phys. Lett. **49B**, 52 (1974); "Supergauge transformations in four-dimensions," Nucl. Phys. **B70** (1974) 39.

21. H.P. Nilles, "Supersymmetry, supergravity and particle physics," Phys. Rept. **110** (1984) 1.

22. H.P. Nilles, "Beyond the Standard Model," *Lectures given at Theoretical Study Inst. in Elementary Particle Physics (TASI90), Boulder, CO, Jun 3-29, 1990*. In Boulder 1990, Proceedings, Testing the Standard Model, p. 633-718.

23. H.P. Nilles, "Minimal supersymmetric Standard Model and grand unification," *Lectures given at Theoretical Advanced Study Inst. in Elementary Particle Physics (TASI93), Boulder, CO, Jun 6 - Jul 2, 1993*. In Boulder 1993, Proceedings, The building blocks of creation, p. 291-346.

24. D.R.T Jones, "Introduction to supersymmetry," *Lectures given at Theoretical Advanced Study Inst. in Elementary Particle Physics (TASI93), Boulder, CO, Jun 6 - Jul 2, 1993*. In Boulder 1993, Proceedings, The building blocks of creation, p. 259-290.

25. J.D. Lykken, "Introduction to supersymmetry," *Lectures given at The-oretical Advanced Study Institute in Elementary Particle Physics (TASI 96): Fields, Strings, and Duality, Boulder, CO, 2-28 Jun 1996*, hep-th/9612114.

26. S.P. Martin, "A Supersymmetry primer," hep-ph/9709356.

27. J. Wess and J. Bagger, *Supersymmetry and supergravity* (Princeton, NJ, 1991).

28. H.E. Haber and G.L. Kane, "The search for supersymmetry: Probing physics beyond the standard model," Phys. Rept. **117** (1985) 75.

29. H.E. Haber, "Introductory low-energy supersymmetry," *Lectures given at Theoretical Advanced Study Institute in Elementary Particle Physics (TASI 92): From Black Holes and Strings to Particles, Boulder, CO, 3-28 Jun 1992*, hep-ph/9306207.

30. J.L. Feng, N. Polonsky and S. Thomas, "The light higgsino - gaugino window," Phys. Lett. **B370**, 95 (1996).

31. G.R. Farrar, "Status of light gaugino scenarios," Nucl. Phys. Proc. Suppl. **62**, 485 (1998), hep-ph/9710277;
"Experimental and cosmological implications of light gauginos," hep-ph/9710395.

32. S. Raby and K. Tobe, "The Phenomenology of SUSY models with a gluino LSP," Nucl. Phys. **B539**, 3 (1999).

33. L. Girardello and M.T. Grisaru, "Soft breaking of supersymmetry," Nucl. Phys. **B194**, 65 (1982).

34. L.J. Hall and L. Randall, "Weak scale effective supersymmetry," Phys. Rev. Lett. **65**, 2939 (1990).

35. F. Borzumati, G.R. Farrar, N. Polonsky and S. Thomas, "Soft Yukawa couplings," hep-ph/9902443, Nucl. Phys. B, in press.

36. I. Jack and D.R.T. Jones, "Non-standard soft supersymmetry breaking," hep-ph/9903365.

37. H. Dreiner, "An introduction to explicit R-parity violation," hep-ph/9707435.

38. G.R. Farrar and P. Fayet, "Phenomenology of the production, decay, and detection of new hadronic states associated with supersymmetry," Phys. Lett. **76B**, 575 (1978).

39. L.E. Ibanez and G.G. Ross, "Discrete gauge symmetries and the origin of baryon and lepton number conservation in supersymmetric versions of the standard model," Nucl. Phys. **B368**, 3 (1992).

40. D. Kapetanakis, P. Mayr and H.P. Nilles, "Discrete symmetries and solar neutrino mixing," Phys. Lett. **B282**, 95 (1992).

41. L.J. Hall and M. Suzuki, "Explicit R-Parity Breaking In Supersymmetric

Models," Nucl. Phys. **B231**, 419 (1984).

42. L. Covi, J.E. Kim and L. Roszkowski, "Axinos as cold dark matter," Phys. Rev. Lett. **82**, 4180 (1999);
    E.J. Chun and H.B. Kim, "Nonthermal axino as cool dark matter in supersymmetric standard model without R-parity," hep-ph/9906392.

43. X. Tata, "Supersymmetry: Where it is and how to find it," *Lectures given at Theoretical Advanced Study Institute in Elementary Particle Physics (TASI 95): QCD and Beyond, Boulder, CO, 4-30 Jun 1995*, hep-ph/9510287.

44. A. Bottino and N. Fornengo, "Dark matter and its particle candidates," *Lectures given at 5th ICTP School on Nonaccelerator Astroparticle Physics, Trieste, Italy, 29 Jun - 10 Jul 1998*, hep-ph/9904469.

45. M. Drees, "Particle dark matter physics: An update," Pramana **51**, 87 (1998), hep-ph/9804231.

46. The 1998 Review of Particle Physics, C. Caso *et al.* [Particle Data Group], The Eur. Phys. J. **C3**, 1 (1998), http://pdg.lbl.gov/pdg.html.

47. D.M. Pierce, "Renormalization of supersymmetric theories," *Lectures given at the Theoretical Advanced Study Institute in Elementary Particle Physics (TASI 97): Supersymmetry, Supergravity and Supercolliders, Boulder, CO, 1-7 Jun 1997*, hep-ph/9805497;
    J. Erler and D.M. Pierce, "Bounds on supersymmetry from electroweak precision analysis," Nucl. Phys. **B526**, 53 (1998).

48. J.A. Bagger, "Weak scale supersymmetry: Theory and practice," *Lectures given at Theoretical Advanced Study Institute in Elementary Particle Physics (TASI 95): QCD and Beyond, Boulder, CO, 4-30 Jun 1995*, hep-ph/9604232.

49. S. Dawson, "The MSSM and why it works," *Lectures given at the Theoretical Advanced Study Institute in Elementary Particle Physics (TASI 97): Supersymmetry, Supergravity and Supercolliders, Boulder, CO, 1-7 Jun 1997*, hep-ph/9712464.

50. R.N. Mohapatra, "Supersymmetric grand unification," *Lectures given at the Theoretical Advanced Study Institute in Elementary Particle Physics (TASI 97): Supersymmetry, Supergravity and Supercolliders, Boulder, CO, 1-7 Jun 1997*, hep-ph/9801235.

51. K.R. Dienes, "String theory and the path to unification: A Review of recent developments," Phys. Rept. **287**, 447 (1997).

52. P. Langacker and N. Polonsky, "Uncertainties in coupling constant unification," Phys. Rev. **D47**, 4028 (1993);
    'The Bottom mass prediction in supersymmetric grand unification: Uncertainties and constraints," Phys. Rev. **D49**, 1454 (1994);

"The Strong coupling, unification, and recent data," Phys. Rev. **D52**, 3081 (1995);

N. Polonsky, "On supersymmetric b - tau unification, gauge unification, and fixed points," Phys. Rev. **D54**, 4537 (1996).

53. R. Arnowitt and P. Nath, "SUSY mass spectrum in SU(5) supergravity grand unification," Phys. Rev. Lett. **69**, 725 (1992);
"Radiative breaking, proton stability and the viability of no scale supergravity models," Phys. Lett. **B287**, 89 (1992);
"Predictions in SU(5) supergravity grand unification with proton stability and relic density constraints," Phys. Rev. Lett. **70**, 3696 (1993);
"Supersymmetry and supergravity: Phenomenology and grand unification," hep-ph/9309277;
"Supergravity unified models," hep-ph/9708254;
J. Hisano, H. Murayama and T. Yanagida, "Nucleon decay in the minimal supersymmetric SU(5) grand unification," Nucl. Phys. **B402**, 46 (1993);
K. Hagiwara and Y. Yamada, "GUT threshold effects in supersymmetric SU(5) models," Phys. Rev. Lett. **70**, 709 (1993);
V. Barger, M.S. Berger and P. Ohmann, "Supersymmetric grand unified theories: Two loop evolution of gauge and Yukawa couplings," Phys. Rev. **D47**, 1093 (1993);
M. Carena, S. Pokorski and C.E. Wagner, "On the unification of couplings in the minimal supersymmetric Standard Model," Nucl. Phys. **B406**, 59 (1993);
W.A. Bardeen, M. Carena, S. Pokorski and C.E. Wagner, "Infrared fixed point solution for the top quark mass and unification of couplings in the MSSM," Phys. Lett. **B320**, 110 (1994);
L.J. Hall, R. Rattazzi and U. Sarid, "The Top quark mass in supersymmetric SO(10) unification," Phys. Rev. **D50**, 7048 (1994);
P.H. Chankowski, Z. Pluciennik and S. Pokorski, "$\sin^2 \theta_W(M(Z))$ in the MSSM and unification of couplings," Nucl. Phys. **B439**, 23 (1995);
D.M. Pierce, J.A. Bagger, K. Matchev and R. Zhang, "Precision corrections in the minimal supersymmetric standard model," Nucl. Phys. **B491**, 3 (1997);
J.A. Bagger, K.T. Matchev, D.M. Pierce and R. Zhang, "Gauge and Yukawa unification in models with gauge mediated supersymmetry breaking," Phys. Rev. Lett. **78**, 1002 (1997);
T. Goto and T. Nihei, "Effect of RRRR dimension five operator on the proton decay in the minimal SU(5) SUGRA GUT model," Phys. Rev. **D59**, 115009 (1999).

54. B. Pendleton and G.G. Ross, "Mass and mixing angle predictions from infrared fixed points," Phys. Lett. **98B**, 291 (1981);
C.T. Hill, "Quark and lepton masses from renormalization group fixed points," Phys. Rev. **D24**, 691 (1981).

55. L.E. Ibanez and G.G. Ross, "low-energy predictions in supersymmetric grand unified theories," Phys. Lett. **105B**, 439 (1981);
L.E. Ibanez and C. Lopez, "N=1 Supergravity, the breaking Of SU(2) X U(1) and the top quark mass," Phys. Lett. **126B**, 54 (1983);
L.E. Ibanez, C. Lopez and C. Munoz, "The low-energy supersymmetric spectrum according to $N = 1$ supergravity guts," Nucl. Phys. **B256**, 218 (1985);
H.P. Nilles, "Dynamically broken supergravity and the hierarchy problem," Phys. Lett. **115B**, 193 (1982);
K. Inoue, A. Kakuto, H. Komatsu and S. Takeshita, "Low-energy parameters and particle masses in a supersymmetric grand unified model, "Prog. Theor. Phys. **67**, 1889 (1982);
"Aspects of grand unified models with softly broken supersymmetry," Prog. Theor. Phys. **68**, 927 (1982); *Erratum ibid.* **70**, 330 (1983);
"Renormalization of supersymmetry breaking parameters revisited," Prog. Theor. Phys. **71**, 413 (1984);
L. Alvarez-Gaume, J. Polchinski and M.B. Wise, "Minimal low-energy supergravity," Nucl. Phys. **b221**, 495 (1983);
P.H. Chankowski, "Radiative $SU(2) \times U(1)$ breaking in the supersymmetric standard model and Decoupling Of Heavy Squarks And Gluino," Phys. Rev. **D41**, 2877 (1990);
M. Drees and M.M. Nojiri, "Radiative symmetry breaking in minimal N=1 supergravity with large Yukawa couplings," Nucl. Phys. **B369**, 54 (1992).

56. J.L. Feng, C. Kolda and N. Polonsky, "Solving the supersymmetric flavor problem with radiatively generated mass hierarchies," Nucl. Phys. **B546**, 3 (1999);
J. Bagger, J.L. Feng and N. Polonsky, "Naturally heavy scalars in supersymmetric grand unified theories," hep-ph/9905292.

57. S.P. Martin and M.T. Vaughn, "Two loop renormalization group equations for soft supersymmetry breaking couplings," Phys. Rev. **D50**, 2282 (1994);
Y. Yamada, "Two loop renormalization group equations for soft SUSY breaking scalar interactions: Supergraph method," Phys. Rev. **D50**, 3537 (1994);
I. Jack and D.R. Jones, "Soft supersymmetry breaking and finiteness,"

Phys. Lett. **B333**, 372 (1994);
I. Jack, D.R. Jones, S.P. Martin, M.T. Vaughn and Y. Yamada, "Decoupling of the epsilon scalar mass in softly broken supersymmetry," Phys. Rev. **D50**, 5481 (1994).

58. R. Arnowitt and P. Nath, "Loop corrections to radiative breaking of electroweak symmetry in supersymmetry," Phys. Rev. **D46**, 3981 (1992);
S. Kelley, J.L. Lopez, D.V. Nanopoulos, H. Pois and K. Yuan, "Aspects of radiative electroweak breaking in supergravity models," Nucl. Phys. **B398**, 3 (1993);
V. Barger, M.S. Berger and P. Ohmann, "The supersymmetric particle spectrum," Phys. Rev. **D49**, 4908 (1994);
G.L. Kane, C. Kolda, L. Roszkowski and J.D. Wells, "Study of constrained minimal supersymmetry," Phys. Rev. **D49**, 6173 (1994);
M. Olechowski and S. Pokorski, "Bottom - up approach to unified supergravity models," Nucl. Phys. **B404**, 590 (1993);
M. Carena, M. Olechowski, S. Pokorski and C.E. Wagner, "Radiative electroweak symmetry breaking and the infrared fixed point of the top quark mass," Nucl. Phys. **B419**, 213 (1994); M. Carena, M. Olechowski, S. Pokorski and C.E. Wagner, "Electroweak symmetry breaking and bottom - top Yukawa unification," Nucl. Phys. **B426**, 269 (1994);
N. Polonsky and A. Pomarol, "GUT effects in the soft supersymmetry breaking terms," Phys. Rev. Lett. **73**, 2292 (1994);
"Nonuniversal GUT corrections to the soft terms and their implications in supergravity models," Phys. Rev. **D51**, 6532 (1995);
S.A. Abel and B. Allanach, "The quasifixed MSSM," Phys. Lett. **B415**, 371 (1997);
M. Carena, P. Chankowski, M. Olechowski, S. Pokorski and C.E. Wagner, "Bottom - up approach and supersymmetry breaking," Nucl. Phys. **B491**, 103 (1997).

59. J.E. Kim and H.P. Nilles, "The μ-problem and the strong CP problem," Phys. Lett. **138B**, 150 (1984);
"Symmetry principles toward solutions of the mu problem," Mod. Phys. Lett. **A9**, 3575 (1994).

60. G.F. Giudice and A. Masiero, "A natural solution to the μ-problem in supergravity theories," Phys. Lett. **B206**, 480 (1988).

61. C. Kolda, S. Pokorski and N. Polonsky, "Stabilized singlets in supergravity as a source of the μ-parameter," Phys. Rev. Lett. **80**, 5263 (1998).

62. R. Barbieri and L. Maiani, "Renormalization of the electroweak rho parameter from supersymmetric particles," Nucl. Phys. **B224**, 32 (1983).

63. H.E. Haber and A. Pomarol, "Constraints from global symmetries on radiative corrections to the Higgs sector," Phys. Lett. **B302**, 435 (1993).
64. M. Cvetic, in this volume.
65. H.E. Haber, "How well can we predict the mass of the Higgs boson of the minimal supersymmetric model?," hep-ph/9901365.
66. H.E. Haber, "Higgs boson masses and couplings in the minimal supersymmetric model," hep-ph/9707213.
67. M.A. Diaz and H.E. Haber, "Can the Higgs mass be entirely due to radiative corrections?," Phys. Rev. **D46**, 3086 (1992);
    H.E. Haber and R. Hempfling, "Can the mass of the lightest Higgs boson of the minimal supersymmetric model be larger than M_Z?," Phys. Rev. Lett. **66**, 1815 (1991);
    H.E. Haber and R. Hempfling, "The renormalization group improved Higgs sector of the minimal supersymmetric model," Phys. Rev. **D48**, 4280 (1993);
    H.E. Haber, R. Hempfling and A.H. Hoang, "Approximating the radiatively corrected Higgs mass in the minimal supersymmetric model," Z. Phys. **C75**, 539 (1997);
    Y. Okada, M. Yamaguchi and T. Yanagida, "Upper bound of the lightest Higgs boson mass in the minimal supersymmetric standard model," Prog. Theor. Phys. **85**, 1 (1991);
    J. Ellis, G. Ridolfi and F. Zwirner, "Radiative corrections to the masses of supersymmetric Higgs bosons," Phys. Lett. **B257**, 83 (1991);
    "On radiative corrections to supersymmetric Higgs boson masses and their implications for LEP searches," Phys. Lett. **B262**, 477 (1991);
    R. Barbieri, M. Frigeni and F. Caravaglios, "The Supersymmetric Higgs for heavy superpartners," Phys. Lett. **B258**, 167 (1991);
    J.R. Espinosa and M. Quiros, "Two loop radiative corrections to the mass of the lightest Higgs boson in supersymmetric standard models," Phys. Lett. **B266**, 389 (1991);
    J.R. Espinosa and M. Quiros, "Higgs triplets in the supersymmetric standard model," Nucl. Phys. **B384**, 113 (1992);
    V. Barger, M.S. Berger, P. Ohmann and R.J. Phillips, "Phenomenological implications of the m_t RGE fixed point for SUSY Higgs boson searches," Phys. Lett. **B314**, 351 (1993);
    P. Langacker and N. Polonsky, "Implications of Yukawa unification for the Higgs sector in supersymmetric grand unified models," Phys. Rev. **D50**, 2199 (1994);
    M. Carena, J.R. Espinosa, M. Quiros and C.E. Wagner, "Analytical expressions for radiatively corrected Higgs masses and couplings in the

MSSM," Phys. Lett. **B355**, 209 (1995);

M. Carena, M. Quiros and C.E. Wagner, "Effective potential methods and the Higgs mass spectrum in the MSSM," Nucl. Phys. **B461**, 407 (1996);

M. Carena, P.H. Chankowski, S. Pokorski and C.E. Wagner, "The Higgs boson mass as a probe of the minimal supersymmetric standard model," Phys. Lett. **B441**, 205 (1998).

68. J. Ellis, S. Ferrara and D.V. Nanopoulos, "CP violation and supersymmetry," Phys. Lett. **114B**, 231 (1982);

W. Buchmuller and D. Wyler, "CP violation and R invariance in supersymmetric models of strong and electroweak interactions," Phys. Lett. **121B**, 321 (1983);

J. Polchinski and M.B. Wise, "The electric dipole moment of the neutron in low-energy supergravity," phys. lett. **125b**, 393 (1983);

W. Fischler, S. Paban and S. Thomas, "Bounds on microscopic physics from P and T violation in atoms and molecules," Phys. Lett. **B289**, 373 (1992);

Y. Kizukuri and N. Oshimo, "The Neutron and electron electric dipole moments in supersymmetric theories," Phys. Rev. **D46**, 3025 (1992);

S. Bertolini and F. Vissani, "On soft breaking and CP phases in the supersymmetric standard model," Phys. Lett. **B324**, 164 (1994);

S. Dimopoulos and S. Thomas, "Dynamical relaxation of the supersymmetric CP violating phases," Nucl. Phys. **B465**, 23 (1996);

T. Moroi, "Electric dipole moments in gauge mediated models and a solution to the SUSY CP problem," Phys. Lett. **B447**, 75 (1999).

69. T. Ibrahim and P. Nath, "The Neutron and the lepton EDMs in MSSM, large CP violating phases, and the cancellation mechanism," Phys. Rev. **D58**, 111301 (1998);

M. Brhlik, G.J. Good and G.L. Kane, "Electric dipole moments do not require the CP violating phases of supersymmetry to be small," Phys. Rev. **D59**, 115004 (1999).

70. J.A. Bagger, K.T. Matchev and R. Zhang, "QCD corrections to flavor changing neutral currents in the supersymmetric standard model," Phys. Lett. **B412**, 77 (1997);

M. Cìuchìnì et al., "$\Delta M(K)$ and $\epsilon(K)$ in SUSY at the next-to-leading order," JHEP **10**, 008 (1998).

71. J.A. Grifols and A. Mendez, "Constraints on supersymmetric particle masses from $(g-2)_\mu$," Phys. Rev. **D26**, 1809 (1982);

J. Ellis, J. Hagelin and D.V. Nanopoulos, "Spin 0 leptons and the anomalous magnetic moment of the muon," Phys. Lett. **116B**, 283 (1982);

R. Barbieri and L. Maiani, "The muon anomalous magnetic moment in broken supersymmetric theories," Phys. Lett. **117B**, 203 (1982); D.A. Kosower, L.M. Krauss and N. Sakai, "Low-energy supergravity and the anomalous magnetic moment of the muon," Phys. Lett. **133B**, 305 (1983).

72. S. Bertolini, F. Borzumati, A. Masiero and G. Ridolfi, "Effects of supergravity induced electroweak breaking on rare B decays and mixings," Nucl. Phys. **B353**, 591 (1991); F.M. Borzumati, "The Decay $b \to s\gamma$ in the MSSM revisited," Z. Phys. **C63**, 291 (1994).

73. F. Gabbiani, E. Gabrielli, A. Masiero and L. Silvestrini, "A complete analysis of FCNC and CP constraints in general SUSY extensions of the standard model," Nucl. Phys. **B477**, 321 (1996); A. Masiero and L. Silvestrini, "Two lectures on FCNC and CP violation in supersymmetry," hep-ph/9711401; R. Contino and I. Scimemi, "The supersymmetric flavor problem for heavy first two generation scalars at next-to-leading order," hep-ph/9809437; A. Masiero and H. Murayama, "Can ϵ'/ϵ be supersymmetric?," hep-ph/9903363; R. Barbieri, L. Hall and A. Strumia, "Violations of lepton flavor and CP in supersymmetric unified theories," Nucl. Phys. **B445**, 219 (1995); P. Ciafaloni, A. Romanino and A. Strumia, "Lepton flavor violations in SO(10) with large $\tan\beta$," Nucl. Phys. **B458**, 3 (1996); F. Borzumati and A. Masiero, "Large muon and electron number violations in supergravity theories," Phys. Rev. Lett. **57**, 961 (1986); J. Hisano, T. Moroi, K. Tobe and M. Yamaguchi, "Lepton flavor violation via righthanded neutrino Yukawa couplings in supersymmetric standard model," Phys. Rev. **D53**, 2442 (1996); U. Chattopadhyay and P. Nath, "Probing supergravity grand unification in the Brookhaven g-2 experiment," Phys. Rev. **D53**, 1648 (1996); T. Moroi, "The Muon anomalous magnetic dipole moment in the minimal supersymmetric standard model," Phys. Rev. **D53**, 6565 (1996); *Erratum ibid.* **D56**, 4424 (1997); M. Carena, G.F. Giudice and C.E. Wagner, "Constraints on supersymmetric models from the muon anomalous magnetic moment," Phys. Lett. **B390**, 234 (1997).

74. N. Hata, in this volume.

75. K. Choi, J.S. Lee and C. Munoz, "Supergravity radiative effects on soft terms and the μ term," Phys. Rev. Lett. **80**, 3686 (1998).

76. H.P. Nilles and N. Polonsky, "Gravitational divergences as a mediator of supersymmetry breaking," Phys. Lett. **B412**, 69 (1997).
77. V.S. Kaplunovsky and J. Louis, "Model independent analysis of soft terms in effective supergravity and in string theory," Phys. Lett. **B306**, 269 (1993).
78. S. Dimopoulos and H. Georgi, "Softly broken supersymmetry and $SU(5)$," Nucl. Phys. **B193**, 150 (1981);
    R. Barbieri, S. Ferrara and C.A. Savoy, "Gauge models with spontaneously broken local supersymmetry," Phys. Lett. **119B**, 343 (1982);
    P. Nath, R. Arnowitt and A.H. Chamseddine, "Gravity induced symmetry breaking and ground state of local supersymmetric GUTs," Phys. Lett. **121B**, 33 (1983);
    H.P. Nilles, "Supergravity generates hierarchies," Nucl. Phys. **B217**, 366 (1983);
    H.P. Nilles, M. Srednicki and D. Wyler, "Weak interaction breakdown induced by supergravity," Phys. Lett. **120B**, 346 (1983);
    L. Hall, J. Lykken and S. Weinberg, "Supergravity as the messenger of supersymmetry breaking," Phys. Rev. **D27**, 2359 (1983).
79. L. Randall and R. Sundrum, "Out of this world supersymmetry breaking," hep-th/9810155;
    G.F. Giudice, M.A. Luty, H. Murayama and R. Rattazzi, "Gaugino mass without singlets," JHEP **12**, 027 (1998).
80. G.D. Kribs, "Disrupting the one loop renormalization group invariant M/α in supersymmetry," Nucl. Phys. **B535**, 41 (1998).
81. M. Dine, A.E. Nelson and Y. Shirman, "Low-energy dynamical supersymmetry breaking simplified," Phys. Rev. **D51**, 1362 (1995);
    M. Dine, A.E. Nelson, Y. Nir and Y. Shirman, "New tools for low-energy dynamical supersymmetry breaking," Phys. Rev. **D53**, 2658 (1996);
    J.A. Bagger, K. Matchev, D.M. Pierce and R. Zhang, "Weak scale phenomenology in models with gauge mediated supersymmetry breaking," Phys. Rev. **D55**, 3188 (1997);
    S. Dimopoulos, S. Thomas and J.D. Wells, "Sparticle spectroscopy and electroweak symmetry breaking with gauge mediated supersymmetry breaking," Nucl. Phys. **B488**, 39 (1997);
    S.P. Martin, "Generalized messengers of supersymmetry breaking and the sparticle mass spectrum," Phys. Rev. **D55**, 3177 (1997);
    G.F. Giudice and R. Rattazzi, "Theories with gauge mediated supersymmetry breaking," hep-ph/9801271.
82. S. Raby and K. Tobe, "Dynamical SUSY breaking with a hybrid messenger sector," Phys. Lett. **B437**, 337 (1998).

83. M. Dine and A.E. Nelson, "Dynamical supersymmetry breaking at low-energies," Phys. Rev. **D48**, 1277 (1993);
G. Dvali, G.F. Giudice and A. Pomarol, "The μ-problem in theories with gauge mediated supersymmetry breaking," Nucl. Phys. **B478**, 31 (1996);
T. Yanagida, "A Solution to the μ-problem in gauge mediated supersymmetry breaking models," Phys. Lett. **B400**, 109 (1997);
H.P. Nilles and N. Polonsky, "Gravitational divergences as a mediator of supersymmetry breaking," Phys. Lett. **B412**, 69 (1997);
K. Agashe and M. Graesser, "Improving the fine tuning in models of low-energy gauge mediated supersymmetry breaking," Nucl. Phys. **B507**, 3 (1997);
A. de Gouvea, A. Friedland and H. Murayama, "Next-to-minimal supersymmetric standard model with the gauge mediation of supersymmetry breaking," Phys. Rev. **D57**, 5676 (1998);
E.J. Chun, "Strong CP and μ-problem in theories with gauge mediated supersymmetry breaking," Phys. Rev. **D59**, 015011 (1999);
P. Langacker, N. Polonsky and J. Wang, "A Low-energy solution to the μ-problem in gauge mediation," hep-ph/9905252.
84. M. Drees, "$N = 1$ Supergravity guts with noncanonical kinetic energy terms," Phys. Rev. **D33**, 1468 (1986);
S. Dimopoulos and G.F. Giudice, "Naturalness constraints in supersymmetric theories with nonuniversal soft terms," Phys. Lett. **B357**, 573 (1995);
A. Pomarol and D. Tommasini, "Horizontal symmetries for the supersymmetric flavor problem," Nucl. Phys. **B466**, 3 (1996);
A.G. Cohen, D.B. Kaplan and A.E. Nelson, "The More minimal supersymmetric standard model," Phys. Lett. **B388**, 588 (1996).
85. G. Dvali and A. Pomarol, "Anomalous U(1) as a mediator of supersymmetry breaking," Phys. Rev. Lett. **77**, 3728 (1996);
A.E. Nelson and D. Wright, "Horizontal, anomalous U(1) symmetry for the more minimal supersymmetric standard model," Phys. Rev. **D56**, 1598 (1997).
86. D.E. Kaplan, F. Lepeintre, A. Masiero, A.E. Nelson and A. Riotto, "Fermion masses and gauge mediated supersymmetry breaking from a single U(1)," hep-ph/9806430.
87. H. Cheng, J.L. Feng and N. Polonsky, "Superoblique corrections and nondecoupling of supersymmetry breaking," Phys. Rev. **D56**, 6875 (1997).
88. M. Dine, R. Leigh and A. Kagan, "Flavor symmetries and the problem of squark degeneracy," Phys. Rev. **D48**, 4269 (1993);

Y. Nir and N. Seiberg, "Should squarks be degenerate?," Phys. Lett. **B309**, 337 (1993);

M. Leurer, Y. Nir and N. Seiberg, "Mass matrix models: The Sequel," Nucl. Phys. **B420**, 468 (1994);

S. Dimopoulos, G.F. Giudice and N. Tetradis, "Disoriented and plastic soft terms: A Dynamical solution to the problem of supersymmetric flavor violations," Nucl. Phys. **B454**, 59 (1995).

89. N. Arkani-Hamed, H. Cheng, J.L. Feng and L.J. Hall, "Probing lepton flavor violation at future colliders," Phys. Rev. Lett. **77**, 1937 (1996); N. Arkani-Hamed, J.L. Feng, L.J. Hall and H. Cheng, "CP violation from slepton oscillations at the LHC and NLC," Nucl. Phys. **B505**, 3 (1997).

90. M. Gell-Mann, P. Ramond and R. Slansky, in *Supergravity, Proceedings of the Workshop, Stony Brook, New York, 1979*, eds. P. van Nieuwenhuizen and D. Freedman (North Holland, Amsterdam, 1979) p. 315; T. Yanagida, in *Proceedings of the Workshop on Unified Theories and Baryon Number in the Universe, Tsukuba, Japan, 1979*, eds. A. Sawada and A. Sugamoto (KEK Report No. 79-18, Tsukuba, 1979).

91. K.S. Babu, J.C. Pati and F. Wilczek, "Fermion masses, neutrino oscillations, and proton decay in the light of SuperKamiokande," hep-ph/9812538.

92. H. Nilles and N. Polonsky, "Supersymmetric neutrino masses, R symmetries, and the generalized μ problem," Nucl. Phys. **B484**, 33 (1997); M.A. Diaz, J.C. Romao and J.W. Valle, "Minimal supergravity with R-parity breaking," Nucl. Phys. **B524**, 23 (1998).

93. R.M. Godbole, P. Roy and X. Tata, "τ signals of R-parity breaking at LEP200," Nucl. Phys. **B401**, 67 (1993).

94. J. Erler, J.L. Feng and N. Polonsky, "A Wide scalar neutrino resonance and $b\bar{b}$ production at LEP," Phys. Rev. Lett. **78**, 3063 (1997).

95. F. Borzumati, J. Kneur and N. Polonsky, "Higgs-Strahlung and Slepton-Strahlung at hadron colliders," hep-ph/9905443.

1998 TASI LECTURES: NEUTRINOS FROM STRINGS
A Practical Introduction to String Theory, String Model-Building, and String Phenomenology
— Part I: Ten Dimensions —

KEITH R. DIENES

CERN Theory Division, CH-1211 Geneva 23, Switzerland
and
Department of Physics, University of Arizona, Tucson, AZ 85721 [a]

This is the written version of an introductory self-contained course on string model-building and string phenomenology given at the 1998 TASI summer school. No prior knowledge of string theory is assumed. The goal is to provide a practical, "how-to" manual on string theory, string model-building, and string phenomenology with a minimum of mathematics. These notes consist of Part I of these lectures, which cover the construction of bosonic strings, superstrings, and heterotic strings prior to compactification. These notes also develop the ten-dimensional free-fermionic construction. Part II (not published here) covers compactification, semi-realistic four-dimensional heterotic string models, and general features of low-energy string phenomenology. The complete set of lectures can be found on the Los Alamos hep-ph archive located at http://xxx.lanl.gov.

Introduction: An explanation of the title

Perhaps it is best to begin by explaining the title of these lectures.

These lectures were delivered at the 1998 Theoretical Advanced Study Institute (TASI), to an audience of graduate students whose interests were primarily oriented towards high-energy phenomenology. Indeed, this school had a particularly interesting title, one notable both for its breadth as well as its sharp focus: "Neutrinos in Particle Physics and Astrophysics: From 10^{-33} to 10^{+28} Centimeters". My job was to give a set of lectures on the phenomenological aspects of string theory within the context of this school. It may seem, at first glance, that this topic is incompatible with the title of the school. However, given the title, one can judiciously extract the subset of words "Neutrinos in Physics from 10^{-33} Centimeters". Defining "Physics from 10^{-33} Centimeters" as "String Theory" then yields the title "Neutrinos from String Theory", and so that is the title of these lectures. Of course, string theory contains a lot more than neutrino physics (and also, in some ways, a lot less!), and in the course of these lectures I will not really focus so much

---

[a] Permanent address. E-mail: dienes@physics.arizona.edu

on neutrinos as on string theory as a whole. Nevertheless, I will continue to keep neutrinos as a running theme throughout these lectures as a way of reminding ourselves that our discussion of string theory is ultimately aimed at understanding something real and observable, such as an actual neutrino. Therefore, one of the highlights of these lectures will be the actual derivation of a bona-fide neutrino as a low-energy state of string theory in Lecture #9, and everything prior to that point will aim in that direction.

The remainder of the title states that these lectures are meant to serve as a practical introduction to string theory, string model-building, and string phenomenology. Let me explain, in a rough sense, what each of these words is meant to convey. We are all familiar with quantum field theory, which is a *language* through which we might construct particular *models* of physics (such as the Standard Model or the Minimal Supersymmetric Standard Model). Such models then have certain physical characteristics, certain *phenomenologies*. String theory, at least as I shall try to present it, can likewise be considered as a *language* for discussing physics: in this sense it replaces quantum field theory (a language based on point-particle physics) with a new language suitable for theories whose fundamental objects are the one-dimensional extended objects known as *strings*. However, from this perspective, string theory is still only a language: it is still necessary to take the next step and *use* this language to construct *models* that describe the everyday world. Therefore, although I will attempt to give a self-contained introduction to the language of string theory, these lectures will primarily focus on the model-building aspects of string theory and on the resulting phenomenologies that these models have. While there already exist many excellent reviews of string theory, there are relatively few that focus on its model-building and phenomenological aspects. These lecture notes will therefore hopefully help to fill the gap, especially for those readers who might care less for the formal aspects of string theory and more for their phenomenological implications.

Finally, I should explain the word "practical" which also appears in the title. The word "practical" refers to actual *practice* — the things that practitioners actually need to know in order to build *bona-fide* string models and/or comprehend their low-energy properties. Of course, string theory is a rich and beautiful subject, with many mathematical aspects that are compelling and ultimately essential for a deep understanding of the subject. However, the goal of these lectures is simply to present the basic features of string theory with a minimum of mathematics — as stated in the abstract, I am seeking to provide a "how-to" manual which cuts the subject to the bone and conveys only that information which will be important for phenomenology. Therefore, in many places the omissions will be substantial. Certainly they do not do justice to the subject. However, these lectures were designed for phenomenologically-oriented graduate students whose desire (I hope) was to learn something of string theory without being deluged by mathematical formalism. It is with them in mind that I designed these lectures to be as elementary as feasible, and to "get to the physics" as rapidly as possible. Therefore, I now issue the following

> **Warning:** These lectures are meant to cover a considerable amount of introductory material very rapidly and without mathematical sophistication. The purpose is to advance quickly to the model-building and phenomenological aspects of string theory, while still conveying an intuitive flavor of the essential issues. The target audience consists of people who have had no prior exposure to string theory, and who wish to understand the basic concepts from a purely phenomenological perspective.

Hopefully, the students came away with a sense that string theory is a real part of physics, one with direct relevance for the real world. Perhaps the reader will too. If so, then these lectures will have served their purpose.

1 Lecture #1: Why strings? — an overview

Why should we be interested in string theory? In this lecture, we shall review our present state of knowledge about the underlying constituents of matter, and discuss how string theory has the potential to extend that knowledge in a profoundly new direction. Since this lecture is meant only as an overview, we shall keep the discussion at an extremely superficial level and seek to present the intuitive *flavor* of string theory rather than its substance. We shall deal with the substance in subsequent lectures.

1.1 From atoms to the Standard Model: A quick review

Certainly we do not need to understand string theory in order to appreciate modern high-energy particle physics, or to understand or interpret the results of collider experiments. Why then should one study string theory, a subject whose connections to observable phenomena are usually considered rather tenuous at best?

The primary reason, of course, is that the goal of high-energy physics has always been to uncover the fundamental "elements" or building-blocks of the natural world. These consist of both the fundamental *particles* that make up the *matter*, and the fundamental *forces* that describe their *interactions*. In this way, we hope to expose the underlying laws of physics in their simplest forms.

But what is "fundamental"? Clearly, the answer depends on the energy scale, or equivalently the inverse length scale, at which these constituents are being probed. In order to establish our frame of reference, recall that 1 eV $\approx 1.6 \times 10^{-19}$ Joules $\approx (10^{-7}$ meters$)^{-1}$. At the eV scale, the fundamental objects are atoms, or nuclei plus electrons. But it turns out that there are many different types of atoms or nuclei — indeed, they fill out an entire periodic table, the complexity but regularity of which suggests a deeper substructure. And indeed such a deeper substructure exists: at the keV to MeV scale, the nuclei are no longer fundamental, but decompose into new fundamental objects — protons and neutrons. Thus, at this energy scale, the fundamental objects are protons, neutrons, and electrons. But once again, it is found that there are many different "types" of protons and neutrons — collectively they are called *hadrons*, and include not only the proton (p) and neutron (n), but also the pions (π), kaons (K), rho (ρ), omega (Ω), and so forth. Indeed, the "periodic table of the elements" at this energy scale is nothing but the Particle Properties Data Book! But once again, the complexity and regularity of these "elementary" particles suggests a deeper substructure, and indeed such a substructure is found, this time at the GeV scale: the proton and neutron are just made of two kinds of *quarks*, the so-called up and down quarks. Thus, at the GeV scale, the fundamental objects are up quarks, down quarks, and electrons. But once again complexity emerges: it turns out that there are many different "types" (flavors) of quarks: up, down, strange, charm, top, and bottom. Likewise, there are many different "types" of electrons (collectively called *leptons*): the electron, the muon, the tau, and their associated neutrinos. And indeed, once again there is a mysterious pattern, usually referred to as a family or generational structure. This once again suggests a deeper substructure.

Unfortunately, this is as far as we've come. Indeed, all of our present-day knowledge down to this energy scale is gathered together into the so-called *Standard Model* of particle physics. The primary features of the Standard Model are as follows. The fundamental *particles* are the quarks and leptons. They are all fermions, and are arranged into three generations of doublets:

$$\text{quarks}: \quad \begin{pmatrix} u \\ d \end{pmatrix}, \quad \begin{pmatrix} c \\ s \end{pmatrix}, \quad \begin{pmatrix} t \\ b \end{pmatrix}$$

$$\text{leptons}: \quad \begin{pmatrix} \nu_e \\ e \end{pmatrix}, \quad \begin{pmatrix} \nu_\mu \\ \mu \end{pmatrix}, \quad \begin{pmatrix} \nu_\tau \\ \tau \end{pmatrix}. \tag{1.1}$$

The fundamental *forces* also come in three varieties. First, there is the strong (or "color") force, associated with the non-abelian Lie group $SU(3)$. Its fine-structure constant is $\alpha_3 \approx 1/8$ (as measured at energy scales of approximately 100 GeV), and it is responsible for binding quarks together to form hadrons and nuclei. As such, it is felt only by quarks. Its mediators or carriers are called *gluons*. Second, there is the electroweak force, associated with the non-abelian Lie group $SU(2)$. Its fine-structure constant is $\alpha_2 \approx 1/30$ (indeed, weaker than the strong force!), and it is responsible for β-decay. Unlike the strong force, it is felt by *all* of the fundamental particles. Finally, there is the "hypercharge" force, associated with the abelian Lie group $U(1)$, with fine-structure constant $\alpha_1 \approx 1/59$. Once again, this force is felt by essentially *all* particles, both quarks and leptons. The carriers of the latter two forces are the photon as well as the $W^{\pm}$ and Z particles. Indeed, ordinary electromagnetism is a combination of the electroweak and hypercharge forces, and is the survivor of electroweak symmetry breaking. This breaking is induced by the one remaining particle of the Standard Model, a boson called the *Higgs particle*. An excellent introduction to the physics of the Standard Model can be found in the TASI lectures of G. Altarelli (this volume).

1.2 Beyond the Standard Model: Two popular ideas

Is that all there is? Clearly, there are lots of reasons to believe in something deeper! First, the Standard Model contains many *arbitrary parameters*, such as the masses and "mixings" of fundamental particles. All of these must ultimately be *fit* to data rather than *explained*. Second, there are many *conceptual* questions. *Why* are there three generations? *Why* are there three kinds of forces? *Why* do these forces have different strengths and ranges? A fundamental theory should explain these features. Finally, there is also another force which we have not yet mentioned: the *gravitational* force. How do we incorporate the gravitational force into this framework? In other words, how do we "quantize" gravity?

There is only one conclusion we can draw from this state of affairs. Just as in each previous case, there must still be a deeper underlying principle. It is important to stress that this is not simply an issue of academic interest. Rather, it is one of practical importance, because the next generation of particle accelerators are being built right now! (Two of the most prominent that will be exploring physics beyond the Standard Model are Fermilab, where upgrades to the TeVatron are being implemented, and CERN, where construction of the Large Hadron Collider (LHC) is already underway.) The pressing question, therefore, is: What do we expect to see at these machines? What will high-energy physics be focusing on over the next ten to twenty years? It turns out that there are two very popular sets of ideas, both of which are thoroughly reviewed in the TASI lectures of N. Polonsky (this volume).

Low-energy supersymmetry

The first idea is *supersymmetry* (SUSY). This refers to a new kind of symmetry in physics, one which relates bosons (particles with integer spin) to fermions (particles with half-integer spin). Thus, for every known particle, there is a predicted new particle, its so-called superpartner:

$$
\begin{aligned}
\text{quarks} &\Longleftrightarrow \textit{squarks} \\
\text{leptons} &\Longleftrightarrow \textit{sleptons} \\
\text{gauge bosons} &\Longleftrightarrow \textit{gauginos} \ .
\end{aligned}
\tag{1.2}
$$

Clearly, this implies the existence of a *lot* of new particles and a *lot* of new interactions! Why then go through all this trouble?

Well, it turns out that supersymmetry can provide a number of striking benefits. First, through supersymmetry, we can explain the relative strengths of the forces ("gauge coupling unification"). Second, we can explain the origin of electroweak symmetry breaking. Third, supersymmetry has a number of favorable cosmological implications (for example, supersymmetry provides a natural set of dark-matter candidates). Finally, it turns out that supersymmetry is the only known answer to certain difficult theoretical puzzles in the Standard Model (chief among them the so-called "gauge hierarchy problem", *i.e.*, the difficulty of explaining the lightness of the Higgs particle, or equivalently to difficulty of explaining the stability of the scale of electroweak symmetry breaking against radiative corrections). In order to serve as an explanation of the gauge hierarchy problem, the energy scale associated with supersymmetry must not be too much higher than the scale of electroweak symmetry breaking. This is therefore called "low-energy supersymmetry", which refers to the common expectation that superparticles should exist at or near the TeV-scale.

Supersymmetry is a beautiful theory, both phenomenologically and mathematically. But it is *not* observed in nature. Therefore, supersymmetry must be broken. The problem, however, is that supersymmetry is very robust! It turns out to be quite hard to find mechanisms that can easily ("spontaneously") break supersymmetry at the expected energy scales. Therefore, we are faced with a major unsolved problem: *How do we break supersymmetry?* Indeed, we often have to resort to introducing SUSY-breaking by hand, which requires the introduction of *many* additional unknown parameters. This is quite unpleasant, not only from an aesthetic point of view but also a phenomenological (predictive) point of view. However, it is often possible to consider only a minimal supersymmetric extension to the Standard Model (the so-called MSSM) where a minimal number of supersymmetry-breaking parameters are chosen.

Grand unification

The second popular idea for physics beyond the Standard Model concerns so-called *Grand Unified Theories* (GUTs). This refers to an attempt to realize the different forces and particles in nature as different "faces" or "aspects" of a single GUT force and a single GUT particle. An electromagnetic analogy here might be useful. Recall that the electric force is felt or caused by static charges, and that the magnetic force is felt or caused by moving charges. Are these therefore different forces? As we know, the answer is most definitely "no": we can Lorentz-boost from a rest frame to a moving frame, whereupon the distinction between the electric and magnetic forces melts away and these forces become intertwined. Thus, we conclude that the electric and magnetic forces are merely different aspects of *one* force, the "electromagnetic" force.

Is the same true for the strong, electroweak, and hypercharge forces? Is there a single "strong-weak-hypercharge" GUT force?

At first glance, this doesn't seem possible, because these different forces have different strengths. Recall their fine-structure constants: $\alpha_1 \approx 1/59$, $\alpha_2 \approx 1/30$, and $\alpha_3 \approx 1/8$. However, also recall that in quantum field theory, the strengths of forces ultimately depend on the energy scale through which they are measured. To see why this is so, let us think of placing a positive charge next to a dielectric. The positive charge draws some negative charge from within the dielectric towards it, so that the dielectric medium partially screens the positive charge. Therefore, in a rough sense, the less of the dielectric we see (*i.e.*, the more finely resolved our experimental apparatus to probe the original positive charge), the stronger our original positive charge seems to be. Thus, we see that at shorter distances (corresponding to higher energies), our electric charges (and therefore the corresponding electric forces) appear to be stronger. If this dielectric analogy serves as a good model for the results of a true quantum field-theoretic calculation (and in this case it does), we conclude that the electric force appears to grow stronger with increasing energy.

Of course, this is just a mechanical analogy. However, in the supersymmetric Standard Model, it turns out that the quantum field-theoretic vacuum itself indeed behaves like a dielectric for the hypercharge and weak forces. However, for the strong force, it behaves as an *anti*-dielectric. Thus, while the hypercharge and electroweak forces become *stronger* at higher energies, the strong force becomes *weaker* at higher energies. (This latter feature is the celebrated phenomenon of *asymptotic freedom*.) Together, these observations imply that these three forces have a chance of *unifying* at some energy scale if their strengths become equal, and indeed, carrying out the appropriate calculations, one finds the results shown in Fig. 1. We see from this figure that the forces appear to unify at the scale

$$M_{GUT} \approx 2 \times 10^{16} \text{ GeV} . \tag{1.3}$$

This would then be the natural energy scale for grand unification. Note that this unification also requires the existence of weak-scale supersymmetry in the form of weak-scale superpartners. Without such superpartners, the evolution of these fine-structure constants as a function of the energy scale is different, and they fail to unify at any scale. This then serves as another motivation for weak-scale supersymmetry.

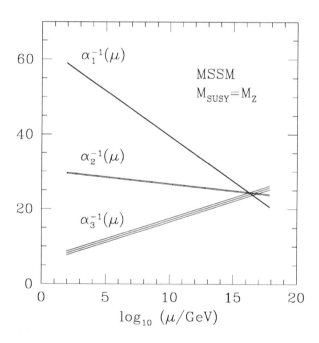

Figure 1: One-loop evolution of the gauge couplings within the Minimal Supersymmetric Standard Model (MSSM), assuming supersymmetric thresholds at the Z scale. Here $\alpha_1 \equiv (5/3)\alpha_Y$, where α_Y is the hypercharge coupling in the conventional normalization. The relative width of each line reflects current experimental uncertainties.

GUTs would have numerous important effects on particle physics. First, by their very nature, they would imply new interactions that can mix the three fundamental forces. Second, this in turn implies that GUTs naturally lead to new, rare decays of particles. The most famous example of this is proton decay, the rate for which is experimentally known to be exceedingly small (since the proton lifetime is $\tau_p \gtrsim 10^{32}$ years). Third, GUTs would naturally explain the quantum numbers of all of the fundamental particles. Along the way, GUTs would also explain charge quantization. GUTs might also explain the origins of fermion mass. Finally, because they generally lead to baryon-number violation, GUTs even have the potential to explain the cosmological baryon/anti-baryon asymmetry. By combining GUTs with supersymmetry in the context of SUSY GUTs, it might then be possible to realize the attractive features of GUTs simultaneously with those of supersymmetry in a single theory.

Both the SUSY idea and the GUT idea are very compelling. Certainly, the SUSY idea (and indirectly the GUT idea, through measurements of proton decay and other rare decays) will be the focus of experimental high-energy physics over the next 20 years. But high-energy theorists also have plenty of work to do — we must build theories in order to interpret the data. But *how* do we build realistic SUSY theories? *How* do we build realistic GUT theories? *How* do we incorporate gravity?

Clearly, the possibilities seem endless. And even the SUSY or GUT ideas have not answered our most fundamental questions, such as why there are three gauge forces, or why there are three generations. Therefore, it is natural to hope that there is yet a deeper principle that can provide some theoretical guidance. And that's where string theory comes in.

1.3 So what is string theory?

The basic premise of string theory is very simple: all elementary particles are really closed vibrating loops of energy called *strings*. The length scale of these loops of energy is on the order of 10^{-35} meters (corresponding to 10^{19} GeV), so it is not possible to probe this stringy structure directly.

This idea has great power, because it provides a way to unify all of the particles and forces in nature. Specifically, each different elementary particle can be viewed as corresponding to a different vibrational mode of the string. A pictorial representation of this idea is given in Fig. 2, where we are schematically associating higher vibrational string modes with string loops containing more "wiggles". From the point of view of a low-energy observer who cannot make out this stringy structure, the different excitations each appear to be point particles. However, to such an observer, the states with more underlying "wiggles" appear to have higher spin. Thus, in this way we find that string

210

theory predicts not only spin-1/2 and spin-1 states (which can be associated with the fermions and gauge bosons of the Standard Model respectively), but also a spin-2 state (which can naturally be associated with the graviton). Thus, through string theory, we see that the gauge interactions, particles, *and also gravity* are unified into a common quantized description as corresponding to different excitation modes of a single fundamental entity, the string itself.

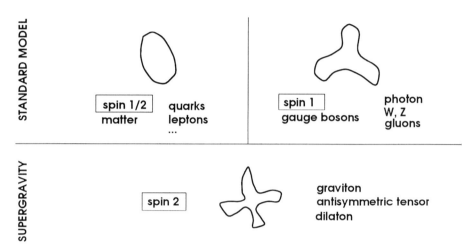

Figure 2: The basic hypothesis of string theory is that the different elementary particles correspond to the different vibrational modes of a single fundamental entity, a closed loop of energy called a string. In this way one obtains not only spin-1/2 and spin-1 states which can be associated with the matter and gauge bosons of the Standard Model, but also a spin-2 state which can be identified with the graviton. Thus, string theory provides a way of unifying the Standard Model with gravity.

Of course, this is not the end of the story. Just as a violin string has an infinite number of harmonics, so too does a string give rise to an infinite tower of states corresponding to higher and higher vibrational modes. Since it takes more and more energy to excite these higher vibrational string modes, such states are increasingly massive. Indeed, because the fundamental string scale is on the order of $M_{string} \approx 10^{18}$ GeV, these string states are quantized in units of M_{string}. The states which we have illustrated in Fig. 2 are all *massless* with respect to M_{string}, and correspond, in some sense, to the ground states of the string. These are the so-called "observable states", and include not only the (supersymmetric) Standard Model and (super)gravity, but also may include various additional states (often called "hidden-sector states" which contain their own matter and gauge particles). However, there also exists an infinite tower of massive states with masses $M_n \approx \sqrt{n} M_{\text{string}}$, $n \in \mathbb{Z}^+$. In most discussions of the phenomenological properties of string theory, these massive states are ignored (since they are so heavy), and one concentrates on the phenomenology of the massless states. One then presumes that they accrue (relatively small) masses through other means, such as through radiative corrections.

Nevertheless, the passage from point particles to strings has tremendous consequences. Not only have we replaced the physics of zero-dimensional objects (elementary point particles) with the physics of one-dimensional objects (strings), but we have also replaced the physics of the one-dimensional world-lines that they sweep out with the physics of two-dimensional so-called *world-sheets*. Likewise, we have replaced the physics of Feynman diagrams with the physics of two-dimensional *manifolds*, so that a tree diagram corresponds to a genus-zero manifold (a sphere) and a one-loop diagram corresponds to a genus-one manifold (a torus). These comparisons are illustrated in Fig. 3. Note that the latter descriptions as spheres and tori correspond to shrinking the external strings to points, essentially "pinching off" the external legs. This is a valid description for reasons to be discussed in Lecture #2.

This is clearly a new language for doing physics. However, as we have seen, because string theory also includes gravity (which is exceedingly weak compared with the other forces), its fundamental mass scale is very high. Indeed, since the fundamental energy scale for gravity is the Planck mass

$$M_{\text{Planck}} \equiv \sqrt{\frac{\hbar c}{G_N}} \approx 10^{19} \text{ GeV} \approx (10^{-33} \text{ cm.})^{-1} \, , \qquad (1.4)$$

the string scale must also be very high. Indeed, to a first approximation, it turns out that

$$M_{\text{string}} \approx g_{\text{string}} M_{\text{Planck}} \qquad (1.5)$$

where g_{string} is the string coupling constant, typically assumed to be $\sim \mathcal{O}(1)$. Thus, we see that string theory is ultimately a theory of Planck-scale physics.

There are lots of "formal" reasons for being excited about string theory. First, it turns out that string theory requires the existence of extra space-time dimensions in order to be consistent, and consequently we now have to consider physics in different numbers of dimensions as well as all sorts of geometric questions pertaining to different possible "compactification" scenarios. Second, string theory gives us a new perspective on the structure of spacetime itself. For example, string theory gives rise to many novel Planck-scale effects. One of these is called T-duality: the physics of a closed string in a spacetime one of whose dimensions is compactified on a circle of radius R turns out to be

212

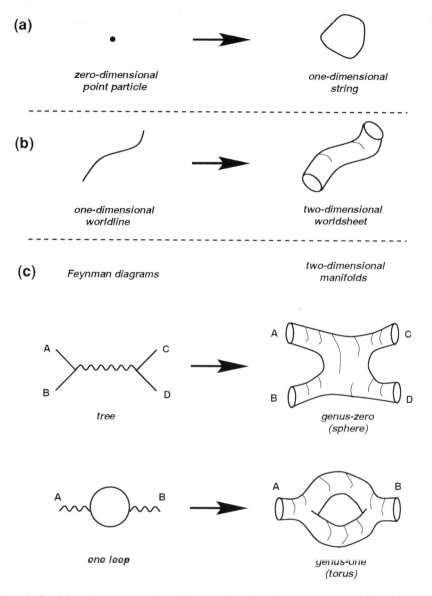

Figure 3: In string theory, we replace (a) zero-dimensional elementary particles with one-dimensional strings; (b) one-dimensional worldlines with two-dimensional worldsheets; and (c) Feynman diagrams with two-dimensional manifolds. For example, tree diagrams correspond to genus-zero manifolds (spheres), and one-loop diagrams correspond to genus-one manifolds (tori).

equivalent to the physics of the same string in a spacetime in which the radius is M^2_{string}/R. Thus, T-duality interchanges large radii and small radii, and suggests that our naïve view of spacetime and its linear hierarchy of energy and length scales cannot ultimately be correct. Third, string theory also provides new types of strong/weak coupling dualities. These have proven useful for elucidating the strong-coupling dynamics of not only string theory, but also *field* theory. Finally, there have even been novel applications to black-hole physics. The most famous example of this is the fact that various non-perturbative string structures called D-branes have provided the first *statistical* (*i.e.*, microscopic) derivation of the Bekenstein-Hawking entropy formula $S = A/4$ that relates the entropy S of a black hole to its surface area A. Indeed, the above list only begins to scratch the surface of all of the many exciting recent formal developments in string theory.

But we are phenomenologists, so it is natural to ask about the rest of high-energy physics. How does string theory connect with the rest of particle physics?

Some of the answers to this question have already been given above. We have seen, in particular, that string theory is capable of reproducing the Standard Model as its low-energy limit. Moreover, as we have also seen, the Standard Model naturally emerges coupled with gravity. Furthermore, in many cases this entire structure is also joined with *supersymmetry*. Finally, this entire structure is also often joined with many properties of GUTs (such as gauge coupling unification). All of this comes out of the low-energy limit of string theory, in some sense automatically.

There are also many other benefits to considering the application of string theory to particle physics. First, string theory provides us with new kinds of symmetries (so-called "worldsheet symmetries") which lead to powerful new constraints on the resulting low-energy phenomenology. Second, in principle[a]

---

[a] In this connection, we hasten to emphasize the phrase "in principle". Unfortunately, our relative inability to understand the non-perturbative structure of string theory often means that the pragmatic consequences of having no free parameters cannot be realized, and in practice one is often forced to introduce many parameters to reflect our ignorance of the underlying dynamics. This will be discussed in subsequent lectures. This situation is rather analogous to one that arises in the MSSM: we do not know how supersymmetry is broken, so we typically parametrize our ignorance through the introduction of various supersymmetry-breaking parameters. Likewise, in string theory, there are analogous questions which come under the heading of "vacuum selection": we do not know how the non-perturbative dynamics of string theory selects a particular vacuum state. Thus, in order to proceed to make phenomenological predictions, we are often forced to *assume* a certain vacuum state, or to parametrize the vacuum via the introduction of essentially unfixed parameters. The important point, however, is that string theory is a complete theory in that it should *in principle*, by virtue of its dynamics, uniquely fix the values of all of its fundamental parameters.

string theory has *no free parameters*, which leads to a very predictive theory. Third, string theory has no divergences — in some sense, string theory is a completely *finite* theory in which many of the troublesome divergences associated with field theory are simply absent. Finally, it turns out that string theory can even give rise to a new perspective on the Standard Model itself, and often provides new and simpler ways to perform calculations.

These last three points (absence of free parameters, absence of divergences, and new ways to perform calculations) are truly remarkable. Therefore, let us pause to explain in an intuitive way why these features arise. First, let us explain why string theory has fewer free parameters. To do this, let us consider a Feynman diagram for a typical tree-level decay $A \to B + C$, as shown in Fig. 4(a). In field theory, such a process depends on many separate parameters ultimately associated with the separate propagators and vertices. Specifically, even though the propagators are determined once the masses and spins of the particles are specified, there still remains an *independent* choice as to the form of the vertex interaction. Thus, in a given field theory, there still remain many independent parameters to choose. In string theory, by contrast, there is no sharp distinction between propagators and vertices; they melt into each other, and are essentially the same. Thus, once the propagators are determined, the vertices are also intrinsically determined. This is one of the underlying reasons why string theory contains fewer free parameters than field theory.

Next, let us discuss why string theory is more finite than field theory. To do this, let us consider a typical one-loop Feynman diagram, as shown in Fig. 4(b). In field theory, the virtual interactions occur at sharp spacetime locations x and y. This is ultimately the origin of the ultraviolet (*i.e.*, short-distance) divergence as $x \to y$. In string theory, by contrast, we have seen that there are no such sharp interaction points — essentially the interaction is "smoothed out" by the presence of the string. Thus, there is no sense in which the dangerous $x \to y$ limit exists, for there are no precise means by which one can define such interaction locations x and y. It is in this manner that string theory automatically removes ultraviolet divergences: the string itself, through its extended geometry, acts as a (Planck-scale) ultraviolet regulator.

Finally, let us discuss why string theory can often give us simpler ways to perform calculations than in field theory. To do this, let us consider the total tree-level amplitude for a typical process $A + B \to C + D$, as illustrated in Fig. 4(c). As we know, in field theory there are two separate topologies of Feynman diagram that must be separately considered: the s-channel diagram and the t-channel diagram. In general, at any given order, there are many separate diagrams to evaluate, and one often finds that great simplifications

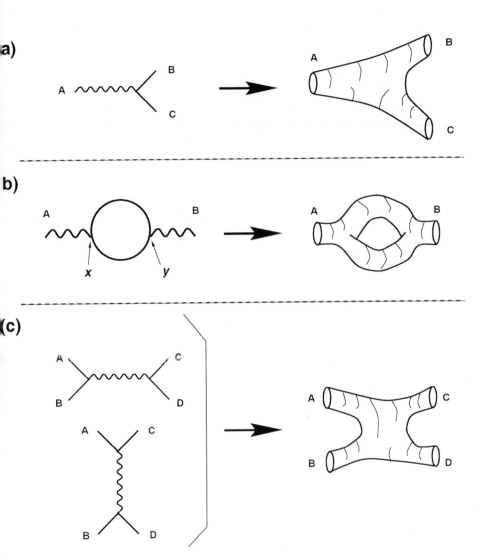

Figure 4: (a) Illustration of the fact that string propagators and string vertices are not independent. (b) Illustration of the fact that string theory lacks many of the ultraviolet divergences that arise in field theory from the short-distance limit $x \to y$. (c) Illustration of the fact that one string diagram often comprises many field-theoretic diagrams.

and cancellations occur only when these individual contributions are added together. In string theory, by contrast, there is only *one* corresponding diagram to evaluate at any given order. Thus, the sorts of simplifications or cancellations that might occur in field theory are automatically "built into" string theory from the very beginning. In some sense, string theory manages to find a way to reorganize the field-theory diagrams in a perturbative expansion in a useful and potentially profitable way. Indeed, this observation has even led to the development of many new techniques for evaluating complicated field-theoretic processes, particularly in QCD where the number of diagrams and the number of terms in each diagram can easily grow to otherwise unmanageable proportions.

We thus see that in a number of ways, string theory is a very useful language in which we might consider thinking about particle physics. Indeed, in various aspects (such as finiteness, fewer parameters, *etc.*) it is superior to field theory. But overall, the fundamental fact remains that if we are thinking about strings, we are abandoning our usual four-dimensional point of view of particle physics. Specifically, since each different particle in spacetime is now interpreted as a different quantum mode excitation of an underlying string, we see that four-dimensional (spacetime) physics is now ultimately the consequence of two-dimensional (worldsheet) physics. Thus, everything we ordinarily focus on in field theory (such as the four-dimensional particle spectrum, the gauge symmetries, the couplings, *etc.*) are now all ultimately determined or constrained by worldsheet symmetries.

And this brings us to string phenomenology.

1.4 So what is string phenomenology?

In order to understand what string phenomenology is, we can draw a useful analogy. Just as we are replacing the *language* of high-energy physics from field theory to string theory, we likewise replace field-theory phenomenology with string-theory phenomenology. The goals of string phenomenology are of course the same as those of ordinary field-theory phenomenology: both seek to reproduce, explain, and predict observable phenomena, and both seek to suggest or constrain new physics at even higher energy scales. Indeed, only the language in which we will carry out this procedure has changed. Thus, in some sense, string phenomenology is the "art" of using the new insights from string theory in order to understand, explain, and predict what physics at the next energy scale is going to look like. Or, recalling that string theory is ultimately a theory of Planck-scale physics, we can say that string phenomenology is the "interplay" or "meeting-ground" between Planck-scale physics and GeV-scale physics.

It is important to understand that we are not abandoning field theory completely. Nor would we want to. Field theory automatically incorporates many desirable features such as causality, spin-statistics relations, and CPT invariance (which in turn implies the existence of antiparticles). These are all generic predictions of field theory, and are the underlying reasons why field theory is the appropriate language for particle physics. However, since string theory ultimately reduces to field theory in its low-energy limit, all of these features will still be retained in string theory. Moreover, as we have seen, string theory *additionally* predicts or explains gravity, supersymmetry, and the absence of ultraviolet divergences. Furthermore, as we shall see, string theory also automatically predicts the existence of gauge symmetry, and even incorporates features such as gauge coupling unification. These are all generic predictions of string theory. It is for these reasons to believe that a change in language from field theory to string theory might be useful.

String theory will also provide us with new tools for model-building, new mechanisms and new guiding principles. Let us give some examples. In field theory, there are many well-known ideas that are part and parcel of the model-building game: one must enforce ABJ anomaly cancellation (to preserve gauge symmetries); one can employ the Higgs mechanism (to generate spontaneous symmetry breaking and give masses to particles); one has the GIM mechanism (to preserve flavor symmetries); and one has supersymmetry (to cancel quadratic divergences). Likewise, in string theory there are analogous sets of ideas, many of which are extensions of their field-theory counterparts. For example, one has the so-called "Green-Schwarz" mechanism for anomaly cancellation (to preserve gauge symmetries); one has string vacuum shifting via pseudo-anomalous $U(1)$ gauge symmetries (to generate spontaneous symmetry breaking and generate particle masses); one has spacetime compactification (to generate gauge symmetries); one has hidden string sectors (to break supersymmetry and impose selection rules); and one has massive towers of string states (to enforce finiteness). Thus, model-building proceeds, but with a different set of principles.

There is also a much more subtle effect of changing our language from field theory to string theory. Ultimately, since four-dimensional physics is now derived from an underlying two-dimensional (worldsheet) theory, string phenomenology is ultimately much more constrained than field-theory phenomenology. One given worldsheet symmetry, which might serve as an "input", can have various seemingly unrelated effects in the resulting spacetime phenomenological "output". Thus, string theory not only leads to unexpected connections or correlations between seemingly disparate spacetime phenomena, but can also give rise to entirely new phenomenological scenarios that could not have been anticipated within field theory alone. We will see many examples of this in the coming lectures.

Thus, we see that string phenomenology does many things and has many goals:

- to provide a new framework for addressing and answering numerous phenomenological questions;

- to provide a rigorous test of string theory as a theory of physics;

- to explore the interplay between worldsheet physics and spacetime physics (*i.e.*, to ultimately determine which "patterns" of low-energy phenomenology are allowed or consistent with being realized as the low-energy limit of an underlying string theory); and

- to augment field theories of "low-energy" physics into the string framework so as to give them the full benefits of the language of string theory.

Because of these different roles, string phenomenology occupies a rather central position in high-energy physics: it allows the transmission of ideas from high-scale string theory to guide "low"-scale particle physics, and vice versa. This situation is illustrated in Fig. 5. At the lowest energies (lower left), string phenomenology has direct relevance for the Standard Model, where it can potentially explain features such as the choice of the gauge group, the number of generations, and numerous other parameters such as the masses and mixings of Standard-Model particles. At slightly higher energies (lower right), we see that string phenomenology can also suggest or constrain various extensions to the Standard Model, such as SUSY and SUSY-breaking, grand unification, and hidden-sector physics. At the highest energies (upper left), string phenomenology is also concerned with the more formal aspects of string theory: such important questions include string vacuum selection, nonperturbative string dynamics, string duality, and new mathematical structures and techniques. And string phenomenology even has relevance outside the strict confines of particle physics. For example, string theory should have a profound impact on cosmology (upper right), where important stringy issues include the role of the dilaton, the effects of many other light degrees of freedom (the so-called *moduli*), the possibility of extra spacetime dimensions, the cosmological constant problem, and even more exotic ideas such as topology change. As illustrated in Fig. 5, string phenomenology sits at the center of this web of ideas. Exploring the connections between the different corners of this figure is, therefore, the job of the string phenomenologist. Indeed, through

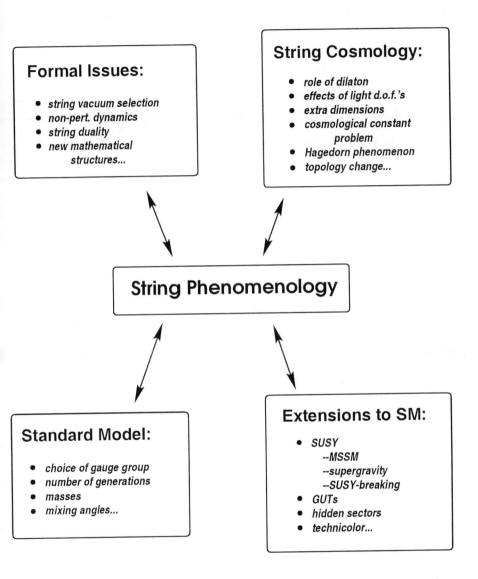

Figure 5: String phenomenology is the central "meeting-ground" between Standard-Model physics, extensions to the Standard Model, formal string issues, and string cosmology.

string phenomenology, one "uses" string theory in order to open a window into the possibilities for physics beyond the Standard Model.

1.5 Plan of these lectures

For much of the past decade, string phenomenology has been practiced assuming a particular type of underlying string theory, the so-called *perturbative heterotic string*. Therefore, this string will be the focal point of most of these lectures. However, it turns out that the heterotic string is built directly on the foundations of two other kinds of strings, the *bosonic string* and *Type II superstring*. Indeed, in a sense to be made more precise in Lecture #5, one can view the heterotic string as the "sum" of the bosonic string and the superstring string. Therefore, in these lectures, we will have to start at the beginning by studying first the bosonic string, then the Type II string, and finally the heterotic string. Indeed, this situation is analogous to the way in which one often studies quantum field theory: first one learns how to quantize the Klein-Gordon field, then the Dirac field, and finally the gauge field. In a certain sense, the bosonic string is the analogue of the Klein-Gordon field, while the Type II superstring is the analogue of the Dirac field and the heterotic string is the analogue of the gauge field. Of course, this analogy is only a pedagogical organizational one, since the heterotic string itself will ultimately contain *all* of the phenomenological properties (*e.g.*, scalars, fermions, and gauge symmetries) that we desire.

In Lecture #2, we will therefore give a brief introduction to the bosonic string, stopping only long enough to develop the ideas and techniques we will need for later applications. In Lectures #3 and #4, we will then proceed to develop the Type II superstring, once again focusing on only those aspects that will be useful for later applications. Finally, in Lecture #5, we will arrive at our destination: the heterotic string. In Lecture #6 we will construct some ten-dimensional heterotic string models, and in Lecture #7 we will develop a useful set of rules for heterotic string model-building.

The above material comprises Part I of these lectures. In Part II (not published here), we will consider the physics that emerges after compactification. Specifically, in Lecture #8 we will learn how to compactify these strings down to four dimensions, and in Lecture #9 we will finally construct some semi-realistic string models. In keeping with the title of these lectures, it is here that we will finally learn how to "derive" a neutrino from string theory.

It is important to note, however, that all of string phenomenology is not based on the heterotic string. Particularly over the past several years, there has been a profound shift in our understanding of both string theory and its phenomenological implications. One of the consequences of this so-called "second superstring revolution" has been a new emphasis on yet another class of strings, the *Type I (open) strings*. Indeed, there has even emerged a new superstructure which promises to relate all of these strings to each other: this structure is called *M-theory*, and is deeply tied to many non-perturbative aspects of string theory which are still being understood. Needless to say, these recent developments have the potential to completely change the way we think about string theory and string phenomenology. We will therefore discuss some of these modern developments in the final Lecture #10. Nevertheless, the bulk of these lectures will primarily be focused on the more traditional aspects of string phenomenology that concern the weakly coupled heterotic string. Indeed, this affords the best introduction to string theory and string phenomenology, regardless of the future directions that string theory and string phenomenology might ultimately take.

We also remind the reader that our goal here is to provide an introduction to string theory that avoids mathematical complications wherever possible, and which "gets to the physics" as rapidly as possible. Therefore, in many places, we will simply assert a mathematical result to be true, leaving its derivation to be found in various textbooks on the subject. For this purpose, we recommend Volume I of the textbook *Superstring Theory*, by M.B. Green, J.H. Schwarz, and E. Witten (henceforth to be referred to as GSW[b]). In fact, our initial approach will be very similar to that of GSW, and we will continually refer back to this textbook as we proceed. Another recommended textbook with a more modern mathematical perspective is *Introduction to String Theory*, by J. Polchinski. Finally, we remind the reader that although only Part I of these lectures are published in this volume, the full set of lectures (including Part II) can be found on the Los Alamos hep-ph archive located at http://xxx.lanl.gov.

2 Lecture #2: Strings and their spectra: The bosonic string

2.1 The action

We begin by studying the simplest string of all: the bosonic string. As we discussed in Lecture #1, the physics of a string is ultimately described by the shape it takes (*e.g.*, its vibrational mode of oscillation) as it propagates through

---

[b]Not to be confused with another great GSW trio, namely Glashow, Salam, and Weinberg. One can only hope that someday string theory will be as well-established, both theoretically and experimentally, as the GSW electroweak theory. This may sound a bit optimistic, but a possible new *experimental* direction for string theory and string phenomenology will be discussed in Lecture #10.

an external spacetime and thereby sweeps out a two-dimensional worldsheet. Therefore, we must first have a way of describing the shape of this worldsheet. To this end, we parametrize the worldsheet by two worldsheet coordinates (σ_1, σ_2) as illustrated in Fig. 6, and describe the embedding of this worldsheet into the external spacetime by giving the spacetime coordinates X^μ of any location (σ_1, σ_2) on the worldsheet. Thus, the physics of the string is ultimately encapsulated in the embedding functions $X^\mu(\sigma_1, \sigma_2)$, where $\mu = 0, 1, ..., D - 1$. Here D is the total spacetime dimension, which we shall keep arbitrary for now.

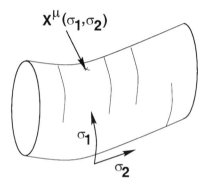

Figure 6: The string worldsheet can be parametrized by two worldsheet coordinates (σ_1, σ_2). Thus, the location in the external spacetime of any point on the string worldsheet is described by a set of functions $X^\mu(\sigma_1, \sigma_2)$. It is convenient to think of σ_1 as a spacelike worldsheet coordinate, and σ_2 as a timelike worldsheet coordinate.

Given these embedding functions, we can attempt to write down an appropriate action for the string. To do this, we first note that as we might expect, strings have *tension* — *i.e.*, strings generically have a non-zero energy per unit length. In other words, it takes energy to stretch a string and to give the worldsheet a larger area. Thus, as the string propagates along in spacetime, we expect on physical grounds that this string should choose a configuration that minimizes the area of the worldsheet. This leads us to identify the string action with the area of the corresponding worldsheet. Indeed, this results in the so-called *Nambu-Goto action*, which involves a non-trivial square root of the X^μ coordinates. For certain calculational purposes, however, this square root is often problematic. Fortunately, however, there exists an alternative action, the so-called *Polyakov action*, which is classically equivalent to the Nambu-Goto action but which does not involve fractional powers of the X

coordinates. This action is given by

$$S = -\frac{1}{4\pi\alpha'} \int d^2\sigma \sqrt{h}\, h^{\alpha\beta}\, g_{\mu\nu}\, \partial_\alpha X^\mu \partial_\beta X^\nu \ . \tag{2.1}$$

Here $g_{\mu\nu}$ is the metric of the external spacetime, $h_{\alpha\beta}$ is the metric of the worldsheet, the worldsheet derivative is given by $\partial_\alpha \equiv \partial/\partial\sigma^\alpha$, and $h \equiv \det h_{\alpha\beta}$. In the prefactor, α' is a dimensionful constant (called the *Regge slope*) with units of (length)^2. Since these units are equivalent to length/energy, we see that α' is an inverse tension, and indeed the string tension T turns out to be related to α' via $T = (2\pi\alpha')^{-1}$. We shall discuss the numerical value of α' below. Note that the action (2.1) is manifestly spacetime Lorentz-invariant.

Before proceeding further, it may be useful to draw an analogy between this action and the analogous action for a *point* particle propagating through spacetime and sweeping out a *worldline* rather than a worldsheet. The worldline can be parametrized by a single coordinate σ, which functions as a proper time along the worldline. The point-particle action can then be written in the form

$$S_{\text{point particle}} = \frac{1}{2} \int d\sigma \left(e^{-1} g_{\mu\nu}\partial_\sigma X^\mu \partial_\sigma X^\nu - e\hat{m}^2\right) \tag{2.2}$$

where $\hat{m}$ is the mass of the point particle and where $e(\sigma)$ is an auxiliary field (a so-called *einbein*). Solving for $e(\sigma)$ through its equation of motion and substituting back into (2.2) yields an action proportional to the length of the worldline and involving a square root. Thus, we see that the string action (2.1) is nothing but the generalization of the point-particle action (2.2), where we have associated

$$e^{-1}(\sigma) \iff h^{\alpha\beta}(\sigma_1, \sigma_2) \ , \qquad \hat{m} = 0 \ . \tag{2.3}$$

In other words, the string action (2.1) is the two-dimensional generalization of the action of a *massless* point particle, where the worldsheet metric functions as an auxiliary field (a "zweibein"). This masslessness property will be crucial shortly.

It is now possible to make some simplifications. Perhaps the most obvious is to restrict our attention to a flat spacetime and take $g_{\mu\nu} = \eta_{\mu\nu}$. We shall do this throughout these lectures. A much more subtle simplification, however, is to simplify the worldsheet metric. Let us therefore pause to discuss how this can be done.

One of the first things we realize is that the ultimate physics of the string should not depend on the particular choice of coordinate system (σ_1, σ_2) on the string worldsheet. After all, on purely physical grounds, we know that the particular choice of worldsheet coordinate system cannot have a physical effect, for the same worldsheet geometry can ultimately be described using an infinite variety of coordinate systems which differ from each other through relative reparametrizations or rescalings. (Indeed, in the point-particle case, we are likewise free to reparametrize our proper-time variable along the particle worldline.) Therefore, the string action should have a symmetry that makes it invariant under reparametrizations and rescalings of the worldsheet coordinates. Note, in particular, that the invariance under *rescalings* follows from the fact that we chose our string action (2.1) to generalize that of a *massless* point particle. In other words, we have taken $\hat{m} = 0$ in (2.3). While it is possible to add terms to the action of the bosonic string which mimic the effects of possible mass terms and which explicitly break the scale invariance of the bosonic string, we shall not need to consider such theories in these lectures.

The symmetry that comprises both reparametrizations and rescalings of the worldsheet coordinates is called *conformal symmetry*, and the bosonic string action (2.1) is thus said to be "conformally invariant". Clearly, this symmetry must hold not only at the classical level, but also at the quantum level, for we would not have a consistent theory if this symmetry were broken by quantum anomalies. Conformal invariance of the action is a very powerful physical tool which will play an important role throughout these lectures, and indeed the mathematical structure underlying conformal symmetry and its implications is a deep and beautiful subject which we will not have time or space to discuss here. A recommended starting point is *Applied Conformal Field Theory* (Proceedings of Les Houches, Session XLIX, 1988), by P. Ginsparg. Therefore, in order to proceed, we will have to make the first of many "great leaps", and take certain results on faith. Our first great leap will therefore be the following:

Great Leap #1: Conformal invariance of the string action allows us to replace the string metric $h_{\alpha\beta}$ with the two-dimensional Minkowski metric $\eta_{\alpha\beta}$ without loss of generality.

This then results in the simplified bosonic string action

$$S = -\frac{1}{4\pi\alpha'} \int d^2\sigma \, \partial_\alpha X^\mu \partial^\alpha X_\mu \, , \qquad (2.4)$$

Looking at the action (2.4), we see that it has two possible interpretations. The first interpretation is the one that we have already been following: minimizing this action is classically equivalent to minimizing the worldsheet area. This follows directly from the interpretation of $X^\mu(\sigma_1, \sigma_2)$ as the *spacetime coordinates* of a given worldsheet position (σ_1, σ_2). Note that this action is invariant under $SO(D-1,1)$ Lorentz transformations of the spacetime coordinates, with the index μ interpreted as a spacetime vector index relative to the Lorentz group. We shall refer to this as the *spacetime interpretation*.

There is, however, a completely different interpretation of (2.4): this is the action of a *two-dimensional quantum field theory* where the two dimensions refer to the worldsheet coordinates and where the "fields" are nothing but the functions $X^\mu(\sigma_1, \sigma_2)$, $\mu = 0, 1, ..., D-1$. Indeed, we see that these spacetime coordinate functions are simply a collection of D different massless bosonic Klein-Gordon fields which happen to exhibit an internal $SO(D-1,1)$ rotation symmetry (analogous to a gauge symmetry) between them. In such a case, the index μ is simply an internal symmetry index which tells us that the X^μ fields transform as vectors with respect to the internal $SO(D-1,1)$ symmetry. We shall refer to this as the *worldsheet interpretation*. Indeed, it is because this string action contains only bosonic worldsheet fields that we call this the *bosonic string*. In such a description, spacetime is not a fundamental concept but rather a "derived" concept: it results from the interpretation of various worldsheet fields as spacetime coordinates, and from the interpretation of an internal symmetry as a spacetime Lorentz symmetry. It is indeed remarkable that such different interpretations can be made of the same physics, and we shall often go back and forth between these different worldsheet and spacetime points of view.

Given these two descriptions of the action, we can also understand the origin of the Regge slope parameter α' on dimensional grounds. Let us first take the worldsheet point of view, so that our length dimensions are determined with respect to the coordinates (σ_1, σ_2). In such a case, we know that the ordinary Klein-Gordon action does not require any dimensionful prefactor, for $\int d^2\sigma (\partial_\alpha X^\mu)^2$ is indeed dimensionless when the Klein-Gordon field X^μ is itself dimensionless. However, from the spacetime point of view, we see that X^μ cannot be dimensionless, for we ultimately need to interpret this field as a spacetime coordinate with units of length. Thus, we are forced to compensate by inserting a dimensionful prefactor α' in front of the action. In other words, *the need for the dimensionful prefactor α' arises from the need to interpret our dimensionless (scale-free) worldsheet theory as a dimensionful (spacetime) theory*. Or, to put it slightly differently, the parameter α' is the dimensionful conversion factor that describes the overall scale of the embedding of the dimensionless worldsheet physics into the dimensionful spacetime. We shall see this phenomenon very often throughout these lectures: the worldsheet physics is by itself scale-invariant (since it generalizes the physics of a massless point particle with $\hat{m} = 0$), and it is only in the conversion to dimensionful *spacetime* quantities that the overall scale α' plays a role. Thus, α' sets the overall spacetime mass scale of string theory, often called the *string scale*:

$$M_{\text{string}} \equiv \frac{1}{\sqrt{\alpha'}}. \qquad (2.5)$$

A priori, this mass scale is unfixed, but we shall see shortly how this scale is ultimately determined.

Now that we have established the worldsheet picture and the spacetime picture, it is easy to see how they are related to each other: each quantum excitation of the Klein-Gordon worldsheet fields X^μ corresponds to a different particle in spacetime. Thus, the study of string theory can be reduced to the study of a two-dimensional quantum field theory! For example, particle scattering amplitudes in spacetime can be re-interpreted as the correlation functions of our two-dimensional worldsheet fields, evaluated on various two-dimensional manifolds. Of course, as we have stated above, this is not just *any* two-dimensional quantum field theory, for physical consistency also requires the presence of conformal symmetry. Thus, from this point of view, string theory is the study of two-dimensional *conformal* field theories. In two dimensions, it turns out conformal symmetry is extremely powerful, for it gives rise to an infinite number of conserved currents. Indeed, two-dimensional conformal symmetry is often sufficiently powerful to permit the *exact* evaluation for many scattering amplitudes.

In the case in question, the particular conformal field theory that concerns us is that of D free massless bosonic fields X^μ, $\mu = 0, 1, ..., D-1$. However, just as with any symmetry, there is always the danger of quantum anomalies. Nevertheless, it is straightforward to show that

> **Great Leap #2:** Conformal invariance of the string action is preserved at the quantum level (*i.e.*, all quantum anomalies are cancelled) if and only if $D = 26$.

This is clearly a big result, and we will not have space to provide a proper mathematical derivation of this fact. At the very least, however, we can give a guide as to the most useful way of thinking about this result. Note that our D bosonic fields are identical to each other and essentially decoupled from each other. Therefore, each contributes the same amount to any potential anomaly. This amount is called the *central charge*, and the central charge c of each bosonic field X will be denoted c_X. It turns out that $c_X = 1$, and therefore the total central charge from the D bosonic fields is $c_{\text{fields}} = D$. However, it can be shown that there also exists a "background" central charge (*i.e.*, a background quantum anomaly) of magnitude $c_{\text{background}} = -26$. Thus, the total anomaly is cancelled only if $D = 26$. Clearly, the most mysterious part of this discussion is the origin of this "background" central charge. In technical terms, it reflects the contributions of the conformal ghosts that arose when we used the conformal symmetry to set (or "gauge-fix") the worldsheet metric $h_{\alpha\beta} \to \eta_{\alpha\beta}$. However, all we will need to know for the future is that the value of the "background" anomaly $c_{\text{background}}$ depends on only the particular symmetry of the worldsheet action that we are dealing with. In the present case, this worldsheet symmetry is simply conformal invariance, and the corresponding background central charge corresponding to conformal invariance is $c_{\text{background}} = -26$. Therefore, we see that the total conformal anomaly is cancelled only if $D = 26$. This is typically called the *critical dimension* of the bosonic string.

We see, then, that string theory is able to determine the spacetime dimension as the result of an *anomaly cancellation argument*! It is worth reflecting on how this happened by considering an analogous situation in field theory, namely the cancellation of the triangle axial anomaly. We know that this anomaly is cancelled only for very particular combinations of particle representations (*e.g.*, we require complete generations of Standard-Model fields, with three colors of quark for every lepton). So we are used to the idea that anomalies are extremely sensitive to the field content of the theory. In string theory, however, we have seen that the analogous worldsheet field content is parametrized by the spacetime dimension. More worldsheeet fields correspond to more spacetime dimensions. Therefore, just as triangle anomaly cancellation requires three colors, conformal anomaly cancellation requires 26 dimensions.

Of course, our world does not consist of 26 flat spacetime dimensions, and we shall ultimately need to find a way of reducing this to a four-dimensional theory. We will discuss various ways of doing this in Lecture #8 when we discuss the more realistic heterotic string. For now, however, we can just think of the present bosonic string as a 26-dimensional toy model.

2.2 Quantizing the bosonic string

Let us now quantize this theory. Having already noted that the action (2.4) is nothing but the action of a set of 26 Klein-Gordon fields X^μ, we already know how to proceed: in the usual fashion, we introduce a Fourier-expansion of the fields X^μ, and interpret the coefficients of this expansion as creation and annihilation operators obeying canonical quantization relations.

Because we ultimately wish to interpret the fields X^μ as spacetime coordinates, we must first impose the constraint

$$X^\mu(\sigma_1 + \pi, \sigma_2) = X^\mu(\sigma_1, \sigma_2) \qquad (2.6)$$

where we have chosen to normalize the length of the closed string as π. In other words, the spacetime coordinates must be single-valued as we make one complete circuit around the closed string. This is the first place where we have essentially incorporated the requirement that we are dealing with closed strings whose topology is that of a circle. Moreover, because of this topology (and because of the linear nature of the wave equation resulting from the action (2.4)), we know that we can also decompose any possible quantum excitation of the wiggling string into a superposition of modes that travel clockwise around the string (in the direction of, say, decreasing σ_1) and those that travel counter-clockwise (in the direction of increasing σ_1). These are respectively called *left-movers* and *right-movers*. We can therefore decompose each of our Klein-Gordon fields into the form

$$X^\mu(\sigma_1, \sigma_2) = X_L^\mu(\sigma_1 + \sigma_2) + X_R^\mu(\sigma_1 - \sigma_2) . \tag{2.7}$$

The most general mode-expansion consistent with the boundary condition (2.6) is then

$$X^\mu(\sigma_1, \sigma_2) = x^\mu + \ell^2 p^\mu \sigma_2 + \frac{i}{2}\ell \sum_{n \neq 0} \left[\frac{\alpha_n^\mu}{n} e^{-2in(\sigma_1 + \sigma_2)} + \frac{\tilde{\alpha}_n^\mu}{n} e^{+2in(\sigma_1 - \sigma_2)} \right] ,$$
$$\tag{2.8}$$

which decomposes into

$$X_L^\mu(\sigma_1 + \sigma_2) = \tfrac{1}{2}x^\mu + \frac{\ell^2}{2}p^\mu(\sigma_1 + \sigma_2) + \frac{i}{2}\ell \sum_{n \neq 0} \frac{\alpha_n^\mu}{n} e^{-2in(\sigma_1 + \sigma_2)}$$

$$X_R^\mu(\sigma_1 - \sigma_2) = \tfrac{1}{2}x^\mu - \frac{\ell^2}{2}p^\mu(\sigma_1 - \sigma_2) + \frac{i}{2}\ell \sum_{n \neq 0} \frac{\tilde{\alpha}_n^\mu}{n} e^{+2in(\sigma_1 - \sigma_2)} \tag{2.9}$$

Here $\ell \equiv \sqrt{2\alpha'}$ is a fundamental length that has been inserted on dimensional grounds.

It is easy to interpet the different terms in (2.8) and (2.9). Clearly the final terms in each line represent the internal quantum vibrational oscillations of the string, where α_n^μ and $\tilde{\alpha}_n^\mu$ are the left-moving and right-moving creation/annihilation operators corresponding to vibrational modes of a given frequency n. We shall discuss these operators shortly. Note that the contribution from the "zero-mode" has been separated out and written explicitly in the form $x^\mu \pm \frac{1}{2}\ell^2(\sigma_1 \pm \sigma_2)$ for the left- and right-movers respectively. In the case when there are no quantum excitations (so that we can ignore the final exponential terms), these "zero-modes" are all that remain of the mode-expansion, whereupon we see from (2.8) that the total X^μ field takes the form

$X^\mu = x^\mu + \ell^2 p^\mu \sigma_2$. Interpreting σ_2 as the timelike coordinate on the string worldsheet, we thus see that x^μ is nothing but the center-of-mass position of the string, and p^μ its center-of-mass momentum.

Let us now consider the quantization rules that we must impose. The first one (for the zero-modes) is easy: we simply impose the usual commutation relation $[x^\mu, p^\nu] = i\hbar\eta^{\mu\nu}$. We shall henceforth set $\hbar = 1$. The excited modes also have a similar commutation relation. First, note that because the X fields are interpreted as spacetime coordinates, they are necessarily *real*. This implies that we must identify $\alpha^\mu_{-n} = (\alpha^\mu_n)^\dagger$, with a similar result for the right-moving oscillator modes. In other words, the negative modes *create* excitations, while the positive modes *annihilate* the same excitations. Given this, we then can immediately write down the commutation relation for the creation/annihilation operators:

$$[\alpha^\mu_m, \alpha^\nu_n] = m\,\delta_{m+n}\,\eta^{\mu\nu}\,, \qquad [\tilde{\alpha}^\mu_m, \tilde{\alpha}^\nu_n] = m\,\delta_{m+n}\,\eta^{\mu\nu}\,. \qquad (2.10)$$

Here we have introduced the notation $\delta_x = \delta_{x,0} \equiv 1$ if $x = 0$, and $\equiv 0$ if $x \neq 0$. Note that these are exactly the harmonic oscillator commutation relations, except that we have rescaled each mode α_n by its corresponding frequency n in (2.9). Thus, $a_n \equiv \alpha_n/\sqrt{n}$ obey the usual harmonic oscillator commutation relations. This rescaling has become conventional in string theory, and we shall retain it here. Likewise, it is often conventional to define the zero-mode $\alpha^\mu_0 = \frac{1}{2}\sqrt{\alpha'}p^\mu$.

Given this mode-expansion, we can now construct the corresponding number operators

$$n > 0: \qquad N_n = \frac{1}{n}\alpha^\mu_{-n}\alpha_{n\mu}\,, \qquad \tilde{N}_n = \frac{1}{n}\tilde{\alpha}^\mu_{-n}\tilde{\alpha}_{n\mu} \qquad (2.11)$$

which count the number of excitations of the nth frequency modes of the string. Once again, this is completely analogous to the harmonic-oscillator creation/annihilation modes, after we take into account the rescaling $\alpha_n \equiv \sqrt{n}a_n$ and the hermiticity condition $\alpha_{-n} = \alpha^\dagger_n$.

Likewise, we can also write down the total *energy* of the system. To do this, let us consider the different contributions to the total energy. First, there is the energy associated with the internal quantum vibrational oscillations of the string. As we might expect, this is given by

$$L^{(\text{osc})}_0 \equiv \sum_{n=1}^{\infty} n N_n = \sum_{n=1}^{\infty} \alpha^\mu_{-n}\alpha_{n\mu}$$

$$\bar{L}^{(\text{osc})}_0 \equiv \sum_{n=1}^{\infty} n \tilde{N}_n = \sum_{n=1}^{\infty} \tilde{\alpha}^\mu_{-n}\tilde{\alpha}_{n\mu}\,. \qquad (2.12)$$

For convenience, we are defining these energy operators in such a way that they are dimensionless numbers (*i.e.*, they are *worldsheet* energies). These L_0 operators are often called *Virasoro generators*, which are more generally defined $L_m \equiv \sum_n \alpha^\mu_{m-n} \alpha_{n\mu}$. These generators are nothing but the different frequency modes of the total worldsheet stress-energy tensor, and together they satisfy the so-called *Virasoro algebra*. We shall only consider L_0 in these lectures.

Next, there is the energy of the zero-modes, which correspond to the net center-of-mass motion of the string. This is given by

$$L_0^{(\text{com})} \equiv \alpha_0^\mu \alpha_{0\mu} = \frac{\alpha'}{4} p^\mu p_\mu$$

$$\bar{L}_0^{(\text{com})} \equiv \tilde{\alpha}_0^\mu \tilde{\alpha}_{0\mu} = \frac{\alpha'}{4} p^\mu p_\mu . \tag{2.13}$$

Note that factors of α' must appear in order to counter-balance the fact that the center-of-mass momentum p^μ is a spacetime quantity, and hence dimensionful.

Finally, there is the possibility of an overall non-zero vacuum energy for both the left-movers and the right-movers. In other words, there is no reason to assume that the vacuum state (the state without any excitations) is exactly at zero energy. This is important, of course, since string theory is ultimately a theory which will contain gravity, and it is precisely in theories containing gravity that the overall zero of energy becomes important. Indeed, mathematically, one can imagine that due to the commutation relations (2.10), there can be an overall normal-ordering ambiguity in the definitions in (2.12), and this overall normal-ordering constant would be our "vacuum energy".

Thus, denoting the left- and right-moving vacuum energies as $a_{L,R}$, we have the total left- and right-moving energies

$$H \equiv L_0^{(\text{com})} + L_0^{(\text{osc})} + a_L , \qquad \bar{H} \equiv \bar{L}_0^{(\text{com})} + \bar{L}_0^{(\text{osc})} + a_R . \tag{2.14}$$

These are the total worldsheet Hamiltonians.

Clearly, the important thing to do at this stage is to determine the vacuum energies $a_{L,R}$. Of course, the symmetry between left-movers and right-movers requires $a_L = a_R$. Calculating this vacuum energy can be done in numerous ways, each of which would take too much space for our purposes. Once again, we refer the reader to Chapter 2 of GSW, where a full calculation is given. Therefore, it is time for another

Great Leap #3: Conformal invariance of the string action implies that $a_L = a_R = -1$.

Finally, in order to determine the total *spacetime mass* of a given string state, we must have a *mass-shell condition* for the string. Rather than provide a rigorous derivation (for which we again refer the curious reader to GSW), we can instead give an intuitive argument which suggests the proper answer. In a quantum field theory of point particles, the mass $\hat{m}$ is a parameter that appears in the Lagrangian through an explicit mass term that might be generated in some separate manner, *e.g.*, through the Higgs mechanism. Since a point particle has no internal degrees of freedom beyond those associated with its center-of-mass motion, such a mass parameter $\hat{m}$ would then be directly identified with M, the resulting physical mass of the particle. Such a physical mass M is the quantity satisfying the condition $p^\mu p_\mu = -M^2$, or equivalently the condition $L_0^{(\mathrm{com})} = \bar{L}_0^{(\mathrm{com})} = -\alpha' M^2/4$. In the special case of a massless particle (for which $\hat{m} = M = 0$), this mass-shell condition then takes the simple form $L_0^{(\mathrm{com})} = \bar{L}_0^{(\mathrm{com})} = 0$.

A similar condition emerges in string theory. We have already seen that our string action (2.1) generalizes that of a massless particle, which again suggests that our effective Lagrangian mass parameter $\hat{m}$ vanishes. Indeed, as we have discussed, this is the root of the scale invariance of the string action (2.1). However, unlike the point-particle case, a string *does* have additional, purely internal degrees of freedom — these are the oscillations of the string itself, whose additional energy contributions are represented by $L_0^{(\mathrm{osc})}$, $\bar{L}_0^{(\mathrm{osc})}$, and $a_{L,R}$. Thus, even though $\hat{m} = 0$, the resulting string state can still have a non-zero physical mass M in spacetime. Indeed, just as the mass-shell condition for massless point particles is given by $L_0^{(\mathrm{com})} = \bar{L}_0^{(\mathrm{com})} = 0$, the mass-shell condition for our scale-invariant string is generalized to $H = \bar{H} = 0$. This then becomes our scale-invariant mass-shell condition in string theory. Of course, spacetime Lorentz invariance still allows us to identify the physical spacetime mass M of a given string state via the relations $L_0^{(\mathrm{com})} = \bar{L}_0^{(\mathrm{com})} = -\alpha' M^2/4$. Thus, the string mass-shell conditions $H = \bar{H} = 0$ lead to the identifications

$$\frac{1}{4}\alpha' M^2 = L_0^{(\mathrm{osc})} - 1 \,, \qquad \frac{1}{4}\alpha' M^2 = \bar{L}_0^{(\mathrm{osc})} - 1 \,. \qquad (2.15)$$

Note that these two conditions can also be written in the form

$$\alpha' M^2 = 2\left(L_0^{(\mathrm{osc})} + \bar{L}_0^{(\mathrm{osc})} - 2\right) \qquad (2.16)$$

where we must obey the constraint

$$L_0^{(\mathrm{osc})} = \bar{L}_0^{(\mathrm{osc})} \,. \qquad (2.17)$$

Interpreting the conditions (2.16) and (2.17) is easy. The condition (2.16) simply tells us that the physical spacetime mass M of a given string state (and thus the square of its center-of-mass momentum) is generated *solely* from its internal left- and right-moving vibrational excitations. The condition (2.17), by contrast, tells us that the mass of the string must come *equally* from left-moving and right-moving excitations. The latter condition (2.17) is often referred to as the *level-matching condition*, since it implies that a given string oscillator state is considered to be "on shell" (or "physical") only if the total excitation level of the left-movers matches the total excitation level of the right-movers. This condition implies that the string does not have an unbalanced "wobbling", for if such a wobbling existed, it could ultimately be used to determine a preferred coordinate system on the worldsheet (thereby breaking conformal invariance). Indeed, demanding invariance under shifts in the σ_1 variable leads directly to the condition (2.17). We remark, however, that states not satisfying (2.17) are nevertheless important for understanding the "off-shell" or "virtual" structure of string theory. Such "virtual" states contribute, for example, within loop amplitudes. In these lectures, however, we shall focus on only the so-called "tree-level" string spectrum for which the level-matching constraint (2.17) is imposed and the corresponding physical masses are given by (2.16).

2.3 The spectrum of the bosonic string

Having discussed the quantization of the bosonic string, we can now examine its spectrum. The procedure is simple: we simply consider all possible combinations of left- and right-moving mode excitations of the string worldsheet, subject to the level-matching constraint (2.17), and then we tensor these left- and right-moving states together to form the total resulting string state. The spacetime mass of this string state is then given by (2.16), and the properties of the state are deduced directly from the underlying vibrational configuration of the string.

The simplest state, of course, is the string vacuum state

$$|0\rangle_R \otimes |0\rangle_L \qquad (2.18)$$

in which the right- and left moving vacuum states are tensored together. This state trivially satisfies (2.17), which indicates that this state is indeed part of the physical string spectrum. Unfortunately, we see from (2.16) that this state has a negative squared mass — *i.e.*, the spacetime mass of this state is imaginary! This state is thus a *tachyon*. Making sense of this string state is problematic, and is one of the reasons that we shall not ultimately be interested in the bosonic string.

Let us continue, however. The first excited string state is

$$\tilde{\alpha}^\mu_{-1}|0\rangle_R \otimes \alpha^\nu_{-1}|0\rangle_L \ . \tag{2.19}$$

This state has $L_0^{(\text{osc})} = \bar{L}_0^{(\text{osc})} = 1$, and according to (2.16) is therefore massless. As evident from its Lorentz index structure, this state transforms under the spacetime Lorentz group as the tensor product of two spin-one Lorentz vectors. We can therefore decompose this tensor product into a spin-two state (the symmetric traceless component), a spin-one state (the antisymmetric component), and a spin-zero state (the trace). Mathematically, this is equivalent to the tensor-product rule for Lorentz $SO(8)$ vector representations:

$$\mathbf{V}_8 \otimes \mathbf{V}_8 \ = \ \mathbf{1} \ \oplus \ \mathbf{28} \ \oplus \ \mathbf{35} \tag{2.20}$$

where $\mathbf{V}_8$ is the eight-dimensional vector representation, and where the $\mathbf{1}$ representation is the spin-zero state, the $\mathbf{28}$ representation is the spin-one state, and the $\mathbf{35}$ representation is the spin-two state.

How can we interpret these states? A massless spin-two state must, by Lorentz invariance, have equations of motion which are equivalent to the Einstein field equations of general relativity. Thus, we are forced to identify the spin-two (traceless symmetric) component of the state (2.19) as the *graviton* $g_{\mu\nu}$, which is the spin-two mediator of the gravitational interactions. The spin-one (antisymmetric) state within (2.19) is an antisymmetric tensor field, often denoted $B_{\mu\nu}$, and the spin-zero (trace) component is the so-called *dilaton*, denoted ϕ. Together, $(g_{\mu\nu}, B_{\mu\nu}, \phi)$ are called the *gravity multiplet*.

By identifying (2.19) with the gravity multiplet, we see that string theory becomes a theory that contains gravity! This in turn allows us to determine the value of our previously unfixed mass scale α'. We shall now sketch how this happens (with details available in GSW). It turns out that if one calculates loop amplitudes in string theory, one finds that $e^{-\phi}$ serves as a loop expansion parameter (*i.e.*, higher-loop amplitudes come multiplied by more powers of $e^{-\phi}$). Given this observation, it is natural to identify the string coupling constant as the vacuum expectation value of the dilaton:

$$g_{\text{string}} \ = \ e^{-\langle\phi\rangle} \ . \tag{2.21}$$

This string coupling constant describes the strength of string interactions. Given this definition, we then find that the graviton state couples to matter with the expected gravitational strength only if we choose

$$\alpha' \ = \ \frac{G_{\text{Newton}}}{g_{\text{string}}^2} \tag{2.22}$$

where G_{Newton} is Newton's constant. Substituting this result into (2.5), we then find

$$M_{\text{string}} = g_{\text{string}} M_{\text{Planck}} , \qquad (2.23)$$

where $M_{\text{Planck}} \equiv 1/\sqrt{G_{\text{Newton}}}$. Thus, because it contains gravity, string theory becomes a theory whose fundamental mass scale is related to the Planck scale.

We can also construct more and more massive string states. Ultimately, these fill out an infinite tower of string states. It is clear that such additional states all have $\alpha' M^2 > 0$. Given the above value for α', this implies that these additional states all have Planck-scale masses. Such Planck-scale excited states are therefore not of direct relevance for string phenomenology. Let us note, however, one interesting fact about these states. For any given spacetime mass level M, the string state with maximum spin is achieved by exciting only the lowest vibrational modes α^{μ}_{-1} and $\tilde{\alpha}^{\mu}_{-1}$. We thus find that for a given spacetime mass M, the maximum spin J_{max} that can be realized is

$$\alpha' M^2 = 2J_{\text{max}} - 4 . \qquad (2.24)$$

For example, we see that the maximum spin that can be realized for a massless state is $J = 2$ (the graviton). The relation (2.24) was originally observed for hadron resonances, and historically gave rise to the so-called "dual resonance models" (which eventually became modern string theory). In such dual resonance models, the relation (2.24) describes a so-called "Regge trajectory", with α' serving as the so-called "Regge slope". It is for this reason that in modern string theory, we continue to refer to α' as the Regge slope.

Before concluding, let us briefly mention one further important issue. In ordinary four-dimensional quantum field theory, we know that a massless spin-one state (e.g., a photon) naïvely has four distinct states (corresponding to the four components of a vector field A^{μ}). However, the underlying gauge invariance allows us to make a unitary gauge choice wherein only two of these states (the two helicity states) are truly physical. The timelike and longitudinal states decouple, leaving only the transverse components. In the above description of the string spectrum, however, we have taken a covariant approach analogous to the description of a photon as a four-component vector. One might then wonder which of these states are truly physical. This issue is an important one in string theory, and once again we cannot here provide a proper proof. We shall therefore make recourse to another

Great Leap #4: The physical string states are those which are realized by exciting the oscillator modes of only the *transverse* coordinates X^i $(i = 1, ..., 24)$.

Proving this statement requires showing that even after we have used conformal invariance to set the string worldsheet metric to $\eta_{\alpha\beta}$, there still remains sufficient freedom to make a further "gauge" choice wherein we set the oscillator modes of the timelike and longitudinal spacetime coordinates to zero. This gauge choice, which is called *light-cone gauge*, is thus the analogue of unitary gauge in quantum field theory, and essentially tells us that only the 24 transverse coordinates correspond to physical degrees of freedom in the string worldsheet action. An important by-product of this fact is that every remaining string state has a non-negative norm. This is non-trivial. For example, if our metric signature is chosen such that $\eta^{00} = -1$, then the state $\alpha_{-n}^{\mu=0}|0\rangle$ has a negative norm. However, one can demonstrate that in light-cone gauge all resulting states are physical and have non-negative norm.

2.4 Summary

Let us quickly review those features of the bosonic string that we shall need to bear in mind in subsequent lectures. We shall separate these features into worldsheet features and spacetime features.

Worldsheet: The worldsheet fields consist of D copies of the left- and right-moving spacetime coordinates X_L^μ and X_R^μ (the worldsheet bosons). The fact that these X coordinates are periodic as we traverse the closed string loop implies that they have integer modings α_n and $\tilde{\alpha}_n$, where $n \in \mathbb{Z}$. The relevant worldsheet symmetry is conformal invariance, which tells us that the number of these X^μ fields is $D = 26$ and also tells us that the vacuum energy corresponding to these fields is $a_L = a_R = -1$. As we have stated above, a useful way to think about these results is to imagine that there is a "background" conformal anomaly $c_{\text{background}} = -26$, and that each X^μ field makes a contribution $c_X = 1$. In general, the "background" conformal anomaly is only a function of the relevant worldsheet symmetry (in this case conformal invariance), and it will always remain true that $c_X = 1$. Thus, cancellation of the conformal anomaly requires $D = 26$. A similar interpretation can also be given to the vacuum energy. When calculating the vacuum energies, only the physical (*i.e.*, transverse) fields are relevant. It is a general result that each X field contributes $a_X = -1/24$ to the vacuum energy. Therefore, we find $a_L = a_R = 24a_X = -1$.

Spacetime: The above worldsheet theory leads to the following features in spacetime. We find that the *spacetime dimension* (often called the *critical* spacetime dimension) is 26. The spectrum consists of a spinless tachyon, as well as a massless gravity multiplet consisting of the graviton $g_{\mu\nu}$, the antisymmetric tensor $B_{\mu\nu}$, and the dilaton ϕ. There is also an infinite tower of massive (Planck-scale) string states.

Comments: Two remarkable things have happened. First, we have a theory of quantized gravity! The graviton has emerged as the quantum excitation of a closed string. This alone is very exciting, but also somewhat mysterious. We started by assuming a closed string propagating through an external, fixed, flat spacetime. But this string itself includes a graviton mode, which implies a distortion in that background spacetime. This then acts back to change the worldsheet theory. Thus, in some sense, the string itself not only "creates" the spacetime in which it propagates, but is then affected by this change in the spacetime geometry. This coupling or interplay between the string and its spacetime is not fully understood, and is clearly at the heart of the many mysterious features of string theory as a theory of quantum gravity.

A second remarkable thing has also happened, although we have not demonstrated it explicitly. As indicated in (2.21), a coupling constant has been determined *not* as a free parameter, but rather *dynamically* as the vacuum expectation of a string field. It is in this sense that string theory contains no free parameters, and that all parameters such as coupling constants are determined dynamically.

There are, however, a number of drawbacks to this bosonic string theory. First, it contains a tachyonic state. We must somehow find a way to eliminate this. Second, all string excitations are spacetime *bosons* (*i.e.*, they have integer spin). We must find a way to obtain spacetime fermions. Third, there are no massless spin-one states (which we would wish to associate with gauge fields). Thus, there are no gauge symmetries. It is for these reasons that we shall go on to consider more complicated string theories.

And finally, there is another major drawback that we need to be aware of. Although it is compelling that the string coupling g_{string} is in principle determined dynamically, as the vacuum expectation value of the dilaton scalar field, in practice we do not understand how to calculate the potential of the dilaton field and thereby deduce its vacuum expectation value. In the bosonic string we are considering here, the dilaton potential $V(\phi)$ is actually divergent for all $\phi < \infty$, and so this question cannot be meaningfully addressed. However, even in the more realistic string theories to be discussed, this potential is either completely flat (as happens in a supersymmetric context), or generally taken a shape that sends $\langle \phi \rangle \to \infty$. This is the famous *dilaton runaway problem*. Solving this problem is perhaps one of the most important (unsolved) problems in string phenomenology.

How can we remedy these features? One possibility is prompted by the appearance of the tachyon. In ordinary quantum field theory, the existence of a tachyon (a state with a negative mass-squared) signals that the vacuum has been misidentified (as in the Higgs mechanism); the theory then "rolls" to a different vacuum configuration in which the tachyon is eliminated. So it is natural to speculate that perhaps the bosonic string theory also "rolls" to a new vacuum in such a way that the tachyon is no longer present and the dilaton is stabilized. Perhaps fermions and gauge fields might also appear in this new vacuum, as desired. However, as we have already indicated, it is not known how the bosonic string behaves in this context. We do not know if there exists a new ("stable") vacuum to roll to, and if so, what its properties might be. Of course, knowing the potential $V(\phi)$ would be extremely useful, yet as we indicated this potential is naïvely divergent and therefore requires some knowledge of the non-perturbative structure of string theory. So (at least for the time being) this option does not appear promising.

A second possibility, then, is simply to abandon the bosonic string and attempt to construct a new string theory altogether. And this is what we shall now do.

3 Lecture #3: Neutrinos are fermions: The superstring

As we saw in the last lecture, the bosonic string has two glaring failures: it contains a tachyon, and it does not give rise to spacetime fermions. Both of these features are troubling, especially since the announced goal of these lectures is to derive a neutrino from string theory, and we know that the neutrino is a fermionic object. We therefore seek to construct a new string theory which can give rise to excitations with half-integer spins.

3.1 The action

We have already seen that string theories are defined by their two-dimensional worldsheet actions. Thus, in order to construct a new string theory, we must construct a new worldsheet action. At the very least, this action should contain that of the bosonic string, since we still wish to retain the spacetime interpretatation that we had previously. Thus, our only option is to *introduce additional worldsheet fields* into the action:

$$ S = -\frac{1}{4\pi\alpha'} \int d^2\sigma \ (\partial_\alpha X^\mu \partial^\alpha X_\mu + \dots) \ . \tag{3.1} $$

What new fields can we add? If our goal is to produce spacetime fermions, a natural guess would be to add worldsheet fermions! These would complement the worldsheet bosonic fields X^μ that are already present. For the moment, let us denote such fermionic fields schematically as ψ. We would then attempt to consider an action of the form

$$S = -\frac{1}{4\pi\alpha'} \int d^2\sigma \left(\partial_\alpha X^\mu \partial^\alpha X_\mu + \bar\psi i\rho^\alpha \partial_\alpha \psi\right) . \tag{3.2}$$

Here $\psi(\sigma_1, \sigma_2)$ represents our two-dimensional fermionic fields, and ρ^α are an appropriate set of two-dimensional Dirac matrices (the analogues of the γ^μ matrices in four dimensions).

We then face a number of questions. First, how many ψ fields must we add? Second, what kinds of worldsheet fermions should these be? Should they be Dirac fermions, or Majorana fermions, or Majorana-Weyl fermions? Third, how should these two-dimensional spinors ψ transform under the (internal) $SO(D-1,1)$ spacetime Lorentz symmetry? We already know that the X^μ fields, for example, transform as vectors under this symmetry. Note that it is not obvious that the ψ fields should necessarily transform as spinors under $SO(D-1,1)$ and carry a spacetime spinor index. In particular, all we know thus far is that the ψ fields transform as spinors under worldsheet *two-dimensional* Lorentz transformations. This does not *a priori* give us any information about their *spacetime* transformation properties.

There is also another potential worry that appears if we try to add new worldsheet fields. We have already seen in the bosonic string that worldsheet conformal invariance was sufficiently powerful a symmetry to allow us to choose a light-cone gauge and thereby eliminate all negative-norm states. However, the presence of new worldsheet fields implies the existence of new quantum excitation modes in the resulting string spectrum, and some of these new states may also have negative norm. Thus, conformal symmetry may no longer be sufficient (and indeed would not be sufficient) to allow us to eliminate these states as well.

It turns out that all of these questions have a common answer: we can impose an extra symmetry beyond simple worldsheet conformal invariance. Indeed, the extra symmetry that we shall impose is nothing but worldsheet (*i.e.*, two-dimensional) *supersymmetry*. Specifically, we shall require that the ψ fields be the two-dimensional superpartners of the X fields, so that the resulting action has a manifest worldsheet (two-dimensional) supersymmetry.[a] This new theory will be called the *superstring*.

---

[a] We remark that this is only one possible choice, and will ultimately lead us to the so-called Ramond/Neveu-Schwarz (RNS) formalism. Another possible choice would be to demand *spacetime* supersymmetry, and to imagine that the ψ fields are the Grassmann coordinates θ of a super-spacetime. This possibility would then lead to the so-called Green-Schwarz (GS) formalism. It turns out that these two formalisms are ultimately equivalent, however, and both provide suitable descriptions of the resulting superstring theory. This

It is important to stress that this supersymmetry that we will be discussing is *not* the spacetime supersymmetry that might be seen in the next round of accelerator experiments. Instead, this is a *worldsheet* supersymmetry which stems directly from the worldsheet interpretation of the original Polyakov action (2.4), and which relates the worldsheet bosons X to worldsheet fermions ψ via a worldsheet supercurrent J.

Imposing this worldsheet supersymmetry then answers all of the questions we previously raised. How many ψ fields? The answer is D, one for each boson X^μ. What kind of ψ spinor? The answer is a Majorana (two-dimensional) spinor. How does ψ field transform under the $SO(D-1,1)$ spacetime Lorentz symmetry? The answer is that the ψ field must transform as a *vector* under the Lorentz symmetry, since the X^μ field (for which it is the worldsheet superpartner) also transforms as a vector. In other words, the *worldsheet* supersymmetry commutes with the *spacetime* Lorentz symmetry, and thus does not change the Lorentz index structure. Thus, the ψ fields transform as spacetime vectors, and carry a spacetime vector index: $\psi^\mu(\sigma_1, \sigma_2)$.

This last point may initially seem confusing, so we reiterate: the ψ fields are worldsheet fermions, but spacetime bosons! They transform as spinors under worldsheet Lorentz transformations, but as vectors under the spacetime Lorentz transformations.

Given this, we can now explicitly write down the superstring action:

$$S = -\frac{1}{4\pi\alpha'} \int d^2\sigma \left(\partial_\alpha X^\mu \partial^\alpha X_\mu - i\bar{\psi}_\mu \rho^\alpha \partial_\alpha \psi^\mu \right) . \tag{3.3}$$

Our worldsheet fields are $X^\mu(\sigma_1, \sigma_2)$ and $\psi^\mu(\sigma_1, \sigma_2)$, and the μ index (with $\mu = 0, 1, 2, ..., D-1$) is a vector index with respect to the internal symmetry $SO(D-1,1)$. From the worldsheet perspective, each X^μ is a scalar field (containing one component), while each ψ^μ is a two-component spinor. The ρ^α are two-dimensional Dirac matrices satisfying the two-dimensional Clifford algebra $\{\rho^\alpha, \rho^\beta\} = -2\eta^{\alpha\beta}$, and $\bar{\psi} \equiv \psi^\dagger \rho^0$. One can then show that the action (3.3) is invariant under the worldsheet supersymmetry transformations

---

equivalence is possible because the RNS superstring ultimately also has spacetime supersymmetry (as we shall discover below). In these lectures, however, we shall restrict our attention to the RNS formulation in which the ψ fields are *worldsheet* (rather than spacetime) superpartners of the X^μ fields. Aside from being more useful for string phenomenology, the RNS formalism has the philosophical advantage that it treats the *string* as the fundamental object, with the spacetime structure emerging as a *derived consequence*. The RNS formalism thus reinforces one of the central themes of these lectures, namely that we define a string theory by its worldsheet properties alone, and then deduce the spacetime effects of these properties as consequences. The GS formalism, on the other hand, has the benefit of being manifestly spacetime supersymmetric from the very beginning.

$\delta X^\mu = \bar{\epsilon}\psi^\mu$, $\delta\psi^\mu = -i\rho^\alpha\epsilon\partial_\alpha X^\mu$, where ϵ is a constant anticommuting spinor that parametrizes the "magnitude" of the supersymmetry transformation. The corresponding generator of this worldsheet supersymmetry transformation is the worldsheet supercurrent $J_\alpha = \frac{1}{2}\rho^\beta\rho_\alpha\psi^\mu\partial_\beta X_\mu$.

It is convenient to choose a particular Weyl (chiral) representation for the two-dimensional ρ^α matrices:

$$\rho^0 = \begin{pmatrix} 0 & -i \\ i & 0 \end{pmatrix}, \quad \rho^1 = \begin{pmatrix} 0 & i \\ i & 0 \end{pmatrix} \quad \Longrightarrow \quad \rho^0\rho^1 = \begin{pmatrix} 1 & 0 \\ 0 & -1 \end{pmatrix}. \quad (3.4)$$

Here the product $\rho^0\rho^1$ plays the role of the chirality operator (the analogue of γ_5 in four dimensions), and thus in this basis we can identify the upper and lower components of the two-component Majorana spinor ψ as being left-moving and right-moving respectively. Our worldsheet action (3.3) then decomposes into the form

$$S = -\frac{1}{4\pi\alpha'} \int d^2\sigma \; (\partial_\alpha X^\mu \partial^\alpha X_\mu - \psi_{\mu R}\partial_-\psi_R^\mu - \psi_{\mu L}\partial_+\psi_L^\mu) \quad (3.5)$$

where $\partial_\pm$ are derivatives with respect to the left- and right-moving worldsheet coordinates $\sigma_1 \pm \sigma_2$. The worldsheet content of this theory therefore consists of D left-moving worldsheet bosons X_L, D right-moving worldsheet bosons X_R, D left-moving worldsheet Majorana-Weyl (one-component) fermions ψ_L, and D right-moving worldsheet Majorana-Weyl (one-component) fermions ψ_R. There are two worldsheet supercurrents in this theory:

$$J_L = \psi_{\mu L} \, \partial_+ X_L^\mu \,, \qquad J_R = \psi_{\mu R} \, \partial_- X_R^\mu \,. \quad (3.6)$$

Note that our original goal in constructing the superstring had been to obtain spacetime fermions. However, it may seem from the above that we have failed in this regard, since we have only introduced new fields ψ which themselves are spacetime vectors. How then are we to obtain spacetime fermions? It turns out that this will happen in a surprising way.

Let us proceed to analyze this string following the same steps as we used for the bosonic string. First, we see that our worldsheet symmetry has been enlarged: rather than simply have conformal invariance, we now have conformal invariance *plus* worldsheet supersymmetry. Together, this is called *superconformal invariance*, which is a much larger symmetry than conformal invariance alone.

This enlargement of the worldsheet symmetry changes many of the features of the resulting string. The most profound is the value of the spacetime dimension D. Recall from our discussion of the bosonic string that associated with each worldsheet symmetry there is a particular "background" conformal (central charge) anomaly, and that it is necessary to choose a sufficient number of worldsheet fields so as to cancel this anomaly and ensure that conformal invariance is maintained even at the quantum level. The same argument applies here as well, except that

> **Great Leap #5:** The "background" conformal anomaly associated with *superconformal* invariance is not $c = -26$ but rather $c = -15$. Likewise, the conformal anomaly contribution from each worldsheet Majorana fermion is $c = 1/2$.

We can understand the origin of the "background" conformal anomaly $c = -15$ as follows. Just as in the bosonic string, a certain contribution $c = -26$ is attributable to the conformal ghosts resulting from conformal gauge fixing. The new feature here is that we now have an additional contribution $+11$ which is attributable to the worldsheet *superpartners* of these ghosts. Together, this produces a background anomaly $c = -15$. What this means is that we must choose the number D of worldsheet bosons and fermions such that this "background" anomaly is cancelled. We have already seen that the anomaly contribution from each worldsheet boson X^μ is $c_X = 1$. Since the anomaly contribution from each Majorana fermion is $c_\psi = 1/2$, we must satisfy

$$D\left(1 + \tfrac{1}{2}\right) - 15 = 0 \qquad \Longrightarrow \qquad D = 10 \ . \tag{3.7}$$

Thus, we see that the critical dimension of the superstring is $D = 10$ rather than $D = 26$. Moreover, just as for the bosonic string, the *super*conformal symmetry of the superstring worldsheet action again allows us to choose a light-cone gauge in which only *eight* transverse bosons and *eight* transverse fermions represent the truly physical propagating worldsheet fields.

3.2 Quantizing the superstring

Let us now quantize the superstring, just as we did for the bosonic string. The boundary conditions (2.6) for the X^μ fields remain valid even for the superstring, since the X^μ continue to have the interpretation of spacetime coordinates. Therefore the mode-expansions (2.9) continue to apply.

The only new feature, then, is the mode-expansion for the fermionic fields ψ^μ. However, unlike the bosonic fields X^μ which must be periodic because of their interpretation as spacetime coordinates, these fermionic fields ψ^μ do not have any immediate interpretation in spacetime. Therefore, the only boundary conditions that might be imposed on these fields are those that are required directly from the symmetries of the action. In particular, we must choose boundary conditions for the ψ^μ fields so as to maintain the single-valuedness of the action as we traverse the closed string (*i.e.*, as $\sigma_1 \to \sigma_1 + \pi$), and so as to maintain the worldsheet supersymmetry of the action. It turns out that are only two choices of boundary conditions that satisfy these requirements. One possibility is that the ψ^μ fields are periodic under $\sigma_1 \to \sigma_1 + \pi$:

$$\text{Ramond:} \quad \psi^\mu(\sigma_1 + \pi, \sigma_2) \;=\; +\,\psi^\mu(\sigma_1, \sigma_2) \,. \tag{3.8}$$

Such periodic boundary conditions are typically called "Ramond" (R) boundary conditions, after P. Ramond (who introduced these fermionic boundary conditions in 1971). The second possibility is that the ψ^μ fields are *anti*periodic under $\sigma_1 \to \sigma_1 + \pi$:

$$\text{Neveu-Schwarz:} \quad \psi^\mu(\sigma_1 + \pi, \sigma_2) \;=\; -\,\psi^\mu(\sigma_1, \sigma_2) \,. \tag{3.9}$$

Such periodic boundary conditions are typically called "Neveu-Schwarz" (NS) boundary conditions, after A. Neveu and J. Schwarz (who introduced these fermionic boundary conditions in 1971). As we shall see in Lecture #4, both of these boundary conditions are ultimately required for the self-consistency of the superstring.

In the case of periodic (Ramond) boundary conditions, the mode-expansion of the ψ^μ field resembles that of the X^μ field:

$$\text{Ramond:} \quad \psi^\mu_L(\sigma_1 + \sigma_2) \;=\; \sum_{n \in \mathbb{Z}} b^\mu_n\, e^{-2in(\sigma_1 + \sigma_2)}$$

$$\psi^\mu_R(\sigma_1 - \sigma_2) \;=\; \sum_{n \in \mathbb{Z}} \tilde{b}^\mu_n\, e^{+2in(\sigma_1 - \sigma_2)} \,. \tag{3.10}$$

Here $b^\mu_n, \tilde{b}^\mu_n$ are the (fermionic) creation and annihilation operators, satisfying the *anti*-commutation relations

$$\{b^\mu_m, b^\nu_n\} \;=\; \eta^{\mu\nu} \delta_{m+n} \tag{3.11}$$

where we recall the hermiticity condition $b^\mu_{-n} = (b^\mu_n)^\dagger$. The same relations hold for the right-moving modes as well. This hermiticity condition follows from the fact that the ψ fields are Majorana (*i.e.*, real) fields. Note that unlike the bosonic mode-expansion (2.9), we have joined the zero-modes together with the excited modes in (3.10).[b] There is also no "center-of-mass" term

---

[b] We are cheating slightly here, since the treatment of Ramond zero-modes for Majorana worldsheet fermions is actually quite subtle. In some sense, each Majorana fermion has only "half" a zero-mode. We will provide a rigorous discussion of this fact in Lecture #5. In the meantime, it will suffice to ignore this subtlety.

in the mode-expansion (a fermionic analogue of x^μ) because the ψ fields are Grassmann variables and thus lack a classical limit. Finally, also note that unlike the bosonic α_n^μ modes, which are rescaled relative to the usual harmonic oscillator modes by powers of the mode frequency n, the fermionic b_n^μ modes are defined without this rescaling and hence satisfy the usual harmonic-oscillator commutation relations (3.11) directly. This too is traditional in string theory.

In the case of anti-periodic (Neveu-Schwarz) boundary conditions, the mode-expansion of the ψ^μ field involves *half-integer* rather than integer modes:

$$\text{Neveu-Schwarz:} \quad \psi_L^\mu(\sigma_1 + \sigma_2) = \sum_{r \in \mathbb{Z}+1/2} b_r^\mu \, e^{-2ir(\sigma_1+\sigma_2)}$$

$$\psi_R^\mu(\sigma_1 - \sigma_2) = \sum_{r \in \mathbb{Z}+1/2} \tilde{b}_r^\mu \, e^{+2ir(\sigma_1-\sigma_2)} \, . \quad (3.12)$$

Once again, $b_r^\mu, \tilde{b}_r^\mu$ are the (fermionic) creation and annihilation operators, satisfying the *anti*-commutation relations

$$\{b_r^\mu, b_s^\nu\} = \eta^{\mu\nu} \delta_{r+s} \quad (3.13)$$

where we have the hermiticity condition $b_{-r}^\mu = (b_r^\mu)^\dagger$.

The expressions for the total energy of a given string configuration now receive contributions from not only the bosonic oscillator modes, as in (2.12), but also the fermionic oscillator modes. These new contributions are given by

$$\text{R:} \quad L_0^{(\text{osc})} = \sum_{n=0}^{\infty} n \, b_{-n}^\mu b_{n\mu}$$

$$\text{NS:} \quad L_0^{(\text{osc})} = \sum_{r=1/2}^{\infty} r \, b_{-r}^\mu b_{r\mu} \, , \quad (3.14)$$

with similar expressions for the right-movers.

Finally, we must consider the vacuum energies a_L and a_R for the superstring. Recall that for the bosonic string, each of the 24 transverse X^μ fields contributed $a_X = -1/24$, yielding a total of $a_L = a_R = -1$. This contribution from each bosonic field remains the same for the superstring, so we continue to have $a_X = -1/24$. It therefore only remains to determine the vacuum-energy contributions from the worldsheet Majorana fermions, and it is found that

Great Leap #6: Each Ramond fermion contributes vacuum energy $a_\psi = +1/24$, whereas each Neveu-Schwarz fermion contributes vacuum energy $a_\psi = -1/48$.

We thus see that like the bosons, the Neveu-Schwarz fermions contribute negative vacuum energies, while Ramond fermions contribute positive vacuum energies.

Given these mode-expansions and commutation relations, it is instructive to consider the Fock space of an individual Ramond (R) or Neveu-Schwarz (NS) fermion. It turns out to be simplest to consider the Fock space of an individual (left- or right-moving) NS fermion first. The two lowest-lying states are

$$\text{vacuum:} \qquad |0\rangle \qquad L_0^{(\text{osc})} = 0$$
$$\text{first-excited state:} \qquad b_{-1/2}|0\rangle \qquad L_0^{(\text{osc})} = 1/2 \ . \qquad (3.15)$$

Note that relative to the vacuum, all further excited states are reached through only half-integer excitations. Also note that the vacuum of the NS Fock space is unique, just like that of the bosons X^μ. What this means is that from the spacetime perspective, the vacuum is spinless (and hence a spacetime bosonic state), and that all subsequent excitations of the vacuum are also spacetime bosons. Recall, in this connection, that the fermion mode operators b are only fermionic from the worldsheet perspective; they are still bosonic operators (just like the fields ψ^μ themselves) relative to *spacetime* Lorentz symmetries.

Let us now consider the corresponding Fock space for the Ramond fermions with periodic boundary conditions. Once again, we have a tower of states

$$\text{vacuum:} \qquad |0\rangle \qquad L_0^{(\text{osc})} = 0$$
$$\text{first-excited state:} \qquad b_{-1}|0\rangle \qquad L_0^{(\text{osc})} = 1 \qquad (3.16)$$

which now continues upwards through integer, rather than half-integer, steps. However, in this case it is important to observe that we also have a *zero-mode* in the theory. The existence of this zero-mode means that it is possible to excite this zero-mode without increasing the overall energy of the state. We therefore have the additional tower of states

$$\text{vacuum:} \qquad b_0^\dagger|0\rangle \qquad L_0^{(\text{osc})} = 0$$
$$\text{first-excited state:} \qquad b_{-1}b_0^\dagger|0\rangle \qquad L_0^{(\text{osc})} = 1 \ . \qquad (3.17)$$

(Note that b_0 and $b_0^\dagger$ are equivalent.) In other words, combining (3.16) and (3.17), we see that the Ramond vacuum consists of *two degenerate states*,

$$|0\rangle \qquad \text{and} \qquad b_0^\dagger|0\rangle \ , \qquad (3.18)$$

and that all further excitations maintain this two-fold degeneracy.

How can we interpret this two-fold degeneracy of the Ramond vacuum? It may seem, at first, that both of the states in (3.18) cannot be considered as the true vacuum, because the second state in (3.18) appears to be realized as a zero-mode excitation of the first. However, let us define the first state in (3.18) as $|V_0\rangle$ and let us also define $|V_1\rangle \equiv \sqrt{2}b_0^\dagger|0\rangle$, which is a rescaling of the second state in (3.18). Then using (3.11), it is easy to show that

$$|V_1\rangle = \sqrt{2}b_0^\dagger|V_0\rangle , \qquad |V_0\rangle = \sqrt{2}b_0^\dagger|V_1\rangle . \tag{3.19}$$

Thus, we see that neither state in (3.18) is more fundamental than the other, and there exists an unbroken symmetry between them — they are realized as zero-mode excitations of each other. The interpretation of this fact is that the true Ramond vacuum state is a two-component object, a spacetime spinor! It then follows that all of the excited states in the Ramond spectrum are also spacetime spinors, since they are realized as non-zero-mode excitations of a spinorial ground state.

Of course, the above discussion is only suggestive, since we have not proven that these two vacuum states actually form a Lorentz spinor representation with respect to the spacetime Lorentz algebra. However, it is easy to see that this is indeed the case. Observe from (3.11) that the zero-modes satisfy the algebra $\{b_0^\mu, b_0^\nu\} = \eta^{\mu\nu}$. Thus, if we define $\Gamma^\mu \equiv \sqrt{2}ib_0^\mu$, then we see that $\{\Gamma^\mu, \Gamma^\nu\} = -2\eta^{\mu\nu}$, which is nothing but the spacetime Clifford algebra. In other words, the zero-modes act as spinorial gamma-matrices. This implies that all states built upon such a vacuum state will transform in spinor representations of the spacetime Lorentz symmetry group $SO(D-1,1)$, and hence will be spacetime fermions.

This is a remarkable result. Even though we have introduced worldsheet ψ^μ fields which are spacetime bosons and which carry a spacetime Lorentz *vector* index, the algebra of zero-modes in the case of Ramond boundary conditions has managed to change these vector indices into spinor indices and thereby produce spacetime fermions. Of course, this is completely analogous to what happens in the usual four-dimensional Dirac equation, where the γ^μ matrices are matrices in a spinor space but nevertheless carry vector indices. Thus, we see that by choosing Ramond boundary conditions for worldsheet fermions, string theory affords us with the same possibility. We therefore now see that string theory can indeed give rise to spacetime fermions: while excitations of worldsheet Neveu-Schwarz fermions give rise to spacetime bosons, excitations of worldsheet Ramond fermions give rise to spacetime fermions.

4 Lecture #4: Some famous superstrings

The next step is to determine the spectrum of the full superstring, just as we did for the bosonic string. However, the presence of two possibilities (Neveu-Schwarz and Ramond) for the modings of the fermions introduces several new complications relative to the bosonic string, and enables us to make different choices for what kind of superstring we wish to construct. These different choices are typically called different "string models", and so we are finally in a position to begin to discuss string model-building. That is the subject of the present lecture.

4.1 String sectors

Recall from the previous lecture that in light-cone gauge, the worldsheet field content of the ten-dimensional superstring consists of eight right-moving bosons X_R, eight right-moving Majorana-Weyl (one-component) fermions ψ_R, and a similar set of left-moving fields X_L and ψ_L. The bosons X_L and X_R must have periodic (integer) modings because of their interpretation as spacetime coordinates, but their worldsheet fermionic superpartners ψ_L and ψ_R can have either Ramond (periodic, integer) or Neveu-Schwarz (anti-periodic, half-integer) modings. The question then immediately arises: What rules govern the possible self-consistent choices of fermion modings? A priori, the appearance of 16 distinct fermions would seem to lead to 2^{16} different choices.

It is easy to see that not all possibilities are allowed, however. One quick way to see this is to realize that if some of the right-moving fermions had different periodicities than other right-moving fermions, then these different periodicities would necessarily break spacetime Lorentz invariance because these fermions carry a spacetime vector index μ. A similar situation would also hold for the left-moving fermions. This would then imply that all of the right-moving fermions should have the same periodicity as each other, and that all of the left-moving fermions should have the same periodicity as each other (though not necessarily the same as that of the right-moving fermions). However, this argument is not really satisfactory because we do not necessarily wish to preserve the full *ten*-dimensional Lorentz invariance (or even its eight-dimensional transverse subgroup); after all, our sole phenomenological requirement is that *four*-dimensional Lorentz invariance must be maintained. Moreover, it goes against the spirit of string theory (as we have been presenting it) that we should demand a certain phenomenological property of the resulting *spacetime* physics when formulating our worldsheet theory. In string theory the spacetime physics is a *consequence* of the worldsheet physics, and we would ultimately like to base our worldsheet choices directly on worldsheet symmetries.

Fortunately, it is easy to find a worldsheet argument that leads to the same constraint. Recall that the worldsheet symmetry that we must maintain is superconformal invariance. The worldsheet supersymmetry that makes up superconformal invariance is generated by the two worldsheet supercurrents given in (3.6). Because these two supercurrents are also worldsheet fermionic, they may also be either periodic or anti-periodic as we traverse the closed string. Indeed, each individual term $\psi^\mu \partial X_\mu$ in these supercurrents will have the periodicity property of the fermion ψ^μ. However, in order for each of these supercurrents J_R and J_L to have a unique, well-defined periodicity as we traverse the closed string, we see that it is necessary that all right-moving fermions have the same periodicity as each other, and that all left-moving fermions have the same periodicity as each other. This is required in order to preserve worldsheet supersymmetry. Thus, we have our first constraints on fermion modings:

- All right-moving fermions ψ_R^μ must have the same periodicity as each other, either Ramond or Neveu-Schwarz.

- All left-moving fermions ψ_L^μ must have the same periodicity as each other, either Ramond or Neveu-Schwarz.

Note that there is no requirement that the right- and left-moving periodicities be the same.

#	$\psi_R^{i=1,\ldots,8}$	$\psi_L^{i=1,\ldots,8}$	a_R	a_L
1	NS	NS	$-1/2$	$-1/2$
2	R	R	0	0
3	R	NS	0	$-1/2$
4	NS	R	$-1/2$	0

Table 1: The four possible sectors of the ten-dimensional superstring, numbered 1 through 4. Here 'NS' and 'R' respectively indicate Neveu-Schwarz (anti-periodic) and Ramond (periodic) boundary conditions for worldsheet fermions, and a_R and a_L respectively denote the corresponding right- and left-moving vacuum energies.

Given these constraints, we see that we are left with four distinct periodicity choices for our sixteen Majorana-Weyl worldsheet fermions, as shown in Table 1. Each individual choice is called a *sector* or *spin structure* of the superstring, so we see that the ten-dimensional superstring has four possible sectors. For future convenience, these sectors have been numbered in Table 1. We have also indicated the corresponding right- and left-moving vacuum energies of these sectors. Recall from the previous lecture (in particular, Great

Leap #6) that the vacuum-energy contribution of each Ramond fermion is +1/24, while that of each Neveu-Schwarz fermion is −1/48 and that of each worldsheet boson is −1/24. Therefore, generally assuming n_{NS} Neveu-Schwarz fermions and n_R Ramond fermions, we can add these individual contributions to find

$$a = -\frac{1}{24}\left(8 - n_R - \tfrac{1}{2}n_{NS}\right) = -\frac{n_{NS}}{16} . \tag{4.1}$$

The second equality results from setting $n_R = 8 - n_{NS}$. Of course, as discussed above, in the ten-dimensional superstring we are restricted to the cases $n_{NS} = 0, 8$ for both the right- and left-moving fermions.

4.2 Modular invariance and GSO projections

The next question that arises is whether we are free to pick any one of these sectors to construct our superstring theory, or whether we must consider all of them together, superposing the spectrum from each sector separately in order to construct the full superstring spectrum. What rules govern the choices of sectors?

Ultimately, it turns out that a special form of conformal invariance known as *modular invariance* will give us the answer. In keeping with the spirit of these lectures, we will not be able to provide a proper mathematical discussion of modular invariance. (Indeed, doing so would require a preliminary discussion of string partition functions and the modular group.) However, we can discuss the relevance and implications of modular invariance at a conceptual level.

Recall from Lecture #2 that our string actions always have a certain symmetry known as conformal invariance, which reflects the fact that the action should be invariant under *local* reparametrizations and rescalings of the coordinates (σ_1, σ_2) that parametrize the string worldsheet. For *tree*-level string interactions, demanding this *local* symmetry is sufficient to ensure that the resulting physics is indeed invariant under arbitrary coordinate reparametrizations. This is because any tree-level string interaction has the topology of a sphere (a genus-zero surface, with no handles), and on a sphere it can be shown that any possible net coordinate reparametrization can be generated or "built up" in small steps as the cumulative effect of small, local coordinate reparametrizations. Geometrically, this is equivalent to saying that any closed loop on the surface of a sphere can be continuously shrunk to a point, as illustrated in Fig. 7(a), by sliding the loop along the surface of the sphere towards one side. Thus, demanding invariance under *local* coordinate reparametrizations (*i.e.*, conformal invariance) by itself is sufficient to guarantee consistency for tree-level string amplitudes.

(a)   **(b)**

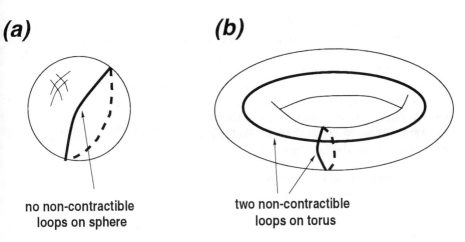

no non-contractible   two non-contractible
loops on sphere   loops on torus

Figure 7: (a) On a sphere, all closed loops can be continuously shrunk to a point. (b) On a torus, there exist two topologically distinct non-contractible loops.

However, this situation changes drastically if we now consider one-loop amplitudes. As discussed in Lecture #1, these amplitudes have the world-sheet topology of a *torus* (a genus-*one* surface), and we see from Fig. 7(b) that on a torus there exist *two* types of closed loops that cannot be continuously shrunk to a point. Such loops are said to be *non-contractible*, which is indeed the defining property of such higher-genus surfaces. The presence of these non-contractible loops means that for torus diagrams, there exist possible coordinate reparametrizations that *cannot* be built up from local coordinate reparametrizations alone. Indeed, these reparametrizations non-trivially involve "large", discrete mappings around these non-contractible loops. Thus, we see that demanding conformal invariance alone is not sufficient to ensure that one-loop string amplitudes are truly invariant under worldsheet coordinate reparametrizations: we must also demand an invariance under these "large" discrete mappings around these non-contractible loops. This additional global invariance is called "modular invariance", and just like conformal invariance, it too stems from our need to maintain the overall invariance of the string under reparametrizations and rescalings of the worldsheet coordinates.

One might wonder, at this stage, why we are suddenly worrying about modular invariance, whereas we did not need to consider modular invariance in Lecture #2 when we discussed the bosonic string. The truth of the matter is that we must *always* consider modular invariance in addition to conformal invariance, regardless of the type of (closed) string we are discussing. However, in the simple case of the 26-dimensional bosonic string, it turns out that all amplitudes are trivially modular-invariant, so we did not need to make recourse to modular invariance in order to distinguish between different possibilities. However, for the superstring (and particularly for the heterotic string to be discussed later), the possible sector choices become quite numerous, and it turns out that modular invariance is the powerful tool by which we are able to narrow down the self-consistent possibilities.

What, then, are the effects of modular invariance? It turns out that at the level of string model-building, modular invariance has two primary effects:

- it forces us to consider only certain selected *sets* or *combinations* of underlying sectors, and

- it produces new *constraints* (beyond the level-matching constraint $L_0 = \bar{L}_0$) that govern which Fock-space excitations are allowed in each sector.

These new constraints are called *GSO constraints*, after F. Gliozzi, J. Scherk, and D. Olive who first imposed some of these constraints in 1977. The important point is that these conditions stem directly from modular invariance, and thus they follow from the worldsheet physics of the string and do not represent any additional arbitrary input. We will provide many explicit examples of such combinations and constraints shortly.

In order to construct a fully consistent string model, therefore, our procedure is as follows. First, we must determine which are the allowed sectors that need to be considered as part of our set. For each of these sectors in our allowed set, we then determine the corresponding Fock space of physical states by applying not only the usual level-matching constraint, but also the GSO constraints appropriate for that sector. In this way each underlying sector then gives rise to a different Fock space of states, and the full Hilbert space of states for the full string theory (*i.e.*, for the resulting string "model") is nothing but the direct sum of these different Fock spaces corresponding to each of the underlying sectors in the specified set. This then yields a fully self-consistent (and in particular, modular-invariant) theory.

This is an important point, so it is worth repeating: the full Hilbert space of string states is given by the direct sum of the different Fock spaces corresponding to different underlying boundary conditions for worldsheet fields. In order to better understand this fact, an analogy with QCD may be useful. Recall that Yang-Mills quantum field theory contains non-perturbative instanton solutions, and therefore one can imagine doing quantum field theory in an n-instanton background $|n\rangle$. Of course, as we know, the full vacuum state of QCD is not composed of any one of these $|n\rangle$ vacua by itself, but rather by an appropriately weighted *combination* of these vacua:

$$|\theta\rangle = \sum_n e^{in\theta} |n\rangle . \qquad (4.2)$$

This is the famous θ-vacuum of QCD. The situation that we now face in string theory is somewhat analogous. The fact that the string worldsheet fermions can have different boundary conditions (thereby giving rise to different sectors) is in some sense analogous to the fact that QCD can have different instanton backgrounds. Indeed, each underlying string sector is analogous to a different n-instanton vacuum state $|n\rangle$, and the different "combinations of sectors" that we are now being forced to consider are analogous to the different QCD θ-vacua. In this sense, then, each different "string model" that we will be constructing can be viewed as a different θ-vacuum of string theory! Of course, this analogy with the QCD θ-vacuum can take us only so far. One important difference is that whereas the θ-vacuum necessarily involves *all* of the $|n\rangle$ states regardless of the value of θ, in string theory our "vacuum" may consist of more complicated combinations of sectors which may or may not include all possible sectors. In fact, the more sectors that are included in our "combination of sectors", the more GSO constraints there are for each sector. But the important lesson that emerges from all of this is that no single sector by itself forms a consistent string vacuum; rather, we must select an appropriate combination of sectors and add together their corresponding Fock spaces in order to produce the fully self-consistent string model.

In Lecture #7, we shall provide an explicit set of rules which will enable us to quickly determine the appropriate sector combinations and GSO constraints that can be chosen in order to yield self-consistent theories. For the time being, however, we shall defer a discussion of these rules and proceed directly with the construction of actual string models in order to deduce their physical properties. Therefore, even though we shall simply assert certain sector combinations and GSO projections to be required by modular invariance, we stress that all of these features can (and ultimately will) be derived using the rules to be presented in Lecture #7.

4.3 Ten-dimensional superstring models

In the case of the ten-dimensional superstring, we have already seen that the four possible sectors are listed in Table 1. It then only remains to determine the particular sector combinations and GSO constraints that are required by modular invariance. In this case, it turns out that there are only two possible combinations or sets of sectors that can be considered:

- we consider the contributions from only Sectors #1 and #2, or

- we consider the contributions from *all* Sectors #1 through #4.

Moreover, for each of the above cases, it turns out that there are two possible choices of GSO projections that may be imposed in each sector. Thus, combining all of these possibilities, we see that there are four distinct possible superstring "models" that can be constructed in ten dimensions. We shall therefore now turn to a construction of these models.

The Type 0 strings

Let us begin by considering the first option, taking our set of sectors to consist only of Sectors #1 and #2. For each of these sectors, we need to determine the appropriate GSO constraints that must be applied in addition to the usual level-matching constraint. In order to write down these GSO constraints, let us first recall that for a given left-moving worldsheet fermion (with either Ramond or Neveu-Schwarz boundary conditions), the corresponding number operator is defined by

$$\text{R}: \qquad N^{(i)} = \sum_{n=0}^{\infty} b^i_{-n} b_{ni}$$

$$\text{NS}: \qquad N^{(i)} = \sum_{r=1/2}^{\infty} b^i_{-r} b_{ri} \ . \qquad (4.3)$$

Here the index $i = 1, ..., 8$ labels the individual fermion. For right-moving fermions, the analogous number operators $\bar{N}^{(i)}$ are constructed using the right-moving mode operators $\tilde{b}_n, \tilde{b}_r$. Let us also define N_L and N_R respectively as the total left- and right-moving number operators, *i.e.*,

$$N_L \equiv \sum_{i=1}^{8} N^{(i)} \ , \qquad N_R \equiv \sum_{i=1}^{8} \bar{N}^{(i)} \ . \qquad (4.4)$$

Note that these number operators are defined to include only the contributions of the worldsheet *fermions*, and in particular do not include the contributions of the worldsheet bosons. It then turns out (and we shall see in Lecture #7) that if we choose our set of sectors to consist only of Sectors #1 and #2, then the appropriate GSO constraints in each sector are as follows:

$$\text{Sector \#1}: \qquad N_L - N_R = \text{even}$$

$$\text{Sector \#2}: \qquad N_L - N_R = \left\{ \begin{array}{c} \text{odd} \\ \text{even} \end{array} \right\} \ . \qquad (4.5)$$

In the second line, we have used a brace notation to indicate a further choice: we can choose to impose *either* the 'odd' constraint, or the 'even' constraint. As we shall see, this is a residual choice that is not fixed by modular invariance (or by any other worldsheet symmetry), leading to two equally valid possibilities. Thus, we see that if we choose our set of sectors to consist of only Sectors #1 and #2, then this leads to *two different string models* depending on our subsequent choice of which GSO constraint we choose to impose in (4.5).

Let us now determine the spectra of these two models, beginning with the states that arise from Sector #1. Note that in this sector, both models have the same states (because both models have the same GSO constraint for Sector #1). As with the bosonic string, our procedure is to consider all possible excitations of the worldsheet fields (in this case, the worldsheet fermions as well as the worldsheet bosons). These excitations are subject to the level-matching constraint $L_0 = \bar{L}_0$ (which ensures that the total bosonic and fermionic worldsheet *energy* is distributed equally between left- and right-moving excitations) and the GSO constraint $N_L - N_R =$ even (which is a constraint on the worldsheet *number* operators of the worldsheet fermions only). In general, the mass-shell condition for the superstring is

$$\alpha' M^2 = 2(L_0 + \bar{L}_0 + a_L + a_R) \qquad (4.6)$$

where a_L and a_R are the individual left- and right-moving vacuum energies, and where L_0 and $\bar{L}_0$ include the contributions from not only the worldsheet bosons, but also the worldsheet fermions. Note from Table 1 that the left- and right-moving vacuum energies in Sector #1 are $a_L = a_R = -1/2$.

We see that the tachyonic vacuum state $|0\rangle_R \otimes |0\rangle_L$ satisfies both constraints, and thus it remains in the spectrum. However, unlike the tachyon in the bosonic string (which has spacetime mass $\alpha' M^2 = -4$), we see from (4.6) that the tachyonic state in the superstring has spacetime mass $\alpha' M^2 = -2$. This is the result of the smaller (less negative) vacuum energy of the superstring compared to that of the bosonic string.

Because the vacuum energies in Sector #1 are $a_L = a_R = -1/2$, we see that massless states cannot be obtained by exciting the quantum modes of worldsheet bosons, for each of these excitations would add a full unit of energy. Instead, massless states can be obtained only by adding a half-unit of energy. Fortunately, this is possible in Sector #1 because in this sector, all worldsheet fermions have Neveu-Schwarz boundary conditions and therefore have half-integer modings. The first excited states in Sector #1 are therefore

$$\tilde{b}^\mu_{-1/2}|0\rangle_R \otimes b^\nu_{-1/2}|0\rangle_L \,. \qquad (4.7)$$

Note that these states satisfy both the level-matching constraint (since $L_0 = \bar{L}_0 = 1/2$) as well as the GSO constraint (since $N_L = N_R = 1$). The interpretation of these states is precisely the same as in the bosonic string: these states give us the gravity multiplet, consisting of the graviton $g_{\mu\nu}$, dilaton ϕ, and anti-symmetric tensor $B_{\mu\nu}$. It is indeed a general principle that *all* weakly coupled closed strings contain at least these massless states, and this is a useful cross-check of the GSO constraints.

Let us now turn to the states from Sector #2. Before concerning ourselves with the implication of the GSO constraints in (4.5), let us first understand the general structure of the states from this sector. In this sector, the vacuum energy (according to Table 1) is $(a_R, a_L) = (0, 0)$, so we see immediately that this sector contains no tachyons. Indeed, the ground state is already massless, so all that will concern us here is the nature of this ground state. As we discussed at the end of Lecture #3, the left- and right-moving ground states in this sector are each spacetime *spinors* since all worldsheet fermions in this sector have Ramond boundary conditions. Because the nature of these spinors will be important to us, let us pause to review some properties of these spinors.

Since we are considering these ten-dimensional strings in light-cone gauge, the Lorentz group that concerns us here is the transverse ("little") Lorentz group $SO(8)$. In general, the groups $SO(2n)$ share a number of properties. Their smallest representations, of course, are simply the identity representations. These are singlets, which will be denoted **1**. The next representations are the vector representations, which are $(2n)$-dimensional, and which will be denoted $\mathbf{V}_{2n}$. Along with these are the spinor representations, which are (2^{n-1})-dimensional. In general, there are two types of spinor representations, **S** and **C**, the so-called "spinor" and "conjugate spinor" representations. In the special case of $SO(8)$, the vector, spinor, and conjugate spinor representations are all eight-dimensional, and will be denoted $\mathbf{V}_8$, $\mathbf{S}_8$, and $\mathbf{C}_8$ respectively. The distinction between $\mathbf{S}_8$ and $\mathbf{C}_8$ is one of spacetime *chirality*, but the choice of which is to be associated with a given physical chirality is a matter of convention.

The ground state of Sector #2 has the structure

$$\{\tilde{b}_0^\mu\} \, |0\rangle_R \, \otimes \, \{b_0^\mu\} \, |0\rangle_L \qquad (4.8)$$

where the notation $\{b_0^\mu\}$ (and similarly for the right-movers) indicates that each of the individual Ramond zero-modes can be either excited or not excited.

How can we interpret (4.8) physically? This issue is actually quite subtle, and we shall not have the space to give a proper discussion. Moreover, as we have already indicated, we are not giving a fully rigorous treatment of Ramond zero-modes in these lectures, since our aim is to focus more on the physics than the formalism. However, it is possible to understand the appropriate physical interpretation intuitively. First, let us *count* the number of states in (4.8). *A priori*, it would seem that we have 2^{16} individual states, since each Ramond fermion zero-mode can either be excited or not excited. However, this is not correct because (as we shall discuss more completely in Lecture #5, and as we have already hinted in the footnote in Sect. 3.2), one should really count only one zero-mode per *pair* of Ramond Majorana-Weyl fermions. Thus, we can imagine that there are only four independent zero-modes for the right-movers, and four for the left-movers. Therefore, (4.8) consists of only $2^8 = 128$ states.

All combinations of these zero-mode excitations already satisfy the level-matching constraint (since $L_0 = \bar{L}_0 = 0$). Imposing either of the GSO constraints for Sector #2 in (4.5) then reduces the number of allowed states by a factor of two. Specifically, if we impose the constraint $N_L - N_R = $ odd, then we can choose only an even number of right-moving zero-mode excitations together with an odd number of left-moving zero-mode excitations, or an odd number of right-moving excitations together with an even number of left-moving excitations. Choosing the constraint $N_L - N_R = $ even has the opposite effect, pairing even numbers of excitations for left- and right-movers with each other, and likewise pairing odd numbers with each other.

Interpreting these results is therefore quite simple. As we discussed at the end of Lecture #3, the left-moving states and right-moving states are spacetime spinors, and we have already seen that there are two possible spinors, $\mathbf{S_8}$ and $\mathbf{C_8}$. At this stage, the names assigned to each are arbitrary, so we shall now establish the following convention: spinors realized by an even number of zero-mode excitations will be identified with $\mathbf{C_8}$, and those realized by an odd number of zero-mode excitations will be identified with $\mathbf{S_8}$. Of course, only the relative difference between these two spinors is physically significant (having the interpretation of spacetime chirality).

Given these definitions, we see that if we choose the first GSO constraint $N_L - N_R = $ odd, the 128 states in (4.8) decompose into

$$\left(\bar{\mathbf{C}}_8 \otimes \mathbf{S}_8\right) \; \oplus \; \left(\bar{\mathbf{S}}_8 \otimes \mathbf{C}_8\right) , \qquad\qquad (4.9)$$

whereas if we choose the second GSO constraint $N_L - N_R = $ even, these states instead decompose into

$$\left(\bar{\mathbf{C}}_8 \otimes \mathbf{C}_8\right) \; \oplus \; \left(\bar{\mathbf{S}}_8 \otimes \mathbf{S}_8\right) . \qquad\qquad (4.10)$$

If we wish to further decompose these states into representations of the Lorentz group, we can use the $SO(8)$ tensor-product relations

$$\mathbf{S}_8 \otimes \mathbf{S}_8 \;\; = \;\; \mathbf{1} \oplus \mathbf{28} \oplus \mathbf{35'}$$

$$\mathbf{C}_8 \otimes \mathbf{C}_8 = \mathbf{1} \oplus \mathbf{28} \oplus \mathbf{35}''$$
$$\mathbf{S}_8 \otimes \mathbf{C}_8 = \mathbf{V}_8 \oplus \mathbf{56} . \tag{4.11}$$

Here the **28** representation is the anti-symmetric component of the spinor tensor product (spin-one), while the **35**′ and **35**″ representations are the symmetric components of the spinor tensor product (also spin-one). (These latter representations are not to be confused with the spin-two **35** graviton representation in (2.20).) Likewise, the **56** is a certain vectorial (spin-one) higher-dimensional representation.[a] However, for our present purposes it will be sufficient to think of these states in the tensor-product forms (4.9) and (4.10). Note that in each case, the tensor product of two spacetime fermionic (spinor) states produces a spacetime bosonic state. Thus, just as in Sector #1, the states emerging in Sector #2 are spacetime bosons.

Thus, summarizing, we see that the spectra of our two resulting superstring models are as follows. First, from Sector #1, we have the tachyonic state $|0\rangle_R \otimes |0\rangle_L$. In the notation of $SO(8)$ Lorentz representations, this state may be denoted $\bar{\mathbf{1}} \otimes \mathbf{1}$; this tachyon is a Lorentz singlet. Next, we have the massless gravity multiplet. In the notation of $SO(8)$ Lorentz representations, this state takes the form $\bar{\mathbf{V}}_8 \otimes \mathbf{V}_8$. Finally, from Sector #2, we have massless states whose form depends on the particular choice of the GSO projection. In the first case, we have the states given in (4.9), while in the second case, we have the states given in (4.10). There are then, as usual, an infinite tower of massive (Planck-scale) states above these.

The string model produced by the first GSO projection is called the *Type 0A* string model, and the second is called the *Type 0B* string model. Collectively, these are sometimes simply called the Type 0 strings. As we see, both of these strings are tachyonic, and moreover they contain only bosonic states. Furthermore, as is evident from (4.9) and (4.10), both of these strings are non-chiral. In other words, they are invariant under the transposition $\mathbf{S}_8 \leftrightarrow \mathbf{C}_8$ for the left- and right-movers. These string theories were first constructed by N. Seiberg and E. Witten in 1985. Although not relevant for phenomenology,

---

[a] For the mathematically inclined reader, we can succinctly describe all of these states as follows. Recall that a given representation is called a p-form if it can be realized as the totally anti-symmetric combination within the tensor product of p different vector indices of $SO(8)$, with resulting dimension $8 \times 7 \times 6 \times \ldots \times (9-p)/p!$. Using this language, we see that singlet states are zero-forms, the **28** representations are two-forms, and the **35**′ and **35**″ representations are "self-dual" four-forms. (The self-duality condition eliminates exactly half of the degrees of freedom in the four-form.) Likewise, the $\mathbf{V}_8$ state is a one-form, and the **56** representation is a three-form. These different forms (and the so-called *D-branes* whose existence they imply) are important when considering the non-perturbative structure of these string theories.

they are currently proving to have an important role in understanding certain non-perturbative aspects of non-supersymmetric string theory.

The Type II strings

Let us now turn to the second choice outlined at the beginning of Sect. 4.3, namely the case in which we consider the contributions from *all* of the sectors in Table 1. This will result in the so-called *Type II strings*. As we discussed at the end of Sect. 4.2, it is a general property that the larger the set of sectors that we consider, the more GSO constraints there are that must be imposed in each sector. Thus, the introduction of new sectors generally leads to new GSO constraints in each of the sectors (old and new), and likewise the introduction of new GSO constraints in a given sector requires the introduction of entire new sectors to compensate.

It turns out (and we shall see explicitly in Lecture #7) that if we consider the full set of sectors in Table 1, then the appropriate GSO constraints in each sector are given as follows:

$$
\begin{aligned}
\text{Sector } \#1: \quad & N_L - N_R = \text{odd} \ , \quad N_R = \text{odd} \\
\text{Sector } \#2: \quad & N_L - N_R = \left\{ \begin{array}{c} \text{odd} \\ \text{even} \end{array} \right\} \ , \quad N_R = \text{odd} \\
\text{Sector } \#3: \quad & N_L - N_R = \left\{ \begin{array}{c} \text{odd} \\ \text{even} \end{array} \right\} \ , \quad N_R = \text{odd} \\
\text{Sector } \#4: \quad & N_L - N_R = \text{even} \ , \quad N_R = \text{odd} \ .
\end{aligned}
\tag{4.12}
$$

Note that in each case where a choice is possible, these choices are correlated: we simultaneously choose either the top lines within all braces, or the bottom lines. Thus, once again there are two sets of GSO conditions that can be imposed, resulting in two distinct string models.

Before proceeding further, it is useful to note the *pattern* of these GSO projections. In the case of the Type 0 strings, we considered only Sectors #1 and #2; as shown in Table 1, these were the sectors for which the right-moving fermions were always identical to the left-moving fermions and shared the same boundary conditions. The corresponding GSO projections in (4.5) likewise did not distinguish between right- and left-moving fermions. (In this context, note that the GSO projections in (4.5) can equivalently be written with minus signs replaced by plus signs.) Thus, in some sense, the Type 0 strings are symmetric under exchange of left- and right-movers. However, for the Type II strings, we have now introduced two additional sectors (Sectors #3 and #4) whose structure explicitly breaks this symmetry between left- and right-movers. No longer does each sector individually exhibit this left/right symmetry. As we see from (4.12), the effect of this breaking is to introduce additional GSO conditions which mirror this broken symmetry by becoming sensitive to right- or left-moving number operators *by themselves*. The technical word for this breaking of symmetry is "twisting" or "orbifolding", for by including Sectors #3 and #4, we see that we have twisted the left-movers relative to the right-movers by allowing them to have oppositely moded boundary conditions. Thus, the Type II strings that will result can be viewed as twisted (or orbifolded) versions of the Type 0 strings. This twisting procedure ultimately serves as the means by which more and more complicated (and more and more phenomenologically realistic) string models may be constructed, and will be discussed more fully in Lecture #7.

Given the GSO constraints in (4.12), we can proceed to determine the resulting spectrum just as we did for the Type 0 strings. Let us begin with Sector #1 (this is often called the "NS-NS sector"). Because the boundary conditions of the worldsheet fermions are the same in this sector as they were for the Type 0 strings, the possible states that arise are the same as they were for the Type 0 strings, and consist of the tachyon $|0\rangle_R \otimes |0\rangle_L$ as well as the gravity multiplet (4.7). The only difference is that we must now impose the additional GSO constraint $N_R = $ odd. It is immediately clear that the effect of this new GSO constraint is that *the tachyon is projected out of the spectrum*, while the gravity multiplet is retained. Thus, by "twisting" the Type 0 strings in just this way, we have succeeded in curing one of the major problems of the bosonic and Type 0 strings, namely the appearance of tachyons. Moreover, we have done this *without* eliminating the desirable gravity multiplet.

Let us now consider the states from Sector #2 (this is often called the "Ramond-Ramond" sector). Once again, if we impose only the first GSO constraint in (4.12), we obtain the states in either (4.9) or (4.10). Imposing the additional GSO constraint in (4.12) then enables us to project out half of these states, so that we retain only the states

$$\bar{\mathbf{S}}_8 \otimes \left\{ \begin{matrix} \mathbf{C}_8 \\ \mathbf{S}_8 \end{matrix} \right\} . \tag{4.13}$$

These states are spacetime bosons.

Finally, let us consider the states that arise in the new Sectors #3 and #4. In Sector #4, the vacuum energy is $(a_R, a_L) = (-1/2, 0)$. Therefore, in order to have level-matching ($L_0 = \bar{L}_0$), we see that we are immediately forced to excite a half-unit of energy for the right-movers while not increasing the energy of the left-movers. This is the only way to produce a massless state. This also ensures that this sector does not give rise to tachyons. Fortunately, since the right-moving fermions have Neveu-Schwarz boundary conditions in this sector, these fermions have half-integer modings, and thus by exciting their lowest modes we can indeed introduce a half-unit of energy. The left-moving fermions have Ramond boundary conditions in this sector, and hence their ground state is the Ramond zero-mode state. The massless states in Sector #3 therefore take the form

$$\tilde{b}^\mu_{-1/2} |0\rangle_R \otimes \{b^\nu_0\} |0\rangle_L . \tag{4.14}$$

At this stage, of course, these states satisfy only the level-matching constraint. Imposing the GSO constraints then leaves us with the state in which we excite only an odd (or even) number of left-moving Ramond zero modes.

How can we interpret this state? First, we notice that this state is a space-time *fermion* because it results from tensoring a right-moving Neveu-Schwarz state with a left-moving Ramond state. Thus, we now have a string theory that contains spacetime fermions! This is yet another benefit of performing the "twist" that takes us from the Type 0 strings to the Type II strings. However, let us examine this state a bit more closely. Clearly, it has the Lorentz structure

$$\mathbf{V_8} \otimes \left\{ \begin{matrix} \mathbf{C_8} \\ \mathbf{S_8} \end{matrix} \right\} \tag{4.15}$$

where we have retained the spinor-labelling conventions that we employed for the Type 0 strings. The relevant tensor-product decompositions in this case are given by

$$\mathbf{V_8} \otimes \mathbf{C_8} = \mathbf{S_8} \oplus \mathbf{56'}$$
$$\mathbf{V_8} \otimes \mathbf{S_8} = \mathbf{C_8} \oplus \mathbf{56''} \tag{4.16}$$

where the $\mathbf{S_8}$ and $\mathbf{C_8}$ representations are spin-1/2 and where the $\mathbf{56'}$ and $\mathbf{56''}$ representations are spin-3/2. Thus, we see that the Type II strings contain a massless, spin-3/2 object! Just as a massless spin-two object satisfies the Einstein field equations and must be interpreted as the graviton, a massless spin-3/2 object must be interpreted as a *gravitino* — i.e., a superpartner of the graviton. This implies that this string not only gives rise to spacetime bosons *and* fermions, but actually gives rise a spectrum which exhibits *spacetime supersymmetry*! This is yet another phenomenologically compelling feature.

Finally, let us now consider Sector #3. This sector has vacuum energies $(a_R, a_L) = (0, -1/2)$, so now we must excite right-moving zero-modes and left-moving $b^\mu_{-1/2}$ modes. This then leads to states of the form

$$\{\tilde{b}^\mu_0\} |0\rangle_R \otimes b^\nu_{-1/2} |0\rangle_L , \tag{4.17}$$

and imposing the GSO projections results in states with the Lorentz structure $\bar{\mathbf{S}}_8 \otimes \mathbf{V}_8$. Once again, this also contains a gravitino!

So what do we have in the end? The first choice of GSO projections results in the so-called *Type IIA string*, while the second choice results in the *Type IIB string*. Both of these strings are tachyon-free, and their spectra contain both bosons and fermions. Moreover, these strings exhibit *spacetime* supersymmetry. This is most easily seen in the following suggestive way. Let us collect together the states from all four sectors, retaining our Lorentz-structure tensor-product notation:

$$
\bar{\mathbf{V}}_8 \otimes \mathbf{V}_8 \,, \qquad \bar{\mathbf{S}}_8 \otimes \left\{ \begin{array}{c} \mathbf{C}_8 \\ \mathbf{S}_8 \end{array} \right\} \,, \qquad \bar{\mathbf{V}}_8 \otimes \left\{ \begin{array}{c} \mathbf{C}_8 \\ \mathbf{S}_8 \end{array} \right\} \,, \qquad \bar{\mathbf{S}}_8 \otimes \mathbf{V}_8 \,. \tag{4.18}
$$

Together, this collection of states can be written in the factorized form

$$
\left(\bar{\mathbf{V}}_8 \oplus \bar{\mathbf{S}}_8 \right) \; \otimes \; \left(\mathbf{V}_8 \oplus \left\{ \begin{array}{c} \mathbf{C}_8 \\ \mathbf{S}_8 \end{array} \right\} \right) \,. \tag{4.19}
$$

We thus see that there are *two* spacetime supersymmetries exhibited in this massless spectrum: the first exchanges $\bar{\mathbf{V}}_8 \leftrightarrow \bar{\mathbf{S}}_8$ amongst the right-movers, while the second exchanges

$$
\mathbf{V}_8 \leftrightarrow \left\{ \begin{array}{c} \mathbf{C}_8 \\ \mathbf{S}_8 \end{array} \right\} \tag{4.20}
$$

amongst the left-movers. Thus, the massless spectrum exhibits $N = 2$ supersymmetry. This is, of course, consistent with the appearance of two gravitinos in the massless spectrum (one from Sector #3 and one from Sector #4). Another way to understand this $N = 2$ supersymmetry is to realize that the first supersymmetry relates the bosonic states in Sector #1 to the fermionic states in Sector #3 (and the bosons in Sector #2 to the fermions in Sector #4), while the second supersymmetry relates the bosons in Sector #1 to the fermions in Sector #4 (and the bosons in Sector #2 to the fermions in Sector #3). In either case, we thus see that we have two independent spacetime supersymmetries.

It is important to note that we did not demand spacetime supersymmetry when constructing the superstring. We merely introduced *worldsheet* supersymmetry, and found that spacetime supersymmetry emerged naturally as the result of certain GSO projections. This further illustrates the fact that in string theory, spacetime properties such as supersymmetry emerge only as the consequences of deeper, more fundamental *worldsheet* symmetries. Another important point is that the same "twist" which eliminated the tachyon has introduced spacetime supersymmetry. While this is certainly an interesting phenomenon that arises for ten-dimensional superstrings, it is certainly *not* a general property that the elimination of the tachyon requires spacetime supersymmetry. In particular, we shall see in Lecture #6 that it is possible to construct string theories whose tree-level spectra lack spacetime supersymmetry but nevertheless are tachyon-free.

One might question whether we have really demonstrated the existence of $N = 2$ supersymmetry, since we have examined only the massless spectrum. However, it can be shown that any unitary theory which contains a massless spin-3/2 state necessarily exhibits supersymmetry, and hence must be supersymmetric at all mass levels (*i.e.*, for all massive, excited states as well). Of course, this is still not a proof, since we do not *a priori* know (and would therefore need to verify) that string theory is a consistent theory in this sense. However, it is possible to construct (two) explicit spacetime supercurrent operators and to demonstrate that they commute with the full (massless and massive) spectrum of the string. Another approach (as indicated in the footnote in Sect. 3.1) is to develop an alternative formulation of the superstring in which *spacetime* (rather than worldsheet) supersymmetry is manifest at the level of the string action, and to demonstrate the equivalence of the two formulations. Indeed, both approaches have been successfully carried out, thereby demonstrating that the Type II spectrum is indeed $N = 2$ supersymmetric. It is for this reason that these strings are referred to as Type II strings.

One important distinction between these two strings is their chirality. The Type IIA string, as we see, contains two supersymmetries of opposite chiralities, interchanging $\bar{V}_8 \leftrightarrow \bar{S}_8$ for the right-movers and $V_8 \leftrightarrow C_8$ for the left-movers. Equivalently, the two gravitinos associated with these supersymmetries are of opposite chiralities (because the $56'$ and $56''$ representations in (4.16) are of opposite chiralities). Because it contains supersymmetries of both chiralities, this string is ultimately non-chiral, and its low-energy (field-theoretic) limit consists of so-called *Type IIA* supergravity (whose discovery predates that of the Type IIA string). It is for this reason that this string is called the Type IIA string. The Type IIB string, by contrast, contains two supersymmetries (or two gravitinos) of the *same* chirality, exchanging $\bar{V}_8 \leftrightarrow \bar{S}_8$ and $V_8 \leftrightarrow S_8$ respectively. Thus, this string theory is *chiral*, and has a low-energy field-theoretic limit consisting of Type IIB supergravity.

We conclude, then, that by introducing a twist relative to the Type 0 strings, we have constructed a set of strings (the Type IIA and Type IIB strings) that exhibit a number of compelling features: they are tachyon-free, they contain both bosons and fermions in their spacetime spectra, they contain gravity, and they are spacetime $N = 2$ supersymmetric. Despite this success, however, there is still something that we lack: we do not, as yet, have gauge symmetries. Specifically, there are no gauge bosons (such as photons, gluons, or W and Z particles). Likewise, there are no states which carry gauge charges. Therefore, once again, we shall need to construct a new kind of string.

5 Lecture #5: Neutrinos have gauge charges: The heterotic string

5.1 Motivation and alternative approaches

Thus far in these lectures, we have shown how string theory can give rise to quantized gravity, spacetime bosons and fermions, spacetime supersymmetry, and tachyon-free spectra. There is, however, one important phenomenological feature that is still missing: *gauge symmetry*. In other words, we wish to have massless gauge bosons, *i.e.*, spacetime vectors that transform in the adjoint representation of some internal symmetry group. As a side issue, we would also like to find a way of breaking $N = 2$ supersymmetry to $N = 1$ supersymmetry (if our goal is to reproduce the MSSM) or even to $N = 0$ supersymmetry (if our goal is to reproduce the Standard Model).

It is worth considering why such gauge-boson states fail to appear for the ten-dimensional Type II strings discussed in the previous lecture. The problem is the following. In order to produce worldsheet bosons, we are restricted to considering only the NS-NS or Ramond-Ramond sectors (Sectors #1 and #2 in Table 1). In the NS-NS sector (Sector #1), the vacuum energy is $(a_R, a_L) = (-1/2, -1/2)$, so we must excite the half-energy fermionic mode oscillators $\tilde{b}^\mu_{-1/2}, b^\mu_{-1/2}$ for the both the left- and right-movers. This produces a state with *two* vector indices rather than one, and as we see from the vector-vector tensor-product decomposition in (2.20), this does not contain a vectorial state. In the Ramond-Ramond sector (Sector #2), by contrast, the vacuum energy is $(a_R, a_L) = (0, 0)$, which implies that our massless states comprise the tensor product of two Ramond spinors as in (4.9) for the Type IIA string, or as in (4.10) for the Type IIB string. In the case of the Type IIB string, we see from (4.11) that the tensor product $\bar{S}_8 \otimes S_8$ does not contain a vector state V_8. Thus, the Type IIB string contains no massless vectors. In the case of the Type IIA string, we observe from (4.16) that indeed $\bar{S}_8 \otimes C_8 \supset V_8$, and thus the Type IIA string does contain a massless vector. (This state is often called a "Ramond-Ramond gauge boson".) However, the $U(1)$ "gauge" symmetry associated with this state is too small to contain the Standard-Model gauge group, and moreover it can be shown that no states in the perturbative spectrum of the Type IIA string spectrum can carry this Ramond-Ramond charge.[a]

---

[a] Despite this fact, Ramond-Ramond charge plays a crucial role in recent developments concerning string duality. While none of the states in the *perturbative* Type IIA string spec-

In each case, the fundamental obstruction that we face is that we need to generate representations of a *gauge group* (*i.e.*, an internal symmetry group) that is *different* from the Lorentz group. Until now, all of our worldsheet fields (such as $X^\mu_{L,R}$ and $\psi^\mu_{L,R}$) have carried Lorentz indices associated with the $SO(D-1,1)$ Lorentz symmetry. In order to produce a separate gauge symmetry, we therefore need fields which do *not* carry a Lorentz index but which carry a purely internal index. (Note that these fields cannot carry a Lorentz index because we ultimately want our gauge symmetries to commute with the Lorentz symmetries.)

How can we do this? One idea is to *compactify* the Type II strings that we constructed in the previous lecture. Although this approach ultimately fails for phenomenological reasons, it will be instructive to briefly explain this idea. Recall that for the superstring, the critical dimension $D = 10$ emerges as the result of an anomaly cancellation argument: each worldsheet boson X contributes $c_X = 1$, each Majorana fermion ψ^μ contributes $c_\psi = 1/2$, and thus ten copies of each are necessary in order to cancel the "background" central charge associated with the worldsheet superconformal symmetry. But, even though we require ten bosons and ten fermions, there is no reason why we must endow *all* of them with Lorentz vector indices μ. Since we are ultimately interested in four-dimensional string theories, one natural idea is to consider these ten bosons and ten fermions in two groups, four with indices $\mu = 0, 1, 2, 3$, and the remaining six with purely internal indices $i = 1, ..., 6$. This internal symmetry could then be interpreted as a gauge symmetry.

This idea is in fact reminiscent of the original Kaluza-Klein idea whereby gauge symmetries are realized from higher-dimensional gravitational theories upon compactification. Moreover, this idea does succeed in producing gauge bosons (and gauge symmetries) in dimensions $D < 10$. However, the problem is that this idea fails to produce *enough* gauge symmetry. Specifically, although we obtain gauge symmetries that are large enough to contain the Standard Model gauge symmetry $SU(3) \times SU(2) \times U(1)$, we cannot obtain massless representations that simultaneously transform as triplets of $SU(3)$ and doublets of $SU(2)$. Such "quark" representations are required phenomenologically. Thus, even though this compactification idea is interesting as a way of generating certain amounts of gauge symmetry (and we shall return to this idea in Lecture #8), it cannot be used in order to save the superstring.

What we require, then, is a different way of introducing worldsheet fields without Lorentz vector indices. Since we will (temporarily) abandon the idea

---

trum carry Ramond-Ramond charge, these strings also contain non-trivial *solitonic* states (so-called *D-branes*) which do carry Ramond-Ramond charge. We shall briefly discuss D-branes in Lecture #10.

of removing Lorentz indices from our ten worldsheet bosons and fermions, what this means is that we require a way of obtaining *even more worldsheet fields* in ten dimensions. In other words, if we want bigger gauge symmetries in $D = 4$, then we require more than six extra fields with internal indices i, which in turn means that we already want extra fields even in the original ten-dimensional interpretation.

But how can we introduce extra worldsheet fields without violating our previous conformal anomaly cancellation arguments? Just adding extra fields will reintroduce the conformal anomaly at the quantum level.

5.2   The heterotic string: Constructing the action

The idea, of course, is to abandon the Type II string and proceed to construct a new kind of string that can accomplish the goal. This string is called the *heterotic* string, and it is this string that will be our focus for the remainder of these lectures. This string was first introduced by D. Gross, J. Harvey, E. Martinec, and R. Rohm in 1985, and for more than a decade dominated (and still continues to play a pivotal role in) discussions of string phenomenology.

Let us begin by recalling the action of the bosonic string:

$$S_{\text{bosonic}} \;=\; -\frac{1}{4\pi\alpha'} \int d^2\sigma \; \left\{ (\partial_- X_R^\mu)^2 \;+\; (\partial_+ X_L^\mu)^2 \right\} \;. \tag{5.1}$$

Here the worldsheet symmetry is simply conformal invariance, which requires that we take $\mu = 0, 1, ..., 25$ in order to cancel the conformal anomaly. Clearly, this action contains lots of worldsheet fields. However, we saw in Lecture #2 that this string does not give rise to spacetime bosons.

Next, we considered the superstring, whose action is given by:

$$S_{\text{super}} = -\frac{1}{4\pi\alpha'} \int d^2\sigma \; \left\{ (\partial_- X_R^\mu)^2 \;-\; \psi_R^\mu \partial_- \psi_{R\mu} \;+\; (\partial_+ X_L^\mu)^2 \;-\; \psi_L^\mu \partial_+ \psi_{L\mu} \right\} \;.$$
$$\tag{5.2}$$

Here the worldsheet symmetry is *superconformal* invariance, which requires that we take $\mu = 0, 1, ..., 9$ in order to cancel the superconformal anomaly. Unlike the bosonic string, this string gives rise to spacetime fermions. But as we have just explained, this string does not contain enough worldsheet fields to give rise to appropriate gauge symmetries.

Clearly, each of these strings has an advantage lacked by the other. The natural solution, then, is to attempt to "weld" them together, to "cross-breed" them in such a way as to retain the desirable attributes of each. But how can this be done?

The fundamental observation is that we are always dealing with *closed* strings, and for closed strings, we have seen that the left- and right-moving modes are essentially independent of each other and form separate theories. Indeed, only the level-matching constraint $L_0 = \bar{L}_0$ serves to relate these two halves to each other, but even this constraint applies at the level of the physical Fock space rather than the level of the action. Therefore, since these two halves are essentially independent, a natural idea is to construct a new hybrid string whose left-moving half is the left-moving half of the bosonic string, but whose right-moving half is the right-moving half of the superstring. As we shall see, this fundamental idea is just what we need. The resulting string is therefore called a *heterotic* string, where the prefex *hetero-* indicates the joining of two different things.

Given this idea, let us now see how the action for the heterotic string can be constructed. We shall do this in three successive attempts. Our first attempt would be to write an action of the form

$$S = -\frac{1}{4\pi\alpha'} \int d^2\sigma \left\{ (\partial_- X_R^\mu)^2 - \psi_R^\mu \partial_- \psi_{R\mu} + (\partial_+ X_L^\mu)^2 \right\} . \qquad (5.3)$$

In this case, the worldsheet symmetry would be conformal invariance for the left-movers, but superconformal invariance for the right-movers.

But what is the spacetime dimension of such a string? If we consider the right-moving sector, then just as in the superstring we would require $D = 10$, so that $\mu = 0, 1, ..., 9$. But given this, how do we interpret the left-moving side of the heterotic string? On the left-moving side, cancellation of the *conformal* (rather than superconformal) anomaly requires that we still retain 26 X_L fields! But if only ten of these fields are spacetime coordinates, then the remaining sixteen must be mere internal scalar fields. In other words, rather than carry the μ index (which would imply that these X fields would transform as vectors under the spacetime Lorentz group $SO(9,1)$), these sixteen extra fields must instead carry a purely internal index $i = 1, ..., 16$. So our second attempt at writing a heterotic string action would result in an action of the form

$$S = -\frac{1}{4\pi\alpha'} \int d^2\sigma \left\{ (\partial_- X_R^\mu)^2 - \psi_R^\mu \partial_- \psi_{R\mu} + (\partial_+ X_L^\mu)^2 + (\partial_+ X_L^i)^2 \right\}$$
$$(5.4)$$

where we have explicitly separated the left-moving bosons into two groups, with $\mu = 0, 1, ..., 9$ and $i = 1, ..., 16$.

But there still remains a subtlety. We cannot simply *decide* to remove the μ index from the X fields and make no other changes, because these X^i fields would continue to have a mode-expansion of the form (2.9) with the μ index replaced by an internal index i. While the interpretation of the oscillation exponential terms in (2.9) is not problematic, how would we interpret the "zero-mode" terms $x^i + \ell^2 p^i(\sigma_1 + \sigma_2)$? In the case of the spacetime coordinate fields X^μ, recall that these "zero-mode" quantities x^i and p^i are interpreted as the center-of-mass position and momentum of the string. But for purely internal fields X^i, this interpretation is problematic. To clarify this difficulty, let us consider the worldsheet energy $L_0^{(\mathrm{com})}$ associated with these degrees of freedom, as in (2.13). Just as in the case of the spacetime coordinates X^μ, these worldsheet energies for the X^i fields would *a priori* take *continuous* values, thereby leading to a continuous spectrum even in $D = 10$. A continuous spectrum, of course, indicates nothing but the appearance of extra spacetime dimensions, so even though we may have replaced the index μ with the index i, we have not really solved the fundamental problem that there are too many uncompactified degrees of freedom amongst the left-movers.

Therefore, we still must find a way to replace this continuous spectrum with a discrete one. Because the following discussion is slightly technical and outside the main line of the development of the heterotic string action, we shall separate it from the main flow of the text. The reader uninterested in the following details can skip them completely and proceed directly to the resumption of the main text.

In order to eliminate this continuous spectrum, we must compactify these extra sixteen dimensions. This is analogous to discretizing the continuous spectrum of a free particle (plane wave) by localizing it in a box. In the present case, we can choose to compactify each of these extra spacetime "coordinates" X^i on a circle of radius R_i. What this means, operationally, is that we make the following topological identification in *spacetime*:

$$X^i \iff X^i + 2\pi R_i \ . \tag{5.5}$$

For simplicity (and as we shall see, without loss of generality), we shall take $R_i = R$ for all i. Thus, rather than demand simple periodicity of the X^i "coordinates" as in (2.6) as we traverse the closed string worldsheet, we must allow for the more general possibility

$$X^i(\sigma_1 + \pi, \sigma_2) = X^i(\sigma_1, \sigma_2) + 2\pi n_i R \ , \qquad n_i \in \mathbb{Z} \tag{5.6}$$

where the integer n_i is called the "winding number". The interpretation of this condition is that as we traverse the closed string once on the *worldsheet* (*i.e.*, as $\sigma_1 \to \sigma_1 + \pi$), the spacetime "coordinate" field X^i traverses the compactified spacetime circle n_i times.

In other words, the closed string "winds" around the ith compactified *spacetime* circle n_i times. Because of this compactification, we see that the momentum p^i is now quantized (as we would expect for any particle in a periodic box of length R), and is restricted to take the values $p^i = m_i/R$, $m_i \in \mathbb{Z}$. Indeed, working out the most general mode-expansion consistent with (5.6), we find that a given such coordinate X^i takes the form

$$X(\sigma_1, \sigma_2) = x + 2nR\sigma_1 + \ell^2 \frac{m}{R}\sigma_2 + \text{oscillators} , \qquad (5.7)$$

where $\ell \equiv \sqrt{2\alpha'}$ is our fundamental length scale and where 'oscillators' generically denotes the higher frequency modes. This decomposes into left- and right-moving components

$$X_{L,R}(\sigma_1 \pm \sigma_2) = \tfrac{1}{2}x + \left(\frac{\alpha' m}{R} \pm nR \right)(\sigma_2 \pm \sigma_1) + \text{oscillators} . \quad (5.8)$$

Comparing (5.8) with (2.9) enables us to identify the left- and right-moving compactified momenta

$$p_{L,R} \equiv \frac{m}{R} \pm \frac{nR}{\alpha'} . \qquad (5.9)$$

We would then simply keep X_L in our heterotic theory.

Let us pause here to note an interesting phenomenon: this mode-expansion is invariant under the simultaneous exchange $R \leftrightarrow \alpha'/R$, $m \leftrightarrow n$. This is a so-called *T-duality*. What this means is that unlike point particles, strings cannot distinguish between extremely large spacetime compactification radii and extremely small spacetime compactification radii. Indeed, although the usual momentum m/R is extremely small in the first case and extremely large in the second, we see from the above mode-expansions that there is another contribution to the momentum, a "winding-mode momentum" nR/α', which compensates by growing large in the first case and small in the second. Since there is no physical way of distinguishing between these two types of momenta, the string spectrum is ultimately invariant under this T-duality symmetry. This duality underlies many of the unexpected physical properties of strings relative to point particles, and has important (and still not well-understood) implications for string cosmology. More importantly, however, this duality dramatically illustrates the breakdown of the traditional (field-theoretic) view of the linearly ordered progression of length scales and energy scales as we approach the string scale.

Having succeeded in avoiding the consequences of a continuous momentum p^i, our final question is the size of the radius R. It would certainly be aesthetically undesirable if we were forced to incorporate a new, fundamental, unfixed parameter R into our string theory. Fortunately, it turns out that in $D = 10$, there are only a very restricted set of possibilities that lead to consistent theories, and these restrictions imply that we can restrict our attention to the simple case $R = \ell = \sqrt{2\alpha'}$ *without loss of generality.* Thus, we see that R can be taken to be at the string scale, and hence essentially unobservable to "low-energy" measurements.

In order to see what is special about this radius, recall that the conformal anomaly contribution for each worldsheet boson is $c_X = 1$, while the conformal anomaly contribution for each worldsheet Majorana (real) fermion is $c_\psi = 1/2$. This suggests that the spectrum of a single compactified boson X might somehow be related to the spectrum of two Majorana fermions ψ_1, ψ_2, and this is indeed the case. Such a relation is typically referred to as a "boson-fermion equivalence" (which is possible in two dimensions because the usual spin-statistics distinction between bosons and fermions does not apply in two dimensions). In general, the spectrum of a compactified boson is identical to the spectrum of two Majorana fermions which are *coupled* to each other in a radius-dependent manner, and $R = \sqrt{2\alpha'}$ is the only value of the radius for which this coupling vanishes. Thus, if X is compactified on a circle of radius $R = \sqrt{2\alpha'}$, then the spectrum of quantum excitations of X is identical to the spectrum of quantum excitations of two *free* Majorana fermions ψ_1, ψ_2 (or equivalently those of one *complex* fermion $\Psi \equiv \psi_1 + i\psi_2$).[b] In fact, at a mathematical level, it turns out that this equivalence takes the form of an actual *equality*

---

[b]We are again cheating slightly here. The rigorous statement is that we must compactify the X boson on a so-called $\mathbb{Z}_2$ *orbifold* with this radius in order for the spectrum of X to be identical to that of two free Majorana fermions. The equivalence between these bosonic and fermionic systems can be demonstrated explicitly at the level of their full underlying left/right two-dimensional conformal field theories, and will be exploited further in Lecture #8. By contrast, compactifying X on a *circle* of this radius yields the spectrum of a single *complex* fermion, and the full left/right conformal field theory corresponding to a single complex fermion actually differs from that corresponding to two real fermions. These distinctions between circles and orbifolds, and likewise between a single complex fermion and two real fermions, will not be relevant for what follows.

between the *product* $\psi_1\psi_2$ and the partial derivative ∂X. Note, however, that while this specific radius is special from the point of view of boson/fermion equivalence, this is *not* the self-dual radius with respect to the T-duality transformation $R \leftrightarrow \alpha'/R$.

The upshot, then, is that in the action (5.4), we are free to replace the worldsheet bosons X^i ($i = 1, ..., 16$) with *complex* worldsheet fermions Ψ^i ($i = 1, ..., 16$). For ten-dimensional heterotic strings, we shall see that this replacement can be made *without loss of generality*. This replacement suffices to make the center-of-mass "momenta" associated with the X^i fields discrete rather than continuous, as we require. Given this, the final action for the heterotic string takes the form:

$$S_{\text{heterotic}} = -\frac{1}{4\pi\alpha'} \int d^2\sigma \left\{ (\partial_+ X_L^\mu)^2 - \bar{\Psi}_L^i \partial_+ \Psi_L^i + (\partial_- X_R^\mu)^2 - \psi_R^\mu \partial_- \psi_{R\mu} \right\}$$
$$(5.10)$$

where ψ_R are Majorana-Weyl (real) right-moving worldsheet fermions, where Ψ_L are complex Weyl left-moving fermions, and where $\mu = 0, 1, ..., 9$ and $i = 1, ..., 16$.

5.3 Quantizing the heterotic string

The next step, then, is to quantize the worldsheet fields of the heterotic string. The quantization of the bosonic fields X^μ and worldsheet Majorana fermions ψ_R^μ was discussed in previous lectures, and does not change in this new setting. The only new feature, then, are the mode-expansion and quantization rules for the *complex* fermions Ψ_L^i.

Once again, there are two possible mode expansions for the left-moving complex fermions Ψ, depending on whether we choose Neveu-Schwarz (anti-periodic) or Ramond (periodic) boundary conditions.[c] In the case of anti-periodic boundary conditions, recall that our mode-expansion (3.12) for left-moving *real* (Majorana) fermions can be written in the form

$$\psi(\sigma_1 + \sigma_2) = \sum_{r=1/2}^{\infty} \left[b_r e^{-ir(\sigma_1+\sigma_2)} + b_r^\dagger e^{+ir(\sigma_1+\sigma_2)} \right] \qquad (5.11)$$

---

[c] Because there is no worldsheet supersymmetry that relates these left-moving fermions to corresponding left-moving bosons X^μ, more general boundary conditions may actually be imposed in this case. However, for heterotic strings in ten dimensions, it turns out that we can restrict our attention to periodic or anti-periodic boundary conditions without loss of generality. Fermions with generalized worldsheet boundary conditions will be discussed further in Lecture #7.

270

where we recall the hermiticity condition $b_{-r} = b_r^\dagger$. Thus, for a left-moving *complex* fermion, our analogous mode-expansion takes the form

$$\Psi(\sigma_1 + \sigma_2) = \sum_{r=1/2}^{\infty} \left[b_r e^{-ir(\sigma_1+\sigma_2)} + d_r^\dagger e^{+ir(\sigma_1+\sigma_2)} \right] \qquad (5.12)$$

which of course implies

$$\Psi^\dagger(\sigma_1 + \sigma_2) = \sum_{r=1/2}^{\infty} \left[b_r^\dagger e^{+ir(\sigma_1+\sigma_2)} + d_r e^{-ir(\sigma_1+\sigma_2)} \right] . \qquad (5.13)$$

For $r > 0$, b_r destroys fermionic excitations and $b_r^\dagger$ creates them, while d_r destroys *anti-fermionic* excitations and $d_r^\dagger$ creates them. Thus, as expected, the only new feature is the presence of twice as many mode degrees of freedom, one set associated with fermionic excitations and the other with their anti-fermionic counterparts. These modes satisfy the usual anti-commutation relations

$$\{b_r^\dagger, b_s\} = \{d_r^\dagger, d_s\} = \delta_{rs} . \qquad (5.14)$$

The corresponding number operator and worldsheet energy contributions are then given by

$$N = \sum_{r=1/2}^{\infty} \left(b_r^\dagger b_r - d_r^\dagger d_r \right)$$

$$L_0 = \sum_{r=1/2}^{\infty} r \left(b_r^\dagger b_r + d_r^\dagger d_r \right) . \qquad (5.15)$$

Note that the anti-particle excitations *subtract* from the number operator yet *add* to the total energy. Finally, as expected, the vacuum energy contribution from each complex Neveu-Schwarz fermion is twice that for each real Neveu-Schwarz fermion: $a_\Psi = 2a_\psi = -1/24$.

The Ramond case, of course, is more subtle because of the zero-mode. It turns out that the complex-fermion mode-expansion is given by

$$\Psi(\sigma_1 + \sigma_2) = \sum_{n=1}^{\infty} \left[b_n e^{-in(\sigma_1+\sigma_2)} + d_n^\dagger e^{+in(\sigma_1+\sigma_2)} \right] + b_0$$

$$\Psi^\dagger(\sigma_1 + \sigma_2) = \sum_{n=1}^{\infty} \left[b_n^\dagger e^{+in(\sigma_1+\sigma_2)} + d_n e^{-in(\sigma_1+\sigma_2)} \right] + b_0^\dagger , \qquad (5.16)$$

with the anti-commutation relations

$$\{b_m^\dagger, b_n\} = \{d_m^\dagger, d_n\} = \delta_{mn} . \tag{5.17}$$

In (5.16), we have explicitly separated out the zero-mode from the higher-frequency modes. The number operator and worldsheet energy conributions are given by

$$N = \sum_{r=1/2}^{\infty} \left(b_r^\dagger b_r - d_r^\dagger d_r\right) + b_0^\dagger b_0$$

$$L_0 = \sum_{r=1/2}^{\infty} r \left(b_r^\dagger b_r + d_r^\dagger d_r\right) . \tag{5.18}$$

Note that there is no worldsheet energy contribution from the zero-modes. Finally, the vacuum energy contribution from each complex Ramond fermion is twice that for each real Ramond fermion: $a_\Psi = 2a_\psi = +1/12$.

One might wonder, at first, why there is no *anti-particle* zero-mode d_0. However, such an anti-particle zero-mode d_0 would be *equivalent* to the *particle* zero-mode b_0. The easiest way to see this is to realize that ultimately (5.16) represents a Fourier-decomposition of the $\Psi(\sigma_1 + \sigma_2)$ into different harmonic frequencies (exponentials). By its very nature, the zero-mode is the constant term in such a decomposition (since it corresponds to zero frequency), and this constant term is nothing but b_0. However, there can only be *one* degree of freedom associated with a given constant term. Having an additional zero-mode d_0 would thus represent a redundant (non-independent) degree of freedom. Of course, whether we associate b_0 or d_0 with the constant term is purely a matter of convention.

Given this observation, we are finally in a position to explain our counting of zero-mode states in Lectures #3 and #4. Since there is only one zero-mode degree of freedom for each *complex* worldsheet fermion, there can really be only "half" a zero-mode for each *real* worldsheet fermion. This explains the footnote in Sect. 3.2, and also explains why (in the paragraph following (4.8)) we counted only one zero-mode excitation per *pair* of Majorana fermions. This also explains why, ultimately, the treatment of the Ramond zero-mode for a *real* worldsheet fermion is rather subtle: essentially we must take a "square root" of the complex Ramond zero-mode b_0. There does exist a consistent method for taking this square root, but this is beyond the scope of these lectures. For our purposes, it will simply be sufficient to recall that there is only one zero-mode state for each complex worldsheet fermion, or for each pair of real worldsheet fermions.

6 Lecture #6: Some famous heterotic strings

Our next step is to construct actual heterotic string *models*, just as we did for the superstring. This will be the subject of the present lecture.

6.1 *General overview*

Before plunging into details, it is worthwhile to consider the general features that will govern the construction of our heterotic string models. Recall from the previous lecture that the worldsheet fields of the heterotic string in light-cone gauge consist of eight right-moving worldsheet bosons X_R^μ, eight left-moving worldsheet bosons X_L^μ, eight right-moving Majorana (real) worldsheet fermions ψ_R^μ, and sixteen left-moving complex worldsheet fermions Ψ_L^i ($i = 1, ..., 16$).

The role of the right-moving fermions ψ_R^μ is the same as in the superstring: if they have Neveu-Schwarz modings, the corresponding states are spacetime bosons, and if they Ramond modings, the corresponding states are spacetime fermions. Indeed, by properly stitching these sectors together, it may also be possible to obtain spacetime supersymmetry (as in the superstring). Note that unlike the superstring, however, these boson/fermion identifications hold *regardless* of the modings of the left-moving complex fermions Ψ_L^i. This is because only the right-moving fermions carry spacetime Lorentz indices μ, and hence only these fermions determine the representations of the spacetime Lorentz algebra.

The role of the left-moving complex fermions Ψ_L^i is analogous. Because they carry internal indices rather than spacetime Lorentz indices, the symmetries they carry are also internal, and as we shall see, they can be interpreted as gauge symmetries. Indeed, these Ψ_L^i fields are precisely the internal fields we were hoping to obtain in Sect. 5.1. When they have Neveu-Schwarz modings, these fermions provide "vectorial" (scalar, vector, tensor) representations of the internal gauge symmetry. When they have Ramond modings, by contrast, they provide "spinorial" representations of the internal gauge symmetry. Thus, we expect a rich gauge representation structure in these models as well.

As with the superstring, different models can be constructed depending on how the different modings are joined together to form our set of underlying sectors, and how the corresponding GSO constraints are implemented. We shall construct explicit models below. But it is already apparent that the heterotic string contains all the ingredients we require for successful phenomenology. By choosing certain combinations of right-moving fermionic modings with left-moving fermionic modings, we can control which gauge-group representations are bosonic and which are fermionic. Moreover, by choosing the relative modings *amongst* the left-moving complex fermions, we can even control the gauge group that is ultimately produced.

6.2 Sectors and GSO constraints

Just as in the superstring, we begin the process of model-building by choosing an appropriate set of underlying sectors and corresponding GSO constraints. Moreover, just as in the superstring, we know that preservation of the right-moving worldsheet supersymmetry (or equivalently spacetime Lorentz invariance) requires that we choose our eight right-moving fermions ψ_R^μ to all have the same boundary condition in each sector. This implies that we can, if we wish, combine these right-moving fermions to form four complex right-moving fermions which we can denote Ψ_R^μ. (We retain the index μ to remind ourselves that these fields carry indices with respect to the spacetime Lorentz algebra, even though strictly speaking only the real fields ψ_R^μ carry such vectorial indices.) However, unlike the superstring, there is no longer any such restriction on the boundary conditions of the left-moving fermions Ψ_L^μ. Thus, there remains substantial freedom in choosing the boundary conditions of these left-moving fermions. Ultimately this choice becomes the choice of the gauge group for the particular model in question.

In the next lecture, we shall provide a detailed discussion of the rules by which one can choose these boundary conditions and determine their associated GSO constraints. Therefore, for the time being, we shall simply restrict our attention to the sectors listed in Table 2. Note that the corresponding vacuum energies are also listed in Table 2. In order to compute these energies, we can continue to use the middle expression in (4.1) where we recall that n_R and n_{NS} count the number of *real* worldsheet fermions. Thus, for complex fermions, these numbers are doubled.

Before proceeding further, we can immediately deduce some physical properties of the string states that would emerge in each sector. First, we see that Sector #1 is the only sector from which tachyons can possibly emerge. This is because the level-matching constraints prevent tachyons in any other sector (*i.e.*, there is no other sector which for which both a_L and a_R are negative). Second, we observe that Sectors #2 and #3 cannot give rise to massless states. This again follows from the level-matching constraints, and implies that (for phenomenological purposes) we will not need to consider the states arising in these sectors. Finally, we observe that Sectors #1,3,5,6 give rise to spacetime bosons, while Sectors #2,4,7,8 give rise to spacetime fermions.

In some sense, Sectors #1–4 are the direct analogues of the four possible sectors in Table 1 for the superstring. Thus, the heterotic models that result

#	$\psi_R^{i=1,\ldots,8}$	$\Psi_L^{i=1,\ldots,8}$	$\Psi_L^{i=9,\ldots,16}$	a_R	a_L
1	NS	NS		$-1/2$	-1
2	R	R		0	$+1$
3	NS	R		$-1/2$	$+1$
4	R	NS		0	-1
5	NS	NS	R	$-1/2$	0
6	NS	R	NS	$-1/2$	0
7	R	NS	R	0	0
8	R	R	NS	0	0

Table 2: Eight possible sectors for ten-dimensional heterotic strings, numbered 1 through 8. Here 'NS' and 'R' respectively indicate Neveu-Schwarz (anti-periodic) and Ramond (periodic) boundary conditions for worldsheet fermions, and a_R and a_L respectively denote the corresponding right- and left-moving vacuum energies.

from these sectors will be the analogues of the Type 0 and Type II superstring models. However, the additional Sectors #5–8 represent new sectors that arise only for heterotic strings. We hasten to add that these sectors are not unique, and others could equally well have been chosen. We will discuss these possibilities in the next lecture.

The next issue we face is to determine which *combinations* of sectors form self-consistent sets. It turns out (following the rules to be discussed in Lecture #7) that there are three different possibilities:

- Case A: we consider Sectors #1 and #2 by themselves;

- Case B: we consider Sectors #1 through #4 by themselves; or

- Case C: we consider *all* Sectors #1 through #8.

For each of these cases, there is then a different set of GSO constraints for each sector. As we have seen in our discussion of the superstring, the more sectors we have in our model, the more GSO constraints there are in each sector. In particular, each time the number of sectors doubles, the number of GSO constraints in each sector increases by one. For completeness, Table 3 lists the GSO constraints that apply in each sector for each of these three cases.

Once again, observe the *pattern* of the GSO constraints. In Case A, we have only Sectors #1 and #2, for which all right-moving and left-moving boundary conditions are identical. Thus, the GSO constraints that apply in Case A combine N_L and N_R together. (Recall that since $N_{L,R} \in \mathbb{Z}$, we can just as easily write the GSO constraint for Case A as $N_L + N_R = $ odd.) When we move from Case A to Case B, we introduce two new sectors (Sectors #3

Sector #	Case A	Case B	Case C
1	$N_L - N_R = $ odd	$N_L - N_R = $ odd $N_L = $ even	$N_L - N_R = $ odd $N_L = $ even $^{(8)}N_L = $ even
2	$N_L - N_R = $ odd	$N_L - N_R = $ odd $N_L = $ even	$N_L - N_R = $ odd $N_L = $ even $^{(8)}N_L = $ even
3	—	$N_L - N_R = $ odd $N_L = $ even	$N_L - N_R = $ odd $N_L = $ even $^{(8)}N_L = \left\{ \begin{array}{c} \text{odd} \\ \text{even} \end{array} \right\}$
4	—	$N_L - N_R = $ odd $N_L = $ even	$N_L - N_R = $ odd $N_L = $ even $^{(8)}N_L = \left\{ \begin{array}{c} \text{odd} \\ \text{even} \end{array} \right\}$
5	—	—	$N_L - N_R = $ odd $N_L = \left\{ \begin{array}{c} \text{odd} \\ \text{even} \end{array} \right\}$ $^{(8)}N_L = \left\{ \begin{array}{c} \text{odd} \\ \text{even} \end{array} \right\}$
6	—	—	$N_L - N_R = $ odd $N_L = \left\{ \begin{array}{c} \text{odd} \\ \text{even} \end{array} \right\}$ $^{(8)}N_L = $ even
7	—	—	$N_L - N_R = $ odd $N_L = \left\{ \begin{array}{c} \text{odd} \\ \text{even} \end{array} \right\}$ $^{(8)}N_L = $ even
8	—	—	$N_L - N_R = $ odd $N_L = \left\{ \begin{array}{c} \text{odd} \\ \text{even} \end{array} \right\}$ $^{(8)}N_L = \left\{ \begin{array}{c} \text{odd} \\ \text{even} \end{array} \right\}$

Table 3: GSO constraints for each of the eight heterotic string sectors in Table 2. Here the notation $^{(8)}N_L \equiv \sum_{i=1}^{8} N^{(i)}$ indicates the total left-moving number operator for only the first *eight* left-moving complex fermions. As before, the braces indicate different *correlated* choices of GSO projections, so that we simultaneously choose either the upper choice or the lower choice for all sets.

and #4 in Table 2) which "twist" the boundary conditions of the right-movers relative to those of the left-movers. This has the effect of introducing a new GSO constraint in each sector, one which distinguishes separately between N_L and N_R. Finally, when we move from Case B to Case C, we introduce four new sectors (Sectors #5 through #8) which introduce an additional "twist" that distinguishes between the first eight left-moving fermions $\Psi_L^{i=1,\ldots,8}$ and the second eight left-moving fermions $\Psi_L^{i=9,\ldots,16}$. The corresponding new GSO constraint in each sector is then one which is sensitive only to $^{(8)}N_L \equiv \sum_{i=1}^8 N^{(i)}$. This suggests (and we shall see explicitly in Lecture #7) that the set of sectors is deeply correlated with the set of GSO constraints that are applied in each sector: each new "twist" introduces both a new set of sectors and a new GSO constraint in each sector. The fact that we are considering only Ramond or Neveu-Schwarz boundary conditions for our left-moving complex fermions Ψ_L^i means that each successive twist doubles the number of sectors and introduces one new GSO constraint in each sector. These are called $\mathbb{Z}_2$ twists. If we were to consider more general "multi-periodic" boundary conditions for the left-moving fermions (which is possible because they are not related to the left-moving worldsheet bosons by worldsheet supersymmetry), then we could introduce so-called "higher-order" twists that would result in more complicated GSO constraints. However, it turns out that in ten dimensions, we lose no generality by restricting our attention to such $\mathbb{Z}_2$ twists.

6.3 Four ten-dimensional heterotic string models

It is apparent from Table 3 that Case A and Case B each correspond to one heterotic string model, while Case C corresponds to two separate heterotic string models. Thus, the GSO constraints in Table 3 together give rise to four distinct heterotic string models. In the remainder of this lecture, we shall work out the physical properties of these four models.

The non-supersymmetric $SO(32)$ string

Let us begin by considering Case A, which consists of only Sectors #1 and #2. Only Sector #1 (the so called "NS-NS sector") can contain massless states. As indicated in Table 1, the vacuum energy in this sector is $(a_R, a_L) = (-1/2, -1)$. Thus, at the bare minimum, the level-matching constraint $L_0 = \bar{L}_0$ forces us to excite at least a half-unit of energy on the left-moving side. This can be accomplished by exciting any of the left-moving half-unit fermionic modes, since in this sector the left-moving fermions all have Neveu-Schwarz boundary conditions and thus contain half-integer modings. This produces the 32 possible states

$$|0\rangle_R \otimes b^i_{-1/2}|0\rangle_L \quad \text{and} \quad |0\rangle_R \otimes d^i_{-1/2}|0\rangle_L \ . \tag{6.1}$$

Note that these states also satisfy the single applicable GSO constraint $N_L - N_R =$ odd, so they remain in the spectrum. From (4.6), we see that these states are tachyonic with $\alpha' M^2 = -2$.

Further states are realized by exciting higher worldsheet modes. Because our worldsheet modes are quantized in minimum half-integer steps, we see that the next excited states in this model are massless. These states come in two varieties:

$$\tilde{b}^\mu_{-1/2}|0\rangle_R \otimes \alpha^\nu_{-1}|0\rangle_L \tag{6.2}$$

and

$$\tilde{b}^\mu_{-1/2}|0\rangle_R \otimes \begin{cases} b^i_{-1/2}\, b^j_{-1/2}\, |0\rangle_L \\ b^i_{-1/2}\, d^j_{-1/2}\, |0\rangle_L \\ d^i_{-1/2}\, b^j_{-1/2}\, |0\rangle_L \\ d^i_{-1/2}\, d^j_{-1/2}\, |0\rangle_L \ . \end{cases} \tag{6.3}$$

In (6.2), we have excited the lowest mode of the left-moving worldsheet boson X^μ_L, whereas in (6.3) we have excited two of the lowest modes of the left-moving fermions $\Psi^{i,j}_L$. Note that it is possible to excite both the particle and anti-particle modes from the same fermion Ψ^i, and thus there is no restriction that $i \neq j$. Also note that all of these states in (6.2) and (6.3) satisfy the GSO constraint $N_L - N_R =$ odd. While $N_R = 1$ in all cases, we have $N_L = 0$ in (6.2) (since the number operators are defined not to include the contributions from worldsheet bosons), and $N_L = 2$ in (6.3).

How do we interpret these states? Once again, the states (6.2) are easily recognized as our gravity multiplet, consisting of the spin-two graviton $g_{\mu\nu}$, the spin-one anti-symmetric tensor $B_{\mu\nu}$, and the spin-zero dilaton ϕ. It is interesting to note that this state (6.2) is realized as a hybrid of the gravity multiplet state in the bosonic string (2.19) and in the superstring (4.7). This reflects the underlying construction of the heterotic string, and ensures that the heterotic string, like its predecessors, is also a theory of quantized gravity. Once again, the appearance of the gravity multiplet is a useful cross-check of the GSO constraints.

The states in (6.3) have a different interpretation, however. Clearly, their Lorentz structure indicates that they are massless Lorentz vectors. Thus, they are to be interpreted as spacetime *gauge bosons*. Thus, we see that the heterotic string has succeeded in providing us with spacetime gauge symmetry, just as we had originally hoped.

But what is the gauge group? Of course, the gauge group is ultimately determined from the i, j indices, and since (in Cases A and B) we have not destroyed the rotational symmetry in the space of the 16 complex left-moving fermions Ψ_L^i (or the 32 real left-moving fermions into which they can be decomposed), we immediately suspect that the gauge symmetry should be $SO(32)$. There are number of ways to deduce that this is correct. Perhaps the easiest way is simply to *count* the gauge boson states in (6.3). If we restrict our attention to the cases $i \neq j$, then there are $(2 \cdot 16)(2 \cdot 15)/2$ states. The first factor $(2 \cdot 16)$ reflects the fact that for each of the 16 possible choices of Ψ_L^i, we can excite either the fermion or anti-fermion mode. The second factor $(2 \cdot 15)$ reflects the same set of options for the second fermion Ψ_L^j, and we divide by two as the interchange symmetry factor. There are also the cases with $i = j$: from such cases we obtain 16 possible states, reflecting the 16 different fermions Ψ_L^i whose fermion and anti-fermion modes are jointly excited. The total number of states is then

$$\frac{(2 \cdot 16)(2 \cdot 15)}{2} \; + \; 16 \; = \; 496 \; = \; \dim SO(32) \, . \tag{6.4}$$

Of course, the above counting method for determining the gauge group is hardly precise, for there are a number of gauge groups with the same overall dimension (and we shall come across another such gauge group very soon). We therefore require a more sophisticated method which also generalizes to more complicated cases. By definition, of course, the gauge group can be determined by explicitly examining the charges of the gauge boson states and determining which Lie algebra (*i.e.*, which root system) they fill out. We therefore need a way of determining the charges of the gauge boson states. Since our gauge symmetry is ultimately associated with the left-moving worldsheet fermions Ψ_L^i, the relevant current in this case is simply the worldsheet current $J^i \equiv \overline{\Psi}_L^i \Psi_L^i$. From this, we can deduce the associated charge Q_i. It turns out that

Great Leap #7: The charge associated with each worldsheet fermion Ψ_L^i for a given string state with fermionic excitation number $N^{(i)}$ is given by $Q_i \equiv N^{(i)} + q_i$. Here q_i is a "background" charge which is 0 if Ψ_L^i is a Neveu-Schwarz fermion and $-1/2$ if Ψ_L^i is a Ramond fermion.

Given this result, we can easily deduce the gauge group for the case in question. For simplicity, let us first imagine that there are only *two* left-moving fermions $\Psi_L^{i=1,2}$. In this case, (6.3) reduces to six states:

$$b_{-1/2}^1 b_{-1/2}^2 |0\rangle_L \, , \quad b_{-1/2}^1 d_{-1/2}^2 |0\rangle_L \, , \quad d_{-1/2}^1 b_{-1/2}^2 |0\rangle_L \, ,$$
$$d_{-1/2}^1 d_{-1/2}^2 |0\rangle_L \, , \quad b_{-1/2}^1 d_{-1/2}^1 |0\rangle_L \, , \quad b_{-1/2}^2 d_{-1/2}^2 |0\rangle_L \, . \tag{6.5}$$

For each of these states, there are two charges, Q_1 and Q_2, associated with each of the two complex fermions. If we denote these states as A through F respectively, we can plot the charges of these six states as in Fig. 8. The resulting diagram is easily recognized as the root system (or equivalently the weight system of the adjoint representation) of the Lie group $SO(4)$. Generalizing from two complex fermions to n complex fermions analogously yields the gauge group $SO(2n)$, provided that all n complex fermions have the same modings. Thus, in the case of 16 complex fermions, we find the gauge group $SO(32)$.

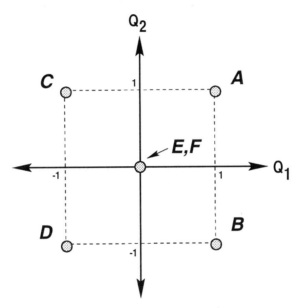

Figure 8: The two-dimensional "charge lattice" associated with the six string states A through F in (6.5). Note that the two states E and F fill out the Cartan subalgebra of the root system. For a ten-dimensional heterotic string, the charge lattice is always sixteen-dimensional (generally implying a gauge group of rank 16), with a Cartan subalgebra consisting of sixteen gauge boson states.

Note that this argument suffices to show that the gauge bosons fill out the adjoint representation of $SO(32)$. However, it does not demonstrate that all other string states in the model fall into representations of this gauge group. Of course, this is required for the consistency of the string. However, such a result can indeed be proven mathematically by constructing the current operators associated with the gauge group in question (as discussed above), and demonstrating that all states surviving the appropriate GSO constraints transform appropriately under these currents. For example, the 32 tachyonic states in (6.1) transform in the vector representation of $SO(32)$, and the gravity multiplet (6.2) transforms as a singlet of $SO(32)$ (as it must). However, a proof that this holds for all states in both the massless and massive string spectrum is beyond the scope of these lectures.

We should also point out that what emerges in such closed string theories is not simply the algebra associated the gauge symmetry in question, but rather an infinite-dimensional extension (or "affinization") of it. Such affine Lie algebras are discussed in Ginsparg (reference given at the end of Lecture #1), and play an important role in the consistency and phenomenology of such heterotic string theories.

To summarize, then, we see that Case A results in a tachyonic string model with quantum gravity and $SO(32)$ gauge symmetry. In addition to 32 scalar tachyons transforming in the vector representation of $SO(32)$, this model contains massless gauge bosons transforming in the adjoint representation of $SO(32)$ as well as the usual gravity multiplet. This non-supersymmetric $SO(32)$ heterotic string model is the heterotic analogue of the Type 0 string models in Lecture #4.

The supersymmetric $SO(32)$ string

Let us now proceed to Case B. In this case there are four sectors (#1 through #4 in Table 2), and we must impose the GSO constraints listed in the second column of Table 3.

Let us begin by considering the states from Sector #1. These are the same as those considered in Case A, except that we must now impose the additional GSO constraint $N_L = $ even. This projects out the tachyonic states (6.1), but preserves the gravity multiplet as well as the gauge bosons.

As we discussed previously, Sectors #2 and #3 contain no massless states. Therefore, all that remains is to consider the states from Sector #4. Here the vacuum energy is $(a_R, a_L) = (0, -1)$. The right-moving ground state in this sector is the Ramond zero-mode ground state, which we have previously denoted $\{\tilde{b}_0^\mu\}|0\rangle_R$, and thus massless states are realized only through non-zero excitations of the left-movers. The possible states are

$$\{\tilde{b}_0^\mu\}|0\rangle_R \otimes \begin{cases} \alpha_{-1}^\nu |0\rangle_L \\ b_{-1/2}^i \, b_{-1/2}^j \, |0\rangle_L \\ b_{-1/2}^i \, d_{-1/2}^j \, |0\rangle_L \\ d_{-1/2}^i \, b_{-1/2}^j \, |0\rangle_L \\ d_{-1/2}^i \, d_{-1/2}^j \, |0\rangle_L \; . \end{cases} \tag{6.6}$$

In each case, the GSO constraints imply that we can excite only an odd number of right-moving zero-modes. According to our previous conventions, this indicates that the right-moving ground state corresponds to the spacetime Lorentz spinor $\bar{\mathbf{S}}_8$ (rather than the conjugate spinor $\bar{\mathbf{C}}_8$).

It is, by now, easy to interpret the states in (6.6). The first state provides the superpartner states to the gravity multiplet, and contains a gravitino. This implies that the model has spacetime supersymmetry. Likewise, the remaining states correspond to the superpartners of the $SO(32)$ gauge bosons, and contain the $SO(32)$ gauginos. The chirality of these spinor states is fixed by the GSO constraint and the right-moving ground state $\bar{\mathbf{S}}_8$.

Summarizing, we see that this model therefore consists of the following states. We shall describe these states using the notation $\bar{R}_1 \otimes (R_2; R_3)$ where R_1, R_2 are representations of the spacetime Lorentz group, and where R_3 is a representation of the $SO(32)$ gauge group. These states consist of

$$\bar{\mathbf{V}}_8 \otimes (\mathbf{V}_8; \mathbf{1}) \ , \quad \bar{\mathbf{V}}_8 \otimes (\mathbf{1}; \mathbf{adj}) \ , \quad \bar{\mathbf{S}}_8 \otimes (\mathbf{V}_8; \mathbf{1}) \ , \quad \bar{\mathbf{S}}_8 \otimes (\mathbf{1}; \mathbf{adj}) \ , \qquad (6.7)$$

where the first and third states form the $N = 1$ supergravity multiplet and the second and fourth states form the $SO(32)$ gauge boson supermultiplet. Together these states can be written in the factorized form

$$(\bar{\mathbf{V}}_8 \oplus \bar{\mathbf{S}}_8) \ \otimes \ \{(\mathbf{V}_8; \mathbf{1}) \oplus (\mathbf{1}; \mathbf{adj})\} \ , \qquad (6.8)$$

thereby explicitly exhibiting the supersymmetry $\mathbf{V}_8 \leftrightarrow \mathbf{S}_8$.

This string is the famous supersymmetric $SO(32)$ heterotic string. Although not directly relevant for string phenomenology, this string plays a vital role in recent developments in string duality (to be discussed briefly in Lecture #10).

The $SO(16) \times SO(16)$ and $E_8 \times E_8$ strings

Let us now proceed to Case C. As discussed in Sect. 6.2, this case differs from Case B because we have now "twisted" the second group of eight left-moving complex worldsheet fermions relative to the first set. A priori, it is easy to imagine that this twist will break the gauge symmetry $SO(32) \rightarrow SO(16) \times SO(16)$. However, there a few surprises still in store for us.

We begin in Sector #1, which previously gave rise to the states given in (6.3). Introducing the third GSO constraint [8] $N_L \equiv \sum_{i=1}^{8} N^{(i)} =$ even does not affect the gravity multiplet (top line of (6.3)), but has a drastic effect on the remaining gauge boson states. We now see that we cannot excite arbitrary combinations of (i, j) fermions; instead we must choose either $(i, j) = 1, ..., 8$ or $(i, j) = 9, ..., 16$. In string-theory parlance, all of the other states have been "projected out of the spectrum". It is in this manner that we remove gauge boson states and break gauge symmetries in string theory. (There are other methods for doing this in string theory, but this is the only method at tree-level.) It is easy to see (following the arguments given above) that the remaining gauge boson states fill out the adjoint representation of two copies of $SO(16)$, and thus the gauge group is *a priori* $SO(16) \times SO(16)$. Therefore, we shall henceforth denote our string states in the notation $\bar{R}_1 \otimes (R_2; R_3, R_4)$ where $\bar{R}_1, R_2$ are the representations of the Lorentz group from the right- and left-movers, and where R_3, R_4 are the representations with respect to the two gauge group factors of $SO(16)$ respectively. Thus, we see that Sector #1 gives rise to the states

$$\bar{\mathbf{V}}_8 \otimes (\mathbf{V}_8; \mathbf{1}, \mathbf{1}) \ , \quad \bar{\mathbf{V}}_8 \otimes (\mathbf{1}; \mathbf{adj}, \mathbf{1}) \ , \quad \bar{\mathbf{V}}_8 \otimes (\mathbf{1}; \mathbf{1}, \mathbf{adj}) \ , \tag{6.9}$$

where the first states form the gravity multiplet and the second and third states are the $SO(16) \times SO(16)$ gauge bosons.

As before, Sectors #2 and #3 do not give rise to massless states. Let us now consider what happens in Sector #4. The states that previously emerged in Sector #4 are given in (6.6). We now must impose the remaining GSO constraint $^{(8)}N_L = \begin{Bmatrix} \text{odd} \\ \text{even} \end{Bmatrix}$. Let us consider each case separately. If we impose the odd choice, then the gravitino state in (6.6) is projected out of the spectrum, indicating that *supersymmetry is broken*. Likewise, we find that the gaugino states are also affected: we can now excite only those states for which $i = 1, ..., 8$ and $j = 9, ..., 16$. This spinor state transforms in the $(\mathbf{16}, \mathbf{16})$ representation of $SO(16) \times SO(16)$ (*i.e.*, as the vector-vector bifundamental). By contrast, if we impose the even choice, then the gravitino state in (6.6) remains in the spectrum, indicating that *supersymmetry is preserved*. Likewise, the gaugino states are affected only by the new requirement that either $i, j = 1, ..., 8$ or $i, j = 9, ..., 16$. Thus, the new GSO projection projects our $SO(32)$ gauginos down to $SO(16) \times SO(16)$ gauginos, as expected. Summarizing, we find that in the "even" case, the states from Sector #4 are

$$\bar{\mathbf{S}}_8 \otimes (\mathbf{V}_8; \mathbf{1}, \mathbf{1}) \ , \quad \bar{\mathbf{S}}_8 \otimes (\mathbf{1}; \mathbf{adj}, \mathbf{1}) \ , \quad \bar{\mathbf{S}}_8 \otimes (\mathbf{1}; \mathbf{1}, \mathbf{adj}) \ . \tag{6.10}$$

Let us now consider Sector #5. As indicated in Table 2, in this sector the vacuum energy is $(a_R, a_L) = (-1/2, 0)$ and the first eight left-moving complex fermions are Neveu-Schwarz while the second eight are Ramond. Choosing the "odd" GSO constraints projects all possible massless states out of the spectrum (because there is no simultaneous solution to all three GSO constraints in the "odd" case). By contrast, choosing the "even" GSO constraints yields the states

$$\tilde{b}^{\mu}_{-1/2}|0\rangle_R \otimes \{b_0^i\}|0\rangle_L \qquad (i = 9, ..., 16) \tag{6.11}$$

where we must choose an even number of zero-mode excitations on the left-moving side. This produces a massless vector state which transforms in a (128-dimensional) *spinorial* representation of the second $SO(16)$ gauge group factor. Following our previous conventions, we shall refer to this spinor as $\mathbf{C}_{128}$ rather than its conjugate $\mathbf{S}_{128}$. This state can therefore be denoted as

$$\bar{\mathbf{V}}_8 \otimes (1; 1, \mathbf{C}_{128}) . \tag{6.12}$$

We shall discuss the physical interpretation of this state shortly.

Sector #6 is similar to Sector #5, except that now the first eight left-moving complex fermions are Ramond and the second eight are Neveu-Schwarz. In a similar way we then find that there are no states in the "odd" case, while in the "even" case we find the states

$$\bar{\mathbf{V}}_8 \otimes (1; \mathbf{C}_{128}, 1) . \tag{6.13}$$

We now turn to Sector #7. Here the vacuum energy is $(a_R, a_L) = (0, 0)$, which implies that if we restrict our attention to massless states, we can tolerate only zero-mode excitations amongst both the left- and right-movers. In the "odd" case, we find the states

$$\{\tilde{b}_0^{\mu}\}|0\rangle_R \otimes \{b_0^i\}|0\rangle_L \qquad (i = 9, ..., 16) \tag{6.14}$$

where the GSO projections restrict us to an even number of zero-mode excitations on the right-moving side and an odd number on the left-moving side. According to our conventions, this produces the state $\bar{\mathbf{C}}_8 \otimes (1; 1, \mathbf{S}_{128})$. In the "even" case, by contrast, we are restricted to (6.14) where now we must have an even number of zero-mode excitations on the right-moving side and an odd number of the left-moving side. This produces the state

$$\bar{\mathbf{S}}_8 \otimes (1; 1, \mathbf{C}_{128}) . \tag{6.15}$$

Finally, in Sector #8, we similiarly find the states $\bar{\mathbf{C}}_8 \otimes (1; \mathbf{S}_{128}, 1)$ in the "odd" case and

$$\bar{\mathbf{S}}_8 \otimes (1; \mathbf{C}_{128}, 1) \tag{6.16}$$

in the "even" case.

What are we to make of these results? Collecting our states for the "odd" case, we find a string model with the following massless spectrum:

$$\bar{\mathbf{V}}_8 \otimes (\mathbf{V}_8; 1, 1) , \quad \bar{\mathbf{V}}_8 \otimes (1; \mathbf{adj}, 1) , \quad \bar{\mathbf{V}}_8 \otimes (1; 1, \mathbf{adj})$$
$$\bar{\mathbf{S}}_8 \otimes (1; \mathbf{V}_{16}, \mathbf{V}_{16}) , \quad \bar{\mathbf{C}}_8 \otimes (1; \mathbf{S}_{128}, 1) , \quad \bar{\mathbf{C}}_8 \otimes (1; 1, \mathbf{S}_{128}) . \tag{6.17}$$

This is clearly a non-supersymmetric spectrum consisting of a gravity multiplet, vector bosons transforming of the adjoint of $SO(16) \times SO(16)$, one spinor transforming as a vector-vector bifundamental with respect to the gauge group, and two additional spinors of opposite chirality transforming in the spinor representations of the gauge group. This is the non-supersymmetric $SO(16) \times SO(16)$ heterotic string model, first constructed in 1986. Note that this spectrum configuration is anomaly-free, as required for a self-consistent string theory. Also note that this string is tachyon-free even though it is non-supersymmetric. This example thus proves that *not all non-supersymmetric strings have tachyons* (although it is certainly true that all supersymmetric strings lack tachyons). While this is the only non-supersymmetric tachyon-free heterotic string in ten dimensions, there exist a plethora of such strings in lower dimensions. We shall discuss some of the properties of such strings in Lecture #10, but this raises an interesting issue: Does string theory *predict* spacetime supersymmetry? As this example makes clear, string theory certainly does not predict spacetime supersymmetry on the basis of tachyon-avoidance. However, the general answer to this question is unknown, and will be discussed more fully in Lecture #10.

Even more interesting is the model that results in the "even" case. Collecting our states from (6.9), (6.10), (6.12), (6.13), (6.15), and (6.16), we find that the total massless spectrum of this string can be written in the factorized form

$$(\bar{\mathbf{V}}_8 \oplus \bar{\mathbf{S}}_8) \otimes \left\{ (\mathbf{V}_8; 1, 1) \oplus \left(1; \left\{ \mathbf{adj} \oplus \mathbf{C}_{128} \right\}, 1 \right) \oplus \left(1; 1, \left\{ \mathbf{adj} \oplus \mathbf{C}_{128} \right\} \right) \right\}.$$

$$(6.18)$$

The appearance of the right-moving factor $\bar{\mathbf{V}}_8 \oplus \bar{\mathbf{S}}_8$ indicates that this model has $N = 1$ supersymmetry, as expected from the appearance of a single gravitino in the massless spectrum. The left-moving factor, by contrast, contains three terms. The first term combines with the right-moving factor to produce the supergravity multiplet. The second two terms formerly gave rise to the $SO(16) \times SO(16)$ gauge supermultiplet. However, we now see that for each $SO(16)$ gauge group factor, the massless vector states transform in the $\mathbf{adj} \oplus \mathbf{C}_{128}$ representation rather than simply in the $\mathbf{adj}$ representation. While the $\mathbf{adj}$ contribution is easy to interpret (giving rise to the usual gauge bosons of $SO(16)$), the extra massless vector states transforming in the $\mathbf{C}_{128}$ representation of each gauge group factor appear to cause an inconsistency, for we know that all massless vector states must be interpreted as gauge bosons, and hence such states can only transform in the adjoint representation. Thus, the only possible way that this string can be consistent is if the massless vector states in this model somehow combine to fill out the adjoint representation of some *other* group G:

$$\mathbf{adj}_{SO(16)} \; \oplus \; \mathbf{C}_{128} \; \overset{?}{=} \; \mathbf{adj}_G \; . \tag{6.19}$$

Remarkably, this is precisely what occurs: the group G is nothing but the exceptional Lie group E_8! Indeed, the 120 states of the adjoint representation of $SO(16)$ together with the 128 states of the spinor representation of $SO(16)$ combine to produce the 248 states of the adjoint representation of E_8! In string parlance, we thus say that the presence of the "twisted" states (6.12), (6.13), (6.15), and (6.16) has *enhanced* the total gauge group from $SO(16) \times SO(16)$ to $E_8 \times E_8$. This, then, is the famous supersymmetric $E_8 \times E_8$ heterotic string.

Unlike the supersymmetric $SO(32)$ string, this string is generally considered to have excellent phenomenological prospects. It has $N = 1$ spacetime supersymmetry, quantum gravity, and an $E_8 \times E_8$ gauge symmetry. E_8 is a compelling gauge group for phenomenology because it contains E_6 as a subgroup, and E_6 is a group that contains chiral representations which can be associated with grand unification and which thereby contain all of the particle content of the Standard Model. (Of course, it is still necessary to obtain actual *matter* representations from this string, but these can arise upon compactification.) Moreover, while we can imagine the Standard Model to reside entirely within one of the E_8 gauge group factors, the other factor may be interpreted as a "hidden" sector which can also have important phenomenological uses (such as triggering supersymmetry breaking, providing dark-matter candidates, and enforcing string selection rules). Thus, historically, much of the original work in string phenomenology began with a study of the compactification of this model down to four dimensions. However, as we shall see in Lectures #8 and #9, it is possible to construct heterotic string models directly in four dimensions, and to obtain models which do not necessarily have an interpretation as arising via the compactification of any particular string model in ten dimensions. Thus, as we shall see, the prospects for phenomenological heterotic string model-building are broader than merely studying the compactifications of the $E_8 \times E_8$ heterotic string.

6.4 More ten-dimensional heterotic strings

So far, we have constructed four heterotic string models in ten dimensions. Of these, two have spacetime supersymmetry, and two do not. However, it is readily apparent that further models can be constructed by introducing further "twists" which further enlarge the set of sectors in Table 2 and which further break the gauge group into smaller factors (or which break the original

$SO(32)$ gauge group in entirely different ways). The question that arises, then, is whether there exist other ten-dimensional heterotic strings with spacetime supersymmetry, or whether there exist other non-supersymmetric strings in ten dimensions that are tachyon-free. The answer to both questions turns out to be "no". A complete list of ten-dimensional heterotic strings is given in Table 4.

gauge group	spacetime SUSY?	tachyon-free?
$SO(32)$	yes	yes
$E_8 \times E_8$	yes	yes
$SO(16) \times SO(16)$	no	yes
$SO(32)$	no	no
$SO(16) \times E_8$	no	no
$SO(8) \times SO(24)$	no	no
$(E_7)^2 \times [SU(2)]^2$	no	no
$U(16)$	no	no
E_8	no	no

Table 4: The complete set of ten-dimensional heterotic string models. Two have spacetime supersymmetry, one is non-supersymmetric but tachyon-free, and the remaining six are non-supersymmetric and tachyonic.

The presence of the last string in Table 4 might seem surprising. After all, the rank of the gauge group for this string is only eight rather than sixteen, which implies that its construction must differ substantially from that of the previous strings. It turns out that this is indeed the case.[a] We briefly indicate in Lecture #7 how such strings may be constructed.

7 Lecture #7: Rules for string model-building

In the last several lectures, we constructed many different string models. Amongst the superstring models, we constructed the Type 0A, Type 0B, Type IIA, and Type IIB models, while amongst the heterotic string models, we constructed the non-supersymmetric $SO(32)$ model, the non-supersymmetric $SO(16) \times SO(16)$ model, and the supersymmetric $SO(32)$ and $E_8 \times E_8$ models. In each

---

[a] Unlike the other ten-dimensional heterotic strings, this string involves splitting each complex worldsheet fermion into a pair of two real worldsheet fermions and then introducing relative "twists" within each pair. In technical language, this results in a gauge group whose rank is reduced but whose so-called *affine level* is increased relative to those of the other strings. This increase in the affine level is important for string GUT model-building, and will be discussed in subsequent lectures.

case, we simply *asserted* a set of sectors (combinations of Neveu-Schwarz and Ramond modings) and a set of GSO constraints in each sector. Of course, each of these sets of sectors and GSO constraints conspires to yield a self-consistent string model, and occasionally it is even possible to see intuitively which choices can lead to self-consistent string models. However, we ultimately wish to construct semi-realistic string models where the groups are broken down to much smaller pieces than we have been dealing with thus far (*e.g.*, $SU(3) \times SU(2) \times U(1)$, or even $SU(5)$ or $SO(10)$), and this is going to require more complicated twists than we have thus far been using. Furthermore, all of our string models thus far have been in ten dimensions, yet we are ultimately going to wish to compactify our string models to four dimensions. It turns out that this will introduce even further choices for modings, twists, and their associated GSO projections. (In geometric language, these further choices amount the choice of compactification manifold.)

The question that arises, then, is to determine the minimal set of parameters that govern these choices. What we require is a way to *systematize* the whole process of string model-construction, so that we will know precisely which choices govern the construction of a string model and guarantee its internal self-consistency. In other words, we require *rules for string model-building*. This is the subject of the present lecture.

Once we learn the rules for the construction of ten-dimensional string models, it will be relatively straightforward to generalize these rules for the construction of models in four dimensions. This will be the subject of Lecture #8. We will then have the tools whereby we may finally construct semi-realistic four-dimensional string models. This will be the subject of Lecture #9.

7.1 Generating the sector combinations: The 20-dimensional lattice

The first issue we face is that of choosing the appropriate sector combinations. For example, let us recall the possible heterotic string sectors in Table 2. As we discussed in Sect. 6.2, this set of sectors permits only three distinct sector combinations: either we choose Sectors #1 and #2 only, or we choose Sectors #1 through #4 only, or we choose Sectors #1 through #8. How can we know which combinations are allowed, and which sectors are required in each grouping? In Sect. 6.2, we discussed how modular invariance ultimately governs these choices. Here, however, we shall develop a rule which we can use in order to deduce these sector combinations rather quickly and which can easily be generalized to more complicated situtions.

First, let us introduce some notation. Since it is rather awkward to consider left-moving complex fermions Ψ_L^i at the same time as right-moving *real* (Majorana) fermions ψ_R^μ, let us "complexify" our right-moving Majorana fermions so that *all* of our worldsheet fermions are complex. This means that instead of having eight left-moving real fermions ψ_R^μ in light-cone gauge, we have instead four complex ones Ψ_R^μ formed by pairing the left-moving real fermions in groups of two. (We retain the index μ to remind ourselves that these fields carry indices with respect to the spacetime Lorentz algebra, even though strictly speaking it is only their real component fields ψ_R^μ that carry such vectorial indices.)

We also need a more general notation for discussing the possible boundary conditions and modings that any such complex worldsheet fermion can take. In general, we can parametrize any possible worldsheet boundary condition in the form

$$\Psi(\sigma_1 + \pi, \sigma_2) \;=\; -\, e^{-2\pi i v}\, \Psi(\sigma_1, \sigma_2) \tag{7.1}$$

where $-\frac{1}{2} \leq v < \frac{1}{2}$. Thus the quantity v parametrizes the boundary condition of the individual fermion, with

$$
\begin{array}{llll}
v = 0: & \text{anti-periodic} & \text{(Neveu-Schwarz)} & \\
v = -1/2: & \text{periodic} & \text{(Ramond)}. &
\end{array}
\tag{7.2}
$$

General values of v correspond to so-called "multi-periodic fermions". For example, the general moding of a multi-periodic left-moving complex fermion is given by

$$\Psi_L(\sigma_1 + \sigma_2) \;=\; \sum_{n=1}^{\infty} \left[b_{n+v-1/2}\, e^{-i(n+v-1/2)(\sigma_1+\sigma_2)} + d^\dagger_{n-v-1/2}\, e^{+i(n-v-1/2)(\sigma_1+\sigma_2)} \right]$$

$$\tag{7.3}$$

and the corresponding number operator and worldsheet energy are defined accordingly. Note that these modings generalize those given in Sect. 5.3. Likewise, the vacuum energy contribution from such a fermion is given by

$$a_\Psi \;=\; \frac{1}{2}\left(v^2 - \frac{1}{12} \right). \tag{7.4}$$

This too generalizes our previous results

In ten dimensions, it turns out that we lose no generality by considering only the specific cases $v = 0, -\frac{1}{2}$ for all worldsheet fermions. What this means is that all self-consistent ten-dimensional string models can ultimately be realized using worldsheet fermions with only Neveu-Schwarz or Ramond boundary conditions. In lower dimensions, by contrast, other choices are possible. Therefore, even though we shall primarily focus our attention on the cases $v \in \{0, -\frac{1}{2}\}$, we shall develop our formalism in such a way that it holds for arbitrary values of v.

Given this parametrization, we can describe the boundary conditions within any sector rather succinctly by specifying twenty v-values, four for the complex right-movers Ψ_R^μ and sixteen for the complex left-movers Ψ_L^i. We can group these twenty v-values to form a "boundary-condition" vector

$$\mathbf{V} = [\bar{v}_1, \bar{v}_2, \bar{v}_3, \bar{v}_4 \,|\, v_1, ..., v_{16}] \,, \qquad (7.5)$$

and thus we may associate a vector with each underyling string sector. For example, the sectors in Table 2 now correspond to the vectors shown in Table 5. Note that in Table 5, we have used a shorthand notation in which superscripts indicate repeated components. We have also dropped the minus signs from the Ramond entries $v = -\frac{1}{2}$. We stress, however, that even though we shall no longer explicitly indicate the Ramond minus sign, it should continue to be implicitly understood for all Ramond boundary conditions. (This minus sign can play an important role for string models in lower dimensions.)

Sector #	$\mathbf{V}$
1	$[(0)^4 \,|\, (0)^{16}]$
2	$[(\frac{1}{2})^4 \,|\, (\frac{1}{2})^{16}]$
3	$[(0)^4 \,|\, (\frac{1}{2})^{16}]$
4	$[(\frac{1}{2})^4 \,|\, (0)^{16}]$
5	$[(0)^4 \,|\, (0)^8 (\frac{1}{2})^8]$
6	$[(0)^4 \,|\, (\frac{1}{2})^8 (0)^8]$
7	$[(\frac{1}{2})^4 \,|\, (0)^8 (\frac{1}{2})^8]$
8	$[(\frac{1}{2})^4 \,|\, (\frac{1}{2})^8 (0)^8]$

Table 5: The eight possible sectors for ten-dimensional heterotic strings from Table 2, written in the boundary-condition vector notation of (7.5). Here the superscripts indicate repeated components, and we have dropped the minus sign for Ramond boundary conditions.

What, then, are the self-consistent combinations of sectors? Recall from the previous lecture that the first self-consistent combination of sectors comprises Sectors #1 and #2 only. Let us therefore study this simplest combination. Sector #1 (the so-called NS-NS sector) corresponds to the *zero-vector* $\mathbf{0}$, the vector whose entries all vanish. Thus, in this sense, we might associate the NS-NS sector with the *origin* in a twenty-dimensional vector space. Sector #2 (the so-called Ramond-Ramond sector) then corresponds to some other point in the vector space which is some distance away from the origin. Let us call this other location $\mathbf{V}_0 \equiv [(\frac{1}{2})^4 | (\frac{1}{2})^{16}]$.

If we were to consider $\mathbf{V}_0$ to be a lattice basis vector, a natural question would be to determine the lattice that is generated by this basis vector. Because there is only one such non-zero vector, this would clearly be a one-dimensional "lattice". Since $\mathbf{V}_0 \equiv [(\frac{1}{2})^4 | (\frac{1}{2})^{16}]$, the next point in the lattice would be $2\mathbf{V}_0 \equiv [(1)^4 | (1)^{16}]$. How can we interpret this point? Recall from (7.1) that the components of such vectors (*i.e.*, the values of v) are defined only modulo 1 (*i.e.*, they are restricted to the unit interval $-\frac{1}{2} \leq v < \frac{1}{2}$). Thus, we see that $v = 1$ is physically the same as $v = 0$, once again implying a Neveu-Schwarz boundary condition. In other words, we should only add our vectors *modulo 1*. Given this, we find that $2\mathbf{V}_0 \overset{1}{=} \mathbf{0}$, where we have introduced the notation $\overset{1}{=}$ to indicate equality modulo 1. Likewise, $3\mathbf{V}_0 \overset{1}{=} \mathbf{V}_0$, and so forth. Thus, we see that $\mathbf{V}_0$ generates a "lattice" consisting of only two physically distinct "points":

$$\{\mathbf{0}, \mathbf{V}_0\} \,. \tag{7.6}$$

However, these are precisely the two "points" that comprised our first self-consistent set of sectors (Case A in Lecture #6), and which led to our first string model!

It turns out that this is a general property: *All self-consistent choices of string sectors are those that correspond to the "points" in a twenty-dimensional lattice generated by a set of basis vectors.* To illustrate this principle, let us consider the next case (Case B in Lecture #6). In this case, we included only Sectors #1 through #4. This indicates that we need a larger lattice, which in turn implies the existence of not just the single lattice-generating basis vector $\mathbf{V}_0$, but also an additional basis vector $\mathbf{V}_1$. One choice is:

$$\begin{aligned}
\mathbf{V}_0 &= [(\tfrac{1}{2})^4 \,|\, (\tfrac{1}{2})^{16}] \\
\mathbf{V}_1 &= [(0)^4 \,|\, (\tfrac{1}{2})^{16}] \,.
\end{aligned} \tag{7.7}$$

Using these choices, we can see that indeed all four of these sectors can be generated as the different "points" in the resulting lattice: Sector #1 corresponds to the origin $\mathbf{0}$, Sector #2 corresponds to $\mathbf{V}_0$ itself, Sector #3 corresponds to $\mathbf{V}_1$ itself, and Sector #4 corresponds to the remaining lattice point $\mathbf{V}_0 + \mathbf{V}_1$. Note that no other points exist in this lattice, since $2\mathbf{V}_0 \overset{1}{=} 2\mathbf{V}_1 \overset{1}{=} \mathbf{0}$. Thus, we see that the introduction of the additional basis vector $\mathbf{V}_1$ is physically equivalent to the "twist" that shifts the boundary conditions of the left-moving fermions relative to those of the right-moving fermions in Sectors #3 and #4.

Finally, let us consider the full set (Case C) consisting of Sectors #1 through #8. It is easy to see that this set is generated by the *three* basis vectors:

$$\mathbf{V}_0 = [(\tfrac{1}{2})^4 \,|\, (\tfrac{1}{2})^{16}]$$

$$\mathbf{V}_1 = [(0)^4 \,|\, (\tfrac{1}{2})^{16}]$$
$$\mathbf{V}_2 = [(0)^4 \,|\, (\tfrac{1}{2})^8 (0)^8] \,. \tag{7.8}$$

Once again, the introduction of the new basis vector $\mathbf{V}_2$ implements the "twist" that separates the boundary conditions of the first set of eight left-moving fermions from those of the second set.

This procedure can be continued. Each additional basis vector introduces a new twist, increases the size of the resulting lattice, and leads to the introduction of new physical string sectors (so-called "twisted sectors"). For example, one further basis vector that might be introduced is $\mathbf{V}_3 \equiv [(0)^4 |(\tfrac{1}{2})^4 (0)^4 (\tfrac{1}{2})^4 (0)^4]$. This vector would have the effect of introducing a further twist amongst the left-moving fermions within each group of eight.

Clearly, given a set of N basis vectors $\mathbf{V}_i$ $(i = 0, ..., N-1)$, the procedure for generating the full set of resulting string sectors is to consider all possible lattice vectors $\sum_{i=0}^{N-1} \alpha_i \mathbf{V}_i$ where $\alpha_i \in \{0, 1\}$. Note that this restriction on the values of α_i assumes that we are considering only Neveu-Schwarz or Ramond boundary conditions for the worldsheet fermions; generalizations to multi-periodic fermions will be discussed shortly. We shall henceforth denote a given string sector as $\alpha\mathbf{V} \equiv \sum_i \alpha_i \mathbf{V}_i$. For example, the NS-NS sector (*i.e.*, Sector #1) always corresponds to $\alpha = (0, 0, ...)$ and the Ramond-Ramond sector (*i.e.*, Sector #2) corresponds to $\alpha = (1, 0, ...)$.

At this stage, we now know how to generate the full set of underlying string sectors once we are given a "primordial" set of basis vectors $\mathbf{V}_i$. The next issue that arises is to determine the rules that govern the allowed choices of these basis vectors. Of course, we have already derived one such rule: each basis vector $\mathbf{V}_i$ must take the form

$$\mathbf{V}_i = [(\bar{v})^4 \,|\, v_1, ..., v_{16}] \tag{7.9}$$

where the right-moving fermions all have *same* moding $\bar{v} \in \{0, -\tfrac{1}{2}\}$. Indeed, as we saw in Lectures #5 and #6, this requirement is necessary for the preservation of the right-moving worldsheet supersymmetry (so that the right-moving worldsheet supercurrent has a unique moding in each sector). This is also necessary for the preservation of spacetime Lorentz invariance, since the right-moving worldsheet fermions carry Lorentz spacetime indices.

As we might expect, there are still several additional conditions that our basis vectors $\mathbf{V}_i$ must satisfy. But before we can discuss these conditions, we must turn to the generation of the GSO constraints in each sector.

7.2 Generating the GSO constraints

We have already seen in previous lectures that the appearance of new string sectors is correlated with the appearance of new GSO constraints in each sector. We are now in a position to formulate this correlation more precisely: *in each string sector, there is one GSO projection for each basis vector.* Our task, then, is to find a simple way to generate the exact forms of these GSO projections.

Let us return to Case A, and consider the model consisting of only Sectors #1 and #2. As we have seen above, this model is generated by the single basis vector $V_0 \equiv [(\frac{1}{2})^4 | (\frac{1}{2})^{16}]$, resulting in the two sectors **0** (Sector #1) and V_0 (Sector #2). In each of these sectors, recall from Table 3 that we then had the single GSO constraint $N_L - N_R = $ odd, or equivalently

$$\sum_{i=1}^{16} N^{(i)} - \sum_{j=1}^{4} \bar{N}^{(j)} = \text{odd} . \tag{7.10}$$

(Here we have used the j-index to span our four complex right-moving fermions, while the i-index spans our sixteen complex left-moving fermions.) It is this GSO constraint that we now wish to write in a more transparent manner.

Given our success in using the lattice idea and modular arithmetic in order to generate the complete set of string sectors, let us attempt to write (7.10) in a form that makes use of both ideas. Let us first concentrate on the modular arithmetic idea. Since all of our basis vectors are defined only modulo one, let us cast (7.10) into the form of a modulo-one relation. Since (7.10) is already a modulo-two relation, this can be achieved by dividing by two:

$$\frac{1}{2}\sum_{i=1}^{16} N^{(i)} - \frac{1}{2}\sum_{j=1}^{4} N^{(j)} \stackrel{1}{=} \frac{1}{2} \tag{7.11}$$

where we have used the notation $\stackrel{1}{=}$ to indicate equality modulo 1.

Let us now try to incorporate the lattice idea. To do this, let us make a *vector* out of our twenty number operators:

$$\mathbf{N} \equiv [\bar{N}^{(1)}, \bar{N}^{(2)}, \bar{N}^{(3)}, \bar{N}^{(4)} \mid N^{(1)}, \quad , N^{(16)}] \tag{7.12}$$

Clearly, each different possible string state in a given sector corresponds to a different **N**-vector, and the physical (surviving) string states are those satisfying (7.11). Let us now attempt to write (7.11) in a vector notation. Neglecting the minus sign in (7.11) for the moment, we see that (7.11) involves a sum of vector components, which reminds us of a vector dot product. Thus, if we define the "signature" of our twenty-dimensional lattice to be $[(-)^4 \,|\, (+)^{16}]$, we can write (7.11) in the form of a vector dot product:

$$[(\tfrac{1}{2})^4 \,|\, (\tfrac{1}{2})^{16}] \cdot \mathbf{N} \stackrel{1}{=} \tfrac{1}{2} \tag{7.13}$$

where we have introduced a vector each of whose components is equal to $\tfrac{1}{2}$. However, this vector is nothing but $\mathbf{V}_0$, the basis vector that generates the lattice for this model! Thus, we see that if our model is generated by the basis vector $\mathbf{V}_0$, then in each of the resulting sectors $\{0, \mathbf{V}_0\}$ the GSO projections take the form

$$\mathbf{V}_0 \cdot \mathbf{N} \stackrel{1}{=} \tfrac{1}{2} \,. \tag{7.14}$$

This produces the non-supersymmetric $SO(32)$ string model from Lecture #6!

Let us now consider Case B, consisting of Sectors #1 through #4. As we saw in Lecture #6, this produces the *supersymmetric* $SO(32)$ heterotic string model, and is generated by the set of two basis vectors given in (7.7). In each of the four resulting sectors $\{0, \mathbf{V}_0, \mathbf{V}_1, \mathbf{V}_0 + \mathbf{V}_1\}$, the *two* GSO projections were $N_L - N_R = \text{odd}$ and $N_L = \text{even}$. (Recall Table 3.) These now take the form

$$\mathbf{V}_0 \cdot \mathbf{N} \stackrel{1}{=} \tfrac{1}{2} \,, \qquad \mathbf{V}_1 \cdot \mathbf{N} \stackrel{1}{=} \tfrac{1}{2} \,. \tag{7.15}$$

Similarly, Case C is generated by the *three* basis vectors in (7.8), and the three GSO constraints in each sector take the general form

$$N_L - N_R = \dots \,, \qquad N_L = \dots, \qquad {}^{(8)}N_L = \dots \,. \tag{7.16}$$

Here ${}^{(8)}N_L \equiv \sum_{i=1}^{8} N^{(i)}$, and we shall momentarily defer a discussion of the values of the right sides of these constraint equations. We then find that these three GSO constraints take the general forms

$$\mathbf{V}_0 \cdot \mathbf{N} \stackrel{1}{=} \dots \,, \qquad \mathbf{V}_1 \cdot \mathbf{N} \stackrel{1}{=} \dots \,, \qquad \mathbf{V}_2 \cdot \mathbf{N} \stackrel{1}{=} \dots \,. \tag{7.17}$$

Depending on the right sides of these equations, this generates either the supersymmetric $E_8 \times E_8$ string or the non-supersymmetric $SO(16) \times SO(16)$ string.

The final question, then, is to determine what appears on the right sides of these GSO constraint equations. In general, this will be some value x which satisfies $-\tfrac{1}{2} \le x < \tfrac{1}{2}$. This x-value is called a *GSO projection phase*, and is generally different for each sector. Thus, we know that x must itself depend on α, where (as discussed in Sect. 7.1) α parametrizes the particular sector in question. We also know from our prior experience (in particular, from Table 3) that x must also contain some additional *free* parameters because we occasionally still had the freedom to make choices such as $\left\{\begin{array}{c}\text{even}\\\text{odd}\end{array}\right\}$ when constructing our GSO constraints.

It turns out the final result is the following. Within any given string sector $\alpha\mathbf{V} \equiv \sum_{i=0}^{N-1} \alpha_i \mathbf{V}_i$, the states that survive are those whose number operator vectors $\mathbf{N}$ satisfy the equations

$$\mathbf{V}_i \cdot \mathbf{N} \stackrel{1}{=} \sum_{j=0}^{N-1} k_{ij}\alpha_j + s_i - \mathbf{V}_i \cdot (\alpha\mathbf{V}) , \qquad 0 \le i \le N - 1 . \tag{7.18}$$

This is therefore the full set of GSO constraint equations for the sector $\alpha\mathbf{V}$. In (7.18), the notation is as follows. There are N different equations here, depending on the value of i. In the last term, the dot product $\mathbf{V}_i \cdot (\alpha\mathbf{V})$ is the dot product between $\mathbf{V}_i$ and the sector $\alpha\mathbf{V}$ for which the GSO constraint is being applied. In the second-to-last term, s_i is defined as the first component (*i.e.*, the first of the right-moving components) of the vector $\mathbf{V}_i$:

$$s_i \equiv \mathbf{V}_i^{(1)} . \tag{7.19}$$

Thus s_i parametrizes the *spacetime statistics* of the sector $\mathbf{V}_i$, with $s_i = 0$ indicating spacetime bosons and $s_i = -\frac{1}{2}$ indicating spacetime fermions. Likewise, the sum $\sum \alpha_i s_i$ (mod 1) indicates the statistics of the sector $\alpha\mathbf{V}$. In the remaining term, k_{ij} denotes a certain $N \times N$ matrix of numbers (so-called *GSO projection phases*) satisfying $-\frac{1}{2} \le k_{ij} < \frac{1}{2}$. These are therefore the remaining degrees of freedom that enter into our GSO constraints. In the case of $\mathbb{Z}_2$ twists (for which all fermionic boundary conditions have either Neveu-Schwarz or Ramond boundary conditions), one has $k_{ij} \in \{0, -\frac{1}{2}\}$ only. The case of multi-periodic fermions will be discussed shortly.

Thus, if we are given a set of parameters $\{\mathbf{V}_i, k_{ij}\}$, we can now generate the resulting string model and the entire corresponding spectrum! These parameters are ultimately the parameters that physically describe a given string model.

7.3 Self-consistency constraints

We finally turn to the remaining question: what determines how the parameters $\{\mathbf{V}_i, k_{ij}\}$ are to be chosen? What are the rules that guarantee a self-consistent choice?

Clearly, as we have discussed earlier, modular invariance is one of many symmetries that govern these choices. Other requirements for self-consistency include proper spacetime spin-statistics relations (so that all Ramond states are indeed anti-commuting spacetime fermions, and all Neveu-Schwarz states are commuting spacetime bosons) and physically sensible GSO projections (so that unitarity is not violated, among other things). It is important to stress that these are not *additional* constraints that need to be imposed in order to guarantee the consistency of the string in spacetime; rather these constraints are intrinsic to string theory itself at the worldsheet level, emerging as string self-consistency constraints, and together imply these features in spacetime.

We have already discussed the first contraint that governs the choices of the basis vectors: they must all have the form (7.9), with all right-moving fermions sharing the same boundary condition. Second, these vectors must all be linearly independent with respect to addition (modulo 1); otherwise, at least one of these vectors is redundant. The third constraint also turns out to be quite simple: among our set of basis vectors, we must always start with the vector

$$\mathbf{V}_0 \equiv [(\tfrac{1}{2})^4 \,|\, (\tfrac{1}{2})^{16}] \,. \tag{7.20}$$

The presence of this vector ensures that the resulting string model contains at least a Ramond-Ramond sector in addition to a NS-NS sector.

The remaining constraints serve to correlate the $\mathbf{V}_i$ vectors with the GSO projection phases k_{ij}, and take the form:

$$k_{ij} + k_{ji} \overset{1}{=} \mathbf{V}_i \cdot \mathbf{V}_j$$
$$k_{ii} + k_{i0} \overset{1}{=} \tfrac{1}{2}\mathbf{V}_i \cdot \mathbf{V}_i - s_i \,. \tag{7.21}$$

Note that given a set of boundary condition vectors $\mathbf{V}_i$, the constraints (7.21) imply that only the elements k_{ij} with $i > j$ are independent parameters. The first equation in (7.21) then enables us to uniquely determine k_{ij} with $i < j$, and the second equation in (7.21) enables us to uniquely determine the diagonal elements k_{ii}.

7.4   Summary, examples, and generalizations

Let us now summarize the rules for heterotic string model-building in $D = 10$. We begin by choosing a set of linearly independent basis vectors $\mathbf{V}_i$ ($i = 0, ..., N - 1$) and a corresponding matrix of GSO projection phases k_{ij} ($i, j = 0, ..., N - 1$). Our set of basis vectors may be as large as we desire; since each vector corresponds to an additional twist, larger sets of vectors lead to more complicated string models. Among our choice of basis vectors must always appear the vector $\mathbf{V}_0$ defined in (7.20), and every basis vector is required to have the form (7.9). We must also ensure that our choices of basis vectors $\mathbf{V}_i$ and GSO projection phases k_{ij} are properly *correlated* according to (7.21). If there does not exist a solution for k_{ij}, then our original choice of $\mathbf{V}_i$ must be discarded or repaired. These are the only constraints that govern the choices of the parameters $\{\mathbf{V}_i, k_{ij}\}$.

Given such a self-consistent choice of parameters $\{\mathbf{V}_i, k_{ij}\}$, we are then guaranteed to have a self-consistent string model. The different sectors of this model are generated as all combinations $\sum_i \alpha_i \mathbf{V}_i$ that fill out the twenty-dimensional lattice, where $\alpha_i \in \{0, 1\}$. In each sector $\alpha \mathbf{V} \equiv \sum_i \alpha_i \mathbf{V}_i$, the allowed states are then those whose number operator vectors $\mathbf{N}$ simultaneously satisfy the constraints (7.18) for $i = 0, ..., N - 1$. This is often called the *spectrum-generating formula*.

It is straightforward to see how this formalism can be applied in practice. We shall leave it as an exercise to verify that the choice

$$\mathbf{V}_0 \equiv [(\tfrac{1}{2})^4 \,|\, (\tfrac{1}{2})^{16}] \,, \qquad k_{00} = (0) \tag{7.22}$$

generates the non-supersymmetric $SO(32)$ heterotic string model; that the choice

$$\begin{cases} \mathbf{V}_0 \equiv [(\tfrac{1}{2})^4 \,|\, (\tfrac{1}{2})^{16}] \\ \mathbf{V}_1 \equiv [(0)^4 \,|\, (\tfrac{1}{2})^{16}] \end{cases} \qquad k_{ij} = \begin{pmatrix} 0 & 0 \\ 0 & 0 \end{pmatrix} \tag{7.23}$$

generates the *supersymmetric* $SO(32)$ heterotic string model; and that the choices

$$\begin{cases} \mathbf{V}_0 \equiv [(\tfrac{1}{2})^4 \,|\, (\tfrac{1}{2})^{16}] \\ \mathbf{V}_1 \equiv [(0)^4 \,|\, (\tfrac{1}{2})^{16}] \\ \mathbf{V}_2 \equiv [(0)^4 \,|\, (\tfrac{1}{2})^8 (0)^8] \end{cases} \qquad k_{ij} = \begin{pmatrix} 0 & 0 & 0 \\ 0 & 0 & k \\ 0 & k & 0 \end{pmatrix} \tag{7.24}$$

generate the supersymmetric $E_8 \times E_8$ string model if we choose $k = 0$, and the non-supersymmetric $SO(16) \times SO(16)$ string model if we choose $k = 1/2$. Indeed, it is a general property that if we choose our vector $\mathbf{V}_1$ as above, then spacetime supersymmetry is preserved if $k_{i0} = k_{i1}$ for all $i = 0, 1, ..., N - 1$, and broken otherwise. Thus, we see that we now have a very compact notation and procedure for generating and analyzing ten-dimensional heterotic string models! We should also stress that these are not the only parameter choices of $\{\mathbf{V}_i, k_{ij}\}$ that will lead to these models. In fact, there is often a great redundancy in this procedure, so that a given physical string model can have many different representations in terms of the worldsheet parameters $\{\mathbf{V}_i, k_{ij}\}$. However, a given set of parameters always corresponds to a single, unique, self-consistent string model in spacetime.

The formalism that we have presented in this lecture is called the "free-fermionic construction", and was developed in 1986 by H. Kawai, D.C. Lewellen, and S.-H.H. Tye and by I. Antoniadis, C. Bachas, and C. Kounnas. The name stems from the fact that the fundamental degrees of freedom on the string worldsheet (in addition to the spacetime coordinate fields X^μ) are taken to be the free fermionic fields Ψ. Even though we have presented this formalism for the case of ten-dimensional heterotic strings, we shall see in Lecture #8 that there also exists a straightforward generalization of this formalism to *four-dimensional* heterotic string models.

As we have indicated, this formalism also carries over directly to the case of multi-periodic complex fermions for which the boundary condition parameter v in (7.1) can be an arbitrary rational number in the range $-\frac{1}{2} \leq v < \frac{1}{2}$. For each resulting boundary-condition vector $\mathbf{V}_i$, let us define m_i to be the smallest integer such that if we multiply each element in $\mathbf{V}_i$ by m_i, we obtain a vector of integer entries. In general, m_i is called the "order" of the vector $\mathbf{V}_i$, and is also the order of the corresponding physical twist introduced by that vector. For example, in the case of only Neveu-Schwarz or Ramond fermions, we have $m_i = 2$ for all i, implying only $\mathbb{Z}_2$ twists. Nevertheless, even for general multi-periodic boundary conditions, the above constraints continue to apply exactly as written. Indeed, the only small change is that we now must take $\alpha_i \in \{0, 1, ..., m_i - 1\}$ when generating our lattice of corresponding string sectors. Likewise, each GSO projection phase k_{ij} must now also be chosen such that $m_j k_{ij} \in \mathbb{Z}$.

In this regard, it is important to note that the only fermions which can possibly have such generalized boundary conditions are those which are *not* the worldsheet superpartners of worldsheet bosons. This restriction arises because the structure of the worldsheet supersymmetry algebra itself restricts the corresponding fermions to have only Neveu-Schwarz or Ramond boundary conditions. For example, in the case of the ten-dimensional heterotic string, only the left-moving worldsheet fermions are *a priori* permitted to have generalized boundary conditions. By contrast, the right-moving fermions are restricted by the right-moving worldsheet supersymmetry algebra to have either Neveu-Schwarz or Ramond boundary conditions. This in turn implies that $s_i \in \{0, -\frac{1}{2}\}$, so that a given string sector continues to give rise to only spacetime bosons or spacetime fermions. Also note that although we are capable *in principle* of utilizing multi-periodic fermions while constructing ten-dimensional heterotic string models, in practice it turns out that this does not lead to new models which are physically distinct from those using only Ramond or Neveu-Schwarz fermions. It is for this reason that we can ultimately restrict ourselves to these simpler boundary conditions in ten dimensions without loss of generality. In lower dimensions, by contrast, this is no longer true, and the number of possible models grows dramatically.

This formalism can also be carried over to the case of ten-dimensional *superstrings* (rather than heterotic strings). For superstrings, the boundary-condition vectors take the simpler form

$$\mathbf{V}_i = [(\bar{v})^4 \mid (v)^4] \tag{7.25}$$

where $v, \bar{v} \in \{0, \frac{1}{2}\}$. Our mandatory vector $\mathbf{V}_0$ then takes the form $[(\frac{1}{2})^4|(\frac{1}{2})^4]$, and we define $s_i \equiv v + \bar{v} \pmod 1$ as our new spacetime statistics parameter, replacing (7.19). The results (7.21) and (7.18) then continue to apply directly. Of course, this formalism is fairly trivial in the case of *ten-dimensional* superstrings, for the maximal set of linearly independent basis vectors of the form (7.25) consists of only $\mathbf{V}_0$ and $\mathbf{V}_1 \equiv [(0)^4|(\frac{1}{2})^4]$. As we have seen in Lecture #4, this results in only four distinct superstring models in ten dimensions: omitting $\mathbf{V}_1$ from our basis set generates the Type 0 models, while including $\mathbf{V}_1$ in our basis set generates the Type II models. However, just as for the heterotic strings, we shall see in Lecture #8 that this formalism can also be generalized to the case of four-dimensional superstring models where the possibilities become much richer.

It turns out that the free-fermionic formalism can be extended still further. For example, one can also extend this formalism to compactifications of the *bosonic* string. Moreover, one can even extend this formalism to special types of superstring and heterotic string models whose worldsheet actions must be represented in terms of real rather than complex fermions. Likewise, there even exist generalizations to string models involving non-free worldsheet fermions (*i.e.*, models whose worldsheet actions involve additional Thirring-type interactions between the worldsheet fermions). In fact, even though there exist alternative model-construction formalisms that do not involve free worldsheet fermions at all, the free-fermionic construction can often yield models that are physically equivalent to those that are constructed through these other means.

How general, then, is the free-fermionic construction? It turns out that for *ten-dimensional* string models, this construction is completely general. What this means is that all known physically consistent superstring and heterotic string models in ten dimensions can be realized via this construction (*i.e.*, as stemming from an underlying set of free-fermionic parameters $\{\mathbf{V}_i, k_{ij}\}$). In lower dimensions, by contrast, this construction is *not* completely general — there exist self-consistent lower-dimensional string models which cannot be written or constructed in this manner. However, as we shall illustrate in Lecture #9, the free-fermionic construction does comprise a *vast set* of semi-realistic string models. Moreover, the free-fermionic construction has the great advantage that the rules for construction are relatively simple, and that they enable one to *systematically* construct many string models and examine their phenomenological properties. Indeed, many computer programs have been written that use this formalism in order to scan the space of string models and analyze their low-energy phenomenologies. Thus, for these reasons, the free-fermionic construction has played a very useful role as the underlying method through which the majority of string model-building has historically been pursued.

7.5 Assessment: Conclusion to Part I

At this point, it is perhaps useful to assess the position in which we now find ourselves. Clearly, through these constructions, we are able to produce *many* string models. In fact, as we shall see, the number of self-consistent string models in $D < 10$ is virtually infinite, and there exists a whole space of such models. This space of models is called a *moduli space*, where the so-called moduli are various continuous parameters which can be adjusted in order to yield different models. (Of course, we have seen that we have only discrete parameter choices in ten dimensions, but these parameters can become continuous in lower dimensions.) Moreover, each of these models has a completely different spacetime phenomenology. What, then, is the use of string theory as an "ultimate" theory, if it does not lead to a single, unique model with a unique low-energy phenomenology?

To answer this question, we should recall our discussion at the beginning of these lectures. Just as field theory is a language for building certain models (one of which, say, is the Standard Model), string theory is a new and deeper language by which we might also build models. The advantages of using this new language, as discussed in Lecture #1, include the fact that our resulting models incorporate quantum gravity and Planck-scale physics. Of course, in field theory, many parameters enter into the choice of model-building. These parameters include the choice of fields (for example, the choice of the gauge group, and whether or not to have spacetime supersymmetry), the number of fields (for example, the number of generations), the masses of particles, their mixing angles, and so forth. These are all *spacetime* parameters. In string theory, by contrast, we do not choose these spacetime parameters; we instead choose a set of *worldsheet* parameters. For example, in the free-fermionic construction, we choose the parameters $\{\mathbf{V}_i, k_{ij}\}$. All of the phenomenological properties in spacetime are then derived as consequences of these more fundamental choices. But still, just as in field theory, we are faced with the difficult task of model-building.

Is this progress, then? While opinions on this question may differ, one can argue that the answer is still definitely "yes". Recall that quantum gravity is automatically included in these string models. This is one of the benefits of model-building on the worldsheet rather than in spacetime. Also recall that string theory is a finite theory, and does not contain the sorts of ultraviolet divergences that plague us in field theory. This is another benefit of worldsheet, rather than spacetime, model-building. Moreover, worldsheet model-building ultimately involves choosing *fewer* parameters than we would have to choose in field theory — for example, we have seen that an entire infinite tower of string states, their gauge groups and charges and spins, are all ultimately encoded in a few underlying worldsheet parameters such as $\{\mathbf{V}_i, k_{ij}\}$. Furthermore, because of this drastic reduction in the number of free parameters, string phenomenology is in many ways more tightly constrained than ordinary field-theoretic phenomenology. Thus, it is in this way that string theory can guide our choices and expectations for physics beyond the Standard Model. Indeed, from a string perspective, we see that we should favor only those patterns of spacetime physics that can ultimately be derived from an underlying set of worldsheet parameters such as $\{\mathbf{V}_i, k_{ij}\}$. These would then serve as a "minimal set" of parameters which would govern all of spacetime physics!

Of course, at a theoretical or philosophical level, this state of affairs is still somewhat unsatisfactory. After all, we still do not know *which* self-consistent choice of string parameters ultimately corresponds to reality. However, *in principle*, string theory should be able to predict this dynamically. Indeed, even though there exists a whole moduli space of self-consistent string models, there should exist an energy or potential function in this space (*i.e.*, some function $V(\{\phi\})$ of all the moduli $\{\phi\}$) which should dynamically select a particular point in moduli space (*e.g.*, as a local or global minimum of V). This would then fix all of the moduli to specific values, or equivalently (in the language of the free-fermionic construction) tell us which choices of parameters $\{\mathbf{V}_i, k_{ij}\}$ are preferred dynamically.

Unfortunately, we do not understand the dynamics of string theory well enough to carry out such an ambitious undertaking. Certainly, at the level of perturbative (weakly coupled) string theory, we have no way to distinguish amongst the possible low-energy models by calculating such a function $V(\{\phi\})$. This is particularly true for string models exhibiting spacetime supersymmetry, for which $V = 0$ exactly to all orders in perturbation theory. Even if the spacetime supersymmetry is broken, the resulting potential $V(\{\phi\})$ often turns out not to have a stable minimum. This is the so-called "runaway problem", to be discussed further in Lecture #10. Of course, one might hope that recent advances in understanding the *non-perturbative* structure of string theory will ultimately be able to provide guidance in this direction. However, as we shall discuss briefly in Lecture #10, although these non-perturbative insights (particularly those concerning string duality) have thus far changed our understanding of the size and shape of this moduli space, they have not yet succeeded in leading us to an explanation of which points in this moduli space are dynamically selected.

So where do we stand? As string phenomenologists, we can do two things. First, we can pursue *model-building*: we can search through the moduli space of self-consistent string models in order to determine how close to realistic spacetime physics we can come. This is, in some sense, a direct test of string theory as a phenomenological theory of physics. Of course, this approach to string phenomenology is ultimately limited by many factors: we have no assurance that our model-construction techniques are sufficiently powerful or general to include the "correct" string model (assuming that one exists); we have no assurance that our model-construction techniques will not lead to physically distinct models which nevertheless "agree" as far as their testable low-energy predictions are concerned; and we have no assurance that the most important phenomenological features that describe our low-energy world (such as the pattern of supersymmetry-breaking) are to be found in perturbative string theory rather than in non-perturbative string theory. For example, it may well be (and it has indeed been argued) that the true underlying string theory that describes nature is one which is intrinsically non-perturbative, and which would therefore be beyond the reach of the sorts of approaches typically followed in studies of string phenomenology.

Another option, then, is to temporarily abandon string model-building somewhat, and to seek to extract general phenomenological theorems or corre lations about spacetime physics that follow directly from the general structure of string theory itself. Clearly, we would wish such information to be *model-independent*, *i.e.*, independent of our particular location in moduli space or the values of particular string parameters such as $\{V_i, k_{ij}\}$. For example, if some particular configuration of spacetime physics (some pattern of low-energy phenomenology) can be shown to be inconsistent with being realized from an underlying set of $\{V_i, k_{ij}\}$ parameters, and if such a demonstration can be made to transcend the particular free-fermionic construction so that it relies on only the primordial string symmetries themselves, then such patterns of phenomenology can be ruled out. In this way, one can still use string theory in order to narrow the list of possibilities for physics at higher energies, and to correlate various seemingly disconnected phenomenological features with each other. Such correlations would then be viewed as "predictions" from string theory, and we shall see many examples of this phenomenon in subsequent lectures.

In summary, then, we have seen that there exist powerful ways of constructing string models and surveying their low-energy phenomenologies, but that this leads to the problem of selecting the true model (*i.e.*, the true "ground state" or "vacuum") of string theory. Despite recent advances in understanding various non-perturbative aspects of string theory, our inability to answer the fundamental question of vacuum selection persists. Until this challenge is overcome, string phenomenology therefore must content itself with answering questions of a *relative* nature (such as questions concerning relative *patterns* of phenomenology) rather than the sorts of absolute questions (such as calculating the mass of the electron) that one would also ideally like to ask. Nevertheless, as we shall see, string theory can still provide us with considerable guidance for physics beyond the Standard Model.

Acknowledgments

First, I wish to thank the organizers of the 1998 TASI school, especially P. Langacker and K.T. Mahanthappa, for giving me the opportunity to spend two weeks in Boulder, Colorado, and to deliver these lectures in such a pleasant and stimulating environment. I am particularly grateful to P. Langacker for the infinite patience he exhibited while waiting for these lectures to appear in written form. Second, I wish to thank the TASI students themselves for their questions and sustained interest which made my efforts worthwhile. Third, I would like to thank my physics colleagues and collaborators for their insights and explanations, many of which have found their way into these lectures. Finally, I would like to thank T. Gherghetta, D. Marfatia, and J. Wells, each of whom has provided numerous helpful comments and suggestions pertaining to earlier drafts of these notes.

COLLIDER PHYSICS

DIETER ZEPPENFELD

Department of Physics, University of Wisconsin,
1150 University Avenue, Madison, WI 53706, USA
E-mail: dieter@pheno.physics.wisc.edu

These lectures are intended as a pedagogical introduction to physics at e^+e^- and hadron colliders. A selection of processes is used to illustrate the strengths and capabilities of the different machines. The discussion includes W pair production and chargino searches at e^+e^- colliders, Drell-Yan events and the top quark search at the Tevatron, and Higgs searches at the LHC.

1 Introduction

Over the past two decades particle physics has advanced to the stage where most, if not all observed phenomena can be described, at least in principle, in terms of a fairly economic $SU(3) \times SU(2) \times U(1)$ gauge theory: the Standard Model (SM) with its three generations of quarks and leptons[1]. Still, there are major short-comings of this model. The spontaneous breaking of the electroweak sector is parameterized in a simple, yet ad-hoc manner, with the help of a single scalar doublet field. The smallness of the electroweak scale, $v - 246$ GeV, as compared to the Planck scale, requires incredible fine-tuning of parameters. The Yukawa couplings of the fermions to the Higgs doublet field, and thus fermion mass generation, cannot be further explained within the model. For all these reasons there is a strong conviction among particle physicists that the SM is an effective theory only, valid at the low energies presently being probed, but eventually to be superseded by a more fundamental description.

Extensions of the SM include models with a larger Higgs sector,[2,3] e.g. in the form of extra doublets giving rise to more than the single scalar Higgs resonance of the SM, models with extended gauge groups which predict e.g. extra Z-bosons to exist at high energies,[4] or supersymmetric extensions of the SM with a doubling of all known particles to accommodate the extra bosons and fermions required by supersymmetry[5]. In all these cases new heavy quanta are predicted to exist, with masses in the 100 GeV region and beyond. One of the main goals of present particle physics is to discover unambiguous evidence for these new particles and to thus learn experimentally what lies beyond the SM.

While some information on physics beyond the SM can be gleaned from precision experiments at low energies,[1] via virtual contributions of new heavy particles to observables such as the anomalous magnetic moment of the muon, rare decay modes of heavy fermions (such as $b \rightarrow s\gamma$), or the so-called oblique corrections to electroweak observables, a complete understanding of what lies beyond the SM will require the direct production and detailed study of the decays of new particles.

The high center of mass energies required to produce these massive objects can only be generated by colliding beams of particles, which in practice means electrons and protons[a]. e^+e^- colliders have been operating for almost thirty years, the highest energy machine at present being the LEP storage ring at CERN with a center of mass energy close to 200 GeV. Proton-antiproton collisions have been the source of the highest energies since the early 1980's, with first the CERN collider and now the Tevatron at Fermilab, which will continue operation with a center of mass (c.m.) energy of 2 TeV in the year 2000. In 2005 the Large Hadron Collider (LHC) in the LEP tunnel at CERN will raise the maximum energy to 14 TeV in proton-proton collisions. Somewhat later we may see a linear e^+e^- collider with a center of mass energy close to 1 TeV.

The purpose of these lectures is to describe the research which can be performed with these existing, or soon to exist machines at the high energy frontier. No attempt will be made to provide complete coverage of all the questions which can be addressed experimentally at these machines. Books have been written for this purpose.[7] Rather I will use a few specific examples to illustrate the strengths (and the weaknesses) of the various colliders, and to describe a variety of the analysis tools which are being used in the search for new physics. Section 2 starts with an overview of the different machines, of the general features of new particle production processes, and of the ensuing implications for the detectors which are used to study them. This section includes a theorist's picture of the structure of modern collider detectors.

Specific processes at e^+e^- and hadron colliders are discussed in the three main Sections of these lectures. Section 3 first deals with W^+W^- production in e^+e^- collisions, and how this process is used to measure the W-mass. It then discusses chargino pair production as an example for new particle searches in the relatively clean environment of e^+e^- colliders. Hadron collider experiments are discussed in Sections 4 and 5. After consideration of the basic structure of cross sections at hadron colliders and the need for non-perturbative input in the form of parton distribution functions, the simplest new physics process, single gauge boson (Drell-Yan) production will be considered in some detail. The top quark search at the Tevatron is used as an example for a successful

---

[a]The feasibility of a $\mu^+\mu^-$ collider is under intense study as well.[6] For the purpose of these lectures, the physics investigations at such machines would be very similar to the ones at e^+e^-colliders.

search for heavy quanta. Section 5 then looks at LHC techniques for finding the Higgs boson. Some final conclusions are drawn in Section 6.

2 Overview

Some features of production processes for new heavy particles are fairly general. They are important for the design of colliding beam accelerators and detectors, because they concern the angular and energy dependence of pair-production processes.

In order to extend the reach for producing new heavy particles, the available center of mass energy, at the quark, lepton or gluon level, is continuously being increased, in the hope of crossing production thresholds. In turn this implies that new particles will be discovered close to pair-production threshold, where their momenta are still fairly small compared to their mass. Small momenta, however, imply little angular dependence of matrix elements because all $\cos\theta$ or $\sin\theta$ dependence is suppressed by powers of $\beta = |\mathbf{p}|/p^0$. As a result the production of heavy particles is fairly isotropic in the center of mass system. This is simple quantum mechanics, of course; higher multipoles L in the angular distribution are suppressed by factors of β^{2L+1}.

Another feature follows from the fact that pair production processes have a dimensionless amplitude, which can only depend on ratios like $M^2/\hat{s}$, where M is the mass of the produced particles and $\sqrt{\hat{s}}$ is the available center of mass energy. At fixed scattering angle, the amplitude approaches a constant as $M^2/\hat{s} \to 0$. The effect is most pronounced when the pair production process is dominated by gauge-boson exchange in the s-channel as is the case for a large class of new particle searches, be it $q\bar{q} \to t\bar{t}$ at the Tevatron, $e^+e^- \to ZH$ at LEP2, or $q\bar{q}$ annihilation to two gluinos at the LHC (see Fig. 1). s-channel production allows $J = 1$ final states only, which limits the angular dependence of the production cross section to a low order polynomial in $\cos\theta$ and $\sin\theta$. We thus have little variation of the production amplitude, i.e.

$$\mathcal{M}(\hat{s}) \approx \text{constant} , \tag{1}$$

where the constant is determined by the coupling constants at the vertices of the Feynman graphs of Fig. 1.

Since the production cross section is given by

$$\frac{d\hat{\sigma}}{d\Omega} = \frac{1}{2\hat{s}} \frac{\beta}{32\pi^2} \overline{\sum_{\text{pol}}} |\mathcal{M}|^2 , \tag{2}$$

approximately constant $\mathcal{M}$ implies that the production cross section drops like $1/\hat{s}$ and becomes small fast at high energy, being of order $\alpha^2/\hat{s} \sim 1$ pb

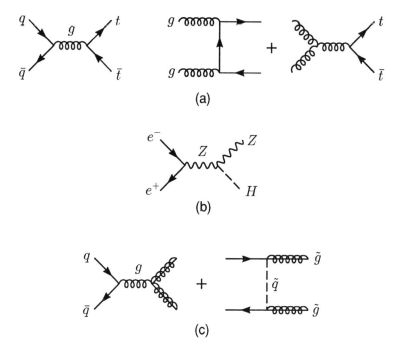

Figure 1: Representative Feynman graphs for new particle production processes, (a) top quark production, $q\bar{q} \to t\bar{t}$ and $gg \to t\bar{t}$, (b) Higgs production, $e^+e^- \to ZH$, and (c) gluino pair production, $q\bar{q} \to \tilde{g}\tilde{g}$.

at $\sqrt{\hat{s}} = 200$ GeV for electroweak strength cross sections. This $1/s$ fall-off of production cross sections is clearly visible in Fig. 2, above the peak at 91.2 GeV, produced by the Z resonance. The rapid decrease of cross sections with energy implies that in order to search for new heavy particles one needs both high energy and high luminosity colliders, with usable center of mass energies of order several hundred GeV and luminosities of order 1 fb^{-1} per year or higher.

The most direct and cleanest way to provide these high energies and luminosities is in e^+e^- collisions. For most of the nineties, LEP at CERN and the SLAC linear collider (SLC) have operated on the Z peak, at $\sqrt{s} \approx m_Z = 91.187$ GeV, collecting some 10^7 Z decays at LEP and several hundred thousand at SLC. The high counting rate provided by the Z resonance (see Fig. 2) has allowed very precise measurements of the couplings of the Z to the SM quarks and leptons. At the same time searches for new particles were conducted, and the fact that nothing was found excludes any new particle which

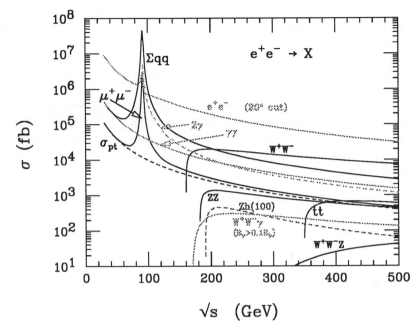

Figure 2: Representative SM production cross sections in e^+e^- annihilation. For Bhabha scattering and final states involving photons, the differential cross sections $d\sigma/d\cos\theta$ have been integrated with a cut of $20^o < \theta < 160^o$, in order to avoid the singular forward region. Note the Z peak and the characteristic $1/s$ fall-off of cross sections at high energies.

could be pair-produced in Z decays, i.e. which have normal gauge couplings to the Z and have masses $M \lesssim m_Z/2$. This includes fourth generation quarks and leptons and the charginos, sleptons and squarks of supersymmetry.[1]

Since 1996 the energy of LEP has been increased in steps, via 161, 172, and 183 GeV, to 189 GeV in 1998. About 250 pb^{-1} of data have been collected by each of the four LEP experiments over this period, mostly at the highest energy point. The experiments have mapped out the W^+W^- production threshold (see Fig. 2), measured the W mass and W couplings directly, searched for the Higgs and set new mass limits on other new particles (see Section 3).

Why has progress been so incremental? The main culprit is synchrotron radiation in circular machines. The centripetal acceleration of the electrons on their circular orbit leads to an energy loss which grows as γ^4/R, where R is the bending radius of the machine, approximately 4.2 km at LEP, and $\gamma = E_e/m_e$ is the time dilatation factor for the electrons, which at LEP exceeds 10^5. The small electron mass requires very large bending radii, which can be achieved with very modest bending magnets in the synchrotron. Scaling up LEP to higher energies soon leads to impossible numbers. A 1 TeV electron synchrotron with the same synchrotron radiation loss as LEP (about 2.8 GeV per turn at $\sqrt{s} = 200$ GeV)[8] would have to be over 600 times larger. The future of e^+e^- machines belongs to linear colliders, which do not reuse the accelerated electrons and positrons, but rather collide the beams once, in a very high intensity beam spot. The SLC has been the first successful machine of this type. The next linear collider (NLC) would be a 400 GeV to 1 TeV e^+e^- collider with a luminosity in the 10–100 fb^{-1} per year range. With continued cooperation of physicists worldwide, such a machine might become a reality within a decade.

The easiest way to get to larger center of mass energies is to use heavier beam particles, namely protons or anti-protons. Their 2000 times larger mass makes synchrotron radiation losses negligible, even for much higher beam energies. Thus, the energy of proton storage rings is limited by the maximum magnetic fields which can be achieved to keep the particles on their circular orbits, i.e. the beam momentum p is limited by the relation

$$p = eBR . \tag{3}$$

The Tevatron at Fermilab is the highest energy $\bar{p}p$ collider at present, and so far has accumulated about 120 pb^{-1} of data, at a $\bar{p}p$ center of mass energy of 1.8 TeV, in each of two experiments, CDF and D0. It was this data taking period in the mid-nineties, called run I, which led to the discovery of the top-quark (see Section 4). Note that the pair production of top-quarks, with a mass of $m_t = 175$ GeV, will be possible at e^+e^- colliders only in the NLC era. A higher luminosity run at 2 TeV, run II, is scheduled to collect 1–2 fb^{-1} of data, starting in the spring of 2000. And before an NLC will be constructed, the LHC at CERN will start with pp collisions at a c.m. energy of 14 TeV and with a luminosity of 10^{33}–10^{34} cm^{-2}sec^{-1}, corresponding to 10–100 fb^{-1} per year.

Given these much higher energies available at hadron colliders, why do we still invest in e^+e^- machines? The problem with hadron colliders is that protons are composite objects, made out of quarks and gluons, and these partons only carry a fraction of the proton energy. In order to produce new heavy particles the c.m. energy in a parton-parton collision must be larger than the sum of the masses of the produced particles, and this becomes increasingly unlikely as the required energy exceeds some 10–20% of the collider energy. Most pp or $\bar{p}p$ collisions are collisions between fairly low energy partons. Since the proton is a composite object, of finite size, the total pp or $\bar{p}p$ cross section does not decrease with energy, in fact it grows logarithmically and reaches about

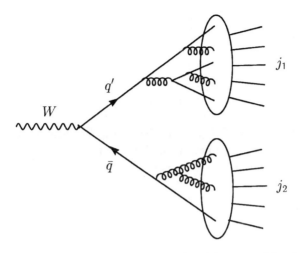

Figure 3: Schematic drawing representing decay of a $W^+ \to \bar{q}q$ with subsequent parton showering and hadronization, leading to a dijet signature for the W in the detector.

100 mb at LHC energies. The high energy parton-parton collisions, however, suffer from the $1/s$ suppression discussed before. Production cross sections for new particles are of order 1 pb (with large variations), i.e. 10^{-11} times smaller. At hadron colliders, one thus needs to identify one interesting event in a background of 10^{11} bad ones, which poses a daunting task to the experimentalists and their detectors. The much cleaner situation at e^+e^- colliders can be appreciated from Fig. 2: backgrounds to new physics searches arise form processes like $e^+e^- \to q\bar{q}$, $e^+e^- \to W^+W^-$, or $e^+e^- \to Z\gamma$, with cross sections in the 1–100 pb region and thus not much larger than the expected signal cross sections.

Whether new particles are produced in e^+e^- or pp collisions, they are never expected to be seen directly in the detector. Expected lifetimes, τ, and ensuing decay lengths, $\gamma c\tau$, are of nuclear scale and, therefore, only the decay products can be observed. A typical example is shown in Fig. 3. A W boson has a decay width of $\Gamma_W \approx 2.1$ GeV and is depicted to decay into a $u\bar{d}$ quark pair (other decay modes are $e^+\nu_e$, $\mu^+\nu_\mu$, $\tau^+\nu_\tau$ and $c\bar{s}$). The produced quarks, of course, are not observable either, due to confinement, rather they emit gluons and quark-antiquark pairs, which eventually hadronize and form jets of hadrons containing pions, kaons, and so on. The detectors have to observe these hadrons, measure the directions and energies of the jets, and deduce from here the four-momenta of the original W^+ or u and $\bar{d}$ quarks.

The situation is very similar for other new heavy quanta. A Higgs boson, of mass $m_H = 120$ GeV say, is expected to decay into $b\bar{b}$, $\tau^+\tau^-$ or $\gamma\gamma$, among others, with expected branching ratios

$$B(H \to b\bar{b}) = 74\% \tag{4}$$

$$B(H \to \tau^+\tau^-) = 6\% \tag{5}$$

$$B(H \to \gamma\gamma) = 0.2\% \ . \tag{6}$$

At LEP2, ZH production is searched for in the $b\bar{b}$ decay mode of the Higgs, i.e. b-quark jets need to be observed and distinguished from lighter quark jets.

Supersymmetric particles are expected to produce an entire decay chain before they can be observed in the detector, an example being the decay of a gluino to a squark and a quark, where the squark in turn decays to the lightest neutralino, $\tilde{\chi}^0$ or chargino, $\tilde{\chi}^\pm$,

$$\tilde{g} \to \tilde{q}\bar{q} \to \bar{q}q\tilde{\chi}^0 \to jj\not{E}_T \ , \tag{7}$$

$$\tilde{g} \to \tilde{q}\bar{q} \to \bar{q}q'\tilde{\chi}^\pm \to \bar{q}q'W^\pm\tilde{\chi}^0 \to \bar{q}q'\ell^\pm\nu\tilde{\chi}^0 \to jj\ell^\pm\not{E}_T \ . \tag{8}$$

In the last step of the decay chain, hadronization of the quarks leads to jets, and the neutrino and the neutralino escape the detector, leading to an imbalance in the measured momenta of observable particles transverse to the beam, i.e. to a missing E_T signature.

Collider detectors, which are to discover these new particles, must be designed to observe the decay products, positively identify them and measure their direction and their energy in the lab, i.e. determine the momenta of electrons, muons, photons and jets. In addition, the identification of heavy quarks, in particular the b quarks arising in Higgs boson and top quark decays, is very important. Many technical solutions have been developed for this purpose, variations in the type of detector doing the reconstruction of charged particle tracks, or the energy measurement of electrons and photons (via electromagnetic showers) or jets (in a hadronic calorimeter). The basic layout of modern collider detectors is remarkably uniform, however, for fundamental physics reasons. For the basic level addressed in these lectures, it is sufficient to have a brief look at these common, global features.

The schematic drawing of Fig. 4 shows a cross section through a typical collider detector. The detector has a cylindrical structure, which is wrapped around the beam pipe in which the electrons or protons collide, at the center of the detector. The standard components of the detector are as follows.

- Central Tracking: The innermost part of the detector, closest to the interaction region, records the tracks of the produced charged particles, $e^\pm$,

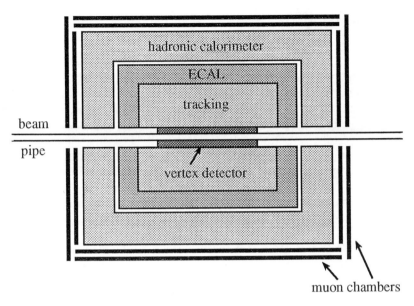

Figure 4: Schematic cross section through a typical collider detector. Shown, from the beam pipe out, are the locations of vertex detector, tracker, electromagnetic calorimeter, hadronic calorimeter, and muon chambers.

$\mu^{\pm}$, $\pi^{\perp}$, $K^{\perp}$ and so on. Via the bending of the tracks in a strong magnetic field, pointing along the beam direction, it measures the charged particle momenta.

- Electromagnetic Calorimeter: Neutral particles leave no tracks. They need to be measured via total absorption of their energy in a calorimeter. Electrons and photons loose their energy relatively quickly, via an electromagnetic shower. Since a shower develops randomly, statistical fluctuations limit the accuracy of the energy measurement. Excellent results, such as for the CMS detector at the LHC,[10] are

$$\frac{\Delta E}{E} = \frac{0.02}{\sqrt{E}} \quad \text{to} \quad \frac{0.05}{\sqrt{E}} \, , \tag{9}$$

where the energy E is measured in GeV. Note that neutral pions decay into two photons before they leave the beam pipe, and these photons will stay very close to each other for large π^0 momentum. Hence, π^0s will be recorded in the electromagnetic calorimeter and they can fake photons.

- Hadronic Calorimeter: Other hadrons are absorbed in the hadron calorimeter where their energy deposition is measured, with a statistical error which may reach[9]

$$\frac{\Delta E}{E} = \frac{0.4}{\sqrt{E}} . \tag{10}$$

Typical hadron calorimeters have a thickness of some 25 absorption lengths and normally only muons, which do not interact strongly with the heavy nuclei of the hadron calorimeter, will penetrate it. Outside the calorimeter one therefore places the

- Muon Chambers: They record the location where a penetrating particle leaves the inner detector, and in several layers follows the direction of the track. Together with the central tracking information, this allows to measure the curvature of the muon track in the known magnetic fields inside the detector and determines the muon momentum.

- Vertex Detector: Bottom and charm quarks can be identified by the finite lifetime, of order 1 ps, of the hadrons which they form. This lifetime leads to a decay length of up to a few mm, i.e. the b or c decay products do not point back to the primary interaction vertex, which is much smaller, but to a secondary vertex. The decay length can be resolved with very high precision tracking. For this purpose modern collider detectors possess a solid state micro-vertex detector, very close to the beam-pipe, which provides information on the location of tracks with a resolution of order 10 μm. This technique now allows to identify centrally produced b-quarks, i.e. those that are produced at angles of more than a few degrees with respect to the beams, with efficiencies above 50% and with high purity, rejecting non-b jets with more than 95% probability.

The various elements of the detector work together to identify the components of an event. An electron would deposit its energy in the electromagnetic calorimeter and produce a central track, which distinguishes it from a neutral photon. A charged pion or kaon produces a track also, but only a fraction of its energy ends up in the electromagnetic calorimeter: most of it leaks into the hadronic calorimeter. A muon, finally, deposits little energy in either calorimeter, rather it leads to a central track and to hits in the muon chambers.

If this muon originates from a b-decay, it will be traveling in the same direction as other hadrons belonging to the b-quark jet. This muon is not isolated, as opposed to a muon from a decay $W \to \mu\nu$ which only has a small probability to travel into the same direction as the twenty or thirty hadrons of a typical jet. One thus obtains a very detailed picture of the entire event, and from this picture one needs to reconstruct what happened at the parton level.

3 e^+e^- Colliders

A full event reconstruction is most easily done at an e^+e^- collider where the beam particles are elementary objects, i.e. the entire energy of the collision can go into the production of heavy particles. In contrast, at hadron colliders, the additional partons in the parent protons lead to a spray of hadrons in the detector which obscure the parton-parton collision we are interested in.

In e^+e^- collisions at energies $\sqrt{s} \lesssim 160$ GeV, the dominant hard process is fermion pair production, $e^+e^- \to \bar{f}f$. For $f = e, \mu$ this leads to extremely clean events with two particles in the final state only, and even the bulk of $\bar{q}q$ production events are easily recognized as dijet events. With the advent of LEP2 the situation has become more complicated. As can be seen in Fig. 2, W^+W^- and ZZ events are as copious as fermion pair production, but they lead to a more complex 4-fermion final state, via the decay of the two Ws or Zs into a pair of quarks or leptons each. Let us start our survey of e^+e^- physics with this new class of events, which will be important in all new high energy e^+e^- colliders. They show interesting features and provide information on fundamental parameters of the SM in their own right, but also they form an important background to new particle searches, such as charginos for example, and we need to analyze them in some detail.

3.1 W^+W^- production

At tree level, three Feynman graphs contribute to the production of W pairs.[11,12] They are s-channel γ- and Z-exchange, and t-channel neutrino exchange, and are shown in Fig. 5. The dominant features of the production cross section can be read off the propagator structure of the individual graphs. While the two s-channel amplitudes show a modest dependence on scattering angle only, the neutrino exchange graph produces a strong peaking at small scattering angles: the propagator factor for this graph is $1/t$ with $t = (p-k)^2 = -2p \cdot k + m_W^2$. In the c.m. system we choose the initial electron momentum, p, along the

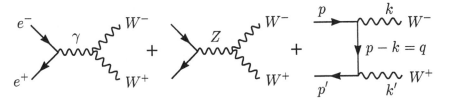

Figure 5: Feynman graphs for the process $e^+e^- \to W^+W^-$. The labels on the neutrino exchange graph give the particle momenta.

314

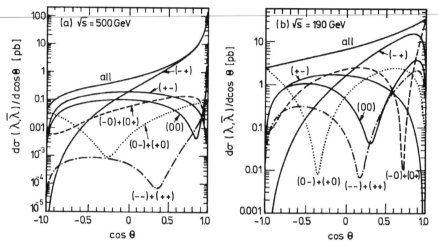

Figure 6: Angular distribution $d\sigma/d\cos\theta$ for $e^+e^- \to W^+W^-$ production at (a) $\sqrt{s} = 500$ GeV and (b) $\sqrt{s} = 190$ GeV, within the SM. Also shown are the cross sections for the various W^-W^+ helicity combinations $(\lambda, \bar{\lambda})$. Where a sum is shown, both helicity combinations give the same result. From Ref. [11].

z-axis and define the scattering angle as the angle between the W^- and the e^- direction, i.e.

$$p = \frac{\sqrt{s}}{2}(1,0,0,1) , \qquad k = \frac{\sqrt{s}}{2}(1,\beta\sin\theta,0,\beta\cos\theta) , \qquad (11)$$

where $\beta = \sqrt{1 - 4m_W^2/s}$ is the velocity of the produced W. The denominator of the propagator then becomes

$$|t| = \frac{s}{4}\left(2 - 2\beta\cos\theta - \frac{4m_W^2}{s}\right) = \frac{s}{4}\left(1 + \beta^2 - 2\beta\cos\theta\right) \geq \frac{s}{4}(1-\beta)^2 . \quad (12)$$

Close to threshold, i.e. for $\beta \approx 0$, there is little angular dependence. At high energies, however, as $\beta \to 1$, $|t|$ becomes very small near $\theta = 0$ and the $1/t$ pole induces a strongly peaked W pair cross section near forward scattering angles. This effect is shown in Fig. 6, where the angular distribution $d\sigma/d\cos\theta$ for W^+W^- production in the SM is shown, including the contributions from different W^- and W^+ helicities, λ and $\bar{\lambda}$. Angular momentum conservation and the fact that left-handed electrons only contribute to the ν-exchange graph, lead to a strong polarization of the produced W's in the forward region: the produced W^- helicity is mostly -1 in this region, i.e. it picks up the electron helicity.

The Ws are only observed via their decay products, $W \to \bar{f}f'$, where three $\ell\nu$ combinations ($\ell = e, \mu, \tau$) and six quark combinations can be produced ($u\bar{d}$ and $c\bar{s}$ and counting $N = 3$ different colors). Since all quarks and leptons couple equally to Ws (when neglecting Cabibbo mixing), and because any CKM effects exactly compensate in the decay widths, the branching ratios to leptons and hadrons are simply given by

$$B(W \to e\nu_e) = B(W \to \mu\nu_\mu) = B(W \to \tau\nu_\tau) = \frac{1}{9} \,, \tag{13}$$

$$B(W \to \text{hadrons}) = \frac{1}{9} \cdot 2 \cdot 3 = \frac{2}{3} \,. \tag{14}$$

QCD effects induce corrections of a few percent to these relations. In lowest order, a decay $W \to \bar{q}q'$ leads to two jets in the final state, and this, combined with the branching ratios of Eq. (13,14), fixes the probabilities for the various classes of events to be observed for W^+W^- production,

$$B(W^+W^- \to 4\text{jets}) = 46\% \,, \tag{15}$$

$$B(W^+W^- \to jj + e\nu_e, \mu\nu_\mu) = 29\% \,, \tag{16}$$

$$B(W^+W^- \to \ell^+\nu\ell^-\bar{\nu}) = 10.5\% \,, \tag{17}$$

$$B(W^+W^- \to jj + \tau\nu_\tau) = 14.5\% \,. \tag{18}$$

Thus, it is most likely to observe the two Ws in 4-jet events, followed by the 'semileptonic' channel, where one W decays into either electrons or muons. The remaining channels have at least two neutrinos in the final state (the τ decays inside the beam-pipe!) and hence a substantial fraction of the final state particles cannot be observed, which limits the reconstruction of the event. Fortunately, these more difficult situations comprise only one quarter of the W pair sample.

As we saw when discussing production angular distributions, the Ws are strongly polarized. Fortunately, the $V - A$ structure of the W-fermion couplings provides a very efficient polarization analyzer for the Ws, via their decay distributions. Consider the decay of a right-handed W^-, i.e. of helicity $\lambda = +1$, as depicted in Fig. 7 in the W rest frame. Because of the $V - A$ coupling, the decay $W \to f_1\bar{f}_2$ always leads to a left-handed fermion and a right-handed anti-fermion, in the massless fermion limit. The fermion spins therefore always line up as shown in the figure. Taking the decay polar angle $\theta_1^* \to 0$, the combined fermion spins point opposite to the spin of the parent W^-. Angular momentum conservation does not allow this, which means that the decay amplitude vanishes at $\cos\theta_1^* = 1$. The same argument shows that for a left-handed W^-, of helicity $\lambda = -1$, the decay amplitude must vanish

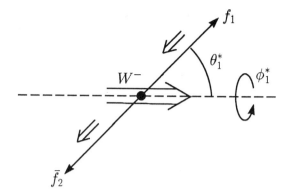

Figure 7: Orientation of spins for the decay of a right-handed W^- into a pair of fermions, $f_1 \bar{f}_2$. Because of the $V - A$ structure of charged currents, the fermion helicities are fixed as shown.

for $\cos \theta_1^* = -1$. A quick calculation shows that the decay amplitudes for $W^- \to f_1 \bar{f}_2$ are proportional to

$$D(W^- \to f_1 \bar{f}_2)(\lambda = +1) = \frac{1}{\sqrt{2}} (1 - \cos \theta_1^*) \, e^{i\phi_1^*} , \qquad (19)$$

$$D(W^- \to f_1 \bar{f}_2)(\lambda = 0) = -\sin \theta_1^* , \qquad (20)$$

$$D(W^- \to f_1 \bar{f}_2)(\lambda = -1) = \frac{1}{\sqrt{2}} (1 + \cos \theta_1^*) \, e^{-i\phi_1^*} , \qquad (21)$$

where ϕ_1^* is the decay azimuthal angle, as indicated in the figure. Analogous results are obtained, of course, for the W^+ decay angles θ_2^* and ϕ_2^*.

By measuring the decay angular distributions, i.e. by distinguishing the $(1 - \cos \theta^*)^2$, $\sin^2 \theta^*$, and $(1 + \cos \theta^*)^2$ distributions for right-handed, longitudinal and left-handed polarized Ws, we can measure the average W polarization. The full polarization information is contained in the 5-fold production and decay angular distributions,[11,12,13]

$$\frac{d^5 \sigma}{d \cos \theta \, d \cos \theta_1^* \, d\phi_1^* \, d \cos \theta_2^* \, d\phi_2^*} , \qquad (22)$$

Here the decay angles are defined in the W^- and W^+ rest frames, respectively, and are measured against the direction of the parent W in the lab.

A major reason to study this 5-fold angular distribution is the experimental determination of the WWZ and $WW\gamma$ couplings which enter in the first two Feynman graphs of Fig. 5. This is analogous to the measurement of vector and axial vector couplings of the various fermions to the Z and the W. The experiments at LEP2 so far confirm the SM predictions for these triple-gauge-boson couplings at about the 10% level.[13,14]

Why would one consider W decay angular distributions if one does not want to measure polarizations or triple-gauge-boson couplings? As we saw when discussing W production, the produced W^- is very strongly left-handed polarized in the forward direction. This polarization has important conse-quences for the energy distribution of the decay products, and therefore for the way the event appears in the detector. To be definite, let us consider the decay $W^- \to \ell^- \bar{\nu}$. In the W rest frame, the four-momentum of the charged lepton is given by

$$p^* = \frac{m_W}{2} \left(1, \sin\theta_1^* \cos\phi_1^*, \sin\theta_1^* \sin\phi_1^*, \cos\theta_1^*\right) . \tag{23}$$

The charged lepton energy in the lab frame is obtained from here via a boost of its four-momentum, with a γ-factor $\gamma = E_W/m_W = \sqrt{s}/2m_W$,

$$p^0 = \gamma\left(p^{0*} + \beta p_z^*\right) = \gamma \frac{m_W}{2}\left(1 + \beta\cos\theta_1^*\right) = \frac{\sqrt{s}}{4}\left(1 + \beta\cos\theta_1^*\right) . \tag{24}$$

Thus, the polar angle of the lepton in the W rest frame can be measured in terms of the lepton energy in the lab frame, and the two observables directly correspond to each other. This also implies that the energy distributions of the leptons in the lab are determined by their angular distributions in the W rest frame, and these are fixed by the polarization of the parent W.

As a concrete example, consider the average energy of the charged lepton, for the decay of a left-handed W^-. We need to average the result of Eq. (24) over the normalized decay angular distribution, $3/8(1 + \cos\theta^*)^2$, for a left-handed W^-:

$$\langle p^0(\ell^-)\rangle = \frac{E_W}{2} \int_{-1}^{1} d\cos\theta^* \frac{3}{8}(1 + \cos\theta^*)^2(1 + \beta\cos\theta^*) = \frac{E_W}{2}\left(1 + \frac{\beta}{2}\right) . \tag{25}$$

Energy conservation fixes the average neutrino momentum to

$$\langle p^0(\nu)\rangle = \frac{E_W}{2}\left(1 - \frac{\beta}{2}\right) . \tag{26}$$

In the relativistic limit, $\beta \to 1$, the neutrino receives only $1/3$ of the energy of the charged lepton, on average, which has important consequences for detection and energy measurement of the leptons as well as for the consideration of W pair production as backgrounds to new physics searches.

Polarization effects can have dramatic effects and one therefore needs predictions for W pair production and decay which consider the full chain $e^+e^- \to W^+W^- \to 4$ fermions. In this full $2 \to 4$ process the Ws merely appear as resonant propagators, which are treated as Breit-Wigner resonances. Away from the peak of the resonance, seven additional Feynman graphs contribute, beyond the three shown in Fig. 5, even for the simplest case, $e^+e^- \to \mu^- \bar{\nu}_\mu u\bar{d}$. These calculations have been performed [12] and are being used in the actual data analysis.

Another application, which nicely demonstrates the advantages of e^+e^- collisions, is the W-mass measurement in W^+W^- production at LEP2.[15] Let us consider the decay $W^+W^- \to \ell\nu jj$ as an example. A full reconstruction of the Breit-Wigner resonances, and a measurement of its center, at m_W, is possible with the two jet momenta. However, the measurements of the jets' energies have large errors, of order $\pm 15\%$, and such a direct approach would lead to fairly large errors on the extracted W-mass. One can do much better by making use of the known kinematics of the event.

The 3-momentum of the neutrino in the event can be reconstructed from momentum conservation, as

$$\mathbf{p}_\nu = -\mathbf{p}_\ell - \mathbf{p}_{j_1} - \mathbf{p}_{j_2} , \qquad (27)$$

where we have used the fact that the lab frame is the c.m. frame, i.e. the sum of all the final state momenta in the lab must add up to zero. The energy of the massless neutrino is then given by $E_\nu = |\mathbf{p}_\nu|$. Energy conservation and the equal masses of the two Ws now imply the constraints

$$E_{W_1} = E_{j_1} + E_{j_2} = E_b , \qquad E_{W_2} = E_\ell + E_\nu = E_b , \qquad (28)$$

where $E_b = \sqrt{s}/2$ is the beam energy. Even when considering the finite widths of the W resonances and the possibility of initial state radiation, i.e. emission of photons along the beam direction, which effectively lowers the c.m. energy $\sqrt{s}$, the constraint of Eq. (28) is satisfied to much higher accuracy than the precision of the jet energy measurement. One can thus drastically improve the W-mass resolution by using the two constraints to solve for the two unknowns E_{j_1} and E_{j_2} and use these values to calculate the W^+ and W^- mass. The expected improvement is illustrated in Fig. 8.

First measurements of the W-mass, using both $\ell\nu jj$ and 4-jet events, have already been performed with the 172 and 183 GeV data and resulted in [16]

$$m_W = 80.36 \pm 0.09 \text{ GeV} , \qquad (29)$$

where the results from all four LEP experiments have been combined. Further improvements are expected in the near future from the four times larger event

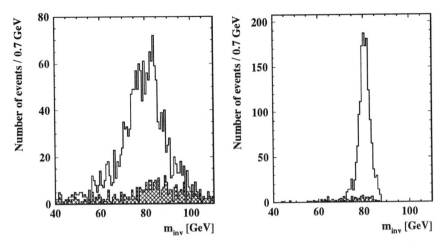

Figure 8: Expected W-mass reconstruction in $W^+W^- \to \ell\nu jj$ events at LEP2. The two plots show the invariant mass distribution of the W decay products before and after using the kinematic constraints described in the text. From Ref.[15].

sample already collected in 1998. The LEP value agrees well with the one extracted from order 10^5 leptonic W-decays observed at the Tevatron, $m_W = 80.41 \pm 0.09$ GeV[17]

3.2 Chargino pair production

So far we have considered W^+W^- production as a signal. However, W-pairs can be a serious background in the search for other new particles. Let us consider one example in some detail, the production of charginos at LEP2.

Charginos arise in supersymmetric models[5] as the fermionic partners of the $W^\pm$ and the charged Higgs, $H^\pm$. Since they carry electric and weak charges, they couple to both the photon and the Z and can be pair-produced in e^+e^- annihilation. The relevant Feynman graphs are shown in Fig. 9(a). Note that the graphs for chargino production are completely analogous to the ones in Fig. 5 for W-pair production, a reflection of the supersymmetry of the couplings.

Two charginos are predicted by supersymmetry, and the lighter one, $\tilde{\chi}_1^\pm$, might be light enough to be pair-produced at LEP2.[18] Production cross sections can be sizable, ranging from 2 to 5 pb at LEP2 for typical parameters of SUSY models. However, the s-channel γ and Z exchange graphs, and t-channel sneutrino exchange in Fig. 9(a), interfere destructively. For small sneutrino masses, of order 100 GeV or less, this can lead to a drastic reduction in the

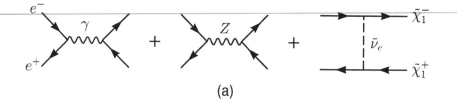

(a)

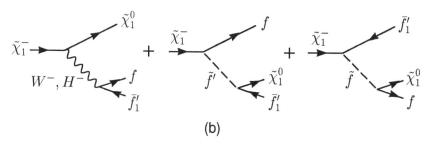

(b)

Figure 9: Feynman graphs for (a) chargino production in e^+e^- annihilation and (b) chargino decay into the lightest neutralino, $\tilde{\chi}_1^0$ and a pair of quarks or leptons.

chargino pair production cross section, in particular near threshold. As a result, one must be prepared for production cross sections well below 1 pb at LEP2 also. In any case, the expected chargino pair cross section can be smaller than the cross section for the dominant background, $\sigma(e^+e^- \to W^+W^-) \approx 17$ pb, by one order of magnitude or more.

A chargino, once it is produced, is expected to decay into the lightest neutralino, $\tilde{\chi}_1^0$, and known quarks and leptons, as shown in Fig. 9(b). The lightest neutralino is stable in most SUSY models, due to conserved R-parity, and does not interact inside the detector, thus leading to a missing momentum signature. Over large regions of parameter space, where the squarks and sleptons entering the chargino decay graphs are quite heavy, the chargino effectively decays into the neutralino and a virtual W. This results in a signature which is quite similar to W-pair production, since only the W^* decay products are seen inside the detector. Thus, LEP looks for the charginos via the production and decay chain

$$e^+e^- \to \tilde{\chi}_1^+\tilde{\chi}_1^- \to \tilde{\chi}_1^0\tilde{\chi}_1^0 W^{+*}W^{-*}\,, \tag{30}$$

which, just like W^+W^- production, leads to a $jjjj$ signature 46% of the time and to an $\ell\nu jj$ final state (including τ's) in 43.5% of all cases. The only difference is the additional presence of two massive neutralinos, which needs to be exploited.

Since the neutralinos escape detection, the energy deposited in the detector typically is less than for W^+W^- events. In addition, the missing neutralinos spoil the momentum balance of the visible particles, which leads to a more spherically symmetric event than encountered in W pair production. The most effective cut arises from the fact, however, that the four-momentum of the missing neutrinos and neutralinos can be reconstructed due to the beam constraint. Four-momentum conservation for the $\ell \nu j j \tilde{\chi}^0 \tilde{\chi}^0$ final state state reads

$$p_{e^-} + p_{e^+} = p_\ell + p_{j_1} + p_{j_2} + \not{p} \tag{31}$$

from which the missing mass, $M^2_{\mathrm{miss}} = \not{p}^2$ can be reconstructed. For a W^+W^- event the missing momentum corresponds to a single neutrino and, thus, $M_{\mathrm{miss}} = 0$. For the signal, however, one has

$$\not{p}^2 > 4m^2_{\tilde{\chi}^0} . \tag{32}$$

Not only does the consideration of the missing mass allow for effective background reduction, to a level below 0.05–0.1 pb,[18] but it would also provide a measurement of the lightest neutralino mass, once charginos are discovered.

So far, no chargino signal has been observed at LEP. This pushes the chargino mass bound above 90 GeV, provided the chargino-neutralino mass difference is sufficiently large to allow enough energy for the visible chargino decay products.[19] The LEP experiments have searched for other super-partners also, like squarks and sleptons, and have not yet discovered any signals. The sfermions, since they are scalars, have a softer threshold turn-on, proportional to β^3, and hence have a very low pair production cross section if their mass is close to the beam energy. As a result, squark and slepton mass bounds currently are somewhat weaker than for charginos, but even here, sfermions with masses below 70–85 GeV (depending on flavor) are excluded.[19]

3.3 Future e^+e^- and $\mu^+\mu^-$ colliders

An exciting search presently being conducted at LEP is the hunt for the Higgs boson, in $e^+e^- \to ZH$.[22] The mass of the Higgs boson does influence radiative corrections to 4-fermion amplitudes, via ZH and WH loops contributing to the Z and W propagator corrections. Precise measurements of asymmetries in $e^+e^- \to \bar{f}f$, of partial Z widths to leptons and quarks, of atomic parity violation etc. allow to extract the expected Higgs mass within the SM. These measurements point to a relatively small Higgs mass, of about 100 GeV, albeit with a large error of about a factor of two.[1,20,21] LEP is exactly searching in this region.

In $e^+e^- \to ZH$, the large mass of the accompanying Z limits the reach of the LEP experiments, to about [22]

$$m_H \lesssim \sqrt{s} - m_Z - 5 \text{ GeV}. \tag{33}$$

With an eventual c.m. energy of $\sqrt{s} \approx 200$ GeV, this allows discovery of the Higgs at LEP, provided its mass is below about 105 GeV. However, measurements at energies up to 189 GeV have not discovered anything yet, setting a lower Higgs mass bound, within the SM, of 95 GeV.[21] We need luck to still find a Higgs signal at LEP, before the LEP tunnel needs to be cleared for installing the LHC.

Given the indications for a relatively light Higgs from electroweak precision data, an expectation which is shared by supersymmetric models,[5] an e^+e^- collider with higher c.m. energy than LEP2 is called for. As explained in Section 2 this cannot be a circular machine, due to excessive synchrotron radiation, but rather should be a linear e^+e^- collider.[23] A 500 GeV NLC, with a yearly integrated luminosity of 10–100 fb^{-1}, would be a veritable Higgs factory. At such a machine, the Higgs production cross section is of order 0.1 pb in both the $e^+e^- \to ZH$ production channel (for $m_H \lesssim 300$ GeV) and also in the weak boson fusion channel, where the Higgs boson is radiated off a t-channel W ($\sigma(e^+e^- \to H\nu\bar{\nu}) \gtrsim 0.03$ pb for $m_H \lesssim 200$ GeV). In this mass range, the 1000 to 10000 produced Higgs bosons per year would allow for detailed investigations of Higgs boson properties, in the clean environment of an e^+e^- collider. Somewhat higher Higgs boson masses, up to $m_H \approx \sqrt{s} - 100$ GeV (see Eq. (33)) are accessible as well, albeit with lower production rates.

The NLC would greatly extend the search region for other new heavy particles as well, like the charginos and neutralinos of the MSSM, or its squarks and sleptons. Even if theses particles are first discovered at a hadron collider like the LHC, the cleaner environment of e^+e^- collisions, the more constrained kinematics, and the observability of most of the decay channels give linear e^+e^- colliders great advantages for detailed studies of the properties of any new particles.

This is true also for the latest new particle that has been discovered already, the top quark.[24,25] A scan of the top production threshold in e^+e^- collisions, at $\sqrt{s} = 2m_t \approx 350$ GeV, would give an unprecedented precision in the measurement of the top quark mass. The simultaneous direct measurement of the top quark width would determine the V_{tb} CKM matrix element and thus provide a significant test of the electroweak sector.

All these e^+e^- collider measurements could also be performed at a $\mu^+\mu^-$ collider.[6] Such a machine would have the added advantage of an excellent beam energy resolution, of order 10^{-4} or better, while beam-strahlung in the tight focus of an e^+e^- linear collider leads to a significant smearing of the c.m. energy. The very precisely determined beam energy can then be used for a scan of the $\bar{t}t$ production threshold, which resolves detailed features like the location of the (extremely short-lived) first $\bar{t}t$ bound state, QCD binding effects and the value of the strong coupling constant, or even Higgs exchange effects on the shape of the $\bar{t}t$ production threshold.

Another advantage of a $\mu^+\mu^-$ collider is the larger coupling of the Higgs boson to the muon as compared to the electron, due to the muon's 200 times larger mass. This allows the direct s-channel production of the Higgs resonance in muon collisions, $\mu^+\mu^- \to H$. Because of the excellent energy resolution of a muon collider, an energy scan of the Higgs resonance would provide us with a very precise measurement of the Higgs boson mass, with an error of order MeV, and if dedicated efforts are made to keep the energy spread as small as possible, even the full width of the Higgs resonance can be determined directly.[6]

These examples clearly show that a linear e^+e^- collider or a muon collider would be a terrific experimental tool and would greatly advance our understanding of particle interactions. Unfortunately, no such machine has been approved for construction yet. And it may be argued that we first need to establish the existence of new heavy particles before investing several billion dollars or euros into a machine to search for them and then study their detailed properties. For many of the particles predicted by supersymmetry, or for the Higgs boson, the machines needed for discovery already exist or are under construction, namely the Tevatron at Fermilab and the LHC at CERN.

4 Hadron Colliders

The highest center of mass energies and, hence, the best reach for new heavy particles is provided by hadron colliders, the Tevatron with its 2 TeV $\bar{p}p$ collisions at present, and the LHC with 14 TeV pp collisions after 2005. At the Tevatron the top quark has been discovered in 1994, and Higgs and supersymmetry searches will resume in run II. The LHC is expected to do detailed investigations of the Higgs sector, and should answer the question whether TeV scale supersymmetry is realized in nature. Before discussing how these studies can be performed in a hadron collider environment, we need to consider the general properties of production processes at these machines in some detail.

4.1 Hadrons and partons

A typical hard hadronic collision is sketched in Fig. 10: one of the subprocesses contributing to Z+jet production, namely $ug \to uZ$. The up-quark and the gluon carry a fraction of the parent proton momenta only, x_1 and x_2,

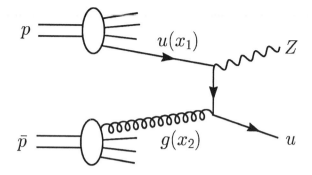

Figure 10: One of the subprocesses contributing to Zj production in $p\bar{p}$ collisions. The two initial partons carry momentum fractions x_1 and x_2 of the incoming proton and anti-proton, respectively.

respectively. Thus, the incoming parton momenta are given by

$$p_u = x_1 p = x_1 \frac{\sqrt{s}}{2}(1,0,0,1)\,,$$

$$p_g = x_2 \bar{p} = x_2 \frac{\sqrt{s}}{2}(1,0,0,-1)\,, \tag{34}$$

and the available center of mass energy for the Z+jet final state is given by the root of $\hat{s} = (p_u + p_g)^2 = 2p_u \cdot p_g = x_1 x_2 s$, i.e. it is only a fraction $\sqrt{x_1 x_2}$ of the collider energy.

In order to calculate observable production rates, for a process $p\bar{p} \to X$, we first need to identify all the parton level subprocesses $a_1 + a_2 \to \hat{x} = b_1 + b_2 + \cdots + b_n$ which give the desired signature. In the above example of Z+jet production this includes, at tree level, $u + \bar{u} \to Z + g$, $\bar{u} + u \to Z + g$ (i.e. the subprocess where the anti-u comes from the proton), $g + u \to Z + u$, $u + g \to Z + u$, $g + \bar{u} \to Z + \bar{u}$, $\bar{u} + g \to Z + \bar{u}$, and all the corresponding subprocesses with the up quark replaced by down, strange, charm, and bottom quarks, which are all treated as massless partons inside the proton. The full cross section for the process $p\bar{p} \to X$ is then given by

$$\upsilon = \int dx_1 dx_2 \sum_{\text{subprocesses}} f_{a_1/p}(x_1)\, f_{a_2/\bar{p}}(x_2)$$

$$\frac{1}{2\hat{s}} \int d\Phi_n(x_1 p + x_2 \bar{p};\ p_1 \ldots p_n)\Theta(\text{cuts})$$

$$\sum |\mathcal{M}|^2 (a_1 a_2 \to b_1 b_2 \ldots b_n)\,. \tag{35}$$

Here $f_{a_1/p}(x_1)$ is the probability to find parton a_1 inside the proton, carrying a fraction x_1 of the proton momentum, i.e. the a_1 parton distribution function. Similarly, $f_{a_2/\bar{p}}(x_2)$ is the a_2 parton distribution function (pdf) inside the antiproton. In the second line of (35) $1/2\hat{s}$ is the flux factor for the partonic cross section,

$$d\Phi_n(P; \, p_1 \ldots p_n) = \prod_{i=1}^{n} \left(\frac{d^3 \mathbf{p}_i}{(2\pi)^3 2E_i} \right) (2\pi)^4 \, \delta^4(P - \sum_i p_i) \tag{36}$$

is the Lorentz invariant phase space element, and $\Theta(\text{cuts})$ is the acceptance function, which summarizes the kinematical cuts on all the final state particles, i.e. $\Theta = 1$ if all the partons a_i satisfy all the acceptance cuts and $\Theta = 0$ otherwise. Finally, $\mathcal{M}$, in the third line of (35) is the Feynman amplitude for the subprocess in question, squared and summed/averaged over the polarizations and colors of the external partons.

Compared to the calculation of cross sections for e^+e^- collision, the new features are integration over pdf's and the fact that a much larger number of partonic subprocesses must be considered for a given experimental signature. The introduction of pdf's introduces an additional uncertainty since they need to be extracted from other data, like deep inelastic scattering, W production at the Tevatron, or direct photon production. The extraction of pdf's is continuously being updated and refined, and in practice, hadron collider cross sections are calculated by using numerical interpolations which are provided by the groups improving the pdf sets.[26] These pdf determinations have been dramatically improved over time and typical pdf uncertainties now are below the 5% level, at least in the range $10^{-3} \lesssim x \lesssim 0.2$, which is most important for our discussion of new particle production processes.

More important than the appearance of pdf's are the kinematic effects which result from the fact that the hard collision is between partons. Since momenta of the incoming partons are not known a priori, we cannot make use of a beam energy constraint as in the case of e^+e^- collisions. The missing information on the momentum parallel to the beam axis affects the analysis of events with unobserved particles in the final state, like neutrinos or the lightest neutralino. Momentum conservation can only be used for the components transverse to the beam axis, i.e. only the missing transverse momentum vector, $\not{p}_T$, can be reconstructed.

Another effect is that the lab frame and the c.m. frame of the hard collision no longer coincide. Rather the partonic c.m. system receives a longitudinal boost in the direction of the beam axis, which depends on x_1/x_2. This longitudinal boost is most easily taken into account by describing four-momenta in terms of rapidity y instead of scattering angle θ. For a momentum vector

$$p = (E, p_x, p_y, p_z) = E(1, \beta \sin \theta \cos \phi, \beta \sin \theta \sin \phi, \beta \cos \theta) , \qquad (37)$$

rapidity is defined as

$$y = \frac{1}{2} \log \frac{E + p_z}{E - p_z} , \qquad (38)$$

which, in the massless limit ($\beta \to 1$), reduces to pseudo-rapidity,

$$\eta = \frac{1}{2} \log \frac{1 + \cos \theta}{1 - \cos \theta} . \qquad (39)$$

The advantage of using rapidity is that under an arbitrary boost along the z-axis, rapidity differences remain invariant, i.e. they directly measure relative scattering angles in the partonic c.m. frame. Using Eq. (34), the c.m. momentum is given by $P = \sqrt{s}/2(x_1 + x_2, 0, 0, x_1 - x_2)$ which results in a c.m. rapidity $y_{c.m.} = 1/2 \log(x_1/x_2)$. As a result, rapidities y^* in the partonic c.m. system and rapidities y in the lab frame are connected by

$$y = y^* + y_{c.m.} = y^* + \frac{1}{2} \log \frac{x_1}{x_2} , \qquad (40)$$

a relation which can be used to determine all scattering angles in the theoretically simpler partonic rest frame whenever all final state momenta can be measured and, hence, the c.m. momentum P is known.

4.2 Z and W Production

One of the early highlights of $\bar{p}p$ colliders was the discovery of the W and Z bosons at the CERN S$\bar{p}p$S, a 630 GeV $\bar{p}p$ collider.[27] W and Z production provide a nice example which demonstrates the use of some of the transverse observables discussed in the previous subsection. In addition, the story might repeat, since nature might have an additional neutral or charged heavy gauge boson in store, a Z' or an W' which might appear at the LHC. Let us thus consider Z production in some detail.

The prototypical Drell-Yan process is $\bar{p}p \to Z \to \ell^+ \ell^-$, where, to lowest order in the strong coupling constant, the Z can be produced by annihilation of a quark-antiquark pair . The partonic subprocess $\bar{q}q \to \ell^+ \ell^-$ leads to two leptons with balancing transverse momenta, which can be parameterized in terms of their lab frame p_T, pseudo-rapidity η, and azimuthal angle ϕ,

$$\ell = p_T \left(\cosh \eta, \cos \phi, \sin \phi, \sinh \eta \right) , \qquad (41)$$
$$\bar{\ell} = p_T \left(\cosh \bar{\eta}, - \cos \phi, - \sin \phi, \sinh \bar{\eta} \right) . \qquad (42)$$

Since transverse momentum is invariant under a boost along the z-axis, we may as well determine it in the partonic c.m. frame, where the ℓ^- momentum is given by

$$\ell^* = \frac{m_Z}{2}\left(1, \sin\theta^* \cos\phi, \sin\theta^* \sin\phi, \cos\theta^*\right) . \tag{43}$$

One finds $p_T = m_Z/2 \sin\theta^*$ by equating the transverse momenta in the two frames. This implies

$$|\cos\theta^*| = \sqrt{1 - \sin^2\theta^*} = \sqrt{1 - \frac{4p_T^2}{m_Z^2}} . \tag{44}$$

Using this relation we obtain the transverse momentum spectrum in terms of the lepton angular distribution in the c.m. frame,

$$\frac{d\sigma}{dp_T^2} = \frac{d\sigma}{d\cos\theta^*}\left|\frac{d\cos\theta*}{dp_T^2}\right| = \frac{d\sigma}{d\cos\theta^*}\frac{1}{m_Z\sqrt{m_Z^2/4 - p_T^2}} . \tag{45}$$

The p_T distribution diverges at the maximum transverse momentum value, $p_T = m_Z/2$. This Jacobian-peak, so called because it arises from the Jacobian factor in Eq. (45), is smeared out in practice by finite detector resolution, the finite Z-width and QCD effects. Nevertheless, it is an excellent tool to determine the W mass in the analogous $W \to \ell\nu$ decay, which has the Jacobian peak at half the W mass.

The Jacobian peak is smeared out considerably by QCD effects, namely the emission of additional partons in Drell-Yan production. Only at lowest order, in $\bar{q}q \to \ell^+\ell^-$ or $\bar{q}q' \to \ell\nu$, do the transverse momenta of the two decay leptons exactly balance. Taking QCD effects into account, we must consider gluon radiation or subprocesses like $qg \to qZ$ as depicted in Fig. 10. The lepton pair now obtains a transverse momentum, which balances the transverse momentum of the additional parton(s) in the final state. In fact, multiple soft gluon emission renders a zero probability to lepton pairs with $p_T(Z) = 0$. In real life, their transverse momentum distribution peaks at a few GeV, as can be seen in Fig. 11, which shows the $p_T(e^+e^-)$ distribution as observed at the Tevatron.

When trying to determine the kinematics of an event to better than some 10% ($= \mathcal{O}(\alpha_s)$), we need to take soft parton emission into account. One way to do this is to use transverse mass instead of transverse momentum of the lepton. For $W \to \ell\nu$ decay the transverse mass is defined as

$$m_T(\ell, \not{\boldsymbol{p}}_T) = \sqrt{(E_{T\ell} + \not{E}_T)^2 - (\mathbf{p}_{T\ell} + \not{\boldsymbol{p}}_T)^2} , \tag{46}$$

328

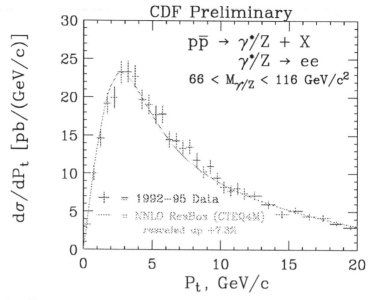

Figure 11: Transverse momentum distribution of the produced lepton pair in Drell-Yan production, $\bar{q}q \to \gamma^*/Z \to e^+e^-$, as measured by the CDF Collaboration at the Tevatron.

i.e. it is determined from the difference of squares of the transverse energy and the transverse momentum vector of the $\ell\nu$ pair. This is analogous to the definition of invariant mass, which in addition includes the contributions from the longitudinal momentum, along the beam axis. The transverse mass retains a Jacobian peak, at $m_T(\ell, \boldsymbol{p}_T) = m_W$, even in the presence of QCD radiation.

4.3 Extra W and Z bosons

The SM is a gauge theory based on the gauge group $G_{SM} = SU(3) \times SU(2)_L \times U(1)_Y$ and each of these factors is associated with a set of gauge bosons, the eight gluons of $SU(3)$, the photon and the $W^\pm$ and Z of the electroweak sector. It is possible, however, that the gauge symmetry of nature is larger, which in turn would predict the existence of extra gauge bosons. The apparent symmetry of the SM would arise because the extra gauge bosons are too heavy to have been observed as yet, made massive by the spontaneous breaking of the extra gauge symmetry. Examples of such extended gauge sectors are left-right symmetric models [28] with

$$G = SU(3) \times SU(2)_R \times SU(2)_L \times U(1)_{B-L} , \qquad (47)$$

where the new $SU(2)_R$ factor gives rise to an extra charged W' with $V + A$ couplings to quarks and leptons and an additional Z', or extensions with extra $U(1)$ factors,[4,29]

$$G = SU(3) \times SU(2)_L \times U(1) \times U(1) \,, \tag{48}$$

which lead to the existence of an extra Z'. Present indirect limits on such extra gauge bosons are relatively weak, and allow extra W' are Z' bosons to exist with masses above some 500 GeV.[30] These indirect bounds are obtained form the apparent absence of additional contact interactions, similar to Fermi's four-fermion couplings, in low energy data. Given mass bounds of a few hundred GeV only, additional gauge bosons, with masses up to several TeV, could readily be observed at the LHC.

The cleanest method for discovering extra Z' bosons would be through a repetition of the historic CERN experiments [27] which lead to the discovery of the Z in 1983, i.e. by searching for the Z' resonance peak in

$$\bar{q}q \to Z' \to \ell^+\ell^- \,, \qquad \ell = e, \mu \,. \tag{49}$$

The reach of this search depends on the coupling $g_{L,R}^{\prime f}$ of the extra Z' to left- and right-handed fermions. The production cross section is proportional to $(g_L^{\prime q})^2 + (g_R^{\prime q})^2$. The decay branching fraction, $B(Z' \to \ell^+\ell^-)$, depends on the relative size of this combination of left- and right-handed couplings for lepton ℓ to the same combination, summed over all fermions. If the product of production cross section times leptonic branching ratio,

$$\sigma \cdot B = \sigma(pp \to Z') \, B(Z' \to \ell^+\ell^-) \,, \tag{50}$$

is the same as for the SM Z-boson, scaled up to Z' mass, i.e. if the couplings of the Z' are SM-like, the LHC experiments can observe a Z' with a mass up to $m_{Z'} = 5$ TeV. Smaller (larger) couplings would decrease (increase) this reach, of course.[9]

While the LHC will not be capable of repeating the precision experiments of LEP/SLC, for a Z', a lot of additional information can be obtained by more detailed observations of leptonic Z' decays.[29] One measurement which would be of particular importance is the determination of the lepton charge asymmetry. At the parton level, the forward-backward charge asymmetry measures the relative number of $\bar{q}q \to Z' \to \ell^+\ell^-$ events where the ℓ^- goes into the same hemisphere as the incident quark, as compared to events where the ℓ^+ goes into the quark direction. In terms of the pseudo-rapidity of the ℓ^-, η^*, as measured relative to the incident quark direction, i.e.

$$\eta^* = \frac{1}{2} \log \frac{1 + \cos\theta^*}{1 - \cos\theta^*} \,, \tag{51}$$

where θ^* is the c.m. frame angle between the incident quark and the final state ℓ^- (or the angle between the incident anti-quark and the final state ℓ^+), the forward-backward asymmetry, at the parton level, is given by

$$\hat{A}^\ell_{FB} = \frac{\hat{\sigma}(\eta^* > 0) - \hat{\sigma}(\eta^* < 0)}{\hat{\sigma}(\eta^* > 0) + \hat{\sigma}(\eta^* < 0)} = \frac{3}{4} \frac{(g_R^{\prime q})^2 - (g_L^{\prime q})^2}{(g_R^{\prime q})^2 + (g_L^{\prime q})^2} \frac{(g_R^{\prime \ell})^2 - (g_L^{\prime \ell})^2}{(g_R^{\prime \ell})^2 + (g_L^{\prime \ell})^2} \tag{52}$$

One sees that a measurement of the forward-backward asymmetry gives a direct comparison of left-handed and right-handed couplings of the Z' to leptons and quarks.

Unfortunately, at a pp-collider, the two proton beams have equal probabilities to originate the quark or the antiquark in the collision, and, therefore, the forward backward asymmetry averages to zero when considering all events. One can make use of the different Feynman x distributions of up and down quarks as opposed to anti-quarks, however. At small x, quarks and anti-quarks have roughly equal pdf's, $q(x) \approx \bar{q}(x)$, while at large x the valence quarks dominate by a sizable fraction, $q(x) \gg \bar{q}(x)$. In the experiment one measures the rapidities y_+ and y_- of the ℓ^+ and ℓ^-, respectively. Since the leptons are back-to-back in the c.m. frame, their c.m. rapidities cancel, and the sum of the lab frame rapidities gives the rapidity $y = y_{c.m.}$ of the c.m. frame,

$$y = \frac{1}{2}(y_+ + y_-) = \frac{1}{2} \log \frac{x_1}{x_2} , \tag{53}$$

while their difference measures η^*,

$$\eta^* = \frac{1}{2} \log \frac{1 + \cos \theta^*}{1 - \cos \theta^*} = \frac{1}{2}(y_- - y_+) . \tag{54}$$

At $y > 0$ we have $x_1 > x_2$, and therefore it is more likely that the quark came from the left, while at $y < 0$ an anti-quark from the left is dominant. Measuring both η^* and y, the charge asymmetry,

$$A(y) = \frac{\frac{d\sigma}{dy}(\eta^* > 0) - \frac{d\sigma}{dy}(\eta^* < 0)}{\frac{d\sigma}{dy}(\eta^* > 0) + \frac{d\sigma}{dy}(\eta^* < 0)} , \tag{55}$$

can be determined. Of course, we have $A(y) = A(-y)$ at a pp collider, where the two sides are equivalent, and the average over all y vanishes. At fixed y, however, $A(y)$ is analogous to the forward-backward charge asymmetry at e^+e^- colliders, and it measures the relative size of left-handed and right-handed Z' couplings.

Different extra gauge groups predict substantially different sizes for left- and right-handed couplings of the Z' to quarks and leptons. Thus, the lepton charge asymmetry, $A(y)$, is a very powerful tool to distinguish different models, once a Z' has been discovered.

Similar to the Z' search, the search for a heavy charged gauge boson, W', would repeat the W search at CERN in the early eighties. One would study events consisting of a charged lepton and a neutrino, signified by missing transverse momentum opposite to the charged lepton direction. The mass of the W' is then determined by the Jacobian peak in the transverse mass distribution, at $m_T(\ell, \mathbf{p}_T) = m_{W'}$. For a W' with SM strength couplings, the LHC can find it and measure its mass, up to W' masses of order 5 TeV, by searching for a shoulder in the transverse mass distribution and measuring its cutoff at $m_{W'}$.[9]

4.4 Top search at the Tevatron

The leptonic decay of a W or Z produces a fairly clean signature at a hadron collider. Perhaps more typical for a new particle search was the discovery of the top quark [24,25] at the Fermilab Tevatron. A much more complex signal needed to be isolated from large QCD backgrounds. At the same time the top discovery provides a beautiful example for the use of hadronic jets as a tool for discovering new particles. Let us have a brief, historical look at the top quark search at the Tevatron, from this particular viewpoint.

In $p\bar{p}$ collisions at the Tevatron, the top quark is produced via quark anti-quark annihilation, $q\bar{q} \to t\bar{t}$, and, less importantly, via $gg \to t\bar{t}$. Production cross sections have been calculated at next-to-leading order, and are expected to be around 5 pb for a top mass of 175 GeV.[31] The large top decay width which is expected in the SM,

$$\Gamma(t \to W^+ b) \approx 1.6 \text{ GeV} , \qquad (56)$$

implies that the t and $\bar{t}$ decay well before hadronization, and the same is true for the subsequent decay of the W bosons. Thus, a parton level simulation for the complete decay chain, including final parton correlations, is a reliable means of predicting detailed properties of the signal. The top quark signal, $t\bar{t} \to bW^+ \bar{b}W^-$, is determined by the various decay modes of the W^+W^- pair, whose branching ratios were discussed in Sec. 3.1. In order to distinguish the signal from multi-jet backgrounds, the leptonic decay $W \to \ell\nu$ ($\ell = e, \mu$) of at least one of the two final state Ws is extremely helpful. On the other hand, the leptonic decay of both Ws into electrons or muons has a branching ratio of $\approx 4\%$ only, and thus the prime top search channel is the decay chain

$$t\bar{t} \to bW^+ \bar{b}W^- \to \ell^\pm \nu \, q\bar{q} \, b\bar{b} . \qquad (57)$$

332

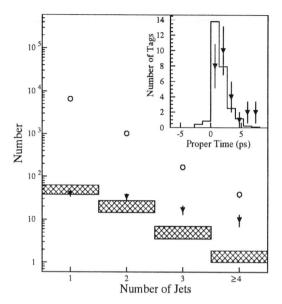

Number of Jets

Figure 12: Number of $W + n$ jet events in the CDF top quark search as a function of jet multiplicity. Number of observed events are given without b-tagging (open circles) and with an SVX tag (triangles). The expected background, mainly from QCD $W + n$ jet events, is given by the cross-hatched bars. From the original CDF top quark discovery paper, Ref.[24].

Within the SM, this channel has an expected branching ratio of $\approx 30\%$. After hadronization each of the final state quarks in (57) may emerge as a hadronic jet, provided it carries enough energy. Thus the $t\bar{t}$ signal is expected in $W+3$ jet and $W + 4$ jet events[b].

Events with leptonic W decays and several jets can also arise from QCD corrections to the basic Drell-Yan process $q\bar{q} \rightarrow W^{\pm} \rightarrow \ell^{\pm}\nu$. The process $ug \rightarrow dggW^+$, for example, will give rise to $W + 3$ jet events and its cross section and the cross sections for all other subprocesses with a W and three partons in the final state need to be calculated in order to assess the QCD background of $W + 3$ jet events, at tree level. $W + n$ jet cross sections have been calculated for $n = 3$ jets[32] and $n = 4$ jets.[33] As in the experiment, the calculated $W + n$ jet cross sections depend critically on the minimal transverse energy of a jet. CDF, for example, requires a cluster of hadrons to carry $E_T > 15$ GeV to be identified as a jet,[34] and this observed E_T must then be translated into the corresponding parton transverse momentum in order to get

---
[b]Gluon bremsstrahlung may increase the number of jets further and thus all $W+ \geq 3$ jet events are potential $t\bar{t}$ candidates.

a prediction for the $W + n$ jet cross sections.

At this level the QCD backgrounds are still too large to give a viable top quark signal. The situation was improved substantially by using the fact that two of the four final state partons in the signal are b-quarks, while only a small fraction of the $W + n$ parton background events have b-quarks in the final state. These fractions are readily calculated by using $W + n$ jet Monte Carlo programs. There are several experimental techniques to identify b-quark jets, all based on the weak decays of the produced b's. One method is to use the finite b lifetime of about $\tau = 1.5$ ps which leads to b-decay vertices which are displaced by $\gamma c\tau = $ few mm from the primary interaction vertex. These displaced vertices can be resolved by precision tracking, with the aid of their Silicon VerteX detector in the case of CDF, and the method is, therefore, called SVX tag. In a second method, b decays are identified by the soft leptons which arise in the weak decay chain $b \rightarrow W^*c$, $c \rightarrow W^*s$, where either one of the virtual Ws may decay leptonically.[25,34]

The combined results of using jet multiplicities and SVX b-tagging to isolate the top quark signal are shown in Fig. 12. A clear excess of b-tagged 3 and 4 jet events is observed above the expected background. The excess events would become insignificant if all jet multiplicities were combined or if no b-tag were used (see open circles). Thus jet counting and the identification of b-quark jets are critical for identification of top quark events.

Beyond counting the number of jets above a certain transverse energy, the more detailed kinematic distributions, their summed scalar E_T's [25] and multi-jet invariant masses, have also been critical in the top quark search. The top quark mass determination, for example, relies on a good understanding of these distributions. Ideally, in a $t\bar{t} \rightarrow (\ell^+\nu b)(q\bar{q}\bar{b})$ event, for example, the two subsystem invariant masses should be equal to the top quark mass,

$$m_t \approx m(\ell^+\nu b) \approx m(q\bar{q}\bar{b}) . \tag{58}$$

Including measurement errors, wrong assignment of observed jets to the two clusters, etc. one needs to perform a constrained fit to extract m_t. The 1995 CDF result of this fit [24] is shown in Fig. 13. In addition, the figure demonstrates that the observed b-tagged $W + 4$ jet events (solid histogram) are considerably harder than the QCD background (dotted histogram). On the other hand the data agree very well with the top quark hypothesis (dashed histogram).

By now, the top quark has been observed in all three decay channels of the W^+W^- pair, purely leptonic W decays, $W^+W^- \rightarrow \ell\nu jj$, and 4-jet decays, and all channels have been used to extract the top-quark mass. Results at present are

$$m_t = 172.1 \pm 7.1 \text{ GeV} , \tag{59}$$

334

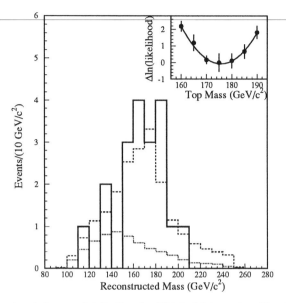

Figure 13: Reconstructed mass distribution for $W+ \geq 4$ jet events with a b-tag. The solid histogram represents the CDF data. Also included are the expected background (dotted histogram) and the expected signal+background for $m_t = 175$ GeV. The insert shows the likelihood fit to determine the top quark mass, which yielded $m_t = 176 \pm 8 \pm 10$ GeV at the time. From Ref. [24].

from the D0 Collaboration,[35] and

$$m_t = 176.0 \pm 6.5 \text{ GeV} , \qquad (60)$$

from the CDF Collaboration.[36]

5 Higgs search at the LHC

With the top-quark discovery at the Tevatron, the elementary fermions of the SM have all been observed. The missing ingredient, as far as the SM is concerned, is the Higgs boson. LEP2 is likely to find it if its mass is below ≈ 105 GeV. The Tevatron has a chance to discover the Higgs boson in the processes $\bar{q}q \to WH$, $H \to \bar{b}b$ (for Higgs masses below ≈ 130 GeV) [37] or $gg \to H \to WW^*$ (for $130 \text{ GeV} \lesssim m_H \lesssim 180$ GeV) [38] if sufficient luminosity can be collected within the next few years. (Between 10 and 30 fb^{-1} are required for this purpose.) The best candidate for Higgs discovery and detailed Higgs studies within the next ten years is the LHC, however.

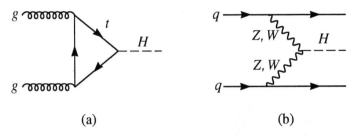

| | (a) | | (b) |

Figure 14: Feynman graphs for the two dominant Higgs production processes at the LHC, (a) gluon-gluon fusion via a top-quark loop and (b) weak boson fusion.

5.1 Higgs production channels

A Higgs boson can be produced in a variety of processes at the LHC. The machine has sufficient energy to excite heavy quanta, and since the Higgs boson couples to other particles proportional to their mass, this leads to efficient Higgs production modes. The two dominant processes are shown in Fig. 14, gluon fusion, $gg \to H$, which proceeds via a top quark loop, and weak boson fusion, $qq \to qqH$, where the two incoming quarks radiate two virtual Ws or Zs which then annihilate to form the Higgs. The expected cross sections for both are in the 1–30 pb range, and are shown in Fig. 15 as a function of the Higgs mass.

Beyond these two, a variety of heavy particle production processes may radiate a relatively light Higgs boson at an appreciable rate. These include WH (or ZH) associated production,

$$\bar{q}q \to WH \,, \tag{61}$$

which is the analogue of ZH production at e^+e^- colliders, and $t\bar{t}H$ (or $b\bar{b}H$) associated production,

$$\bar{q}q \to \bar{t}tH \,, \qquad gg \to \bar{t}tH \,. \tag{62}$$

As can be seen in Fig. 15, these associated production cross sections are quite small for large Higgs boson masses, but can become interesting for $m_H \lesssim$ 150 GeV, because the decay products of the additional Ws or top quarks provide characteristic signatures of associated Higgs production events which allow for excellent background suppression.

In the most relevant region, 400 GeV $\gtrsim m_H \gtrsim$ 110 GeV, which will not be accessible by LEP, the total SM Higgs production cross section is of the order 10–30 pb, which corresponds to some 10^5 events per year at the LHC, even at the initial 'low' luminosity of $\mathcal{L} = 10^{33}$ cm^{-2}sec^{-1}. This already

336

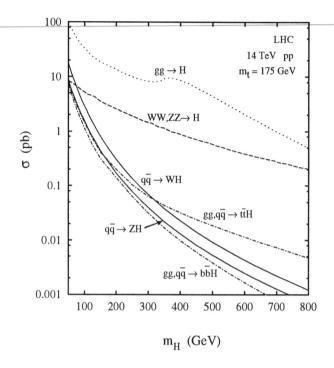

Figure 15: Production cross sections for the SM Higgs boson at the LHC. From Ref.[39].

indicates that the main problem at the LHC is the visibility of the signal in an environment with very large QCD backgrounds, and this visibility critically depends on the decay mode of the Higgs. The decays $H \to b\bar{b}$, $c\bar{c}$, gg all lead to a dijet signature and are very difficult to identify, because dijet production at a hadron collider is such a common-place occurrence. More promising are $H \to ZZ \to \ell^+\ell^-\ell^+\ell^-$ and $H \to \gamma\gamma$ which have much smaller backgrounds to contend with.

5.2 Higgs search in the $H \to ZZ$ mode

The expected branching ratios of the SM Higgs to the various final states are shown in Fig. 16. For $m_H > 180$ GeV, the $H \to ZZ$ threshold, Higgs decay to WW and ZZ dominates, and of the various decay modes of the two weak bosons, $ZZ \to \ell^+\ell^-\ell^+\ell^-$ gives the cleanest signature.[2] Not only can the invariant mass of the two lepton pairs be reconstructed, and their arising from Z decay be confirmed, also the invariant mass of the four charged leptons reconstructs the Higgs mass. Thus, the Higgs signal appears as a resonance in

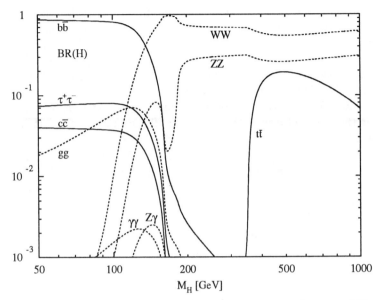

Figure 16: Decay branching fractions of the SM Higgs boson, as a function of Higgs mass. From Ref. [40].

the $m_{4\ell}$ invariant mass distribution. ZZ backgrounds are limited at the LHC, they mostly arise from $\bar{q}q \to ZZ$ processes, i.e. continuum ZZ production. The Higgs resonance needs to be observed on top of this irreducible background. Estimates are that with 100 fb^{-1} the LHC detectors can see $ZZ \to 4\ell$ for 180 GeV$< m_H \lesssim 600$ GeV.[9,10]

The only disadvantage of this 'gold-plated' Higgs search mode is the relatively small branching ratio of $B(H \to ZZ \to \ell^+\ell^-\ell^+\ell^-) \approx 0.15\%$. For large Higgs masses the gold-plated mode becomes rate limited, and additional Higgs decay modes must be searched for. $H \to ZZ \to \nu\bar{\nu}\ell^+\ell^-$, the 'silver-plated' mode, has about a six times larger rate, but because of the unobserved neutrinos it does not provide for a direct Higgs mass reconstruction. The large missing E_T and the two observed leptons allow a measurement of the transverse mass, however, with a Jacobian peak at m_H, analogous to the $W \to \ell\nu$ example. $H \to ZZ \to \nu\bar{\nu}\ell^+\ell^-$ events allow an extension of the Higgs search to $m_H \approx 0.8\text{--}1$ TeV.

Another promising search mode for a heavy Higgs boson is $H \to WW \to \ell\nu jj$, where the Higgs boson is produced in the weak boson fusion process, as depicted in Fig. 14(b). The two quarks in the process $qq \to qqH$ result in two additional jets, which have a large mutual invariant mass, and the presence of these two jets is a very characteristic signature of weak boson fusion processes. These two jets are typically emitted at forward angles, corresponding to pseudo-rapidities between ± 2 to ± 4.5. Requiring the observation of these two 'forward tagging jets' substantially reduces the backgrounds and leads to an observable signal for Higgs boson masses in the 600 GeV to 1 TeV range and above.[9,10,41] This technique of forward jet-tagging can more generally be used to search for any weak boson scattering processes.[3,42] However, it is as useful for the study of a weakly interacting Higgs sector at the LHC, and we shall consider it below in some detail in that context.

Fits to electroweak precision data, from LEP and SLC, from lower energy data as well as from the Tevatron, are increasingly pointing to a relatively small Higgs boson mass,[21] between 100 and 200 GeV, at least within the context of the SM. Thus, interest at present is focused on search strategies for a relatively light Higgs boson. The search in the gold-plated channel, $H \to \ell^+\ell^-\ell^+\ell^-$, can be extended well below the Z-pair threshold.[9,10] As can be read off Fig. 16, the branching ratio $H \to ZZ^*$, into one real and one virtual Z, remains above the few percent level for Higgs boson masses as low as 130 GeV. With an integrated luminosity of around 100 fb^{-1} the SM Higgs resonance can be observed the LHC, above $m_H \approx 130$ GeV. A problematic region is the 160 GeV$< m_H <$180 GeV range, however, where the branching ratio $H \to \ell^+\ell^-\ell^+\ell^-$ takes a serious dip: this is the region where the Higgs boson can decay into two on-shell Ws, while the ZZ channel is still below threshold for two on-shell Zs. While sufficiently large amounts of data will yield a positive Higgs signal in this region, the observation in the dominant decay channel, via $H \to W^+W^- \to \ell^+\nu\ell^-\bar{\nu}$ has a much higher rate and can in fact be distinguished from backgrounds as well.[43]

5.3 Search in the $H \to \gamma\gamma$ channel

For a relatively light SM Higgs boson, of mass $m_H \lesssim 150$ GeV, the cleanest search mode is the decay $H \to \gamma\gamma$. Photon energies can be measured with high precision at the LHC, resulting in a very good $m_{\gamma\gamma}$ invariant mass resolution, of order 1 GeV for Higgs masses around 100–150 GeV. In fact, the design of the LHC detectors has been driven by a good search capability for these events. Since the natural Higgs width is only a few MeV in this mass range, the Higgs would appear as as a very narrow peak in the $\gamma\gamma$ invariant mass distribution.

The simplest search is for all $H \to \gamma\gamma$ events, irrespective of the Higgs production mode. Backgrounds arise from double photon bremsstrahlung processes like

$$gq \to gq\gamma\gamma\,, \qquad qq \to qq\gamma\gamma\,, \quad \text{etc.} \tag{63}$$

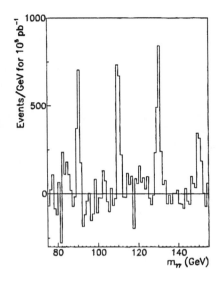
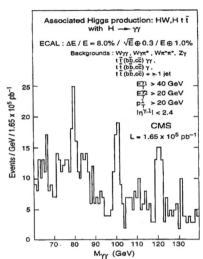

Figure 17: Expected $H \to \gamma\gamma$ signal in the CMS detector at the LHC, after one year of running at a luminosity of 10^{34} cm^{-2}sec^{-1}. Shown are the expected Higgs resonance peaks in the $\gamma\gamma$ invariant mass distributions (a) in the inclusive $H \to \gamma\gamma$ search after background subtraction for Higgs masses of 90, 110, 130, and 150 GeV, and (b) for WH and $\bar{t}tH$ associated production with $m_H = 80$, 100, and 120 GeV. From Ref. [10].

pair production of photons in $\bar{q}q$ annihilation and in gluon fusion (via quark loops),

$$\bar{q}q \to \gamma\gamma \,, \qquad gg \to \gamma\gamma \,, \qquad (64)$$

and from reducible backgrounds, in particular from jets where the bulk of the jet's energy is carried by a single π^0, whose decay into two nearby photons may not be resolved and may mimic a single photon in the detector.

Bremsstrahlung photons tend to be emitted close to the parent quark direction, i.e. they are close in angle to a nearby jet. The same is true for photons from π^0 decays, these photons are usually embedded into a hadronic jet. One therefore requires the signal photons to be well isolated, i.e. to have little hadronic activity at small angular separations

$$\Delta R = \sqrt{(\eta_\gamma - \eta_j)^2 + (\phi_\gamma - \phi_j)^2} \,. \qquad (65)$$

Here $(\eta_\gamma, \ \phi_\gamma)$ give the photon direction and $(\eta_j, \ \phi_j)$ denotes the direction of hadronic activity, be it hadrons, partons in a perturbative calculation, or jets. A practical requirement may be that the total energy carried by hadrons within a cone of radius $\Delta R = 0.3$ be less that 10% of the photon energy and that no hard jet is found with a separation $\Delta R < 0.7$. Photon isolation requirements drastically reduce bremsstrahlung and QCD (π^0) backgrounds and are absolutely crucial for identifying any signal.

Another characteristic feature of the backgrounds is their soft photon spectrum: background rates drop quite fast with increased photon transverse momentum, $p_{T\gamma}$. This plays into the features of the signal. Since a Higgs boson decays isotropically, with $E_\gamma = m_H/2$ in the Higgs rest frame, photon transverse momenta in the range $m_H/4 \lesssim p_{T\gamma} \lesssim m_H/2$ are favored. In practice, when searching for a Higgs boson in the 100 GeV $< m_H <$ 150 GeV range, one requires $p_{T\gamma_1} > 40$ GeV, $p_{T\gamma_2} > 25$ GeV for the two photons, which together with the isolation requirement reduces the background to a level of order $d\sigma/dm_{\gamma\gamma} = $100–200 fb/GeV.[10] This needs to be compared to the SM Higgs signal, which has a cross section, after cuts, of $\sigma \cdot B(H \to \gamma\gamma) \approx 40$ fb for masses around $m_H = 120$ GeV.

The visibility of the signal crucially depends on the mass resolution of of the detector. For CMS (ATLAS) one expects a resolution of order $\sigma_m = \pm 0.8$ GeV (± 1.5 GeV). Taking the better CMS resolution and an integrated luminosity of 50 fb^{-1} as an example, one would see $S = 0.683 \cdot 2000 = 1400$ signal events in a mass bin of full width 1.6 GeV, on top of a background of $B = 13000 \cdot 1.6 = 21000$ events, giving a statistical significance $S/\sqrt{B}$ of almost 10 standard deviations, a very significant discovery! The expected two-photon mass spectrum, after background subtraction, is shown in Fig. 17(a). The above example only conveys the rough size of the signal and background in the inclusive $H \to \gamma\gamma$ search. More detailed estimates for a range of Higgs masses, including detection efficiencies, the decline of background rates with increasing $m_{\gamma\gamma}$, and variations in the signal rate as a function of m_H can be found in Refs.[9,10].

The main disadvantage of the inclusive $H \to \gamma\gamma$ search is the relatively small signal size as compared to the background, $S : B \approx 1 : 15$ for CMS and even smaller for ATLAS. A much cleaner signal can be found by looking for Higgs production in association with other particles, in particular the isolated leptons arising from W decays in WH and $\bar{t}tH$ associated production. A high p_T isolated lepton is very unlikely to be produced in most background processes with two photons, and, thus, a signal to background ratio of about 1:1 or even better can be achieved. The results of a simulation of the anticipated $m_{\gamma\gamma}$ spectrum are shown in Fig. 17(b), again for the CMS detector.[10] While $S : B$ is very good, cross sections are much lower than in the inclusive $H \to \gamma\gamma$ search and integrated luminosities of order 100 fb^{-1} or larger are needed for a significant signal in these associated production channels.

5.4 Weak boson fusion

There is a danger in relying too much on the $H \to \gamma\gamma$ decay channel for a light Higgs boson, of course: it implicitly assumes that the Higgs partial decay widths are indeed as large as predicted by the SM. Approximately, the two-photon branching ratio is given by

$$B(H \to \gamma\gamma) = \frac{\Gamma(H \to \gamma\gamma)}{\Gamma(H \to \bar{b}b) + \ldots}, \qquad (66)$$

i.e. a strongly increased partial width for $H \to \bar{b}b$ can render the $H \to \gamma\gamma$ channel unobservably small. This is what happens in the MSSM, with its two Higgs doublets, which lead to two CP-even scalars, the light h and a heavier state, H. For large $\tan\beta = v_1/v_2$, the b-quark Yukawa coupling is enhanced, leading to a suppressed $h \to \gamma\gamma$ branching ratio over large regions of parameter space.[c] One thus needs to prepare for a search in other decay channels as well. And even if the $H \to \gamma\gamma$ mode is observed first, the other channels will be needed to learn about the various couplings of the Higgs boson, to weak bosons, quarks and leptons, i.e. they need to be studied in order to understand the dynamics of the symmetry breaking sector.

In any model which treats lepton and quark mass-generation symmetrically, the $H \to \bar{b}b$ and $H \to \tau^+\tau^-$ decay widths move in unison because both represent the isospin $-1/2$ component of a third generation doublet. Thus, the $h \to \tau^+\tau^-$ branching ratio is fairly stable, staying at the 8–9% level in e.g. the MSSM over large regions of parameter space where the $h \to \gamma\gamma$ branching ratio may be suppressed by large factors. Interestingly, the tau decay mode is observable in the most copious of the associated Higgs production processes, weak boson fusion as depicted in Fig. 14.

Traditionally, weak boson fusion has been considered mainly as a method for studying a strongly interacting symmetry breaking sector,[?] where one encounters either a very heavy Higgs boson or non-Higgs dynamics such as in technicolor models. However, as is evident from Fig. 15, the weak boson fusion cross section, $\sigma(qq \to qqH)$, is as large as a few pb also for a Higgs boson in the 100 GeV range, which in the SM corresponds to 10–20% of the total Higgs production rate.

A characteristic feature of weak boson fusion events are the two accompanying quarks (or anti-quarks) from which the "incoming" Ws or Zs have been

---

[c]In the following, no distinction will be made between different scalar states. H generically denotes the Higgs resonance which is being searched for.

radiated (see Fig. 14(b)). In general these scattered quarks will give rise to hadronic jets. By tagging them, i.e. by requiring that they are observed in the detector, one obtains a powerful background rejection tool.[44,45] Whether such an approach can be successful depends on the properties of the tagging jets: their typical transverse momenta, their energies, and their angular distributions.

Similar to the emission of virtual photons from a high energy electron beam, the incoming weak bosons tend to carry a small fraction of the incoming parton energy. At the same time the incoming weak bosons must carry substantial energy, of order $m_H/2$, in order to produce the Higgs boson. Thus the final state quarks in $qq \to qqH$ events typically carry very high energies, of order 1 TeV. This is to be contrasted with their transverse momenta, which are of order $p_T \approx m_W$. This low scale is set by the weak boson propagators in Fig. 14(b), which introduce a factor

$$D_V(q^2) = \frac{-1}{q^2 - m_V^2} \approx \frac{1}{p_T^2 + m_V^2} \tag{67}$$

into the production amplitudes and suppress the $qq \to qqH$ cross section for quark transverse momenta above m_V. The modest transverse momentum and high energy of the scattered quark corresponds to a small scattering angle, typically in the $1 < \eta < 5$ pseudo-rapidity region.

These general arguments are confirmed by Fig. 18, where the transverse momentum and pseudo-rapidity distributions of the two potential tagging jets are shown for the production of a $m_H = 120$ GeV Higgs boson at the LHC. One finds that one of the two quark jets has substantially lower median p_T (≈ 35 GeV) than the other (≈ 70 GeV), and therefore experiments must be prepared to identify fairly low p_T forward jets. A typical requirement would be[48]

$$p_{T_{j(1,2)}} \ge 40, 20 \text{ GeV}, \qquad |\eta_j| \le 5.0, \qquad m_{j1,j2} > 1 \text{ TeV}, \tag{68}$$

where, in addition, the tagging jets are required to be in opposite hemispheres, with the Higgs decay products between them.

While these requirements will suppress backgrounds substantially, the most crucial issue is identification of the $H \to \tau\tau$ decay products and the measurement of the $\tau^+\tau^-$ invariant mass. The τ's decay inside the beam pipe and only their decay products, an electron or muon in the case of leptonic decays, $\tau^- \to \ell^- \bar{\nu}_\ell \nu_\tau$, and an extremely narrow hadronic jet for $\tau^\pm \to h^\pm \nu_\tau$ ($h = \pi, \rho, a_1$) are seen inside the detector. The presence of a charged lepton, of $p_T > 20$ GeV, is crucial in order to trigger on an $H \to \tau^+\tau^-$ event. Allowing the other τ to decay hadronically then yields the highest signal rate. Even

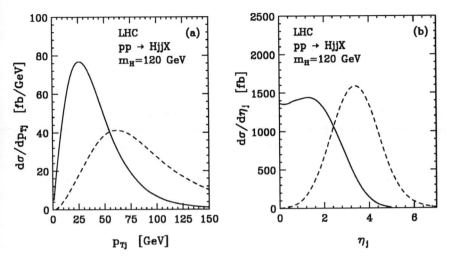

Figure 18: Transverse momentum and pseudo-rapidity distributions of the two (anti)quark jets in $qq \rightarrow qqH$ events at the LHC. Shown are (a) $d\sigma/dp_{Tj}$ for the highest (dashed curve) and lowest p_T jet (solid curve) and (b) $d\sigma/d|\eta_j|$ for the most forward (dashed curve) and the most central jet (solid curve).

though the hadronic τ decay is seen as a hadronic jet, this jet is not a typical QCD jet. Rather, the τ-jet is extremely well collimated and it normally contains a single charged track only (so called 1-prong τ decay). An analysis by ATLAS[9,46] has shown that hadronically decaying τ's, of $p_T > 40$ GeV, can be identified with an efficiency of 26% while rejecting QCD jets at the 1:400 level.

At first sight the two or more missing neutrinos in $H \rightarrow \tau^+\tau^-$ decays seem to preclude a measurement of the $\tau^+\tau^-$ invariant mass. However, because of the small τ mass, the decay products of the τ^+ or the τ^- all move in the same direction, i.e. the directions of the unobserved neutrinos are known. Their energy can be inferred by measuring the missing transverse momentum vector of the event. Denoting by x_{τ_ℓ} and x_{τ_h} the fractions of the parent τ carried by the observed lepton and decay hadrons, the transverse momentum vectors are related by

$$\not{p}_T = (\frac{1}{x_{\tau_l}} - 1)\, \mathbf{p}_{T\ell} + (\frac{1}{x_{\tau_h}} - 1)\, \mathbf{p}_{Th}\,. \tag{69}$$

As long as the the decay products are not back-to-back, Eq. (69) gives two conditions for x_{τ_i} and provides the τ momenta as $\mathbf{p}_\ell/x_{\tau_l}$ and $\mathbf{p}_h/x_{\tau_h}$, respectively. As a result, the $\tau^+\tau^-$ invariant mass can be reconstructed,[47] with an accuracy of order 10–15%.

Backgrounds to $H \rightarrow \tau\tau$ events in weak boson fusion arise from several sources. First is the production of real $\tau^+\tau^-$ pairs in "Zjj events", where the real or virtual Z or photon, which decays into a τ pair, is produced in association with two jets. In addition, any source of isolated leptons and three or more jets gives a background since one of the jets may be misidentified as a τ hadronic decay. Such reducible backgrounds can arise from $W + 3$ jet production or heavy flavor production, in particular $b\bar{b}jj$ events, where the W or one of the b-quarks decay into a charged lepton. Identifying the two forward jets of the signal, with a large separation, $|\eta_{j_1} - \eta_{j_2}| > 4.4$, and large invariant mass, substantially limits these backgrounds, as do the τ-identification requirements. Additional background reduction is achieved by asking for consistent values of the reconstructed τ momentum fractions carried by the central lepton and τ-like jet,

$$x_{\tau_l} < 0.75, \qquad x_{\tau_h} < 1. \tag{70}$$

Finally, a characteristic difference between the weak boson fusion signal and the QCD backgrounds is in the amount and angular distribution of gluon radiation in the central region.[49] The $qq \to qqH$ signal proceeds without color exchange between the scattered quarks. Similar to photon bremsstrahlung in Rutherford scattering, gluon radiation will be emitted in the very forward and very backward directions, between the tagging jets and the beam direction. Gluons giving rise to a soft jet are a rare occurrence in the central region. The background processes, on the other hand, proceed by color exchange between the incident partons, and here, fairly hard gluon radiation in the central region is quite common. A veto on any additional jet activity between the two tagging jets, of $p_{Tj} > 20$ GeV, is expected to reduce the QCD backgrounds by about 80% while reducing the signal by 20-30% only.[48] Combining these various techniques, forward jet tagging, τ-identification, τ-pair mass reconstruction, and the central jet veto, one obtains a very low background signal (S:B $\approx$ 7:1 for $m_H = 120$ GeV, significantly worse only for a Higgs which is degenerate with the Z) which is large enough to give a highly significant signal with an integrated luminosity of 30–50 fb^{-1}. The expected τ-pair invariant mass distribution, for a Higgs mass of 120 GeV, is shown in Fig. 19.

The forward jet tagging and central jet vetoing techniques described above can be used for isolating any weak boson fusion signal. Another example is the search for $H \to \gamma\gamma$ events, where the Higgs has been produced via $qq \to qqH$. About 20-30 fb^{-1} of integrated luminosity are sufficient to observe the SM Higgs boson in the 110-150 GeV mass range this way,[50] which is comparable to the inclusive $H \to \gamma\gamma$ search described earlier. By observing both production channels, however, $gg \to H$ and $qq \to qqH$, separate information is obtained on the $H t\bar{t}$ and HVV couplings which determine the production cross sections.

As should be clear from the preceding examples, a veritable arsenal of

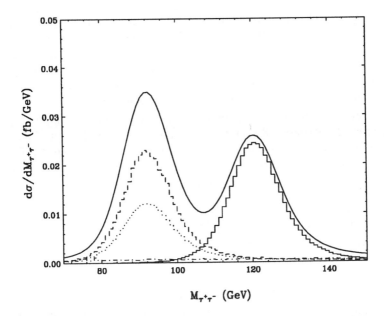

Figure 19: Reconstructed $\tau^+\tau^-$ invariant mass distribution expected in weak boson fusion events at the LHC, for a SM Higgs boson of mass $m_H = 120$ GeV. The solid line represents the sum of the signal and all backgrounds. Individual components are shown as histograms: the $qq \to qqH$ signal (solid), the irreducible QCD Zjj background (dashed), the irreducible EW Zjj background (dotted), and the combined $Wj + jj$ and $b\bar{b}jj$ reducible backgrounds (dash-dotted). From Ref. [48].

methods is available at the LHC to search for the Higgs boson and to analyze its properties. A single one of the methods may be sufficient for discovery of the Higgs and measurement of its mass. However, discovery will only start a much more important endeavor, for which all the tools will be needed: determining the various couplings of the Higgs boson to gauge bosons and heavy fermions, answering the question whether additional particles arise from the symmetry breaking sector, like the H^+ and pseudo-scalar A of two Higgs doublet models, and, thus, finding the dynamics which is responsible for $SU(2) \times U(1)$ breaking in nature. New methods are still being developed for this purpose, and all will be needed to answer these fundamental questions.

6 Conclusions

Our discussion has touched on a number of the investigations which can be conducted at e^+e^- and hadron colliders, or at a future muon collider. It is by no means complete, however. The strategies for identifying a supersymmetry signal at the LHC, for example, have not been discussed in any detail. The reader is referred to other TASI lectures for this purpose.[5] Some general properties of experimental possibilities should have emerged, however. Both e^+e^- and hadron colliders provide the necessary energy and luminosity to search for new particles, and to investigate their properties, once they have been discovered. In these goals, the two types of machines complement each other.

For the foreseeable future, hadron colliders produce the highest parton center of mass energies and therefore have the longest reach in producing very heavy objects. Their disadvantage is the large pp cross section, i.e. the fact that any new physics signal needs to be extracted from backgrounds whose rates are larger by many orders of magnitude. We have studied several examples of how this can be done. Electroweak decays, with their resulting isolated photons and leptons, are crucial to identify top quarks, heavy gauge bosons or the Higgs. But precise features of the hadronic part of new particle production events provide equally important information. Examples are the multiple jets and tagged b-quarks in top decays, and the two forward tagging jets in weak boson fusion events.

e^+e^- colliders benefit from much better signal to background ratios, lack of an underlying event as encountered in hadron collisions, and the constrained kinematics which results from the point-like character of the beam particles: beam constraints are extremely useful tools in the reconstruction of events. The disadvantage of e^+e^- colliders is their limited energy reach, of course, as compared to hadron colliders.

Only time will show which of these machines, the Tevatron, LEP2, the LHC, an NLC, or a muon collider, will give us the most important clues to what lies beyond the standard model. But given their capabilities, exciting times ahead of us are virtually assured.

Acknowledgments

I would like to thank the organizers of TASI 98 for bringing students and lecturers together in a most pleasant and inspiring environment. The help of T. Han and Y. Pan in obtaining several figures and other information is greatfully acknowledged. Many thanks go to D. Rainwater and K. Hagiwara for a most enjoyable collaboration which led to some of the results reported in Section 5.4. This work was supported in part by the University of Wisconsin Research Committee with funds granted by the Wisconsin Alumni Research Foundation and in part by the U. S. Department of Energy under Contract No. DE-FG02-95ER40896.

References

1. For a full discussion see the lectures of G. Altarelli, these proceedings [hep-ph/9811456].
2. See e.g. J. F. Gunion, H. E. Haber, G. Kane, and S. Dawson, *The Higgs Hunter's Guide*, Addison-Wesley (1990).
3. M. E. Peskin, *Beyond the Standard Model*, Lectures given at *The 1996 European School of High-Energy Physics*, SLAC-PUB-7479 (1997) [hep-ph/9705479].
4. For details and further references see M. Cvetic, these proceedings.
5. See the lectures by N. Polonsky, these proceedings; Xerxes Tata, Supersymmetry: Where it is and how to find it, in *TASI 95, QCD and Beyond*, ed. by D.E. Soper, World Scientific (1996), p. 163 [hep-ph/9510287].
6. For recent reviews see V. Barger, preprint MADPH-98-1040 (1998) [hep-ph/9803480]; J.F. Gunion, preprint UCD-98-5 (1998) [hep-ph/9802258]; V. Barger et al., The physics capabilities of $\mu^+\mu^-$ colliders, in *Santa Barbara 1996, Future high energy colliders*, p. 219 [hep-ph/9704290].
7. See e.g. V. Barger and R. Phillips, *Collider Physics*, Addison-Wesley (1987); R.K. Ellis, W.J. Stirling and B.R. Webber, *QCD and Collider Physics*, Cambridge Univ. Press (1996).
8. E. Keil, in Proceedings of the *ECFA Workshop on LEP 200*, ed. by A. Böhm and W. Hoogland, report CERN 87-08 (1987), Vol. I, p. 17.
9. W. W. Armstrong et al., Atlas Technical Proposal, report CERN/LHCC/94-43 (1994).
10. G. L. Bayatian et al., CMS Technical Proposal, report CERN/LHCC/94-38 (1994).
11. K. Hagiwara, K. Hikasa, R.D. Peccei and D. Zeppenfeld, Nucl. Phys. **B282**, 253 (1987).
12. W. Beenakker et al., *WW* cross sections and distributions, in *Physics at LEP2*, ed. by G. Altarelli, T. Sjöstrand and F. Zwirner, CERN Yellow report CERN-96-01 (1996) vol. 1, p. 79 [hep-ph/9602351] and references therein.
13. G. Gounaris et al., Triple gauge boson couplings, in *Physics at LEP2*, ed. by G. Altarelli, T. Sjöstrand and F. Zwirner, CERN Yellow report CERN-96-01 (1996) vol. 1, p. 525 [hep-ph/9601233].
14. ADLO and D0 TGC Combination Group, report LEPEWWG/TGC/98-02, summer 1998.
15. Z. Kunszt et al., Determination of the mass of the *W* boson, in *Physics at LEP2*, ed. by G. Altarelli, T. Sjöstrand and F. Zwirner, CERN Yellow report CERN-96-01 (1996) vol. 1, p. 141 [hep-ph/9602352].
16. LEP Electroweak Working Group, *LEP WW cross section and W*

mass for '98 summer conferences, report LEPEWWG/MW/98-02 at http://www.cern.ch/LEPEWWG/wmass/.

17. CDF Collaboration, F. Abe et al., Phys. Rev. Lett. **75**, 11 (1995); D0 Collaboration, B. Abbot et al., Phys. Rev. Lett. **80**, 3000 (1998).

18. For details see G.F Giudice et al., Searches for new physics, in *Physics at LEP2*, ed. by G. Altarelli, T. Sjöstrand and F. Zwirner, CERN Yellow report CERN-96-01 (1996) vol. 1, p. 463 [hep-ph/9602207] and references therein.

19. P. Rebecchi, talk given at ICHEP98, Vancouver, Canada, July 23-29, 1998.

20. J. Erler and P. Langacker, to appear in the Proceedings of the *5th International WEIN Symposium*, Santa Fe, NM, June 14-21, 1998 [hep-ph/9809352].

21. For a recent summary see W. Marciano, talk given at the 1999 DPF meeting, Los Angeles, Jan.5-9, 1999.

22. M. Carena et al., Higgs Physics, in *Physics at LEP2*, ed. by G. Altarelli, T. Sjöstrand and F. Zwirner, CERN Yellow report CERN-96-01 (1996) vol. 1, p. 351 [hep-ph/9602250] and references therein.

23. For recent reviews of physics at linear e^+e^- colliders see e.g. E. Accomando et al., Phys. Rept. **299**, 1 (1998) [hep-ph/9705442]; S. Kuhlman et al., Physics and Technology of the NLC, report SLAC-R-0485 (1996) [hep-ex/9605011].

24. CDF Collaboration, F. Abe et al., Phys. Rev. Lett. **74**, 2626 (1995).

25. D0 Collaboration, S. Abachi et al., Phys. Rev. Lett. **74**, 2422 (1995); Phys. Rev. Lett. **74**, 2632 (1995); Phys. Rev. **D52**, 4877 (1995).

26. See e.g. H.L. Lai et al., Phys. Rev. **D55**, 1280 (1997); A.D. Martin, R.G. Roberts, W.J. Stirling, and R.S. Thorne, Eur. Phys. J. **C4**, 463 (1998).

27. UA1 Collaboration, G. Arnison et al., Phys. Lett. **122B**, 103 (1983); UA2 Collaboration, M. Banner et al., Phys. Lett. **122B**, 476 (1983).

28. D. Chang, R. Mohapatra, and M. Parida, Phys. Rev. Lett. **50**, 1072 (1984); Phys. Rev. **D30**, 1052 (1984).

29. For a recent review see e.g. M. Cvetic and P. Langacker, preprint hep-ph/9707451, to appear in *Perspectives in Supersymmetry*, World Scientific, ed. by G. Kane.

30. G.-C. Cho, K. Hagiwara, and S. Matsumoto, Eur. Phys J. **C5**, 155 (1998) [hep-ph/9707334]; G.-C. Cho, K. Hagiwara, and Y. Umeda, hep-ph/9805448.

31. P. Nason, S. Dawson, and R. K. Ellis, Nucl.Phys. **B303**, 607 (1988); E. Laenen, J. Smith, and W. L. van Neerven, Nucl.Phys. **B369**, 543

(1992); Phys. Lett. **B321**, 254 (1994); E. Berger and H. Contopanagos, Phys. Rev. **D54**, 3085 (1996); S. Catani et al., Phys. Lett. **B378**, 329 (1996).

32. V. Barger, T. Han, J. Ohnemus, and D. Zeppenfeld, Phys. Rev. **D40**, 2888 (1989); F. A. Berends, W. T. Giele, and H. Kuijf, Nucl. Phys. **B321**, 39 (1989); F. A. Berends et al., Phys. Lett. **B224**, 237 (1989).

33. F. A. Berends, H. Kuijf, B. Tausk, and W. T. Giele, Nucl. Phys. **B357**, 32 (1991).

34. CDF Collaboration, F. Abe et al., Phys. Rev. **D50**, 2966 (1994); Phys. Rev. **D52**, 2605 (1995).

35. D0 Collaboration, B. Abbot et al., report FERMILAB-PUB-98-261-E (1998), [hep-ex/9808029].

36. CDF Collaboration, F. Abe et al., Phys. Rev. Lett. **82**, 271 (1999).

37. S. Kim, S. Kuhlmann, and W.M. Yao, report CDF-ANAL-EXOTIC-PUBLIC-3904 (1996), presented at 1996 DPF/DPB Summer Study on New Directions for High-energy Physics (Snowmass 96), Snowmass, CO, 25 Jun - 12 Jul 1996.

38. Tao Han and Ren-Jie Zhang, Phys. Rev. Lett. **82**, 25 (1999) [hep-ph/9807424].

39. J. F. Gunion, A. Stange, and S. S. D. Willenbrock, in *Electroweak symmetry breaking and new physics at the TeV scale*, ed. by T.L.Barklow et al., p. 23-145 [hep-ph/9602238].

40. M. Spira and P. Zerwas, Electroweak symmetry breaking and Higgs Physics, in Schladming 1997 *Computing particle properties*, p. 161-225, [hep-ph/9803257].

41. K. Iordanidis and D. Zeppenfeld, Phys. Rev. **D57**, 3072 (1998), [hep-ph/9709506].

42. J. Bagger et al., Phys. Rev. **D49**, 1246 (1994); Phys. Rev. **D52**, 3878 (1995).

43. M. Dittmar and H. Dreiner, Phys. Rev. **D55**, 167 (1997) [hep-ph/9608317].

44. R. N. Cahn et al., Phys. Rev. **D35**, 1626 (1987); V. Barger et al., Phys. Rev. **D37**, 2005 (1988); R. Kleiss and W. J. Stirling, Phys. Lett. **200B**, 193 (1988).

45. U. Baur and E. W. N. Glover, Nucl. Phys. **B347**, 12 (1990); U. Baur and E. W. N. Glover, Phys. Lett. **B252**, 683 (1990); D. Froideveaux, in *Proceedings of the ECFA Large Hadron Collider Workshop*, Aachen, Germany, 1990, edited by G. Jarlskog and D. Rein (CERN report 90-10, Geneva, Switzerland, 1990), Vol II, p. 444; M. H. Seymour, ibid, p. 557; V. Barger, K. Cheung, T. Han, J. Ohnemus, and D. Zeppenfeld, Phys.

Rev. **D44**, 1426 (1991).

46. D. Cavalli *et al.*, ATLAS Internal Note PHYS-NO-051, Dec. 1994.
47. R. K. Ellis *et al.*, Nucl. Phys. **B297**, 221 (1988).
48. D. Rainwater, D. Zeppenfeld, and K. Hagiwara, Phys. Rev. **D59** (1999) 014037.
49. Y. L. Dokshitzer, V. A. Khoze, and S. Troyan, in *Proceedings of the 6th International Conference on Physics in Collisions*, (1986) ed. M. Derrick (World Scientific, Singapore, 1987) p. 365; J. D. Bjorken, Int. J. Mod. Phys. **A7**, 4189 (1992); Phys. Rev. **D47**, 101 (1993); J. D. Bjorken, in *Proceedings of the International Workshop on Photon-Photon Collisions*, (1992) ed. D. O. Caldwell and H. P. Paar (World Scientific, Singapore, 1992) p. 502.
50. D. Rainwater and D. Zeppenfeld, J. High Energy Physics 12 (1997) 005.

THE EXPERIMENTAL SEARCH FOR FINITE NEUTRINO MASS

THOMAS J. BOWLES

Physics Division
Los Alamos National Laboratory, Los Alamos, New Mexico 87545 USA
E-mail: tjb@lanl.gov

A finite neutrino mass would have broad implications for our understanding of the Universe. Thus, a very wide range of experiments in high energy and nuclear physics has been mounted to search for evidence for finite mass neutrinos. These range from direct kinematic tests to double beta decay to neutrino oscillation searches. At present, there is evidence reported for neutrino oscillations in an accelerator experiment studying pion beta and muon beta decay (LSND), by several experiments studying atmospheric neutrinos, and in a number of measurements of solar neutrinos. Additional experiments are under way to test these claims and the next few years promise to be an exciting time in neutrino physics. In this paper I discuss the present status and future plans for experimental searches for a finite neutrino mass.

1 Impact of Neutrinos in Different Fields

Neutrinos play a dominant role in particle physics, astrophysics, and cosmology. In our present understanding of the strong, weak, and electromagnetic forces, the group structure of the Standard Model is $SU(3)_C \otimes SU(2)_L \otimes U(1)_{EM}$. In the Weinberg-Salam-Glashow (W-S-G) Standard Electroweak Model [1,2], left-handed neutrinos sit in a doublet, while right-handed neutrinos are in a singlet, and therefore do not interact with the other known particles. However, while the W-S-G model provides an amazingly accurate picture of our present cold Universe, it has a number of deficits. The Standard Model does not explain the origin of the group structure, it does not reduce the number of coupling constants required, nor does it offer any prediction for the physical masses of the particles. Thus, it is generally assumed that the Standard Model is but a subset of some larger gauge theory. A wide variety of Grand Unified field Theories (GUTs), Super Symmetric Models (SUSY), and Superstring models have been proposed as the model for this larger structure. In general, these models predict nonzero neutrino masses and contain mechanisms that provide for lepton-number violation. Thus, a variety of new phenomena are predicted, including finite neutrino masses and the possibility that neutrinos can oscillate from one type to another.

2 Neutrino Properties

In the Standard Weinberg-Salam-Glashow Model (SM), the neutrinos are contained in the gauge structure of $SU(3)_C \otimes SU(2)_L \otimes U(1)_{EM}$. The electroweak fields γ are chiral, i.e., $\gamma_L = (1 + \gamma_5)/2$, $\gamma_R = (1 + \gamma_5)/2$, and $\gamma = \gamma_L + \gamma_R$. Under $SU(2)_L$, the right-handed projections γ_R are singlets and the left-handed projections γ_L are doublets. Since the left- and right-handed components sit in different representations and the right handed components do not interact, we observe maximal parity violation. Thus, the SM provides us with left-handed neutrinos and right-handed antineutrinos (as observed in nature) but without any underlying physical explanation as to the cause of this structure. In the SM, neutrinos are massless and lepton number is conserved.

There are three known flavors of neutrinos: electron, muon, and tau (together with their corresponding antineutrinos). In addition, the existence of a fourth neutrino that is sterile (i.e., is of right-handed helicity and thus does not participate in the normal weak interactions) has been postulated in order to accommodate all of the experimental results reported to date. Neutrinos may also be classified as Dirac, in which case the right- and left-handed neutrinos v_R and v_L and their antiparticles $\bar{v}_R$ and $\bar{v}_L$ are distinct states, or Majorana, in which case $v \equiv \bar{v}$, so that there are only two states v_R and v_L. The electron and muon neutrinos have been directly observed in many experiments. The tau neutrino has not yet been directly observed, but is assumed to exist as it is required in the Standard Model and from LEP results that show [3] that the number of neutrinos with mass less than 45 GeV is 2.993 ± 0.011.

3   Direct Searches for Finite Neutrino Mass

Searches for a finite neutrino mass can be undertaken with a variety of experimental techniques. I define direct neutrino mass measurements as those in which the only physical requirement is that the neutrino has a finite mass. Experiments that observe beta decay and some measurements of supernova fall within this category. Indirect measurements are those in which in addition to having a finite mass, the neutrino must also possess other properties that are not predicted by the Standard Weinberg-Salam-Glashow Electroweak Model. One class includes those experiments that require lepton number violation in addition to a finite neutrino mass. Neutrino oscillation experiments fit into this category. Experiments that indirectly probe for neutrino mass include neutrino oscillation experiments at reactors and accelerators, atmospheric neutrino measurements, studies of heavy neutrino branches in beta decay, solar neutrino experiments, and some measurements of supernova. Other classes of indirect measurements require additional new physics, such as requiring that the neutrino be a Majorana neutrino or that in addition to being Majorana neutrinos that new particles, such as Majorons, exist. Zero neutrino double beta decay and double beta decay with Majoron emission fall into these classes, respectively.

In this paper, I will focus primarily on experiments that have reported evidence for a finite and stable (on cosmological time scales) neutrino mass – in particular tritium beta decay and neutrino oscillation experiments (both terrestrial and astrophysical).

3.1 Beta Decay

Kinematic searches for a finite neutrino mass are made in the beta decay spectra of nuclei, muons, and tau particles, by carrying out measurements in which one can kinematically reconstruct the events. The existence of a finite neutrino mass will produce a well-defined distortion in the spectra in the region of the endpoint of the decay.

3.2 Tau Neutrino

The best means to search for a finite tau neutrino mass in the laboratory is to study the kinematics of tau decay. One measures the momenta of the decay products and forms an invariant mass plot. Any missing mass can be attributed to a finite tau neutrino mass. By observing the decay into multiple charged and neutral particles, the uncertainties in the measurements of the particle momenta are minimized (since the masses of the particles are known with much higher precision than the measurements of the particle momenta). Thus, searches have been made for tau decay into either 3- or 5-pion final states. The best limit on the ν_τ mass comes from the ALEPH collaboration at CERN. By observing the kinematics in the decay of 2939 $\tau^- \rightarrow 2\pi^- \pi^+ \nu_\tau$ and 52 $\tau^- \rightarrow 3\pi^- 2\pi^+ \nu_\tau$ final states, one can constrain the ν_τ mass by looking for events very close to the endpoint. This provides an upper limit on the ν_τ mass of $m_{\nu_\tau} < 18.2$ MeV (95% CL). [4] With higher statistics and improved resolution, one might hope to achieve limits of 10 MeV in the future. The only real hope for obtaining sensitivity in the cosmologically interesting region (tens of eV) is by observation of a supernova in our galaxy with one of the large underground detectors now in operation or under construction.

3.3 Muon Neutrino

In the same manner as for tau decay, one searches for a finite muon neutrino mass by constructing the invariant mass for the decay of $\pi^+ \rightarrow \mu^+ + \nu_\mu$. There are two modes in which this can be done: either in pion decay at rest or pion decay in flight. In the former, one stops a π^+ (stopped π^- undergo rapid nuclear capture before they can decay) in a target and then measures the momentum of the muon emitted during the beta decay. In the latter, one has to measure the momenta of both the pion and the muon.

The most stringent limit on the muon neutrino mass comes from measurements of the muon momentum following the decay of stopped pions. In 1988, measurements of the mass of the muon (m_{μ^+}), coupled with the muon decay data of Abela et al., [5-6] gave $m_{\nu_\mu} < -0.097(72)$ MeV, which, using the Bayesian prescription, [7] yields an upper limit of 270 keV (90% CL).

However, a problem arose in 1993 when new measurements at the Paul Scherrer Institute (PSI) of m_{π^+} caused a reevaluation [8] of the Abela et al. This resulted in a new value of the neutrino mass of $m_{\nu_\mu} < -0.127(25)$ MeV with a 5 σ negative central value. The difficulty is that m_{π^+} is determined using pionic X-rays, and the precision is limited principally by theoretical uncertainties such as electron screening and strong interactions (e.g., absorption from the 3d state). Thus, the measurement using stopped pions had a severe systematic problem and cannot be used.

The situation again changed with new data on the mass of the pion from PSI. [10] The mass value of the pion is determined using the 4f – 3d transition which in turn depends on the average number of K electrons present in the atom. That number is deduced using the relative intensity of radiative transitions that pass through the 3d state, as Auger transition intensities depend on the population of the K shell. One has to correct the data for direct strong-interaction absorption from the 3d state. From the 1985 data, one found that the K-shell occupancy was 0.22. However, with more precise data taken in 1992, it was found that a misidentification of the 0-K, 1-K, and 2-K electron occupancies of the K shell had been made. Using the proper identification, the value of the K-shell occupancy was found to be 0.86. This resulted in a reevaluation of the π^+ mass, as indicated in Table 1. The revised value for the muon neutrino mass is now $m_{\nu_\mu}^2 = -0.016[23]$ MeV^2, resulting in an upper limit of $m_{\nu_\mu} < 170$ keV (90% CL) [11].

Table 1. Data on $\pi^+ \rightarrow \mu^+ + \nu_\mu$ at rest.

Reference	m_μ (MeV)	p_μ (MeV/c)	m_π (MeV)	$m_{\nu_\mu}^2$(MeV^2)
Abela et al. 84 [5]	105.65932(29)	29.79239(83)	139.56761(77)	-0.163(80)
Jeckelman et al. 86 [6]			139.56871(53)	-0.097(72)
PDG 88, 92 [7]	105.658389(34)		139.56752(37)	
Daum et al. 91 [9]		29.79206(68)	139.56996(67)	
Jeckelman et al. 94 [10]		29.79177(74)		0.143(24)
Assamagan et al 96 [11]			139.56995(35)	-0.016(23)

As the above limit is derived from data in which the momentum resolution is already 10^{-5}, it is difficult to improve substantially on this limit. A renewed effort is currently underway at PSI to improve the accuracy of the pion mass [12]. While this will not substantially reduce the uncertainty in the quoted limit, it may lead to a change in the central value of the pion mass and thus change the limit on the muon neutrino mass.

Another method of determining the v_μ mass is using π^+ decay in flight in which both the pion and muon momenta are directly measured. This method is relatively insensitive to m_π and m_μ. The measurement of Anderhub et al. [13] at PSI thus provides the most reliable limit with $m_{v_\mu}^2 = -0.14(20)$ MeV^2, resulting in an upper limit of $m_{v_\mu} < 500$ keV (90% CL). There is also a proposal to use the spectrometer of the muon g-2 measurement at Brookhaven to measure pion decay in flight. Both the pion and muon momenta would be measured in the spectrometer ring. As the magnetic fields within the spectrometer are known to a fraction of a ppm, it might be possible to improve the accuracy of the measurement to be able to search for a finite muon neutrino mass in the 10-20 keV range [12]. However, as in the case of the tau neutrino, it is probably only through observation of a supernova in our galaxy that one can reach a sensitivity of better than 100 eV.

3.4 Electron Antineutrino

Tritium beta decay offers an almost ideal means of searching for a neutrino mass. It is a superallowed decay, the endpoint energy is quite low (18.6 keV), and the atomic final-state effects can be well understood. In these experiments, the beta decay spectrum is measured over a wide region far below the endpoint and then the expected spectral shape in the endpoint region is extrapolated assuming a zero neutrino mass and including all effects (energy loss in the source, spectrometer resolution, decay to different atomic final states, backscattering, background, variation of spectrometer acceptance efficiency with energy, ...) that can distort the spectrum. One then measures the spectrum in the endpoint region and compares it with the extrapolated spectrum. A deviation between the extrapolated and measured spectra can be indicative of a finite neutrino mass, as shown in the Kurie plot in Figure 1 (in which the effects of phase space and Coulomb effects are divided out). The tritium beta decay experiments are very challenging. First, the region of the spectrum that is affected by a finite neutrino mass extends only over about 3 times the neutrino mass. In the region of the endpoint, the fractional amount of the total beta decay strength in a region ΔE below the endpoint to the endpoint is proportional to ΔE^3. Thus, if one is searching for a neutrino mass of order 10 eV, the fraction of the beta decay strength in the last 30 eV is only about 7 x 10^{-9} of the total decay rate. Second, any systematic effect that affects the beta spectrum at the 1 eV level must be accurately taken into account. As the binding energy of the first excited state in the

356

^{3}He atom is at 40 eV, one must very accurately know the final state distribution to both bound and continuum final atomic and molecular states. Similarly, for a source with any finite thickness (which is required in order to get enough count rate), the energy loss in the source is appreciable and must be accurately known. The resolution response of the spectrometer must either be so narrow with negligible tails that it does not introduce any uncertainty (this is essentially unachievable in any current experiment) or must be extremely well characterized. Finally, as count rates in the region of the endpoint of the spectrum are low (the highest count rate experiment has achieved a rate of 1 count/30 seconds 10 eV below the endpoint), the backgrounds must be kept to extremely low levels.

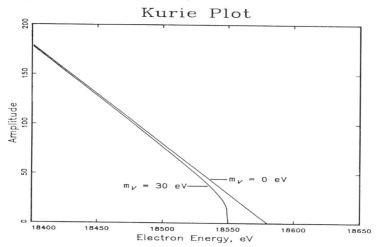

Figure 1. Effect of finite neutrino mass in tritium beta decay spectrum.

In 1980 a Russian group from the Institute of Theoretical and Experimental Physics (ITEP) initially claimed to see evidence for a $m_{\bar{\nu}_e}$ mass of 35 eV, which was later reduced to 26 eV [14]. This experiment used a solid source in the form of a tritiated amino acid (valine) as the source material and a toroidal magnetic spectrometer with a resolution of about 20 eV to measure the beta spectrum. The use of a complex molecule as the source required extensive (and somewhat uncertain) theoretical calculations to take into account the atomic and molecular final-state effects. The data showed a statistically significant effect that was consistent with a finite neutrino mass, as shown in Figure 2. The real concern in this experiment is systematic effects, in particular the determination of the resolution function of the spectrometer and the final state excitations in the tritiated amino acid.

The data is generally fit to an expression of the following form:

$$W(E,Z) = A\,F(E,Z)\,E\,p\,\Sigma\,W_i(E_{0i} - E)\,[(E_{0i} - E)^2 - m_\nu^2 c^4]^{1/2}$$

where the decay probability W is given by a constant A, the Fermi function $F(E,Z)$, the energy E and momentum p of the beta, the branching ratio W_i to each of the atomic or molecular final states with energy E_{0i}, and the neutrino mass m_ν . This expression applies only to the ideal case in which one measures the spectrum with perfect resolution and perfect linearity, no energy loss in the source, and no background. In practice of course the extraction of m_ν is complicated due to the presence of such effects.

In order to determine the neutrino mass using this expression to fit the observed beta decay spectrum, one must *a priori* know the spectrometer resolution function, the energy loss and backscattering in the source, the final state distribution of the source molecule, and the energy dependence of the spectrometer acceptance. One then fits for the amplitude, end-point energy, background, and the square of the neutrino mass. Assuming that the square of the neutrino mass is positive, one then can take the square root to obtain the neutrino mass.

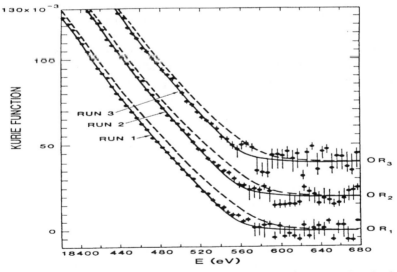

Figure 2. The ITEP data taken as evidence for a finite neutrino mass. Results from 3 runs are shown where the slid line is the best fit to the data with a 35 eV neutrino mass and the dashed lines are the fits for zero neutrino mass.

The ITEP result led to a number of groups around the world working to make precision measurements of the tritium beta decay spectrum in order to test the ITEP claim. Three groups (Zurich, INS Tokyo, and Beijing) used solid sources coupled to magnetic spectrometers to measure the spectrum. The form of the tritium source in these experiments had a simpler molecular form but there were still significant uncertainties in the result associated with final state interactions in the source.

Our group at Los Alamos initiated an experiment using a differentially pumped gaseous molecular tritium source coupled to a toroidal magnetic spectrometer. While this led to a rather large and complex experimental setup as shown in Figure 3, it greatly simplified the interpretation of the data as very accurate calculations of the final state effects in the tritium molecule have been carried out. In the Los Alamos experiment, gaseous tritium is introduced at the center of a long tube that is inside a superconducting solenoid. A small fraction of the atoms in the source tube beta decay before they exit the tube where they are pumped away and recirculated back to the source. The betas spiral along the field lines and are then accelerated by a constant electrostatic potential that raises the entire beta spectrum uniformly by about 6 keV. This ensures that only betas from the source region have enough energy to be transported through the spectrometer to the detector resulting in a substantial reduction in the background. The betas then pass through a differential pumping region and then are focused by nonadiabatic fast passage through a rapidly falling magnetic field region to a collimator at the entrance to the spectrometer. The collimator ensures that only those betas that are in the central region of the source tube (and thus cannot have interacted with the walls of the source tube) are transmitted to the spectrometer. The betas are then momentum analyzed in a five-pass toroidal magnetic spectrometer. The betas are then detected in a position-sensitive silicon strip detector at the end of the spectrometer. An experiment at Livermore used a similar, but substantially improved spectrometer design to measure the tritium beta decay spectrum.

Two other groups (at the University of Mainz and the Institute for Nuclear Research (INR) in Moscow, used pure molecular tritium sources coupled to electrostatic integral spectrometers. The Mainz experiment uses molecular tritium frozen onto a substrate at temperatures of a few degrees Kelvin. The INR experiment uses a differentially pumped gaseous tritium source. Both experiments use integral spectrometers in which the tritium source is in a high magnetic field. The electrons from the beta decay are transported adiabatically along the magnetic field lines as the magnetic field decreases to a low value. There is an adiabatic invariant (B^2/A) (where B is the magnetic field strength and A is the area enclosed by the orbit of the beta particle in the magnetic field) requiring that as the field strength falls the transverse component of the electron momentum is converted into the longitudinal component. No work is done in the system, so the total kinetic energy of the beta is unchanged but the phase space distribution of the electrons is changed. Figure 4 shows the Mainz spectrometer in order to demonstrate the principle. The INR spectrometer is very similar in design, but twice as large. The field in the source region is a few Tesla while the field in the middle of the spectrometer is a few gauss. In the low field region all of the betas have a very small component of transverse momentum. However, the cross-sectional area of the beta in this region scales as the source area x B(source) / B(spectrometer) and thus is quite large in the center of the spectrometer. By applying an electrostatic field (parallel to the magnetic field) in the center of the spectrometer, the betas are slowed down and only those that have a kinetic energy higher than the electrostatic field potential are transmitted across the electrostatic field. Those that pass through experience an increasing magnetic field strength and so the cross sectional area of the betas in decreased and the betas are transported along the magnetic field lines to a detector.

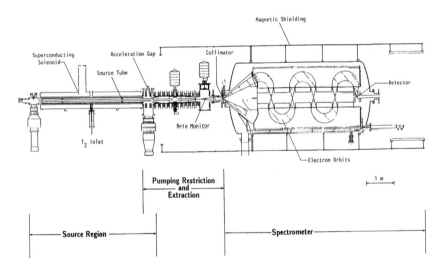

Figure 3. Design of the Los Alamos tritium beta decay experiment.

The advantage of these spectrometers is that they have very large acceptance and can achieve a resolution of a few eV. However, as the betas are decelerated to almost zero energy, any process that generates a low energy electron in the central region of the spectrometer (such as field emission, ionization of residual gas, etc.) can represent a serious background. It is necessary to essentially eliminate field emission, obtain extremely high vacuums, and prevent any tritium from the source from entering into the spectrometer region. After a great deal of effort, very low backgrounds have been achieved in both of these experiments.

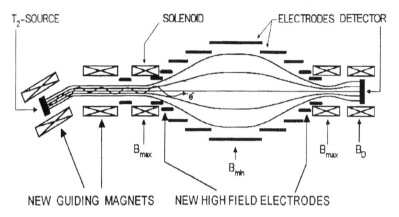

Figure 4. Diagram of the Mainz electrostatic spectrometer.

As of 1995, seven experiments have reported results using source materials that are much simpler than the ITEP source, ranging from pure molecular tritium to tritiated molecular compounds. As shown in Table 2, all seven experiments rule out the ITEP result. However, it must also be noted that all of the experiments find a best fit for $m_{\bar{\nu}_e}^2$ that is negative. Physically, this corresponds to an observed excess of events in the endpoint region, rather than a deficit, which would be indicative of a finite neutrino mass. Using the Bayesian method, these combined data result in a limit on $m_{\bar{\nu}_e}$ of about 5 eV (95% CL). However, as the combined result is more than 5 σ negative, the experiments obviously are subject to some unresolved systematic problem. This has led the Particle Data Group to increase the combined uncertainty by a scale factor of 4.2 resulting in an average value of -27 ± 20 eV^2.

The origin of the negative central value could be due to either an undetermined systematic error in the experiments, difficulties with the theoretical description of the spectrum, or some new physics that is being manifested in the shape of the beta spectrum. An independent check for possible systematic problems has been made by comparing the endpoint measurements from the experiments with the known T-^{3}He mass difference. One finds very good agreement at the few-eV level, making it less likely (but not ruling out) that the explanation is a systematic error. Clearly additional studies of possible effects are warranted and are under way.

Table 2. Limits on the mass of $m_{\bar{\nu}_e}^2$.

Group	$m_{\bar{\nu}_e}^2$ (eV^2)	Limit (eV)	CL
Los Alamos [15]	$-147 \pm 68 \pm 41$	9.3	95%
Zurich [16]	$-24 \pm 48 \pm 61$	11.5	95%
INS Tokyo [17]	$-65 \pm 85 \pm 65$	13	95%
Livermore [18]	$-130 \pm 20 \pm 15$	7.0	95%
Mainz [19]	$-39 \pm 34 \pm 15$	7.2	95%
Beijing [20]	$-31 \pm 75 \pm 48$	12.4	95%
INR [21]	-22 ± 4.8	4.35	95%
PDG [22]	$-27 + 20$	15	--%

There are two possible uncertainties in the theoretical description of the spectrum. The first is that the effect of decays populating different atomic and molecular final states comes entirely from theory. In the case of tritium, it is believed that these final-state distributions can be calculated with high accuracy, and in fact several different calculations agree quite well. The other possibility is that some new physics is involved; for example, tachyonic neutrinos, capture of relic neutrinos from the Big Bang, the existence of new particles, etc. While most such ideas are ruled out by other data, one should not preclude such a possibility and further theoretical work is under way.

Both the INR and Mainz experiments published results with a significantly negative $m_{\bar{\nu}_e}^2$. Subsequently, both experiments have discovered problems with their tritium sources. In the Mainz experiment, it was discovered that at 3 K the frozen tritium film became nonuniform over a period of a day or so and anomalous energy loss resulted [23]. This effect was found to disappear at lower source temperatures (the surface mobility of the tritium decreases exponentially with temperature) and that by running the source at 1.8 K no effect was observed over a period of weeks. The Mainz group also redesigned the source, which reduced the background due to migration of tritium into the spectrometer region. With these changes incorporated, the new data produced results with a $m_{\bar{\nu}}^2$ essentially consistent with zero.

In the INR experiment, they discovered that their Monte Carlo calculation of the trapping of electrons in the source region (in which the magnetic field configuration serves to act as a magnetic bottle) was incorrectly done. After correcting the calculation, their results also yielded a $m_{\bar{\nu}}^2$ essentially consistent with zero. However, the INR group observed another anomaly – a distortion in the spectrum that is consistent with a line about 12 eV below the endpoint that has an amplitude corresponding to about 10^{-11} of the total beta decay rate, as shown in Figure 5 [24]. Further, this distortion changes both in position and amplitude as a function of time. The change in position as a function of time is shown in Figure 6. The INR group

362

has fit the change in position of the line with time to see if the shift is a periodic function of time and found a best fit of 0.496 ± 0.003 yr. The INR group has proffered an explanation for this phenomenon by invoking new physics. It is possible for a tritium nucleus to capture a neutrino from the relic sea of neutrinos resulting in a monoenergetic electron that appears as a line source above the endpoint of the tritium decay spectrum by an amount that corresponds to the Fermi energy of the degenerate sea of relic neutrinos. If one assumes that there is a local enhancement in the density of the sea of relic neutrinos around our solar system and that this enhancement is asymmetric with respect to the plane of the solar system (i.e., forms a pancake of relic neutrinos that is orthogonal to the plane of the solar system), then the Earth moves through an enhanced density region of relic neutrinos twice per year. One also has to postulate some new interaction between neutrinos and matter that binds the neutrinos into this enhanced density region and that also reduces the capture energy to about 10 eV below the endpoint. The density of relic neutrinos has to be about $10^{14} - 10^{15}$ neutrinos/cm^3 (compared to the density of 110 neutrinos/cm^3 expected to be left over in the Standard Big Bang model). No plausible mechanism for such an enhancement has been put forward.

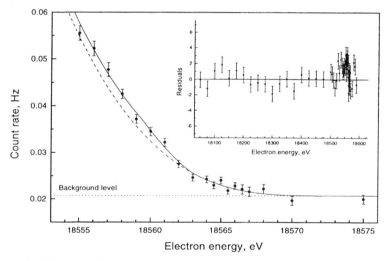

Figure 5. The spectral shape near the endpoint in the INR experiment.

The Mainz group is working to see if they can confirm or reject the line source observed by the INR group. The Mainz group has taken three runs with about 150 days of new data during 1998. In two of the runs, there was no indication for any step function. For the second run, however, there was strong evidence (more than

2σ) for a step function with a best-fit value of about 12 eV down from the endpoint. Of the three new runs, two showed this tendency towards small negative $m\bar{\nu}^2$ while the third run was completely consistent with zero $m\bar{\nu}^2$. The deduced limits on m_ν from each of the runs was about 3 eV (95% CL) with a m_ν^2 value fairly consistent with zero. Since the runs were not consistent in terms of a fitted step in the spectrum, it is difficult to determine how to combine the limits on m_ν from the run with a step and those without a step. Thus, the Mainz group plans to continue to take more data to see if the issue of a step in the function can be resolved. If one uses only one of the 1998 runs (run Q5) that does not show any obvious distortion, then one finds a value of $m\bar{\nu}_e^2 = -3.7 \pm 5.3 \pm 2.1$ eV^2 and using the unified approach of Cousins and Feldman, one derives a limit of $m\bar{\nu}_e < 2.5$ eV (95% CL) [25].

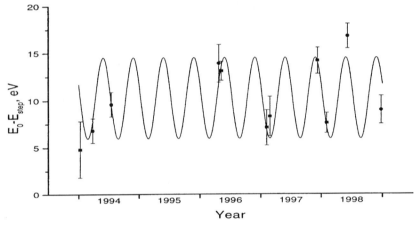

Figure 6. The change in the position of the line source observed in the INR experiment as a function of time.

The tritium beta decay experiment at the INR has also taken more data in 1998 [26]. They continue to see a step in the beta decay spectrum in the region of 10-15 eV below the endpoint. In runs in November and December 1998, the position and amplitude of the step function did not behave in the expected manner (based on the hypothesis of a 6-month variation due to passage through a locally dense cloud of relic neutrinos). The step was significantly different in the two runs, thus indicating a much more rapid variation than expected. The INR group offered an explanation for this effect that involves the Earth moving through a region where the local density of relic neutrinos was much higher. The INR group plans to work further to study the step in the spectrum in future runs.

At present, the INR group uses the data to set a limit on the mass of the electron antineutrino by simply fitting the step in the spectrum to an analytical function and subtracting it from the spectrum. Although the effect of doing this is to shift the value of $m_{\bar{\nu}}^2$ by more than 20 eV^2, no systematic uncertainty due to this subtraction is included. Using this subtraction technique, and combining the 1998 and 1999 data, Lobashev derives a value of $m_{\bar{\nu}}^2 = -1.0 \pm 3.0 \pm 2.1$ eV^2. Using the Bayesian prescription, the INR group arrives at a limit of $m_\nu < 2.5$ eV (95% CL) on the mass of the electron antineutrino. Since it is certainly not clear what the source of the observed step in the spectrum is, it is impossible to tell how covariant this effect is with the derived neutrino mass.

Thus, one is left with a very unsatisfactory situation at present. If one follows the prescription of the Particle Data Group, which gives a combined result of $- 27 \pm 24$ eV^2, then one arrives (using the Bayesian approximation) at a limit of $m_{\bar{\nu}_e} < 15$ eV. Since this involves rescaling the uncertainty by a large amount, no confidence level is quoted for this limit. Until the source of the observed distortion is understood, one cannot determine what the covariance of this effect is with neutrino mass. Therefore, it is impossible to set a model-dependent limit on the mass of the electron antineutrino from laboratory experiments. Hopefully this situation will be resolved with more data from the Mainz and INR experiments. Nonetheless, since the effect is quadratic (as the quantity fit is $m_{\bar{\nu}}^2$), it seems unlikely that the mass of the electron antineutrino can be much larger than 10 eV.

3.5 Supernova

The observation of neutrinos from supernova 1987a by the Kamiokande and IMB underground detectors is certainly a landmark event in modern physics. Not only did it show us that our understanding of the energetics of supernova is quantitatively correct, it also provided valuable information on the properties of neutrinos. As the neutrinos survived the 150,000 year transit from the supernova and arrived at Earth in the expected numbers, we know that the electron antineutrino is essentially stable. From studies of the timing and energies of the observed neutrinos, numerous authors [27] published claims that ranged from setting limits on the electron antineutrino mass of 50 meV to claiming evidence for the existence of a 40 eV neutrino. Careful studies which include the systematic uncertainties in the supernova evolution during neutrino emission provide reasonable estimators for a limit on the mass of the electron antineutrino of 23 eV by the Particle Data Group [28].

With the new generation of large underground detectors (SuperKamiokande and SNO) now on line (and Borexino coming on line in the next few years), the possibility exists that if a supernova occurs within our galaxy that these detectors will have sensitivities for neutrino masses in the 10-100 eV range for all types of neutrinos.

4 Double Beta Decay

The pairing force in nuclei results in a number of cases in which beta decay from a nucleus with (Z, A) is not energetically allowed to a daughter with (Z-1, A), but double beta ($\beta\beta$) decay from the (Z, A) parent is energetically allowed to a daughter with (Z-2, A). This can proceed by three possible mechanisms: two-neutrino $\beta\beta$ decay, zero-neutrino $\beta\beta$ decay, and zero-neutrino $\beta\beta$ decay with the emission of a Majoron. The two-neutrino $\beta\beta$ decay is an allowed second-order process requiring no new physics. Zero-neutrino $\beta\beta$ decay requires that the neutrino has a mass, that lepton number be violated, and that the neutrino is a Majorana neutrino. Zero-neutrino $\beta\beta$ decay with Majoron emission has, in addition, the requirement that a new particle, the Majoron (a massless Goldstone boson) also exists.

The beta spectra that are produced in these different decay modes are shown in Figure 7. In the case of zero-neutrino $\beta\beta$ decay, a sharp line for the sum of the energies of the two electrons is produced. As this is the most interesting case in terms of searching for a finite neutrino mass, I will concentrate on that decay mode here. The decay rate for zero-neutrino double beta decay is given by [29]:

$$[T_{1/2} (0\nu)]^{-1} = G^{0\nu} \mid M^{0\nu} \mid <m_\nu>^2$$

where $T_{1/2}$ (0ν) is the observed zero neutrino half-life, $G^{0\nu}$ is the phase space integral for the decay, $M^{0\nu}$ is the nuclear matrix element, and m_ν is the neutrino mass.

The figure of merit for a given experiment is given by [30]:

$$T_{1/2}(0\nu) \propto \sqrt{Mt / B\Delta E}$$

where a is the isotopic abundance, M is the active mass of the detector, t is the measuring time, B is the average background at the energy of the peak, and ΔE is the energy resolution of the detector.

Thus, it is clear that in order to have the highest sensitivity to neutrino mass, one wants to study cases in which: 1) the mass difference between the mother and daughter nuclei is as large as possible, 2) the matrix element is as large as possible, 3) the source can be made as massive as possible, 4) the energy resolution is as high as possible, and 5) the backgrounds are as low as possible. Thus, while the nucleus chosen for study determines the mass difference and matrix element, the mass,

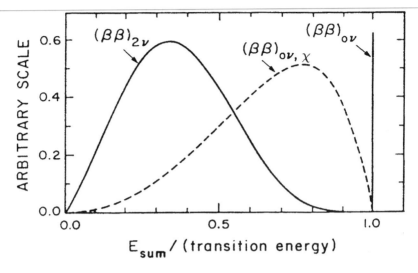

E_{sum} / (transition energy)

Figure 7. Double beta decay spectra for emission with a) zero neutrinos, b) two neutrinos, c) zero neutrinos and a Majoron.

energy resolution, and background can be optimized in each experiment. One means of making the source as massive as possible has been to use isotopically pure sources. This not only maximizes the number of nuclei in the source that is of interest, but also reduces backgrounds since the isotopic enrichment often results in a reduction of other isotopes that may be a source of background. Achieving the best possible energy resolution has invariably meant employing cryogenic detectors. Trying to achieve the lowest possible backgrounds has resulted in a whole research field in itself in which specially selected materials are chosen for use in the experiments. [31] There are generally three sources of backgrounds that must be addressed: long-lived isotopes due primarily to the decays of nuclei in the uranium and thorium chains, prompt activities that are produced by cosmic ray muons, and cosmogenic activities that are a primarily a result of spallation reactions by cosmic ray muons. In order to reduce background due to long-lived isotopes, one generally uses shielding and source materials that are specially prepared (e.g., by zone refining or chemical extraction of U and Th from the materials). Cosmic ray backgrounds are reduced by operating the experiments deep underground where the cosmic ray flux is reduced by several orders of magnitude. Cosmogenic activation can be reduced in some cases by processing of the materials. In other cases, it is simply necessary to store the materials deep underground for long periods of time while the cosmogenic activities die away. The extreme effort that has been undertaken to reduce the level of radioactivity in these experiments is remarkable. For example, ^{210}Pb is a long-lived (22.3 year half-life) activity that is a significant source of background in experiments that typically use Pb shielding. It has been found that in the process of smelting of Pb that U and Th are removed and the decay chain is in disequilibrium since both decay chains have short-lived radon isotopes in them. In the case of the Th decay chain, all of the daughters following the decay of ^{220}Rn decay quickly and after a few weeks there is essentially no radioactivity left in the Pb due to the ^{232}Th that was originally present before the smelting of the Pb. In the case of the U decay chain, after the decay of ^{222}Rn, the only long-lived radioisotope is ^{210}Pb. One can eliminate the ^{210}Pb background if one uses old smelted Pb, as the U was removed in the smelting process and the ^{210}Pb has decayed away. Thus, these experiments have resorted to using Pb that has been recovered from old Spanish galleons and ancient Roman shipping vessels. This sort of care has allowed one to achieve reductions in backgrounds by up to 6 orders of magnitude over that which can be achieved in counting on the surface with normal shielding materials [32].

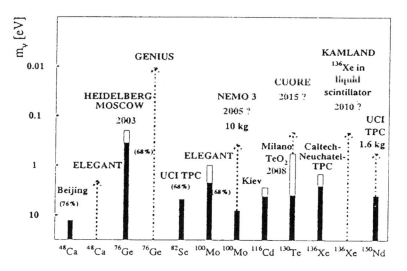

Figure 8. The current limits (black) on a Majorana neutrino mass and the projected sensitivity (open for experiments currently running and dashed for proposed) of various experiments.

The most stringent limits on the neutrino mass in zero-neutrino $\beta\beta$ decay come from the experiments using Ge solid-state detectors that are constructed from highly enriched [76]Ge. In this case, the detector is also the sample. The best limit to date comes from a Moscow-Heidelberg collaboration, which has set a limit on zero neutrino double beta decay of $< 1.6 \times 10^{24}$ yrs (90% CL) [33]. Using calculated nuclear matrix elements, one can then use this result to set a limit on the mass of a Majorana electron neutrino of < 1.2-1.4 eV (90% CL). A variety of new double beta decay experiments have been proposed or are under construction, ranging from large cryogenic detectors to Time Projection Chambers to dissolving one ton of Xe in a large solar neutrino detector. The current limits on neutrino mass and projected sensitivity of possible future experiments is shown if Figure 8 [34].

5 Neutrino Oscillations

If the neutrino has a nonzero mass and lepton number is also violated, then the physical neutrinos that we observe (v_e, v_μ, and v_τ) are not mass eigenstates. Instead there exist three mass eigenstates v_1, v_2, and v_3) with masses m_1, m_2, and m_3. The three physical neutrinos are linear combinations of these mass eigenstates. For simplicity, one can consider the case of only two neutrinos, v_e and v_μ. Then, v_e is predominantly composed of v_1 with a small admixture (determined by a mixing angle θ between v_1 and v_2) of v_2. It is then possible for v_e to oscillate into v_μ as it propagates. The probability for oscillation to occur is [35, 36]:

$$P(v_e \rightarrow v_\mu) = \sin^2 2\theta \ \sin^2\{1.27 \times \Delta m^2(eV^2) \times L(m)/E_v(MeV)\} \ .$$

where $\Delta m^2 = |m_2{}^2 - m_1{}^2|$.

In order for neutrinos to oscillate, at least one of the types of neutrinos must have a finite mass and a mechanism must exist to allow for lepton number violation. Thus, if observed, neutrino oscillations would require new physics beyond the Standard Electroweak Model. A variety of searches, both terrestrial and nonterrestrial, have been carried out. At present, there is evidence reported for $v_\mu \rightarrow v_e$ and $\bar{v}_\mu \rightarrow \bar{v}_e$ oscillations by the LSND experiment and for $v_\mu \rightarrow v_\tau$ or $v_\mu \rightarrow v_s$ in atmospheric neutrino experiments. There is also evidence from measurements of solar neutrinos that electron neutrinos oscillate into another type of neutrino.

5.1 Reactor Neutrino Experiments

Reactors provide an extremely intense source of low-energy electron antineutrinos (of order 10^{20} $\bar{v}_e$/s/GW thermal power) with a continuous energy spectrum peaked at a few MeV, as shown in Figure 9. This provides the basis for a neutrino oscillation disappearance experiment. The sensitivity of these experiments is limited to relatively large mixing angles (typically > 0.1) that is limited by the knowledge of the absolute neutrino intensity and backgrounds. In principle one could reach to somewhat smaller mixing angles by using a detector at a very close distance to the reactor to normalize the neutrino source intensity and spectral shape, but in practice this has not yet been carried out (due primarily to cost considerations). The sensitivity to Δm^2 is limited by the distance between the reactor source and the detector. This in turn is largely limited by the detector size. To date, experiments have been able to probe down to the 10^{-2} eV^2 range. An ambitious new project (KamLAND) plans to use a 1 kiloton detector installed underground at the site of the old Kamiokande detector to observe neutrinos from a number of Japanese reactors that are at distances of 80 – 350 km. KamLAND expects to achieve a sensitivity to Δm^2 down to 10^{-5} eV^2. This would allow one to probe into the Large Mixing Angle solution of the solar neutrino problem.

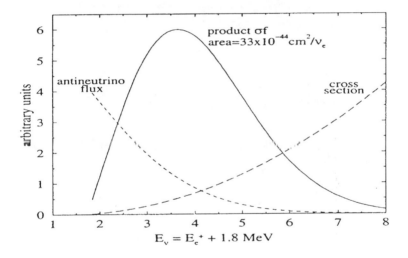

Figure 9. Typical electron antineutrino energy spectrum emitted from a nuclear reactor. Also shown is the cross section for inverse beta decay on the proton and the resultant product of the energy spectrum and cross section.

A number of short baseline (few tens of meters) reactor oscillation experiments have been carried out. While at times evidence for neutrino oscillations has been claimed, invariably as more sensitive experiments have been carried out the evidence for neutrino oscillations has disappeared [37].

Two medium-baseline experiments have been carried out at reactors. One uses the CHOOZ reactor in France and the other uses the Palo Verde reactor in Arizona.

5.2 CHOOZ

This experiment uses the CHOOZ reactor in northern France. It takes advantage of a large underground chamber with 300 meters of water equivalent (mwe) of shielding which exists at a distance of 1.1 and 1.0 km from the two new reactors that have been constructed at the CHOOZ site. This overhead shielding reduces the external cosmic ray flux by a factor of about 300. The two power reactors have a total thermal power of 8.5 GW.

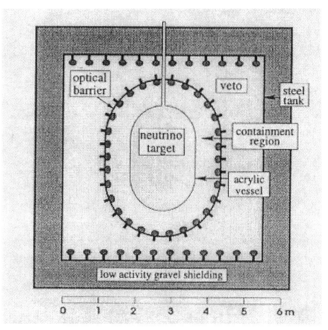

Figure 10. Schematic view of the CHOOZ detector.

The $\bar{v}_e$ from the reactor are detected using the inverse beta decay reaction on the proton:

$$\bar{v}_e + p \rightarrow e^+ + n; \quad E_{e^+} = E\bar{v}_e - 1.804 \text{ MeV}$$

The $\bar{v}_e$ signature is a delayed coincidence between the prompt e^+ signal (together with its two 511 keV annihilation gamma rays) and the signal from neutron capture in Gd-loaded scintillator. CHOOZ is a disappearance experiment, i.e., it compares the measured $\bar{v}_e$ flux at a given distance to the calculated flux at that distance. This provides a search for the oscillations mode $\bar{v}_e \rightarrow v_x$ where v_x can be any type of neutrino, either active or sterile.

A schematic view of the CHOOZ detector is shown in Figure 10. The detector consists of 5 tons of Gd loaded liquid scintillator as the neutrino target that is contained within an acrylic vessel and is surrounded by a containment region with 17 tons of ordinary liquid scintillator, and an outer veto region with 90 tons of scintillator. The inner two regions are viewed by an array of 160 photomultipliers and the outer veto region is viewed by another set of photomultipliers on the bottom of the tank. This coincidence requirement, together with the greatly reduced cosmic ray background, results in a background rate of only 1.2 ± 0.3 events per day in the detector while the signal rate (with both reactors on) was measured to be 25.5 ± 1.0

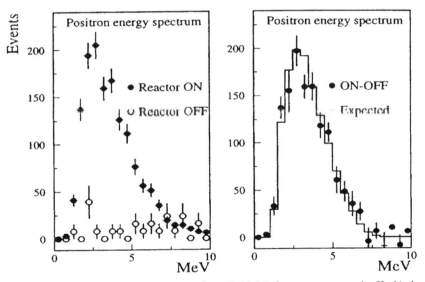

Figure 11. a) Positron energy spectra from CHOOZ for reactor on and off, b) the positron energy spectrum compared to the expected spectrum.
events per day.

CHOOZ was also fortunate in that it was able to measure the background rate before either of the two new reactors came on line. As the reactors were brought on line,

CHOOZ was able to measure the neutrino interaction rate at many different power levels for the reactors, thus allowing an accurate determination of the backgrounds. The positron energy spectra measured by CHOOZ are shown in Figure 11a and Figure 11b that shows the comparison between the data and the expected spectrum. As one can see from the plot of the of the expected and measured rates, the measured rate is in good agreement with that which is expected. Thus, CHOOZ does not observe any evidence for oscillations and is able to exclude a large region of oscillation parameters for the mode $\bar{v}_e \rightarrow v_x$, as is shown in Figure 12 [38].

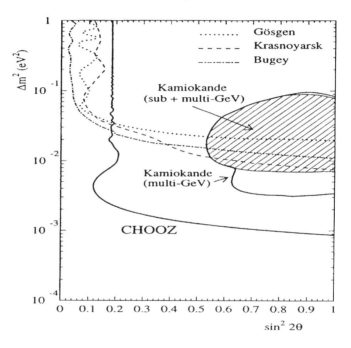

Figure 12. Exclusion plot from the CHOOZ experiment for $\bar{v}_e \rightarrow v_x$.

5.4 The Palo Verde Experiment

The detector is installed in an underground chamber under 32 meters water equivalent (mwe) of shielding at a distance of 890 m from two reactors and 750 m from a third reactor [39]. The total thermal power of the three reactors is 11 GW. The detector, as shown in Figure 13a, has a fiducial volume of 12 tons of Gd-doped liquid scintillator. The inner detector is surrounded by a passive water shield that is in turn surrounded by an active cosmic ray veto. At this shallow depth, the soft hadronic cosmic rays (i.e., cosmic ray neutrons and protons) are ranged out, but the cosmic ray muon flux, which has typical energies of a few GeV, is essentially unattenuated., This experiment observes the $\bar{v}_e + p \rightarrow e^+ + n$ reaction in a liquid scintillator detector in which the neutron is captured on Gd, yielding on the average 4 capture gamma rays with a total energy of about 8 MeV, as is shown schematically in Figure 13b. In order to achieve a signal/background of 1/1, a four-fold coincidence requirement of the prompt e^+ signal, the two annihilation gammas of the e^+, and the neutron capture gamma rays is required. The background can be determined by measuring the rate when one of the three reactors is turned off for refueling and then extrapolating to having all three reactors turned off. The signal to background ratio is measured to be only about 1.2/1, which is reflection of the very high cosmic muon background at this shallow depth. The detector started operation in November 1998 and searches for the same type of neutrino oscillation as the

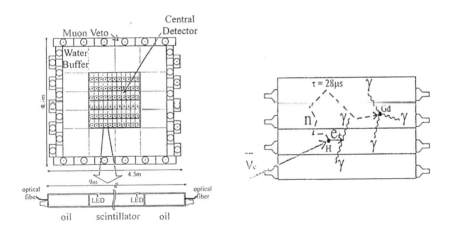

Figure 13. a) Schematic view of the Palo Verde detector, b) Illustration of the neutrino reaction and detected interactions in the Palo Verde detector.

CHOOZ experiment. It has not detected any evidence for neutrino oscillations and has achieved a sensitivity that is roughly comparable to the CHOOZ result, as is shown in Figure 14 [40].

374

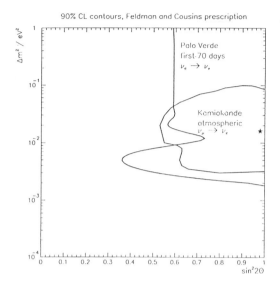

Figure 14. The exclusion plot from the Palo Verde experiment.

5.5 KamLAND

KamLAND is a long baseline oscillation experiment that searches for the oscillation mode $\bar{\nu}_e \to \nu_x$ with a sensitivity down to $\Delta m^2 = 10^{-5}$ eV^2 for $\sin^2 2\theta \geq 0.1$ with reactor antineutrinos. With this degree of sensitivity KamLAND will be able to carry out a terrestrial test for the Large Mixing Angle (LMA) solution to the solar neutrino problem. Additional goals of KamLAND include measurements of geophysical neutrinos from U and Th decay in the Earth, observation of neutrinos from a supernova, and possibly double beta decay of ^{136}Xe and observations of the ^{7}Be solar neutrinos [41].

Japan derives a significant fraction of its total electrical power from nuclear reactors and KamLAND will be able to observe reactions via the inverse beta decay on protons from 9 reactors at distances from 80 km to 350 km from Kamioka. KamLAND is sufficiently large and has sufficiently low backgrounds that it will observe about 450 events a year due to reactor antineutrinos within a 600 ton fiducial volume. KamLAND will provide measurements of the absolute flux of reactor antineutrinos, will search for spectral distortion (due to oscillations) in the antineutrino spectrum, and will also use the temporal variations in the antineutrino flux (as different of the reactors are shut off for refueling) as a means of varying the average distance from the reactor sources and thus to carry out a model-independent search for neutrino oscillations.

KamLAND is being constructed in the cavity at the Kamioka Mine where the Kamiokande III detector was previously situated. The detector consists of an inner liquid scintillator volume of 1200 m^3 contained within a plastic balloon. Surrounding the balloon is 3000 m^3 of nonscintillating mineral oil that is contained within a 18 m diameter stainless steel support shell. Mounted on the shell are 1280 17" photomultipliers that view the 1000 tons of scintillator. Surrounding the stainless steel shell is an active water shield within the 20 m diameter cavity. This veto is used to tag cosmic ray muons which are a serious source of background at the relatively shallow depth of 2700 mwe at the Kamioka site. KamLAND is expected to become operational in early 2001.

5.6 Accelerator experiments

Medium and high energy accelerators can serve as intense sources of neutrinos due to their production of pions and muons. Pions are produced through interactions with the accelerator beam (typically a proton beam) and the pions subsequently beta decay to muons which subsequently decay to electrons. During the beta decay, both muon and electron neutrinos are produced according to the decay scheme shown in the insert in Figure 15a. Thus, each pion produces a muon neutrino, a muon antineutrino, and an electron neutrino. Similarly, negative pions can decay producing a muon neutrino, a muon antineutrino, and an electron antineutrino. If the pions decay in flight (DIF), the energies of the neutrinos depends on the incident beam energy, but are typically 100 – 200 MeV for medium energy accelerators and several GeV to tens of GeV for high energy accelerators. Alternately, the pions can come to rest in a beam stop and then decay at rest (DAR). In that case, essentially all of the negative pions undergo capture by a nucleus before they decay and thus only π^+ produce neutrinos. The spectrum from a beam stop source is shown in Figure 15a. The different sources for neutrino production from both DIF and DAR, together with the relative amplitudes for each neutrino source, for the LSND experiment are shown in Figure 15b. It is the fact that the flux of $\overline{v}_e$ in DAR is only about 7 x 10^{-4} of the v_e flux that allows one to search for oscillations in the $\overline{v}_\mu \rightarrow \overline{v}_e$ channel. It is the residual flux of $\overline{v}_e$ that ultimately limits the sensitivity to small values of $sin^2 2\theta$ in the DAR oscillation search. Neutrino oscillation experiments can be carried out using large detectors at a distance of tens of meters for medium energy accelerators (short baseline oscillation experiments) to hundreds of kilometers for high energy accelerators (long baseline experiments). Two short-baseline accelerator experiments have reported results – one (LSND) that claims evidence for neutrino oscillations and one (KARMEN) that is in some disagreement with the LSND result.

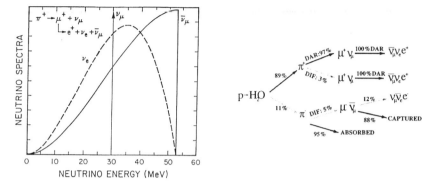

Figure 15a. a) Neutrino spectrum from π^+ decay at rest, b) neutrino production chain in LSND.

5.7 LSND

The Liquid Scintillation Neutrino Detector (LSND) [42] is a short-baseline neutrino oscillation experiment that observes both pion decay at rest and pion decay in flight at the Los Alamos Meson Production Facility (LAMPF). The 1 mA, 800 MeV proton beam from the LAMPF accelerator strikes a beam stop at the end of the nuclear physics experimental area, providing a copious source of neutrinos. While most of the pions are produced in the beam stop and decay at rest, a few percent of the pions are produced in targets upstream from the beam stop. Some of those pions decay in flight with energies of up to 250 MeV. The LSND detector consists of a cylindrical tank 8.3 m long by 5.7 m in diameter located at a distance of 30 m from the LAMPF beam stop. The detector is located under 2 kg/cm^2 of overhead shielding that serves to strongly attenuate the flux of cosmic ray neutrons and protons. The flux of high-energy neutrons from the LAMPF beam stop are reduced to extremely low levels by several meters of Fe shielding between the beam dump and the detector. The tank is filled with 167 tons of liquid scintillator that has a low concentration of scintillator mixed into mineral oil. This allows the detection of both scintillation light and Cherenkov light. The isotropic scintillation light allows an energy measurement of the neutrino interactions with a resolution of 43% /$\sqrt{E(MeV)}$.

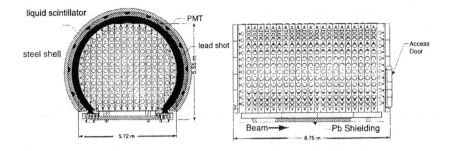

Figure 16. View of the LSND detector.

The Cherenkov light provides a cone of light that allows one to determine the position and direction of the charged particles produced in the neutrino interactions. The light produced is detected in 1220 8-inch photomultipliers. The detector is surrounded on the tops and sides by a liquid scintillator cosmic ray veto shield. The detector is shown in Figure 16. The experiment is designed to search for $\bar{\nu}_\mu \to \bar{\nu}_e$ oscillations from pion decay at rest and $\nu_\mu \to \nu_e$ from pion decay in flight. The signature for a $\bar{\nu}_e$ interaction is $\bar{\nu}_e + p \to n + e^+$ followed by the $np \to d\gamma$ capture reaction that produces a 2.2 MeV gamma ray. In order to separate signal from backgrounds, a maximum likelihood fitter is employed that takes into account the number of phototubes that detect a photon, the reconstructed distance between the positron and the capture gamma, and the relative time between the positron and the capture gamma. Any event in which the veto shield observes any activity is eliminated. It is necessary to subtract neutral cosmic ray backgrounds (typically events in which a cosmic ray neutron enters the detector without being vetoed) that can mimic the signature for a neutrino interaction. Since the LAMPF beam is pulsed, it is on only 6% of the time, and one can take data both during the time that beam is on and beam is off. One can then subtract a normalized background spectrum of the beam-off data from the beam-on data. The experiment has taken data from 1993 to 1997. After applying the maximum likelihood fitter, LSND observes a total of 61 beam-on events as possible candidates for neutrino interactions, of which 15.6 are expected to be cosmic ray background events (based on the beam-off data) and 11.5 are expected to arise from other neutrino-induced events other than the $\bar{\nu}_e + p \to n + e^+$ reaction. This leaves an excess of events over that expected in the case of no oscillations of 33.9 ± 8.0 events. A number of systematic checks of the experiment have been carried out and the data seems to be consistent with the hypothesis of neutrino oscillations. The oscillation probability based on this number of events is $(0.31 \pm 0.12 \pm 0.05)$ % [43]. The energy

378

distribution of the excess events is shown in Figure 17 that also shows the spectrum expected for different values of the oscillation parameters. The data set has been fit with different values of Δm^2 and $\sin^2 2\theta$ in order to map out the allowed parameter space for oscillations. More recently, LSND has analyzed the data from pion decay in flight in which the v_e is detected through the v_e C $\rightarrow$ e⁻ X reaction. The backgrounds in this reaction are much higher since there is no coincidence required. Thus, the sensitivity is not as good for the decay in flight measurement as in the decay at rest. The allowed regions of parameter space for both the decay at rest and the decay in flight experiments are shown in Figure 18 [44].

One possible concern about the LSND data is that there is some difference in the data depending on the type of beam stop used to produce the pions. In the initial data set the beam stop was water-cooled copper while in the second data set the beam stop was water-cooled tungsten. The copper beam stop has a high pion production rate and a fairly low neutron production rate. In contrast, the tungsten beam stop was used as a neutron spallation target and produces more than 15 than neutrons for every incident 800 MeV proton and the pion yield is down roughly a factor of two from that in the copper beam stop. The initial LSND data showed an excess of events for energies greater 36 MeV and no excess for energies between 20 and 36 MeV. The second LSND data set showed no excess for energies greater than 36 MeV and an excess for energies between 20 and 36 MeV. While both data sets can be fit with an oscillation probability of about 0.3%, the allowed regions of Δm^2 –

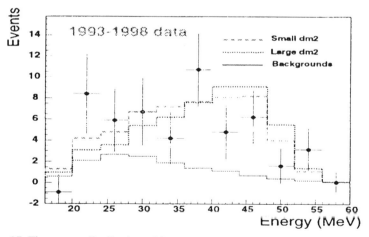

Figure 17. The energy distribution of the excess events from LSND.

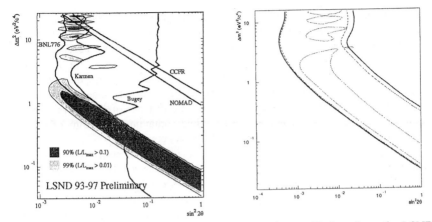

Figure 18. a) Plot of the allowed regions for neutrino oscillations from the LSND decay at rest data at the 90% and 99% confidence levels. Also shown are the 90% CL limits from the KARMEN, BNL E776, and Bugey reactor experiments. b) Plot of the allowed regions for neutrino oscillations from the LSND decay in flight data (solid curve). The dotted line shows the 90% CL limits from the LSND decay at rest data.

$\sin^2 2\theta$ parameter space are somewhat different for the two data sets. This may be just a statistical fluctuation at the 2 σ level, or it may indicate a systematic problem in understanding backgrounds. It is clearly important to verify the LSND claim for observation of neutrino oscillations. Thus, a new collaboration has been formed to carry out another oscillation experiment miniBooNE) at Fermilab using a larger version of the LSND detector [45]. If the LSND result is correct, this new experiment should observe about 5000 excess events in two years of data acquisition.

5.8 KARMEN

Another short-baseline accelerator experiment has taken data at the ISIS accelerator in England. The 200 μA, 800 MeV proton beam from the ISIS accelerator is stopped in a Ta-D$_2$O neutron spallation target. The pions produced decay at rest in the beam stop, producing 6.4 x 10^{13} neutrinos/second of ν_μ, $\bar{\nu}_\mu$, and ν_e. The Karlsruhe Rutherford Medium Energy Neutrino (KARMEN) experiment observes neutrinos produced from pion decay at rest via the $\bar{\nu}_e\, p \to e^+\, n$ reaction to search for the oscillation of $\bar{\nu}_\mu \to \bar{\nu}_e$, $\nu_\mu \to \nu_e$, and $\nu_e \to \nu_x$. The detector consists of 56 tons of liquid scintillator that is designed in a cellular form with air gaps between the cells.

380

Thus, light produced in a single cell undergoes total internal reflection at the walls of the cell and is detected by two photomultipliers at each end of the cell. In the air gaps between the cells there is Gd_2O_3 coated paper that acts as an absorber for the neutron produced in the inverse beta decay. A schematic view of the KARMEN detector is shown in Figure 19. The neutron capture gamma rays thus provide a delayed coincidence signal (similar to LSND) for the inverse beta decay reaction on the proton. The KARMEN detector is located at a distance of 18 meters from the beam stop and is housed inside a passive shield containing 7000 tons of steel in combination with two layers of active scintillator veto counters.

Another very important aspect of KARMEN is the timing structure of the proton beam. The accelerator produces two proton pulses that are 100 ns wide and are spaced apart by 225 ns. This set of double pulses is repeated at a rate of 50 Hz. Thus, the duty factor of the accelerator is dominated by the muon lifetime and beam-associated neutrino interactions must occur within a few microseconds following the beam pulse. This provides an effective beam-on duty factor of only 5 x 10^{-4} (compared to 6% for LSND). This very low duty factor provides a very strong suppression of cosmic-ray induced backgrounds and also allows discrimination against target-related fast neutron backgrounds by using time-of-flight cuts. Thus,

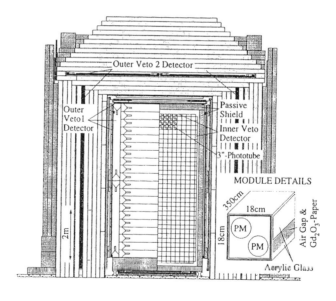

Figure 19. The KARMEN detector inside the 7000-ton steel shield. The insert shows one of the scintillator modules.

the backgrounds expected in KARMEN are much lower than in LSND. The dominant background in KARMEN was due to cosmic ray muons that did not pass through the active scintillator veto counters but produced energetic neutrons in the steel shielding around the detector that subsequently entered the detector and produced a delayed coincidence in the detector. This background was sufficiently large that KARMEN was initially not able to reach sufficient sensitivity to test the LSND claims.

In 1996 the KARMEN detector was upgraded by the addition of a third anti counter system with a total area of 300 m^2 that was installed within the 3-m thick steel roof and the 2- to 3-m thick walls of the steel shielding. This reduced the muon-induced fast neutron background by a factor of 50. A second data set (KARMEN2) was taken with the additional veto system in place and does not show any excess events above the expected background rate. This allows KARMEN to set limits [46] on $\bar{\nu}_\mu \to \bar{\nu}_e$ and $\nu_e \to \nu_x$ oscillations. Figure 20 shows the excluded regions of $\Delta m^2 - \sin^2 2\theta$ parameter space from the KARMEN2 $\bar{\nu}_\mu \to \bar{\nu}_e$ data [47]. There are two excluded regions plotted: one is the 90% CL limit determined by a maximum likelihood fit and the second is the sensitivity of the experiment, i.e., the

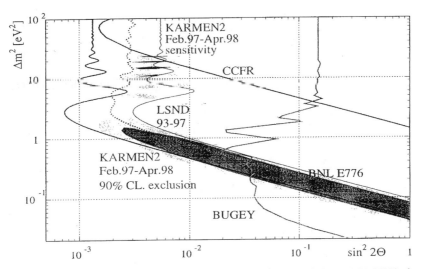

Figure 20. The 90% CL exclusion limit and sensitivity of the KARMEN2 data compared to other experiments.

most stringent upper limit on $\sin^2 2\theta$ assuming that there are no neutrino oscillations in the sensitive Δm^2 range of KARMEN. The use of sensitivity to characterize the result of a null experiment has a well-defined meaning even when backgrounds fluctuate to mimic a negative counting rate for the expected signal. The mathematical definition of sensitivity has been worked out [48] for a wide range of cases and the use of sensitivity to characterize experimental results has now been adopted by the Particle Data Group as the preferred analysis method.

Thus, while KARMEN cannot totally exclude all of the range allowed by the LSND results, the KARMEN results certainly do not provide confirmation of the LSND results. It is clear that a new experiment with greatly improved sensitivity is required to either confirm or reject the LSND claims. As mentioned earlier, that is the goal of the miniBooNE experiment now under construction at Fermilab.

5.9 High Energy Oscillation Experiments

Two high energy experiments, CHORUS and NOMAD, have been carried out to search for the oscillation mode $v_\mu \rightarrow v_\tau$. The muon neutrino beam is a wide band neutrino beam produced by 450 GeV protons from the CERN proton synchrotron striking a target in which pions are produced. The pions are focused by a pulsed magnet so that either π^+ or π^- can be preferentially selected to be focused into a long decay channel. The pions decay in flight producing muon neutrinos with an average energy of 23.5 GeV. There is also a low flux of $\bar{v}_\mu$, v_e, and $\bar{v}_e$ and a negligibly small flux of v_τ. Thus, it is possible to search for $v_\mu \rightarrow v_\tau$ with high sensitivity in $\sin^2 2\theta$.

The CHORUS and NOMAD detectors are located at a distance of 630 m from the neutrino source. Both detectors are designed to observe v_τ (which actually has never been observed yet) by identifying the tau that is produced in the neutrino interaction in the detector. The tau is very short lived and decays into one of several channels. All of these decays have a signature of a very short track (a few hundred microns) of the tau followed by a kink in which other particles are produced when the tau decays. The two detectors are designed specifically to be able to measure the track of the tau before it decays as well as the tracks of the decay particles. However, the two experiments approach the detection of the tau in quite different ways.

CHORUS uses a 770 kg stack of nuclear emulsion plates in which tracks from charged particles can be visually identified after the plates are developed. Nuclear emulsions have extremely good spatial resolution ($\approx 1\ \mu$) and thus can be used to identify the kink in the τ decay. In order to reduce the amount of the emulsion that must be scanned, the vertex at which the tau decays in the emulsion is identified by tracking the decay products in an array of 150 μ thick scintillating fibers. This array is followed by an air-core hexagonal magnet, high resolution electromagnetic and hadron calorimeters and muon spectrometer, as shown in Figure 21. The vertex position in the emulsion, as determined by the tracking spectrometer, is then scanned automatically to search for a kink. CHORUS took data for about 150 days each year from May 1994 to November 1997. Only about 20% of the data has been analyzed so far and no evidence for $v_\mu \to v_\tau$ oscillations has been found [49].

NOMAD uses an active target with a fiducial mass of 2.7 tons made of 44 drift chambers. The target region is followed by a transition radiation detector that provides electron identification that is followed by a preshower and an electromagnetic calorimeter. The entire tracking part of the detector is housed inside a 0.4 T dipole magnet, as is shown in Figure 22. NOMAD uses a kinematical method to identify tau candidates. Using the tracking and momentum information, NOMAD is able to reconstruct the vertex and momenta of the all of the particles in an interaction. In the case of a neutrino-induced interaction, the missing transverse momentum is associated with the neutrino. The challenge in NOMAD is to reduce badly reconstructed tracks (due to finite position and momentum resolution) that mimic neutrino interactions to a very low level. NOMAD took data during the period of 1995 – 1997 and did not find any evidence for $v_\mu \to v_\tau$ oscillations [50]. The exclusion plot for both CHORUS and NOMAD is shown in Figure 23 [51].

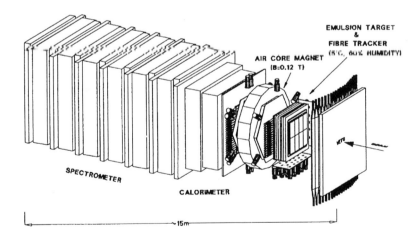

Figure 21. Schematic view of the CHORUS detector.

384

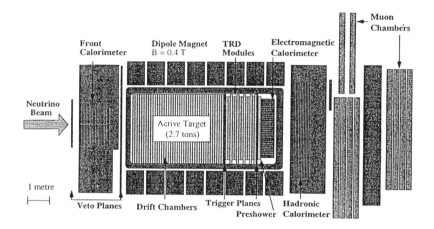

Figure 22. Schematic view of the NOMAD detector.

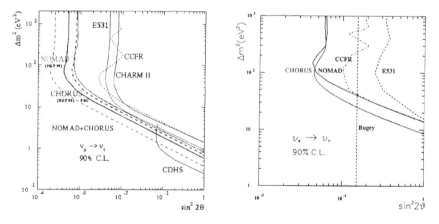

Figure 23. Exclusion region (90% CL) for $\nu_\mu \to \nu_\tau$ oscillations from CHORUS and NOMAD. Also shown are limits from previous experiments.

6. Atmospheric Neutrinos

Cosmic rays (primarily protons and gammas) striking the upper atmosphere produce showers of pions and muons that decay, yielding v_e and v_μ with typical energies close to 1 GeV. The flux of these neutrinos can be calculated to about 20% accuracy and is sufficiently intense that they can be observed in large underground detectors. The flux of atmospheric neutrinos also depends on the location of the detector as the Earth's magnetic field affects the distribution of incoming charged cosmic rays that strike the atmosphere. Both the shape and the amplitude of the atmospheric neutrinos also depends on solar activity and thus varies with the 11-year solar cycle, as is shown in Figure 24 [52]. The spectrum of the atmospheric neutrinos, as measured in the SuperKamiokande detector, is shown in Figure 25 [53].

By measuring the ratio of electron- to muon-neutrinos, one can search for neutrino oscillations in a manner that is relatively free of the individual flux uncertainties, as the ratio can be calculated to about 5% accuracy. As the distance traveled by the neutrinos ranges from 10 to 10,000 km, the oscillation parameter L(km) / E(GeV) ranges from 1 to 10^5, providing a sensitivity to values of Δm^2 down into the 10^{-4} eV^2 range.

Measurements of atmospheric neutrinos have been carried out by a number of large underground detectors: Kamiokande, Superkamiokande, and IMB (which are large water Cherenkov detectors); Frejus, NUSEX, Baksan, MACRO, and Soudan II (which are either scintillator based detectors or tracking calorimeters). In their analyses, they compare the observed ratio of v_μ / v_e divided by the Monte Carlo (MC) calculated ratio of v_μ / v_e. The results as of 1996 are given in Table 3. The most sensitive of the detectors (Kamiokande and IMB) observe a significant deficit of the relative number of v_μ compared to v_e. A possible explanation of this deficit may be attributed to neutrino oscillations. IMB does not make any claim to observe neutrino oscillations, due to their much larger systematic uncertainties.

Table 3. Atmospheric neutrino data as of 1996.

Group	v_μ/v_e(Data) / v_μ/v_e(MC)
Kamiokande [54]	0.60 +0.07/-0.06 ± 0.05
IMB-3 [55]	0.54 ± 0.05 ± 0.12
	1.01 ± 0.03 ± 0.11
Frejus [56]	1.06 ± 0.18 ± 0.15
	0.87 ± 0.16 ± 0.08
NUSEX [57]	0.99 +0.35/-0.25 ± ?

However, using the ratio of ratios could be susceptible to systematic effects rather than new physics. Two major concerns were raised. First, it is quite difficult to separate electrons from muons at low energies (a few hundred MeV) based on the differences in the observed Cherenkov rings. However, IMB and Kamiokande have checked this by building a 1-kiloton water Cherenkov detector at KEK by using beams of charged particles (π, μ, and e) to check the accuracy of the identification.

Second, there is possible concern that the cross sections used in the Monte Carlo calculations may be incorrect. These cross sections are calculated using the Fermi Gas Model (FGM), as the momentum transfer is low at these energies, and nuclear effects are important. However, data recently published of measurements made at LAMPF of ν_μ cross sections on ^{12}C are at some variance with the FGM predictions [58]. Preliminary theoretical work apparently does not find any large differences with improved nuclear models. One way to check this concern is by using one detector at a close (1 km) distance to the neutrino beam at KEK and another detector at a long distance (several hundred kilometers). Then comparing the rates in the close and far detectors allows one to make a model-independent test for neutrino oscillations in a long baseline experiment.

6.1 SuperKamiokande

The situation changed substantially in 1998 with results from SuperKamiokande, Soudan II, and MACRO. In particular, the high data rates at SuperKamiokande have allowed them to carry out a very sensitive search for neutrino oscillations from atmospheric neutrinos.

SuperKamiokande is an improved and enlarged version of the Kamiokande detector [59]. It is located at the Kamiokande site at a depth of 2700 mwe. It consists of a detector 39.3-m diameter and 41.4-m high with a 2-m-thick water shield that also serves as an active anti-counter to veto cosmic-ray muons. The inner detector contains 50 kilotons of H_2O with a 22-kton fiducial volume viewed by 11,146 50-cm diameter PMTs. There is an outer (veto) detector that is optically separated from the inner detector by a pair of opaque sheets and the outer detector is viewed by 1885 20-cm diameter PMTs. The PMTs are fitted with reflectors to increase the effective photocathode coverage to about 40%. SuperKamiokande began operation in April, 1996. A schematic view of the SuperKamiokande detector is shown in Figure 24.

SuperKamiokande is able to search for atmospheric neutrino oscillations in a number of ways. The first is by measuring the ratio R of ν_μ/ν_e(Data)$/\nu_\mu/\nu_e$(MC), which SuperKamiokande has measured this ratio for different types of events in the detector. SuperKamiokande defines fully contained (FC) events as those in which the interaction vertex occurs within the fiducial volume of the detector and all of the energy is deposited within the inner detector, partially contained (PC) events are those in which the interaction vertex occurs within the fiducial volume of the detector but some of the energy is deposited in the outer detector, upward-going muons are those events in which the neutrino interaction occurred in the rock below the detector and the muon enters the detector from outside the detector (SupeKamiokande cannot observe down-going muons produced by neutrino interactions in the rock around the detector since these events are overwhelmed by the high rate of down-going cosmic ray muons). SuperKamiokande also breaks up the data set into sub-GeV and multi-GeV since the fraction of charged-current (CC) events (those events mediated by the exchange of a charged intermediate vector boson W^+ or W^-) depends on the energy of the interaction, with 88 (96)% of the sub-

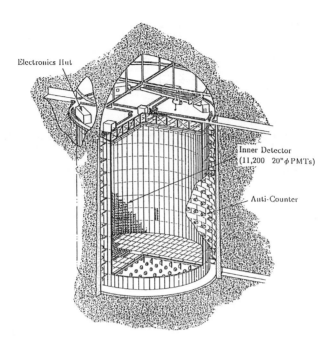

Figure 24. Schematic view of the SuperKamiokande detector.

GeV e-like (μ-like) events are ν_e (ν_μ) CC interactions and 84% (99%) of the multi-GeV e-like (μ-like) are ν_e (ν_μ) CC interactions. The results of the measurements by SuperKamiokande are shown in Table 4 together with the Monte Carlo predictions. Combining the results of the FC and PC data providing evidence for neutrino oscillations at the 7 σ level [60, 62]. This confirms the earlier Kamiokande measurement and provides substantially reduced statistical and systematic uncertainties. However, while the ratio is relatively robust measurement, any evidence for neutrino oscillations from a measurement of R is based on the assumption that the model for the production of atmospheric neutrinos is correct. This assumption has been called into question and measurements of the high-altitude primary muon flux indicate that there may be some deviations from the model predictions [61].

An essentially model-independent test for neutrino oscillations can be carried out by measuring the azimuthal dependence. Neutrino are produced within a fairly limited region in the upper atmosphere (of order 20 km). One can vary the distance at which one observes atmospheric neutrinos by using the azimuthal information from the interaction in the detector. Thus, if one looks at neutrinos produced in the atmosphere directly above the detector, the average distance from the production of the neutrino in the atmosphere to the detector is about 15 km. Alternately, if one looks at neutrinos coming from directly below the detector, those neutrinos are

388

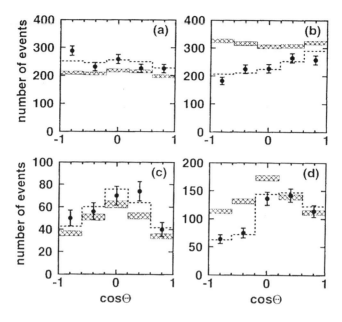

Figure 25. The azimuthal dependence of the SuperKamiokande data.

produced on the opposite side of the globe at a distance of about 13,000 km. Thus, one can vary the distance L from 15 to 13,000 km and see how the oscillation probability changes by measuring the event rate in the detector as a function of azimuthal angle. One needs to know how the cosmic ray flux varies with azimuth, but there is considerably less uncertainty in that in the absolute neutrino production rate. SuperKamiokande has made this measurement and finds that there is a deviation from that predicted for no oscillations, as is shown in Figure 25 [60, 62].

Table 4. Summary of the up-down ratios for SuperKamiokande and Kamiokande.

		SuperKamiokande	Kamiokande
	Monte Carlo	Data	Data
e-like			
Sub-GeV(< 400 MeV/c)	1.00±0.04±0.03	1.20+0.11/-0.10±0.03	1.29+0.27/-0.22
Sub-GeV(>400 MeV/c)	1.02±0.04±0.03	1.10+0.11/-0.10±0.03	0.76 +0.22/-0.18
Multi-GeV	1.01±0.06±0.03	0.93+0.13/-0.12±0.02	1.38 +0.39/-0.30
μ-like			
Sub-GeV(<400 MeV/c)	1.05±0.03±0.02	1.03+0.11/-0.10±0.02	1.18 +0.31/-0.24
Sub-GeV(>400 MeV/c)	1.00±0.03±0.02	0.65+0.10/-0.05±0.01	1.09 +0.22/-0.18
Multi-GeV(FC + PC)	0.98±0.03±0.02	0.54+0.06/-0.05±0.01	0.58 +0.13/-0.11

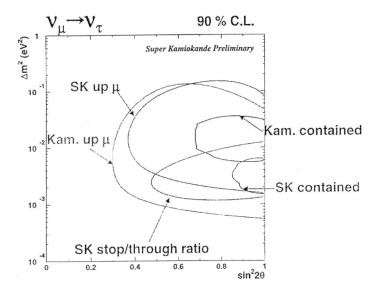

Figure 26. The allowed regions from the different SuperKamiokande data sets.

One can characterize the azimuthal dependence by defining an up-down ratio U/D in which U is the number of upward-going events ($-1 < \cos\theta < -0.2$) and D is the number of downward-going events ($0.2 < \cos\theta < 1$). Table 4 shows the summary of the U/D ratio for both SuperKamiokande and Kamiokande. One sees that there is a very significant deficit ($> 5\,\sigma$) of u-like events above 400 MeV/c.

SuperKamiokande has fit the data to the hypothesis of neutrino oscillations. Since all of the e-like data is consistent with expectations, oscillations of ν_e is not supported by the data. In addition, the CHOOZ and Palo Verde results rule out oscillations of $\nu_\mu \to \nu_e$ in the region of $\Delta m^2 - \sin^2 2\theta$ parameter space that SuperKamiokande is sensitive to. Thus, the evidence for ν_μ oscillations must be either $\nu_\mu \to \nu_\tau$ or $\nu_\mu \to \nu_s$ (where ν_s is a sterile neutrino). The χ^2/DOF fit to the data is 62/67 for $\nu_\mu \to \nu_\tau$ oscillations and 64/67 for $\nu_\mu \to \nu_s$ oscillations. The range of parameter space that is allowed by the different SuperKamiokande measurements for the oscillation mode is shown in Figure 26 [62].

Two other experiments have reported results that confirm the SuperK claim of the observation of atmospheric neutrino oscillations: MACRO and Soudan II.

6.2 MACRO

MACRO is a very large detector located in Hall B of the Gran Sasso National Laboratory that was constructed to search for magnetic monopoles in addition to neutrino studies. It is 76.6 m x 12 m x 9.3 m, divided longitudinally in six similar supermodules and vertically into two parts, as shown in Figure 27. The active

390

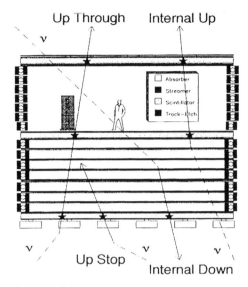

Figure 27. Schematic view of the MACRO detector.

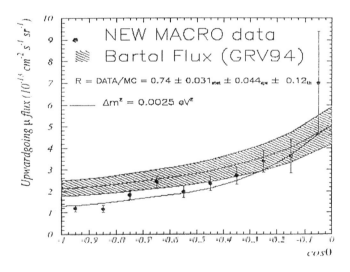

Figure 28. The azimuthal dependence of the MACRO data.

detection elements are planes of streamer tubes for tracking and of liquid scintillation counters for fast timing. The lower half of the detector is filled with trays of crushed rock absorbers alternating with streamer tube planes, while the upper part is open and contains the electronics and work areas. Scintillator planes and streamer tube planes are located on the sides, top, and bottom of the detector as well as at the midplane between the upper and lower halves of the detector. The detector has very good timing characteristics and thus is able to resolve upward-going muons (produced by neutrinos in the rock around the detector) from down-going cosmic ray muons with very high separation. Figure 27 shows the zenith angle distribution of upward going through muons in MACRO along with that predicted by the Bartol model [63]. MACRO observes a significant deficit compared to that expected. For the upward throughgoing muons with an energy greater than 1 GeV, MACRO finds the ratio of expected to Monte Carlo rate to be R = 0.74 ± 0.031 ± 0.044 ± 0.12 where the last number is that due to the uncertainty in the theoretical calculation of the predicted muon flux [64].

6.3 Soudan II

The Soudan II experiment is located at the Soudan Mine Underground State Park in northern Minnesota at a depth of 2100 mwe. The detector is a 963 ton fine-grained gas tracking calorimeter. It consists of 224 1 m x 2.7 m iron modules weighing 3.4 tons each. Ionization deposited in the plastic tubes of a module drifts in an electric field to the faces of the module where it is detected by vertical anode wires and horizontal cathode strips. The third coordinate is determined by the drift time in the module. The calorimeter operates in the proportional mode and pulse height measurements are used for particle identification [65].

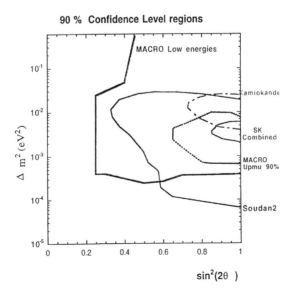

Figure 29. Allowed region for a combined analysis of atmospheric neutrino results.

The Soudan II experiment also observes a significant deficit of u-like events in the detector from atmospheric neutrinos with a ratio of R = 0.66 ± 0.11 ± 0.06 based on 4.2 kty (kiloton years) of data [66]. This result is based on a data set of 144 e-like and 116 μ-like events. With additional data acquisition the result will become more significant and Soudan II might possibly be able to provide some insight as to the type of oscillation involved. At present, they analyze the data for $v_\mu \rightarrow v_x$ and find that the 90% confidence level (CL) allowed region is $0.68 < \sin^2 2\theta < 1.0$ and $2 \times 10^{-3} < \Delta m^2 < 3 \times 10^{-2}$ eV^2, in agreement with the region allowed by SuperKamiokande.

A combined analysis of all of the atmospheric neutrino data has been carried out [67] and the results for the allowed region of oscillation parameters is shown in Figure 29. The plot shown is for the oscillation mode $v_\mu \rightarrow v_\tau$ but the allowed region for $v_\mu \rightarrow v_s$ oscillations is similar.

6.4 Long Baseline Experiments

It is clearly desirable to confirm the atmospheric neutrino results in a terrestrial experiment. In order to reach into the $\Delta m^2 = 10^{-3}$ eV^2 range, the oscillation probability $P(v_\mu \rightarrow v_x) = \sin^2 2\theta \sin^2\{1.27\Delta m^2(eV^2)L(m)/E_V(MeV)\}$ must have $\Delta m^2(eV^2)L(m)/E_V(MeV) \approx 1$. In a disappearance experiment, the v_μ beam must have enough energy to produce muons in the final state and a beam energy of order 1 GeV is reasonable. If we then set $\Delta m^2 = 10^{-3}$ eV^2 we find that L $\approx 10^3$ km. Two long baseline experiments to check the atmospheric neutrino results (K2K and MINOS) are already under way and others are planned.

6.5 K2K

The K2K experiment (called K2K for KEK To Kamiokande) has been constructed with a monitor detector at 1 km from the neutrino beam at KEK. The near detector is comprised of a tracking calorimeter and a small water Cherenkov detector with characteristics similar to SuperKamiokande and the far detector is the SuperKamiokande detector at a distance of 250 km from KEK. The near detector will be able to characterize the neutrino beam as well as providing a normalization of the muon neutrino event rate for the SuperKamiokande detector [68].

The muon neutrino beam is produced using the 12 GeV proton synchrotron at KEK. Pions are produced by a two interaction length aluminum target and double horn system that focuses pions with a pulsed current of 250 kA. The pions decay in a 200 m-long decay pipe and produce a neutrino beam that peaks between 1 and 2 GeV and extends up to 5 GeV. The beam also contains a contamination of about 10% $\bar{v}_\mu$ and v_e and about 1% $\bar{v}_e$. Running of K2K began in early 1999 and will require 2-3 years of data in order to reach a sensitivity of the best value for the atmospheric neutrinos oscillation parameters determined by SuperKamiokande. K2K will have a sensitivity for $v_\mu \rightarrow v_x$ oscillations for $\Delta m^2 \geq 3 \times 10^{-3}$ eV^2. This only covers about half of the allowed range for oscillations from the

SuperKamiokande results. Thus, in order to fully test the atmospheric neutrino results one must carry out more sensitive experiments.

6.6 MINOS

The MINOS experiment [69] uses the 120 GeV proton beam from the Main Injector at Fermilab on a small diameter target segmented target. The neutrino beam is sign selected using a parabolic focusing horn that can be run in several different configurations. This allows one to change the average energy of the beam from one that peaks at about 3 GeV to one that peaks at about 15 GeV. This provides the ability to optimize the neutrino beam properties for a specific range of oscillation parameters.

The MINOS experiment uses a near detector with a mass of about 1 ton that is segmented with layers of steel between planes of 4-cm wide scintillator strips. The near detector consists of four functional parts: a veto system to ensure that there are no interactions induced by neutrons, a 1-m thick steel target, a 1.5-m thick hadron catcher that contains the showers induced by neutrino interactions in the target, and a muon spectrometer. The near detector employs a magnetic field that allows a determination of the muon momentum.

The far detector, located 732 km away from Fermilab at the Soudan site, is a 5400 ton magnetized iron calorimeter with scintillator strips as active elements. There are 242 modules composed of 8-m diameter, 2.54-cm thick steel planes followed by 1 cm thick scintillator planes. The average magnetic field in the detector is 1.5 Tesla.

Using the low-energy neutrino beam, MINOS expects to reach a sensitivity for $v_\mu \to v_x$ oscillations $\Delta m^2 \geq 6 \times 10^{-4}$ eV^2 (90% CL). Thus, MINOS should be able to cover (at 90% CL) all of the region of oscillation parameters allowed by the SuperKamiokande data. MINOS is now under construction and plans to start taking data in 2003.

A number of other experiments are under consideration as long-baseline experiments between CERN and the Gran Sasso National Laboratory [70]. As the distance from CERN to Gran Sasso is virtually identical to that between Fermilab and Soudan, the experiments at Gran Sasso are likely to have sensitivity comparable to that of MINOS.

7   The Solar Neutrino Problem

7.1 Theory of Neutrino Production in the Sun

The Standard Solar Model (SSM) is based upon the nucleosynthesis reactions that occur in the Sun [71]. The primary energy-producing reactions are the p-p chain in which two protons fuse to form deuterium. The p-p chain extends out through nucleosynthesis chain through the production of ^{8}B, as shown in Figure 30. In addition, there are other reactions that produce neutrinos (the CNO cycle) but it is the p-p reactions that account for most of the energy production in the Sun. It is important to note that the amplitude of any of the fusion reactions is a prediction of the SSM but the shape of any individual component is determined solely by the Q-value and the phase space in nuclear beta decay. Thus, the SSM cannot affect (at least down to the 10^{-5} level [71]) cannot affect the spectral shape of the neutrinos emitted in any single nucleosynthesis reaction.

The solar neutrino spectrum, together with the thresholds for various detectors, is shown in Figure 31. The SSM calculations are made under the assumptions [71] that:

1. The interior of the Sun is in hydrostatic equilibrium.
2. Energy is transported from the interior by radiative transfer except in the outer convective zone.
3. There is spherical symmetry and no significant rotation.
4. Chemical homogeneity exists at formation and is changed only by nuclear reactions.
5. No unknown physics plays any significant role in the Sun.

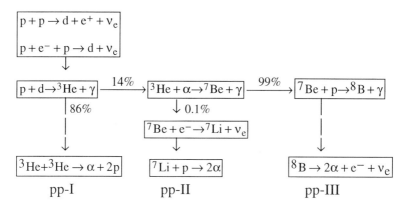

Figure 30. The nucleosynthesis chain reactions that form the basis of the Standard Solar Model.

The SSM calculations use as input the assumed initial elemental abundances, nuclear cross sections, solar luminosity, solar age, equation of state, and calculated radiative opacities. The SSMs then track the Sun as it evolves in time to the present period. A number of checks are applied to ensure self consistency (e.g., that the assumed initial He abundance is consistent with that presently measured).

A number of SSMs that make somewhat different predictions for the fluxes have been published in the last few years [72-78]. The SSM predictions may vary from model to model depending on the input parameters used and different approximations for the physics involved. The range of SSM predictions, as well as the uncertainties in them, becomes very important in trying to resolve the origin of the "solar neutrino problem." I use the predictions of two particular SSMs (Bahcall-Pinsonneault [72] and Turck-Chieze [74]) as being representative of the spread of predicted values for the solar neutrino fluxes and capture rates. The predictions of the capture rates for different experiments in the two representative SSMs are shown in Table 5.

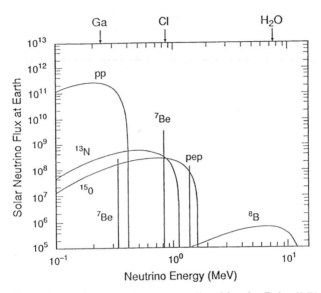

Figure 31. The solar neutrino spectrum as predicted by the Bahcall-Pinsonneault Standard Solar Model. The thresholds for the Ga, Cl, and Kamiokande experiments are shown at the top.

Table 5. Predictions of solar neutrinos fluxes in the Bahcall-Pinsonneault and Turck-Chieze Standard Solar Models.

Flux ($v/cm^2/s$))	Bahcall-Pinsonneault	Turck-Chieze
p-p	$(5.94 \pm 0.59) \times 10^{10}$	5.98×10^{10}
pep	$(1.39 \pm 0.01) \times 10^8$	1.40×10^8
^7Be	$(4.80 \pm 0.43) \times 10^9$	4.70×10^9
^8B	$(5.15 +0.98/-0.72) \times 10^6$	$(4.82 \pm 1.1) \times 10^6$
^{13}N	$(6.05 +1.15/-1.09) \times 10^8$	4.66×10^8
^{15}O	$(5.32 +1.17/-0.80) \times 10^8$	3.99×10^8
^{17}F	$(6.33 +0.76/-0.70) \times 10^6$	
hep	2.1×10^3	

An important point concerning the SSMs is the temperature dependence of the neutrino fluxes. The different stages of nucleosynthesis occur at different temperatures (and therefore at different solar radii). The temperature dependences have been found [71] to depend on the core temperature (T_C) as $\phi(^8B) \propto T_C^{18}$, $\phi(^7Be) \propto T_C^8$, and $\phi(p\text{-}p) \propto T_C^{-1.2}$. These effective temperature dependences are not the same as the temperature dependences of the nuclear reactions involved, but rather come from modeling variations in the production rates with temperature under the boundary conditions that the SSM reproduce the measured physical properties of the Sun (e.g., luminosity).

Finally, it is critical to note that the flux of the p-p neutrinos is directly related to the observed solar luminosity in an essentially model-independent manner and is thus determined to an accuracy of about 1-2% [71]. This is of particular importance to the results of the gallium experiments, as they are primarily sensitive to the p-p neutrinos.

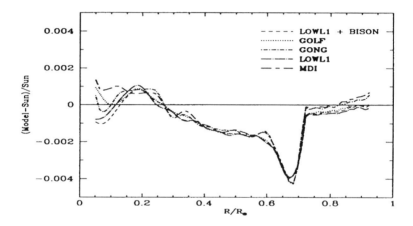

Figure 32. Residuals spectrum showing the difference between different helioseismological measurements and the Bahcall-Pinsonneault Standard Solar Model.

Recent improvements in helioseismology have now provided another probe into the interior of the Sun. The Sun acts as a resonant cavity and as a result, there are solar oscillations in which patches of the surface of the Sun oscillate intermittently with periods of the order of five minutes and velocity amplitudes of order 0.5 km/s [71]. The oscillation modes are of two types: pressure waves (p-modes) are oscillations that are largely trapped between the solar surface and the lower boundary of the convection zone, and gravity waves (g-modes) that penetrate deeply into the solar interior. Some 3500 p-modes have now been identified [79], while the claims for observations of the g-modes are still controversial. From the helioseismology measurements, it is possible to deduce the speed of sound as a function of solar radius, as this determines the propagation of the acoustic modes [80-81]. One is then able to check the SSM inputs of the equation of state, radiative opacities, and microscopic diffusion by comparing the SSM predictions with the helioseismology measurements. The agreement that the Bahcall-Pinsonneault model [72] provides with several different helioseismology measurements can be seen in Figure 32 [82].

While previous predictions disagreed at about the 2% level, the most recent Bahcall-Pinsonneault predictions now agree with the measurements in the region of 0.2 to 1.0 solar radius to within 0.4%. This lends a great deal of support to the validity of the SSM calculations. Unfortunately, the helioseismology measurements are not yet sensitive to the g-mode waves that probe into the solar core, where all of the neutrino producing reactions occur. With continued satellite-based helioseismology experiments now underway and it is hoped that this will soon be possible.

7.2 Introduction to the Solar Neutrino Problem

The original solar neutrino problem is based on the observation that the flux of high-energy solar neutrinos measured by the chlorine experiment of Davis et al. is a factor of about three below that predicted by Standard Solar Model (SSM) calculations. Although many checks of the operation of the experiment have been made, this discrepancy has persisted for more than twenty years and is referred to as the classical solar neutrino problem. The deficit of high-energy ^{8}B solar neutrinos was corroborated by the Kamiokande water Cherenkov experiment. The extreme energy dependence of the ^{8}B flux means that T_C must only be reduced by 4–8% to suppress the ^{8}B flux by a factor of 2–4. However, all Nonstandard Solar Models [71, 83-84] proposed so far that decrease T_C run into problems in reproducing other measured properties of the Sun (such as the solar luminosity). Nonetheless, the extreme temperature dependence of the ^{8}B flux has led many to preclude the possibility of determining the source of the solar neutrino problem from observations of high-energy solar neutrinos. Thus, the need for a measurement of the low-energy p-p solar neutrinos (which occur in the primary energy producing reaction in the Sun) has been clear, as the flux of the p-p neutrinos is directly related to the observed solar luminosity and is insensitive to alterations in the solar models.

Two gallium radiochemical experiments, SAGE (originally the Soviet-American Gallium Experiment and now the Russian-American Gallium Experiment) and GALLEX (the GALLium EXperiment) have reported measurements that are primarily sensitive to the p-p neutrinos. Their results also indicate a significant deficit (by a factor of about two) of the solar neutrino flux.

More recently, the SuperKamiokande experiment has come on line. The much higher statistics from SuperKamiokande has verified the Kamiokande results and has provided new sensitivity for a finite neutrino mass by searching for time variations in the flux of ^{8}B neutrinos and a possible spectral distortion of the ^{8}B neutrino spectrum.

The most likely interpretation of the combined results of the solar neutrino experiments is that the p-p flux is about that predicted by the SSM, there is no observable flux of ^{7}Be neutrinos, and the ^{8}B flux is about half of that predicted by the SSM. Thus, the observation of ^{8}B neutrinos requires ^{7}Be to be present in the Sun as ^{8}B is produced from the ^{7}Be(p,γ)^{8}B reaction. However, the flux of ^{7}Be neutrinos is apparently strongly suppressed. The conclusion is essentially that ^{7}Be is present in the Sun but there are no ^{7}Be neutrinos. This is referred to as the ^{7}Be solar neutrino problem. This observation rules out any astrophysical explanation invoking a cooler Sun is ruled out at the 99.99% CL [85].

All of the experimental evidence to date in favor of a neutrino oscillation solution to the solar neutrino problem is model dependent in that the measured rate in all of the detectors show a deficit compared to the SSM predictions. Model independent experiments can also be carried out. In particular, one can look for possible time variations in the observed rates, distortions of the observed neutrino spectrum, and the appearance of muon or tau neutrinos from the Sun. A new round of solar neutrino experiments now on line offers the promise of definitively resolving the source of the solar neutrino problem by carrying out these model-independent measurements.

7.3 Solar Neutrinos and Neutrino Properties

In addition to Nonstandard Solar Models, numerous explanations of the observed deficit of solar neutrinos have been proffered involving particle physics explanations (neutrino oscillations, heavy neutrinos decaying into lighter neutrinos, Weakly Interacting Massive Particles (WIMPs), etc.) [86-87]. Most of the particle physics explanations have also been ruled out by either laboratory measurements or astrophysical observations. The apparent sole surviving explanation requires that neutrinos can oscillate from one type to another. In this case, the flux of electron neutrinos produced in the Sun may be accurately predicted by the SSMs, but the flux measured by solar neutrino detectors at the Earth may be depleted, as the neutrinos may oscillate into other types of neutrinos (muon or tau) during the transit from the Sun to the Earth and thereby escape detection at the Earth, as the detectors are predominantly sensitive to electron-type neutrinos. Thus, the solar neutrino spectrum that we observe on Earth may be different from the spectrum of electron neutrinos emitted within the Sun. In some extensions of the Electroweak Model, the neutrinos can also oscillate between different helicity states into so-called sterile neutrinos (right-handed neutrinos or left-handed antineutrinos) that do not interact with quarks or leptons in the low-energy limit of the Standard Model (i.e., our present Universe).

It was observed in 1985 [88] that the Sun also provides sensitivities to very small values of the mixing angle θ. This is due to the resonant amplification of the mixing in the high-matter densities of the Sun, which are called (Mikheyev-Smirnov-Wolfenstein) MSW oscillations. The physical process is due to the fact that the electron neutrinos that are produced in the core of the Sun can scatter from electrons as they traverse the Sun by both the charged and neutral weak currents (i.e., by exchange of either a W^- or a Z^0, respectively). However, muon and tau neutrinos can scatter only by the neutral weak current. As a result, as the neutrinos propagate, the index of refraction (or forward-scattering amplitude) depends on neutrino flavor. Thus, the phase of the electron neutrino can evolve differently than the phase of a muon or tau neutrino. Under proper circumstances, the phases of the electron neutrino and muon or tau neutrinos will be such that full mixing can occur between the different flavors. This condition occurs when [36,71] $L_{osc}/L_0 = -\cos 2\theta$ where the L_{osc} is the oscillation length for the neutrinos in vacuum ($L_{osc} = 2\pi\ 2p\_/\Delta m^2$) and L_0 is the oscillation length in matter and is given by:

$$L_0 = 2\pi/(\sqrt{2}G_F N_e) \approx 1.7 \times 10^7/[\rho(g/cm^3)\ Z/A]\ m\ ,$$

where G_F is the Fermi constant, $N_e = \rho N_0 Z/A$, and N_0 is Avagadro's number. Thus, in the Sun, where $\rho \approx 150$ g/cm^3 and $Z/A \approx 2/3$, $L_0 \approx 200$ km, compared to the radius of the Sun of 7×10^5 km.

It is important to note two things about MSW oscillations. First, the probability for matter oscillations is dependent on the neutrino energy and therefore can affect neutrinos from different fusion reactions in the Sun differently, resulting in a distortion of the energy spectrum from that predicted by the SSMs. The effects of MSW distortions on the solar neutrino spectrum have been detailed by numerous authors [89]. Second, one finds that the solar neutrino experiments have sensitivities to mixing angles as small as $\sin^2 2\theta = 10^{-4}$, compared to sensitivities for oscillations in vacuum of $\sin^2 2\theta \geq 0.2$.

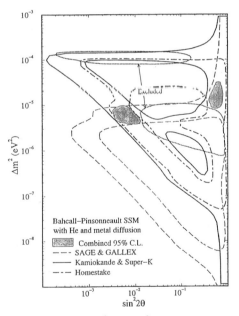

Figure 33. Allowed regions within Δm^2 and $\sin^2 2\theta$ parameter space.

The results of all of the operating solar neutrino experiments are consistent with neutrino oscillations within a fairly restricted range of possible mass and mixing parameters. Each of the experimental results can be described by bands of Δm^2 and $\sin^2 2\theta$, as shown in Figure 33 [90] that shows each of the bands allowed by the

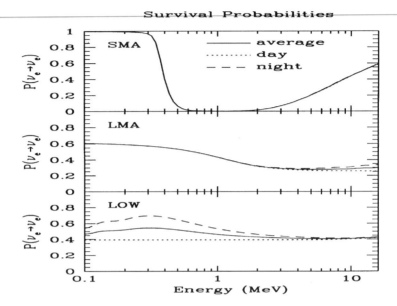

Figure 34. Energy dependence of the oscillation probabilities for different solutions to the solar neutrino problem, including the effect of day-night variations.

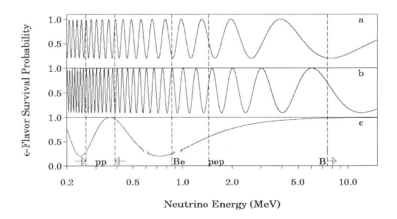

Figure 35. Energy dependence of different vacuum oscillation solutions.

chlorine, gallium, and water Cherenkov results. The bands differ for each experiment due to the different thresholds of the detectors and thus the combination provides complementary information. The overlap of the allowed bands from the different experiments results in four regions of allowed parameters, as is shown in Figure 44. The adiabatic branch of the MSW solution is the horizontal branch at $\Delta m^2 = 10^{-4}$ and $\sin^2 2\theta$ near unity [called the Large Mixing Angle (LMA) solution], while the nonadiabatic branch [called the Small Mixing Angle (SMA) solution] is the one that has a slope extending from $\Delta m^2 = 10^{-4}$, $\sin^2 2\theta = 3 \times 10^{-3}$ to $\Delta m^2 = 7 \times 10^{-8}$, $\sin^2 2\theta = 0.5$. There is also a small region [called the Large Mixing Angle with LOW Δm^2) with $\Delta m^2 \approx 10^{-7}$ and $\sin^2 2\theta$ near unity. Finally, there are several allowed regions called the "just-so" or vacuum oscillations with $\Delta m^2 \approx 10^{-10}$ eV^2 and large values of $\sin^2 2\theta$ ranging from 0.6 to 1.0.

The different oscillation solutions all exhibit different energy dependences [91] and it is this fact that provides an explanation for the observed difference from the SSM model predictions. The energy dependence of the oscillation probability is shown in Figures 34 and 35. Since the oscillation probability depends on L/E, for some of the regions there may be quite sizeable temporal effects due to the eccentricity of the Earth's orbit and due to regeneration effects in the Earth (in which neutrinos that had oscillated from v_e to v_μ or v_τ in the Sun oscillate back into v_e in passing through the Earth due to the MSW effect). In the case of the LMA solution, there is only a gentle energy dependence and no (small) temporal variations for the p-p and ^{7}Be (^{8}B) neutrinos. In the case of the SMA solution, we expect little suppression of the p-p neutrinos, almost complete suppression of the ^{7}Be neutrinos, a 50% suppression of the ^{8}B neutrinos, and no observable temporal variations. For the LOW case, we expect uniformly large suppression for the p-p, ^{7}Be, and ^{8}B neutrinos, and appreciable day-night temporal variations of the p-p and ^{7}Be neutrinos. For the vacuum oscillation case, there is a very large distortion of the SSM spectrum that varies rapidly with small changes in the oscillation parameters and very large and rapid temporal variations are also expected.

7.4 Radiochemical Experiments

Radiochemical experiments involve large passive targets in which solar neutrinos produce a particular daughter atom by inverse beta decay. The daughter atom is then chemically extracted and its subsequent decay is observed in a small electronic detector (typically a miniature proportional counter). The signal and backgrounds in the detector are separated by carrying out an energy and time analysis by fitting the data to an exponential decay of the daughter and a constant background. All of the radiochemical experiments to date also employ some form of additional background rejection in addition to an energy cut of the region of interest. This usually takes the form of measuring the rise time of the signal in the proportional counter. The signals of interest are usually K (and L) X-rays and Auger electrons, all of which have a short range in the proportional counter gas. This results in a fast rise time signal compared to Compton electrons which typically have an extended path length and therefore longer rise time. Some of the experiments employ active vetos consisting of NaI detectors around the proportional counters.

The primary advantage of radiochemical experiments is that they are insensitive to many backgrounds. Only those backgrounds that result in the formation of the daughter atom of interest are of concern. This greatly reduces the radiopurity level that must be achieved in the detectors and thus the first experiment to begin operation (the chlorine experiment) was a radiochemical experiment.

7.5 The Chlorine Experiment

The chlorine radiochemical experiment employs 610 tons of perchlorethylene (C_2Cl_4) housed in a steel tank located 4900 mwe (meters water equivalent) underground at the Homestake Gold Mine in Lead, South Dakota. Solar neutrinos can produce ^{37}A by inverse beta decay reaction $^{37}Cl(v_e,e^-)^{37}A$ on the ^{37}Cl in the tank with a threshold of 814 keV. Thus, the chlorine experiment is predominantly sensitive to the ^8B neutrinos (77% of the rate) while ^7Be neutrinos contribute 15% of the rate, and smaller contributions occur from pep, and CNO solar neutrinos. Originally there was a fairly large (25%) difference between the Bahcall-Pinsonneault [73] and Turck-Chieze [76] SSM predictions for the ^{37}Cl capture rate. However, improvements in the nuclear cross sections, solar opacities, and the inclusion of diffusion in the models has brought them into good agreement with the Bahcall-Pinsonneault SSM predicting 7.7 +1.2/-1.0 SNU [72] and the Turck-Chieze SSM predicting 7 ± 1.35 SNU [74].

The chemical extraction procedure [92] in the chlorine experiment is very simple, as it only requires the extraction and separation of a noble gas (^{37}A) and is shown schematically in Figure 36. The ^{37}A atoms are chemically extracted about every 60 days by sweeping the detector with a flow of helium. After each extraction, a few tenths of a cc of either ^{36}A or ^{38}A are added to the tank. This carrier gas is removed during the next extraction (along with any ^{37}A atoms) and serves as a means to measure the extraction efficiency. The helium and perchlorethylene are circulated through the tank by a set of eductors. The helium is also circulated through a charcoal trap held at LN temperature where the ^{37}A atoms are absorbed. The fraction of the carrier gas removed provides a measure of the extraction efficiency, which is typically 94%. At the end of the extraction cycle, the charcoal trap is heated and the trapped gas is swept out by helium. The extracted gas is purified to remove non-noble gases and by gas chromatography to remove other noble gases. The purified argon is then mixed with methane (to make a good counting gas) and then inserted into a small proportional counter. The proportional counter is placed in the well of a NaI detector inside a large passive shield. The NaI acts to veto events with any associated gamma activity. ^{37}A is detected in the proportional counter as it decays by electron capture with a 35-day half-life. The only way to observe this decay is to detect the low-energy 2.82-keV Auger electrons from K-shell capture produced during electron shell relaxation in the resulting ^{37}Cl atom.

Pulse shape discrimination based on rise-time measurements is used to separate the ^{37}A decays from background. In contrast to the spatially localized ionization produced by Auger electrons during ^{37}A decay, background radioactivity primarily produces fast electrons in the counter, which result in extended ionization. Pulses from the counter are differentiated with a time constant of 10 ns. The amplitude of the differentiated pulse is proportional to the product of the amplitude and the inverse rise time of the pulse. For every event in the counter, the time of occurrence, energy, amplitude of the differentiated pulse, and any associated NaI signal are recorded. Candidate ^{37}A decay events are selected by taking events that fall within the K-peak energy and rise-time acceptance windows that do not have any associated NaI activity.

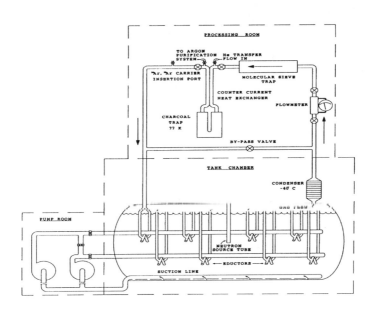

Figure 36. Schematic operation of the chlorine experiment.

Each extraction is counted for about one year in order to allow an accurate determination of the background after the ^{37}A decays away. The candidate events are fit to an exponential decay with a 35-day half-life and a time-independent background using a maximum-likelihood method.

Runs have been made with rise-time information since mid-1970 on a regular basis (except for periods during 1985–86 when both extraction pumps failed and 1996-99 when funding for the experiment was not available). The published results through 1995 [92] are shown in Figure 37. The analysis was made using a 1-FWHM energy cut and a 95% acceptance rise-time cut. The average production rate of ^{37}A from 1970 to 1995 was $0.478 \pm 0.030 \pm 0.029$ atoms per day after a background due to cosmic-ray muon interactions of 12% is subtracted from this rate. This results in a value of $2.56 \pm 0.016 \pm 0.016$ SNU attributed to solar neutrinos for the chlorine experiment for the period from 1970 to 1995. This corresponds to a total number of about 875 ^{37}A atoms observed during 20 years of operation.

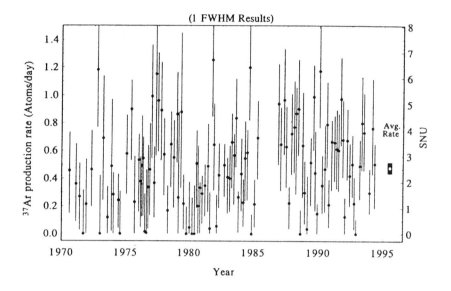

Figure 37. Results of each run of the Cl solar neutrino experiment.

It was noted in the 1980s that an apparent anticorrelation between the production rate of ^{37}A and the number of sunspots existed in the Cl data [93], although various analyses differ in the statistical significance of the effect [94]. Such an inverse correlation was attributed to the possible existence of a magnetic moment of the neutrino, causing a spin flip of the neutrino as it traverses the magnetic fields of the convection zone [95]. The magnitude of the magnetic moment required (about 10^{-10} Bohr magnetons with 10-kGauss fields in the convection zone) is much larger than that anticipated in conventional particle physics models. While this value is just slightly below laboratory limits [96], it appears to be ruled out by astrophysical observations [97], in particular from observed values of the flash point of red giants in globular clusters [98]. More recent analyses still indicate a statistically significant variation of the Cl data [99]. However, there is no corroborating evidence for any time variations from any of the other solar experiments and thus it is most likely that any variation in the Cl data is just a statistical fluctuation.

7.6 SAGE

The two gallium experiments (SAGE and GALLEX) both observe the reaction $^{71}Ga(\nu_e,e^-)^{71}Ge$ produced by solar neutrinos by chemically extracting and observing the subsequent decay of the ^{71}Ge atoms. The threshold for this reaction is 233 keV, so that the gallium experiments are sensitive to all of the neutrino-producing reactions in the Sun [100]. In particular, of the 129 +8/-6 (127 ± 8.6) SNU of the total capture rate predicted by the SSMs, 55% is due to the p-p neutrinos. The largest second contribution is due to ^{7}Be neutrinos with 34 SNU in the Bahcall-Pinsonneault SSM. Thus, the prospect exists that the gallium experiments may be able to unravel solar physics from particle physics effects, as the flux of p-p neutrinos is essentially independent of solar modeling.

The Russian-American Gallium Experiment [originally the Soviet-American Gallium Experiment (SAGE)] uses the Gallium-Germanium Neutrino Telescope situated in an underground laboratory specially built at the Baksan Neutrino Observatory of the Institute for Nuclear Research of the Russian Academy of Sciences in the Northern Caucasus Mountains [101]. It is located 3.5 km from the entrance of a horizontal adit driven into the side of Mount Andyrchi and has an overhead shielding of 4700 mwe.

SAGE initially used 30 tons and later increased the amount of gallium to 57 tons in the form of the liquid metal. The gallium is held in eight reactors of 2 m^3 each, with about 7 tons of gallium in each reactor. The tanks are lined with Teflon, have Teflon vanes around the inside walls of the tank, and Teflon-coated stirrer paddles.

The chemical extraction process from metallic gallium was first worked out in the US and later fully tested in a 7.5-ton pilot experiment in the USSR. The extraction is based on the fact that Ga melts at 29.8 C, which makes it possible to keep the gallium in its liquid form, and is central to the extraction. Each measurement of the solar neutrino flux begins by adding approximately 700 μg of natural Ge carrier in the form of a solid Ga-Ge alloy to the reactors holding the Ga. After a typical exposure interval of 4 weeks, the Ge carrier and any ^{71}Ge atoms that have been produced by neutrino capture are chemically extracted from the Ga.

The extraction process involves mixing HCl, H$_2$O$_2$, and H$_2$O in each reactor, which forms an emulsion from which the Ge is extracted as GeCl$_4$. The extracts from the separate reactors are vacuum siphoned off, combined, and the Ge is concentrated, extracted into CCl$_4$ and then back-extract it into H$_2$O. The resulting solution is used to synthesize the counting gas GeH$_4$ (germane). A measured quantity of xenon is added, and this mixture is inserted into a sealed proportional counter. The overall extraction efficiency has been measured to be typically 80 ± 6%. The extraction efficiency is limited by the amount of gallium one is willing to dissolve. At 80% efficiency, less than 0.1% of the gallium is dissolved in each extraction.

The SSM predicts a production rate of 1.2 ^{71}Ge atoms/day in 30 tons of Ga. At the end of a 4-week exposure period, an average of 16 ^{71}Ge atoms will be present. Taking into account all efficiencies, one would expect that SAGE would detect only about 4 ^{71}Ge atoms from 30 tons of gallium in each run, assuming the SSM flux. Thus, the counting backgrounds must be kept to a small fraction of a count/day.

^{71}Ge decays with an 11.4-day half life by electron capture to the ground state of ^{71}Ga. The only way to observe this decay is to detect the low-energy (10.4 keV) Auger electrons and (1.2 and 10.4 keV) X-rays produced during electron-shell relaxation in the resulting ^{71}Ga atom. The low-energy electrons and X-rays are detected in a small-volume (0.75 cm^3) proportional counter that is placed in the well of a NaI detector (used as a veto) inside a large passive shield and counted for 6 months. Pulse-shape discrimination based on rise-time measurements is used to separate the ^{71}Ge decays from background by digitizing the waveform of the proportional counter pulses with a 1 GHz transient digitizer. A rectangular acceptance window is then calculated around the 5.9-keV ^{55}Fe calibration peak, and the K and L peak acceptance windows are then determined by extrapolation. This procedure was verified by filling a counter with ^{71}GeH$_4$, together with the standard counter gas and the resultant spectrum is shown in Figure 38.

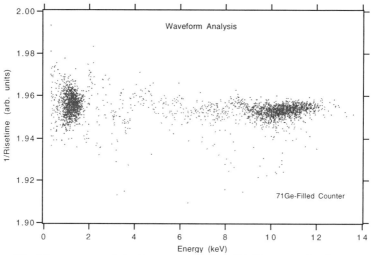

Figure 38. The energy and risetime spectrum in a SAGE proportional counter filled with a ^{71}GeH$_4$-Xe gas mixture.

A maximum likelihood analysis is then carried out on these events by fitting the time distribution to an 11.4-day half-life exponential decay plus a constant rate background.

The systematic uncertainties in the chemical extraction and counting efficiencies are typically ±3.4% and +2.3/-3.9%, respectively, resulting in a systematic uncertainty of +3.5/-2.7 SNU for the combined 1990–97 data.

The final possible systematic effect is due to possible background reactions that could produce ^{71}Ge and the possible presence of radon, which can mimic a ^{71}Ge signal. The main source of ^{71}Ge other than from solar neutrinos is from protons arising as secondary particles produced by external neutrons, internal radioactivity, and cosmic-ray muons. These protons can initiate the reaction ^{71}Ga(p,n)^{71}Ge. Radon can also produce backgrounds but the radon levels are greatly suppressed in the chemical extraction stages and eliminating any data within a few hours of any saturation pulse in the proportional counters (indicating an alpha decay in the counter) reduces the effect of radon to very low levels. The total uncertainty assigned to backgrounds is +0/-1.2 SNU.

SAGE was the first experiment to measure the integral flux of solar neutrinos and initially observed quite a low signal. Thus a great deal of effort was expended to check that the experiment was working correctly. Several measurements of the chemical extraction and counting efficiencies were carried out, all of which indicated that the experiment was performing as expected. The best possible test of the experiment can be carried out by using an artificial neutrino source. A suitable neutrino calibration source can be made using ^{51}Cr, which decays with a 27.7-day half-life by electron capture, emitting monoenergetic neutrinos of 751 keV (90.2% BR) and 426 keV (9.8% BR). An engineering test run was made in 1991 and a final ^{51}Cr experiment was carried out in 1994-95 using a 517 kCi ^{51}Cr source [102]. This source produced a neutrino flux in the SAGE detector that was 50 times brighter than the Sun (of course it was a great deal closer than the Sun!). In total, 130 ^{71}Ge atoms were observed that were produced in 13 tons of gallium metal by the ^{51}Cr source. This production rate can be compared with that expected based on calculations of the neutrino cross sections on ^{71}Ga. While the cross section for the ground state in ^{71}Ga is well known, there is some uncertainty in the cross sections for the excited states. Two calculations of the cross sections have been carried out. The Cr measurement showed that the combined extraction and counting efficiencies was 95 ± 12 (experimental) +3.5/-2.7 (theoretical) % of that expected from the Bahcall cross sections [103] and 87 ± 11 (exp) ± 9 (theor)% of that expected from the Haxton cross sections [104]. Thus, SAGE concluded that the measured value of the integral flux of solar neutrinos reflects a deficit of solar neutrinos and is not due to some experimental artifact.

SAGE
1990 - 1997

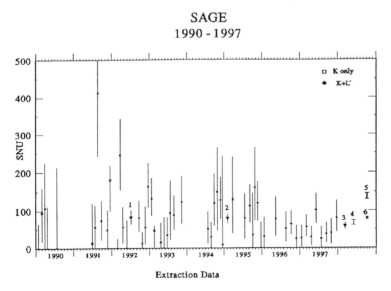

Extraction Data

Figure 39. SAGE results.

The measured rates in SAGE from 1990 to 1998 are shown in Figure 39. The data is consistent with a statistical distribution with a 58% probability. The result of the combined fit to the data gives a capture rate of:

$$^{71}\text{Ga Capture Rate} = 67.2 +7.2/-7.0 + 3.5/-3.0 \text{ SNU}.$$

Thus, SAGE observes an integral flux of solar neutrinos that is only 52% of the Bahcall-Pinsonneault SSM prediction and 53% of the Turck-Chieze SSM prediction. SAGE does not observe any statistically significant temporal variation in the rate [105] and is able to place some (rather loose) constraints on a vacuum oscillation solution to the solar neutrino problem.

SAGE is continuing to operate with extractions now being made on a monthly basis. In order to improve the overall accuracy of the measurement of the integral flux of solar neutrinos, to search for possible time variations in the solar neutrino flux, to test the model-independent upper limit of 79 SNU [71] for a Ga experiment, and to overlap with the operation of SuperKamiokande, SNO, and Borexino, SAGE plans to continue runs until 2006.

7.7 GALLEX

The GALLium EXperiment (GALLEX) [106] experiment is located in Hall A of the Gran Sasso National Laboratory in central Italy. The laboratory is located 6.3 km from the entrance of the tunnel at a depth of 3300 mwe. The experiment is housed in two buildings, one containing the tanks holding the gallium solution and one containing the counting electronics and computers, as shown in Figure 40. Many of the physical principles involved in the GALLEX experiment have been treated in the previous section on SAGE (^{71}Ge decay, background sources, ^{51}Cr neutrino source) and need not be repeated here.

GALLEX employs 30.3 tons of gallium in 101 tons of 8.13 molar aqueous gallium chloride solution. The gallium solution is contained in one of two 7-m-high 70-m^3 target tanks. One of the tanks is equipped with a central elongated thimble designed to hold an intense ^{51}Cr artificial neutrino source, as well as to accept a 470-liter vessel containing a calcium-nitrate solution used to monitor the fast neutron background.

GALLEX begins a run by adding about 1 mg of a stable Ge isotope to the target solution. After an exposure interval of 3 weeks, the Ge carrier and any ^{71}Ge atoms are recovered by purging the solution with 1900 m^3 of nitrogen gas through eductors located within the tank. The nitrogen gas is scrubbed in 50 liters of counter-flowing water in three large absorber columns packed with glass helices. The extraction takes about 20 hours and typically 99% of the Ge is recovered. The resulting aqueous solution of Ge is then concentrated by acidifying it with the addition of HCl gas, and the solution is purged again with a stream of nitrogen and reabsorbed in 1 liter of water. This solution is then extracted into CCl$_4$ and back extracted into tritium-free water. Germane is then synthesized from this solution and purified by gas chromatography. The germane is then mixed with xenon (70% by volume) and inserted into a small proportional counter at a slight overpressure. GALLEX has gone to great lengths to maximize the counting efficiency and reduce the backgrounds of these counters. The counters are made from hyperpure Suprasil quartz and have either zone-refined iron or ultrapure silicon cathodes. Two locations within a large passive shield have been provided for counting. The passive shield is hermetically sealed and purged with dry nitrogen to eliminate radon. The pulses from the counters are amplified and multiplexed into a GHz transient digitizer to provide rise-time information. In addition, the pulse waveform is also recorded at slower speeds to check for delayed coincidences in the counters and to allow identification of noise events.

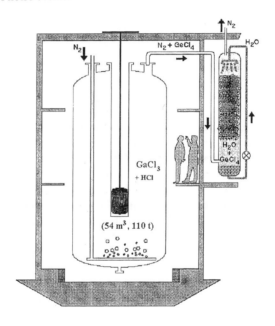

Figure 40. Layout of the GALLEX solar neutrino experiment.

GALLEX carried out measurements of the solar neutrino flux from 1991 to 1997. Due to shallower depth of GALLEX (with an attendant increase in cosmic ray rates) and the use of an aqueous solution (the presence of hydrogen atoms in the water makes it more sensitive to fast neutron backgrounds), GALLEX must subtract a background of 6.5 ± 1.7 SNU. Systematic uncertainties other than those due to backgrounds include uncertainties in the chemical yield (± 2.2%), the counting efficiency (± 4.5%), and the correction for the ^{68}Ge correction (+0.9/-2.6%).

The results of the runs, after correcting for backgrounds, are shown in Figure 41 and the combined fit yields [105]:

$$^{71}\text{Ga Capture Rate} = 77.5 \pm 6.3 +4.3/-4.7 \text{ SNU}.$$

410

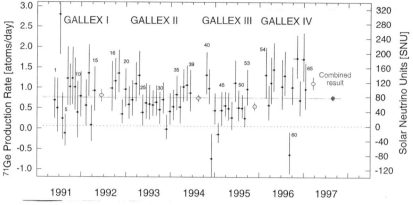

Figure 41. GALLEX results.

A number of consistency checks have been carried out by GALLEX. In one test, additional runs were made by extracting "blank" samples immediately after a solar neutrino extraction. This tests for any effects that do not depend on exposure time. The result was $-1.5 \pm 5.1 \pm 1.4$ SNU, consistent with no additional sources of ^{71}Ge in the chemical extraction. GALLEX has also looked at the statistical distribution of the data in time. While the result for the GALLEX III data set was low and the result for the GALLEX IV data set was high, GALLEX concluded that the overall distribution agreed well with Monte Carlo simulations. GALLEX also carried out a number of tests in which the gallium solution was spiked with a large number ($\approx 10^6$) of ^{71}As atoms, letting them decay to ^{71}Ge in situ. The ^{71}Ge atoms were extracted and counted. A number of systematic variations of the procedures were carried out and the result indicated that the extraction and counting efficiencies were as expected [107].

GALLEX carried out two measurements [108, 109] with a ^{51}Cr source with total activities of 63.4 PBq and 68.9 PBq. These are the most intense artificial neutrino sources ever produced. However, although the source intensities were more than 3 times larger than in SAGE, the larger physical size of the source and the lower density of gallium in the aqueous solution in GALLEX produced a counting rate in GALLEX that was slightly lower than in SAGE. The results of the two measurements, compared to the Bahcall prediction, 99 ± 11% and 83 ± 10%, giving a combined result of 91 ± 8%, in good agreement with the expected value of 100%.

Thus, while the basic principles of SAGE and GALLEX are very similar, there are important systematic differences that provide valuable cross checks between the two experiments. It is reassuring that SAGE and GALLEX agree well within the 1-σ level.

GALLEX formally stopped running 1997, but a revised collaboration has been formed as the Gallium Neutrino Observatory (GNO) [110]. After a number of upgrades to the chemistry and counting systems used by GALLEX, GNO began running in 1998 and plans to run for one full solar cycle of 11 years. While initial operation is with 30 tons of Ga, in order to improve the counting statistics, GNO plans to upgrade to first to 60 tons of Ga in aqueous and finally to 100 tons of Ga with the last 40 tons being in the metallic form. This will allow GNO to search for possible time variations in the integral flux of solar neutrinos with a sensitivity to variations at about the 3% level.

7.8 Kamiokande

The Kamiokande experiment is a water Cherenkov detector originally designed to search for proton decay, which was later upgraded to be sensitive to ^{8}B neutrinos [111]. The detector is located at the Kamioka mine in the Japanese Alps at a depth of 1000 m (2700 mwe). The signature for solar neutrinos in this experiment is by the elastic scattering (ES) reaction: $v_X + e^- \rightarrow v_X + e^-$ and observing the Cherenkov light cone radiated by the scattered electron. Timing and pulse-height information from the photomultipliers (PMTs) that register the Cherenkov photons allows one to determine the position, direction, and energy of the recoil electron, thereby providing a measure of the neutrino energy and direction.

The elastic scattering cross section can proceed via both charged and neutral current interactions. In the charged current (CC) interaction, a charged W^- intermediate vector boson is exchanged, and this can occur only in the case of electron neutrinos scattering from electrons. In the neutral current (NC) interaction, a neutral Z^0 intermediate vector boson is exchanged and this can occur with any flavor neutrino. However, as the ratio of the cross sections for the neutral current to charged current at these low energies is $\sigma(NC)/\sigma(CC) \approx 1/6$, Kamiokande is predominately sensitive to electron neutrinos [112].

The Kamiokande detector is similar in design, but smaller, than the SuperKamiokande detector that was described earlier. It was initially constructed to search for proton decay and consists of a cylinder 14.4 m diameter by 13.1 m high holding 4,500 tons of pure water. The inner volume is viewed by 948 50-cm-diameter photomultipliers (PMTs) providing 20% photocathode coverage. A 2-m-thick outer water volume is used as a veto for cosmic-ray muons and is viewed by 123 PMTs.

The Kamiokande proton decay experiment was upgraded to Kamiokande II by reducing the threshold to below 10 MeV in order to be sensitive to ^{8}B solar neutrinos. In order to make such a substantial improvement in the threshold, a fiducial volume cut of 680 tons was made, the amount of heavy radioactive elements in the water was reduced by recirculating the water through ion-exchange columns, a degasification system to eliminate radon was added, the detector was hermetically sealed with radon-free gas, and spallation products (which subsequently beta decay) due to cosmic-ray muons were vetoed. These techniques enabled Kamiokande to initially achieve a threshold of 9 MeV. In 1990, the threshold was lowered to 7.5 MeV by doubling the gain of the photomultipliers. Finally, the experiment was upgraded to Kamiokande III [113] by replacing about 100 failed photomultipliers, upgrading the fast electronics, and placing wavelength-shifting plates around the PMTs to increase the fraction of light collected, resulting in a threshold of 7 MeV.

412

The results from Kamiokande II are shown in Figure 42. A clear signature for events correlated with the direction of the Sun ($\cos\theta = 1$) is observed, but the flux is less than that predicted by the SSM. As the Kamiokande experiment can point back to the direction the neutrinos come from, this capability has provided the first direct observation of neutrinos coming from the Sun, and confirms the long-held belief that fusion reactions are occurring in the Sun.

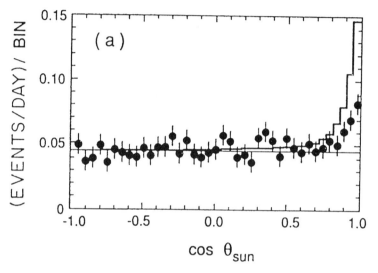

Figure 42. The electron recoil angular distribution measured in the Kamiokande II experiment. The observed forward scattering peak at $0°$ is in line with the direction of the Sun.

Kamiokande measured the ^{8}B solar neutrino flux for a total of 2079 days until February 1995 when the experiment was completed. The data from Kamiokande II and III yields a flux at Earth of ^{8}B neutrinos (assuming the standard ^{8}B neutrino spectrum) of $(2.80 \pm 0.19 \pm 0.33) \times 10^6$ v/cm^2/s [114]. This can be compared to the predictions of the SSMs:

$$\phi(\text{Kamiokande})/ \phi(\text{Bahcall-Pinsonneault}) = 0.54 +0.20/-0.16$$
$$\phi(\text{Kamiokande})/ \phi(\text{Turck-Chieze}) = 0.58 \pm 0.24$$

where the experimental and theoretical errors have been added in quadrature. Kamiokande does not observe any statistically significant variations on a seasonal, annual, or solar cycle time scale [114].

The Kamiokande results have been confirmed by SuperKamiokande. With it's much higher statistical accuracy and better measurements of systematic effects, the SuperKamiokande results discussed in the next section have superceded the Kamiokande results.

Kamiokande is able to constrain the region of allowed parameter space for MSW oscillations in three ways. First, the absolute rate compared to that predicted by the SSMs provides a constraint [115]. Second, Kamiokande can further constrain the parameters by looking at the energy dependence of the observed spectrum, which rules out the adiabatic branch. Third, the effect of matter oscillations in the earth, in which muon or tau neutrinos are regenerated back into electron neutrinos, can be determined by looking at the difference in rates between day and night (in which case the neutrinos do not or do pass through the Earth, respectively). The day-night Kamiokande data [116] are able to rule out a region of the large mixing-angle solution with $\Delta m^2 = 3 \times 10^{-6}$. The resultant allowed regions, together with the region allowed by the Cl, Ga, and SuperKamiokande data, are shown in Figure 33.

7.9 SuperKamiokande

SuperKamiokande [59] provides rates that are almost a factor of 100 times higher than in the Kamiokande detector. In addition, the quality of the data is substantially improved: 1) the threshold is currently 5.5 MeV and is expected to reach 5 MeV, compared to 7.5 MeV for KII, 2) the energy resolution is about 14% (at 10 MeV), compared to 20% for KII, and 3) the vertex resolution is about 50 cm (for 10-MeV e$^-$), compared to 1.1 m for KII. Reductions by a factor of several for U and Th, and by a factor of more than 100 for ^{222}Rn have been achieved by improved water purification equipment [117]. This results in an improvement in the S/N for SuperKamiokande of a factor of about 16 compared to KII. In addition, SuperKamiokande has installed a small electron linac that can inject low energy (from 5 to 15 MeV) electrons into the SuperKamiokande detector at different positions. This is very important in determining the absolute energy calibration and detector resolution.

The impressive statistics that SuperKamiokande has achieved permits precision searches for possible time variations of the ^{8}B solar neutrino flux, with an accuracy of $\pm 3\%$ for day/night or seasonal variations and ± 1-2% for yearly variations. The most exciting prospect for SuperKamiokande comes from its improved performance, which provides a much more accurate measurement of the energy spectrum of the scattered electrons. This permits a determination of the spectral distortion caused by MSW oscillations free from any ambiguities in the SSMs. In the ES reaction, the MSW distortions are diluted somewhat by kinematics, so that the difference in spectral shapes between the nonadiabatic and large mixing-angle MSW solutions is only about 10% at 10 MeV and almost zero at 5 MeV. Thus, the ability to discriminate between these two solutions requires a great deal of attention to systematic effects in the detector and very careful attention to energy and efficiency calibrations.

SuperKamiokande has reported a measurement of the flux of ^{8}B solar neutrinos based on 825 days of data [118] The results (which are essentially unchanged from earlier measurements [119], compared to the Bahcall-Pinsonneault and Turck-Chieze SSMs, are:

414

$$\phi(^8B)\ (SK)\ /\ \phi(^8B)\ (B\text{-}P) = 0.475 \pm 0.008/\ 0.007 \pm 0.013$$
$$\phi(^8B)\ (SK)\ /\ \phi(^8B)\ (T\text{-}C) = 0.508 \pm 0.008 \pm 0.014$$

SuperKamiokande has also reported on a measurement of the spectral shape of the recoil spectrum of electrons, as shown in Figure 43 [120, 121]. The spectral shape at high energies deviates significantly from that expected from the known ^8B recoil spectrum. The SuperKamiokande collaboration notes that there may be several possible explanations for this distortion. First, it may just be a statistical fluctuation and that with additional data the effect may disappear. Second, there is an additional source of high energy neutrinos from the hep reaction, namely those neutrinos produced in the fusion reaction $^3He + p \rightarrow {}^4He + e^+ + \nu_e$, that extend up to 18.8 MeV. If the flux of the hep neutrinos is significantly higher (by about a factor of 16) than the SSM prediction, this could account for the observed spectral shape. The prediction of the hep flux in the SSM is in fact highly uncertain [122]. This is a forbidden transition, and involves cancellation of matrix elements. The difficulty in calculating this rate is so great that no uncertainty is quoted in the SSM predictions for the hep flux. Thus, it may well be plausible that the SuperKamiokande result simply reflects the fact that the hep flux is much larger than predicted. Third, it could indicate the existence of new physics. If vacuum oscillations occur, then the ^8B spectrum would be distorted and there are certain regions in the $\Delta m^2 - \sin^2 2\theta$ parameter space that could account for the distortion. In this case, there may also be observable temporal variations.

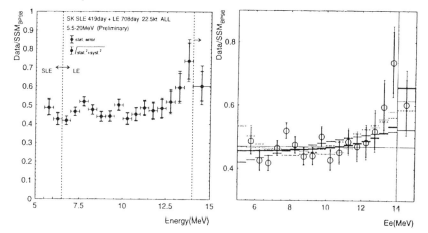

Figure 43. SuperKamiokande recoil electron spectral shape measurement on the left shown as the ratio of data to the SSM prediction. The fits shown on the right are for the 0.48 x SSM (thin horizontal line), typical SMA (solid), vacuum (dots and dashes), and 14 times the hep rate with no oscillations (thick solid).

SuperKamiokande has also carried out an analysis of the data by binning the count rate during each 24 hour period into 6 bins. One of the bins corresponds to the period during the day when the neutrinos from the Sun do not pass through the Earth. The other 5 bins correspond to times of the day when the neutrinos from the Sun pass through different lengths of the Earth. For certain values of $\Delta m^2 - \sin^2 2\theta$, neutrinos may be regenerated as they pass through the Earth. Such an effect would be manifested by an increased count rate during the night. The present SuperKamiokande result does indicate a difference at the 1.9 σ level: (N − D) / [(N + D)/2] = 0.065 ± 0.031 ± 0.013 where N is the number of counts during the day and D is the number of counts during the night [122]. While this observation is intriguing (since the positive sign of the ratio is what is required for regeneration), it is certainly premature to claim observation of neutrino oscillations from this data. With continued running of SuperKamiokande and a reduced statistical uncertainty, it may prove that in fact regeneration within the Earth is occurring (or alternately the effect may just prove to be a statistical fluctuation). The region of parameter space that is allowed by a combined analysis of the SuperKamiokande, Kamiokande, chlorine, and gallium experiments is shown Figure 44 [121, 123].

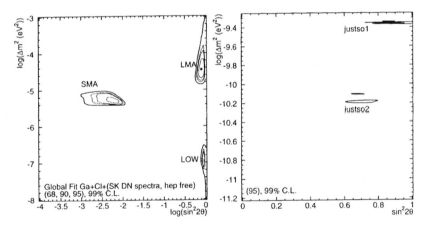

Figure 44. Allowed regions for the combined fit of all experimental data.

9   The Next Generation Solar Neutrino Experiments

9.1 Sudbury Neutrino Observatory (SNO)

In order to be certain that the solution to the solar neutrino problem involves new physics, it is essential to observe new phenomena directly, such as neutrinos oscillating from electron to muon or tau neutrinos. It is with this goal in mind that SNO was conceived. By measuring both charged- and neutral-current interactions on deuterium, SNO will be able to measure directly both the flux of electron neutrinos, as well as the flux of all neutrinos, independent of their flavor. In addition, SNO will be able to make accurate measurements of the spectral shape of

416

the ^{8}B neutrinos. The sensitivity of SNO to distortions of the ^{8}B spectrum caused by neutrino oscillations is shown in Figure 45 [124]. Finally, SNO will be sensitive to temporal variations in the flux of ^{8}B neutrinos. These three measurements all provide model-independent means to search for neutrino oscillations.

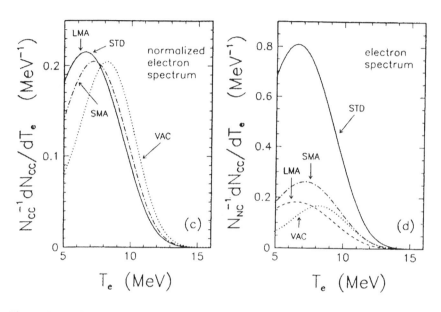

Figure 45. Effect of neutrino oscillations on the shape of the ^{8}B spectrum for the SNO detector: a) normalized SNO spectra (area under curves = 1.0), b) unnormalized SNP spectra (area under curves = CC rate / NC rate). Cases are shown for STD (no oscillation), SMA (small mixing angle), LMA (large mixing angle), and VAC (vacuum oscillation).

The SNO detector (shown in Figure 46) has been constructed at the Creighton #9 nickel mine at Sudbury, Ontario, Canada [125]. A 22-m-diameter × 34-m-high chamber was excavated in norite rock at the 6800-ft level (5900 mwe) to house the detector. One kiloton of D_2O is housed inside a 6-cm-thick acrylic sphere 12 m in diameter. The D_2O sits within a 5-m-thick H_2O shield to provide shielding of photons and neutrons from the rock. A photomultiplier support assembly (PSUP) holds 9500 20-cm-diameter PMTs on a 17-m diameter. The PMTs are fitted with reflectors to increase the light collection surface, providing 56% effective photocathode coverage.

SNO will be sensitive to and able to discriminate electron neutrinos from muon and tau neutrinos by the following reactions:

1) $\nu_e + d \rightarrow p + p + e^-$ [Charged Current reaction (CC)]
2) $\nu_x + e^- \rightarrow \nu_x + e^-$ [Elastic Scattering reaction (ES)]
3) $\nu_x + d \rightarrow \nu_x + p + n$ [Neutral Current reaction (NC)]

The CC reaction is sensitive only to electron neutrinos, provides very good spectral information, as $E_e = E_\nu - 1.44$ MeV, and has some directionality as the angular distribution is $1 + 1/3 \cos\theta$. The ES reaction is primarily sensitive to electron neutrinos, provides good directionality, and contains some spectral information (as discussed in the section on Kamiokande). The NC reaction is equally sensitive to all types of neutrinos, has a 2.2-MeV threshold, but has no directionality or spectral information, as the signature comes from the capture of the thermalized neutron.

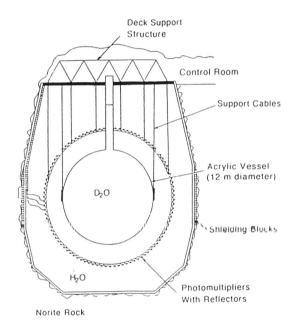

Figure 46. Schematic view of the SNO detector.

The rates in SNO are almost two orders of magnitude higher than in the current generation of solar neutrino experiments. Assuming a 5-MeV threshold, the rates per year will be about 3600 for the CC, 2200 for the NC, and 400 for the ES reactions (assuming 0.5 x Bahcall-Pinsonneault SSM fluxes). These rates are sufficient to get good spectra to allow analysis for MSW energy distortions, and are also sufficient to observe the secular variation of the NC signal with 2-σ accuracy in 4 years, thereby confirming the solar origin of the NC signal.

The primary difficulty in SNO is the reduction of backgrounds necessary to observe the CC reaction with a low threshold and to observe the NC signal. Extensive Monte Carlo calculations have been carried out to determine maximum permissible levels of radioimpurities in the detector. The activities expected indicate that the threshold should be about 5 MeV for the CC and ES reactions, as is shown in

418

Figure 47a [126]. For the NC reaction, U and Th in the detector present significant backgrounds. The decay chains of U and Th result in the emission of 2.44- and 2.61-MeV gammas, which can photodisintegrate deuterium, thus providing a source of free neutrons that cannot be distinguished from those produced by neutrino disintegration. The design goals (in pg/g) are levels of U and Th in the detector of about 10^{-14} in the D_2O and H_2O, and about 10^{-12} in the acrylic.

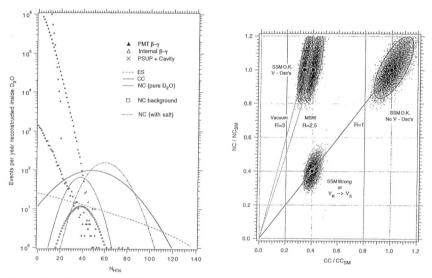

Figure 47a) Simulation of the components that contribute to the Cherenkov spectrum in SNO. The lines show the expected shape for the CC, ES, and NC (for pure D2O and with MgCl2) while the symbols show the expected backgrounds, b) simulation of the NC and CC sensitivity in SNO, showing how the combined NC and CC data will provide a clear separation of different physics solutions to the solar neutrino problem.

The CC and ES reactions are directly observable from the Cherenkov light, allowing a determination of the energy, position, and direction of the track. However, the neutrons from the NC reaction need to be detected by some other means, as the capture reaction n(d,t)γ has such a small cross section. The nominal means to detect the neutron is by adding 2.5 tons of salt (Mg_2Cl) to the D_2O and capturing the neutron on ^{35}Cl, yielding an 8.6-MeV gamma ray that is detected by observing the Cherenkov light from Compton scattering. The detector will be run for 6 months without Mg_2Cl, then 6 months with Mg_2Cl (or alternately by varying the concentration of Mg_2Cl to modulate the NC efficiency). The two spectra will then be subtracted from each other, leaving a pure NC spectra due to capture on the NaCl. One must be certain that no additional radioimpurities have been added along with the NaCl, which might mimic a NC signal by increasing the rate. One must also be sure that the detector gain, resolution, and efficiency do not change between the runs with and without salt. While this appears feasible (albeit difficult), there are two disadvantage to this method: 1) one must assume that the NC rate is independent of time, and 2) the detector is sensitive to NC reactions only half of the time. With a duty factor of only 50%, it would be particularly unfortunate if a supernova were to occur when the detector was not sensitive to NC.

An alternate method is has been developed that will resolve these difficulties that involves deploying an array of 96 proportional counters filled with ^{3}He in the D_2O [127]. The neutrons are detected by the ^{3}He(n,p)t reaction. Measurements of the radioactivity of the components of the proportional counters has shown that backgrounds introduced due to the presence of the proportional counters in the D_2O will be less than 10% of the SSM rate. This method has the advantage that it provides 100% duty factor for the NC reaction, an effective increase in the NC counting rate of a factor of four, and identification of NC reactions on an event-by-event basis.

The great advantage of having both NC and CC information in resolving the origin of the solar neutrino problem is shown in Figure 47b. Thus, with information on the shape of ^{8}B spectrum and a measurement of the NC rate SNO will provide two model-independent measurements that will allow a resolution of the solar neutrino problem.

Construction of the main SNO detector has been completed, all H_2O and D_2O has been filled, and the experiment is now on line. The construction of the ^{3}He counters is well advanced. Initial measurements will be made with pure D_2O followed by adding $MgCl_2$ for an initial NC measurement. The ^{3}He counters will be used for long-term NC measurements.

9.2 Borexino

Borexino [128] is envisaged as a 300-ton (100-ton fiducial) volume detector using liquid scintillator in an acrylic vessel surrounded by a 2-m-thick water shield. The scintillator is composed primarily of 1-2-4 trimethylbenzene doped with fluors to provide efficient light output. The fiducial volume will be viewed by 1650 20-cm PMTs fitted with reflectors, providing an effective photocathode coverage of 50%. The scintillator, PMTs, and support structure will be inside a 16.4-m-diameter steel vessel filled with ultrapure water that serves as a passive shield. The detector will be installed in the Gran Sasso National Laboratory at a depth of 3700 mwe. A view of the Borexino detector is shown in Figure 48.

The detector will have a very low threshold of 250 keV and will be sensitive to the ES reaction:

$$\nu_X + e^- \rightarrow \nu_X + e^- \qquad \text{[Elastic Scattering reaction (ES)]}$$

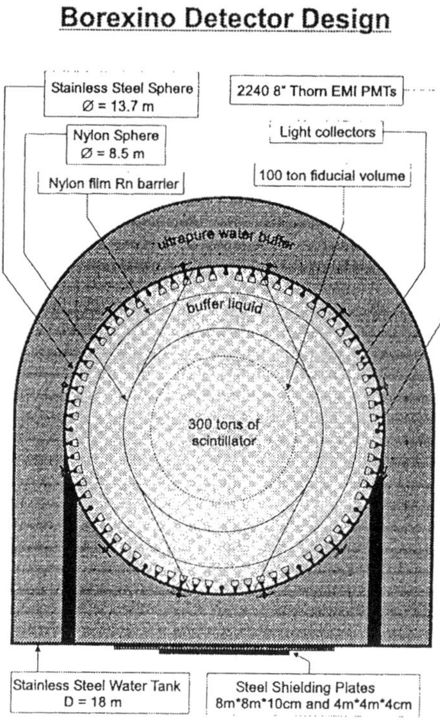

Borexino Detector Design

Figure 48. The Borexino detector.

with a rate of 18,000 events/yr (B-P SSM) from ^{7}Be neutrinos. The CC and NC reactions also occur at much lower rates:

The important contribution that Borexino will make is the direct detection of ^{7}Be neutrinos. The signature comes from a combination of the Compton edge observed in the scintillator from the monoenergetic ^{7}Be line and from the 7% modulation in count rate due to the Earth's orbital eccentricity. As the ^{7}Be flux is much less temperature dependent than the ^{8}B flux, and because the ^{7}Be neutrinos are intermediate in energy between the p-p and ^{8}B neutrinos, their observation provides a strong means of verifying the origin of the solar neutrino problem and in determining which of the allowed MSW solutions is the correct one.

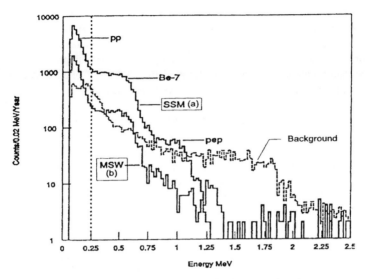

Figure 49. The expected signal and backgrounds in the Borexino detector.

The primary difficulty facing Borexino is to achieve extremely low levels of radioactivity in the scintillator. Unlike SNO and Kamiokande, which are water Cherenkov detectors and thus have an intrinsic Cherenkov threshold, Borexino is sensitive to all radioimpurities. The levels of U and Th must be at the 10^{-16} level, while K must be at the 10^{-14} level. Other possible background sources include ^{14}C, ^{7}Be, and ^{10}Be. A great deal of development effort has gone into demonstrating these purity levels, and it seems likely that they will be reached. A 0.2-ton Counting Test Facility (CTF) test detector has been installed at the Gran Sasso Laboratory and the data taken indicate that it appears feasible to achieve the radiopurity levels required for Borexino. The full Borexino detector is now under construction and it is expected that it will become operational in 2001.

9.3 Other Possible Future Experiments

Although it is hoped that the current generation of solar neutrino detectors now running or under construction will resolve the origin of the solar neutrino problem, it is clear that additional measurements, particularly at low energies, are desired. Thus, a number of ideas for a new generation of detectors are being pursued. Table 6 gives a list of efforts currently under way.

Table 6. Concepts for the next generation of solar neutrino detectors that are under development. ES refers to the elastic scattering reaction $e(v_e, v_e')e$, xtal denotes crystal, rad chem denotes radiochemical experiment, scint denotes a scintillator detector, TPC denotes a Time Projection Chamber, cryo denotes a cryogenic detector, and * denotes that the daughter nucleus is left in an excited state and that the decay of the excited state is used as a delayed coincidence.

Name	Reaction	Sensitivity	Detector Type	Group
Hellaz	ES	p-p	He TPC	College de France [129]
Heron	ES	p-p	Rotons in LHe	Brown [130]
Super MuNu	ES	p-p	CF_4 TPC	Neuchatel [131]
GaAs	$^{71}Ga(v_e,e)^{71}Ge$	p-p	Semiconductor	LANL/INR [132]
In	$^{115}In(v_e,e)^{115}Sn^*$	p-p	Tunnel junction	Oxford [133]
LENS	$^{176}Yb(v_e,e)^{176}Lu^*$	p-p	Liquid scint	LENS [134]
LiF xtal	$^7Li(v_e,e)^7Be$	pep	Cryo bolometer	Lucent [135]
LiF xtal	$^7Li(v_e,e)^7Be$	pep	Crystal scint	U. Maryland [136]
I rad chem	$^{127}I(v_e,e)^{127}Xe^*$	^7Be	NaI solution	U. Penn [137]
Br	$^{81}Br(v_e,e)^{81}Kr^*$	^7Be	Cryo bolometer	Milano [138]
Li rad chem	$^7Li(v_e,e)^7Be$	^8B	Cryo calorimeter	INR/Genova [139]
Xe rad chem	$^{131}Xe(v_e,e)^{131}Cs$	^7Be	Ionization	Kiev [140]
F scint	$^{19}F(v_e,e)^{19}Ne^*$	^8B	Liquid scint	INR [141]
Icarus	$^{40}A(v_e,e)^{40}K^*$	^8B	Liquid TPC	CERN [142]

10. Implications of Experimental Evidence for Neutrino Mass

The experimental search for a finite neutrino mass has been a global, large-scale effort for more than two decades. In the past, a number of claims for observation of a finite neutrino mass have been made, but they were all eventually refuted. In particular, a claim for the existence of a finite neutrino mass in tritium beta decay has been refuted. However, distortions of the tritium beta decay spectrum have been observed that are not yet understood. Until this problem is resolved, one cannot set a model-independent limit on the mass of the electron antineutrino from tritium beta decay. Nonetheless, it appears fairly certain that the tritium beta decay experiments provide a limit on the mass of the electron antineutrino of 10 eV.

Although there is no evidence for a finite neutrino mass in any direct kinematic tests or from double beta decay, there is evidence from the reported observation of neutrino oscillations in three areas.

LSND has reported evidence for $\bar{\nu}_\mu \rightarrow \bar{\nu}_e$ oscillations from pion decay at rest and $\nu_\mu \rightarrow \nu_e$ from pion decay in flight with best fit values of $\Delta m^2 \approx 1$ eV^2 and $\sin^2 2\theta = 0.03$. This claim appears to be at least somewhat inconsistent with the results from the KARMEN experiment. A new experiment (miniBooNE) is now under construction at FermiLab and should be able to definitively resolve the validity of the LSND claim.

A number of atmospheric neutrino experiments reported possible evidence for neutrino oscillations as much as ten years ago. With the much higher statistics and reduced systematics available from the SuperKamiokande, there is now a solid claim for the existence of either oscillations with $\Delta m^2 = 3\text{-}4 \times 10^{-3}$ eV^2 and large $\sin^2 2\theta$. The most convincing evidence for oscillations is from the azimuthal dependence of the ν_μ flux from SuperKamiokande. This claim is also supported by new data from MACRO and Soudan II. A long-baseline accelerator experiment (K2K) from KEK to SuperKamiokande is now underway to provide a terrestrial test of this claim. While K2K will not have enough sensitivity to cover the entire allowed region, another long-baseline experiment (Minos) from FermiLab to Soudan should start operation in 2002 with enough sensitivity to cover virtually the entire allowed region.

Finally, five experiments have reported a significant deficit of the flux of solar neutrinos. No valid astrophysical explanation of the results has been found and it appears that the most likely solution to the solar neutrino problem involves neutrino oscillations of either $\nu_e \rightarrow \nu_\mu$ (or to ν_τ) or $\nu_e \rightarrow \nu_s$. All of the data is consistent with oscillations occurring with four different ranges of parameters. In order to confirm that oscillations are occurring, it is essential to carry out model independent tests such as temporal variations in the solar neutrino flux, distortion of the ^{8}B neutrino spectrum, or direct observation of ν_μ or ν_τ from the Sun. A new experiment (SNO) is just now starting data acquisition with the goal of providing a model-independent resolution of the solar neutrino problem by searching for all three of the effects listed above. SNO should also be able to determine if sterile neutrinos play a role in solar neutrinos.

If one assumes that there is a natural hierarchy to neutrino mass (so that the neutrinos follow the same ordering in mass as their associated leptons), then one can approximate $\sqrt{\Delta m_{12}^2} = \sqrt{m_2^2 - m_1^2} \approx m_2$ and similarly for Δm^2_{23}, etc. Then one can conclude the following from the data:

The solar neutrino experiments observe only the disappearance of ν_e and allow three regions of Δm^2. Assuming that the mass of the electron neutrino is small compared to the other neutrino masses, then the mass of ν_μ (or ν_τ or ν_s) is in the

424

range of 10 μeV < m_v < 10 meV. Thus, the largest neutrino mass allowed by the solar neutrino data is too small to be of importance in cosmology.

The atmospheric neutrino experiments observe the disappearance of v_μ. However, the long-baseline reactor experiments (CHOOZ and Palo Verde) rule out oscillations to v_e. Thus, either $v_\mu \to v_\tau$ or $v_\mu \to v_s$ and the allowed mass range for v_τ or v_s is 10 meV < m_v < 100 meV. Again, a neutrino mass in this range will not significantly affect cosmology.

Finally, the LSND data indicates $v_\mu \to v_e$ with a range of the mass of v_μ of 0.5 < m_v < 5 eV. A neutrino in this mass range, while not sufficient to close the Universe, is large enough to affect the large scale structure of the Universe, and thus is of cosmological importance [143].

If the LSND result is correct, then one does not necessarily have a natural hierarchy of neutrino masses as v_μ may be as (or more) more massive than v_τ. In fact, if one accepts the LSND, atmospheric, and solar neutrino data at face value, then the implication for neutrino mass is that there must be a fourth, sterile neutrino. This is a result of the fact that $\Delta m^2_{12} + \Delta m^2_{23} = \Delta m^2_{13}$. Thus, with only three types of neutrinos, one can only have two independent values of Δm^2.

A number of authors have made different models for the neutrino mass matrix under varying assumptions about which set of the experimental data should be included [144]. Many of these models have mass hierarchies that do not mirror the lepton masses. Thus, it is impossible to make any general statements about constraints on the neutrino mass matrix given the available data.

The resolution of the existence of a finite neutrino mass is under direct attack on many fronts and one may expect that this question will be definitively resolved in the next several years. This upcoming set of experiments, which are either starting operation or are under construction, should be able to resolve the issue of the existence of a sterile neutrino and determine (to first order) the neutrino mass matrix and mixing parameters. Thus, some 70 years after the postulation of the existence of the neutrino, we appear to be finally on the brink of resolving the question of neutrino mass.

Acknowledgements

The author is indebted to many people for useful discussions as well as to insightful papers in this field. In particular, I would like to thank Frank Avignone, John Bahcall, Baha Dalantenkin, Milla Baldo-Ceolin, Barry Barish, Eugene Beier, Enrico Belotti, Venia Berezinsky, Hans Bethe, Felix Boehm, Jochen Bonn, Steve Brice, Jacques Bouchez, Dave Caldwell, Bruce Cleveland, Raman Cowsik, Bob Cousins, Michel Cribier, Alain de Bellefon, Ray Davis, Guido Drexlin, Steve Elliott, Klaus Eitel, Ettore Fiorini, Vladimir Gavrin, Terry Goldman, Maury Goodman, Giorgio

Gratta, Dick Hahn, Naoya Hata, Wick Haxton, Peter Herczeg, Andrew Hime, Eugen Holzschuh, Boris Kayser, Till Kirsten, Hans Klapdor-Kleingrothaus, Anatoli Kopylov, Ken Lande, Paul Langacker, Bob Lanou, John Learned, Vladimir Lobashev, Bill Louis, Al Mann, Reinhard Maschuw, Art McDonald, Mike Moe, Rabindu Mohapatra, Jeff Nico, Serguey Petcov, Andreas Piepke, Mark Pinsonneault, Raju Raghavan, Hamish Robertson, Peter Rosen, Keith Rowley, Bernard Sadoulet, Hank Sobel, Dick Slansky, Alexei Smirnov, Michel Spiro, Jerry Stephenson, Wolfgang Stoeffl, Yoichiro Suzuki, Charling Tao, Yoji Totsuka,Sylvaine Turck-Chieze, Jose Valle, Evgeny Veretenkin, Daniel Vignaud, Petr Vogel, Franz von Feilitzsch, Hywel White, Paul Wildenhain, John Wilkerson, Klaus Winter, George Zatsepin, Bernard Zeitnitz, and Kai Zuber.

References

1. Kolb E.W., Turner M.S., *The Early Universe*, Addison-Welsey Publishing. Co. (1990) 507.
2. Ramond P., *Field Theory: A Modern Primer*, Benjamin/Cummings (1984).
3. Partricle Data Group, *European Physical Journal* **C3** (1998) 319.
4. Barate R. *et al.*, (ALEPH collaboration), *Euro. Phys. Jour.* **C2** (1998) 395.
5. Abela R. *et al.*, *Phys. Lett.* **B146** (1984) 431.
6. Jeckelman B. *et al.*, *Phys. Rev. Lett.* **56** (1986) 1444; *Nucl. Phys.* **A457** (1986) 709.
7. Particle Data Group, *Phys. Lett.* **B170** (1986) 1.
8. Frosch R., (1992) *private communication*.
9. Daum M. *et al.*, *Phys. Lett.* **B265** (1991) 425.
10. Jeckelman B. *et al.*, *Phys. Lett.* **B335** (1994)326.
11. Assamagan *et al.*, *Phys. Rev.* **D53** (1996) 6065.
12. Kirch K. (1999), *private communication.*
13. Anderhub H.B. *et al.*, *Phys. Lett.* **B114** (1982) 76.
14. Boris S. *et al.*, *JETP Lett.* **45** (1987) 333.
15. Robertson, R.G.H. *et al.*, *Phys. Rev. Lett.* **67** (1991) 957.
16. Holzschuh E. *et al.*, *Phys. Lett.* **B287** (1992) 381.
17. Kawakami H. *et al.*,. *Phys. Lett.* **B256** (1991) 105.
18. Stoeffl W. and Decman D., *Phys. Rev. Lett.* **75** (1995) 3237.
19. Weinheimer C. *et al.*, *Phys. Lett.* **B300** (1993) 210.
20. Sun *et al.*, *Chinese Jour. Nucl. Phys.* **15** (1993) 261.
21. Belesev *et al.*, *Phys. Lett.* **B350** (1995) 263.
22. Partricle Data Group, *Euro. Phys. Jour.* **C3** (1998) 313.

23. Barth H. et al., Proc. 18th International Conference on Neutrino Physics and Astrophysics (Neutrino 98), Takayama, Japan, June 4 - 9, 1998, ed. Y. Suzuki and Y. Totsuka, North Holland, Nucl. Phys. B (Proc. Suppl.) 77 (1999) 321.

24. Lobashev V.M., Proc. 8th Int'l. Workshop on Neutrino Telescopes, Venice Italy, Feb 23-26, 1999, ed. M. Baldo-Ceolin, Edizioni Papergraf, (1999) 19.

25. Weinheimer C. et al., Phys. Lett. B460 (1999) 219.

26. Lobashev V., to appear in Proc. of the 6th Int'l. Workshop on Topics in Astroparticle and Underground Physics (TAUP 99), Paris, France, Sept 6-10, 1999.

27. Bethe. H.A., Rev. Mod. Phys. 62 (1990) 801, Trimble V., Rev. Mod. Phys. 60 (1988) 859.

28. Partricle Data Group, European Physical Journal C3 (1998) 313.

29. Moe M., Proc. 16th Int'l Conference on Neutrino Physics and Astrophysics (Neutrino 94), Eilat, Israel, May 29 – June 3, 1994, ed. Dar A., Eilam G., Gronau M., North Holland, Nucl. Phys. B (Proc Suppl) 38 (1995) 37.

30. Beck M. et al., Massive Neutrinos – Tests of Fundamental Symmetries, Les Arcs, France, Jan 26–Feb 2, 1991, ed. Fackler O., Fontaine G., Tran Thanh Van J., Editions Frontieres (1991) 91.

31. Heusser G., Ann. Rev. Nucl. Part. Sci. 45 (1995) 543.

32. Avignone F.T., et al., J. Phys. G: Nucl. Phys. 17 (1991) S181.

33. Klapdor-Kleingrothaus H.V., Int. J. Mod. Phys. A13 (1998) 3953.

34. Klapdor-Kleingrothaus H.V., Proc. of the Fifth Int'l. Weak and Electromagnetic Interactions (WEIN 98) Symposium, Santa Fe, June 14-19, 1998, ed. Herczeg P., Hoffman C.M., Klapdor-Kleingrothaus H.V., World Scientific (1999) 275.

35. Kayser B., Gibrat-Debu F., Perrier F., The Physics of Massive Neutrinos, World Scientific (1989).

36. Boehm F., Vogel P, Physics of Massive Neutrinos, Cambridge Univ. Press (1987).

37. Zacek G. et al., Phys. Rev. D34 (1986) 2621; Vidaykin G.S. et al., J. Moscow Phys. Soc. 1 (1991) 85.

38. Appollonio et al., Phys. Lett. B420 (1998) 397.

39. Boehm F., Proc. 8th Int'l Workshop on Neutrino Telescopes (1999) 311.

40. Boehm, F., to appear in Proc. of TAUP 99.

41. Suzuki, A., Proc. 8th Int'l Workshop on Neutrino Telescopes (1999) 325.

42. Athanassopoulos, C. et al., Nucl. Instrum. Meth. A338 (1997) 149.

43. Athanassopoulos, C., et al., Phys. Rev. Lett. 75 (1995) 2650, Athanassopoulos, C., Phys. Rev. Lett. 77 (1996) 3082, Athanassopoulos, C., et al., Phys. Rev. C54 (1996) 2685, White, D.H., Proc. Neutrino 98, Nucl. Phys. B (Proc. Suppl.) 77 (1999) 207.

44. Athanassopoulos, C., *et al.*, *Phys. Rev.* **C58** (1998) 2489, Athanassopoulos, C., *Phys. Rev. Lett.* **81** (1998) 1774.
45. Louis, W.C. *et al.*, *Proc. WEIN 98* (1999) 112.
46. Zeitnetz, B. *et al.*, *Prog. Nucl. Physics* **40** (1998) 169.
47. Armbruster, B. et al., *Phys. Rev.* **C57** (1998) 3414.
48. Feldman, G.J. and Cousins, R.D., *Phys. Rev.* **D57** (1998) 3873.
49. Osamu, S. *et al.*, *Proc. Neutrino 98, Nucl. Phys. B (Proc. Suppl.)* **77** (1999) 220.
50. Gomez-Cadenas, J.J., *et al.*, *Proc. Neutrino 98, Nucl. Phys. B (Proc. Suppl.)* **77** (1999) 225.
51. Meyer J.P., to appear in *Proc. TAUP 99*.
52. Honda, M., *Proc. Neutrino 98, Nucl. Phys. B (Proc. Suppl.)* **77** (1999) 140.
53. Gaisser, T.K., *Proc. Neutrino 98, Nucl. Phys. B (Proc. Suppl.)* **77** (1999) 133.
54. Kajita, T. *et al.*, *Proc 26th Intl. Conf. on High Energy Physics*, August 6-12, 1992, Dallas, Texas, ed. Sanford J.R., *AIP Conference Proc.* **272** (1992) 1187.
55. IMB Collaboration, *Phys. Rev. Lett.* **66** (1991) 2561; *Phys. Rev. Lett.* **69** (1992) 1010.
56. Frejus Collaboration, *Phys. Lett. B* **245** (1991) 305.
57. NUSEX Collaboration, *Europhys. Lett.* **8** (1989) 611.
58. Auerbach, N. et al., *Phys. Rev.* **C56** (1997) R2368.
59. Nakamura K., *et al.*, (SuperKamiokande Collaboration) *Physics and Astrophysics of Neutrinos*, ed. Fukugita M., Suzuki A., Springer-Verlag (1994) 249.
60. SuperKamiokande Collaboration, *Phys. Rev. Lett.* **81** (1998) 1562; SuperKamiokande Collaboration, *Phys. Rev. Lett.* **82** (1999) 2644; SuperKamiokande Collaboration, *Phys. Lett.* **B467** (1999) 185.
61. Circella, M., *Proc. 8th International Workshop on Neutrino Telescopes*, (1999) 229; Lipari, P., *Proc. 8th International Workshop on Neutrino Telescopes*, (1999) 245.
62. Kajita K. *et al.*, (SuperKamiokande Collaboration) *Proc. Neutrino 98, Nucl. Phys. B (Proc. Suppl.)* **77** (1999) 123; Scholberg K., *et al.*, (SuperKamiokande Collaboration) *Proc. 8th International Workshop on Neutrino Telescopes*, (1999) 183.
63. Agrawal, V. *et al.*, *Phys. Rev.* **D53** (1996) 1314.
64. Spinetti, M. for the MACRO collaboration, *Proc. 8th International Workshop on Neutrino Telescopes*, (1999) 217.
65. Allison, W.W.M. *et al.*, (Soudan 2 collaboration) *Nucl. Instrum. Methods* **A381** (1996) 385.

428

66. Mann, A. et al., (Soudan 2 collaboration) Proc. 8th International Workshop on Neutrino Telescopes, (1999) 203; Allison, W.W.M. et al., (Soudan 2 collaboration) Phys. Lett. **B449** (1999) 137.

67. Nakahata M., et al., (SuperKamiokande Collaboration) to appear in Proc. of TAUP 99.

68. Nishikawa, K., et al., Proc. Neutrino 98, Nucl. Phys. B (Proc. Suppl.) **77** (1999) 198.

69. Bernstein, B. et al., (MINOS collaboration) Proc. 8th Int'l. Workshop on Neutrino Telescopes, (1999) 391.

70. Ereditato, A. and Migliozzi, P., Proc. 8th Int'l. Workshop on Neutrino Telescopes, (1999) 361.

71. Bahcall J.N., Neutrino Astrophysics, Cambridge Univ. Press (1989)

72. Bahcall J.N., Basu S, and Pinsonneault MH, Phys. Lett. **B433**:1 (1998)

73. Bahcall JN, Pinsonneault MH, Rev. Mod. Phys. **64**:885 (1992)

74. Brun A.S., Turck-Chieze S., and Morel P., Astrophys. J. **506** (1998) 913.

75. Brun A.S., Turck-Chieze S., Zahn J.P., Astrophys. J. **525** (1999) 1032.

76. Turck-Chieze S., Lopes I., Astrophys. J. **408** (1993) 347.

77. Iglesias C.A., Rogers F.J., Astrophys. J. **464** (1996) 943.

78. Rogers F.J., Swenson J., and Iglesias C.A., Astrophys. J. **456** (1996) 902.

79. Turck-Chieze S., Proc. 8th Int'l. Workshop on Neutrino Telescopes, (1999) 147.

80. Lazrek M., Baudin F., Bertello F., et al., Sol. Phys. **175** (1997) vol 2.

81. Kosovichev A.G., Schou J., Scherrer P.H., et al., Sol. Phys. **170** (1997) 43.

82. Fiorentini G., Ricci B., Proc. 8th Int'l. Workshop on Neutrino Telescopes, (1999) 79.

83. Dar A., Shaviv G., Astrophys. J. **468** (1996) 933.

84. Cumming A., Haxton W.C., Phys. Rev. Lett. **77** (1996) 4286.

85. Heeger K.M., Robertson R.G.H., Phys. Rev. Lett. **77** (1996) 3720.

86. Bahcall J.N. et al., Phys. Rev. Lett. **28** (1972) 316.

87. Gilliland R.L., Dappen W., Astrophys. J. **324** (1987) 1153; Gilliland R.L., Faulkner J., Press W.H., Spergel D.N., Astrophys. J. **306** (1986) 703; Spergel D., Press W., Astrophys. J. **294** (1985) 663; Faulkner J., Gilliland R.L., Astrophys. J. **299** (1985) 994.

88. Mikheyev S.P., Smirnov A.Yu., Sov. J. Nucl. Phys. **42** (1985) 1441; Wolfenstein L., Phys. Rev. **D17** (1978) 2369; Wolfenstein L., Phys. Rev. **D20** (1979) 2634.

89. Bethe H.A., Phys. Rev. Lett. **56** (1986) 1305; Bahcall J.N., Haxton W.C., Phys. Rev. **D40** (1989) 931; Bahcall J.N., Phys. Rev. **D44** (1991) 1644; Baltz A.J., Weneser J., Phys. Rev. Lett. **66** (1991) 520; Gelb J.M., Kwong W., Rosen S.P., Phys. Rev. Lett. **69** (1992) 1864, Phys. Rev. Lett. **78** (1997) 2296;

Shi X., Schramm D.N., Bahcall J.N., *Phys. Rev. Lett.* **69** (1992) 717; Berezinsky V., *Proc. 8th Int'l. Workshop on Neutrino Telescopes*, (1999) 91; Bilenky S.M., Petcov S.T., *Rev. Mod. Phys.* **59** (1987) 671; Baltz A.J., Goldhaber A.S., Goldhaber. M., *Phys. Rev. Lett.* **81** (1998) 5730; Langacker P., Hata N., *Phys. Rev.* **D56** (1997) 6107; Maris M., Petcov S.T., *hep-ph/9803244*; Bahcall J.N., Krastev P.I., Smirnov A. Yu., *Phys. Rev.* **D60** (1999) 93001-13, *hep-ph/9905220*; Bahcall J.N., Krastev P.I., Smirnov A. Yu., *Phys. Rev.* **D58** (1998) 09016.

90. Hata N., Langacker P., *Phys. Rev.* **D56** (1997) 6107.

91. Bahcall J.N. et al., *Phys. Rev.* **D56** (1998) 6107

92. Cleveland B.T. *et al.*, *Astrophys. J.* **496** (1998) 505

93. Cleveland B. et al., *Proc. 25th International. Conf High Energy Phys.*, ed. Phua K.K., Yamaguchi Y., South Asia Theor. Phys. Assn. and Phys. Soc. Japan (1991) 667

94. Fillipone B.R., Vogel P., *Phys. Lett.* **B246** (1990) 546; Bieber J.W., Seckel D., Stanev T., Steigman G., *Nature* **348** (1990) 407; Krauss L.M., *Nature* **348** (1990) 403; Bahcall J.N., Press W.H., *Astrophys. J.* **370** (1991) 730; Nunokawa H., Minakata H., *International. J. Mod. Phys.* **A6** (1991) 2347.

95. Voloshin M.B., Vysotskii M.I., Okun L.B., *Sov. J. Nucl. Phys.* **44** (1986) 440; Okun L.B., *Sov. J. Nucl. Phys.* **44** (1986) 546; Lim C.S., Marciano W.J., *Phys. Rev.* **D37** (1988) 1368; Akhmedov E.Kh., *Phys. Lett.* **B225** (1991) 84; Babu K.S., Mohapatra R.N., Rothstein I.Z., *Phys. Rev.* **D44** (1991) 2265; Akhmedov E., Phys. Lett. D213 (1998) 64;

96. Krakauer D. et al., *Phys. Lett.* **B252** (1990) 177

97. Raffelt G., *Astrophys. J.* **336** (1989) 61; Barbieri R., Mohapatra R.N., *Phys. Rev. Lett.* **61** (1988) 27; Fukigita M., Notzold D., Raffelt G., Silk J., *Phys. Rev. Lett.* **60** (1988) 879; Goldman I., Aharanov Y., Alexander G., Nussinov S., *Phys. Rev. Lett.* **60** (1988) 1789; Lattimer J.M., Cooperstein J., *Phys. Rev. Lett.* **61** (1988) 23; Lattimer J.M., Cooperstein J., *Phys. Rev. Lett. erratum* **61** (1988) 2633; Noetzold D., *Phys. Rev.* **D38** (1988) 1658

98. Raffelt G., *Phys. Rev. Lett.* **64** (1990) 2856

99. McNutt, R.L. Jr., *Proc. Fourth Int'l. Solar Neutrino Conference*, Heidelberg, Germany, April 8-11, 1997, ed. Hampel W., Max Planck Institute for Kernphysik (1997) 370; Rust D.M., McNutt, R.L. Jr., *Proc. Fourth Int'l. Solar Neutrino Conf.*, (1997) 380; McNutt L. Jr., *Science* **270** (1995) 1635.

100. Kuzmin V.A., *Zh. Eksp. Teor. Fiz.* **49** (1965) 1532 [*Sov. Phys. JETP* **22** (1966) 1051].

101. Abdurashitov J.N. *et al.*, (SAGE Collaboration) *Phys. Rev.* **C60** (1999) 055801.

102. Abdurashitov J.N. *et al.*, (SAGE Collaboration) *Phys. Rev.* **C59** (1999) 2246.

103. Bahcall J.N., *Phys. Rev.* **C56** (1997) 3391.
104. Haxton W., *Phys. Lett.* **B431** (1998) 110.
105. Abdurashitov J.N. *et al.*, (SAGE Collaboration) *Phys. Rev.Lett.*83 (1999) 4686.
106. Hampel W., *et al.*, (GALLEX Collaboration) *Phys. Lett.* **B447** (1999) 127.
107. Hampel W. *et al.*, (GALLEX Collaboration) *Phys. Lett.* **B346** (1998) 158.
108. Anselmann P. *et al.*, (GALLEX Collaboration) *Phys. Lett.* **B342** (1995) 440.
109. Anselmann P. *et al.*, (GALLEX Collaboration) *Phys. Lett.* **B420** (1998) 114.
110. Kirsten T., *Proc. 18th Neutrino 98*, *Nucl. Phys. B (Proc. Suppl.)* **77** (1999) 26.
111. Hirata K.S. *et al.*, (Kamiokande Collaboration) *Phys. Rev.* **D44** (1991) 2241.
112. Rosen S.P., Kayser B., *Phys. Rev.* **D23** (1981) 669; Bahcall J.N., *Rev. Mod. Phys.* **59** (1987) 505.
113. Nakamura K., (Kamiokande Collaboration) *Proc. 15th Int'l. Conference on Neutrino Physics and Astrophysics (Neutrino 92)*, Granada, Spain, June 7-12, 1992, ed. Morales, A., North Holland, *Nucl. Phys. B (Proc. Suppl.)* **31** (1993) 105.
114. Suzuki Y., et al., (Kamiokande Collaboration) *Proc. 17th Int'l. Conference on Neutrino Physics and Astrophysics (Neutrino 96)*, Helsinki, Finland, June 13-19, 1996, ed. Enqvist K., Huitu K., Maalampi J., World Scientific (1997) 73.
115. Hirata K.S. *et al.*, (Kamiokande Collaboration) *Phys. Rev. Lett.* **65** (1990) 1301.
116. Hirata K.S. *et al.*, *Phys. Rev. Lett.* **66** (1991) 9.
117. Takeuchi Y., *et al.*, (SuperKamiokande Collaboration) *Phys. Lett.* **452** (1999) 418.
118. Totsuka Y., *et al.*, (SuperKamiokande Collaboration) to appear in *Proc. of TAUP 99*.
119. SuperKamiokande Collaboration, *Phys. Rev. Lett.* **81** (1998) 1158.
120. SuperKamiokande Collaboration, Phys. Rev. Lett. 82 (1999) 2430.
121. Inoue K., (SuperKamiokande Collaboration) *Proc. 8th Int'l. Workshop on Neutrino Telescopes*, (1999) 53.
122. Bahcall J.N., Krastev P.I., *Phys. Lett.* **B436** (1998) 243.
123. SuperKamiokande Collaboration, *Phys. Rev. Lett.* **82** (1999) 1810.
124. Bahcall J.N., Lisi E., *Phys. Rev.* **D54** (1996) 5417.
125. SNO Collaboration, to appear in Nucl. Instr. Meth. A (2000); nucl-ex/9910016.
126. Meijer Dress R., (SNO Collaboration) *Proc. Fourth Int'l. Solar Neutrino Conf.*, (1997) 210.
127. Hime A., (SNO Collaboration) *Proc. Fourth Int'l. Solar Neutrino Conf.*, (1997) 218.

128. Borexino Collaboration, *Astroparticle Phys.* **8** (1998) 141; Feilitzsch, F. v., *Prog. in Particle and Nucl. Phys.* **40** (1998) 123.

129. Tao C., *Proc. Fourth Int'l. Solar Neutrino Conf.*, Heidelberg, (1997) 238.

130. Lanou R., *Proc. 8th Int'l. Workshop on Neutrino Telescopes*, (1999) 139; Bandler, S.R., *et al.*, *Phys. Rev. Lett.* **74** (1997) 3169.

131. Broggini C., *Nucl. Phys. B (Proc. Suppl.)* **70** (1998) 188.

132. Bowles T., Gavrin V.N., *Proc. Neutrino 96* (1997) 83.

133. Swift A.M.., *et al.*, *Nucl. Phys. B (Proc. Suppl.)* **35** (1994) 405.

134. Bouchez J., *Proc. 8th Int'l. Workshop on Neutrino Telescopes*, (1999) 127; Raghavan R.S., *Phys. Rev. Lett.* **78** (1997) 3618.

135. Raghavan R.S., *Phys. Rev. Lett.* **71** (1993) 4295.

136. Chang C.C., Chang C.Y., Collins G, *Nucl. Phys. B (Proc. Suppl.)* **35** (1994) 464.

137. Cleveland B.T., *et al.*, *Proc. Fourth Int'l. Solar Neutrino Conf.*, (1997) 228.

138. Alessandrello A., *et al.*, *Astroparticle Phys.* **3** (1995) 239.

139. Kopylov A.V., *et al.*, *Proc. Fourth Int'l. Solar Neutrino Conf.*, (1997) 263.

140. Georgadze A.S., *et al.*, *Proc. Fourth Int'l. Solar Neutrino Conf.*, (1997) 283.

141. Barabanov I.R., et al., *Nucl. Phys. B (Proc. Suppl.)* **35** (1994) 461.

142. Rubbia C., *Nucl. Phys. B (Proc. Suppl.)* **48** (1996) 172.

143. Primack J.R., et al., *Phys. Rev. Lett.* **74** (1995) 2160; Caldwell D.O., Mohapatra R.N., *Phys. Rev.* **D48** (1993) 3259; Peltoniemi J.T., Valle J.W.F., *Nucl. Phys.* **B406** (1993) 409; Shi Z. et al., *Phys. Rev.* **D48** (1993) 2563; Primack J., *Science 280* (1998) 1398; Gawiser E., Silk J., *Science* **280** (1998) 1405; Caldwell D.O., *Proc. Neutrino 98, Nucl. Phys. B (Proc. Suppl.)* **77** (1999) 420; Langacker P., *Proc. Neutrino 98, Nucl. Phys. B (Proc. Suppl.)* **77** (1999) 241; Turner M.S., *Proc. WEIN 98* (1999) 626.

144. Hata N., Langacker, P., *Phys. Rev.* **D56** (1997) 6107; Barger V., *et al.*, *Phys. Lett.* **B437** (1998) 107; Fritzsch H., Xing Z.-z., *hep-ph/9808272*, *hep-ph-9807234*; Georgi H., Glashow S.L., *hep-ph/9808293*; Gibbons C. et al., *Phys. Lett.* **B430** (1998) 296; Liu Q.Y., Smirnov A. Yu., *Nucl. Phys.* **B524** (1998) 505; Chun E.J., Kim C.W., Lee U.W., *Phys. Rev.* **D58** (1998), *hep-ph/9802209*; Okada N., Yasuda O., *Int. J. Mod. Phys.* **A12** (1997) 3669; Berezinsky V., Fiorentini G., Lissia M., hep-ph/9811352; Langacker P., *Phys. Rev.* **D58** (1998) 093017, *hep-ph/9805281*; Bilenky S., Giunti C., Grimus W., *Eur. Phys. J.* **C1** (1998) 247, *hep-ph/9805368*; Smirnov A. Yu., Vissani, F., *hep-ph/9710565*; Mohapatra R.*N., Proc. WEIN 98* (1999) 28; Smirnov A. Yu., *Proc. WEIN 98* (1999) 180; Smirnov A., *Proc. Neutrino 98, Nucl. Phys. B (Proc. Suppl.)* **77** (1999) 98.

TOPICS IN NEUTRINO ASTROPHYSICS

W. C. HAXTON

Institute for Nuclear Theory, Box 351550, and Department of Physics, Box 351560
University of Washington, Seattle, WA 98195, USA
E-mail: haxton@phys.washington.edu

In these TASI summer school lectures I discuss three topics in neutrino astrophysics: the solar neutrino problem, stellar cooling by neutrino emission, and the role of neutrinos in the nucleosynthesis that occurs within core-collapse supernovae.

1 Introduction

Part of the interest in neutrino astrophysics has to do with the fascinating interplay between nuclear and particle physics issues — e.g., whether neutrinos are massive and undergo flavor oscillations, whether they have detectable electromagnetic moments, etc. — and astrophysical phenomena, such as the clustering of matter on large scales, the mechanisms responsible for the synthesis of nuclei, and the evolution of stars. The three lectures here are intended to illustrate this interplay. The first lecture reviews the solar neutrino problem which, along with the atmospheric neutrino problem, provides perhaps our strongest direct evidence that new physics lurks beyond the standard model. The second has to with the implications of neutrino properties — e.g., whether neutrinos are Dirac or Majorana particles — for stellar cooling. The final lecture describes the nucleosynthesis we think accompanies a supernova explosion, and why that synthesis is a delicate probe of neutrino oscillations.

These lectures as well of those from several other 1998 TASI speakers share a common subtheme: how the extraordinary technical revolution in astronomy and astrophysics has made the microphysics of the universe more relevant. It is the precise data coming from the new generation of great observatories — maps of the cosmic microwave spectrum, precision measurements of the products of big bang nucleosynthesis, measurements of the solar neutrino spectrum, Hubble Space Telescope (HST) abundance distributions from early, metal-poor stars, detection of gamma ray bursts from cosmological sources — that allow us to form the connections between observations and the underlying microphysics. This is the driving force that is making the field of nuclear and particle astrophysics of such interest to both senior physicists and new students entering the field.

2 Solar Neutrinos [1]

More than three decades ago Ray Davis, Jr. and his collaborators [2] constructed a 0.615 kiloton C_2Cl_4 radiochemical solar neutrino detector in the Homestake Gold Mine, one mile beneath Lead, South Dakota. Within a few years it was apparent that the number of neutrinos detected was considerably below the predictions of the standard solar model, that is, the standard theory of main sequence stellar evolution.

Today the results from the ^{37}Cl detector, which have become quite accurate due to 30 years of careful measurement, have been augmented by results from four other experiments, the SAGE [3] and GALLEX [4] gallium experiments and the Kamiokanda [5] and SuperKamiokande [6] water Cerenkov detectors. It now appears that the combined results are very difficult to explain — some have argued impossible — by any plausible change in the standard solar model (SSM). Thus most believe that the answer to the solar neutrino problem is new particle physics, most likely some effect like solar neutrino oscillations associated with massive neutrinos. With the recent news that SuperKamiokande sees direct evidence for ν_μ oscillations in the azimuthal dependence of atmospheric [7] neutrinos, it seems that we may be on the threshold of a major discovery.

The purpose of this first (and longest) lecture is to summarize the solar neutrino problem and the arguments that it represents new particle physics.

2.1 The Standard Solar Model [8]

Solar models trace the evolution of the sun over the past 4.6 billion years of main sequence burning, thereby predicting the present-day temperature and composition profiles of the solar core that govern neutrino production. Standard solar models share four basic assumptions:
* The sun evolves in hydrostatic equilibrium, maintaining a local balance between the gravitational force and the pressure gradient. To describe this condition in detail, one must specify the equation of state as a function of temperature, density, and composition.
* Energy is transported by radiation and convection. While the solar envelope is convective, radiative transport dominates in the core region where thermonuclear reactions take place. The opacity depends sensitively on the solar composition, particularly the abundances of heavier elements.
* Thermonuclear reaction chains generate solar energy. The standard model predicts that over 98% of this energy is produced from the pp chain conversion of four protons into ^4He (see Fig. 1)

$$4p \rightarrow\,^4He + 2e^+ + 2\nu_e \tag{1}$$

with proton burning through the CNO cycle contributing the remaining 2%. The sun is a large but slow reactor: the core temperature, $T_c \sim 1.5 \cdot 10^7$ K, results in typical center-of-mass energies for reacting particles of ~ 10 keV, much less than the Coulomb barriers inhibiting charged particle nuclear reactions. Thus reaction cross sections are small: in most cases, as laboratory measurements are only possible at higher energies, cross section data must be extrapolated to the solar energies of interest.

* The model is constrained to produce today's solar radius, mass, and luminosity. An important assumption of the standard model is that the sun was highly convective, and therefore uniform in composition, when it first entered the main sequence. It is furthermore assumed that the surface abundances of metals (nuclei with $A > 5$) were undisturbed by the subsequent evolution, and thus provide a record of the initial solar metallicity. The remaining parameter is the initial ^{4}He/H ratio, which is adjusted until the model reproduces the present solar luminosity after 4.6 billion years of evolution. The resulting ^{4}He/H mass fraction ratio is typically 0.27 ± 0.01, which can be compared to the big-bang value of 0.23 ± 0.01. Note that the sun was formed from previously processed material.

The model that emerges is an evolving sun. As the core's chemical composition changes, the opacity and core temperature rise, producing a 44% luminosity increase since the onset of the main sequence. The temperature rise governs the competition between the three cycles of the pp chain: the ppI cycle dominates below about $1.6 \cdot 10^7$ K; the ppII cycle between $(1.7\text{-}2.3) \cdot 10^7$ K; and the ppIII above $2.4 \cdot 10^7$ K. The central core temperature of today's SSM is about $1.55 \cdot 10^7$ K.

The competition between the cycles determines the pattern of neutrino fluxes. Thus one consequence of the thermal evolution of our sun is that the ^{8}B neutrino flux, the most temperature-dependent component, proves to be of relatively recent origin: the predicted flux increases exponentially with a doubling period of about 0.9 billion years.

A final aspect of SSM evolution is the formation of composition gradients on nuclear burning timescales. Clearly there is a gradual enrichment of the solar core in ^{4}He, the ashes of the pp chain. Another element, ^{3}He, is a sort of catalyst for the pp chain, being produced and then consumed, and thus eventually reaching some equilibrium abundance. The timescale for equilibrium to be established as well as the eventually equilibrium abundance are both sharply decreasing functions of temperature, and thus increasing functions of the distance from the center of the core. Thus a steep ^{3}He density gradient is established over time.

The SSM has had some notable successes. From helioseismology the sound

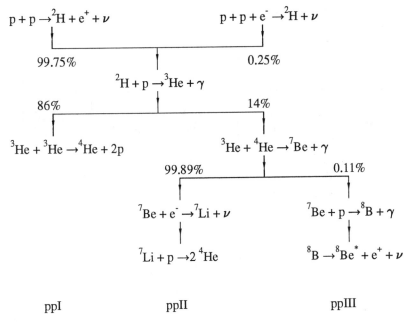

Figure 1: The solar pp chain.

speed profile $c(r)$ has been very accurately determined for the outer 90% of the sun, and is in excellent agreement with the SSM. Such studies verify important predictions of the SSM, such as the depth of the convective zone. However the SSM is not a complete model in that it does not explain all features of solar structure, such as the depletion of surface Li by two orders of magnitude. This is usually attributed to convective processes that operated at some epoch in our sun's history, dredging Li to a depth where burning takes place.

The principal neutrino-producing reactions of the pp chain and CNO cycle are summarized in Table 1. The first six reactions produce β decay neutrino spectra having allowed shapes with endpoints given by E_ν^{max}. Deviations from an allowed spectrum occur for ^8B neutrinos because the ^8Be final state is a broad resonance. The last two reactions produce line sources of electron capture neutrinos, with widths ~ 2 keV characteristic of the temperature of the solar core. Measurements of the pp, ^7Be, and ^8B neutrino fluxes will determine the relative contributions of the ppI, ppII, and ppIII cycles to solar energy generation. As discussed above, and as later illustrations will show more clearly, this competition is governed in large classes of solar models by

Table 1: Solar neutrino sources and the flux predictions of the BP98 and Brun/Turck-Chieze/Morel SSMs in $cm^{-2}s^{-1}$.

Source	E_ν^{max} (MeV)	BP98	BTCM98
$p + p \rightarrow {}^2H + e^+ + \nu$	0.42	5.94E10	5.98E10
${}^{13}N \rightarrow {}^{13}C + e^+ + \nu$	1.20	6.05E8	4.66E8
${}^{15}O \rightarrow {}^{15}N + e^+ + \nu$	1.73	5.32E8	3.97E8
${}^{17}F \rightarrow {}^{17}O + e^+ + \nu$	1.74	6.33E6	
${}^8B \rightarrow {}^8Be + e^+ + \nu$	~ 15	5.15E6	4.82E6
${}^3He + p \rightarrow {}^4He + e^+ + \nu$	18.77	2.10E3	
${}^7Be + e^- \rightarrow {}^7Li + \nu$	0.86 (90%)	4.80E9	4.70E9
	0.38 (10%)		
$p + e^- + p \rightarrow {}^2H + \nu$	1.44	1.39E8	1.41E8

a single parameter, the central temperature T_c. The flux predictions of the 1998 calculations of Bahcall, Basu, and Pinsonneault [8] (BP98) and of Brun, Turck-Chieze and Morel [9] are included in Table 1.

2.2 Solar Neutrino Detection [10]

Let us start with a brief reminder about low energy neutrino-nucleus interactions in detectors. Consider the charged current reaction

$$\nu_e + (A, Z) \rightarrow e^- + (A, Z + 1) \tag{2}$$

Because the momentum transfer to the nucleus is very small for solar neutrinos, it can be neglected in the weak propagator, leading to an effective contact current-current interaction. If we begin with the simplest (though fictitious) case of the free neutron decay $n \rightarrow p$, the corresponding transition amplitude is then

$$S_{fi} = \frac{G_F}{\sqrt{2}} \cos\theta_C \bar{u}(p)\gamma_\mu(1 - g_A\gamma_5)u(n)\bar{u}(e)\gamma^\mu(1 - \gamma_5)u(\nu) \tag{3}$$

where G_F is the weak coupling constant measured in muon decay and $\cos\theta_c$ gives the amplitude for the weak interaction to connect the u quark to its first-generation partner, the d quark. The origin of this effective amplitude is the underlying standard model predictions for the elementary quark and lepton currents. The weak interactions at this level are predicted by the standard model to be exactly left handed. Experiment shows that the effective coupling of the W boson to the nucleon is governed by $\gamma_\mu(1 - g_A\gamma_5)$, as noted above, where $g_A \sim 1.26$. The axial coupling is thus shifted from its underlying value by the strong interactions responsible for the binding of the quarks within the nucleon.

If an isolated nucleon were the target, one could proceed to calculate the cross section from the effective nucleon current given above. The extension to nuclear systems traditionally begins with the observation that nucleons in the nucleus are rather nonrelativistic, $v/c \sim 0.1$. The amplitude $\bar{u}(p)\gamma^\mu(1 - g_A\gamma_5)u(n)$ can be expanded in powers of p/M. The leading vector and axial operators are readily found to be

$$\gamma_0 : \quad 1$$
$$\vec{\gamma} : \quad \vec{p}/M \sim v/c$$
$$\gamma_0\gamma_5 : \quad \vec{\sigma} \cdot \vec{p}/M \sim v/c$$
$$\vec{\gamma}\gamma_5 : \quad \vec{\sigma}$$

Thus it is the time-like part of the vector current and the space-like part of the axial-vector current that survive in the nonrelativistic limit.

(In a nucleus these currents must be corrected for the presence of meson exchange contributions. The corrections to the vector charge and axial three-current, which we just pointed out survive in the nonrelativistic limit, are of order $(v/c)^2 \sim 1\%$. Thus the naive one-body currents are a very good approximation to the nuclear currents. In contrast, exchange current corrections to the axial charge and vector three-current operators are of order v/c, and thus of relative order 1. This difficulty for the vector three-current can be largely circumvented, because current conservation as embodied in the generalized Siegert's theorem allows one to rewrite important parts of this operator in terms of the vector charge operator. In the long-wavelength limit appropriate to solar neutrinos, all terms unconstrained by current conservation do not survive. In effect, one has replaced a current operator with large two-body corrections by a charge operator with only small corrections. In contrast, the axial charge operator is significantly altered by exchange currents even for long-wavelength processes like β decay. Typical axial-charge β decay rates are enhanced by ~ 2 because of exchange currents.)

If such a nonrelativistic reduction is done for our single current one obtains

$$S_{fi} \sim \cos\theta_c \frac{G_F}{\sqrt{2}}(\phi^\dagger(p)\phi(n)\bar{u}(e)\gamma^0(1 - \gamma_5)u(\nu)$$

$$-\phi^\dagger(p)g_A\vec{\sigma}\phi(n)\cdot\bar{u}(e)\vec{\gamma}(1-\gamma_5)u(\nu)) \tag{4}$$

where the ϕ are now two-component Pauli spinors for the nucleons. The above result can be generalized to include $\bar{\nu}_e$ reactions by introducing the isospin operators $\tau_\pm$ where $\tau_+ \mid n\rangle = \mid p\rangle$ and $\tau_- \mid p\rangle = \mid n\rangle$, with all other matrix elements being zero. Thus

$$\phi^\dagger(p)\phi(n) \to \phi^\dagger(N)\tau_\pm\phi(N)$$

$$\phi^\dagger(p)\vec{\sigma}\phi(n) \to \phi^\dagger(N)\vec{\sigma}\tau_\pm\phi(N).$$

This result easily generalizes to nuclear decay. Given our comments about exchange currents, the first step is the replacement

$$\tau_\pm \to \sum_{i=1}^{A}\tau_\pm(i)$$

$$\sigma\tau_\pm \to \sum_{i=1}^{A}\sigma(i)\tau_\pm(i).$$

Plugging S_{fi} into the standard cross section formula (which involves an average over initial and sum over final nuclear spins of the square of the transition amplitude) then yields the allowed nuclear matrix element

$$\frac{1}{2J_i+1}(|\langle f||\sum_{i=1}^{A}\tau_\pm(i)||i\rangle|^2 + g_A^2|\langle f||\sum_{i=1}^{A}\sigma(i)\tau_\pm(i)||i\rangle|^2). \tag{5}$$

Our initial calculation for the nucleon treated that particle as structureless. Implicitly we assumed that the momentum transfer is much smaller than the inverse nucleon size. If we take 10 MeV as a typical solar neutrino momentum transfer, these quantities would be in the ratio 1:20. For a light nucleus, the corresponding result might be 1:10. This long-wavelength approximation in combination with the nonrelativistic approximation yields the allowed result, where only Fermi and Gamow-Teller operators survive. These are the spin-independent and spin-dependent operators appearing above.

The Fermi operator is proportional to the isospin raising/lowering operator: in the limit of good isospin, which typically is good to 5% or better in the description of low-lying nuclear states, it can only connect states in the same isospin multiplet, that is, states with a common spin-spatial structure. If the initial state has isospin (T_i, M_{Ti}), this final state has $(T_i, M_{Ti} \pm 1)$ for (ν, e^-) and $(\bar{\nu}, e^+)$ reactions, respectively, and is called the isospin analog state (IAS).

In the limit of good isospin the sum rule for this operator in then particularly simple

$$\sum_f \frac{1}{2J_i + 1} |\langle f|| \sum_{i=1}^A \tau_+(i) ||i\rangle|^2 = \frac{1}{2J_i + 1} |\langle IAS|| \sum_{i=1}^A \tau_+(i) ||i\rangle|^2 = |N - Z|. \quad (6)$$

The excitation energy of the IAS relative to the parent ground state can be estimated accurately from the Coulomb energy difference

$$E_{IAS} \sim \left(\frac{1.728Z}{1.12A^{1/3} + 0.78} - 1.293\right) \text{MeV}. \quad (7)$$

The angular distribution of the outgoing electron for a pure Fermi $(N, Z) + \nu \to (N - 1, Z + 1) + e^-$ transition is $1 + \beta \cos\theta_{\nu e}$, and thus forward peaked. Here β is the electron velocity.

The Gamow-Teller (GT) response is more complicated, as the operator can connect the ground state to many states in the final nucleus. In general we do not have a precise probe of the nuclear GT response apart from weak interactions themselves. However a good approximate probe is provided by forward-angle (p,n) scattering off nuclei, a technique that has been developed in particular by experimentalists at the Indiana University Cyclotron Facility. The (p,n) reaction transfers isospin and thus is superficially like (ν, e^-). At forward angles (p,n) reactions involve negligible three-momentum transfers to the nucleus. Thus the nucleus should not be radially excited. It thus seems quite plausible that forward-angle (p,n) reactions probe the isospin and spin of the nucleus, the macroscopic quantum numbers, and thus the Fermi and GT responses. For typical transitions, the correspondence between (p,n) and the weak GT operators is believed to be accurate to about 10%. Of course, in a specific transition, much larger discrepancies can arise.

The (p,n) studies demonstrate that the GT strength tends to concentrate in a broad resonance centered at a position $\delta = E_{GT} - E_{IAS}$ relative to the IAS given by

$$\delta \sim \left(7.0 - 28.9 \frac{N - Z}{A}\right) \text{ MeV}. \quad (8)$$

Thus while the peak of the GT resonance is substantially above the IAS for $N \sim Z$ nuclei, it drops with increasing neutron excess. Thus $\delta \sim 0$ for Pb. A typical value for the full width at half maximum Γ is ~ 5 MeV.

The approximate Ikeda sum rule constrains the difference in the (ν, e^-) and $(\bar{\nu}, e^+)$ strengths

$$\sum_f (|M_{GT}^{fi}(\nu, e^-)|^2 - |M_{GT}^{fi}(\bar{\nu}, e^+)|^2) = 3(N - Z) \quad (9)$$

where

$$|M_{GT}^{fi}(\nu, e^-)|^2 = \frac{1}{2J_i + 1}|\langle f|| \sum_{i=1}^{A} \sigma(i)\tau_+(i)||i\rangle|^2. \tag{10}$$

In many cases of interest in heavy nuclei, the strength in the $(\bar{\nu}, e^+)$ direction is largely blocked. For example, in a naive $2s1d$ shell model description of ^{37}Cl, the p $\to$ n direction is blocked by the closed neutron shell at N=20. Thus this relation can provide an estimate of the total β^- strength. Experiment shows that the β^- strength found in and below the GT resonance does not saturate the Ikeda sum rule, typically accounting for $\sim (60 - 70)$ % of the total. Measured and shell model predictions of individual GT transition strengths tend to differ systematically by about the same factor. Presumably the missing strength is spread over a broad interval of energies above the GT resonance. This is not unexpected if one keeps in mind that the shell model is an approximate effective theory designed to describe the long wavelength modes of nuclei: such a model should require effective operators, renormalized from their bare values. Phenomenologically, the shell model seems to require[11] $g_A^{eff} \sim 1.0$ as well as a small spin-tensor term $(\sigma \otimes Y_2(\hat{r}))_{J=1}$ of relative strength ~ 0.1.

The angular distribution of GT $(N, Z)+\nu_e \to (N-1, Z+1)+e^-$ reactions is $3 - \beta \cos \theta_{\nu e}$, corresponding to a gentle peaking in the backward direction.

The above discussion of allowed responses can be repeated for neutral current processes such as (ν, ν'). The analog of the Fermi operator contributes only to elastic processes, where the standard model nuclear weak charge is approximately the neutron number. As this operator does not generate transitions, it is not yet of much interest for solar or supernova neutrino detection, though there are efforts to develop low-threshold detectors (e.g., cryogenic technologies) for recording the modest nuclear recoil energies. The analog of the GT response involves

$$|M_{GT}^{fi}(\nu, \nu')|^2 = \frac{1}{2J_i + 1}|\langle f|| \sum_{i=1}^{A} \sigma(i)\frac{\tau_3(i)}{2}||i\rangle|^2. \tag{11}$$

The operator appearing in this expression is familiar from magnetic moments and magnetic transitions, where the large isovector magnetic moment $(\mu_v \sim 4.700)$ often leads to it dominating the orbital and isoscalar spin operators.

Finally, there is one purely leptonic reaction of great interest, since it is the reaction exploited by Kamiokande and SuperKamiokande. Electron neutrinos can scattered off electrons via both charged and neutral current reactions. The cross section calculation is straight forward and will not be repeated here. Two features of the result are of importance for our later discussions, however.

Because of the neutral current contribution, heavy-flavor (ν_μ and ν_τ) also scatter off electrons, but with a cross section reduced by about a factor of seven at low energies. Second, for neutrino energies well above the electron rest mass, the scattering is sharply forward peaked. Thus this reaction allows one to exploit the position of the sun in separating the solar neutrino signal from a large but isotropic background.

As I mentioned earlier, the first experiment performed was one exploiting the reaction

$$^{37}Cl(\nu, e^-)^{37}Ar.$$

As the threshold for this reaction is 0.814 MeV, the important neutrino sources are the ^{7}Be and ^{8}B reactions. The ^{7}Be neutrinos excite just the GT transition to the ground state, the strength of which is known from the electron capture lifetime of ^{37}Ar. The ^{8}B neutrinos can excite all bound states in ^{37}Ar, including the dominant transition to the IAS residing at an excitation of 4.99 MeV. The strength of excite-state GT transitions can be determined from the β decay $^{37}Ca(\beta^+)^{37}K$, which is the isospin mirror reaction to $^{37}Cl(\nu, e^-)^{37}Ar$. The net result is that, for SSM fluxes, 78% of the capture rate should be due to ^{8}B neutrinos, and 15% to ^{7}Be neutrinos. The measured capture rate [12] 2.56 $\pm 0.16 \pm 0.16$ SNU (1 SNU $= 10^{-36}$ capture/atom/sec) is about 1/3 the standard model value.

Similar radiochemical experiments were done by the SAGE and GALLEX collaborations using a different target, ^{71}Ga. The special properties of this target include its low threshold and an unusually strong transition to the ground state of ^{71}Ge, leading to a large pp neutrino cross section (see Fig. 2). The experimental capture rates are $66 \pm 13 \pm 6$ and 76 ± 8 SNU for the SAGE and GALLEX detectors, respectively. The SSM prediction is about 130 SNU [13]. Most important, since the pp flux is directly constrained by the solar luminosity in all steady-state models, there is a minimum theoretical value for the capture rate of 79 SNU, given standard model weak interaction physics. Note there are substantial uncertainties in the ^{71}Ga cross section due to ^{7}Be neutrino capture to two excited states of unknown strength. These uncertainties were greatly reduced by direct calibrations of both detectors using ^{51}Cr neutrino sources.

The remaining experiments, Kamiokande II/III and SuperKamiokande, exploited water Cerenkov detectors to view solar neutrinos event-by-event. Solar neutrinos scatter off electrons, with the recoiling electrons producing the Cerenkov radiation that is then recorded in surrounding phototubes. Thresholds are determined by background rates; SuperKamiokande is currently operating with a trigger at approximately six MeV. The initial experiment, Kamiokande II/III, found a flux of ^{8}B neutrinos of $(2.91 \pm 0.24 \pm 0.35) \cdot$

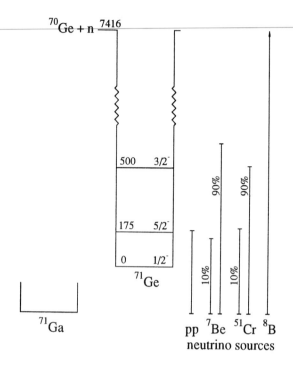

Figure 2: Level scheme for ^{71}Ge showing the excited states that contribute to absorption of pp, ^{7}Be, ^{51}Crm and ^{8}B neutrinos.

10^6/cm^{2}s after about a decade of measurement. Its much larger successor SuperKamiokande, with a 22.5 kiloton fiducial volume, yielded the result $(2.37 \pm 0.06 \pm 0.08) \cdot 10^6$/cm^{2}s after the first 374 days of measurements. This is about 36% of the SSM flux. This result continues to improve in accuracy.

2.3 Uncertainties in Standard Solar Model Parameters

The pattern of solar neutrino fluxes that has emerged from these experiments is

$$\phi(pp) \sim 0.9\, \phi^{\text{SSM}}(pp)$$
$$\phi(^{7}\text{Be}) \sim 0$$

$$\phi(^8\text{B}) \sim 0.43 \, \phi^{\text{SSM}}(^8\text{B}). \tag{12}$$

A reduced ^{8}B neutrino flux can be produced by lowering the central temperature of the sun somewhat, as $\phi(^8\text{B}) \sim T_c^{18}$. However, such an adjustment, either by varying the parameters of the SSM or by adopting some nonstandard physics, tends to push the $\phi(^7\text{Be})/\phi(^8\text{B})$ ratio to higher values rather than the low one of eq. (12),

$$\frac{\phi(^7\text{Be})}{\phi(^8\text{B})} \sim T_c^{-10}. \tag{13}$$

Thus the observations seem difficult to reconcile with plausible solar model variations: one observable ($\phi(^8\text{B})$) requires a cooler core while a second, the ratio $\phi(^7\text{Be})/\phi(^8\text{B})$, requires a hotter one.

An initial question is whether this problem remains significant when one takes into account known uncertainties in the parameters of the SSM. While a detailed summary of standard model uncertainties would take us well beyond the limits of these lectures, a qualitative discussion of pp chain nuclear uncertainties is appropriate. This nuclear microphysics has been the focus of a great deal of experimental work. The pp chain involves a series of nonresonant charged-particle reactions occurring at center-of-mass energies that are well below the height of the inhibiting Coulomb barriers. As the resulting small cross sections preclude laboratory measurements at the relevant energies, one must extrapolate higher energy measurements to threshold to obtain solar cross sections. This extrapolation is often discussed in terms of the astrophysical S-factor

$$\sigma(E) = \frac{S(E)}{E} \exp(-2\pi\eta) \tag{14}$$

where $\eta = \frac{Z_1 Z_2 \alpha}{\beta}$, with α the fine structure constant and $\beta = v/c$ the relative velocity of the colliding particles. This parameterization removes the gross Coulomb effects associated with the s-wave interactions of charged, point-like particles. The remaining energy dependence of S(E) is gentle and can be expressed as a low-order polynomial in E. Usually the variation of S(E) with E is taken from a direct reaction model and then used to extrapolate higher energy measurements to threshold. The model accounts for finite nuclear size effects, strong interaction effects, contributions from other partial waves, etc. As laboratory measurements are made with atomic nuclei while conditions in the solar core guarantee the complete ionization of light nuclei, additional corrections must be made to account for the different electronic screening environments.

Recently a large working group met at a workshop sponsored by the Institute for Nuclear Theory, University of Washington, to review past work on the nuclear reactions of the pp chain and CNO cycle, to recommend best values and appropriate errors, and to identify specific issues in experiment and theory where additional work is needed. The results will soon be published in Reviews of Modern Physics. I will not attempt a summary here, but will give one or two highlights.

The most significant recommend change involves the reaction ^{7}Be(p, γ) ^{8}B, where the standard $S_{17}(0)\sim$ 22.4 evb is that given [14] by Johnson et al. Measurements of $S_{17}(E)$ are complicated by the need to use radioactive targets and thus to determine the areal density of the ^{7}Be target nuclei. Two techniques have been employed, measuring the rate of 478 kev photons from ^{7}Be decay or counting the daughter ^{7}Li nuclei via the reaction ^{7}Li (d,p)^{8}Li. The low-energy data sets for $S_{17}(E)$ disagree by 25%. This is a systematic normalization problem as each data set is consistent with theory in its dependence on E. The energy dependence below $\sim$ 500 keV is believed to be quite simple as it is determined by the asymptotic nuclear wave function.

The Seattle working group on $S_{17}(E)$ found that only one low-energy data set, that of Filippone et al. [15], was described in the published literature in sufficient detail to be evaluated. The target activity in that experiment had been measured by both 478 keV gamma rays and by the (d,p) reaction, with consistent results. The resulting recommended value was thus based on this measurement, yielding

$$S_{17}(0) = 19^{+4}_{-2}\text{eV b}, \quad 1\sigma. \tag{15}$$

The ^{3}He(α, γ)^{7}Be reaction has been measured by two techniques, by counting the capture γ rays and by detecting the resulting ^{7}Be activity. While the two techniques have been used by several groups and have yielded separately consistent results, the capture γ ray value $S_{17}(0) = 0.507 \pm 0.016$ keV b is not in good agreement with the ^{7}Be activity value 0.572 ± 0.026 keV-b. The Seattle working group concluded that the evidence for a systematic discrepancy of unknown origin was reasonably strong and recommended that standard procedures be used in assigning a suitably expanded error. The recommended value S_{34} (0) is 0.53 ± 0.05.

These and other recommended values were recently incorporated into the BP98 solar model calculation. While the workshop's recommended values involve no qualitative changes, there is some broadening of error bars. The downward shift in $S_{17}(0)$ leads to a lower ^{8}B flux. The workshop's Reviews of Modern Physics article summarizes a substantial amount of work on topics not discussed here: screening effects, weak radiative corrections to and exchange current effects on p+p, the atomic physics of ^{7}Be + e$^-$, etc. Much of this discussion was useful in evaluating possible uncertainties in solar microphysics, and in identifying opportunities for reducing these uncertainities.

Are uncertainties in the parameters of the SSM a significant source of uncertainty? The S-factors discussed above comprise one set of parameters, but there are others: the solar lifetime, the opacities, the solar luminosity, etc. In order to answer this question while also taking into account correlations among the fluxes when input parameters are varied, first Bahcall and Ulrich [16] and later Bahcall and Haxton [17] constructed 1000 SSMs by randomly varying five input parameters, the primordial heavy-element-to-hydrogen ratio Z/X and S(0) for the p-p, ^{3}He-^{3}He, ^{3}He-^{4}He, and p-^{7}Be reactions, assuming for each parameter a normal distribution with the mean and standard deviation. (These were the parameters assigned the largest uncertainties.) Smaller uncertainties from radiative opacities, the solar luminosity, and the solar age were folded into the results of the model calculations perturbatively.

The resulting pattern of ^{7}Be and ^{8}B flux predictions is shown in Fig. 3. The elongated error ellipses indicate that the fluxes are strongly correlated. Those variations producing $\phi(^8B)$ below $0.8\phi^{SSM}(^8B)$ tend to produce a reduced $\phi(^7Be)$, but the reduction is always less than 0.8. Thus a greatly reduced $\phi(^7Be)$ cannot be achieved within the uncertainties assigned to parameters in the SSM.

A similar exploration, but including parameter variations very far from their preferred values, was carried out by Castellani et al. [18], who displayed their results as a function of the resulting core temperature T_c. The pattern that emerges is striking (see Fig. 4): parameter variations producing the same value of T_c produce remarkably similar fluxes. Thus T_c provides an excellent one-parameter description of standard model perturbations. Figure 4 also illustrates the difficulty of producing a low ratio of $\phi(^7Be)/\phi(^8B)$ when T_c is reduced.

The 1000-solar-model variations were made under the constraint of reproducing the solar luminosity. Those variations show a similar strong correlation with T_c

$$\phi(pp) \propto T_c^{-1.2} \qquad \phi(^7Be) \propto T_c^8 \qquad \phi(^8B) \propto T_c^{18}. \qquad (16)$$

Figures 3 and 4 offer a strong argument that reasonable variations in the parameters of the SSM, or nonstandard changes in quantities like the metallicity, opacities, or solar age, cannot produce the pattern of fluxes deduced from experiment (eq. (12)). This would seem to limit possible solutions to errors either in the underlying physics of the SSM or in our understanding of neutrino properties.

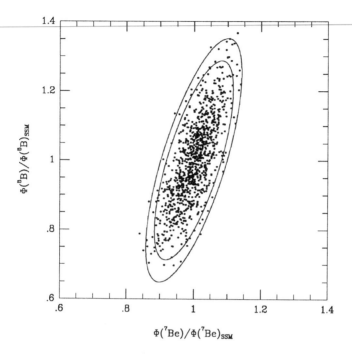

Figure 3: SSM ^{7}Be and ^8B flux predictions. The dots represent the results of SSM calculations where the input parameters were varied according to their assigned uncertainties, as described in the text. The 90% and 99% confidence level error ellipses are shown.

2.4 Nonstandard Solar Models

Nonstandard solar models include both variations of SSM parameters far outside the ranges that are generally believed to be reasonable (some examples of which are given in Figure 4), and changes in the underlying physics of the model. The solar neutrino problem has been a major stimulus to models: in fact, most suggestions were motivated by the hope of producing a cooler sun ($T_c \sim 0.05 T_c$) that would avoid conflict with the results of the ^{37}Cl experiment. The suggestions included models with low heavy element abundances ("low Z" models), in which one abandons the SSM assumption that the initial heavy element abundances are those we measure today at the sun's surface; periodically mixed solar cores; models where hydrogen is continually mixed into the core by turbulent diffusion or by convective mixing; and models where the solar core is partially supported by a strong central magnetic field or by its

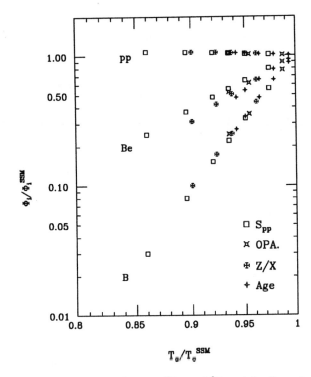

Figure 4: The responses of the pp, ^{7}Be, and ^{8}B neutrino fluxes to the indicated variations in solar model input parameters, displayed as a function of the resulting central temperature T_c. From Castellani et al.

rapid rotation, thereby relaxing the SSM assumption that hydrostatic equilibrium is achieved only through the gas pressure gradient. A larger list is given by Bahcall and Davis [19]. To illustrate the kinds of consequences such models have, two of these suggestions are discussed in more detail below.

In low-Z models one postulates a reduction in the core metallicity from Z ~ 0.02 to Z ~ 0.002. This lowers the core opacity (primarily because metals are very important to free-bound electron transitions), thus reducing T_c and weakening the ppII and ppIII cycles. The attractiveness of low-Z models is due in part to the existence of mechanisms for adding heavier elements to the sun's surface. These include the infall of comets and other debris, as well as the accumulation of dust as the sun passes through interstellar clouds. However, the increased radiative energy transport in low-Z models leads to a thin convective envelope, in contradiction to interpretations of the 5-minute solar surface oscillations. A low Hc mass fraction also results. As diffusion of material from a thin convective envelope into the interior would deplete heavy elements at the surface, investigators have also questioned whether present abundances could have accumulated in low-Z models. Finally, the general consistency of solar heavy element abundances with those observed in other main sequence stars makes the model appear contrived.

Models in which the solar core ($\sim 0.2 \, M_\odot$) is intermittently mixed break the standard model assumption of a steady-state sun: for a period of several million years (the thermal relaxation time for the core) following mixing, the usual relationship between the observed surface luminosity and rate of energy (and neutrino) production is altered as the sun burns out of equilibrium. Calculations show that both the luminosity and the ^{8}B neutrino flux are suppressed while the sun relaxes back to the steady state. Such models have been considered seriously because of instabilities associated with large gradients in the ^{3}He abundance, which in equilibrium varies as $\sim T^{-6}$, where T is the local temperature. The resulting steep profile is unstable under finite amplitude displacements of a volume to smaller r: the energy released by the increased ^{3}He burning at higher T can exceed the energy in the perturbation. For a discussion of the plausibility of such a trigger for core mixing, one can see the original work of Dilke and Gough [20] as well as a more recent critique by Merryfield [21]. The possibility that continuous mixing on time scales of ^{3}He mixing could produce a flux pattern close to that observed (e.g., a suppression in both the ^{8}B neutrino flux and the ^{7}Be/^{8}B flux ratio) was recently discussed by Cumming and Haxton [22].

This discussion of two of the more seriously explored nonstandard model possibilities illustrates how changes motivated by the solar neutrino problem often produce other, unwanted consequences. In particular, many experts feel that the good SSM agreement with helioseismology is likely to be destroyed by changes such as those discussed above.

Figure 5 is an illustration by Hata et al. [23] of the flux predictions of several nonstandard models, including a low-Z model consistent with the ^{37}Cl results. As in the Castellani et al. exploration, the results cluster along a track that defines the naive T_c dependence of the $\phi(^7\text{Be})/\phi(^8\text{B})$ ratio, well separated from the experimental contours.

There is now a popular argument that no such nonstandard model can solve the solar neutrino problem: if one assumes undistorted neutrino spectra, no combination of pp, ^{7}Be, and ^{8}B neutrino fluxes fits the experimental results well [24]. In fact, in an unconstrained fit, the required ^{7}Be flux is unphysical, negative by almost 3σ. Thus, barring some unfortunate experimental error, it appears we are forced to look elsewhere for a solution.

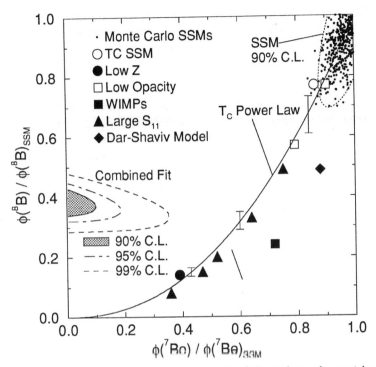

Figure 5: The fluxes allowed by the combined results of the various solar neutrino experiments compared to the results of SSM variations and various nonstandard solar models. The solid line in the naive T_c power law discussed in the text. From Hata et al.

If experimental error, SSM parameter uncertainties, and nonstandard solar physics are ruled out as potential solutions, new particle physics is left as the leading possibility. Suggested particle physics solutions of the solar neutrino problem include neutrino oscillations, neutrino decay, neutrino magnetic moments, and weakly interacting massive particles. Among these, the Mikheyev-Smirnov-Wolfenstein effect — neutrino oscillations enhanced by matter interactions — is widely regarded as the most plausible.

2.5 Neutrino Oscillations

One odd feature of particle physics is that neutrinos, which are not required by any symmetry to be massless, nevertheless must be much lighter than any of the other known fermions. For instance, the current limit on the $\bar{\nu}_e$ mass is $\lesssim 5$ eV. The standard model requires neutrinos to be massless, but the reasons are not fundamental. Dirac mass terms m_D, analogous to the mass terms for other fermions, cannot be constructed because the model contains no right-handed neutrino fields. Neutrinos can also have Majorana mass terms

$$\overline{\nu_L^c} m_L \nu_L \quad \text{and} \quad \overline{\nu_R^c} m_R \nu_R \tag{17}$$

where the subscripts L and R denote left- and right-handed projections of the neutrino field ν, and the superscript c denotes charge conjugation. The first term above is constructed from left-handed fields, but can only arise as a nonrenormalizable effective interaction when one is constrained to generate m_L with the doublet scalar field of the standard model. The second term is absent from the standard model because there are no right-handed neutrino fields.

None of these standard model arguments carries over to the more general, unified theories that theorists believe will supplant the standard model. In the enlarged multiplets of extended models it is natural to characterize the fermions of a single family, e.g., ν_e, e, u, d, by the same mass scale m_D. Small neutrino masses are then frequently explained as a result of the Majorana neutrino masses. In the seesaw mechanism,

$$M_\nu \sim \begin{pmatrix} 0 & m_D \\ m_D^T & m_R \end{pmatrix}. \tag{18}$$

Diagonalization of the mass matrix produces one light neutrino, $m_{\text{light}} \sim \frac{m_D^2}{m_R}$, and one unobservably heavy, $m_{\text{heavy}} \sim m_R$. The factor (m_D/m_R) is the needed small parameter that accounts for the distinct scale of neutrino masses. The masses for the ν_e, ν_μ, and ν_τ are then related to the squares of the corresponding quark masses m_u, m_c, and m_t. Taking $m_R \sim 10^{16}$ GeV, a typical grand unification scale for models built on groups like SO(10), the seesaw mechanism gives the crude relation

$$m_{\nu_e} : m_{\nu_\mu} : m_{\nu_\tau} \leftrightarrow 2 \cdot 10^{-12} : 2 \cdot 10^{-7} : 3 \cdot 10^{-3} \text{eV}. \tag{19}$$

The fact that solar neutrino experiments can probe small neutrino masses, and thus provide insight into possible new mass scales m_R that are far beyond the reach of direct accelerator measurements, has been an important theme of the field.

Now one of the most interesting possibilities for solving the solar neutrino problem has to do with neutrino masses. For simplicity we will discuss just two neutrinos. If a neutrino has a mass m, we mean that as it propagates through free space, its energy and momentum are related in the usual way for this mass.

Thus if we have two neutrinos, we can label those neutrinos according to the eigenstates of the free Hamiltonian, that is, as mass eigenstates.

But neutrinos are produced by the weak interaction. In this case, we have another set of eigenstates, the flavor eigenstates. We can define a ν_e as the neutrino that accompanies the positron in β decay. Likewise we label by ν_μ the neutrino produced in muon decay.

Now the question: are the eigenstates of the free Hamiltonian and of the weak interaction Hamiltonian identical? Most likely the answer is no: we know this is the case with the quarks, since the different families (the analog of the mass eigenstates) do interact through the weak interaction. That is, the up quark decays not only to the down quark, but also occasionally to the strange quark. (This is why we had a $\cos\theta_c$ in our weak interaction amplitude: the amplitude for $u \to s$ is proportional to $\sin\theta_c$.) Thus we suspect that the weak interaction and mass eigenstates, while spanning the same two-neutrino space, are not coincident: the mass eigenstates $|\nu_1\rangle$ and $|\nu_2\rangle$ (with masses m_1 and m_2) are related to the weak interaction eigenstates by

$$|\nu_e\rangle = \cos\theta_v|\nu_1\rangle + \sin\theta_v|\nu_2\rangle$$
$$|\nu_\mu\rangle = -\sin\theta_v|\nu_1\rangle + \cos\theta_v|\nu_2\rangle \tag{20}$$

where θ_v is the (vacuum) mixing angle.

An immediate consequence is that a state produced as a $|\nu_e\rangle$ or a $|\nu_\mu\rangle$ at some time t — for example, a neutrino produced in β decay — does not remain a pure flavor eigenstate as it propagates away from the source. This is because the different mass eigenstates comprising the neutrino will accumulate different phases as they propagate downstream, a phenomenon known as vacuum oscillations (vacuum because the experiment is done in free space). To see the effect, suppose we produce a neutrino in some β decay where we measure the momentum of the initial nucleus, final nucleus, and positron. Thus the outgoing neutrino is a momentum eigenstate[25]. At time $t=0$

$$|\nu(t = 0)\rangle = |\nu_e\rangle = \cos\theta_v|\nu_1\rangle + \sin\theta_v|\nu_2\rangle. \tag{21}$$

Each eigenstate subsequently propagates with a phase

$$e^{i(\vec{k}\cdot\vec{x}-\omega t)} = e^{i(\vec{k}\cdot\vec{x}-\sqrt{m_i^2+k^2}\,t)}. \tag{22}$$

But if the neutrino mass is small compared to the neutrino momentum/energy, one can write

$$\sqrt{m_i^2 + k^2} \sim k(1 + \frac{m_i^2}{2k^2}). \tag{23}$$

Thus we conclude

$$|\nu(t)\rangle = e^{i(\vec{k}\cdot\vec{x} - kt - (m_1^2 + m_2^2)t/4k)}$$
$$\times [\cos\theta_v |\nu_1\rangle e^{i\delta m^2 t/4k} + \sin\theta_v |\nu_2\rangle e^{-i\delta m^2 t/4k}]. \tag{24}$$

We see there is a common average phase (which has no physical consequence) as well as a beat phase that depends on

$$\delta m^2 = m_2^2 - m_1^2. \tag{25}$$

Now it is a simple matter to calculate the probability that our neutrino state remains a $|\nu_e\rangle$ at time t

$$P_{\nu_e}(t) = |\langle \nu_e | \nu(t) \rangle|^2$$
$$= 1 - \sin^2 2\theta_v \sin^2\left(\frac{\delta m^2 t}{4k}\right) \to 1 - \frac{1}{2}\sin^2 2\theta_v \tag{26}$$

where the limit on the right is appropriate for large t. Now $E \sim k$, where E is the neutrino energy, by our assumption that the neutrino masses are small compared to k. Thus we can reinsert the units above to write the probability in terms of the distance x of the neutrino from its source,

$$P_\nu(x) = 1 - \sin^2 2\theta_v \sin^2\left(\frac{\delta m^2 c^4 x}{4\hbar c E}\right). \tag{27}$$

(When one properly describes the neutrino state as a wave packet, the large-distance behavior follows from the eventual separation of the mass eigenstates.) If the the oscillation length

$$L_o = \frac{4\pi\hbar c E}{\delta m^2 c^4} \tag{28}$$

is comparable to or shorter than one astronomical unit, a reduction in the solar ν_e flux would be expected in terrestrial neutrino oscillations.

The suggestion that the solar neutrino problem could be explained by neutrino oscillations was first made by Pontecorvo in 1958, who pointed out the analogy with $K_0 \leftrightarrow \bar{K}_0$ oscillations. From the point of view of particle physics, the sun is a marvelous neutrino source. The neutrinos travel a long distance and have low energies (~ 1 MeV), implying a sensitivity to

$$\delta m^2 \gtrsim 10^{-12} eV^2. \tag{29}$$

In the seesaw mechanism, $\delta m^2 \sim m_2^2$, so neutrino masses as low as $m_2 \sim 10^{-6}$ eV could be probed. In contrast, terrestrial oscillation experiments with accelerator or reactor neutrinos are typically limited to $\delta m^2 \gtrsim 0.1$ eV^2.

From the expressions above one expects vacuum oscillations to affect all neutrino species equally, if the oscillation length is small compared to an astronomical unit. This is somewhat in conflict with the data, as we have argued that the ^{7}Be neutrino flux is quite suppressed. Furthermore, there is a weak theoretical prejudice that θ_v should be small, like the Cabibbo angle. The first objection, however, can be circumvented in the case of "just so" oscillations where the oscillation length is comparable to one astronomical unit. In this case the oscillation probability becomes sharply energy dependent, and one can choose δm^2 to preferentially suppress one component (e.g., the monochromatic ^{7}Be neutrinos). This scenario has been explored by several groups and remains an interesting possibility. However, the requirement of large mixing angles remains.

Below we will see that stars allow us to "get around" this problem with small mixing angles. In preparation for this, we first present the results above in a slightly more general way. The analog of eq. (24) for an initial muon neutrino ($|\nu(t=0)\rangle = |\nu_\mu\rangle$) is

$$|\nu(t)\rangle = e^{i(\vec{k}\cdot\vec{x}-kt-(m_1^2+m_2^2)t/4k)}$$
$$\times[-\sin\theta_v|\nu_1\rangle e^{i\delta m^2 t/4k} + \cos\theta_v|\nu_2\rangle e^{-i\delta m^2 t/4k}] \tag{30}$$

Now if we compare eqs. (24) and (30) we see that they are special cases of a more general problem. Suppose we write our initial neutrino wave function as

$$|\nu(t=0)\rangle = a_e(t=0)|\nu_e\rangle + a_\mu(t=0)|\nu_\mu\rangle. \tag{31}$$

Then eqs. (24) and (30) tell us that the subsequent propagation is described by changes in $a_e(x)$ and $a_\mu(x)$ according to (this takes a bit of algebra)

$$i\frac{d}{dx}\begin{pmatrix} a_e \\ a_\mu \end{pmatrix} = \frac{1}{4E}\begin{pmatrix} -\delta m^2\cos 2\theta_v & \delta m^2\sin 2\theta_v \\ \delta m^2\sin 2\theta_v & \delta m^2\cos 2\theta_v \end{pmatrix}\begin{pmatrix} a_e \\ a_\mu \end{pmatrix}. \tag{32}$$

Note that the common phase has been ignored: it can be absorbed into the overall phase of the coefficients a_e and a_μ, and thus has no consequence. Also, we have equated $x = t$, that is, set $c = 1$.

2.6 The Mikheyev-Smirnov-Wolfenstein Mechanism

The view of neutrino oscillations changed when Mikheyev and Smirnov [26] showed in 1985 that the density dependence of the neutrino effective mass, a phenomenon first discussed by Wolfenstein in 1978, could greatly enhance oscillation probabilities: a ν_e is adiabatically transformed into a ν_μ as it traverses a critical density within the sun. It became clear that the sun was not only an excellent neutrino source, but also a natural regenerator for cleverly enhancing the effects of flavor mixing.

While the original work of Mikheyev and Smirnov was numerical, their phenomenon was soon understood analytically as a level-crossing problem. If one writes the neutrino wave function in matter as in eq. (31), the evolution of $a_e(x)$ and $a_\mu(x)$ is governed by

$$i\frac{d}{dx}\begin{pmatrix} a_e \\ a_\mu \end{pmatrix} = \frac{1}{4E}\begin{pmatrix} 2E\sqrt{2}G_F\rho(x) - \delta m^2\cos 2\theta_v & \delta m^2\sin 2\theta_v \\ \delta m^2\sin 2\theta_v & -2E\sqrt{2}G_F\rho(x) + \delta m^2\cos 2\theta_v \end{pmatrix}\begin{pmatrix} a_e \\ a_\mu \end{pmatrix} \tag{33}$$

where G_F is the weak coupling constant and $\rho(x)$ the solar electron density. If $\rho(x) = 0$, this is exactly our previous result and can be trivially integrated to give the vacuum oscillation solutions of Sec. 2.5. The new contribution to the diagonal elements, $2E\sqrt{2}G_F\rho(x)$, represents the effective contribution to M_ν^2 that arises from neutrino-electron scattering. The indices of refraction of electron and muon neutrinos differ because the former scatter by charged and neutral currents, while the latter have only neutral current interactions. The difference in the forward scattering amplitudes determines the density-dependent splitting of the diagonal elements of the new matter equation.

It is helpful to rewrite this equation in a basis consisting of the light and heavy local mass eigenstates (i.e., the states that diagonalize the right-hand side of the equation),

$$|\nu_L(x)\rangle = \cos\theta(x)|\nu_e\rangle - \sin\theta(x)|\nu_\mu\rangle$$
$$|\nu_H(x)\rangle = \sin\theta(x)|\nu_e\rangle + \cos\theta(x)|\nu_\mu\rangle. \tag{34}$$

The local mixing angle is defined by

$$\sin 2\theta(x) = \frac{\sin 2\theta_v}{\sqrt{X^2(x) + \sin^2 2\theta_v}}$$
$$\cos 2\theta(x) = \frac{-X(x)}{\sqrt{X^2(x) + \sin^2 2\theta_v}} \tag{35}$$

where $X(x) = 2\sqrt{2}G_F\rho(x)E/\delta m^2 - \cos 2\theta_v$. Thus $\theta(x)$ ranges from θ_v to $\pi/2$ as the density $\rho(x)$ goes from 0 to ∞.

If we define

$$|\nu(x)\rangle = a_H(x)|\nu_H(x)\rangle + a_L(x)|\nu_L(x)\rangle, \tag{36}$$

the neutrino propagation can be rewritten in terms of the local mass eigenstates

$$i\frac{d}{dx}\begin{pmatrix} a_H \\ a_L \end{pmatrix} = \begin{pmatrix} \lambda(x) & i\alpha(x) \\ -i\alpha(x) & -\lambda(x) \end{pmatrix}\begin{pmatrix} a_H \\ a_L \end{pmatrix} \tag{37}$$

with the splitting of the local mass eigenstates determined by

$$2\lambda(x) = \frac{\delta m^2}{2E} \sqrt{X^2(x) + \sin^2 2\theta_v} \tag{38}$$

and with mixing of these eigenstates governed by the density gradient

$$\alpha(x) = \left(\frac{E}{\delta m^2}\right) \frac{\sqrt{2}\, G_F \frac{d}{dx}\rho(x) \sin 2\theta_v}{X^2(x) + \sin^2 2\theta_v}. \tag{39}$$

The results above are quite interesting: the local mass eigenstates diagonalize the matrix if the density is constant. In such a limit, the problem is no more complicated than our original vacuum oscillation case, although our mixing angle is changed because of the matter effects. But if the density is not constant, the mass eigenstates in fact evolve as the density changes. This is the crux of the MSW effect. Note that the splitting achieves its minimum value, $\frac{\delta m^2}{2E} \sin 2\theta_v$, at a critical density $\rho_c = \rho(x_c)$

$$2\sqrt{2}EG_F\rho_c = \delta m^2 \cos 2\theta_v \tag{40}$$

that defines the point where the diagonal elements of the original flavor matrix cross.

Our local-mass-eigenstate form of the propagation equation can be trivially integrated if the splitting of the diagonal elements is large compared to the off-diagonal elements,

$$\gamma(x) = \left|\frac{\lambda(x)}{\alpha(x)}\right| = \frac{\sin^2 2\theta_v}{\cos 2\theta_v} \frac{\delta m^2}{2E} \frac{1}{\left|\frac{1}{\rho_c}\frac{d\rho(x)}{dx}\right|} \frac{[X(x)^2 + \sin^2 2\theta_v]^{3/2}}{\sin^3 2\theta_v} \gg 1, \tag{41}$$

a condition that becomes particularly stringent near the crossing point,

$$\gamma_c = \gamma(x_c) = \frac{\sin^2 2\theta_v}{\cos 2\theta_v} \frac{\delta m^2}{2E} \frac{1}{\left|\frac{1}{\rho_c}\frac{d\rho(x)}{dx}\big|_{x=x_c}\right|} \gg 1. \tag{42}$$

The resulting adiabatic electron neutrino survival probability [27], valid when $\gamma_c \gg 1$, is

$$P_{\nu_e}^{\text{adiab}} = \frac{1}{2} + \frac{1}{2}\cos 2\theta_v \cos 2\theta_i \tag{43}$$

where $\theta_i = \theta(x_i)$ is the local mixing angle at the density where the neutrino was produced.

The physical picture behind this derivation is illustrated in Fig. 6. One makes the usual assumption that, in vacuum, the ν_e is almost identical to the

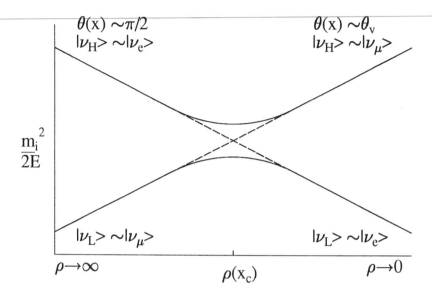

Figure 6: Schematic illustration of the MSW crossing. The dashed lines correspond to the electron-electron and muon-muon diagonal elements of the M^2 matrix in the flavor basis. Their intersection defines the level-crossing density ρ_c. The solid lines are the trajectories of the light and heavy local mass eigenstates. If the electron neutrino is produced at high density and propagates adiabatically, it will follow the heavy-mass trajectory, emerging from the sun as a ν_μ.

light mass eigenstate, $\nu_L(0)$, i.e., $m_1 < m_2$ and $\cos\theta_v \sim 1$. But as the density increases, the matter effects make the ν_e heavier than the ν_μ, with $\nu_e \to \nu_H(x)$ as $\rho(x)$ becomes large. The special property of the sun is that it produces ν_es at high density that then propagate to the vacuum where they are measured. The adiabatic approximation tells us that if initially $\nu_e \sim \nu_H(x)$, the neutrino will remain on the heavy mass trajectory provided the density changes slowly. That is, if the solar density gradient is sufficiently gentle, the neutrino will emerge from the sun as the heavy vacuum eigenstate, $\sim \nu_\mu$. This guarantees nearly complete conversion of ν_es into ν_μs, producing a flux that cannot be detected by the Homestake or SAGE/GALLEX detectors.

But this does not explain the curious pattern of partial flux suppressions coming from the various solar neutrino experiments. The key to this is the behavior when $\gamma_c \lesssim 1$. Our expression for $\gamma(x)$ shows that the critical region for nonadiabatic behavior occurs in a narrow region (for small θ_v) surrounding the crossing point, and that this behavior is controlled by the derivative of the density. This suggests an analytic strategy for handling nonadiabatic crossings: one can replace the true solar density by a simpler (integrable!) two-parameter form that is constrained to reproduce the true density and its derivative at the crossing point x_c. Two convenient choices are the linear ($\rho(x) = a + bx$) and exponential ($\rho(x) = ae^{-bx}$) profiles. As the density derivative at x_c governs the nonadiabatic behavior, this procedure should provide an accurate description of the hopping probability between the local mass eigenstates when the neutrino traverses the crossing point. The initial and ending points x_i and x_f for the artificial profile are then chosen so that $\rho(x_i)$ is the density where the neutrino was produced in the solar core and $\rho(x_f) = 0$ (the solar surface), as illustrated in in Fig. 7. Since the adiabatic result ($P_{\nu_e}^{\mathrm{adiab}}$) depends only on the local mixing angles at these points, this choice builds in that limit. But our original flavor-basis equation can then be integrated exactly for linear and exponential profiles, with the results given in terms of parabolic cylinder and Whittaker functions, respectively.

That result can be simplified further by observing that the nonadiabatic region is generally confined to a narrow region around x_c, away from the end-points x_i and x_f. We can then extend the artificial profile to $x = \pm\infty$, as illustrated by the dashed lines in Fig. 7. As the neutrino propagates adiabatically in the unphysical region $x < x_i$, the exact soluation in the physical region can be recovered by choosing the initial boundary conditions

$$
\begin{aligned}
a_L(-\infty) &= -a_\mu(-\infty) = \cos\theta_i e^{-i\int_{-\infty}^{x_i}\lambda(x)dx} \\
a_H(-\infty) &= a_e(-\infty) = \sin\theta_i e^{i\int_{-\infty}^{x_i}\lambda(x)dx}.
\end{aligned}
\tag{44}
$$

That is, $|\nu(-\infty)\rangle$ will then adiabatically evolve to $|\nu(x_i)\rangle = |\nu_e\rangle$ as x goes from $-\infty$ to x_i. The unphysical region $x > x_f$ can be handled similarly.

With some algebra a simple generalization of the adiabatic result emerges that is valid for all $\delta m^2/E$ and θ_v

$$
P_{\nu_e} = \frac{1}{2} + \frac{1}{2}\cos 2\theta_v \cos 2\theta_i (1 - 2P_{\mathrm{hop}})
\tag{45}
$$

where P_{hop} is the Landau-Zener probability of hopping from the heavy mass trajectory to the light trajectory on traversing the crossing point. For the linear approximation to the density [28,29],

$$
P_{\mathrm{hop}}^{\mathrm{lin}} = e^{-\pi\gamma_c/2}.
\tag{46}
$$

As it must by our construction, P_{ν_e} reduces to $P_{\nu_e}^{\mathrm{adiab}}$ for $\gamma_c \gg 1$. When the crossing becomes nonadiabatic (e.g., $\gamma_c \ll 1$), the hopping probability goes to

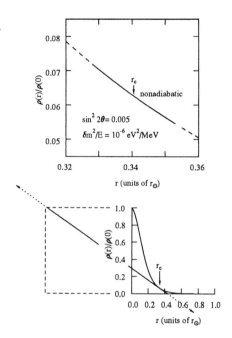

Figure 7: The top figure illustrates, for one choice of $\sin^2 2\theta$ and δm^2, that the region of nonadiabatic propagation (solid line) is usually confined to a narrow region around the crossing point r_c. In the lower figure, the solid lines represent the solar density and a linear approximation to that density that has the correct initial and final values, as well as the correct density and density derivative at r_c. Thus the linear profile is a very good approximation to the sun in the vicinity of the crossing point. The MSW equations can be solved analytically for this wedge. By extending the wedge to $\pm\infty$ (dotted lines) and assuming adiabatic propagation in these regions of unphysical density, one obtains the simple Landau-Zener result discussed in the text.

1, allowing the neutrino to exit the sun on the light mass trajectory as a ν_e, i.e., no conversion occurs.

Thus there are two conditions for strong conversion of solar neutrinos: there must be a level crossing (that is, the solar core density must be sufficient to render $\nu_e \sim \nu_H(x_i)$ when it is first produced) and the crossing must be adiabatic. The first condition requires that $\delta m^2/E$ not be too large, and the second $\gamma_c \gtrsim 1$. The combination of these two constraints, illustrated in Fig. 8, defines a triangle of interesting parameters in the $\frac{\delta m^2}{E} - \sin^2 2\theta_v$ plane, as Mikheyev and Smirnov found by numerically integration. A remarkable feature of this triangle is that strong $\nu_e \to \nu_\mu$ conversion can occur for very small mixing angles $(\sin^2 2\theta \sim 10^{-3})$, unlike the vacuum case.

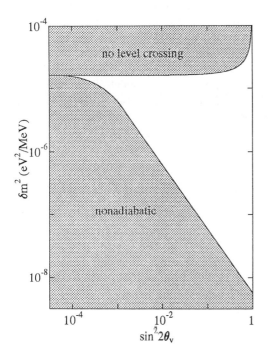

Figure 8: MSW conversion for a neutrino produced at the sun's center. The upper shaded region indices thoses $\delta m^2/E$ where the vacuum mass splitting is too great to be overcome by the solar density. Thus no level crossing occurs. The lower shaded region defines the region where the level crossing is nonadiabatic (γ_c less than unity). The unshaded region corresponds to adiabatic level crossings where strong $\nu_e \to \nu_\mu$ will occur.

One can envision superimposing on Fig. 8 the spectrum of solar neutrinos, plotted as a function of $\frac{\delta m^2}{E}$ for some choice of δm^2. Since Davis sees *some* solar neutrinos, the solutions must correspond to the boundaries of the triangle in Fig. 8. The horizontal boundary indicates the maximum $\frac{\delta m^2}{E}$ for which the sun's central density is sufficient to cause a level crossing. If a spectrum properly straddles this boundary, we obtain a result consistent with the Homestake experiment in which low energy neutrinos (large 1/E) lie above the level-crossing boundary (and thus remain ν_e's), but the high-energy neutrinos (small 1/E) fall within the unshaded region where strong conversion takes place. Thus such a solution would mimic nonstandard solar models in that only the ^{8}B neutrino flux would be strongly suppressed. The diagonal boundary separates the adiabatic and nonadiabatic regions. If the spectrum straddles this boundary, we obtain a second solution in which low energy neutrinos lie within the conversion region, but the high-energy neutrinos (small $1/E$) lie below the conversion region and are characterized by $\gamma \ll 1$ at the crossing density. (Of course, the boundary is not a sharp one, but is characterized by the Landau-Zener exponential). Such a nonadiabatic solution is quite distinctive since the flux of pp neutrinos, which is strongly constrained in the standard solar model and in any steady-state nonstandard model by the solar luminosity, would now be sharply reduced. Finally, one can imagine "hybrid" solutions where the spectrum straddles both the level-crossing (horizontal) boundary and the adiabaticity (diagonal) boundary for small θ, thereby reducing the ^{7}Be neutrino flux more than either the pp or ^{8}B fluxes.

What are the results of a careful search for MSW solutions satisfying the Homestake, Kamiokande/SuperKamiokande, and SAGE/GALLEX constraints? This has been done by several groups: recent results will be discussed by N. Hata in his lectures. One solution, corresponding to a region surrounding $\delta m^2 \sim 6 \cdot 10^{-6} \mathrm{eV}^2$ and $\sin^2 2\theta_v \sim 6 \cdot 10^{-3}$, is the hybrid case described above. It is commonly called the small-angle solution. A second, large-angle solution exists, corresponding to $\delta m^2 \sim 10^{-5} \mathrm{eV}^2$ and $\sin^2 2\theta_v \sim 0.6$. These solutions can be distinguished by their characteristic distortions of the solar neutrino spectrum. The survival probabilities $P_{\nu_e}^{\mathrm{MSW}}(E)$ for the small- and large-angle parameters given above are shown as a function of E in Fig. 9.

The MSW mechanism provides a natural explanation for the pattern of observed solar neutrino fluxes. While it requires profound new physics, both massive neutrinos and neutrino mixing are expected in extended models. The small-angle solution corresponds to $\delta m^2 \sim 10^{-5}$ eV^2, and thus is consistent with $m_2 \sim$ few $\cdot 10^{-3}$ eV. This is a typical ν_τ mass in models where $m_R \sim m_{\mathrm{GUT}}$. This mass is also reasonably close to atmospheric neutrino values. On the other hand, if it is the ν_μ participating in the oscillation, this gives $m_R \sim 10^{12}$ GeV and predicts a heavy $\nu_\tau \sim 10$ eV. Such a mass is of great interest cosmologically as it would have consequences for supernova physics, the dark matter problem, and the formation of large-scale structure.

2.7 Spin-Flavor Oscillations

If the MSW mechanism proves not to be the solution of the solar neutrino problem, it still will have greatly enhanced the importance of solar neutrino physics: the existing experiments have ruled out large regions in the $\delta m^2 - \sin^2 2\theta_v$ plane (corresponding to nearly complete $\nu_e \to \nu_\mu$ conversion) that remain hopelessly beyond the reach of accelerator neutrino oscillation experiments.

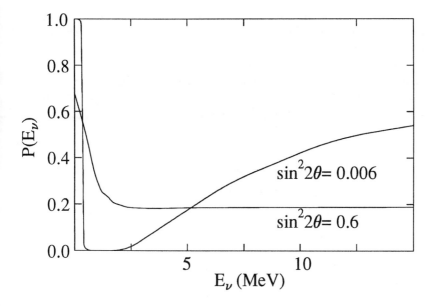

Figure 9: MSW survival probabilities $P(E_\nu)$ for typical small angle and large angle solutions.

A number of other particle physics solutions have been considered, such as neutrino decay, matter-catalyzed neutrino decay, and solar energy transport by weakly interacting massive particles. But perhaps the most interesting possibility, apart from the MSW mechanism, was stimulated by suggestions that the ^{37}Cl signal might be varying with a period comparable to the 11-year solar cycle. While the evidence for this has weakened, the original claims generated renewed interest in neutrino magnetic moment interactions with the solar magnetic field.

The original suggestions by Cisneros and by Okun, Voloshyn, and Vysotsky envisioned the rotation

$$\nu_{e_L} \to \nu_{e_R} \tag{47}$$

producing a right-handed neutrino with sterile interactions in the standard model. With the discovery of the MSW mechanism, it was realized that matter effects would break the vacuum degeneracy of the ν_{e_L} and ν_{e_R}, suppressing the spin precession. Lim and Marciano [30] and Akmedov [31] pointed out that this difficulty was naturally circumvented for

$$\nu_{e_L} \to \nu_{\mu_R} \tag{48}$$

as the different matter interactions of the ν_e and ν_μ can compensate for the vacuum $\nu_e - \nu_\mu$ mass difference, producing a crossing similar to the usual MSW mechanism. Such spin-flavor precession can then occur at full strength due to an off-diagonal (in flavor) magnetic moment.

Quite relevant to this suggestion is the very strong limit on both diagonal and off-diagonal magnetic moments is imposed by studies of the red giant cooling process of plasmon decay into neutrinos

$$\gamma^* \rightarrow \nu_i \bar{\nu}_j. \tag{49}$$

The result is $|\mu_{ij}| \lesssim 3 \cdot 10^{-12} \mu_B$, where μ_B is an electron Bohr magneton [32]. (This can be compared to a simple one-loop estimate [33] of the neutrino magnetic moment of $\sim 10^{-18} \mu_B$, taking a typical dark matter value for the neutrino mass of a few eV.) If a magnetic moment at the red giant limit is assumed, it follows that solar magnetic field strengths of $B_\odot \gtrsim 10^6 G$ are needed to produce interesting effects. Since the location of the spin-flavor level crossing depends on the neutrino energy, such fields have to be extensive to affect an appreciable fraction of the neutrino spectrum. It is unclear whether these conditions can occur in the sun. This constraint from stellar cooling leads us naturally into our next lecture.

3   Dirac and Majorana Neutrinos and Stellar Cooling

3.1   The Neutrino Mass Matrix

Consider a general 4n × 4n neutrino mass matrix where n is the number of flavors

$$(\bar{\Psi}_L^c, \bar{\Psi}^R, \bar{\Psi}_L, \bar{\Psi}_R^c) \begin{pmatrix} 0 & 0 & M_L & M_D^T \\ 0 & 0 & M_D & M_R^\dagger \\ M_L^\dagger & M_D^\dagger & 0 & 0 \\ M_D^* & M_R & 0 & 0 \end{pmatrix} \begin{pmatrix} \Psi_L^c \\ \Psi_R \\ \Psi_L \\ \Psi_R^c \end{pmatrix} \tag{50}$$

where each entry in this matrix is understood to be a n × n matrix operating in flavor space. The entries M_D are the Dirac mass terms, while the M_L and M_R are the Majorana terms. The latter break the invariance of the Dirac equation under the transformation $\psi(x) \rightarrow e^{i\alpha} \psi(x)$ associated with a conserved lepton number. Thus it is these terms that govern lepton-number-violating processes like double beta decay.

One can proceed to diagonalize this matrix

$$\Psi_{\nu(e)}^L = \sum_{i=1}^{2n} U_{ei}^L \tilde{\nu}_i(x) \quad \text{with masses } m_i. \tag{51}$$

The eigenstates are two-component Majorana neutrinos[34], yielding the proper $2 \times 2n = 4n$ degrees of freedom, where n is the number of flavors. We can recover the Majorana and Dirac limits:

• If $M_R = M_L = 0$, the eigenstates of this matrix become pairwise degenerate, allowing the $2n$ two-component eigenstates to be paired to form n four-component Dirac eigenstates.

• If $M_D = 0$, the left- and right-handed components decouple, yielding n left-handed Majorana eigenstates with standard model interactions.

There are interesting physical effects associated with these limits. Dirac neutrinos can have magnetic dipole, electric dipole (CP and T violating), and anapole (P violating) moments, as well as nonzero charge radii. Majorana neutrinos can have anapole moments but only transition magnetic and electric dipole moments. I mentioned in the previous lecture that transition magnetic moments were quite interesting in the context of MSW effects, as well as the use of M_R in the seesaw mechanism.

3.2 Red Giants and Helium Burning

We now consider the evolution off the main sequence of a solar-like star, with a mass above half a solar mass. As the hydrogen burning in the core progresses to the point that no more hydrogen is available, the stellar core consists of the ashes from this burning, ^{4}He. The star then goes through an interesting evolution:

• With no further means of producing energy, the core slowly contracts, thereby increasing in temperature as gravity does work on the core.

• Matter outside the core is still hydrogen rich, and can generate energy through hydrogen burning. Thus burning in this shell of material supports the outside layers of the star. Note as the core contracts, this matter outside the core also is pulled deeper into the gravitational potential. Furthermore, the shell H burning continually adds more mass to the core. This means the burning in the shell must intensify to generate the additional gas pressure to fight gravity. The shell also thickens as this happens, since more hydrogen is above the burning temperature.

• The resulting increasing gas pressure causes the outer envelope of the star to expand by a larger factor, up to a factor of 50. The increase in radius more than compensates for the increased internal energy generation, so that a cooler surface results. The star reddens. Thus this class of star is named a red supergiant.

• This evolution is relatively rapid, perhaps a few hundred million years: the dense core requires large energy production. The helium core is supported by its degeneracy pressure, and is characterized by densities $\sim 10^6$ g/cm^3. This stage ends when the core reaches densities and temperatures that allow helium burning through the reaction

$$\alpha + \alpha + \alpha \to {}^{12}C + \gamma. \tag{52}$$

As this reaction is very temperature dependent (see below), the conditions for ignition are very sharply defined. This has the consequence that the core mass at the helium flash point is well determined.

• The onset of helium burning produces a new source of support for the core. The energy release elevates the temperature and the core expands: He burning, not electron degeneracy, now supports the core. The burning shell and envelope have moved outward, higher in the gravitational potential. Thus shell hydrogen burning slows (the shell cools) because less gas pressure is needed to satisfy hydrostatic equilibrium. All of this means the evolution of the star has now slowed: the red giant moves along the "horizontal branch", as interior temperatures slowly elevate much as in the main sequence.

The 3α process depends on some rather interesting nuclear physics. The first interesting "accident" involves the near degeneracy of the ^{8}Be ground state and two separated αs: The ^{8}Be 0^+ ground state is just 92 keV above the 2α threshold. The measured width of the ^{8}Be ground state is 2.5 eV, which corresponds to a lifetime of

$$\tau_m \sim 2.6 \cdot 10^{-16} \text{s}. \tag{53}$$

One can compare this number to the typical time for one α to pass another. The red giant core temperature is $T_7 \sim 10 \to E \sim 8.6$ keV. Thus v/c ~ 0.002. So the transit time is

$$\tau \sim \frac{d}{v} \sim \frac{5f}{0.002} \frac{1}{3 \cdot 10^{10} \text{cm/sec}} \frac{10^{-13} \text{cm}}{f} \sim 8 \cdot 10^{-21} \text{s}. \tag{54}$$

This is more than five orders of magnitude shorter than τ_m above. Thus when a ^{8}Be nucleus is produced, it lives for a substantial time compared to this naive estimate.

To quantity this, we calculate the flux-averaged cross section assuming resonant capture

$$\langle \sigma v \rangle = (\frac{2\pi}{\mu k T})^{3/2} \frac{\Gamma \Gamma}{\Gamma} e^{-E_r/kT} \tag{55}$$

where Γ is the 2α width of the ^{8}Be ground state. This is the cross section for the $\alpha + \alpha$ reaction to form the compound nucleus then decay by $\alpha + \alpha$. But since there is only one channel, this is clearly also the result for producing the compound nucleus ^{8}Be.

By multiplying the rate/volume for producing ^{8}Be by the lifetime of ^{8}Be, one gets the number of ^{8}Be nuclei per unit volume

$$
\begin{aligned}
N(Be) &= \frac{N_\alpha N_\alpha}{1+\delta_{\alpha\alpha}}\langle \sigma v\rangle \tau_m \\
&= \frac{N_\alpha N_\alpha}{1+\delta_{\alpha\alpha}}\langle \sigma v\rangle \frac{1}{\Gamma} \\
&= \frac{N_\alpha^2}{2}\left(\frac{2\pi}{\mu kT}\right)^{3/2} e^{-E_r/kT}.
\end{aligned}
\tag{56}
$$

Notice that the concentration is *independent* of Γ. So a small Γ is not the reason we obtain a substantial buildup of ^{8}Be. This is easily seen: if the width is small, then the production rate of ^{8}Be goes down, but the lifetime of the nucleus once it is produced is longer. The two effects cancel to give the same ^{8}Be concentration. One sees that the significant ^{8}Be concentration results from two effects: 1) $\alpha+\alpha$ is the only open channel and 2) the resonance energy is low enough that some small fraction of the $\alpha+\alpha$ reactions have the requisite energy. As $E_r = 92$ keV, $E_r/kT = 10.67/T_8$ (where T_8 is the temperature in 10^8K) so that

$$
N(Be) = N_\alpha^2 T_8^{-3/2} e^{-10.67/T_8}(0.94\cdot 10^{-33}\text{cm}^3).
\tag{57}
$$

So plugging in typical values of $N_\alpha \sim 1.5\cdot 10^{28}/\text{cm}^3$ (corresponding to $\rho_\alpha \sim 10^5$ g/cm^3) and $T_8 \sim 1$ yields

$$
\frac{N(^8Be)}{N(\alpha)} \sim 3.2\times 10^{-10}.
\tag{58}
$$

Salpeter suggested that this concentration would then allow $\alpha+{}^8\text{Be}\to{}^{12}\text{C}$ to take place. Hoyle then argued that this reaction would not be fast enough to produce significant burning unless it was also resonant. Now the mass of ^{8}Be $+ \alpha$, relative to ^{12}C, is 7.366 MeV, and each nucleus has $J^\pi = 0^+$. Thus s-wave capture would require a 0^+ resonance in ^{12}C at ~ 7.4 MeV. No such state was then known, but a search by Cook, Fowler, Lauritsen, and Lauritsen revealed a 0^+ level at 7.644 MeV, with decay channels ^{8}Be$+\alpha$ and γ decay to the 2^+ 4.433 level in ^{12}C. The parameters are

$$
\Gamma_\alpha \sim 8.9\text{eV} \qquad \Gamma_\gamma \sim 3.6\cdot 10^{-3}\text{eV}.
\tag{59}
$$

The resonant cross section formula gives

$$
r_{48} = N_8 N_\alpha\left(\frac{2\pi}{\mu kT}\right)^{3/2}\frac{\Gamma_\alpha \Gamma_\gamma}{\Gamma}e^{-E_r/kT}.
\tag{60}
$$

Plugging in our previous expression for $N(^8\text{Be})$ yields

$$r_{48} = N_\alpha^3 T_8^{-3} e^{-42.9/T_8} (6.3 \cdot 10^{-54} \text{cm}^6/\text{s}). \tag{61}$$

If we denote by $\omega_{3\alpha}$ the decay rate of an α in our plasma, then

$$\omega_{3\alpha} = 3N_\alpha^2 T_8^{-3} e^{-42.9/T_8} (6.3 \cdot 10^{-54} \text{cm}^6/\text{sec})$$
$$= (\frac{N_\alpha}{1.5 \cdot 10^{28}/\text{cm}^3})^2 (4.3 \cdot 10^3/\text{sec}) T_8^{-3} e^{-42.9/T_8}. \tag{62}$$

Now the energy release per reaction is 7.27 MeV. Thus we can calculate the energy produced per gram, ϵ:

$$\epsilon = \omega_{3\alpha} \frac{7.27\text{MeV}}{3} \frac{1.5 \cdot 10^{23}}{\text{g}}$$
$$= (2.5 \cdot 10^{21} \text{erg/g sec})(\frac{N_\alpha}{1.5 \cdot 10^{28}/\text{cm}^3})^2 T_8^{-3} e^{-42.9/T_8}. \tag{63}$$

We can evaluate this at a temperature of $T_8 \sim 1$ to find

$$\epsilon \sim (584\text{ergs/g sec})(\frac{N_\alpha}{1.5 \cdot 10^{28}/\text{cm}^3})^2. \tag{64}$$

Typical values found in stellar calculations are in good agreement with this: typical red giant energy production is ~ 100 ergs per gram per second.

To get a feel for the temperature sensitivity of this process, we can do a Taylor series expansion, finding

$$\epsilon(T) \sim (\frac{T}{T_o})^{40} N_\alpha^2. \tag{65}$$

This steep temperature dependence is the reason the He flash is delicately dependent on conditions in the core.

3.3 Neutrino Magnetic Moments and He Ignition

Prior to the helium flash, the degenerate He core radiates energy largely by neutrino pair emission. The process is the decay of a plasmon — which one can think of as a photon "dressed" by electron-hole excitations — thereby acquiring an effective mass of about 10 keV. The photon couples to a neutrino pair through a electron particle-hole pair that then decays into a $Z_o \to \nu\bar{\nu}$.

If this cooling is somehow enhanced, the degenerate helium core would be kept cooler, and would not ignite at the normal time. Instead it would continue to grow until it overcame the enhanced cooling to reach, once again, the ignition temperature.

One possible mechanism for enhanced cooling is a neutrino magnetic moment. Then the plasmon could directly couple to a neutrino pair. The strength of this coupling would depend on the size of the magnetic moment.

A delay in the time of He ignition has several observable consequences, including changing the ratio of red giant to horizontal branch stars. Thus, using the standard theory of red giant evolution, investigators have attempted to determine what size of magnetic moment would produce unacceptable changes in the astronomy. The result is a limit[32] on the neutrino magnetic moment of

$$\mu_{ij} \lesssim 3 \cdot 10^{-12} \text{electron Bohr magnetons} \tag{66}$$

as was mentioned earlier. This limit is more than two orders of magnitude more stringent than that from direct laboratory tests.

This example is just one of a number of such constraints that can be extracted from similar stellar cooling arguments. The arguments above, for example, can be repeated for neutrino electric dipole moments. More interesting, it can be repeated for axion emission from red giants. Axions, the pseudoGolstone bosons associate with the solution of the strong CP problem suggested by Peccei and Quinn, are very light and can be produced radiatively within the red giant by the Compton process, by the Primakoff process off nuclei, or by emission from low-lying nuclear levels, such at from the 14 keV transition in ^{57}Fe. The net result is that axions of mass above a few eV are excluded; if axions have a direct coupling to electrons, so that the Compton process off electrons operates, the constraint is considerably tighter.

A similar argument can be formulated for supernova cooling. During SN1987A the neutrino burst detected by IMB and by Kamiokande was consistent with cooling on a timescale of about 4 seconds. Thus any process cooling the star more efficiently than neutrino emission would have shortened this time, while also reducing the flux in neutrinos. Large Dirac neutrino masses allow trapped neutrinos to scatter into sterile right-handed states. Right-handed neutrinos, lacking standard model interactions, would then escape the star (provided they do not scatter back into interacting left-handed states). Unfortunately the upper bounds imposed on the neutrino mass are quite model dependent, ranging over (1-25) keV.

The supernova cooling argument can also be repeated for axions. The window of sensitive runs from 1 eV (above this mass they are more strongly coupled than neutrinos, and thus cannot compete with neutrino cooling) to about 0.01 eV (below this mass they are too weakly coupled to be produced on the timescale of supernova cooling). It is interesting that the supernova and red giant cooling limit on axions nearly meet: a small window may still exist around a few eV if the axion has no coupling to electrons.

4 Supernovae, Supernova Neutrinos, and Nucleosynthesis

Consider a massive star, in excess of 10 solar masses, burning the hydrogen in its core under the conditions of hydrostatic equilibrium. When the hydrogen is exhausted, the core contracts until the density and temperature are reached where $3\alpha \rightarrow {}^{12}C$ can take place. The He is then burned to exhaustion. This pattern (fuel exhaustion, contraction, and ignition of the ashes of the previous burning cycle) repeats several times, leading finally to the explosive burning of ${}^{28}Si$ to Fe. For a heavy star, the evolution is rapid: the star has to work harder to maintain itself against its own gravity, and therefore consumes its fuel faster. A 25 solar mass star would go through all of these cycles in about 7 My, with the final explosive Si burning stage taking a few days. The result is an "onion skin" structure of the precollapse star in which the star's history can be read by looking at the surface inward: there are concentric shells of H, ${}^{4}He$, ${}^{12}C$, ${}^{16}O$ and ${}^{20}Ne$, ${}^{28}Si$, and ${}^{56}Fe$ at the center.

4.1 The Explosion Mechanism [35]

The source of energy for this evolution is nuclear binding energy. A plot of the nuclear binding energy δ as a function of nuclear mass shows that the minimum is achieved at Fe. In a scale where the ${}^{12}C$ mass is picked as zero:

$$
\begin{array}{ll}
{}^{12}C & \delta/\text{nucleon} = 0.000 \text{ MeV} \\
{}^{16}O & \delta/\text{nucleon} = -0.296 \text{ MeV} \\
{}^{28}Si & \delta/\text{nucleon} = -0.768 \text{ MeV} \\
{}^{40}Ca & \delta/\text{nucleon} = -0.871 \text{ MeV} \\
{}^{56}Fe & \delta/\text{nucleon} = -1.082 \text{ MeV} \\
{}^{72}Ge & \delta/\text{nucleon} = -1.008 \text{ MeV} \\
{}^{98}Mo & \delta/\text{nucleon} = -0.899 \text{ Mev}
\end{array}
$$

Thus once the Si burns to produce Fe, there is no further source of nuclear energy adequate to support the star. So as the last remnants of nuclear burning take place, the core is largely supported by degeneracy pressure, with the energy generation rate in the core being less than the stellar luminosity. The core density is about 2×10^9 g/cc and the temperature is $kT \sim 0.5$ MeV.

Thus the collapse that begins with the end of Si burning is not halted by a new burning stage, but continues. As gravity does work on the matter, the collapse leads to a rapid heating and compression of the matter. As the nucleons in Fe are bound by about 8 MeV, sufficient heating can release αs and a few nucleons. At the same time, the electron chemical potential is increasing. This makes electron capture on nuclei and any free protons favorable,

$$e^- + p \to \nu_e + n. \tag{67}$$

Note that the chemical equilibrium condition is

$$\mu_e + \mu_p = \mu_n + \langle E_\nu \rangle. \tag{68}$$

Thus the fact that neutrinos are not trapped plus the rise in the electron Fermi surface as the density increases, lead to increased neutronization of the matter. The escaping neutrinos carry off energy and lepton number. Both the electron capture and the nuclear excitation and disassociation take energy out of the electron gas, which is the star's only source of support. This means that the collapse is very rapid. Numerical simulations find that the iron core of the star ($\sim$ 1.2-1.5 solar mases) collapses at about 0.6 of the free fall velocity.

In the early stages of the infall the ν_es readily escape. But neutrinos are trapped when a density of $\sim 10^{12}$g/cm^3 is reached. At this point the neutrinos begin to scatter off the matter through both charged current and coherent neutral current processes. The neutral current neutrino scattering off nuclei is particularly important, as the scattering cross section is off the total nuclear weak charge, which is approximately the neutron number. This process transfers very little energy because the mass energy of the nucleus is so much greater than the typical energy of the neutrinos. But momentum is exchanged. Thus the neutrino "random walks" out of the star. When the neutrino mean free path becomes sufficiently short, the "trapping time" of the neutrino begins to exceed the time scale for the collapse to be completed. This occurs at a density of about 10^{12} g/cm^3, or somewhat less than 1% of nuclear density. After this point, the energy released by further gravitational collapse and the star's remaining lepton number are trapped within the star.

If we take a neutron star of 1.4 solar masses and a radius of 10 km, an estimate of its binding energy is

$$\frac{GM^2}{2R} \sim 2.5 \times 10^{53} \text{ergs}. \tag{69}$$

Thus this is roughly the trapped energy that will later be radiated in neutrinos.

The trapped lepton fraction Y_L is a crucial parameter in the explosion physics: a higher trapped Y_L leads to a larger homologous core, a stronger shock wave, and easier passage of the shock wave through the outer core, as will be discussed below. Most of the lepton number loss of an infalling mass element occurs as it passes through a narrow range of densities just before trapping. The reasons for this are relatively simple: on dimensional grounds weak rates in a plasma go as T^5, where T is the temperature. Thus the electron capture rapidly turns on as matter falls toward the trapping radius, and lepton number loss is maximal just prior to trapping. Inelastic neutrino reactions have an important effect on these losses, as the coherent trapping cross section goes as E_ν^2 and is thus least effective for the lowest energy neutrinos. As these neutrinos escape, inelastic reactions repopulate the low energy states, allowing the neutrino emission to continue.

The velocity of sound in matter rises with increasing density. The inner homologous core, with a mass $M_{HC} \sim 0.6 - 0.9$ solar masses, is that part of the iron core where the sound velocity exceeds the infall velocity. This allows any pressure variations that may develop in the homologous core during infall to even out before the collapse is completed. As a result, the homologous core collapses as a unit, retaining its density profile. That is, if nothing were to happen to prevent it, the homologous core would collapse to a point.

The collapse of the homologous core continues until nuclear densities are reached. As nuclear matter is rather incompressible (~ 200 MeV/f^3), the nuclear equation of state is effective in halting the collapse: maximum densities of 3-4 times nuclear are reached, e.g., perhaps $6 \cdot 10^{14}$ g/cm^3. The innermost shell of matter reaches this supernuclear density first, rebounds, sending a pressure wave out through the homologous core. This wave travels faster than the infalling matter, as the homologous core is characterized by a sound speed in excess of the infall speed. Subsequent shells follow. The resulting series of pressure waves collect near the sonic point (the edge of the homologous core). As this point reaches nuclear density and comes to rest, a shock wave breaks out and begins its traversal of the outer core.

Initially the shock wave may carry an order of magnitude more energy than is needed to eject the mantle of the star (less than 10^{51} ergs). But as the shock wave travels through the outer iron core, it heats and melts the iron that crosses the shock front, at a loss of ~ 8 MeV/nucleon. The enhanced electron capture that occurs off the free protons left in the wake of the shock, coupled with the sudden reduction of the neutrino opacity of the matter (recall $\sigma_{coherent} \sim N^2$), greatly accelerates neutrino emission. This is another energy loss. [Many numerical models predict a strong "breakout" burst of ν_es in the few milliseconds required for the shock wave to travel from the edge of the homologous core to the neutrinosphere at $\rho \sim 10^{12}$ g/cm^3 and $r \sim 50$ km. The neutrinosphere is the term from the neutrino trapping radius, or surface of last scattering.] The summed losses from shock wave heating and neutrino emission are comparable to the initial energy carried by the shock wave. Thus most numerical models fail to produce a successful "prompt" hydrodynamic explosion.

Most of the attention in the past decade focused on two explosion scenarios. In the prompt mechanism described above, the shock wave is sufficiently strong to survive the passage of the outer iron core with enough energy to blow off the mantle of the star. The most favorable results were achieved with smaller stars (less than 15 solar masses) where there is less overlying iron, and with soft equations of state, which produce a more compact neutron star and thus lead to more energy release. In part because of the lepton number loss problems discussed earlier, now it is widely believed that this mechanism fails for all but unrealistically soft nuclear equations of state.

The delayed mechanism begins with a failed hydrodynamic explosion; after about 0.01 seconds the shock wave stalls at a radius of 200-300 km. It exists in a sort of equilibrium, gaining energy from matter falling across the shock front, but loosing energy to the heating of that material. However, after perhaps 0.5 seconds, the shock wave is revived due to neutrino heating of the nucleon "soup" left in the wake of the shock. This heating comes primarily from charged current reactions off the nucleons in that nucleon gas; quasielastic scattering also contributes. This high entropy radiation-dominated gas may reach two MeV in temperature. The pressure exerted by this gas helps to push the shock outward. It is important to note that there are limits to how effective this neutrino energy transfer can be: if matter is too far from the core, the coupling to neutrinos is too weak to deposit significant energy. If too close, the matter may be at a temperature (or soon reach a temperature) where neutrino emission cools the matter as fast or faster than neutrino absorption heats it. The term "gain radius" is used to describe the region where useful heating is done.

This subject is still controversial and unclear. The problem is numerically challenging, forcing modelers to handle the difficult hydrodynamics of a shock wave; the complications of the nuclear equation of state at densities not yet accessible to experiment; modeling in two or three dimensions; handling the slow diffusion of neutrinos; etc. Not all of these aspects can be handled reasonably at the same time, even with existing supercomputers. Thus there is considerable disagreement about whether we have any supernova model that succeeds in ejecting the mantle.

However the explosion proceeds, there is agreement that 99% of the $3 \cdot 10^{53}$ ergs released in the collapse is radiated in neutrinos of all flavors. The time scale over which the trapped neutrinos leak out of the protoneutron star is about three seconds. Through most of their migration out of the protoneutron star, the neutrinos are in flavor equilibrium

$$\text{e.g.,} \quad \nu_e + \bar{\nu}_e \leftrightarrow \nu_\mu + \bar{\nu}_\mu. \tag{70}$$

As a result, there is an approximate equipartition of energy among the neutrino flavors. After weak decoupling, the ν_es and $\bar{\nu}_e$s remain in equilibrium with the matter for a longer period than their heavy-flavor counterparts, due to the larger cross sections for scattering off electrons and because of the charge-current reactions

$$\nu_e + n \leftrightarrow p + e^-$$
$$\bar{\nu}_e + p \leftrightarrow n + e^+. \tag{71}$$

Thus the heavy flavor neutrinos decouple from deeper within the star, where temperatures are higher. Typical calculations yield

$$T_{\nu_\mu} \sim T_{\nu_\tau} \sim 8\text{MeV} \quad T_{\nu_e} \sim 3.5\text{MeV} \quad T_{\bar{\nu}_e} \sim 4.5\text{MeV}. \tag{72}$$

The difference between the ν_e and $\bar{\nu}_e$ temperatures is a result of the neutron richness of the matter, which enhances the rate for charge-current reactions of the ν_es, thereby keeping them coupled to the matter somewhat longer.

This temperature hierarchy is crucially important to nucleosynthesis and also to possible neutrino oscillation scenarios. The three-flavor MSW level-crossing diagram is shown in Fig. 10. One very popular scenario attributes the solar neutrino problem to $\nu_\mu \leftrightarrow \nu_e$ transmutation; this means that a second crossing with a ν_τ could occur at higher density. It turns out plausible seasaw mass patterns suggest a ν_τ mass on the order of a few eV, which would be interesting cosmologically. The second crossing would then occur outside the neutrino sphere, that is, after the neutrinos have decoupled and have fixed spectra with the temperatures given above. Thus a $\nu_e \leftrightarrow \nu_\tau$ oscillation would produce a distinctive $T \sim 8$ MeV spectrum of ν_es. This has dramatic consequences for terrestrial detection and for nucleosynthesis in the supernova.

4.2   The Neutrino Process [36]

Core collapse supernovae are one of the major engines driving galactic chemical evolution, producing and ejecting the metals that enrich our galaxy. The discussion of the previous section described the hydrostatic evolution of a pre-supernova star in which large quantities of the most abundant metals (C, O, Ne, ...) are synthesized and later ejected during the explosion. During the passage of the shock wave through the star's mantle, temperature of $\sim (1-3)\cdot 10^9$K and are reached in the silicon, oxygen, and neon shells. This shock wave heating

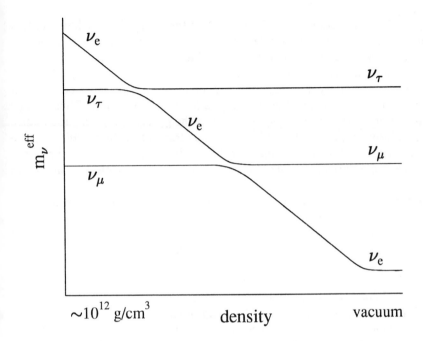

ν_e

ν_τ

ν_τ

ν_e

ν_μ

ν_μ

ν_e

m_ν^{eff}

$\sim 10^{12}$ g/cm^3 density   vacuum

Figure 10: Three-flavor neutrino level-crossing diagram. One popular scenario associates the solar neutrino problem with $\nu_e \leftrightarrow \nu_\mu$ oscillations and predicts a cosmologically interested massive ν_τ with $\nu_e \leftrightarrow \nu_\tau$ oscillations near the supernova neutrinosphere.

induces $(\gamma, \alpha) \leftrightarrow (\alpha, \gamma)$ and related reactions that generate a mass flow toward highly bound nuclei, resulting in the synthesis of iron peak elements as well as less abundant odd-A species. Rapid neutron-induced reactions are thought to take place in the high-entropy atmosphere just above the mass cut, producing about half of the heavy elements above A $\sim$ 80. This is the subject of the Sec. 4.3. Finally, the ν-process described below is responsible for the synthesis of rare species such as ^{11}B and ^{19}F. This process involves the response of nuclei at momentum transfers where the allowed approximation is no longer valid. Thus we will use the ν-process in this section to illustrate some of the relevant nuclear physics.

One of the problems – still controversial – that may be connected with the neutrino process is the origin of the light elements Be, B and Li, elements which are not produced in sufficient amounts in the big bang or in any of the stellar mechanisms we have discussed. The traditional explanation has been cosmic ray spallation interactions with C, O, and N in the interstellar medium. In this picture, cosmic ray protons collide with C at relatively high energy, knocking the nucleus apart. So in the debris one can find nuclei like ^{10}B, ^{11}B, and ^{7}Li. But there are some problems with this picture. First of all, this is an example of a secondary mechanism: the interstellar medium must be enriched in the C, O, and N to provide the targets for these reactions. Thus cosmic ray spallation must become more effective as the galaxy ages. The abundance of boron, for example, would tend to grow quadratically with metalicity, since the rate of production goes linearly with metalicity. But observations, especially recent measurements with the HST, find a linear growth[37] in the boron abundance.

A second problem is that the spectrum of cosmic ray protons peaks near 1 GeV, leading to roughly comparable production of the two isotopes ^{10}B and ^{11}B. That is, while it takes more energy to knock two nucleons out of carbon than one, this difference is not significant compared to typical cosmic ray energies. More careful studies lead to the expectation that the abundance ratio of ^{11}B to ^{10}B might be ~ 2. In nature, it is greater than 4.

Fans of cosmic ray spallation have offered solutions to these problems, e.g., similar reactions occurring in the atmospheres of nebulae involving lower energy cosmic rays. As this suggestion was originally stimulated by the observation of nuclear γ rays from Orion, now retracted, some of the motivation for this scenario has evaporated. Here I focus on an alternative explanation, synthesis via neutrino spallation.

Previously we described the allowed Gamow-Teller (spin-flip) and Fermi weak interaction operators. These are the appropriate operators when one probes the nucleus at a wavelength – that is, at a size scale – where the nucleus responds like an elementary particle. We can then characterize its response by its macroscopic quantum numbers, the spin and charge. On the other hand, the nucleus is a composite object and, therefore, if it is probed at shorter length scales, all kinds of interesting radial excitations will result, analogous to the vibrations of a drumhead. For a reaction like neutrino scattering off a nucleus, the full operator involves the additional factor

$$e^{i\vec{k}\cdot\vec{r}} \sim 1 + i\vec{k}\cdot\vec{r} \tag{73}$$

where the expression on the right is valid if the magnitude of $\vec{k}$ is not too large. Thus the full charge operator includes a "first forbidden" term

$$\sum_{i=1}^{A} \vec{r}_i \tau_3(i) \tag{74}$$

and similarly for the spin operator

$$\sum_{i=1}^{A}[\vec{r}_i \otimes \vec{\sigma}(i)]_{J=0,1,2}\tau_3(i). \tag{75}$$

These operators generate collective radial excitations, leading to the so-called "giant resonance" excitations in nuclei. The giant resonances are typically at an excitation energy of 20-25 MeV in light nuclei. One important property is that these operators satisfy a sum rule (Thomas-Reiche-Kuhn) of the form

$$\sum_{f}|\langle f|\sum_{i=1}^{A}r(i)\tau_3(i)|i\rangle|^2 \sim \frac{NZ}{A} \sim \frac{A}{4} \tag{76}$$

where the sum extends over a complete set of final nuclear states. These first-forbidden operators tend to dominate the cross sections for scattering the high energy supernova neutrinos (ν_μs and ν_τs), with $E_\nu \sim 25$ MeV, off light nuclei. From the sum rule above, it follows that nuclear cross sections per target *nucleon* are roughly constant.

The E1 giant dipole mode described above is depicted qualitatively in Fig. 11a. This description, which corresponds to an early model of the giant resonance response by Goldhaber and Teller, involves the harmonic oscillation of the proton and neutron fluids against one another. The restoring force for small displacements would be linear in the displacement and dependent on the nuclear symmetry energy. There is a natural extension of this model to weak interactions, where axial excitations occur. For example, one can envision a mode similar to that of Fig. 11a where the spin-up neutrons and spin-down protons oscillate against spin-down neutrons and spin-up protons, the spin-isospin mode of Fig. 11b. This mode is one that arises in a simple SU(4) extension of the Goldhaber-Teller model, derived by assuming that the nuclear force is spin and isospin independent, at the same excitation energy as the E1 mode. In full, the Goldhaber-Teller model predicts a degenerate 15-dimensional supermultiplet of giant resonances, each obeying sum rules analogous to the TRK sum rule. While more sophisticated descriptions of the giant resonance region are available, of course, this crude picture is qualitatively accurate.

This nuclear physics is important to the ν-process. The simplest example of ν-process nucleosynthesis involves the Ne shell in a supernova. Because of the first-forbidden contributions, the cross section for inelastic neutrino scattering to the giant resonances in Ne is $\sim 3 \cdot 10^{-41}$ cm^2/flavor for the more energetic heavy-flavor neutrinos. This reaction

$$\nu + A \rightarrow \nu' + A^* \tag{77}$$

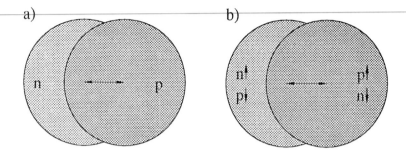

Figure 11: Schematic illustration of a) the E1 giant dipole mode familiar from electromagnetic interactions and b) a spin-isospin giant dipole mode associated with the first-forbidden weak axial response.

transfers an energy typical of giant resonances, ~ 20 MeV. A supernova releases about 3×10^{53} ergs in neutrinos, which converts to about 4×10^{57} heavy flavor neutrinos. The Ne shell in a 20 $M_\odot$ star has at a radius $\sim 20,000$ km. Thus the neutrino fluence through the Ne shell is

$$\phi \sim \frac{4 \cdot 10^{57}}{4\pi (20,000 \text{km})^2} \sim 10^{38}/\text{cm}^2. \tag{78}$$

Thus folding the fluence and cross section, one concludes that approximately 1/300th of the Ne nuclei interact.

This is quite interesting since the astrophysical origin of ^{19}F had not been understood. The only stable isotope of fluorine, ^{19}F has an abundance

$$\frac{^{19}\text{F}}{^{20}\text{Ne}} \sim \frac{1}{3100}. \tag{79}$$

This leads to the conclusion that the fluorine found in toothpaste was created by neutral current neutrino reactions deep inside some ancient supernova.

The calculation of the final ^{19}F/^{20}Ne ratio is more complicated than the simple 1/300 ratio given above:

• When Ne is excited by ~ 20 MeV through inelastic neutrino scattering, it breaks up in two ways

$$^{20}\text{Ne}(\nu, \nu')^{20}\text{Ne}^* \to {}^{19}\text{Ne} + n \to {}^{19}\text{F} + e^+ + \nu_e + n$$
$$^{20}\text{Ne}(\nu, \nu')^{20}\text{Ne}^* \to {}^{19}\text{F} + p \tag{80}$$

with the first reaction occurring half as frequently as the second. As both channels lead to ^{19}F, we have correctly estimated the instantaneous abundance ratio in the Ne shell of

$$\frac{^{19}F}{^{20}Ne} \sim \frac{1}{300}. \tag{81}$$

• We must also address the issue of whether the produced ^{19}F survives. In the first 10^{-8} sec the coproduced neutrons in the first reaction react via

$$^{15}O(n,p)^{15}N \quad ^{19}Ne(n,\alpha)^{16}O \quad ^{20}Ne(n,\gamma)^{21}Ne \quad ^{19}Ne(n,p)^{19}F \tag{82}$$

with the result that about 70% of the ^{19}F produced via spallation of neutrons is then immediate destroyed, primarily by the (n,α) reaction above. In the next 10^{-6} sec the coproduced protons are also processed

$$^{15}N(p,\alpha)^{12}C \quad ^{19}F(p,\alpha)^{16}O \quad ^{23}Na(p,\alpha)^{20}Ne \tag{83}$$

with the latter two reactions competing as the primary proton poisons. This makes an important prediction: stars with high Na abundances should make more F, as the ^{23}Na acts as a proton poison to preserve the produced F.
• Finally, there is one other destruction mechanism, the heating associated with the passage of the shock wave. It turns out the the F produced prior to shock wave passage can survive if it is in the outside half of the Ne shell. The reaction

$$^{19}F(\gamma,\alpha)^{15}N \tag{84}$$

destroys F for peak explosion temperatures exceeding $1.7 \cdot 10^9$K. Such a temperature is produced at the inner edge of the Ne shell by the shock wave heating, but not at the outer edge.

If all of this physics in handled is a careful network code that includes the shock wave heating and F production both before and after shock wave passage, the following are the results:

$[^{19}F/^{20}Ne]/[^{19}F/^{20}Ne]_\odot$	$T_{heavy\ \nu}(MeV)$
0.14	4
0.6	6
1.2	8
1.1	10
1.1	12

where the abundance ratio in the first column has been normalized to the solar value. One sees that the attribution of F to the neutrino process argues that the heavy flavor ν temperature must be greater than 6 MeV, a result theory favors. One also sees that F cannot be overproduced by this mechanism: although the instantaneous production of F continues to grow rapidly with the neutrino temperature, too much F results in its destruction through the (p, α) reaction, given a solar abundance of the competing proton poison ^{23}Na. Indeed, this illustrates an odd quirk: although in most cases the neutrino process is a primary mechanism, one needs ^{23}Na present to produce significant F. Thus in this case the neutrino process is a secondary mechanism.

While there are other significant neutrino process products (^{7}Li, ^{138}La, ^{180}Ta, ^{15}N ...), the most important product is ^{11}B, produced by spallation off carbon. A calculation by Timmes et al. [18] found that the combination of the neutrino process, cosmic ray spallation and big-bang nucleosynthesis together can explain the evolution of the light elements. The neutrino process, which produces a great deal of ^{11}B but relatively little ^{10}B, combines with the cosmic ray spallation mechanism to yield the observed isotope ratio. Again, one prediction of this picture is that early stars should be ^{11}B rich, as the neutrino process is primary and operates early in our galaxy's history; the cosmic ray production of ^{10}B is more recent. There is hope that HST studies will soon be able to descriminate between ^{10}B and ^{11}B: as yet this has not been done.

4.3 The r-process

Beyond the iron peak nuclear Coulomb barriers become so high that charged particle reactions become ineffective, leaving neutron capture as the mechanism responsible for producing the heaviest nuclei. If the neutron abundance is modest, this capture occurs in such a way that each newly synthesized nucleus has the opportunity to β decay, if it is energetically favorable to do so. Thus weak equilibrium is maintained within the nucleus, so that synthesis is along the path of stable nuclei. This is called the s- or slow-process. However a plot of the s-process in the (N,Z) plane reveals that this path misses many stable, neutron-rich nuclei that are known to exist in nature. This suggests that another mechanism is at work, too. Furthermore, the abundance peaks found in nature near masses A $\sim$ 130 and A $\sim$ 190, which mark the closed neutron shells where neutron capture rates and β decay rates are slower, each split into two subpeaks. One set of subpeaks corresponds to the closed-neutron-shell numbers N $\sim$ 82 and N $\sim$ 126, and is clearly associated with the s-process. The other set is shifted to smaller N, $\sim$ 76 and $\sim$ 116, respectively, and is suggestive of a much more explosive neutron capture environment where neutron capture can be rapid.

This second process is the r- or rapid-process, characterized by:
• The neutron capture is fast compared to β decay rates.

• The equilibrium maintained within a nucleus is established by $(n, \gamma) \leftrightarrow (\gamma, n)$: neutron capture fills up the available bound levels in the nucleus until this equilibrium sets in. The new Fermi level depends on the temperature and the relative n/γ abundance.

• The nucleosynthesis rate is thus controlled by the β decay rate: each β^- capture coverting n → p opens up a hole in the neutron Fermi sea, allowing another neutron to be captured.

• The nucleosynthesis path is along exotic, neutron-rich nuclei that would be highly unstable under normal laboratory conditions.

• As the nucleosynthesis rate is controlled by the β decay, mass will build up at nuclei where the β decay rates are slow. It follows, if the neutron flux is reasonable steady over time so that equilibrated mass flow is reached, that the resulting abundances should be inversely proportional to these β decay rates.

Let's first explore the $(n, \gamma) \leftrightarrow (\gamma, n)$ equilibrium condition, which requires that the rate for (n, γ) balances that for (γ, n) for an average nucleus. So consider the formation cross section

$$A + n \to (A + 1) + \gamma. \tag{85}$$

This is an exothermic reaction, as the neutron drops into the nuclear well. Our averaged cross section, assuming a resonant reaction (the level density is high in heavy nuclei) is

$$\langle \sigma v \rangle_{(n,\gamma)} = \left(\frac{2\pi}{\mu kT} \right)^{3/2} \frac{\Gamma_n \Gamma_\gamma}{\Gamma} e^{-E/KT} \tag{86}$$

where $E \sim 0$ is the resonance energy, and the Γs are the indicated partial and total widths. Thus the rate per unit volume is

$$r_{(n,\gamma)} \sim N_n N_A \left(\frac{2\pi}{\mu kT} \right)^{3/2} \frac{\Gamma_n \Gamma_\gamma}{\Gamma} \tag{87}$$

where N_n and N_A are the neutron and nuclear number densities and μ the reduced mass. This has to be compared to the (γ, n) rate.

The (γ, n) reaction requires the photon number density in the gas. This is given by the Bose-Einstein distribution

$$N(\epsilon) = \frac{8\pi}{c^3 h^3} \frac{\epsilon^2 d\epsilon}{e^{\epsilon/kT} - 1}. \tag{88}$$

The high-energy tail of the normalized distribution can thus be written

$$\sim \frac{1}{N_\gamma \pi^2} \epsilon^2 e^{-\epsilon/kT} d\epsilon \tag{89}$$

where in the last expression we have set $\hbar = c = 1$.

Now we need the resonant cross section in the (γ, n) direction. For photons the wave number is proportional to the energy, so

$$\sigma_{(\gamma,n)} = \frac{\pi}{\epsilon^2} \frac{\Gamma_\gamma \Gamma_n}{(\epsilon - E_r)^2 + (\Gamma/2)^2}. \tag{90}$$

As the velocity is c =1,

$$\langle \sigma v \rangle = \frac{1}{\pi^2 N_\gamma} \int_0^\infty \epsilon^2 e^{-\epsilon/kT} d\epsilon \frac{\pi}{\epsilon^2} \frac{\Gamma_\gamma \Gamma_n}{(\epsilon - E_r)^2 + (\Gamma/2)^2}. \tag{91}$$

We evaluate this in the usual way for a sharp resonance, remembering that the energy integral over just the denominator above (the sharply varying part) is $2\pi/\Gamma$

$$\sim \frac{\Gamma_\gamma \Gamma_n}{N_\gamma} e^{-E_r/kT} \frac{2}{\Gamma}. \tag{92}$$

So that the rate becomes

$$r_{(\gamma,n)} \sim 2N_{A+1} \frac{\Gamma_\gamma \Gamma_n}{\Gamma} e^{-E_r/kT}. \tag{93}$$

Equating the (n, γ) and (γ, n) rates and taking $N_A \sim N_{A-1}$ then yields

$$N_n \sim \frac{2}{(\hbar c)^3} \left(\frac{\mu c^2 kT}{2\pi} \right)^{3/2} e^{-E_r/kT} \tag{94}$$

where the $\hbar$s and cs have been properly inserted to give the right dimensions. Now E_r is esssentially the binding energy. So plugging in the conditions $N_n \sim 3 \times 10^{23}/\text{cm}^3$ and $T_9 \sim 1$, we find that the binding energy is ~ 2.4 MeV. Thus neutrons are bound by about 30 times kT, a value that is still small compared to a typical binding of 8 MeV for a normal nucleus. (In this calculation I calculated the neutron reduced mass assuming a nuclear target with A=150.)

The above calculation fails to count spin states for the photons and nuclei and is thus not quite correct. But it makes the essential point: the r-process involves very exotic species largely unstudied in any terrestrial laboratory. It is good to bear this in mind, as in the following section we will discuss the responses of such nuclei to neutrinos. Such responses thus depend on the ability of theory to extrapolate responses from known nuclei to those quite unfamiliar.

The path of the r-process is along neutron-rich nuclei, where the neutron Fermi sea is just $\sim$ (2-3) MeV away from the neutron drip line (where no more bound neutron levels exist). After the r-process finishes (the neutron exposure ends) the nuclei decay back to the valley of stability by β decay. This can involve some neutron spallation (β-delayed neutrons) that shift the mass number A to a lower value. But it certainly involves conversion of neutrons into protons, and that shifts the r-process peaks at N $\sim$ 82 and 126 to a lower N, off course. This effect is clearly seen in the abundance distribution: the r-process peaks are shifted to lower N relative to the s-process peaks. This is the origin of the second set of "subpeaks" mentioned at the start of the section.

It is believed that the r-process can proceed to very heavy nuclei (A $\sim$ 270) where it is finally ended by β-delayed and n-induced fission, which feeds matter back into the process at an A $\sim A_{max}/2$. Thus there may be important cycling effects in the upper half of the r-process distribution.

What is the site(s) of the r-process? This has been debated many years and still remains a controversial subject:

• The r-process requires exceptionally explosive conditions
$$\rho(n) \sim 10^{20} \text{ cm}^{-3} \quad T \sim 10^9 K \quad t \sim 1s.$$

• Both primary and secondary sites proposed. Primary sites are those not requiring preexisting metals. Secondary sites are those where the neutron capture occurs on preexisting s-process seeds.

• Suggested primary sites include the the neutronized atmosphere above the proto-neutron star in a Type II supernova, neutron-rich jets produced in supernova explosions or in neutron star mergers, inhomogeneous big bangs, etc.

• Secondary sites, where $\rho(n)$ can be lower for successful synthesis, include the He and C zones in Type II supernovae, the red giant He flash, etc.

The balance of evidence favors a primary site, so one requiring no preenrichment of heavy s-process metals. Among the evidence:

1) HST studies of very-metal-poor halo stars: The most important evidence are the recent HST measurements of Cowan, Sneden et al. [38] of very metal-poor stars ([Fe/H] $\sim$ -1.7 to -3.12) where an r-process distribution very much like that of our sun has been seen for Z $\gtrsim$ 56. Furthermore, in these stars the iron content is variable. This suggests that the "time resolution" inherent in these old stars is short compared to galactic mixing times (otherwise Fe would be more constant). The conclusion is that the r-process material in these stars is most likely from one or a few local supernovae. The fact that the distributions match the solar r-process (at least above charge 56) strongly suggests that there is some kind of unique site for the r-process: the solar r-process distribution did not come from averaging over many different kinds of r-process events. Clearly the fact that these old stars are enriched in r-process metals also strongly argues for a primary process: the r-process

works quite well in an environment where there are few initial s-process metals.

2) There are also fairly good theoretical arguments that a primary r-process occurring in a core-collapse supernova might be viable[39]. First, galactic chemical evolution studies indicate that the growth of r-process elements in the galaxy is consistent with low-mass Type II supernovae in rate and distribution. More convincing is the fact that modelers have shown that the conditions needed for an r-process (very high neutron densities, temperatures of 1-3 billion degrees) might be realized in a supernova. The site is the last material expelled from the supernova, the matter just above the mass cut. When this material is blown off the star initially, it is a very hot neutron-rich, radiation-dominated gas containing neutrons and protons, but an excess of the neutrons. As it expands off the star and cools, the material first goes through a freezeout to α particles, a step that essentially locks up all the protons in this way. Then the αs interact through reactions like

$$\alpha + \alpha + \alpha \rightarrow ^{12}C$$
$$\alpha + \alpha + n \rightarrow ^{9}Be$$

to start forming heavier nuclei. Note, unlike the big bang, that the density is high enough to allow such three-body interactions to bridge the mass gaps at A = 5,8. The α capture continues up to heavy nuclei, to A $\sim$ 80, in the network calculations. The result is a small number of "seed" nuclei, a large number of αs, and excess neutrons. These neutrons preferentially capture on the heavy seeds to produce an r-process. Of course, what is necessary is to have $\sim$ 100 excess neutrons per seed in order to successfully synthesize heavy mass nuclei. Some of the modelers find conditions where this almost happens.

There are some very nice aspects of this site: the amount of matter ejected is about $10^{-5} - 10^{-6}$ solar masses, which is just about what is needed over the lifetime of the galaxy to give the integrated r-process metals we see, taking a reasonable supernova rate. But there are also a few problems, especially the fact that with calculated entropies in the nucleon soup above the proto-neutron star, neutron fractions appear to be too low to produce a successful A $\sim$ 190 peak. There is some interesting recent work invoking neutrino oscillations to cure this problem: charge current reactions on free protons and neutrons determine the n/p ratio in the gas. Then, for example, an oscillation of the type $\nu_e \rightarrow \nu_{\text{sterile}}$ can alter this ratio, as it would turn off the ν_es that destroy neutrons by charged-current reactions. Unfortunately, a full discussion of such possibilities would take us too far afield today.

The nuclear physics of the r-process tells us that the synthesis occurs when the nucleon soup is in the temperature range of (3-1) $\cdot 10^9$K, which, in the hot bubble r-process described above, corresponds to a freezeout radius of (600-100) km and a time $\sim$ 10 seconds after core collapse. The neutrino fluence after freezeout (when the temperature has dropped below 10^9K and the r-process stops) is then $\sim$ (0.045-0.015) $\cdot 10^{51}$ ergs/(100km). Thus, after completion of the r-process, the newly synthesized material experiences an intense flux of neutrinos. This brings up the question of whether the neutrino flux could have any effect on the r-process.

4.4 Neutrinos and the r-process [40]

Rather than describe the exotic effects of neutrino oscillations on the r-process, mentioned briefly above, we will examine standard-model effects that are nevertheless quite interesting. The nuclear physics of this section – neutrino-induced neutron spallation reactions – is also relevant to recently proposed supernova neutrino observatories such as OMNIS and LAND. In contrast to our first discussion of the ν-process in Sec. 4.2, it is apparent that neutrino effects could be much larger in the hot bubble r-process: the synthesis occurs *much* closer to the star than our Ne radius of 20,000 km: estimates are 600-1000 km. The r-process is completed in about 10 seconds (when the temperature drops to about one billion degrees), but the neutrino flux is still significant as the r-process freezes out. The net result is that the "post-processing" neutrino fluence - the fluence that can alter the nuclear distribution after the r process is completed - is about 100 times larger than that responsible for fluorine production in the Ne zone. Recalling that 1/300 of the nuclei in the Ne zone interacted with neutrinos, and remembering that the relevant neutrino-nucleus cross sections scale as A, one quickly sees that the probability of a r-process nucleus interacting with the neutrino flux is approximately unity.

Because the hydrodynamic conditions of the r-process are highly uncertain, one way to attack this problem is to work backward in time. We know the final r-process distribution (what nature gives us) and we can calculate neutrino-nucleus interactions relatively well. Thus from the observed r-process distribution (including neutrino postprocessing) we can work backward to find out what the r-process distribution looked like at the point of freezeout. In Figs. 12 and 13, the "real" r-process distribution - that produced at freezeout - is given by the dashed lines, while the solid lines show the effects of the neutrino postprocessing for a particular choice of fluence. The nuclear physics input into these calculations is precisely that previously described: GT and first-forbidden cross sections, with the responses centered at excitation energies consistent with those found in ordinary, stable nuclei, taking into account the observed dependence on $|N - Z|$.

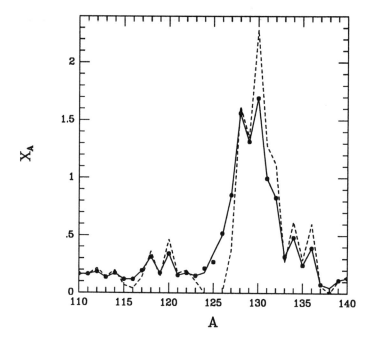

Figure 12: Comparison of the r-process distribution that would result from the freezeout abundances near the A ~ 130 mass peak (dashed line) to that where the effects of neutrino postprocessing have been include (solid line). The fluence has been fixed by assuming that the A = 124-126 abundances are entirely due to the ν-process.

One important aspect of the figures is that the mass shift is significant. This has to do with the fact that a 20 MeV excitation of a neutron-rich nucleus allows multiple neutrons (~ 5) to be emitted. (Remember we found that the binding energy of the last neutron in an r-process neutron-rich nuclei was about 2-3 MeV under typical r-process conditions.) The second thing to notice is that the relative contribution of the neutrino process is particularly important in the "valleys" beneath the mass peaks: the reason is that the parents on the mass peak are abundant, and the valley daughters rare. In fact, it follows from this that the neutrino process effects can be dominant for precisely seven isotopes (Te, Re, etc.) lying in these valleys. Furthermore if an appropriate neutrino fluence is picked, these isotope abundances are produced perfectly (given the abundance errors). The fluences are

$$N = 82 \text{ peak} \qquad 0.031 \cdot 10^{51} \text{ergs}/(100\text{km})^2/\text{flavor}$$

$$N = 126 \text{ peak} \quad 0.015 \cdot 10^{51} \text{ergs}/(100 \text{km})^2/\text{flavor},$$

values in fine agreement with those that would be found in a hot bubble r-process. So this is circumstantial but significant evidence that the material near the mass cut of a Type II supernova is the site of the r-process: there is a neutrino fingerprint.

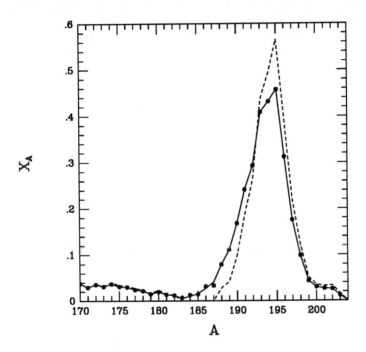

Figure 13: As in Fig. 12, but for the A $\sim$ 195 mass peak. The A = 183-187 abundances are entirely attributed to the ν-process.

4.5 Neutrino Oscillations and the r-process

For the usual seesaw pattern of neutrino masses and a cosmological interesting ν_τ (i.e., a heavy neutrino with a mass in the neighborhood of 10 eV), the full MSW pattern is shown in Fig. 10. If the $\nu_e - \nu_\mu$ crossing is responsible for the solar neutrino problem, a second crossing, $\nu_e - \nu_\tau$, is expected at a density large compared to that of the solar core, but small compared to the location of the supernova neutrinosphere ($\sim 10^{12}$ g/cm^3). For a very large range of

486

mixing angles, this crossing is adiabatic and thus leads to $\nu_e \leftrightarrow \nu_\tau$ conversion. These spectra thus change identities, leading to an anomalously hot ν_e flux from a Type II supernova.

As the ν-nucleon cross section is proportional to E_ν^2, the reaction $\nu_e + n \rightarrow e^- + p$ is enhanced, while $\bar{\nu}_e + p \rightarrow e^+ + n$ is unchanged. For a rather extensive range of $\nu_e \leftrightarrow \nu_\tau$ mixing angles and δm^2, this crossing then destroys the r-process: the hotter ν_es drive the matter proton rich [41]. Thus, if one accepts this location as the site of the r-process, very strong constraints on cosmologically interesting ν_τs are obtained. These limits are truly remarkable for their sensitivity to small mixing angles, extended to $\sin^2 2\theta \sim 10^{-5}$ for neutrino mass differences above a few eV^2.

I thank A. S. Brun and R. E. Shrock for helpful comments, and Paul Langacker for his able organization of the 1998 TASI summer school. This work was supported in part by the US Department of Energy.

References

1. This section taken from W. C. Haxton, *Ann. Rev. Astron. Astrophys.* **33**, 459 (1995).
2. R. Davis, Jr., D. S. Harmer, and K. C. Hoffman, *Phys. Rev. Lett.* **20**, 1205 (1966).
3. J. N. Abdurashitov et al., *Phys. Lett.* B **328**, 234 (1994).
4. P. Anselmann et al., *Phys. Lett.* B **285**, 376 (1992).
5. Y. Suzuki, *Nucl. Phys.* B **38**, 54 (1995).
6. Y. Suzuki, talk presented at Neutrino '98 (Takayama, Japan, June, 1998).
7. T. Kajita, talk presented at Neutrino '98 (Takayama, Japan, June, 1998).
8. J. N. Bahcall, S. Basu, and M. H. Pinsonneault, *Phys. Lett.* B **433**, 1 (1998).
9. A. S. Brun, S. Turck-Chieze, and P. Morel, *Ap. J.* **506**, 913 (1998) and private commun; S. Turck-Chieze and I. Lopez, *Ap. J.* **408**, 347 (1993).
10. This section taken from W. C. Haxton, nucl-th/9901037.
11. S. M. Austin, N. Anantaraman, and W. G. Love, *Phys. Rev. Lett.* **73**, 30 (1994); J. W. Watson et al., *Phys. Rev. Lett.* **55**, 1369 (1985).
12. K. Lande, talk presented at Neutrino '98 (Takayama, Japan, June, 1998).
13. J. N. Bahcall, *Neutrino Astrophysics*, (Cambridge University, Cambridge, 1989).
14. C. W. Johnson, E. Kolbe, S. E. Koonin, and K. Langanke, *Ap. J.* **392**, 320 (1992).
15. B. W. Filippone, A. J. Elwyn, C. N. Davids, and D. D. Koetke, *Phys. Rev. Lett.* **50**, 412 (1983).

16. J. N. Bahcall and R. Ulrich, *Rev. Mod. Phys.* **60**, 297 (1988).
17. J. N. Bahcall and W. C. Haxton, *Phys. Rev.* D **40**, 931 (1989).
18. V. Castellani, S. Degl'Innocenti, G. Fiorentini, M. Lissia, and B. Ricci, *Phys. Rev.* D **50**, 4749 (1994).
19. J. N. Bahcall and R. Davis, Jr., in *Essays in Nuclear Astrophysics*, ed. C. A. Barnes, D. D. Clayton, and D. Schramm (Cambridge Univ. Press, Cambridge) p. 243.
20. F. W. W. Dilke and D. O. Gough, N **240**, 262 (1972).
21. W. J. Merryfield, in *Solar Modeling*, ed. A. B. Balantekin and J. N. Bahcall (World Scientific, Singapore, 1995).
22. A. Cumming and W. C. Haxton, *Phys. Rev. Lett.* **77**, 4286 (1996).
23. N. Hata and P. Langacker, *Phys. Rev.* D **56**, 6107 (1997).
24. K. M. Heeger and R. G. H. Robertson, *Phys. Rev. Lett.* **77**, 3720 (1996).
25. For a proper (wave packet) derivation see M. Nauenberg, submitted to *Phys. Lett.* B.
26. S. P. Mikheyev and A. Smirnov, *Sov. J. Nucl. Phys.* **42**, 913 (1985); L. Wolfenstein, *Phys. Rev.* D **17**, 2369 (1979).
27. H. Bethe, *Phys. Rev. Lett.* **56**, 1305 (1986).
28. W. C. Haxton, *Phys. Rev. Lett.* **57**, 1271 (1986).
29. S. J. Parke, *Phys. Rev. Lett.* **57**, 1275 (1986).
30. C. S. Lim and W. J. Marciano, *Phys. Rev.* D **37**, 1368 (1988).
31. E. Kh. Akhmedov, *Sov. J. Nucl. Phys.* **48**, 382 (1988).
32. G. Raffelt, *Phys. Rev. Lett.* **64**, 2856 (1990).
33. K. Fujikawa and R. E. Shrock, *Phys. Rev. Lett.* **43**, 963 (1980).
34. W. C. Haxton and G. J. Stephenson, Jr., *Prog. Part. Nucl. Phys.* **12**, 409 (1984).
35. A. Mezzacappa et al., *Ap. J.* **495**, 911 (1998); H.-Th. Janka and E. Muller, *Astron. Astrophys.* **306**, 167 (1996); A. Burrows, S. Hayes, and B. A. Fryxell, *Ap. J.* **450**, 830 (1995).
36. S. E. Woosley and W. C. Haxton, *Nature* **334**, 45 (1988); S. E. Woosley, D. H. Hartmann, R. D. Hoffman, and W. C. Haxton, *Ap. J.* **356**, 272 (1990).
37. F. X. Timmes, S. E. Woosley, and T. A. Weaver, *Ap. J. Suppl.* **98**, 617 (1995).
38. J. J. Cowan et al., astro-ph/9808272 (to appear in *Ap. J*).
39. S. E. Woosley, J. R. Wilson, G. J. Mathews, R. D. Hoffman, and B. S. Meyer, *Ap. J.* **433**, 229 (1994).
40. W. C. Haxton, K. Langanke, Y.-Z. Qian, and P. Vogel, *Phys. Rev. Lett.* **78**, 2694 (1997) and *Phys. Rev.* C **55**, 1532 (1997).
41. Y.-Z. Qian et al., *Phys. Rev. Lett.* **71**, 1965 (1993).

HELIOSEISMOLOGY

SARBANI BASU
Institute for Advanced Study, Olden Lane, Princeton, NJ 00540, USA
E-mail: basu@sns.ias.edu

Helioseismology is the study of the Sun through properties of solar oscillations. Helioseismic techniques allow us to probe the interior of the Sun with very high precision, and in the process test the physical inputs to stellar models. This article gives a brief introduction to the theory of solar oscillations and helioseismic inversion techniques to determine solar structure.

1 Introduction

The Sun is oscillating simultaneously in many millions of global normal modes of oscillation. These oscillations are believed to be excited by turbulent convection just below the photosphere. To a good approximation the oscillations can be described as linear and adiabatic. Each mode of solar oscillation has a velocity amplitude of about 5 to 10 cm/s, which is very small compared to the sound speed. They are also adiabatic. The time scale of the oscillations being of the order of 5 min, which is much smaller than the Kelvin-Helmholtz time scales for the Sun ($\simeq 10^7$ years), and hence, heat transfer during each oscillation period can be neglected in most of the solar interior.

Since the Sun is a spherical body (the departures from sphericity have been measured to be very small), it is most natural to describe the angular dependence of the normal-modes in terms of spherical harmonics. Each mode is described by its radial order n, which is the number of nodes in the radial direction, the degree ℓ, where $L = \sqrt{\ell(\ell+1)}$ is roughly speaking the number of wavelengths along the solar circumference, and the azimuthal order m that measures the number of nodes along the equator. The numbers n, ℓ, and m describe the mode completely and determine its frequency ν (or $\omega \equiv 2\pi\nu$). In the absence of rotation or any other agent such as magnetic field to break spherical symmetry, all modes with the same n, and ℓ have the same frequency. Thus each (n, ℓ) multiplet is $2\ell + 1$ fold degenerate. Fig. 1 shows a sample of the spherical harmonics.

If, as in the case of the Sun, rotation, magnetic fields and other large scale flows and asymmetries are small, these can be assumed to be perturbations, and the mean frequency of an (n, ℓ) multiplet is unaffected to the first order. This can be used to probe the structure of the Sun and this article is devoted to showing how that can be done. A brief history of the subject and a straightforward description of observing techniques can be found in the article by Hill

488

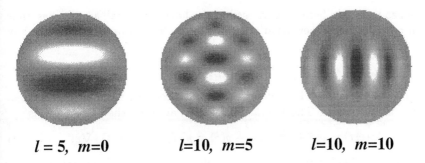

$l = 5, \ m=0$ $l=10, \ m=5$ $l=10, \ m=10$

Figure 1: A few spherical harmonics plotted on the surface of a sphere. An equatorial view is seen, with the polar axis being along the length of the page. The light regions have positive values, the dark have negative value.

et al.[1]

2 The theory of solar oscillations

2.1 The basic equations

Since the Sun is a gaseous object, we begin with the basic equations of hydro-dynamics, i.e., the equations of continuity and motion:

$$\frac{\partial \rho}{\partial t} + \nabla \cdot (\rho \vec{v}) = 0, \tag{1}$$

$$\rho \frac{d\vec{v}}{dt} = -\nabla p + \rho \vec{g}, \tag{2}$$

where,

$$\vec{g} = \nabla \Phi, \quad \text{and,} \quad \nabla^2 \Phi = -4\pi G \rho, \tag{3}$$

together with the energy equation, which can be written in the form

$$\frac{dq}{dt} = \frac{1}{\rho(\Gamma_3 - 1)} \left(\frac{dp}{dt} - \frac{\Gamma_1 p}{\rho} \frac{d\rho}{dt} \right), \tag{4}$$

where

$$\Gamma_1 = \left(\frac{\partial \ln p}{\partial \ln \rho} \right)_{ad}, \quad \text{and,} \quad \Gamma_3 - 1 = \left(\frac{\partial \ln T}{\partial \ln \rho} \right)_{ad}. \tag{5}$$

We assume that the equilibrium structure is static, so all time derivatives vanish. Thus the continuity equation is satisfied trivially, while the equation of motion just becomes the equation of hydrostatic support, i.e.

$$\nabla p = \rho \vec{g} = \rho \nabla \Phi. \tag{6}$$

As has been said before, solar oscillations have very small amplitudes, thus we can consider linear perturbations only. Thus, for example, pressure at any time t can be written as,

$$p(\vec{r}, t) = p(\vec{r}) + p'(\vec{r}, t), \tag{7}$$

where $p(\vec{r})$ is the time-independent equilibrium pressure, and $p'(\vec{r}, t)$ is a small Eulerian perturbation. The Lagrangian perturbation on the other hand is

$$\delta p(r, t) = p(\vec{r} + \vec{\delta r}(\vec{r}, t)) - p(\vec{r}) = p'(\vec{r}, t) + \vec{\delta r}(\vec{r}, t) \cdot \nabla p, \tag{8}$$

where, δr is the displacement from the equilibrium position. The perturbations to the other quantities can be written in exactly the same way. The equilibrium state has no velocities. The velocity $\vec{v}$, in this context, is simply the time derivative of the displacement.

Substituting the (Eulerian) perturbed quantities in the continuity equation, keeping only quantities to the first order in perturbation (and integrating with respect to time), we get

$$\rho' + \nabla \cdot (\rho \vec{\delta r}) = 0. \tag{9}$$

The equation of motion gives us

$$\rho \frac{\partial^2 \vec{\delta r}}{\partial t^2} = -\nabla p' + \rho \vec{g}' + \rho' \vec{g}, \tag{10}$$

where, $\vec{g}' = \nabla \Phi'$, and Φ' is the solution of

$$\nabla^2 \Phi' = -4\pi G \rho'. \tag{11}$$

For the heat equation on the other hand, it is easier to consider the Lagrangian perturbation, since under the assumptions that have been made the time derivative of the various quantities is simply the time derivative of the Lagrangian perturbation in those quantities, e.g., [from. Eq. (8)]

$$\frac{dp}{dt} = \frac{\partial p}{\partial t} + \vec{v} \cdot \nabla p = \frac{\partial p'}{\partial t} + \frac{\partial \delta r}{\partial t} \cdot \nabla p = \frac{\partial}{\partial t}(\delta p). \tag{12}$$

Thus from Eq. (4) one obtains

$$\rho \frac{\partial \delta q}{\partial t} = \frac{1}{\rho(\Gamma_3 - 1)} \left(\frac{\partial \delta p}{\partial t} - \frac{\Gamma_1 p}{\rho} \frac{\partial \delta \rho}{\partial t} \right), \tag{13}$$

or, in the adiabatic limit where energy loss is negligible

$$p' + \vec{\delta r} \cdot \nabla p = \frac{\Gamma_1 p}{\rho}(\rho' + \vec{\delta r} \cdot \nabla \rho). \tag{14}$$

2.2   The equations for solar oscillations

Equations (9), (10) and (14) are enough to describe solar oscillations. However, these equations are still incomplete in the sense that the relationship between solar frequencies and structure is not clear. That becomes clearer once the equations are cast in spherical polar co-ordinates (the Sun is after all a sphere). Once this is done, each vector can be separated into a radial component and a tangential (angular) component. The advantage of this decomposition is the fact that for a spherical body, the tangential gradients of the equilibrium quantities do not exist.

Thus the displacement $\vec{\delta r}$ can be decomposed as

$$\vec{\delta r} = \xi_r \hat{a}_r + \vec{\xi}_t, \tag{15}$$

where, $\hat{a}_r$ is the unit vector in the radial direction. The tangential component of the equation of motion [Eq. (10)] is

$$\rho \frac{\partial^2 \vec{\xi}_t}{\partial t^2} = -\nabla_t p' + \rho \nabla_t \Phi', \tag{16}$$

or (taking the tangential divergence of both sides),

$$\rho \frac{\partial^2}{\partial t^2} (\nabla_t \cdot \vec{\zeta_t}) \quad -\nabla_t^2 p' + \rho \nabla_t^2 \Phi' \tag{17}$$

The continuity equation [Eq. (9)] after decomposition can be used to eliminate the term $\nabla_t \cdot \vec{\xi}_t$ from Eq. (17) to obtain

$$-\frac{\partial^2}{\partial t^2} \left[\rho' + \frac{1}{r^2} \frac{\partial}{\partial r} (\rho r^2 \xi_r) \right] = -\nabla_t^2 p' + \rho \nabla_t^2 \Phi'. \tag{18}$$

The radial component of the equation of motion gives

$$\rho \frac{\partial^2 \xi_r}{\partial t^2} = -\frac{\partial p'}{\partial r} - \rho' g + \rho \frac{\partial \Phi'}{\partial r}, \tag{19}$$

where gravity acts in the negative r direction. Finally, the Poisson's equation becomes

$$\frac{1}{r^2} \frac{\partial}{\partial r} \left(r^2 \frac{\partial \Phi'}{\partial r} \right) + \nabla_t^2 \Phi' = -4\pi G \rho'. \tag{20}$$

Note that in Eqs. (18), (19) and (20) the time and space derivatives appear in a manner such that the solution can be written as a product of a function of time and a function of the spatial co-ordinates (i.e., they are separable in space and time). Note also, that the derivatives with respect to the angular variables θ and ϕ appear only in the combination ∇_t^2. Thus the separation of angular variables can be achieved if the form of the function is an eigenfunction of the tangential Laplacian and such a function is the spherical harmonic function $Y_\ell^m(\theta, \phi)$, where $|m| \leq \ell$. For spherical harmonics

$$\nabla_t^2 Y_\ell^m(\theta, \phi) = -\frac{\ell(\ell + 1)}{r^2} Y_\ell^m(\theta, \phi). \tag{21}$$

Therefore, the perturbed variables can be defined as

$$\xi_r(r, \theta, \phi, t) \equiv \xi_r(r) Y_\ell^m(\theta, \phi) \exp(-i\omega t), \tag{22}$$

$$p'(r, \theta, \phi, t) \equiv p'(r) Y_\ell^m(\theta, \phi) \exp(-i\omega t), \tag{23}$$

and so on.

Substituting the above form of the variables in Eqs. (18), (19), and (20) one obtains

$$\omega^2 \left[\rho' + \frac{1}{r^2} \frac{d}{dr}(r^2 \rho \xi_r) \right] = \frac{\ell(\ell + 1)}{r^2}(p' - \rho \Phi'), \tag{24}$$

$$-\omega^2 \rho \xi_r = -\frac{dp'}{dr} - \rho' g + \rho \frac{d\Phi'}{dr}, \tag{25}$$

and

$$\frac{1}{r^2} \frac{d}{dr}\left(r^2 \frac{d\Phi'}{dr} \right) - \frac{\ell(\ell + 1)}{r^2} \Phi' = -4\pi G \rho'. \tag{26}$$

There is also the heat equation

$$\left(\delta p(r) - \frac{\Gamma_1 p}{\rho} \delta \rho(r) \right) = \rho(\Gamma_3 - 1)\delta q(r). \tag{27}$$

These equations along with a prescription to calculate q give a system of coupled differential equations. Note that the azimuthal harmonic number m does not appear explicitly in the equations, and that is because of spherical symmetry. To define the angles θ and ϕ we need to define an axis and that definition is not unique for a spherically symmetric system.

2.3 Adiabatic oscillations

The oscillation equations derived above are still quite general with respect to heat transfer. In the solar case, which is assumed to be adiabatic, we use the fact that $\delta q(r) = 0$, therefore, Eq. (27) can be written as

$$\rho' = \frac{\rho}{\Gamma_1 p} p' + \rho \xi_r \left(\frac{1}{\Gamma_1 p} \frac{dp}{dr} - \frac{1}{\rho} \frac{d\rho}{dr} \right), \tag{28}$$

and can be used to eliminate ρ' from Eqs. (24) to (26). Eq. (24) gives

$$\frac{d\xi_r}{dr} = -\left(\frac{2}{r} + \frac{1}{\Gamma_1 p}\frac{dp}{dr}\right)\xi_r + \frac{1}{\rho c^2}\left(\frac{S_\ell^2}{\omega^2} - 1\right)p' - \frac{\ell(\ell+1)}{\omega^2 r^2}\Phi', \qquad (29)$$

where $c^2 = \Gamma_1 p/\rho$ is the squared sound speed, and S_ℓ^2 is the Lamb frequency defined by

$$S_\ell^2 = \frac{\ell(\ell+1)c^2}{r^2} = k_t^2 c^2. \qquad (30)$$

Eq. (25) and the equation of hydrostatic equilibrium give

$$\frac{dp'}{dr} = \rho(\omega^2 - N^2)\xi_r + \frac{1}{\Gamma_1 p}\frac{dp}{dr}p' + \rho\frac{d\Phi'}{dr}, \qquad (31)$$

where, N is the *buoyancy frequency* defined as

$$N^2 = g\left(\frac{1}{\Gamma_1 p}\frac{dp}{dr} - \frac{1}{\rho}\frac{d\rho}{dr}\right). \qquad (32)$$

And Eq. (26) becomes

$$\frac{1}{r^2}\frac{d}{dr}\left(r^2\frac{d\Phi'}{dr}\right) = -4\pi G\left(\frac{p'}{c^2} + \frac{\rho\xi_r}{g}N^2\right) + \frac{\ell(\ell+1)}{r^2}\Phi'. \qquad (33)$$

Equations (29), (31) and (33) form a fourth-order system of linear homogeneous differential equations for the four dependent variables ξ_r, p', Φ' and $d\Phi'/dr$.

Four boundary conditions are needed to solve the equations. Two conditions are applied at the surface. One is obtained by demanding the continuity of Φ' and its derivative at the surface $r = R_\odot$. The density perturbation vanishes outside the Sun, and hence, Φ' can be obtained analytically to choose the solution which vanishes at infinity. One gets

$$\Phi' = Ar^{-\ell-1}, \qquad (34)$$

where A is a constant. Thus at $r = R_\odot$, Φ' must satisfy

$$\frac{d\Phi'}{dr} + \frac{\ell+1}{r}\Phi' = 0 \quad \text{at } r = R_\odot. \qquad (35)$$

If the Sun is assumed to have a definite boundary, then the second condition must ensure that there are no forces on the outer surface, even when deformed by the oscillations. This implies that the pressure on the deformed surface vanishes, in other words the Lagrangian perturbation of the pressure is zero. Thus,

$$\delta p = p' + \xi_r \frac{dp}{dr} = 0 \quad \text{at } r = R_\odot. \tag{36}$$

The two inner boundary conditions are derived in order to make the solution non-singular at $r = 0$. The behaviour of the solution near $r = 0$ can be written in terms of an expansion in r. For an acceptable solution, ξ_r is proportional to $r^{\ell-1}$ (expect for the radial mode, $\ell = 0$, where $\xi_r \propto r$) and p',ρ' and Φ' are proportional to r^ℓ. Furthermore as $r \to 0$, $\xi_r \simeq \ell\xi_h$. From the expansion coefficients more precise relations can be obtained between the different variables at small r. These when applied at the innermost mesh-point used in computations determine the two central boundary conditions. In general, non-trivial solutions exist only for specific values of ω^2.

These boundary conditions are perhaps the simplest one can apply. Details of these and more realistic boundary conditions can be found in the books by Unno et al.[2] or Cox[3].

3 A few properties of solar oscillations

Once the equations describing solar oscillations are derived one can go ahead and study the properties of such oscillations. Since the equations are a set of fairly complex equations, detailed analysis of the properties will not be attempted here. We can however, try to find some of the broad properties, particularly those which enable us to use solar oscillations to study the solar interior, under the so called Cowling approximation, where the perturbation to the gravitation field, Φ' is ignored. Detailed analyses of the properties of solar (and stellar) oscillations can be found in Unno et al.[2], Cox[3], the articles by Christensen-Dalsgaard & Berthomieu[4] and Gough[5], etc.

The spherical-harmonic expansion of Φ' shows that it is small when (a)ℓ is large. This implies that the perturbations are restricted to the outer layers where ρ (and hence ρ') is small. Or (b) the radial order $|n|$ is large. In this case the ρ' changes sign rapidly and the positive and negative contributions from ρ' cancel each other, leaving a very small Φ'.

When Φ' is neglected, Eq (29) to (33) reduce to a system of two first order equations,

$$\frac{d\xi_r}{dr} = -\left(\frac{2}{r} - \frac{1}{\Gamma_1}H_p^{-1}\right)\xi_r + \frac{1}{\rho c^2}\left(\frac{S_\ell^2}{\omega^2} - 1\right)p', \tag{37}$$

$$\frac{dp'}{dr} = \rho(\omega^2 - N^2)\xi_r - \frac{1}{\Gamma_1}H_p^{-1}p', \tag{38}$$

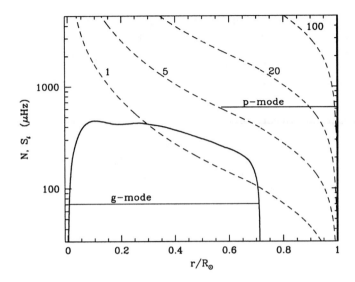

Figure 2: The buoyancy frequency N (Eq. 32) shown by the continuous line, and the characteristic acoustic frequency S_ℓ (Eq. 30) shown by the dashed lines labelled with the value of ℓ, for a standard solar model plotted as a function of radius. The horizontal lines indicate the trapping region of a g-mode and of a p-mode of degree 5.

where, $H_p = -dr/d\ln p$ is the pressure scale height. For high order oscillations, the eigenfunctions vary much more rapidly than the coefficients in the equations. Thus the left hand side of Eq. (37) is much larger than the first term of the right hand side. Hence, in the crudest approximation, we can neglect the first term in Eq. (37) and the second term in Eq. (38) and combine the two resulting equations to obtain a wave equation of the form

$$\frac{d^2\xi}{d\xi^2} = \frac{\omega^2}{c^2}\left(1 - \frac{N^2}{\omega^2}\right)\left(\frac{S_\ell^2}{\omega^2} - 1\right)\xi_r. \tag{39}$$

This is the simplest possible approximation to equations of non-radial oscillations, but is enough to illustrate some of the key properties.

One feature we see immediately is that the equation does not always have an oscillatory solution. The solution is oscillatory when (1) $\omega^2 < S_\ell^2$, and $\omega^2 < N^2$, or (2) $\omega^2 > S_\ell^2$, and $\omega^2 > N^2$. The solution is exponential otherwise.

Fig. 2 shows N^2 and S_ℓ^2 plotted as a function of depth for a standard solar model. The figure shows that for modes for which the first condition is true are trapped mainly in the core (since N^2 is negative in the convection zone). These are the *g-modes* since the restoring force is gravity through buoyancy. Modes that satisfy the second condition are oscillatory in the outer regions, though low-degree modes can penetrate right to the centre. These are the p-modes since the restoring force is predominantly pressure. Only p-modes have been observed so far at the solar surface. Note that p-modes of different degrees penetrate to different depths within the Sun. Thus if we have p-modes for different degrees we can hope to probe different layers of the Sun.

4   Helioseismic inversions

Equations (29), (31) and (33) are not particularly useful when it comes to actually trying to invert the observed solar oscillation frequencies. We start afresh from the perturbed form of the equation of motion, i.e. Eq. (10) to write it in a more useful form. From the previous discussions we know that the displacement vector can be written as $\delta r(\vec{r})\exp(-i\omega t)$. Substituting this in the equation, we get

$$-\omega^2 \rho \vec{\delta r} = -\nabla p' + \rho g' + \vec{g}\rho'. \tag{40}$$

Substituting for ρ' from the continuity equation [Eq. (9)] and for p' from Eq. (28), we get

$$-\omega^2 \rho \vec{\delta r} = \nabla(c^2 \rho \nabla \cdot \vec{\delta r} + \nabla p \cdot \vec{\delta r}) - \vec{g}\nabla \cdot (\rho \vec{\delta r}) - G\rho \nabla \left(\frac{\int_V \nabla \cdot (\rho \vec{\delta r})dV}{|\vec{r} - \vec{r'}|} \right). \tag{41}$$

In Eq. (41), ω is the observed quantity, and we would like to find c^2 and ρ (and hence p assuming hydrostatic equilibrium). However, δr is not observed and therefore the equations cannot be inverted directly. The way out of this is to recognize that Eq. (41) is an eigenvalue equation of the form $L\vec{\delta r} = -\omega^2 \vec{\delta r}$, and, under the specific boundary conditions, namely $\rho(R_\odot) = p(R_\odot) = 0$, Eq. (41) is Hermitian [6]. Thus the variational principle can be used to linearize Eq. (41) around a known solar model to obtain

$$\frac{\delta\omega^2}{\omega^2} = -\frac{\int_V \rho \vec{\delta r}^\star \cdot \delta L \vec{\delta r} dV}{\int_V \rho \vec{\delta r}^\star \cdot \vec{\delta r} dV}, \tag{42}$$

where $\delta\omega$ is the difference in frequency between the known solar model (or the "reference model") and the Sun, δL contains information about the difference between the reference model and the Sun, and δr is the displacement eigenfunction for the known solar model and thus can be calculated. One such equation can be written for each mode, and the set of equations can be used to calculate δL and thus the structure of the Sun.

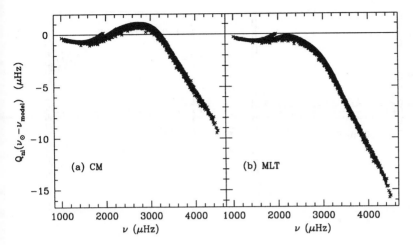

Figure 3: The frequency differences between the Sun and two solar models scaled by the mode inertia. The quantity Q_{nl} is the ratio of the inertia of a mode of given (n, ℓ) to that of a mode with $\ell = 0$ but with the same ν as the given mode. Panel(a) shows the differences for a model constructed with the Canuto-Mazzitelli formulation of convection, Panel (b) shows the differences for a model constructed with the conventional mixing length formalism. The observed frequencies are those obtained by the MDI instrument on board SoHO.[26]

The denominator of the right hand side of Eq. (42) is often called the mode inertia, I, since it can be shown that the time-averaged kinetic energy of a mode is proportional to $\omega^2 I$. Eq. (42) implies that for a given difference in structure, frequencies of modes with a high inertia are changed less than those of modes with lower inertia. For modes of a given frequency, lower degree (i.e., deeply penetrating) modes have higher mode inertias than higher degree (i.e. shallow) modes.

There is an additional complication that comes in the way of inverting Eq. (42) and that is we really do not know how to model the the layers just below the solar surface properly. Eq. (42) implies that we can invert the solar frequencies provided we know how to model the Sun. However, the approximations that are used to calculate the convective flux (e.g., the mixing length formalism) are not enough to model the solar surface properly. Also the adiabatic approximation certainly breaks down in the outer layers. This implies that the RHS of Eq. (42) does not account for the frequency difference $\delta\omega/\omega$. Fortunately, we know the rough form of the frequency differences due to errors in surface layers of the model. For a given mode, the frequency difference is a slowly varying function of the mode frequency divided by the mode inertia[7]. Thus an additional term of the form $F(\omega)/I$ has to be added to

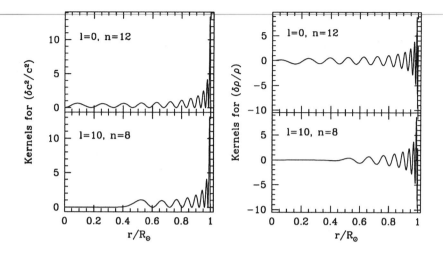

Figure 4: kernels for relative sound-speed difference and relative density difference for two modes.

the RHS of Eq. (42). $F(\omega)$ is generally assumed to be a low degree polynomial. Fig. (3) shows the frequency differences between two standard models and the Sun. The solar data used were obtained by the MDI instrument on board the space-craft SoHO [26]. The difference between the two models lies in the formulation used to calculate convective flux. One uses the standard mixing length theory and the other the formulation by Canuto & Mazzitelli [8]. This difference manifests itself predominantly as a function of frequency — the term $F(\omega)$.

The modified Eq. (42) can be rewritten to isolate the difference between the model and the Sun to obtain

$$\frac{\delta\omega_i}{\omega_i} = \int K^i_{c^2,\rho}(r)\frac{\delta c^2}{c^2}(r)dr + \int K^i_{\rho,c^2}(r)\frac{\delta\rho}{\rho} + \frac{F_{\text{surf}}(\omega_i)}{I_i}, \qquad (43)$$

where, the index i denotes any mode. The terms $\delta c^2/c^2$ and $\delta\rho/\rho$ are the relative squared sound-speed and relative density differences between the reference model and the Sun and the kernels $K_{c^2,\rho}$ and K_{ρ,c^2} are known functions of the reference model. Kernels for a few modes are shown in Fig. (4). Note that kernels for lower-degree modes penetrate deeper in radius than kernels of higher degree modes. Details of the derivations of the kernels can be found in Antia & Basu [9].

There are two complementary methods of using Eq. (43) to determine $\delta c^2/c^2$ and $\delta\rho/\rho$: the regularized least squares (RLS) [7][9] and the optimally localized averages (OLA) [10] [11] [12]. In the former one tries to fit the given data under the constraint that the solution is smooth. The latter involves finding a linear combination of the kernels such that the combination is localized in space. The linear combination is called the averaging kernel. The solution obtained is then an average of the true solution weighted by the averaging kernel. Some details of inversion theory can be found in the article by Gough & Thompson [13].

4.1 Regularized Least Squares Method

In this method, the unknown quantities, $\delta c^2/c^2$, $\delta\rho/\rho$ and F_{surf} are represented over a set of basis functions in r or ω. Thus,

$$\frac{\delta c^2}{c^2} = \sum_{i=1}^{n} b_i\phi_i(r), \qquad \frac{\delta\rho}{\rho} = \sum_{i=1}^{n} c_i\phi_i(r), \qquad F(\omega) = \sum_{i=1}^{m} a_i\psi_i(\omega), \qquad (44)$$

where $\phi_i(r)$ are basis functions, like splines, in r and $\psi_i(\omega)$ are basis functions in ω. It is assumed that n basis functions are enough to represent the radial functions and m are enough to represent the surface term.

The coefficients b_i, c_i and a_i are determined by a least squares fit to minimize

$$\chi^2 = \sum_{i=1}^{N} \left(\frac{\frac{\delta\omega_i}{\omega_i} - \text{RHS}}{\sigma_i}\right)^2 + \alpha^2 \int_0^R \left[\left(\frac{d^2}{dr^2}\frac{\delta\rho}{\rho}\right)^2 + \left(\frac{d^2}{dr^2}\frac{\delta c^2}{c^2}\right)^2\right] dr, \qquad (45)$$

where, RHS is the right hand side of Eq. (43) expressed in terms of the decomposition in Eq. (44), $\sigma_i = \frac{\epsilon_i}{\omega_i}$, ϵ_i being the errors in the observed frequencies, and and α is a trade-off parameter which constrains the obtained results to be smooth.

4.2 Optimally Localized Averages (OLA)

Usually a variant of the OLA [14] method called "Subtractive" Optimally Localized Averages (SOLA) [15] is used. The aim here is to construct explicitly well localized resolution kernels $\mathcal{K}$ such that

$$< f > = \int \mathcal{K}f(r)dr \qquad (46)$$

represents the average of the quantity f over a sufficiently narrow range in r. Thus if

$$\mathcal{K} = \sum_i c_i K_{c^2\rho}^i, \quad \text{then,} \quad \left\langle\frac{\delta c^2}{c^2}\right\rangle = \sum_i c_i \frac{\delta\omega_i}{\omega^i} \qquad (47)$$

From Eq. (43) we see that this is possible only if $\int K dr = 1$, and if $C = \sum_i c_i K^i_{\rho,c^2}$ and $\mathcal{F} = \sum_i c_i F_{\text{surf}}(\omega_i)$ are small. This is achieved by minimizing

$$\int \left(\sum_i c_i K^i_{c^2,\rho} - \mathcal{T} \right)^2 dr + \beta \int \left(\sum_i K^i_{\rho,c^2} \right)^2 dr + \mu \sum_{i,j} c_i c_j E_{ij}, \qquad (48)$$

subject to constraints given below. In the above equation E_{ij} is the error-covariance matrix of the observations, β is a tradeoff parameter designed to minimize C, and μ is a trade-off parameter which controls the propagated error in the solution. $\mathcal{T}$ the target kernel is any localized function, that we would like our averaging kernel to look like. The minimization is done under the constraint that the resulting averaging kernel is unimodular, i.e.,

$$\int \sum_i c_i K^i_{\rho c^2} dr = 1, \qquad (49)$$

To suppress the surface term, we add additional constraints

$$\sum_c c_i \frac{\psi_j(\omega_i)}{I_i} = 0, \quad j = 1, \ldots, m \qquad (50)$$

where $\psi(\omega_i)$ are basis functions such that $F_{\text{surf}}(\omega) = \sum_{j=1}^m a_j \psi_j(\omega)$

Inversions to find the relative sound-speed difference ($\delta c/c$), relative density difference ($\delta\rho/\rho$) between the Sun and any solar model are reasonably straightforward to do. Similarly one can also determine the relative difference in the adiabatic index Γ_1 which can be used to study the solar equation of state [16][17]. These require no assumption that has not been discussed already. However, some of the other interesting quantities that one might want to study are the temperature profile inside the Sun or the composition profile. Unfortunately the equations of stellar oscillations do not directly give us any information about these quantities. To invert for the composition profile etc. once needs to assume that the properties of matter inside the Sun, such as opacity, equation of state and nuclear reaction rates are known, and these can be used to transform the inversion problem of Eq. (43) in terms of the composition or temperature profile. Because of the additional assumption required, these inversions are normally referred to as. "secondary inversions" [18][19][20]

5 Helioseismology and solar neutrinos

One of the reasons helioseismology is of interest to people outside the solar physics community is that helioseismic techniques may enable us to determine

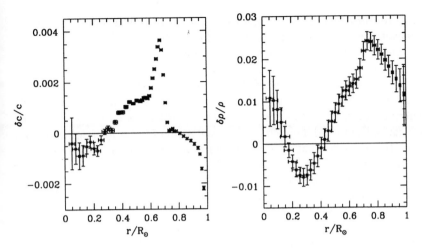

Figure 5: The relative sound speed and density differences between the Sun and a standard solar model.[22] The vertical error-bars are the 1σ error in the results, the horizontal error-bars are a measure of the resolution.

whether or not the solar neutrino problem has an astrophysical solution. In simple terms, the solar neutrino problem is the discrepancy that is found between observed solar neutrino fluxes and those predicted by the standard solar models. Standard solar models predict higher fluxes than is observed. Of course, a solution to the problem does not just mean getting the total neutrino flux right. The neutrino fluxes detectable by the chlorine experiment, the gallium experiments and the flux of Boron neutrinos must be satisfied individually too. Since the oscillation equations do not have any information of the thermal structure of the Sun, inversions do not put any direct constraints on the solar neutrinos. Nevertheless, we find that even indirectly, the constraints put are extremely stringent. [21] [22] [23]

Helioseismic inversions have shown that the difference in sound speed between the Sun and our current, standard solar models is very small. [24] [25] [23] The sound-speed difference in the interior 50% is of the order of a few tenths of percent, and density differences are of the order of a percent (see Fig. 5). So although the structure of the standard models is very close to that of the Sun, the neutrino fluxes are not.

Past suggestions for an astrophysical solution of the solar neutrino problem have included mixing of material inside the core or include particles like WIMPS to provide an additional channel of energy transfer, thereby lowering the mean temperature of the solar core. All such models produced so far have failed to satisfy helioseismic constraints. Although these models lower the temperature, they do not lower the mean-molecular weight enough to keep the sound speed the same as in a standard solar model ($\delta c^2/c^2 = \delta T/T - \delta \mu/\mu$, T being the temperature and μ the mean molecular weight). The difference between the structure of the Sun and these models is much larger than the difference between the structure of the Sun and standard models.

There have been attempts to check whether the solar models can satisfy the neutrino fluxes as well as the helioseismic constraints on the sound-speed and density of the model if opacity (which determines the transfer of energy) is modified. It was found that even allowing for arbitrary variations in the input opacity and relaxing the requirement of thermal equilibrium (which simulated the effect of other sources of energy loss from the Sun, e.g. WIMPS), it is not possible to construct a solar model that is helioseismically consistent and that simultaneously satisfies the neutrino observations from the three different types of experiments[21]. It is therefore unlikely that there is an astrophysical solution to the solar neutrino problem. The solution is more likely to lie in non-standard neutrino physics.

Acknowledgments

The author was supported by an AMIAS fellowship and the Institute for Advanced Study SNS Membership funds.

References

1. Hill, F., Deubner, F.-L., Isaak, G. In *Solar interior and atmosphere*, eds Cox, A.N., Livingston, W.C. & Matthews, M. Space Science Series, (Tucson: University of Arizona Press) 329 (1991)
2. Unno, W., Osaki, Y., Shibahashi, H. *Non-radial Oscillations in stars*, 2nd ed., (Tokyo: Tokyo Univ. Press) (1989)
3. Cox, J.P., *Theory of Stellar Pulsation*, (Princeton: Princeton Univ. Press) (1980)
4. Christensen-Dalsgaard, J. & Berthomieu, G., In *Solar interior and atmosphere*, eds Cox, A.N., Livingston, W.C. & Matthews, M., Space Science Series, (Tucson: University of Arizona Press), 401 (1991)
5. Gough, D.O., In *Astrophysical fluid dynamics, Les Houches Session XLVII*, (eds. Zahn, J.-P. & Zinn-Justin, J.), (Elsevier, Amsterdam), 399 (1993)
6. Chandrasekhar, S., *Astrophys. J.*, **139**, 664 (1964)
7. Dziembowski W.A., Pamyatnykh A.A., Sienkiewicz R., *Mon. Not. Roy. Astron. Soc.*, **244**, 542 (1990)
8. Canuto, V.M., Mazzitelli, I, *Astrophys. J.*, **370**, 295 (1991)

9. Antia, H.M., Basu, S. *Astron. Astrophys. Supp.*, **107**, 421 (1994)

10. Kosovichev, A.G., et al. *Mon. Not. Roy. Astron. Soc.*, **259**, 536 (1992)

11. Christensen-Dalsgaard, J. & Thompson, M.J., In *Proc. GONG'94: Helio- and Astero-seismology from Earth and Space*, eds Ulrich, R.K., Rhodes Jr, E.J. & Däppen, W., ASPCS, **76**, 144 (1995)

12. Basu, S., et al. *Monthly Notices Roy. Astron. Soc.*, **292**, 234 (1997)

13. Gough, D.O. & Thompson, M.J., In *Solar interior and atmosphere*, eds Cox, A.N., Livingston, W.C. & Matthews, M. Space Science Series, (Tucson: University of Arizona Press) 519 (1991)

14. Backus, G.B., & Gilbert J.F. , *Geophys. J. Roy. Astr. Soc.*, **16**, 169 (1968)

15. Pijpers F.P., Thompson M.J., *Astron. Astrophys.*, **262**, L33 (1992)

16. Basu, S., Christensen-Dalsgaard, J. *Astron. Astrophys.*, **322**, L5 (1997)

17. Basu, S., Däppen, W., Nayfanov, A., in *Structure and Dynamics of the Sun and Sun-like Stars*, ESA SP-418 eds. S.G. Korzennik and A. Wilson, (Noordwijk: ESA), in press (1998)

18. Gough, D.O., & Kosovichev, A., 1990, in *Proc. IAU Colloq. 121: Inside the Sun*, eds. G. Berthomieu & M. Cribier, (Dordrecht: Kluwer), p 327

19. Shibahashi, H., Takata, M., Tanuma, S., 1995, in *Proc. Fourth SOHO Workshop, Vol. 2* eds. J. T. Hoeksema, V. Domingo, B. Fleck, B. Battrick (ESA SP-376), p9

20. Antia, H.M., Chitre, S.M., *Astron Astrophys.* in press (astro-ph/9707226)

21. Antia, H.M., Chitre, S.M., *Mon. Not. Roy. Astron. Soc.*, **289**, L1 (1997)

22. Bahcall, J.N., Pinsonneault, M.H., Basu, S., Christensen-Dalsgaard, J., *Phys. Rev. Let.*, **78**, 171 (1997)

23. Bahcall, J.N., Basu, S., Pinsonneault, M.H., *Phys. Let. B*, **433**, 1 (1998)

24. Gough, D.O. et al., *Science*, **272**, 1296 (1996)

25. Kosovichev, A.G. et al., *Sol. Phys.*, **120**, 43 (1997)

26. Rhodes E.J., Kosovichev A.G., Schou J., Scherrer P.H., Reiter J., *Solar Phys.*, **175**, 287 (1997)

NEUTRINOS AND DARK MATTER

CHUNG–PEI MA

Department of Physics and Astronomy, University of Pennsylvania
Philadelphia, PA 19104
E-mail: cpma@strad.physics.upenn.edu

In these lectures I highlight some key features of massive neutrinos in the context of cosmology. I first review the thermal history and the free-streaming kinematics of the uniform cosmic background neutrinos. I then describe how fluctuations in the phase space distributions of neutrinos and other particles arise and evolve after neutrino decoupling according to the linear perturbation theory of gravitational instability. The different clustering properties of massive neutrinos (aka hot dark matter) and cold dark matter are contrasted. The last part discusses the nonlinear stage of gravitational clustering and highlights the effects of massive neutrinos on the formation of cosmological structure.

1 Neutrino Masses

These lectures discuss how the universe serves as a learning ground for massive neutrinos. Before doing so, let us briefly review some experimental measurements of neutrino masses.

Upper bounds on neutrino masses from kinematic measurements in laboratories continue to improve.[1] For the τ-neutrino, $m_{\nu_\tau} < 18.2$ MeV from the decay channel $\tau \to 5\pi + \nu_\tau$. For the μ-neutrino, $m_{\nu_\mu} < 170$ keV from two-body pion decay. For the electron neutrino, the quantity $m_{\nu_e}^2$ is measured in tritium beta decay by fitting the shape of the energy spectrum near the endpoint. Experiments thus far have yielded nonphysical negative values for $m_{\nu_e}^2$, indicating unexplained systematic effects in the measurements. A conservative upper bound is put at $m_{\nu_e} \approx 15$ eV. The spread in arrival times of neutrinos from supernova explosions provides an independent way to constrain the mass of the electron neutrino. Various limits have been reported for SN 1987A; a conservative estimate is $m_{\nu_e} < 23$ eV.[1,2]

2 Properties of Cosmic Background Neutrinos

2.1 Temperature and Density

For a brief 1 second after the big bang, neutrinos enjoy being part of the thermal bath composed of photons, electrons, protons, neutrons, and the associated anti-particles (after the quark-hadron era). The weak interactions at this early time are rapid enough to keep these particles in thermal equilibrium at a single temperature T. After 1 second, when T drops below about 1

MeV, however, the neutrino interaction rate becomes slower than the Hubble expansion, and neutrinos become effectively collisionless and freely-streaming particles whose trajectories are determined by the geodesic equations. This event is commonly referred to as "neutrino decoupling." As the universe expands, the momenta and temperature of neutrinos are simply redshifted, and the neutrino temperature is given by the familiar formulas

$$T_\nu(a) = a^{-1} T_{\nu,0}, \qquad T_{\nu,0} = \left(\frac{4}{11}\right)^{1/3} T_{\gamma,0} = 1.947K, \qquad (1)$$

where a is the cosmic scale factor, the subscripts 0 denote the present-day values, and the cosmic background photon temperature is taken to be $T_{\gamma,0} = 2.728\,K$.[3]

An important feature of the neutrino distribution after decoupling is that, although weak interactions are no longer rapid enough to keep neutrinos in thermal equilibrium with other particle species, neutrinos retain their equilibrium distribution as long as no other physical processes (e.g., gravitational clustering; see Sec. 3) are present to alter it. Therefore, to zeroth order in density and metric perturbations, the phase space distribution f_0 of the cosmic background neutrinos is of the simple Fermi-Dirac form

$$f_0(\epsilon) = \frac{g_s}{h_p^3} \frac{1}{e^{\epsilon/k_B T_{\nu,0}} + 1}, \qquad (2)$$

where $\epsilon = a(p^2 + m_\nu^2)^{1/2}$ is the comoving energy, $T_{\nu,0}$ is the neutrino temperature given by Eq. (1), g_s is the number of spin degrees of freedom, and h_p and k_B are the Planck and the Boltzmann constants.

The situation is further simplified if neutrino masses are $\ll 1$ MeV. Such neutrinos are highly relativistic at decoupling; their energy ϵ, and hence the distribution function f_0, are independent of m_ν to a good approximation. One can easily show that, as long as $m_\nu \ll 1$ MeV, the number density of the cosmic background neutrinos is related to the neutrino temperature by

$$n_\nu(T_\nu) = \frac{7 g_s}{8\pi^2} \zeta(3) \left(\frac{k_B T_\nu}{\hbar c}\right)^3, \qquad (3)$$

where $\zeta(3) \approx 1.202$ is the Riemann zeta function of order 3. This gives a present-day density of ≈ 113 cm^{-3} for every neutrino species independent of their masses. (For comparison, the present-day photon density is ≈ 412 cm^{-3}.) It also follows that the contribution of these neutrinos to the present-day mass density parameter, Ω_ν, is related to their masses by the simple relation

$$\Omega_\nu h^2 = \frac{\Sigma_i m_i}{93\,\text{eV}}, \qquad (4)$$

where the index i runs over all light, stable neutrino species (e.g., ν_e, ν_μ, and ν_τ), and the Hubble constant is $H_0 = 100\,h$ km s^{-1} Mpc^{-1}. One then arrives at the important conclusion that in order for neutrinos not to close universe (i.e. $\Omega_\nu \leq 1$), the sum of neutrino masses must not exceed $93\,h^2$ eV. This value is far below the current laboratory limits (see Sec. 1). Cowsik & McClelland[4] were the first to use such cosmological arguments to place an upper bound on neutrino masses. (Unfortunately, these "hot dark matter" models in which the mass density is dominated by massive neutrinos have been found to produce excessive large voids surrounded by large coherent sheets and filaments that are not seen in the observable universe.[5] Modifications to this model will be discussed below.)

In the high mass regime, $m_\nu \gg 1$ MeV, there exists another window where the neutrino contribution to the mass density parameter Ω of the universe is subcritical. The argument is that neutrinos with $m_\nu \gg 1$ MeV become non-relativistic long before decoupling. Neutrino and anti-neutrino pairs cease to be created in abundance once the thermal temperature drops below m_ν, and the neutrino density is suppressed by the Boltzmann factor $e^{-m_\nu/k_B T}$. This large reduction factor in the relic abundance allows neutrinos to have large masses without overclosing the universe. A more careful calculation[6] shows that an $\Omega \leq 1$ universe implies a *lower* limit of ~ 2 GeV if these heavy neutrinos are Dirac, and ~ 6 GeV if they are Majorana. Since this mass range is well above the current upper mass bounds from laboratory measurements, it is of interest only when one considers more exotic theories for neutrinos.[7]

2.2 Kinematics and Free Streaming

Let us now turn to the kinematics and the streaming properties of neutrinos. In general, neutrinos of mass m_ν become non-relativistic after a redshift of

$$z_{\text{rel}} \approx \frac{m_\nu c^2}{3k_B T_{\nu,0}} = 2 \times 10^3 \left(\frac{m_\nu}{1\,\text{eV}}\right). \tag{5}$$

This redshift has important implications for structure formation because it dictates the time at which massive neutrinos begin to make a transition from being radiation to matter. Note that this transition occurs fairly early, before recombination if $m_\nu \gtrsim 1$ eV. The average momentum of the cosmic background neutrinos at temperature T_ν is given by

$$\langle p \rangle = 3.15\,k_B T_\nu/c. \tag{6}$$

In the non-relativistic regime ($p = m_\nu v$), the average neutrino speed can be written as

$$\langle v \rangle = 160\,\text{km/s}\,\left(\frac{1\,\text{eV}}{m_\nu}\right)\left(\frac{T_\nu}{1.947}\right). \tag{7}$$

Since $T_\nu \propto a^{-1} \propto (1 + z)$, massive neutrinos slow down as time goes on. It is important to keep in mind that neutrinos with a mass of several eV have slowed down to an average velocity below 100 km s^{-1} today.

We also note that at the redshift of matter-radiation equality, $z_{\text{eq}} \sim 24000\,\Omega\,h^2$ (i.e. when the total energy density in radiation in the universe equals that in matter), light neutrinos with $1 < m_\nu < 10$ eV are zooming around with speeds close to c. Such large thermal speeds prevent massive neutrinos from clustering gravitationally during this epoch, and this is why light neutrinos are referred to as hot dark matter (HDM). In contrast, perturbations in cold dark matter (CDM), which by definition has negligible thermal velocities, can grow unimpeded after z_{eq}. I will quantify the different clustering behavior of CDM and HDM further in Sec. 3 and 4.

Since neutrinos cannot cluster appreciably via gravitational instabilities on scales below the free streaming distance, this introduces a characteristic length scale into the problem. This scale is given by the free-streaming wavenumber (in comoving coordinates)

$$k_{\text{fs}}^2 = \frac{4\pi G \rho\, a^2}{\langle v \rangle^2}, \tag{8}$$

which is analogous to the Jeans length for a self-gravitating system of density ρ. For $k < k_{\text{fs}}$ (i.e. large wavelengths), the density perturbation in the neutrinos is Jeans unstable and grows unimpeded in the matter-dominated era. For $k > k_{\text{fs}}$, the density perturbation decays due to neutrino phase mixing. A phase-space interpretation of the free streaming property is that the phase mixing of collisionless particles damps the growth of density perturbations. [8] When the neutrinos are relativistic, $\langle v \rangle \approx c$, and the free-streaming distance is approximately the particle horizon, which scales as $k_{\text{fs}}(a) \propto a^{-1}$ (in the radiation-dominated era). After the neutrinos become non-relativistic the relations $\langle v \rangle \propto a^{-1} m_\nu^{-1}$, $m_\nu \propto \Omega_\nu h^2$, and $\rho \propto a^{-3} h^2$ then imply [9]

$$k_{\text{fs}}(a) \propto a^{1/2} \Omega_\nu h^3. \tag{9}$$

As expected, the free-streaming distance ($\propto k_{\text{fs}}^{-1}$) decreases with time as the neutrinos slow down. This also implies that neutrinos can cluster gravitationally on increasingly small length scales at later times. [9] Such behavior has been seen in cosmological numerical simulations and will be discussed in Sec. 4.

3 Linear Perturbations in Neutrinos and Other Particles

Thus far our discussion has focused on the properties of the smooth cosmic background neutrinos, and Eqs. (1)-(7) were derived under this assumption. The universe today, however, is clearly far from being homogeneous on scales of ~ 100 Mpc and below. Baryons and dark matter in galaxies, clusters, and superclusters show a wide spectrum of overdensities above the cosmic mean. The current theoretical framework for the origin and evolution of these cosmic structures rests upon the assumption that certain primordial fluctuations (perhaps originated from quantum fluctuations of scalar fields during the inflationary era) imprint a perturbation spectrum on all matter and radiation. These fluctuations subsequently grow via gravitational instabilities to give rise to the wide range of observed structures. How are the cosmic relic neutrinos affected by all this?

To understand the growth of density perturbations in neutrinos as well as other forms of matter and radiation, one would need to learn the linear cosmological perturbation theory of gravitational instability. A full description of this theory requires more time than is allocated for these lectures. I will only sketch the theory below with emphasis on the neutrino component. Interested readers should refer to the pioneering work of Lifshitz, later reviewed in Lifshitz & Khalatnikov. [10] More modern treatments of various aspects of this theory can be found in the textbooks by Weinberg and Peebles, [11] in the reviews by Kodama & Sasaki and Mukhanov, Feldman & Brandenberger, [12] and in the Summer School lectures by Efstathiou, Bertschinger, and Bond. [13] A complete description of this theory for all relevant particles is given by Ma & Bertschinger. [14] Here, I will only discuss the essence of the theory and highlight the physical meaning of the key results.

3.1 Neutrino Phase Space

Let us start with the neutrinos. The full phase space distribution function of neutrinos can be written as

$$f(\vec{x}, \vec{p}, t) = f_0(p) + f_1(\vec{x}, \vec{p}, t), \tag{10}$$

where f_1 denotes perturbations to the Fermi-Dirac distribution f_0 given by Eq. (2). Unlike the unperturbed term f_0 that depends only on p, f_1 can have complicated dependence on time as well as positions $\vec{x}$ and the conjugate momenta $\vec{p}$. The equations for neutrino temperature, number density etc. discussed in Sec. 2 were obtained assuming $f = f_0$. A non-vanishing f_1 would lead to perturbations in these quantities. For example, the perturbed neutrino energy density is related to f_1 by

$$\delta\rho(\vec{x}, t) = a^{-4} \int d^3p\, \epsilon\, f_1(\vec{x}, \vec{p}, t).$$ (11)

Other quantities such as perturbations in the pressure and shear can also be related to f_1.

It is in general difficult to compute and sample f_1 directly because at a given time, it depends on six variables. A Monte Carlo technique, or a "general-relativistic N-body" technique, has been developed to evolve f_1 from redshift $z \sim 10^9$ shortly after neutrino decoupling until $z \sim 10$ when nonlinear effects become non-negligible. [15] In this calculation, an ensemble of neutrino simulation particles is initially assigned velocities drawn from the Fermi-Dirac distribution, which is an excellent approximation at $z \sim 10^9$. The trajectory of each neutrino simulation particle is then followed by integrating the geodesic equations in the *perturbed* background spacetime. The metric perturbation gives rise to a nonzero f_1, and the particle positions and velocities at a later time t represent a realization of $f_1(\vec{x}, \vec{p}, t)$. Results from this calculation have revealed that at $z \sim 15$, positive correlations have developed in the rms neutrino velocities and the overdensity, which would be absent if the phase-space distribution were purely Fermi-Dirac (i.e. $f = f_0$). The more spatially clustered neutrinos are found to move faster, possibly resulting from an increase in the kinetic energy during gravitational infalls.

3.2 Evolution of Perturbed Density Fields

The phase-space description above is applicable to all particle species and is used in the full theory. The full, general-relativistic version of the linear perturbation theory is described by a set of coupled and linearized Einstein, Boltzmann, and fluid equations. The variables include the metric perturbations to the homogeneous and isotropic Friedmann-Robertson-Walker spacetime, and the phase-space perturbations in all relevant particle species (e.g., photons, baryons, cold dark matter, massless and massive neutrinos). The Einstein equations describe how the time evolution of the metric perturbations is affected by the perturbations in the density, pressure, shear, and higher-order moments of matter and radiation. The Boltzmann and fluid equations, on the other hand, describe the time evolution of the radiation and matter distribution in the perturbed spacetime. Together, this theory describes the growth of metric and density perturbations throughout the early history of the universe, and it serves as the foundation for all calculations of the linear power spectra for matter and temperature variations imprinted on the cosmic microwave background.

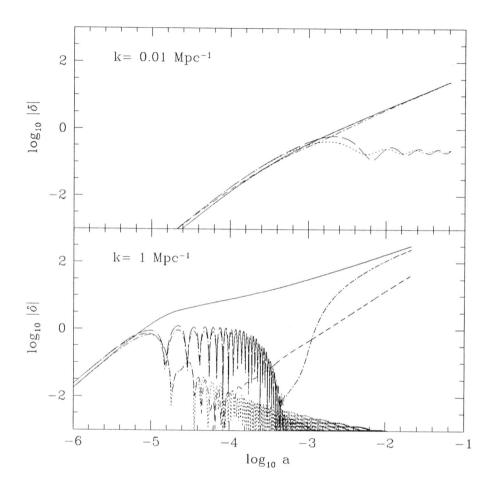

Figure 1: Time evolution of the perturbed energy density field, $\delta = \rho/\bar{\rho} - 1$, for five matter and radiation components in a flat C+HDM cosmological model. (See text for model parameters.) The results are from integration of the coupled Einstein and Boltzmann equations. Since the equations are linearized, each k-mode evolves independently. Two modes are shown here for illustration. In each panel, the five curves represent δ for the cold dark matter (solid), baryons (dash-dotted), photons (long-dashed), massless neutrinos (dotted), and massive neutrinos (short-dashed), respectively.

Figure 1 illustrates a small subset of results that can be obtained from numerical integration of these linearized equations. It shows the time evolution of the perturbed energy density field, $\delta = \delta\rho/\bar{\rho} = \rho/\bar{\rho} - 1$, for the five relevant particle species in a cold+hot dark matter (C+HDM) model. This model assumes an Einstein-de Sitter universe containing a mixture of CDM, HDM (i.e. massive neutrinos), baryons, photons, and massless neutrinos. The density parameters in the first three components in this model are $\Omega_c = 0.75$, $\Omega_\nu = 0.2$, and $\Omega_b = 0.05$, and the Hubble parameter is taken to be $H_0 = 50$ km s^{-1} Mpc^{-1}, or $h = 0.5$. From Eq. (4), these parameters correspond to a neutrino mass of 4.7 eV. (For definiteness, this calculation has assumed that only one type of neutrinos, presumably ν_τ, has a non-negligible mass.) Two wavenumbers, $k = 0.01$ Mpc^{-1} (top) and 1 Mpc^{-1} (bottom), are shown in Figure 1 to demonstrate the intricate dependence of δ on length and time scales. The overall normalization of δ is set arbitrarily.

We observe several salient features in Figure 1. First, the amplitudes of δ for all particles grow monotonically until a critical time, after which different particle species exhibit very different behavior. This critical time is the "horizon crossing" time, and it occurs when the horizon has grown large enough to encompass the wavelength of a given mode of perturbation. A mode of perturbation is therefore not in causal contact until horizon crossing. Naturally, this occurs earlier for smaller wavelengths (i.e. larger k). In Figure 1, one can indeed see that the $k = 1$ Mpc^{-1} mode enters the horizon at $a \sim 10^{-5}$ while the $k = 0.01$ Mpc^{-1} mode enters the horizon a little after $a = 10^{-3}$. A point to keep in mind is that the behavior of δ before horizon crossing is strongly dependent on the choice of gauge. The results shown in Figure 1 are computed in the so-called synchronous gauge, which is a popular choice due to historical precedent. [10] See Mukhanov, Feldman & Brandenberger [12] and Ma & Bertschinger [14] for discussion of a more convenient gauge (the conformal Newtonian gauge).

The second feature in Figure 1 to note is that after horizon crossing, the photons (long-dashed) and baryons (dot-dashed) exhibit rapid, coupled oscillations in the $k = 1$ Mpc^{-1} mode but only the photons oscillate in the $k = 0.01$ Mpc^{-1} mode. This occurs because the former enters the horizon before recombination at $a_{rec} \sim 10^{-3}$, and the photons and baryons are coupled by Thomson scattering and oscillate acoustically. The coupling is not perfect. The friction of the photons dragging against the baryons leads to Silk damping, [16] which is prominent in the bottom panel of Figure 1 at $a \sim 10^{-3.5}$. After recombination, the baryons decouple from the photons and fall quickly into the potential wells formed earlier by the CDM. This results in the rapid growth of the dot-dashed curve in the bottom panel of Figures 1. The mode with $k = 0.01$

Mpc^{-1}, on the other hand, enters the horizon when the universe is neutral. The baryons therefore grow like the CDM and do not oscillate. The critical length scale demarcating these two regimes is the horizon size at recombination a_{rec}: $k_{\text{rec}} \sim 0.03 \text{ Mpc}^{-1}$.

The third feature to note in Figure 1 is the rate of growth of the CDM component (solid curve) after horizon crossing. Close inspection shows that the CDM in the bottom panel grows more slowly at $10^{-5} \lesssim a \lesssim 10^{-4}$ than later on, whereas in the upper panel, the CDM simply grows with a power law after horizon crossing. This is because the shorter wavelength mode ($k = 1 \text{ Mpc}^{-1}$) enters the horizon when the energy density of the universe is dominated by radiation, and fluctuations in matter (e.g. CDM) cannot grow appreciably during this era. The critical scale separating continual and suppressed growth is the horizon size at the time of radiation-matter equality $a_{\text{eq}} \sim 4 \times 10^{-5}(\Omega h^2)^{-1}$: $k_{\text{eq}} \sim 0.1 \text{ Mpc}^{-1}$ for the parameters of this model.

The fourth feature to note in Figure 1 is the behavior of the massive neutrinos (short-dashed). As discussed in Sec. 2, neutrinos of masses within the cosmologically interesting range (~ 1 to 10 eV) are highly non-relativistic today but were relativistic at earlier times. This property is in fact evident in the top panel of Figure 1: Careful inspection shows that at $a \approx 10^{-4}$, the density field δ in massive neutrinos is indeed making a gradual transition from the upper line for the radiation fields to the lower line for the matter fields. (More precisely, the primordial perturbations are assumed to be "isentropic" here, which leads to a perturbation amplitude that is a factor of 4/3 higher for radiation than matter.) The subsequent evolution of δ in massive neutrinos for this mode ($k = 0.01 \text{ Mpc}^{-1}$) is very similar to that of CDM. This is because it enters the horizon when the thermal velocities of the neutrinos have decreased substantially; the free-streaming effect is therefore unimportant. For the $k = 1$ Mpc^{-1} mode, on the other hand, the free streaming effect is evident and the growth of δ in massive neutrinos is suppressed until the characteristic free-streaming scale $k_{\text{fs}}(a)$ given by Eq. (9) grows to $\sim k$. Afterwards, the short-dashed curve for massive neutrinos is seen to grow again and catch up to the CDM. Since $k_{\text{fs}} \propto a^{1/2}$, the larger k modes suffer more free-streaming damping and δ for massive neutrinos can not grow until later times. The damping in the massive neutrino component also affects the growth of the CDM, slowing it down more for models with larger Ω_ν compared to the pure CDM model.

3.3   Growth Rate and Power Spectrum

I have used Figure 1 computed for two particular k-modes in a particular cosmological model to illustrate the physical meaning of many key features in the evolution of the density field for matter and radiation throughout the cosmic history. I will now discuss general descriptions that can be conveniently used to characterize the fluctuation amplitudes over a range of length scales for a variety of cosmological models. For example, it is extremely useful to know the dependence of δ and its time derivative on k at a given time for a wide range of models. The most basic quantity to use is the linear power spectrum, $P(k,t)$, and the growth rate of the density field, $f \equiv d\log\delta/d\log a$. (*Note: I am following the convention of using f to denote the growth rate; it should not be confused with the phase-space distribution function of Sec. 2 and 3.1.*) The power spectrum quantifies the two-point statistics of δ, and for a Gaussian field, $P(k)$ represents its rms fluctuations and completely specifies its statistical properties. The power spectrum is therefore of fundamental importance in cosmology.

Comparing the growth rate and power spectrum for models with and without massive neutrinos is an effective way to illustrate the effects of hot dark matter. In the standard CDM model with $\Omega = 1$ and $h = 0.5$ (neutrino mass is assumed to be zero), the CDM density field grows as the expansion factor a on all scales; therefore $f = 1$. As discussed in Sec. 2.2, massive neutrinos introduce an additional length scale, the free-streaming distance, below which fluctuations are washed out and the growth rate is retarded. The growth rate is therefore more intricate in models with massive neutrinos and is generally a function of the wavenumber k, neutrino density parameter Ω_ν, and time. Figure 2 illustrates such dependence in four different C+HDM models that assume a mixture of CDM and HDM. It shows that the growth is suppressed at large k, and models with a larger fraction of energy density in HDM suffer more. It also shows that the suppression becomes less severe at later times.

One can gain some understanding of the behavior shown in Figure 2 by exploring two asymptotic regions that can be solved analytically for C+HDM models with $\Omega_c + \Omega_\nu = 1$: (1) Since HDM behaves like CDM above the free-streaming distance, $f \to 1$ as $k \to 0$; (2) In the opposite limit of large k, the HDM density field δ_h is severely dampened compared to the CDM density field δ_c because of the neutrino effects. For $\delta_h \ll \delta_c$, the time evolution of the CDM density field is governed by the linearized fluid equation

$$\ddot{\delta}_c + \frac{\dot{a}}{a}\dot{\delta}_c = 1.5H^2a^2\Omega_c\delta_c \, , \tag{12}$$

where the dots denote differentiation with respect to the conformal time τ. Since $Ha = 2/\tau$ in the matter-dominated era, the growing solution in this

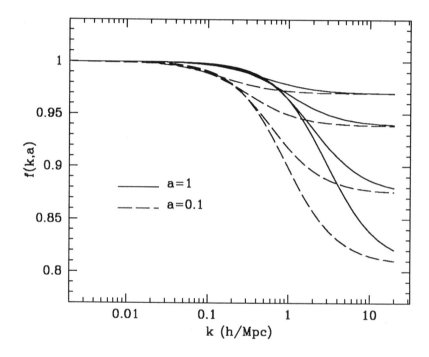

Figure 2: Growth rate of the CDM density field, $f \equiv d\log\delta/d\log a$, in four flat C+HDM models at $a = 1$ (solid) and 0.1 (dashed). The four models assume different neutrino masses: $m_\nu = 1.2$, 2.3, 4.6, and 6.9 eV (from top down), corresponding to $\Omega_\nu = 0.05$, 0.1, 0.2, and 0.3. At small k, the CDM density field in these models grows with the same rate ($\delta \propto a$) as in the standard CDM model. At large k, the growth rate is suppressed because a fraction of the energy density in C+HDM models is in the hot neutrinos that exhibit less gravitational clustering. The suppression at large k becomes less severe at later times because the velocities of the hot neutrinos decrease with time.

regime is easily shown to be [17]

$$f_\infty \equiv f(k \to \infty) = \frac{1}{4}\sqrt{1 + 24\Omega_c} - \frac{1}{4} = \frac{5}{4}\sqrt{1 - \frac{24}{25}\Omega_\nu} - \frac{1}{4}. \tag{13}$$

It is interesting to note that

$$f_\infty \approx \Omega_c^{0.6} \tag{14}$$

is an excellent approximation to the equation above, especially for the cosmologically viable range of $\Omega_\nu \lesssim 0.3$. Using these analytic solutions in the asymptotic regimes and the scaling dependence of the free streaming wavenumber k_{fs} in Eq. (9), one can construct a simple approximation for f for a wide range of model parameters. It is found that the growth rate f is well approximated by [9]

$$f \equiv \frac{d\log\delta}{d\log a} = \frac{1 + \Omega_c^{0.6}\, 0.00681\, x^{1.363}}{1 + 0.00681\, x^{1.363}}, \qquad x \equiv \frac{k}{\Gamma_\nu h}, \tag{15}$$

where Γ_ν is a shape parameter derived from Eq. (9),

$$\Gamma_\nu = a^{1/2}\Omega_\nu h^2, \tag{16}$$

$\Omega_c + \Omega_\nu = 1$, and k is in units of Mpc^{-1}. Note that Eq. (15) depends only on the variable x that characterizes the neutrino free-streaming scale, and Ω_ν (or Ω_c) via f_∞. The fractional error of the fit relative to the numerically computed values is smaller than 0.5% for a wide range of parameters. The seemingly complicated multi-parameter dependence of Figure 2 is succinctly incorporated in Eq. (15).

For the linear power spectrum, the slower time growth of δ at $k > k_{fs}$ for the C+HDM models shown in Figure 2 indicates a suppressed clustering amplitude on these scales. This effect is illustrated in Figure 3, which shows the density-averaged power spectrum (i.e. $P = \{\Omega_c\sqrt{P_c} + \Omega_\nu\sqrt{P_\nu}\}^2$) for three flat C+HDM models. In general, $P_\nu \ll P_c$ on small length scales due to the neutrino thermal velocities, and the models with higher Ω_ν clearly have less power at large k in accordance with Figure 2. A good approximation for the relative $P(k)$ in a flat C+HDM model (with $\Omega_\nu \lesssim 0.3$) and a pure CDM model ($\Omega_\nu = 0$) is given by [9]

$$\frac{P(k, a, \Omega_\nu)}{P(k, a, \Omega_\nu = 0)} = \left(\frac{1 + b_1\, x^{b_4/2} + b_2\, x^{b_4}}{1 + b_3\, x_0^{b_4}}\right)^{\Omega_\nu^{1.05}},$$

$$x = \frac{k}{\Gamma_\nu}, \qquad x_0 = x(a = 1), \tag{17}$$

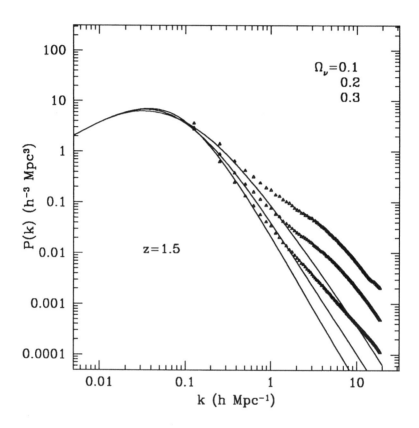

Figure 3: Power spectrum at $z = 1.5$ for the perturbed matter density field for three flat C+HDM models with $\Omega_\nu = 0.1$, 0.2, and 0.3. The solid curves show the linear $P(k)$ computed from the linear perturbation theory; accurate analytical fitting formulas are given by Eq. (17). The triangles show the nonlinear $P(k)$ computed from N-body simulations; accurate analytical approximations are given by Eq. (18).

where the best-fit parameters are $b_1 = 0.004321, b_2 = 2.217 \times 10^{-6}, b_3 = 11.63$, and $b_4 = 3.317$ for k in units of Mpc^{-1}. Analytical approximations for the separate cold and hot spectra P_c and P_ν can be found in the same reference. More complicated approximations for a wider range of parameter space have also been proposed.[18]

4 Nonlinear Gravitational Clustering of Neutrinos

As far as cosmological structure formation is concerned, the main difference between massive and massless neutrinos is that the former can participate in the processes of gravitational clustering and hence serves as a component of the dark matter in the universe. Massless neutrinos, on the other hand, affect cosmology only through their contribution to the radiation energy density. Any primordial perturbations in massless neutrinos are damped out after horizon crossing as a result of phase mixing (see dotted curves in Figure 1), and the only remnant of this component is the elusive $T_{\nu,0} = 1.947$ K background described in Sec. 2.

4.1 Spatial Distribution of Neutrinos

For massive neutrinos in models with a mixture of CDM and HDM, it is interesting to ask: do the neutrinos fall in the CDM potential wells and form a part of dark matter halos? One may naïvely think not because neutrinos are too hot. Our discussion thus far, however, indicates otherwise. We have seen in Eq. (7) that neutrinos slow down with the expansion of the universe. Those with a mass of several eV are travelling with a speed much below the typical velocity dispersions ~ 200 km s^{-1} of stars in galaxies. They can potentially be bound to galactic halos. In general, the extent to which massive neutrinos can cluster gravitationally depends on their mass and speed.

To illustrate this point further, let us examine results from numerical simulations. Figure 4 shows the projected spatial distribution of cold particles (middle), hot particles (bottom), and the sum of the two (top) in a simulated dark matter halo in a flat $\Omega_\nu = 0.2$ C+HDM model ($m_\nu = 4.7$ eV). Three redshifts are shown (from left to right): $z = 2, 1$ and 0 when the universe is 2.5, 4.6, and 13 billion years old, respectively. Each panel is $3.5 \times 3.5\, h^{-1}$ Mpc in *physical* coordinates. The parent simulation is a large N-body run with 23 million simulation particles in a $(100$ Mpc$)^3$ comoving box.[19] The dense halo shown at $z = 0$ is clearly formed from mergers of two smaller halos and their satellites at higher redshifts, demonstrating the "bottom-up" hierarchical pattern of structure formation which is preserved in C+HDM models with $\Omega_c > \Omega_\nu$. Massive neutrinos are visibly clustered in the bottom panels, but their spatial

518

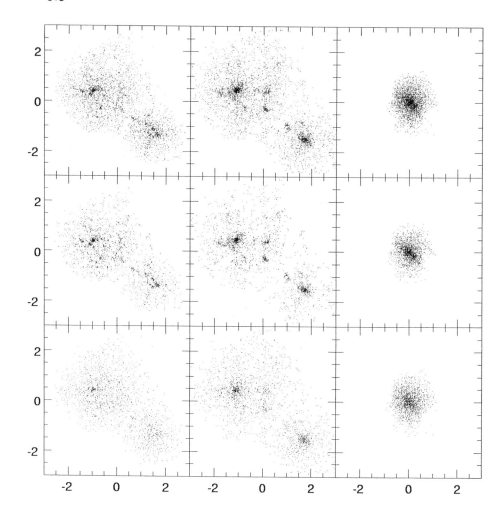

Figure 4: Spatial distribution of cold dark matter (middle) vs. hot dark matter (bottom) particles in a simulated halo formed in a large N-body run for the flat $\Omega_\nu = 0.2$ C I HDM model. The top panels shows the sum of cold and hot particles. Three redshifts are shown: $z = 2$, 1, and 0 (from left to right), corresponding to cosmic times of 2.5, 4.6, and 13 billion years. All boxes have the same physical scale, $3.5 \times 3.5\, h^{-1}$ Mpc. The final halo at $z = 0$ is clearly a merger product of two dominant subhalos and several smaller satellites at higher redshifts. The spatial distribution of the massive neutrinos (4.7 eV) is visibly smoother than that of the CDM.

distribution is smoother than that of the CDM. In the smaller CDM clumps at $z \gtrsim 1$, there are no discernible HDM halos at the same locations. The average thermal speed of an ensemble of 4.7 eV Fermi-Dirac neutrinos at $z \gtrsim 1$ is $\gtrsim 70$ km/s (cf. Eq. 7). As discussed in Sec. 3.1, when linear perturbations are taken into account, neutrinos in overdense regions have even higher velocities. It is therefore not surprising that the shallower potential wells of these small halos cannot trap a substantial number of HDM particles.

The clustering of the 4.7 eV neutrinos shown in Figure 4 should be compared with the constraint derived by Tremaine & Gunn,[20] which states that if cosmic neutrinos were to make up the bulk of galactic and cluster halos, they must be more massive than ~ 10 eV so that the Pauli exclusion principle is not violated. The C+HDM models considered here assume $\Omega_c > \Omega_\nu$ so as to preserve the successful hierarchical formation of structure in pure CDM models. Dark matter halos such as in Figure 4 have a substantial fraction of CDM, which in turn enhances the gravitational infalls of massive neutrinos. The clustering of neutrinos in C+HDM models is therefore more complicated.

4.2 Nonlinear Power Spectrum

The process of nonlinear gravitational clustering can be quantified statistically. Here I will only discuss the lowest-order statistical description given by the nonlinear power spectrum $P(k)$ of the matter density field.

We have already discussed the linear power spectrum in Sec. 3.3. The triangles in Figure 3 show the nonlinear $P(k)$ computed from the particle positions in numerical simulations of three C+HDM models. The hierarchical nature of gravitational collapse in these models is illustrated by the fact that the high-k modes have become strongly nonlinear whereas the low-k modes are still following the linear power spectrum. The fact that the lowest several k modes are still linear at $z = 0$ ensures that our choice of the simulation box size (100 Mpc) is large enough to include all waves that have become nonlinear at present.

The calculation of the fully evolved $P(k)$ for a given model is a laborious task involving the execution of high-resolution N-body simulations. Fortunately, some recent progress has been made in constructing analytical fitting formulas for a wide range of interesting models. The strategy is to examine the mapping between the linear and nonlinear $P(k)$ for a small, selected set of models with N-body data and then to extract systematic behavior for a wider range of parameters. The work carried out thus far has investigated scale-free models with a power-law power spectrum,[21,22] pure CDM and CDM with a cosmological constant Λ (LCDM) models,[22,23,24] and C+HDM models.[24] The proposed formulas typically have a functional form that is motivated by analytical solutions in asymptotic regimes, but in order to obtain accurate approximations, the coefficients are calculated from fits to the nonlinear $P(k)$ computed from the numerical simulations. These approximations have provided physical insight into the process of nonlinear collapse and much practical convenience in incorporating the prominent nonlinear effects illustrated in Figure 3.

The formula applicable for the widest range of cosmological models thus far is given below. It maps the density variance $\Delta(k) \equiv 4\pi k^3 P(k)$ in the linear and nonlinear regimes by [24]

$$\frac{\Delta_{\mathrm{nl}}(k)}{\Delta_{\mathrm{l}}(k_0)} = G\left(\frac{\Delta_{\mathrm{l}}(k_0)}{g_0^{1.5}\sigma_8^\beta}\right),$$

$$G(x) = [1 + \ln(1 + 0.5\,x)]\,\frac{1 + 0.02\,x^4 + c_1\,x^8/g^3}{1 + c_2\,x^{7.5}}. \tag{18}$$

Note that Δ_{l} and Δ_{nl} are evaluated at different wavenumbers, where $k_0 = k\,(1 + \Delta_{\mathrm{nl}})^{-1/3}$ corresponds to the pre-collapsed length scale of k. The parameter σ_8 is the rms linear mass fluctuation on $8\,h^{-1}$ Mpc scale at the redshift of interest, and $\beta = 0.7 + 10\,\Omega_\nu^2$. The functions $g_0 = g(\Omega_{\mathrm{m}}, \Omega_\Lambda)$ and $g = g(\Omega_{\mathrm{m}}(a), \Omega_\Lambda(a))$ are, respectively, the relative growth factor for the linear density field at present day and at a for a model with a present-day matter density Ω_{m} and a cosmological constant Ω_Λ. A good approximation is given by [25]

$$g = \frac{5}{2}\Omega_{\mathrm{m}}(a)[\Omega_{\mathrm{m}}(a)^{4/7} - \Omega_\Lambda(a) + (1 + \Omega_{\mathrm{m}}(a)/2)\,(1 + \Omega_\Lambda(a)/70)]^{-1}, \tag{19}$$

and $\Omega_{\mathrm{m}}(a) = \Omega_{\mathrm{m}}\,a^{-1}/[1 + \Omega_{\mathrm{m}}(a^{-1} - 1) + \Omega_\Lambda(a^2 - 1)]$ and $\Omega_\Lambda(a) = \Omega_\Lambda\,a^2/[1 + \Omega_{\mathrm{m}}(a^{-1} - 1) + \Omega_\Lambda(a^2 - 1)]$. The time dependence is in factors σ_8^β and g. For CDM and LCDM models, a good fit is given by $c_1 = 1.08 \times 10^{-4}$ and $c_2 = 2.10 \times 10^{-5}$. For C+HDM, a good fit is given by $c_1 = 3.16 \times 10^{-3}$ and $c_2 = 3.49 \times 10^{-4}$ for $\Omega_\nu = 0.1$, and $c_1 = 6.96 \times 10^{-3}$ and $c_2 = 4.39 \times 10^{-4}$ for $\Omega_\nu = 0.2$.

4.3 High Redshift Constraints

The C+HDM models discussed thus far are a class of models bridging the much-studied albeit troubled pure CDM and pure HDM models. They are parameterized by the neutrino density parameter $0 < \Omega_\nu < 1$, or equivalently, by the neutrino mass $0 < m_\nu < 93h^2$ eV (see Eq. (4)). The original motivation for examining these mixed models is to study whether the free-streaming effect introduced by the massive neutrinos could suppress the growth of density perturbations below the free-streaming scale, and thereby alleviate the problem of excess small-scale clustering in the standard CDM model. As illustrated in Figure 3, massive neutrinos do indeed reduce the amplitude of clustering at large k, and a larger Ω_ν leads to smaller high-k power.

Any viable cosmological model that provides a good statistical match to the local universe must also reproduce the appropriate evolutionary history out to high redshifts. Although gravitational clustering on galactic scales is indeed reduced in C+HDM models, providing a better match to low-redshift observations,[9,26] this suppression in the clustering power at high redshifts has been found to pose serious problems for some models. Studies based on semi-analytic theories and dissipationless numerical simulations have shown [19,27] that flat C+HDM models with $\Omega_\nu > 0.2$ do not produce enough early structure to explain the statistics of damped Lyα systems at redshift $z \geq 2$. More recent work [28] has included the effects of gas ionization and dissipation in the theoretical calculations and has compared the results to new data for damped Lyα systems at even higher redshift $z \sim 4$. It is found that ionization of hydrogen in the outskirts of halos and gaseous dissipation near the halo centers tend to exacerbate the problem of late galaxy formation. The amount of dense gas associated with the damped systems falls well below that observed, even for the flat $\Omega_\nu = 0.2$ C+HDM model. This has placed an upper bound of ~ 5 eV on the sum of ν_e, ν_μ, and ν_τ masses, which is much more stringent than the upper limits given by current particle experiments (see Sec. 1).

Let us briefly discuss the limitations and uncertainties in these calculations. Still debated is the nature of damped Lyα absorption – whether it is due to intervening large, rapidly-rotating disk galaxies [29] with circular velocities $\gtrsim 200$ km s^{-1}, or infalling and randomly moving protogalactic gas clumps in dark matter halos [30] with virial velocities of ~ 100 km s^{-1}. When kinematic considerations are included, the latter model may have problems balancing the high energy dissipation rate caused by cloud collisions. [31] This uncertainty aside, in either scenario, a host dark halo of velocity at least ~ 100 km s^{-1} is needed to reproduce the large velocity widths and asymmetries of the observed low-ionization lines associated with Lyα systems. Uncertainties associated with the finite resolution of simulations have also been studied in some detail. [32] It is found that even when the contribution from the numerically unresolved halos with velocities $v \lesssim 100$ km s^{-1} is included, the absorption incidence is increased by at most a factor of 2. This is insufficient to erase the discrepancies reported for flat C+HDM models with $\Omega_\nu > 0.2$. The upper bound on neutrino masses from current cosmological studies is therefore ~ 5 eV.

522

Acknowledgments

I thank the hospitality of P. Langacker and K. T. Mahanthappa of the TASI-98 School. This work was supported by the National Scalable Cluster Project at the University of Pennsylvania, the National Center for Supercomputing Applications, and a Penn Research Foundation Award.

References

1. See, e.g., Review of Particle Physics, *Euro. Phys. Jour.* **C3**, 1 (1998); T. Bowles, this volume.
2. See, e.g., A. Mann, this volume.
3. D. J. Fixsen, *Ap. J.* **473**, 576 (1996).
4. R. Cowsik & J. McClelland, *Phys. Rev. Lett.* **29**, 751 (1972).
5. See, e.g., S. D. M. White, C. S. Frenk, & M. Davis, *Ap. J. Lett.* **274**, L1 (1983).
6. B. W. Lee & S. Weinberg, *Phys. Rev. Lett.* **39**, 165 (1977).
7. See, e.g., P. Langacker, this volume.
8. R. Bond & A. Szalay, *Ap. J.* **274**, 443 (1980).
9. C.-P. Ma, *Ap. J.* **471**, 13 (1996).
10. E. M. Lifshitz, *J. Phys. USSR* **10**, 116 (1946); E. M. Lifshitz & I. M. Khalatnikov, *Adv. Phys.* **12**, 185 (1963).
11. S. Weinberg, Gravitation and Cosmology (Wiley, New York, 1972); P. J. E. Peebles, The Large-Scale Structure of the Universe (Princeton University Press, Princeton, 1980).
12. H. Kodama & M. Sasaki. *Prog. Theo. Phys. Suppl.* **78**, 1 (1984); V. F. Mukhanov, H. A. Feldman, & R. Brandenberger, *Phys. Rep.* **215**, 206 (1992).
13. G. Efstathiou, in Physics of the Early Universe: Proceedings of the 36th Scottish Universities Summer School in Physics, ed. J. A. Peacock, A. E. Heavens, & A. T. Davies (New York, Adam Hilger, 1990), p. 361; E. Bertschinger, Proceedings of Les Houches School, Session LX, ed. R. Schaeffer (Elsevier Science, Netherlands, 1995), p. 273; J. R. Bond, *ibid.*, p.469.
14. C.-P. Ma & E. Bertschinger, *Ap. J.* **455**, 7 (1995)
15. C.-P. Ma & E. Bertschinger. *Ap. J.* **429**, 22 (1994).
16. J. Silk, *Ap. J.* **151**, 459 (1968).
17. R. Bond, G. Efstathiou, & J. Silk, *Phys. Rev. Lett.* **45**, 1980 (1980).
18. D. Eisenstein & W. Hu, *Ap. J.* **511**, 5 (1998).
19. C.-P. Ma & E. Bertschinger, *Ap. J. Lett.* **434**, L5 (1994).
20. S. Tremaine, & J. E. Gunn. *Phys. Rev. Lett.* **42**, 407 (1979).

21. A. J. S. Hamilton, P. Kumar, E. Lu, & A. Matthews, *Ap. J. Lett.* **374**, L1 (1991).
22. B. Jain, H. J. Mo, & S. D. M. White, *Mon. Not. Roy. Ast. Soc.* **276**, L25 (1995).
23. J. A. Peacock & S. J. Dodds, *Mon. Not. Roy. Ast. Soc.* **280**, L19 (1996).
24. C.-P. Ma, *Ap. J. Lett.* **508**, L5 (1998).
25. O. Lahav, P. Lilje, J. Primack, & M. Rees, *Mon. Not. Roy. Ast. Soc.* **251**, 128 (1991); S. Carroll, W. Press, & E. L. Turner, *ARAA* **30**, 499 (1992).
26. See, e.g., A. Klypin, J. Holtzman, J. Primack, & E. Regos, *Ap. J.* **416**, 1 (1993); E. Gawiser & J. Silk *Science* **280**, 1405 (1998).
27. H. J. Mo & J. Miralda-Escude, *Ap. J. Lett.* **430**, L25 (1994); G. Kauffmann & S. Charlot, *Ap. J. Lett.* **430**, L97 (1994); A. Klypin, S. Borgani, J. Holtzman, & J. R. Primack, *Ap. J.* **444**, 1 (1995).
28. C.-P. Ma, E. Bertschinger, L. Hernquist, D. Weinberg, & N. Katz, *Ap. J. Lett.* **484**, L1 (1997).
29. J. Prochaska & A. M. Wolfe, *Ap. J.* **487**, 73 (1997).
30. M. Haehnelt, M. Steinmetz, & M. Rauch, *Ap. J.* **495**, 647 (1998).
31. McDonald, P., & J. Miralda-Escudé, astro-ph 9809237 (1998).
32. J. P. Gardner, N. Katz, L. Hernquist, & D. Weinberg, *Ap. J.* **484**, 31 (1997).

LECTURES ON NEUTRINO ASTRONOMY: THEORY AND EXPERIMENT

FRANCIS HALZEN

Department of Physics, University of Wisconsin, Madison, WI 53706

1. Overview of neutrino astronomy: multidisciplinary science.
2. Cosmic accelerators: the highest energy cosmic rays.
3. Neutrino beam dumps: supermassive black holes and gamma ray bursts.
4. Neutrino telescopes: water and ice.
5. Indirect dark matter detection.
6. Towards kilometer-scale detectors.

1   Neutrino Astronomy: Multidisciplinary Science

Using optical sensors buried in the deep clear ice or deployed in deep ocean and lake waters, neutrino astronomers are attacking major problems in astronomy, astrophysics, cosmic ray physics and particle physics by commissioning first generation neutrino telescopes. Planning is already underway to instrument a cubic volume of ice or water, 1 kilometer on the side, as a detector of neutrinos in order to reach the effective telescope area of 1 kilometer squared which is, according to estimates covering a wide range of scientific objectives, required to address the most fundamental questions. This infrastructure provides unique opportunities for yet more interdisciplinary science covering the geosciences and biology.

Among the many problems which high energy neutrino telescopes will address are the origin of cosmic rays, the engines which power active galaxies, the nature of gamma ray bursts, the search for the annihilation products of halo cold dark matter (WIMPS, supersymmetric particles(?)), galactic supernovae and, possibly, even the structure of Earth's interior. Coincident experiments with Earth- and space-based gamma ray observatories, cosmic ray telescopes and gravitational wave detectors such as LIGO can be contemplated. With high-energy neutrino astrophysics we are poised to open a new window into space and back in time to the highest-energy processes in the Universe.

"And the estimate of the primary neutrino flux may be too low, since regions that produce neutrinos abundantly may not reveal themselves in the types of radiation yet detected" Greisen states in his 1960 review.[1] He also establishes that the natural scale of a deep underground neutrino detector is 15 m! The dream of neutrino astronomers is the same today, but experimental techniques are developed with the ultimate goal to deploy kilometer-size instruments.

I will first introduce high energy neutrino detectors as astronomical telescopes using Fig. 1. The figure shows the diffuse flux of photons as a function of their energy and wavelength, from radio-waves to the high energy gamma rays detected with satellite-borne detectors.[2] The data span 19 decades in energy. Major discoveries have been historically associated with the introduction of techniques for exploring new wavelenghts. All of the discoveries were surprises; see Table 1. The primary motivation for commissioning neutrino telescopes is to cover the uncharted territory in Fig. 1: wavelengths smaller than 10^{-14} cm, or energies in excess of 10 GeV. This exploration has already been launched by truly pioneering observations using air Cerenkov telescopes.[3] Larger space-based detectors as well as cosmic ray facilities with sensitivity above 10^7 TeV, an energy where charged protons point back at their sources with minimal deflection by the galactic magnetic field, will be pursuing similar goals. Could the high energy skies in Fig. 1 be empty? No, cosmic rays with energies exceeding 10^8 TeV have been recorded.[4]

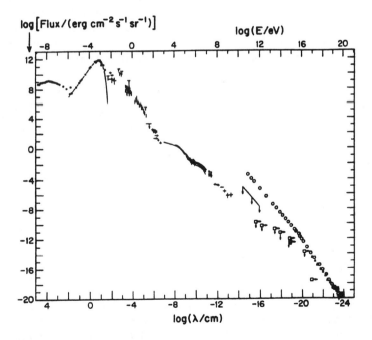

Figure 1: Flux of gamma rays as a function of wavelength and photon energy. In the TeV–EeV energy range the anticipated fluxes are dwarfed by the cosmic ray flux which is also shown in the figure.

526

Telescope	Intended use	Actual results
optical (Galileo)	navigation	moons of Jupiter
radio (Jansky)	noise	radio galaxies
optical (Hubble)	nebulae	expanding Universe
microwave (Penzias-Wilson)	noise	3K cosmic background
X-ray (Giacconi...)	moon	neutron star acc. binaries
radio (Hewish, Bell)	scintillations	pulsars
γ-ray (???)	thermonuclear explosions	γ-ray bursts

Exploring this wide energy region with neutrinos does have the definite advantage that they can, unlike high energy photons and nuclei, reach us, essentially without attenuation in flux, from the largest red-shifts. Gamma rays come from a variety of objects, both galactic (supernova remnants such as the Crab Nebula) and extragalactic (active galaxies), with energies up to at least 30 TeV, but they are absorbed by extragalactic infrared radiation in distances less than 100 Mpc. Photons of TeV energy and above are efficiently absorbed by pair production of electrons on background light above a threshold

$$4E\epsilon > (2m_e)^2 , \qquad (1)$$

where E and ϵ are the energy of the accelerated and background photon in the c.m. system, respectively. Therefore TeV photons are absorbed on infrared light, PeV photons (10^3 TeV) on the cosmic microwave background and EeV (10^{18} eV) on radiowaves. It is likely that absorption effects explain why Markarian 501, at a distance of barely over 100 Mpc the closest blazar on the EGRET list of sources, produces the most prominent TeV signal. Although one of the closest active galaxies, it is one of the weakest; the reason that it is detected whereas other, more distant, but more powerful, AGN are not, must be that the TeV gamma rays suffer absorption in intergalactic space through the interaction with background infra-red light.[5] Absorption most likely provides the explanation why much more powerful quasars such as 3C279 at a redshift of 0.54 have not been identified as TeV sources.

Cosmic rays are accelerated to energies as high as 10^0 TeV. Their range in intergalactic space is also limited by absorption, not by infrared light but by the cosmic microwave radiation. Protons interact with background light by the production of the Δ-resonance just above the threshold for producing pions:

$$4E_p\epsilon > (m_\Delta^2 - m_p^2) . \qquad (2)$$

The dominant energy loss of protons of $\sim$500 EeV energy and above, is photo-production of the Δ-resonance on cosmic microwave photons. The Universe is therefore opaque to the highest energy cosmic rays, with an absorption length of only tens of megaparsecs when their energy exceeds 10^8 TeV. Lower energy protons, below threshold (2), do not suffer this fate. They cannot be used for astronomy however because their direction is randomized in the microgauss magnetic field of our galaxy. Of all high-energy particles, only neutrinos convey directional information from the edge of the Universe and from the hearts of the most cataclysmic high energy processes.

Although Nature is clearly more imaginative than scientists, as illustrated in Table 1, active galactic nuclei (AGN) and gamma ray bursts (GRB) must be considered well-motivated sources of neutrinos simply because they are the sources of the most energetic photons. They may also be the accelerators of the highest energy cosmic rays. If they are, their neutrino flux can be calculated in a relatively model-independent way because the proton beams will photoproduce pions and, therefore, neutrinos on the high density of photons in the source. We have a beam dump configuration where both the beam and target are constrained by observation: by cosmic ray observations for the proton beam and by astronomical observations for the photon target. AGN and GRB have served as the most important "gedanken experiments" by which we set the scale of future neutrino telescopes. We will show that order 100 detected neutrinos are predicted per year in a high energy neutrino telescope with an effective area of 1 km^2. Their energies cluster in the vicinity of 100 TeV for GRB and several 100 PeV for neutrinos originating in AGN jets. For the latter, even larger fluxes of lower energy energy neutrinos may emanate from their accretion disks.

Neutrino telescopes can do particle physics. This is often illustrated by their capability to detect the annihilation into high energy neutrinos of neutralinos, the lightest supersymmetric particle which may constitute the dark matter. This will be discussed in some detail in the context of the AMANDA detector in Section 5. Also, with cosmological sources such as active galaxies and GRB we will be observing ν_e and ν_μ neutrinos over a baseline of 10^3 Megaparsecs. Above 1 PeV these are absorbed by charged-current interactions in the Earth before reaching the opposite surface. In contrast, the Earth never becomes opaque to ν_τ since the τ^- produced in a charged-current ν_τ interaction decays back into ν_τ before losing significant energy. This penetration of tau neutrinos through the Earth above 10^2 TeV provides an experimental signature for neutrino oscillations. The appearance of a ν_τ component in a pure $\nu_{e,\mu}$ would be evident as a flat angular dependence of a source intensity at the highest neutrino energies. Such a flat zenith angle dependence for the farthest sources would indicate tau neutrino mixing with a sensitivity to Δm^2 as low as $10^{-17}\,\mathrm{eV}^2$. With neutrino telescopes we will also search for ultrahigh-energy neutrino signatures from topological defects and magnetic monopoles; for properties of neutrinos such as mass, magnetic moment, and flavor-oscillations; and for clues to entirely new physical phenomena. The potential of neutrino "telescopes" to do particle physics is evident.

We start with a discussion of how Nature may accelerate subnuclear particles to macroscopic energies of more than 10 Joules. We discuss cosmic beam dumps next. We conclude with a status report on the deployment of the first-generation neutrino telescopes.

2 Cosmic Accelerators

2.1 The Machines

Cosmic rays form an integral part of our galaxy. Their energy density is qualitatively similar to that of photons, electrons and magnetic fields. It is believed that most were born in supernova blast waves. Their energy spectrum can be understood, up to perhaps 1000 TeV, in terms of acceleration by supernova shocks exploding into the interstellar medium of our galaxy. Although the slope of the cosmic ray spectrum abruptly increases at this energy, particles with energies in excess of $10^8\,\mathrm{TeV}$ have been observed and cannot be accounted for by this mechanism. The break in the spectrum, usually referred to as the "knee," can be best exhibited by plotting the flux multiplied by an energy dependent power $E^{2.75}$; see Fig. 2.

The failure of supernovae to accelerate cosmic rays above 1000 TeV energy can be essentially understood on the basis of dimensional analysis. It is sensible to assume that, in order to accelerate a proton to energy $E(=pc)$, the size R of the accelerator must be larger than the gyroradius of the particle in the accelerating field B:

$$R > R_{\mathrm{gyro}} = \frac{E}{B}, \tag{3}$$

in units $e = c = 1$. This yields a maximum energy

$$E < BR, \tag{4}$$

or

$$\left[\frac{E_{\mathrm{max}}}{10^5\,\mathrm{TeV}}\right] = \left[\frac{B}{3 \times 10^{-6}\mathrm{G}}\right]\left[\frac{R}{50\,\mathrm{pc}}\right]. \tag{5}$$

Therefore particles moving at the speed of light c reach energies up to a maximum value E_{max} which must be less than $10^5\,\mathrm{TeV}$ for the values of B and

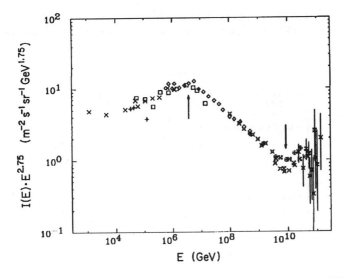

Figure 2: Flux of high energy cosmic rays after multiplication by a factor $E^{2.75}$. Arrows point at structure in the spectrum near 1 PeV, the "knee," and 10 EeV, the "ankle."

R characteristic for a supernova shock. Realistic modelling introduces inefficiencies in the acceleration process and yields a maximum energy which is typically two orders of magnitude smaller than the value obtained by dimensional analysis; see later. One therefore identifies the "knee" with the sharp cutoff associated with particles accelerated by supernovae.

Cosmic rays with energy in excess of 10^8 TeV have been observed, some five orders of magnitude in energy above the supernova cutoff. Where and how they are accelerated undoubtedly represents one of the most challenging problems in cosmic ray astrophysics and one of the oldest unresolved puzzles in astronomy. In order to beat dimensional analysis, one must accelerate particles over larger distances R, or identify higher magnetic fields B.

Although imaginative arguments actually do exist to avoid this conclusion, it is generally believed that our galaxy is too small and its magnetic field too weak to accelerate the highest energy cosmic rays. Furthermore, those with energy in excess of 10^7 TeV have gyroradii larger than our galaxy and should point back at their sources. Their arrival directions fail to show any correlation to the galactic plane, suggesting extra-galactic origin. Searching the sky beyond our galaxy, the central engines of active galactic nuclei and the enigmatic gamma ray bursts stand out as the most likely sites from which particles can be hurled at Earth with joules of energy. The idea is rather compelling because bright AGN and GRB are also the sources of the highest energy photons detected with satellite and air Cherenkov telescopes.

Table 2:

luminosity	B	R	γ	$E = \gamma BR$
• supernova blastwaves 10^{38} erg s^{-1}	10^{-3} G	10^2 pc	1	10^5 TeV knee, efficiency 10^{-2}
• quasar jets 10^{48} erg s^{-1}	10 G	10^{-2} pc $\simeq 1$ day	10	10^6 TeV $\big\}$ 100-PeV neutrinos
• gamma-ray bursts 1 m$_\odot$/msec	$>10^{10}$ G	10^2 km	10^3	10^8 TeV $\big\}$ 100-TeV neutrinos

The jets in AGN consists of beams of electrons and protons accelerated by tapping the rotational energy of the black hole. The black hole is the source of the phenomenal AGN luminosity, emitted in a multi-wavelength spectrum extending from radio up to gamma rays with energies in excess of 20 TeV. In a jet, BR-values in excess of 10^6 TeV can be reached with fields of tens of Gauss extending over sheets of shocked material in the jet of dimension 10^{-2} parsecs; see Table 2. The size of the accelerating region is deduced from the duration of the high energy emission which occurs in bursts lasting days, sometimes minutes; see Figs. 3, 4. Near the supermassive black hole Nature does not only construct a beam, the beam is a beam of smaller accelerating regions. AGN create accelerators more powerful than Fermilab about once a day! The γ-factor in Table 2 reminds us that in the most spectacular sources the jet is beamed in our direction, thus increasing the energy and reducing the duration of the emission in the observer's frame. As previously mentioned, accelerated protons interacting with the ambient light are assumed to be the source of secondary pions which produce the observed gamma rays and, inevitably, neutrinos.[6]

In this context, GRB and AGN jets are similar objects. GRB are somehow associated with neutron stars or solar mass black holes. Characteristic fields in

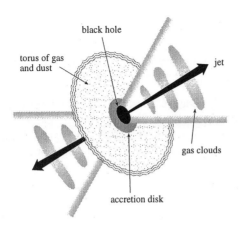

black hole

torus of gas and dust

jet

gas clouds

accretion disk

Figure 3: Active galaxy with accretion disk and a pair of jets. The galaxy is powered by a central super-massive black hole ($\sim 10^9 M_\odot$). Particles, accelerated in shocks in the disk or the jets, interact with the high density of ambient photons ($\sim 10^{14}/\mathrm{cm}^3$).

excess of 10^{10} Gauss are concentrated in a fireball of size 10^2 kilometers, which is opaque to light. The relativistic shock ($\gamma \simeq 10^3$) which dilutes the fireball to the point where the gamma ray display occurs, will also accelerate protons. These interact with the observed light to produce neutrinos;[ ] see Table 2.

2.2 The Blueprint: Shock Acceleration

Cosmic and Earth-based accelerators operate by different mechanisms. In space electric fields of freely moving particles short, and magnetic fields are generally disorganized. Particles gain energy in collisions. A particle of mass m and velocity v colliding head-on with a stellar cloud of mass M and velocity u, gains a kinetic energy:

$$\frac{\Delta E}{E}(u,v) \sim \frac{E_{\text{after}} - E_{\text{before}}}{\frac{1}{2}mv^2} \sim \frac{u}{v}\left(1 + \frac{u}{v}\right). \qquad (6)$$

In order to find the net gain in energy we have to average over all possible directions of the particle-cloud collision. In a one-dimensional (and, note, non-relativistic) world the particle encounters $(v+u)/v$ clouds, head-on, and collides with a smaller number $(v-u)/v$ coming from the opposite direction. The argument is familiar from the Doppler shift. The net gain in energy is given by:

$$\left(\frac{\Delta E}{E}\right)_{\text{net}} \sim \frac{\Delta E}{E}(u,v)\left(\frac{u+v}{v}\right) + \frac{\Delta E}{E}(-u,v)\left(\frac{v-u}{v}\right) \sim \left(\frac{u}{v}\right)^2 . \quad (7)$$

The gain is very small because only the quadratic term $(u/v)^2$ survives. In astrophysical situations u will be typically much smaller than the relativistic particle velocity v. It is clear that averaging over "good" and "bad" collisions is the origin of the cancellation of the linear contribution u/v to the gain in energy.

The key is to find an environment where particles only undergo "good" collisions: a shock, for instance the shock expanding into the interstellar medium produced by a supernova explosion. Acceleration in shocks is referred to as first-order Fermi acceleration, for obvious reasons. In astrophysical shocks the collisions are really magnetic interactions: with magnetic irregularities upstream and turbulence downstream. The particles riding the shockwave collide head-on with both! The highest energy particles will have crossed the shock many times. The details are not simple.[8] Fortunately Nature has provided us with many examples of shocks in action. For example, solar particles produced with MeV energy in nuclear collisions are observed with GeV energy as a result of shock acceleration at the surface.

Unlike the typical mono-energetic beam produced by a synchrotron, shocks produce power law spectra of high energy particles,

$$dN/dE \propto E^{-(\gamma+1)} . \quad (8)$$

The observed high energy cosmic ray spectrum at Earth is characterized by $\gamma \simeq 1.7$. A cosmic accelerator in which the dominant mechanism is first order diffusive shock acceleration, will produce a spectrum with $\gamma \sim 1 + \epsilon$, where ϵ is a small number. The observed cosmic ray spectrum is steeper than the accelerated spectrum because of the energy dependence of their diffusion in the galaxy: high energy ones more readily escape confinement from the magnetic bottle formed by the galaxy. In highly relativistic shocks ϵ can take a negative value.

First-order Fermi acceleration at supernova blast shocks offers a very attractive model for a galactic cosmic accelerator, providing the right power and spectral shape. Acceleration takes time, however, because the energy gain occurs gradually as a particle near the shock scatters back and forth across the front, gaining energy with each transit. The finite lifetime of the shock thus limits the maximum energy a particle can achieve at a particular supernova shock. The acceleration rate is[9]

$$\frac{\Delta E}{\Delta t} = K \frac{u^2}{c} ZeB < ZeBc, \tag{9}$$

where u is the shock velocity, Ze the charge of the particle being accelerated and B the ambient magnetic field. The numerical constant $K \sim 0.1$ is an efficiency factor which depends on the details of diffusion in the vicinity of the shock such as the efficiency by which power in the shock is converted into the actual acceleration of particles. The maximum energy reached is

$$E = \frac{K}{c}(ZeB\,Ru) < ZeB\,R. \tag{10}$$

The crucial time scale used to convert Eq. (9) into this limiting energy is $\Delta t \sim R/u$, where $\Delta t \sim 1000\,\mathrm{yrs}$ for the free expansion phase of a supernova and R is the dimension of the blastwave. This result agrees with Eq. (4). Using Eq. (10) we ascertain that $E_{\max}$ can only reach energies of $\lesssim 10^3\,\mathrm{TeV} \times Z$ for a galactic field $B \sim 3\mu\mathrm{Gauss}$, $K \sim 0.1$ and $u/c \sim 0.1$. Even ignoring all pre-factors the energy can never exceed $10^5\,\mathrm{TeV}$ by dimensional analysis. Cosmic rays with energy in excess of $10^8\,\mathrm{TeV}$ have been observed and the acceleration mechanism leaves a large gap of some five orders of magnitude that cannot be explained by the "standard model" of cosmic ray origin. To reach a higher energy one has to dramatically increase B and/or R. This argument is difficult to beat — it is basically dimensional. Even the details do not matter; elementary electromagnetism is sufficient to identify the EMF of the accelerator or even the Lorentz force in the form of Eq. (10).

First-order Fermi acceleration is also believed to be the origin of the very high energy particles produced near the supermassive black holes in active galaxies and in the explosive release of a solar mass in gamma ray bursts.

3 Neutrinos from Cosmic Beam Dumps

Cosmic neutrinos, just like accelerator neutrinos, are made in beam dumps. A beam of accelerated protons is dumped into a target where they produce pions in collisions with nuclei. Neutral pions decay into gamma rays and charged pions into muons and neutrinos. All this is standard particle physics and, in the end, roughly equal numbers of secondary gamma rays and neutrinos emerge from the dump. In man-made beam dumps the photons are absorbed in the dense target; this may not be the case in an astrophysical system where the target material can be more tenuous. Also, the target may be photons rather than nuclei. For instance, with an ambient photon density a million times larger than the sun, approximately 10^{14} per cm^3, particles accelerated in AGN jets may meet more photons than nuclei when losing energy. Examples of cosmic beam dumps are tabulated in Table 3. They fall into two categories. Neutrinos produced by the cosmic ray beam are, of course, guaranteed and calculable.[9] We know the properties of the beam and the various targets: the atmosphere, the hydrogen in the galactic plane and the CMBR background. Neutrinos from AGN and GRB are not guaranteed, though both represent good candidate sites for the acceleration of the highest energy cosmic rays. That they are also the sources of the highest energy photons reinforces this association.

Table 3: Cosmic Beam Dumps

Beam	Target
cosmic rays	atmosphere
cosmic rays	galactic disk
cosmic rays	CMBR
AGN jets	ambient light, UV
shocked protons	GRB photons

Our main point will be that the rate of neutrinos produced by $p\gamma$ interactions in gamma ray bursts and active galactic nuclei is essentially dictated by the observed energetics of the source. In astrophysical beam dumps, like AGN and GRB, typically one neutrino and one photon is produced per accelerated proton.[9] The accelerated protons and photons are, however, more likely to suffer attenuation in the source before they can escape. So, a hierarchy of particle fluxes emerges with protons < photons < neutrinos. Using these associations, one can constrain the energy and luminosity of the accelerator from the gamma-ray and cosmic ray observations, and subsequently anticipate the neutrino fluxes. These calculations represent the basis for the construction of kilometer-scale detectors as the goal of neutrino astronomy.

Below follow 3 estimates of the luminosity of the Universe in high energy radiation.

1. From the observed injection rate $\dot{F} = 4 \times 10^{44}\,\mathrm{erg\,Mpc^{-3}\,yr^{-1}}$ of GRB and the assumption of equal injection of kinetic energy into electrons (ultimately observed as photons by synchrotron radiation and, possibly, inverse Compton scattering) and protons by the initial fireball, we calculate a proton flux

$$E_p \Phi_p = \frac{c}{4\pi}(t_H \dot{E}) = 2.2 \times 10^{-10}\,\mathrm{TeV\,(cm^2\,s\,sr)^{-1}}\,. \tag{11}$$

Here we assumed injection over a Hubble time t_H of 10^{10} years.

2. From the observed spectrum of the, presumably, extra-galactic cosmic rays beyond the ankle, just above 10^6 TeV energy, in the spectrum:

$$E_{CR}\Phi_{CR} = \int_{10^6\ \text{TeV}} dE \left[E\frac{dN_{CR}}{dE} \right] \cong 1.7 \times 10^{-10}\,\text{TeV}\,(\text{cm}^2\,\text{s\,sr})^{-1}\,.$$

(12)

We fitted the spectrum as $E^{-2.7}$ beyond the ankle to obtain this result. The near equality of the GRB and the cosmic ray flux beyond the ankle supports the speculation that GRB are the source of extra-galactic cosmic rays.

3. A final estimate is based on the luminosity of TeV γ-rays emitted by blazars. Taking the Markarian 421 flux observed by the Whipple collaboration, and the EGRET luminosity function of 130 sources per steradian:

$$(130\,\text{sr}^{-1})(5 \times 10^{-10}\,\text{cm}^{-2}\,\text{s}^{-1}\,\text{TeV}) = 6 \times 10^{-8}\,\text{TeV}\,(\text{cm}^2\,\text{s\,sr})^{-1}\,. \quad (13)$$

This is somewhat less than the observed diffuse γ-ray flux and over an order of magnitude larger than the proton flux, consistent with the expected hierarchy of photons and protons in a beam dump. This result raises the alternative possibility that AGN are the sources of the highest energy cosmic rays.

From this compilation, it is not unreasonable to assign a luminosity of 2×10^{-10} TeV $(\text{cm}^2\,\text{s\,sr})^{-1}$ to the source of the highest energy cosmic rays. This corresponds to an injection rate of several times $\dot{E} = 10^{44}$ erg Mpc^{-3} yr^{-1} and a density of extragalactic cosmic rays of roughly 10^{19} erg cm^{-3}. This is very much in line with estimates made elsewhere.[10] It is important to keep in mind that estimates of neutrino fluxes using this input represent a **lower limit** because of the absorption of the cosmic rays and TeV gamma rays in the interstellar medium and, possibly, in the source itself.

3.1 ν's from AGN

AGN are the brightest sources in the Universe; some are so far away that they are messengers from the earliest of times. Their engines must not only be powerful, but also extremely compact because their luminosities are observed to flare by over an order of magnitude over time periods as short as a day. Only sites in the vicinity of black holes which are a billion times more massive than our sun, will do. It is anticipated that beams accelerated near the black hole are dumped on the ambient matter in the active galaxy, mostly thermal photons with densities of $10^{14}/cm^3$. The electromagnetic spectrum at all wavelengths, from radio waves to TeV gamma rays, is produced in the interactions of the accelerated particles with the magnetic fields and ambient photons in the galaxy. In most models the highest energy photons are produced by Compton scattering of accelerated electrons on thermal UV photons which are scattered up from 10 eV to TeV energy.[11] The energetic gamma rays will subsequently lose energy by electron pair production in photon-photon interactions in the radiation field of the jet or the galactic disk. An electromagnetic cascade is thus initiated which, via pair production on the magnetic field and photon-photon interactions, determines the emerging gamma-ray spectrum at lower energies. The lower energy photons, observed by conventional astronomical techniques, are, as a result of the cascade process, several generations removed from the primary high energy beams.

High energy gamma-ray emission (MeV–GeV) has been observed from at least 40 active galaxies by the EGRET instrument on the Compton Gamma Ray Observatory.[12] Most, if not all, are "blazars". They are AGN viewed from a position illuminated by the cone of a relativistic jet. Of the four TeV gamma-ray emitters conclusively identified by the air Cherenkov technique, two are extra-galactic and are also nearby blazars. The data therefore strongly suggests that the highest energy photons originate in jets beamed at the observer. The cartoon of an AGN, shown in Fig. 4, displays its most prominent features: an accretion disk of stars and gas falling into the spinning black hole as well as a pair of jets aligned with the rotation axis. Several of the sources observed by EGRET have shown variability, by a factor of 2 or so over a time scale of several days. Time variability is more spectacular at higher energies. On May 7, 1996 the Whipple telescope observed an increase of the TeV-emission from the blazar Markarian 421 by a factor 2 in 1 hour reaching eventually a value 50 times larger than the steady flux. At this point the telescope registered 6 times more photons from the Markarian blazar than from the Crab supernova remnant despite its larger distance (a factor 10^5). Recently, even more spectacular bursts have been detected from Markarian 501.[3]

Pion photoproduction may play a central role in blazar jets. If protons are accelerated along with electrons to PeV–EeV energy, they will produce high energy photons by photoproduction of neutral pions on the ubiquitous UV thermal background. Some have suggested that the accelerated protons initiate a cascade which also dictates the features of the spectrum at lower energy. From a theorist's point of view the proton blazar has attractive features. Protons, unlike electrons, efficiently transfer energy in the presence of the magnetic field in the jet. They provide a "natural" mechanism for the en-

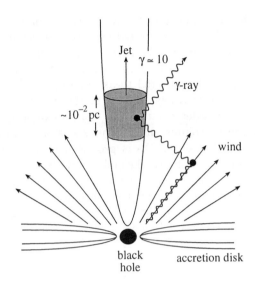

Figure 4: Possible blueprint for the production of high energy photons and neutrinos near the super-massive black hole powering an AGN. Particles, electrons and protons(?), accelerated in sheets or blobs moving along the jet, interact with photons radiated by the accretion disk or produced by the interaction of the accelerated particles with the magnetic field of the jet.

ergy transfer from the central engine over distances as large as 1 parsec as well as for the observed heating of the dusty disk over distances of several hundred parsecs.

Although the relative merits of the electron and proton blazar are hotly debated, it is more relevant that the issues can be settled experimentally. The proton blazar is a source of high energy protons and neutrinos, not just gamma rays. Also, its high energy photon spectrum may exceed the TeV cutoff which is an unavoidable feature of the electron blazar. The opportunities for high energy neutrino astronomy are wonderfully obvious.[9]

Weakly interacting neutrinos can, unlike high energy gamma-rays and high energy cosmic rays, reach us from more distant and much more powerful AGN. Neutrino astronomers anticipate that the high energy neutrino sky will glow uniformly with bright active galaxies outshining our Milky Way. The results may be even more spectacular. As is the case in man-made beam dumps, photons from celestial accelerators may be absorbed in the dump. The most spectacular sources may therefore have no counterpart in high energy photons.

Confronted with the challenge to explain a relatively flat multi-wavelength photon emission spectrum reaching TeV energies which is radiated in bursts of a duration of less than one day, models have converged on the blazar blueprint shown in Fig. 4. Particles are accelerated by Fermi shocks in blobs of matter travelling along the jet with a bulk Lorentz factor of $\gamma \sim 10$ and higher. This factor combines the effects of special relativity and the Doppler effect of the moving source; it is also referred to as the Doppler factor. In order to accommodate bursts lasting a day, or less, in the observer's frame, the size of the blob must be of order $\gamma c\Delta t \sim 10^{-2}$ parsecs or less. The blobs are actually more like sheets, thinner than the jet's size of roughly 1 parsec. The observed radiation at all wavelengths is produced by the interaction of the accelerated particles in the blob and the ambient radiation in the AGN, which has a significant component concentrated in the so-called "UV-bump."

In the following estimate of the neutrino flux from a proton blazar, primes will refer to a frame attached to the blob, which is moving with a Lorentz factor γ relative to the observer. In general, the transformation between blob and observer frame is $R' = \gamma R$ and $E' = \frac{1}{\gamma}E$ for distances and energies, respectively. For a burst of 15 min duration, the strongest variability observed in TeV emission, the size of the accelerator is only

$$R' = \gamma c\Delta t \sim 10^{-4} \text{ to } 10^{-3} \text{ pc} \tag{14}$$

for $\gamma = 10\text{--}10^2$. So, the jet consists of relatively small structures with short lifetime. High energy emission is associated with the periodic formation of these blobs.

Shocked protons in the blob will photoproduce pions on the photons whose properties are known from the observed multi-wavelength emission. From the observed photon luminosity L_γ we deduce the energy density of photons in the shocked region:

$$U'_\gamma = \frac{L'_\gamma \Delta t}{\frac{4}{3}\pi R'^3} = \frac{L_\gamma \Delta t}{\gamma} \frac{1}{\frac{4}{3}\pi(\gamma c\Delta t)^3} = \frac{3}{4\pi c^3} \frac{L_\gamma}{\gamma^4 \Delta t^2} . \tag{15}$$

(Geometrical factors of order unity will be ignored throughout.) The dominant photon density is at UV wavelengths, the UV bump. We will assume that a luminosity $\mathcal{L}_\gamma$ of 10^{45} erg s^{-1} is emitted in photons with energy $E_\gamma = 10$ eV. Luminosities larger by one order of magnitude have actually been observed. The number density of photons in the shocked region is

$$N'_\gamma = \frac{U'_\gamma}{E'_\gamma} = \gamma\frac{U'_\gamma}{E_\gamma} = \frac{3}{4\pi c^3} \frac{L_\gamma}{E_\gamma} \frac{1}{\gamma^3 \Delta t^2} \sim 6.8 \times 10^{14} \text{ to } 6.8 \times 10^{11} \text{ cm}^{-3} . \tag{16}$$

From now on the range of numerical values will refer to $\gamma = 10\text{--}10^2$, in that order. With such high density the blob is totally opaque to photons with

10 TeV energy and above. Because photons with such energies have indeed been observed, one must essentially require that 10 TeV γ's are below the $\gamma\gamma \to e^+ e^-$ threshold in the blob, i.e.,

$$E_{\text{th}} = \gamma E'_{\gamma\,\text{th}} \gtrsim 10 \text{ TeV}, \tag{17}$$

or

$$E_{\text{th}} > \frac{m_e^2}{E_\gamma}\gamma^2 > 10 \text{ TeV}, \tag{18}$$

or

$$\gamma \gtrsim 10. \tag{19}$$

Protheroe et al. find $\gamma \gtrsim 30.$[3] So, we take $10 < \gamma < 10^2$.

The accelerated protons in the blob will produce pions, predominantly at the Δ-resonance, in interactions with the UV photons. The proton energy for resonant pion production is

$$E'_p = \frac{m_\Delta^2 - m_p^2}{4} \frac{1}{E'_\gamma} \tag{20}$$

or

$$E_p = \frac{m_\Delta^2 - m_p^2}{4E_\gamma} \gamma^2 \tag{21}$$

$$E_p = \frac{1.6 \times 10^{17} \text{ eV}}{E_\gamma} \gamma^2 \tag{22}$$

$$= 1.6 \times 10^{18} \text{ to } 1.6 \times 10^{20} \text{ eV}. \tag{23}$$

The jet (hopefully) accelerates protons to this energy, and will definitely do so in models where blazars are the sources of the highest energy cosmic rays.

The secondary ν_μ have energy

$$E_\nu = \frac{1}{4} \langle x_{p\to\pi} \rangle E_p = 7.9 \times 10^{16} \text{ to } 7.9 \times 10^{18} \text{ eV} \tag{24}$$

for $\langle x_{p\to\pi} \rangle \simeq 0.2$, the fraction of energy transferred, on average, from the proton to the secondary pion produced via the Δ-resonance. The $1/4$ is because each lepton in the decay $\pi \to \mu\nu_\mu \to e\nu_e\nu_\mu\bar{\nu}_\mu$ carries roughly equal energy.

The fraction of energy f_π lost by protons to pion production when travelling a distance R' through a photon field of density N'_γ is

$$f_\pi = \frac{R'}{\lambda_{p\gamma}} = R' N'_\gamma \sigma_{p\gamma\to\Delta} \langle x_{p\to\pi} \rangle \tag{25}$$

where $\lambda_{p\gamma}$ is the proton interaction length, with $\sigma_{p\gamma \to \Delta \to n\pi^+} \simeq 10^{-28}$ cm^2. We obtain

$$f_\pi = 3.8\text{–}0.038 \text{ for } \gamma = 10\text{–}10^2 \,. \tag{26}$$

For a total injection rate in high-energy protons $\dot{E}$, the total energy in ν's is $\frac{1}{2}f_\pi t_H \dot{E}$, where $t_H = 10$ Gyr is the Hubble time. The factor $1/2$ accounts for the fact that $1/2$ of the energy in charged pions is transferred to $\nu_\mu + \bar{\nu}_\mu$, see above. The neutrino flux is

$$\Phi_\nu = \frac{c}{4\pi} \frac{\left(\frac{1}{2}f_\pi t_H \dot{E}\right)}{E_\nu} e^{f_\pi} \,. \tag{27}$$

The last factor corrects for the absorption of the protons in the source, i.e., the observed proton flux is a fraction e^{-f_π} of the source flux which photoproduces pions. We can write this as

$$\Phi_\nu = \frac{1}{E_\nu} \frac{1}{2} f_\pi e^{f_\pi} (E_p \Phi_p) \,, \tag{28}$$

For $E_p \Phi_p = 2 \times 10^{-10}$ TeV (cm^2 s sr)$^{-1}$ we obtain

$$\Phi_\nu = 8 \times 10^5 \text{ to } 2 \text{ (km}^2 \text{ yr)}^{-1} \tag{29}$$

over 4π steradians. (Neutrino telescopes are background free for such high energy events and should be able to identify neutrinos at all zenith angles.)

A detailed discussion of how to build high energy neutrino telescopes will be presented further on. For calculational purposes it is sufficient to know that, in order to be detected, i) a ν_μ neutrino must interact in the water or ice near the detector, and ii) the secondary muon must have a sufficient range to reach the detector. The detection probability is thus easily computed from the requirement that the neutrino has to interact within a distance of the detector which is shorter than the range of the muon it produces. Therefore,

$$P_{\nu \to \mu} \simeq \frac{R_\mu}{\lambda_{\text{int}}} \simeq A E_\nu^n \,, \tag{30}$$

where R_μ is the muon range and λ_{int} the neutrino interaction length. For energies below 1 TeV, where both the range and cross section depend linearly on energy, $n = 2$. At TeV and PeV energies $n = 0.8$ and $A = 10^{-6}$, with E in TeV units. For EeV energies $n = 0.47$, $A = 10^{-2}$ with E in EeV.[9]

The observed neutrino event rate in a detector is

$$N_{\text{events}} = \Phi_\nu P_{\nu \to \mu}, \tag{31}$$

with

$$P_{\nu \to \mu} \cong 10^{-2} E_{\nu,\text{EeV}}^{0.4} , \qquad (32)$$

where E_ν is expressed in EeV. Therefore

$$N_{\text{events}} = (3 \times 10^3 \text{ to } 5 \times 10^{-2}) \text{ km}^{-2} \text{ yr}^{-1} = 10^{1 \pm 2} \text{ km}^{-2} \text{ yr}^{-1} \qquad (33)$$

for $\gamma = 10\text{--}10^2$. This estimate brackets the range of γ factors considered. Remember however that the relevant luminosities for protons (scaled to the high energy cosmic rays) and the luminosity of the UV target photons are themselves uncertain.

In summary, for the intermediate value for γ:

$$E_\nu = 7.8 \times 10^{16} \text{ eV} \left(\frac{\gamma}{30} \right) \left(\frac{\Delta t}{15 \text{ min}} \right) , \qquad (34)$$

$$f_\pi = 0.4 \left(\frac{30}{\gamma} \right)^2 \left(\frac{15 \text{ min}}{\Delta t} \right) \left(\frac{\mathcal{L}_\gamma}{10^{45} \text{ erg s}^{-1}} \right) , \qquad (35)$$

$$E_\nu = 7 \times 10^5 \left(\frac{\gamma}{30} \right)^2 \text{ TeV} , \qquad (36)$$

$$N_{\text{events}} = \left(3 \text{ km}^{-2} \text{ yr}^{-1} \right) \left(\frac{f_\pi}{0.4} e^{(f_\pi - 0.4)} \right)$$
$$\times \left(\frac{E_{\text{CR}} \Phi_{\text{CR}}}{2 \times 10^{-10} \text{ TeV cm}^{-2} \text{ s sr}^{-1}} \right) \left(\frac{750 \text{ PeV}}{E_\nu} \right)^{0.6} . \qquad (37)$$

Because of further absorption effects on the input proton flux, e.g. in the CMBR, this result should be interpreted as a lower limit.

3.1.1 The Role of Absorption: Hidden Sources

The large uncertainty in the calculation of the neutrino flux from AGN is predominantly associated with the boost factor γ. The reason for this is clear. The target density of photons in the accelerator is determined by i) the photon luminosity, which is directly observed, and ii) the size of the target which is limited by the short duration of the high energy blazar signals. Large boost factors reduce the photon density of the target because they reduce energy and expand the target size in the accelerator frame. A large γ-factor will render a source transparent to high energy photons and protons despite the high luminosity of photons and the short duration of the burst. Assuming $\gamma = 1$, we would conclude instead that the source is completely opaque to high energy photons and protons. It is a "hidden" source, with reduced or extinct emission of high energy particles, but abundant neutrino production by protons on the high density photon target. The TeV Markarian sources require very large γ-factors. In their absence the sources would be opaque at TeV energy; see Eq. (18). Their neutrino emission is expected to be very low, in the lower range of our prediction. On the other hand, nature presumably made blazars with a distribution of boost factors; observed boost factors are typically less than 10. While uninteresting for high energy gamma ray astronomy, they have the potential to be powerful neutrino emitters, with fluxes near the upper range of our predictions.

3.2 ν's from GRB

Recently, GRB may have become the best motivated source for high energy neutrinos. Their neutrino flux can be calculated in a relatively model-independent way. Although neutrino emission may be less copious and less energetic than from AGN, the predicted fluxes can probably be bracketed with more confidence.

In GRB a fraction of a solar mass of energy ($\sim 10^{53}$ ergs) is released over a time scale of order 1 second into photons with a very hard spectrum. It has been suggested that, although their ultimate origin is a matter of speculation, the same cataclysmic events also produce the highest energy cosmic rays. This association is reinforced by more than the phenomenal energy and luminosity:

- both GRB and the highest energy cosmic rays are produced in cosmological sources, *i.e.*, and, as previously discussed,

- the average rate at which energy is injected into the Universe as gamma rays from GRB is similar to the rate at which energy must be injected in the highest energy cosmic rays in order to produce the observed cosmic ray flux beyond the "ankle" in the spectrum at 10^7 TeV.

There is increasing observational support for a model where an initial event involving neutron stars or black holes deposits a solar mass of energy into a radius of order 100 km. Such a state is opaque to light. The observed gamma ray display is the result of a relativistic shock with $\gamma = 10^2$–10^3 which expands the original fireball by a factor 10^6 over 1 second. Gamma rays are produced by synchrotron radiation by relativistic electrons accelerated in the shock, possibly followed by inverse-Compton scattering. The association of cosmic rays with GRB obviously requires that kinetic energy in the shock is converted into the acceleration of protons as well as electrons. It is assumed that the efficiency with which kinetic energy is converted to accelerated protons is comparable to that for electrons. The production of high-energy neutrinos is a feature of the fireball model because the protons will photoproduce pions and, therefore, neutrinos on the gamma rays in the burst. We have a beam dump configuration where both the beam and target are constrained by observation: of the cosmic ray beam and of the photon fluxes at Earth, respectively.

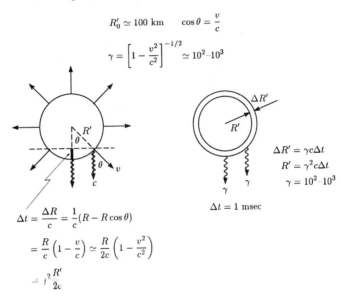

$$R'_0 \simeq 100 \text{ km} \qquad \cos\theta = \frac{v}{c}$$

$$\gamma = \left[1 - \frac{v^2}{c^2}\right]^{-1/2} \simeq 10^2\text{--}10^3$$

$$\Delta t = \frac{\Delta R}{c} = \frac{1}{c}(R - R\cos\theta)$$

$$= \frac{R}{c}\left(1 - \frac{v}{c}\right) \simeq \frac{R}{2c}\left(1 - \frac{v^2}{c^2}\right)$$

$$\simeq \gamma^2 \frac{R'}{2c}$$

$$\Delta R' = \gamma c \Delta t$$
$$R' = \gamma^2 c \Delta t$$
$$\gamma = 10^2\text{--}10^3$$

$$\Delta t = 1 \text{ msec}$$

Figure 5: Kinematics of GRB.

Simple relativistic kinematics (see Fig. 5) relates the radius and width $R', \Delta R'$ to the observed duration of the photon burst $c\Delta t$

$$R' = \gamma^2(c\Delta t) \tag{38}$$
$$\Delta R' = \gamma c\Delta t \tag{39}$$

The calculation of the neutrino flux follows the same path as that for AGN. From the observed GRB luminosity L_γ we compute the photon density in the shell:

$$U'_\gamma = \frac{(L_\gamma \Delta t/\gamma)}{4\pi R'^2 \Delta R'} = \frac{L_\gamma}{4\pi R'^2 c\gamma^2} \tag{40}$$

The pion production by shocked protons in this photon field is, as before, calculated from the interaction length

$$\frac{1}{\lambda_{p\gamma}} = N_\gamma \sigma_\Delta \langle x_{p\to\pi} \rangle = \frac{U'_\gamma}{E'_\gamma} \sigma_\Delta \langle x_{p\to\pi} \rangle \qquad \left(E'_\gamma = \frac{1}{\gamma} E_\gamma\right) \tag{41}$$

As before, σ_Δ is the cross section for $p\gamma \to \Delta \to n\pi^+$ and $\langle x_{p\to\pi}\rangle \simeq 0.2$. The fraction of energy going into π-production is

$$f_\pi \cong \frac{\Delta R'}{\lambda_{p\gamma}} \tag{42}$$

$$f_\pi \simeq \frac{L_\gamma}{E_\gamma}\frac{1}{\gamma^4\Delta t}\frac{\sigma_\Delta\langle x_{p\to\pi}\rangle}{4\pi c^2} \tag{43}$$

$$f_\pi \simeq 0.14\left\{\frac{L_\gamma}{10^{51}\,\text{ergs}^{-1}}\right\}\left\{\frac{1\,\text{MeV}}{E_\gamma}\right\}\left\{\frac{300}{\gamma}\right\}^4\left\{\frac{1\,\text{msec}}{\Delta t}\right\}$$
$$\times\left\{\frac{\sigma_\Delta}{10^{-28}\,\text{cm}^2}\right\}\left\{\frac{\langle x_{p\to\pi}\rangle}{0.2}\right\} \tag{44}$$

The relevant photon energy in the problem is $1\,\text{MeV}$, the energy where the typical GRB spectrum exhibits a break. The number of higher energy photons is suppressed by the spectrum, and lower energy photons are less efficient at producing pions. Given the large uncertainties associated with the astrophysics, it is an adequate approximation to neglect the explicit integration over the GRB photon spectrum. The proton energy for production of pions via the Δ-resonance is

$$E'_p = \frac{m_\Delta^2 - m_p^2}{4E'_\gamma} \tag{45}$$

Therefore,

$$E_p = 1.4\times 10^{16}\,\text{eV}\left(\frac{\gamma}{300}\right)^2\left(\frac{1\,\text{MeV}}{\text{E}_\gamma}\right) \tag{46}$$

$$E_\nu = \frac{1}{4}\langle x_{p\to\pi}\rangle E_p \simeq 7\times 10^{14}\,\text{eV} \tag{47}$$

We are now ready to calculate the neutrino flux:

$$\phi_\nu = \frac{c}{4\pi}\frac{U'_\nu}{E'_\nu} = \frac{c}{4\pi}\frac{U_\nu}{E_\nu} = \frac{c}{4\pi}\frac{1}{E_\nu}\left\{\frac{1}{2}f_\pi t_H\dot{E}\right\} \tag{48}$$

where, as before, the factor $1/2$ accounts for the fact that only $1/2$ of the energy in charged pions is transferred to $\nu_\mu + \bar{\nu}_\mu$. As before, $\dot{E}$ is the injection rate in cosmic rays beyond the ankle ($\sim 4\times 10^{44}\,\text{erg}\,\text{Mpc}^{-3}\,\text{yr}^{-1}$) and t_H is the Hubble time of $\sim 10^{10}\,\text{Gyr}$. Numerically,

$$\phi_\nu = 2\times 10^{-14}\,\text{cm}^{-2}\,\text{s}^{-1}\,\text{sr}^{-1}\left\{\frac{7\times 10^{14}\,\text{eV}}{\text{E}_\nu}\right\}\left\{\frac{f_\pi}{0.125}\right\}\left\{\frac{t_H}{10\,\text{Gyr}}\right\}$$
$$\times\left\{\frac{\dot{E}}{4\times 10^{44}\,\text{erg}\,\text{Mpc}^{-3}\,\text{yr}^{-1}}\right\} \tag{49}$$

The observed muon rate is

$$N_{\text{events}} = \int_{E_{th}}^{E_\nu^{\text{max}}} \Phi_\nu P_{\nu \to \mu} \frac{dE_\nu}{E_\nu}, \tag{50}$$

$$\tag{51}$$

where $P_{\nu \to \mu} \simeq 1.7 \times 10^{-6} E_\nu^{0.8}$ (TeV) for TeV energy. Therefore

$$N_{\text{events}} \cong 26 \text{ km}^{-2} \text{ yr}^{-1} \left\{ \frac{E_\nu}{7 \times 10^{14} \text{ eV}} \right\}^{-0.2} \left\{ \frac{\Delta\theta}{4\pi} \right\} \tag{52}$$

The result is insensitive to beaming. Beaming yields more energy per burst, but less bursts are actually observed. The predicted rate is also insensitive to the neutrino energy E_ν because higher average energy yields less ν's, but more are detected. Both effects are approximately linear.

There is also the possibility that high-energy gamma rays and neutrinos are produced when the shock expands further into the interstellar medium. This mechanism has been invoked as the origin of the delayed high energy gamma rays. The fluxes are produced over seconds, possibly longer. It is easy to adapt the previous calculation to the external shock. Following, for instance, Bottcher et al., the time scale is changed from milliseconds to seconds and the break in the spectrum from 1 to 0.1 MeV, we find that f_π is reduced by two orders of magnitude. In the external shocks higher energies can be reached (a factor 10 higher the for Bottcher et al. model) and this increases the neutrino detection efficiency. In the end, the observed rates are an order of magnitude smaller, but the inherent ambiguities of the estimates are such that it is difficult to establish with confidence the relative rate in internal and external shocks. Again, the result boosts the argument for neutrino telescopes of kilometer scale.

4 Large Natural Cherenkov Detectors

Neutrino telescopes are conventional particle detectors which use natural and clear water and ice as a Cherenkov medium. A three dimensional grid of photomultiplier tubes maps the Cherenkov cone radiated by a muon of neutrino origin. Nanosecond timing provides degree resolution of the muon track which is, at least for high energy neutrinos, aligned with the neutrino direction. The detectors are shielded from the flux of cosmic ray muons by a kilometer, or more, of water and ice. Yet, identifying neutrinos in this down-going muon background is impossible. Cosmic ray muons exceed those of neutrino origin by a factor 10^5, or more, depending on the depth of the instrument. Only up-going muons made by neutrinos reaching us through the Earth can be successfully detected. The Earth is used as a filter to screen cosmic ray muons which makes neutrino detection possible over the lower hemisphere of the detector.

The probability to detect a TeV neutrino is roughly 10^{-6}. As previously discussed, this is easily computed from the requirement that, in order to be detected, the neutrino has to interact within a distance of the detector which is shorter than the range of the muon it produces; see Eq. (30). At PeV energy the cosmic ray flux is of order 1 per m^{-2} per year and the probability to detect a neutrino of this energy is of order 10^{-3}. A neutrino flux equal to the cosmic ray flux will therefore yield only a few events per day in a kilometer squared detector. At EeV energy the situation is worse. With a rate of 1 per km^2 per year and a detection probability of 0.1, one can still detect several events per year in a kilometer squared detector provided the neutrino flux exceeds the proton flux by 2 orders of magnitude or more. For the neutrino flux generated by cosmic rays interacting with CMBR photons and such sources as AGN and topological defects,[13] this is indeed the case. All above estimates are conservative and the rates should be higher because the neutrinos escape the source with a flatter energy spectrum than the protons. In summary, at least where cosmic rays are part of the beam dump, their ray flux and the neutrino cross section and muon range define the size of a neutrino telescope. Needless to say that a telescope with kilometer squared effective area represents a neutrino detector of kilometer cubed volume.

4.1 Baikal and the Mediterranean

First generation neutrino telescopes, launched by the bold decision of the DU-MAND collaboration over 25 years ago to construct such an instrument, are designed to reach a relatively large telescope area and detection volume for a neutrino threshold of 1–100 GeV. This relatively low threshold permits calibration of the novel instrument on the known flux of atmospheric neutrinos. Its architecture is optimized for reconstructing the Cherenkov light front radiated by an up-going, neutrino-induced muon. Up-going muons are to be identified in a background of down-going, cosmic ray muons which are more than 10^5 times more frequent for a depth of 1~2 kilometers. The method is sketched in Fig. 6.

The "landscape" of neutrino astronomy is sketched in Table 4. With the termination of the pioneering DUMAND experiment, the efforts in water are, at present, spearheaded by the Baikal experiment.[14] Operating with 144 optical modules (OM) since April 1997, the *NT-200* detector has been completed in April 1998. The Baikal detector is well understood and the first atmospheric

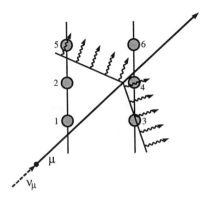

Figure 6: The arrival times of the Cherenkov photons in 6 optical modules determine the direction of the muon track.

neutrinos have been identified; we will discuss this in more detail further on. The Baikal site is competitive with deep oceans although the smaller absorption length of Cherenkov light in lake water requires a somewhat denser spacing of the OMs. This does however result in a lower threshold which is a definite advantage, for instance in WIMP searches. They have shown that their shallow depth of 1 kilometer does not represent a serious drawback. By far the most significant advantage is the site with a seasonal ice cover which allows reliable and inexpensive deployment and repair of detector elements.

In the following years, *NT-200* will be operated as a neutrino telescope with an effective area between $10^3 \sim 5 \times 10^3 \, \text{m}^2$, depending on the energy. Presumably too small to detect neutrinos from AGN and other extraterrestrial sources, *NT-200* will serve as the prototype for a larger telescope. For instance, with 2000 OMs, a threshold of $10 \sim 20 \, \text{GeV}$ and an effective area of $5 \times 10^4 \sim 10^5 \, \text{m}^2$, an expanded Baikal telescope would fill the gap between present underground detectors and planned high threshold detectors of cubic kilometer size. Its key advantage would be low threshold.

The Baikal experiment represents a proof of concept for deep ocean projects. These should have the advantage of larger depth and optically superior water. Their challenge is to design a reliable and affordable technology. Several groups are confronting the problem, both NESTOR and Antares are developing rather different detector concepts in the Mediterranean.

The NESTOR collaboration,[15] as part of an ongoing series of technology tests, are testing the umbrella structure which will hold the OMs. They

Table 4:

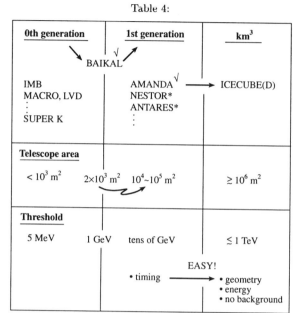

√ taking data
* R & D

deployed two aluminium "floors", 34 m in diameter, to a depth of 2600 m. Mechanical robustness was demonstrated by towing the structure, submerged below 2000 m, from shore to the site and back. The detector will consist of 8 six-legged floors separated by 30 m.

The Antares collaboration[16] is in the process of determining the critical detector parameters at a 2000 m deep, Mediterranean site off Toulon, France. First results on water quality are very encouraging. They have recently demonstrated their capability of deploying and retrieving a string. A deliberate development effort will lead to the construction of a demonstration project consisting of 3 strings with a total of 200 OMs.

For neutrino astronomy to become a viable science several of these, or other, projects will have to succeed. Astronomy, whether in the optical or in any other wave-band, thrives on a diversity of complementary instruments, not on "a single best instrument." When the Soviet government tried out the latter method by creating a national large mirror project, it virtually annihilated the field.

4.2 First Neutrinos from Baikal

The Baikal Neutrino Telescope is deployed in Lake Baikal, Siberia, 3.6 km from shore at a depth of 1.1 km. An umbrella-like frame holds 8 strings, each instrumented with 24 pairs of 37-cm diameter $QUASAR$ photomultiplier tubes (PMT). Two PMTs in a pair are switched in coincidence in order to suppress background from bioluminescence and PMT noise.

They have analysed 212 days of data taken in 94-95 with 36 OMs. Upward-going muon candidates were selected from about 10^8 events in which more than 3 pairs of PMTs triggered. After quality cuts and χ^2 fitting of the tracks a sample of 17 up-going events remained. These are not generated by neutrinos passing the Earth below the detector, but by showers from down-going muons originating below the array. In a small detector such events are expected. In 2 events however the light intensity does not decrease from bottom to top, as expected from invisible showering muons below the detector. A detailed analysis[17] yields a fake probability of 2% for both events.

After the deployment of 96 OMs in the spring of 96, three neutrino candidates have been found in a sample collected over 18 days. This is in agreement with the expected number of approximately 2.3 for neutrinos of atmospheric origin. One of the events is displayed in Fig. 7. In this analysis the most effective quality cuts are the traditional χ^2 cut and a cut on the probability of non-reporting channels not to be hit, and reporting channels to be hit (P_{nohit} and P_{hit}, respectively). To guarantee a minimum lever arm for track fitting, they were forced to reject events with a projection of the most distant channels on the track smaller than 35 meters. This does, of course, result in a loss of threshold.

4.3 The AMANDA South Pole Neutrino Detector

4.3.1 Status of the AMANDA Project

Construction of the first-generation AMANDA detector[18] was completed in the austral summer 96–97. It consists of 300 optical modules deployed at a depth of 1500–2000 m; see Fig. 8. An optical module (OM) consists of an 8 inch photomultiplier tube and nothing else. OM's have only failed when the ice refreezes, at a rate of less than 3 percent. Detector calibration and analysis of the first year of data is in progress, although data has been taken with 80 calibrated OM's which were deployed one year earlier in order to verify the optical properties of the ice below 1 km depth (AMANDA-80).

As anticipated from transparency measurements performed with the shallow strings[19] (see Fig. 8), we found that ice is bubble-free at 1400–1500 meters

NT-96 Array

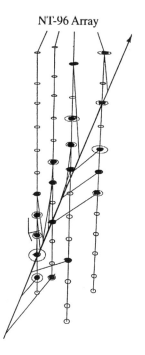

Figure 7: Candidate neutrino event from NT-96 in Lake Baikal.

and below. The performance of the AMANDA detector is encapsulated in the event shown in Fig. 9. Coincident events between AMANDA-80 and four shallow strings with 80 OM's have been triggered for one year at a rate of 0.1 Hz. Every 10 seconds a cosmic ray muon is tracked over 1.2 kilometers. The contrast in detector response between the strings near 1 and 2 km depths is dramatic: while the Cherenkov photons diffuse on remnant bubbles in the shallow ice, a straight track with velocity c is registered in the deeper ice. The optical quality of the deep ice can be assessed by viewing the OM signals from a single muon triggering 2 strings separated by 77.5 m; see Fig. 9b. The separation of the photons along the Cherenkov cone is well over 100 m, yet, despite some evidence of scattering, the speed-of-light propagation of the track can be readily identified.

The optical properties of the ice are quantified by studying the propagation in the ice of pulses of laser light of nanosecond duration. The arrival times of the photons after 20 m and 40 m are shown in Fig. 10 for the shallow and deep

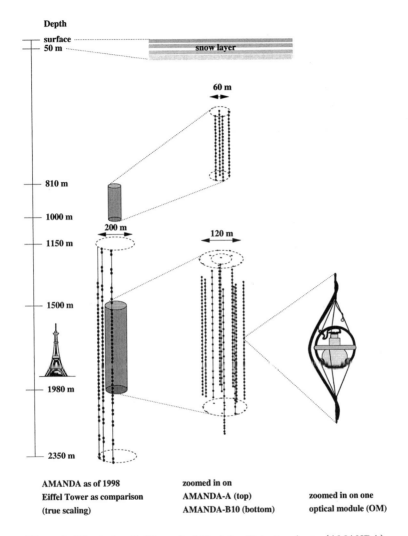

Depth

surface
50 m

snow layer

60 m

810 m

1000 m

200 m

120 m

1150 m

1500 m

1980 m

2350 m

AMANDA as of 1998
Eiffel Tower as comparison
(true scaling)

zoomed in on
AMANDA-A (top)
AMANDA-B10 (bottom)

zoomed in on one
optical module (OM)

Figure 8: The Antarctic Muon And Neutrino Detector Array (AMANDA).

ice.[20] The distributions have been normalized to equal areas; in reality, the probability that a photon travels 70 m in the deep ice is $\sim 10^7$ times larger. There is no diffusion resulting in loss of information on the geometry of the Cherenkov cone in the deep bubble-free ice. These critical results have been verified by the deployment of nitrogen lasers, pulsed LED's and DC lamps in

552

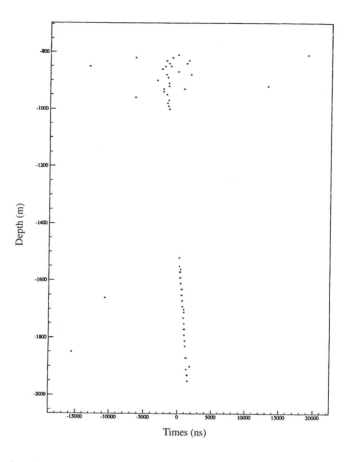

Times (ns)

Figure 9a: Cosmic ray muon track triggered by both shallow and deep AMANDA OM's. Trigger times of the optical modules are shown as a function of depth. The diagram shows the diffusion of the track by bubbles above 1 km depth. Early and late hits, not associated with the track, are photomultiplier noise.

the deep ice. TV cameras have been lowered to 2400 m.

4.3.2 AMANDA: before and after

The AMANDA detector was antecedently proposed on the premise that inferior properties of ice as a particle detector with respect to water could be compensated by additional optical modules. The technique was supposed to be a factor 5~10 more cost-effective and, therefore, competitive. The design

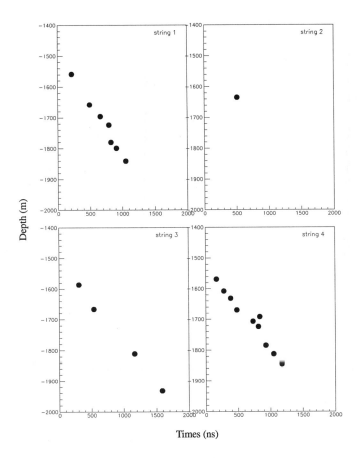

Figure 9b: Cosmic ray muon track triggered by both shallow and deep AMANDA OM's. Trigger times are shown separately for each string in the deep detector. In this event the muon mostly triggers OM's on strings 1 and 2 which are separated by 77.5 m.

was based on then current information:[21]

- the absorption length at 370 nm, the wavelength where photomultipliers are maximally efficient, had been measured to be 8 m;

- the scattering length was unknown;

- the AMANDA strategy would have been to use a large number of closely spaced OM's to overcome the short absorption length. Muon tracks triggering 6 or more OM's are reconstructed with degree accuracy. Taking

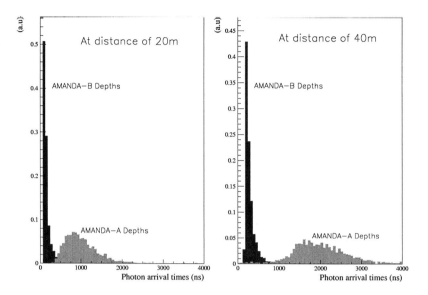

Figure 10: Propagation of 510 nm photons indicate bubble-free ice below 1500 m, in contrast to ice with some remnant bubbles above 1.4 km.

data with a simple majority trigger of 6 OM's or more at 100 Hz yields an average effective area of 10^4 m^2, somewhat smaller for atmospheric neutrinos and significantly larger for the high energy signals previously discussed.

The reality is that:

- the absorption length is 100 m or more, depending on depth;[19]

- the scattering length is ~25 m (preliminary, this number represents an average value which may include the combined effects of deep ice and the refrozen ice disturbed by the hot water drilling);

- because of the large absorption length, OM spacings are similar, actually larger, than those of proposed water detectors. Also, in a trigger 20 OM's report, not 6. Of these more than 5 photons are, on average, "not scattered." A precise definition of "direct" photons will be given further on. In the end, reconstruction is therefore as before, although additional information can be extracted from scattered photons by minimizing a likelihood function which matches measured and expected delays.[22]

The measured arrival directions of background cosmic ray muon tracks, reconstructed with 5 or more unscattered photons, are confronted with their known angular distribution in Fig. 11. There is an additional cut in Fig. 11 which simply requires that the track, reconstructed from timing information, actually traces the spatial positions of the OM's in the trigger. The power of this cut, especially for events distributed over only 4 strings, is very revealing. It can be shown that, in a kilometer-scale detect geometrical track reconstruction using only the positions of triggered OM's is sufficient to achieve degree accuracy in zenith angle. We conclude from Fig. 11 that the agreement between data and Monte Carlo simulation is adequate. Less than one in 10^5 tracks is misreconstructed as originating below the detector.[20] Visual inspection reveals that the remaining misreconstructed tracks are mostly showers, radiated by muons or initiated by electron neutrinos, misreconstructed as up-going tracks of muon neutrino origin. At the 10^{-6} level of the background, up-going muon tracks can be identified; see Fig. 12. Showers can be readily eliminated on the basis of the additional information on the amplitude of OM signals. The rate of tracks reconstructed as up-going is consistent with atmospheric neutrino origin of these events; see next section.

Monte Carlo simulation, based on this exercise, anticipates that AMANDA-300 is a 10^4 m^2 detector for TeV muons, with 2.5 degrees mean angular resolution per event.[22] We have verified the angular resolution of AMANDA-80 by reconstructing muon tracks registered in coincidence with a surface air shower array SPASE.[23] Figure 13 demonstrates that the zenith angle distribution of the coincident SPASE-AMANDA cosmic ray beam reconstructed by the surface array is quantitatively reproduced by reconstruction of the muons in AMANDA.

5 Neutrinos from the Earth's Center: AMANDA-80 events

The capability of neutrino telescopes to discover the particles that constitute the dominant, cold component of the dark matter has been previously mentioned. The existence of the weakly interacting massive particles (WIMPs) in our galactic halo is inferred from observation of their annihilation products. Cold dark matter particles annihilate into neutrinos; *massive* ones will annihilate into *high-energy* neutrinos which can be detected in high-energy neutrino telescopes. This so-called indirect detection is greatly facilitated by the fact that the Earth and the sun represent dense, nearby sources of accumulated cold dark matter particles. Galactic WIMPs, scattering off nuclei in the sun, lose energy. They may fall below escape velocity and be gravitationally trapped. Trapped WIMPs eventually come to equilibrium and accumulate

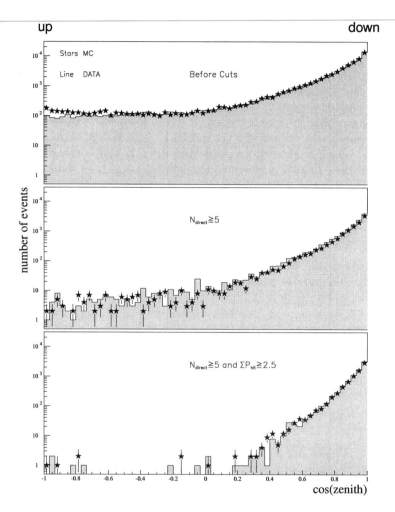

up   down

Figure 11: Reconstructed zenith angle distribution of muons triggering AMANDA-80: data and Monte Carlo. The relative normalization has not been adjusted at any level. The plot demonstrates a rejection of cosmic ray muons at a level of 10^{-5}.

near the center of the sun. While the WIMP density builds up, their annihilation rate into lighter particles increases until equilibrium is achieved where the annihilation rate equals half of the capture rate. The sun has thus become a reservoir of WIMPs which we expect to annihilate mostly into heavy quarks and, for the heavier WIMPs, into weak bosons. The leptonic decays of the

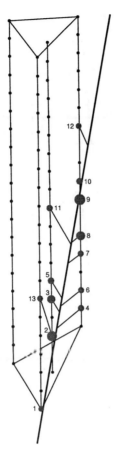

Figure 12: A muon reconstructed as up-going in the AMANDA-80 data. The numbers show the time sequence of triggered OMs, the size of the dots the relative amplitude of the signal.

heavy quark and weak boson annihilation products turn the sun and Earth into nearby sources of high-energy neutrinos with energies in the GeV to TeV range. Existing neutrino detectors have already excluded fluxes of neutrinos from the Earth's center of order 1 event per 1000 m^2 per year. The best limits have been obtained by the Baksan experiment.[24] They are already excluding relevant parameter space of supersymmetric models. We will show that, with data already on tape, the AMANDA detector will have an unmatched discovery reach for WIMP masses in excess of 500 GeV.

We have performed a search[25] for upcoming neutrinos from the center of

558

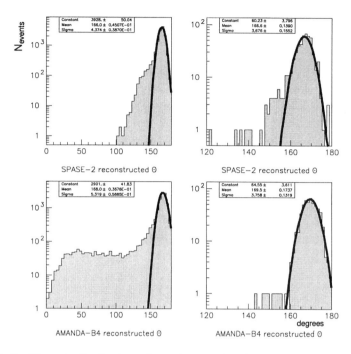

Constant	3928. ±	50.04
Mean	166.0 ±	0.4507E-01
Sigma	4.374 ±	0.3870E-01

Constant	60.23 ±	3.796
Mean	166.6 ±	0.1390
Sigma	3.676 ±	0.1552

SPASE-2 reconstructed Θ   SPASE-2 reconstructed Θ

Constant	2901. ±	41.83
Mean	166.0 ±	0.3676E-01
Sigma	5.319 ±	0.5885E-01

Constant	64.55 ±	3.611
Mean	169.5 ±	0.1737
Sigma	3.758 ±	0.1319

AMANDA-B4 reconstructed Θ   AMANDA-B4 reconstructed Θ

Figure 13: Zenith angle distributions of cosmic rays triggering AMANDA and the surface air shower array SPASE. Reconstruction by AMANDA of underground muons agrees with the reconstruction of the air shower direction using the scintillator array, and with Monte Carlo simulation. The events are selected requiring signals on 2 or more strings (left), and 5 or more direct photons (right).

the Earth. One should keep in mind that the preliminary results are obtained with only 80 OMs, incomplete calibration of the optical modules and only 6 months of data. We nevertheless obtain limits near the competitive level of less than 1 event per $250 \, m^2$ per year for WIMP masses in excess of 100 GeV. Increased sensitivity should result from: lower threshold, better calibration (factor of 3), improved angular resolution (factor of $\sim$2), longer exposure and, finally, an effective area larger by over one order of magnitude. Recall that, because the search is limited by atmospheric neutrino background, sensitivity only grows as the square root of the effective area. First calibration of the full detector is now completed and analysis of the first year of data is in progress. Preliminary results based on the analysis of 1 month of data confirm the performance of the detector derived from the analysis of AMANDA-B4 data. Events reconstructed as going upwards, like the one shown in Fig. 14, are found, as expected.

We reconstructed 6 months of filtered AMANDA-80 events subject to the conditions that 8 OMs report a signal in a time window of 2 microseconds. While the detector accumulated data at a rate of about 20 Hz, filtered events passed cuts[26] which indicate that time flows upwards through the detector. In collider experiments this would be referred to as a level 3 trigger. The narrow, long AMANDA-80 detector (which constitutes the 4 inner strings of AMANDA-300) thus achieves optimal efficiency for muons pointing back towards the center of the Earth which travel vertically upwards through the detector. Because of edge effects the efficiency, which is, of course, a very strong function of detector size, is only a few percent after final cuts, even in the vertical direction. Nevertheless, we will identify background atmospheric neutrinos and establish meaningful limits on WIMP fluxes from the center of the Earth.

That this data set, including prefiltering, is relatively well simulated by the Monte Carlo is shown in Fig. 15. The results reinforce the conclusions, first drawn from Figs. 11 and 13, that we understand the performance of the detector. Cuts are on the number of "direct" photons, i.e. photons which arrive within time residuals of $[-15; 25]$ ns relative to the predicted time. The latter is the time it takes for Cherenkov photons to reach the OM from the reconstructed muon track. The choice of residual reflects the present resolution of our time measurements and allows for delays of slightly scattered photons. The reconstruction capability of AMANDA-80 is illustrated in Fig. 16. Comparison of the reconstructed zenith angle distribution of atmospheric muons and the Monte Carlo is shown in Fig. 16a for 3 cuts in N_{direct}. For $N_{direct} \geq 5$, the resolution is 2.2 degrees as shown in Fig. 16b.

The final cut selecting WIMP candidates requires 6 or more residuals in the interval $[-15, +15]$ ns and $\alpha \geq 0.1$ m/ns. Here α is the slope parameter obtained from a plane wave fit $z_i = \alpha t_i + \beta$, where z_i are the vertical coordinates of hit OMs and t_i the times at which they were hit. The cut selects muons moving vertically along the strings and pointing back towards the center of the Earth.

The two events surviving these cuts are shown in Fig. 17. Their properties are summarized in Table 5. The expected number of atmospheric neutrino events passing the same cuts is $4.8 \pm 0.8 \pm 1.1$. With only preliminary calibration, the systematic error in the time-calibration of the PMTs is ~ 15 ns. This reduces the number of expected events to $2.9 \pm 0.6 \pm 0.6$. The fact that the parameters of both events are not close to the cuts imposed, reinforces their significance as genuine neutrino candidates. Their large tracklengths suggest neutrino energies in the vicinity of 100 GeV which implies that the parent

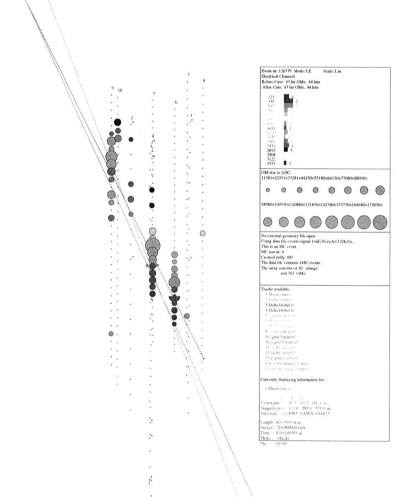

Figure 14:

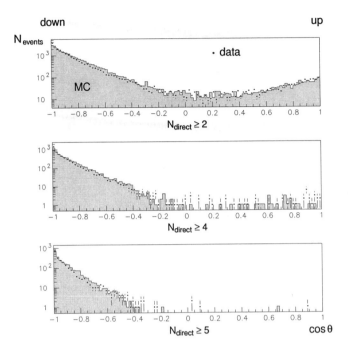

Figure 15: $\cos(\theta_{rec})$ is shown with cuts on the number of residuals in the interval $[-15; 25]$ ns. The histogram represents Monte Carlo simulations with a trigger of 8 or more hits in 2 msec and the dots represent the real data. (Notice that up and down directions are reversed from Fig. 11)

neutrino directions should align with the measured muon track to better than 2 degrees. Conservatively, we conclude that we observe 2 events on a background of 4.8 atmospheric neutrinos. With standard statistical techniques this result can be converted into an upper limit on an excess flux of WIMP origin; see Fig. 13.

In order to interpret this result, we have simulated AMANDA-80 sensitivity to the dominant WIMP annihilation channels:[27] into $b\bar{b}$ and W^+W^-. The upper limits on the WIMP flux are shown in Fig. 18 as a function of the WIMP mass. Limits below 100 GeV WIMP mass are poor because the neutrino-induced muons (with typical energy $\simeq m_\chi/6$) fall below the AMANDA-80 threshold. For the heavier masses, limits approach the limits set by other experiments in the vicinity of $10^{-14}\,\mathrm{cm}^{-2}\,\mathrm{s}^{-1}$. We have previously discussed how data, already on tape from AMANDA-300, will make new incursions into the

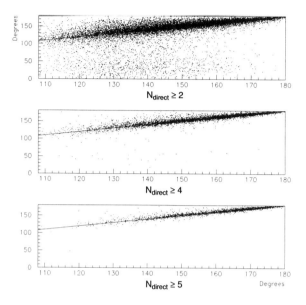

Figure 16a: Scatter-plot showing the AMANDA-reconstructed θ-angle of atmospheric muons versus the MC, at several cut levels.

parameter space of supersymmetric models.

Event ID#	4706879	8427905
α [m/ns]	0.19	0.37
Length [m]	295	182
Closest approach [m]	2.53	1.23
$\theta_{rec}[^\circ]$	14.1	4.6
$\phi_{rec}[^\circ]$	92.0	348.7
Likelihood/OM	5.9	4.2
OM multiplicity	14	8
String multiplicity	4	2

Table 5: Characteristics of the two events reconstructed as up-going muons.

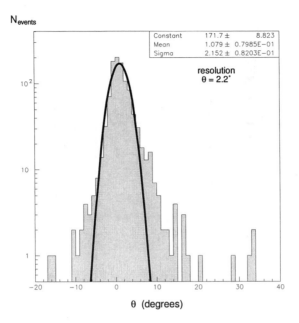

N_events

Constant	171.7 ±	8.823
Mean	1.079 ±	0.7985E−01
Sigma	2.152 ±	0.8203E−01

resolution
θ = 2.2°

θ (degrees)

Figure 16b: $\theta_{\rm rec} - \theta_{\rm MC}$ for reconstructed atmospheric Monte Carlo simulated muons with at least five residuals in the interval [15; 15] ns

6 Kilometer-Scale Detectors

6.1 Towards ICE CUBE(D)

We concluded in previous sections that the study of AGN and GRB is likely to require the construction of a kilometer-scale detector. Other arguments support this conclusion.[28] A strawman detector with effective area in excess of $1\,{\rm km}^2$ consists of 4800 OM's: 80 strings spaced by $\sim 100\,{\rm m}$, each instrumented with 60 OM's spaced by 15 m. A cube with a side of 0.8 km is thus instrumented and a through-going muon can be visualized by doubling the length of the lower track in Fig. 9a. It is straightforward to convince oneself that a muon of TeV energy and above, which generates single photoelectron signals within 50 m of the muon track, can be reconstructed by geometry only. The spatial positions of the triggered OM's allow a geometric track reconstruction with a precision in zenith angle of:

$$\text{angular resolution} \simeq \frac{\text{OM spacing}}{\text{length of the track}} \simeq 15\,{\rm m}/800\,{\rm m} \simeq 1\,\text{degree}; \qquad (53)$$

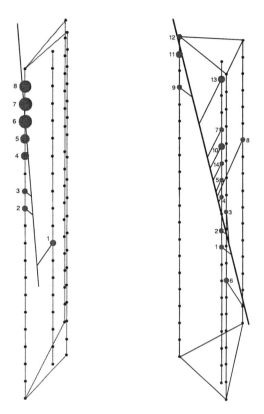

Figure 17: Events reconstructed as up-going satisfying the cuts imposed on the data to search for neutrinos from WIMP annihilation in the center of the Earth.

no timing information is required. Timing is still necessary to establish whether a track is up- or down-going, not a challenge given that the transit time of the muon exceeds 2 microseconds. Using the events shown in Fig. 9, we have, in fact, already demonstrated that we can reject background cosmic ray muons. Once ICE CUBE(D) has been built, it can be used as a veto for AMANDA and its threshold lowered to GeV energy.

In reality, noise in the optical modules and multiple events, crossing the detector in the relatively long triggering window, will interfere with these somewhat over-optimistic conclusions. This is where an ice detector will be at its greatest advantage however. Because of the absence of radioactive potassium, the background counting rate of OMs deployed in sterile ice can be reduced by close to 2 orders of magnitude.

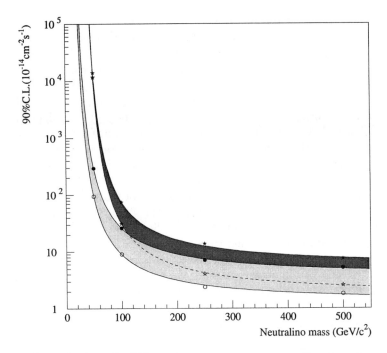

Figure 18: Upper limit at the 90% confidence level on the muon flux from the center of the Earth as a function of neutralino mass. The light shaded band represents the W^+W^- annihilation channel and the dark one represents the $b\bar{b}$ annihilation channel. The width of the bands reflects the inadequate preliminary calculation.

With half the number of OM's and half the price tag of the Superkamiokande and SNO solar neutrino detectors, the plan to commission such a detector over 5 years is not unrealistic. The price tag of the default technology used in AMANDA-300 is $6000 per OM, including cables and DAQ electronics. This signal can be transmitted to the surface by fiber optic cable without loss of information. Given the scientific range and promise of such an instrument, a kilometer-scale neutrino detector must be one of the best motivated scientific endeavors ever.

6.2 Water and Ice

The optical requirements of the detector medium can be readily evaluated, at least to first order, by noting that string spacings determine the cost of the detector. The attenuation length is the relevant quantity because it determines how far the light travels, irrespective of whether the photons are lost by scattering or absorption. Remember that, even in the absence of timing, hit geometry yields degree zenith angle resolution. Near the peak efficiency of the OM's the attenuation length is 25–30 m, larger in deep ice than in water below 4 km. The advantage of ice is that, unlike for water, its transparency is not degraded for blue Cerenkov light of lower wavelength, a property we hope to take further advantage of by using wavelength-shifter in future deployments.

The AMANDA approach to neutrino astronomy was initially motivated by the low noise of sterile ice and the cost-effective detector technology. These advantages remain, even though we know now that water and ice are competitive as a detector medium. They are, in fact, complementary. Water and ice seem to have similar attenuation length, with the role of scattering and absorption reversed; see Table 6. As demonstrated with the shallow AMANDA strings,[29] scattering can be exploited to range out the light and perform calorimetry of showers produced by electron-neutrinos and showering muons. Long scattering lengths in water may result in superior angular resolution, especially for the smaller, first-generation detectors. This can be exploited to reconstruct events far outside the detector in order to increase its effective volume.

Table 6: Optical properties of South Pole ice at 1750 m, Lake Baikal water at 1 km, and the range of results from measurements in ocean water below 4 km.

$\lambda = 385$ nm *	(1700 m) AMANDA	BAIKAL	OCEAN
attenuation	~ 30 m **	~ 8 m	25–30 m ***
absorption	95 ± 5 m	8 m	—
scattering length	24 ± 2 m	150–300 m	—

* peak PMT efficiency
** same for bluer wavelengths
*** smaller for bluer wavelengths

Acknowledgements

We thank Jaime Alvarez for a careful reading of the manuscript. This work was supported in part by the University of Wisconsin Research Committee with funds granted by the Wisconsin Alumni Research Foundation, and in part by the U.S. Department of Energy under Grant No. DE-FG02-95ER40896.

1. K. Greisen, Ann. Rev. Nucl. Science, **10**, 63 (1960); see also F. Reines, Ann. Rev. Nucl. Science, **10**, 1 (1960); M.A. Markov & I.M. Zheleznykh, Nucl. Phys. **27** 385 (1961); M.A. Markov in *Proceedings of the 1960 Annual International Conference on High Energy Physics at Rochester*, E.C.G. Sudarshan, J.H. Tinlot & A.C. Melissinos, Editors (1960).

2. M. T. Ressel and M. S. Turner, Comments Astrophys. **14**, 323 (1990).

3. For a nice overview, see R..J. Protheroe, C.L. Bhat, P. Fleury, E. Lorenz, M. Teshima, T.C. Weekes, astro-ph/9710118. See also M. Punch *et al.*, Nature **358**, 477 (1992); J. Quinn *et al.*, Ap. J. **456**, L83 (1995); Schubnell *et al.*, astro-ph/9602068, Ap. J. (1997 in press); D. J. Macomb , *et al.*, Ap. J. **438**, 59; **446**, 99 (1995).

4. The Pierre Auger Project Design Report, Fermilab report (Feb. 1997) and references therein. Recent references on the origin of the highest energy cosmic rays include: A. A. Watson, Nucl. Phys. B (Proc. Suppl.) **22B**, 116 (1991); D. J. Bird *et al.*, Phys. Rev. Lett. **71**, 3401 (1993); S. Yoshida *et al.*, Astropar. Phys. **3**, 151 (1995); E. Waxman, Phys. Rev. Lett. **75**, 386 (1995); M. Milgrom, and V. Usov, Astrophy. J. **449**, L37 (1995); M. Vietri, Astrophys. J. **453**, 883 (1995); F. Halzen, R.A. Vazquez, T. Stanev, H.P. Vankov, Astropart. Phys. **3**, 151 (1995); T. Stanev *et al.*, Phys. Rev. Lett. **75**, 3056 (1995); J. Lloyd-Evans, A. Watson, Phys. World **9**, No. 9, 47 (1996); G. Burdman, F. Halzen, R. Gandhi, Phys. Lett. **D417**, 107 (1998)

5. F. W. Stecker, O. C. De Jager and M. H. Salamon, Ap. J. **390**, L49 (1992).

6. For a review, see T. K. Gaisser, F. Halzen, and T. Stanev, Phys. Rep. **258**, 173 (1995); see also, P. L. Biermann, & P. A. Strittmatter, Astrophys. J. **322**, 643 (1987); K. Mannheim and P. L. Biermann, Astron. Astrophys. **221**, 211 (1989); F. Stecker, C. Done, M. Salamon, and P. Sommers, Phys. Rev. Lett. **66**, 2697 (1991); erratum Phys. Rev. Lett. **69**, 2738 (1992); K. Mannheim, Astron. Astrophys. **269**, 67 (1993); M. Sikora, M. C. Begelman and M. J. Rees, Ap. J. Lett. **421**, 153 (1994); K. Mannheim, Astropar. Phys. **3**, 295 (1995); K. Mannheim, S. Westerhoff, H. Meyer and H. H. Fink, Astron. Astrophys., in press (1996); R. J. Protheroe, *Gamma Rays and Neutrinos from AGN Jets*, Adelaide preprint ADP-AT-96-4 (astro-ph/9607165) (1996); F. Halzen and E. Zas, Astrophys. J. **488**, 669 (1997); J.P. Rachen and P. Mészáros, submitted to Phys. Rev. D (1998) (astro-ph/9802280); F.W. Stecker and M.H. Salamon, Space Sci. Rev. **75**, 341 (1996); G. C. Hill, Astropart. Phys. **6**, 215 (1997); E. Waxman and J.H. Bahcall, IASSNS-AST-98-38 [hep-ph/9807282] (1998).

7. C. A. Meegan *et al.*, Nature **355**, 143 (1992); B. Paczyński, Nature **355**,

568

521 (1992); T. Piran, Astrophys. J. **389**, L45 (1992); M. Rees, and P. Mészáros, Mon. Not. Roy. Astron. Soc. **258**, 41P (1992); M. Rees, and P. Mészáros, Astrophys. J. **430**, 708 (1994); P. Mészáros, and M. Rees, Mon. Not. Roy. Astron. Soc. **269**, 41P (1994); B. Paczyński, and G. Xu, Astrophys. J. **427**, 708 (1994); J. P. Norris *et al.*, Astrophys. J. **423**, 432 (1994); J.I. Katz, Astrophys. Journal **432**, L27 (1994); E. Waxman, Phys. Rev. Lett. **75**, 386 (1995); F. Halzen and G. Jaczko, Phys. Rev. D **54**, 2774 (1996); E. Waxman and J. N. Bahcall, Phys. Rev. Lett. **78**, 2292 (1997); M. Vietri, Phys. Rev. Lett. **80**, 3690 (1998); M. Botcher and C. D. Dermer, astro-ph/9801027; F. Halzen and D. Saltzberg, hep-ph/9804354, Phys. Rev. Lett. (to be published).

8. M. Longair, *High Energy Astrophysics*, Cambridge University Press.

9. For a review, see T.K. Gaisser, F. Halzen and T. Stanev, Phys. Rep. **258**(3), 173 (1995); R. Gandhi, C. Quigg, M. H. Reno and I. Sarcevic, Astropart. Phys. **5**, 81 (1996).

10. T. K Gaisser, OECD Symposium, Taormina, Italy (1996).

11. M. Sikora, M. C. Begelman and M. J. Rees, Ap. J. Lett. **421**, 153 (1994) and references therein.

12. D. J. Thompson et al., Ap. J. S, **101**, 259.

13. G. Sigl, S. Lee, P. Bhattacharjee, S. Yoshida, hep-ph/9809242 (1998).

14. G. V. Domogatsky, in *Procs. of the 5th International Workshop on "Topics in Astroparticle and Underground Physics (TAUP 97)*, Gran Sasso, Italy, 1997, ed. by A. Bottino, A. di Credico, and P. Monacelli, Nucl. Phys.**B70** (Proc. Suppl.), p. 439 (1998).

15. L. Trascatti, in *Procs. of the 5th International Workshop on "Topics in Astroparticle and Underground Physics (TAUP 97)*, Gran Sasso, Italy, 1997, ed. by A. Bottino, A. di Credico, and P. Monacelli, Nucl. Phys. **B70** (Proc. Suppl.), p. 442 (1998).

16. F. Feinstein, in *Procs. of the 5th International Workshop on "Topics in Astroparticle and Underground Physics (TAUP 97)*, Gran Sasso, Italy, 1997), ed. by A. Bottino, A. di Credico, and P. Monacelli, Nucl. Phys. **B70** (Proc. Suppl.), p. 445 (1998).

17. I.A. Belolaptikov et al., *Proceedings of the 25th International Cosmic Ray Conference*, Durban, South Africa (astro-ph/9705245).

18. S. W. Barwick *et al.*, *The status of the AMANDA high-energy neutrino detector*, in Proceedings of the 25th International Cosmic Ray Conference, Durban, South Africa (1997).

19. The AMANDA collaboration, *Science* **267**, 1147 (1995).

20. S. Tilav et al., *First look at AMANDA-B data*, in Proceedings of the 25th International Cosmic Ray Conference, Durban, South Africa (1997).

21. S. W. Barwick *et al*, *Proceedings of the 22nd International Cosmic Ray Conference*, Dublin (Dublin Institute for Advanced Studies, 1991), Vol. 4, p. 658.
22. C. Wiebusch *et al.*, *Muon reconstruction with AMANDA-B*, in Proceedings of the 25th International Cosmic Ray Conference, Durban, South Africa (1997).
23. T. Miller *et al.*, *Analysis of SPASE-AMANDA coincidence events*, in Proceedings of the 25th International Cosmic Ray Conference, Durban, South Africa (1997).
24. O. Suvorova et al., *Baksan neutralino search*, presented at the First International Workshop on Non-Accelerator New Physics (NANP97), Dubna, Russia, July 7–12, (1997).
25. A. Bouchta, University of Stockholm, PhD thesis (1998)
26. J. E. Jacobsen, University of Wisconsin, PhD thesis (1996)
27. Muflux code written by J. Edsjö, implemented as an event generator by A. Goobar; see also L. Bergström, J. Edsjö and P. Gondolo, *Indirect neutralino detection rates in neutrino telescopes*, Phys. Rev. **D 55** 1765, (1997).
28. F. Halzen, *The case for a kilometer-scale neutrino detector*, in Nuclear and Particle Astrophysics and Cosmology, Proceedings of Snowmass 94, R. Kolb and R. Peccei, eds.; *The Case for a Kilometer-Scale Neutrino Detector, 1996*, in Proc. of the Sixth International Symposium on Neutrino Telescopes, ed. by M. Baldo-Ceolin, Venice (1996).
29. R. Porrata *et al.*, *Analysis of cascades in AMANDA-A*, in Proceedings of the 25th International Cosmic Ray Conference, Durban, South Africa (1997).

SUPERNOVA EXPLOSIONS AND SUPERNOVA NEUTRINOS

A. Burrows

Department of Astronomy, The University of Arizona, Tucson, AZ 85721, USA

E-mail: burrows@as.arizona.edu

We summarize the current status of core–collapse supernova theory, in particular as it relates to multi–dimensional effects and neutrino transport. Highlighted are pulsar kicks, neutrino burst signatures, many–body effects, and observed aspheric-ities that may have a bearing on the mechanism of supernova explosions. Observed systematics of the supernova energy and ^{56}Ni yield with progenitor mass are dis-cussed and attempts are made to put such systematics into the context of the emerging theory of supernova explosions.

1 Introduction

Core–collapse supernova explosions mark the endpoint in the life of a massive star and the birth of either a neutron star or a black hole. The violence of the event belies its fascination, for Nature has contrived an elegant means to create compact objects while at the same time seeding the Galaxy with the elements of existence. Neutrinos plays a key role in the phenomenon of collapse and explosion. They are the major signatures of the inner turmoil of the dense core of the massive star, otherwise shrouded in mystery by its profound opac-ity to photons. Neutrinos of all species play a role, not only in powering the supernova (so we now believe), but in carrying away the binding energy of the young neutron star, a full 10% of its mass–energy. In principle, the detection of collapse neutrinos, their "light curve" and spectra, will allow us to follow in real time the phenomena of stellar death and birth. The supernova, SN1987A, provided a glimpse of what might be possible, but it yielded only 19 events; we can expect the current generation of underground neutrino telescopes to collect thousands of events from a galactic supernova. However, though neutrinos are centrally important, there are many aspects of a supernova explosion and its residues that interest and engage astronomers. Many nascent neutron stars are kicked to high speeds at birth, supernova debris are ejected quite asym-metrically, supernova explosion energies may vary by a factor of ∼5 (they are not all the same!), and the ^{56}Ni yields vary by two orders of magnitude.

I will review the salient features of core–collapse supernova explosions and theory, emphasizing some of what I myself have found most interesting. It is now not unreasonable to imagine a theory that unifies the spins and velocities of neutron stars, the anisotropies observed in supernova ejecta, and stellar collapse and explosion. These may be connected in a given supernova, with the debris asymmetries correlated with the kick directions and the neutrino and gravitational wave emissions related to both. I make some of these connections in this paper and highlight the relevant physics, but apologize in advance for any omissions or confusions. In the spirit of the TASI lectures, some of this paper brings together sections of other papers I have recently penned. These sections, and the glue that binds them, collectively serve as an introduction to some of the technical and academic aspects of the subject. No firm conclusions are drawn. In this way, this paper is a synoptic collection of the current art for the student interested in exploring the issues that still surround supernova theory and supernova neutrinos. The reader is referred to a selection of other works for further details and a mix of views [1,2,3,4,5,6,7].

In §2, I review the status of the theory of explosions. In §3, I discuss the putative role of instabilities and multi–dimensional effects in igniting the explosion. Section 4 contains a useful digression on the observations of young supernova remnants and §5 explores the phenomenon of pulsar "kicks." Section 6 covers aspects of neutrino transfer and cross sections, but is not meant to be complete. Nevertheless, it contains some useful tidbits, including a discussion of many–body correlations effects and the supernova neutrino burst itself. Section 7 summarizes ambiguities in the structures of massive star progenitors and §8 concludes the paper with a discussion of possible systematics with progenitor of the explosion energy and ^{56}Ni yields.

2 Status of Supernova Theory and Explosion Modeling

All groups that do multi–D hydrodynamic modeling of supernovae obtain vigorous convection in the semi–transparent mantle bounded by the stalled shock [1,2,4,3,8,9]. There is a consensus that the neutrinos drive the explosion [10] after a delay whose magnitude has yet to be determined, but that may be between 100 and 1000 milliseconds. Whether any convective motion or hydrodynamic instability is central to the explosion mechanism is not clear, with five groups [1,2,4,8,9] voting yes or maybe and one group [3] voting no. The negative vote is from a group that is taking pains to handle the transport with a minimum of approximations. However, this group opted to do the transport in 1–D, save the result, and impose this history on the 2–D calculation, without feedback. This prescription is suspect, but there are flaws in the calculations of all teams exploring the role of multi–dimensional effects. Herant et $al.$ [2] use a crude variant of flux–limited diffusion and use fewer particles in their SPH simulation than may be warranted. Burrows, Hayes, and Fryxell [1] use another variant of flux–limited diffusion, include the inner core (15 kilometers, but follow it in only 1–D), employ PPM and take pains to resolve the flow with many zones $(500(r) \times 160(\theta))$, but only in one quadrant, and ignore transport in the angular direction. Janka and Müller [4] do not follow collapse, but start with a protoneutron star bounded by a stalled shock, do not follow the hydrodynamics or transport in the inner core, and employ a neutrino light bulb approximation and PPM in the mantle. Their focus is on the character of the multi–dimensional flow and the role of neutrino–driven overturn. Clearly, though some of the multi–D calculations are more self–consistent than others, all leave something to be desired. None yet incorporates general relativity, none adequately handles transport in either the angular or the radial direction, none is three–dimensional, and none is evolved for an adequate duration. Furthermore, none has correctly treated all the known neutrino processes in the core. In sum, though there has been a great deal of conceptual and numerical progress of late and though there is now a consensus that neutrinos mediate the energy transfer to an aspherically exploding mantle, the supernova puzzle in its particulars remains [6]. A definitive calculation in either two or three dimensions has not yet been performed.

3 Hydrodynamic Instabilities and the Supernova Mechanism

The neutrino heating rates just behind a stalled shock exceed the cooling rates and establish the so–called gain region [10,11]. In 1-D, the matter that passes through the shock quickly leaves this region and may settle onto the core unexceptionally. However, in 2–D (and presumably in 3–D), the matter can dwell in the gain region for a longer time. The net result is a higher entropy and a larger gain region, with more matter less bound behind the shock. Burrows, Hayes, and Fryxell [1] and Herant et al. [2] describe this situation differently, but each derive that the core is more unstable to explosion in 2–D than in 1–D, with Herant et al. [2] claiming that overturn is the crucial ingredient. The equilibrium pre–explosion shock radius is larger in the multi–D calculations [1,2,3,4] than in the 1–D calculations, but not all workers agree that this enables the supernova. In particular, Mezzacappa et al. [3] do not obtain an explosion after $\sim$500 milliseconds and declare their run a dud.

A variety of other instabilities and flavors of "convection" have been invoked by theorists over the years to help ignite the supernova explosion. (The study of instabilities in the supernova context can be traced to Epstein [12].) Mayle and Wilson citemaylew suggested that doubly–diffusive ("neutron finger") instabilities in the core enhanced the neutrino luminosity and turned a dud into an explosion after $\sim$0.5 seconds. However, Bruenn and Dineva [14] show that, because the muon neutrinos decouple from matter just at the time they are needed to drive the instability, such a salt–finger instability does not obtain.

Nevertheless, the Mayle/Wilson luminosity enhancement was only ~20%, and this suggested that refinements in the opacity of matter to neutrinos at high and intermediate densities might be important. This is still a fruitful lead to follow.

The idea that core overturn driven by negative entropy and lepton gradients persists during the deleptonization and cooling of the protoneutron star, and that this can boost the driving neutrino luminosities, dates back to Burrows, Mazurek, and Lattimer[15], but was revitalized by Burrows[16] and Keil, Janka, and Müller[17]. Whether this is relevant will be determined only when the best neutrino transport is incorporated in multi–dimensional protoneutron star cooling calculations.

4   Indications of Asphericity in Young Supernova Remnants

There are many observational indications that supernova explosions are indeed aspherical. Fabry–Perot spectroscopy of the young supernova remnant Cas A, formed around 1680 A.D., reveals that its calcium, sulfur, and oxygen element distributions are clumped and have gross back–front asymmetries[18]. No simple shells are seen. Many supernova remnants, such as N132D, Cas A, E0102.2-7219, and SN0540-69.3, have systemic velocities relative to the local ISM of up to 900 km s^{-1}[19]. X–ray data taken by ROSAT of the Vela remnant reveal bits of shrapnel with bow shocks[20]. The supernova, SN1987A, is a case study in asphericity: 1) its X–ray, gamma–ray, and optical fluxes and light curves require that shards of the radioactive isotope ^{56}Ni were flung far from the core in which they were created, 2) the infrared line profiles of its oxygen, iron, cobalt, nickel, and hydrogen are ragged and show a pronounced red–blue asymmetry, 3) its light is polarized, and 4) recent Hubble Space Telescope pictures of its inner debris reveal large clumps and hint at a preferred direction [21]. Furthermore, radio pictures of the supernova SN1993J, which also has polarized optical spectral features, depict a broken shell. One of the most intriguing recent finds is the supernova SN1997X, which is a so–called Type Ic explosion. This supernova shows the greatest optical polarization of any to date (Lifan Wang, private communication). Type Ic supernovae are thought to be explosions of the bare carbon/oxygen cores of massive star progenitors stripped of their envelopes. As such, SN1997X's large polarization implies that the inner supernova cores, and, hence, the explosions themselves, are fundamentally asymmetrical. No doubt, instabilities in the outer envelopes of supernova progenitors clump and mix debris clouds and shatter spherical shells. The observation of hydrogen deep in SN1987A's ejecta[22] strongly suggests the work of such mantle instabilities. All these data collectively and forcefully point to asymmetries in the central engine of explosion. However, nowhere are such asymmetries more indicated than in the large proper motions now being inferred for radio pulsars and neutron stars.

5 Neutron Star Kicks

Strong evidence that neutron stars experience a net kick at birth has been mounting for years. In 1993[23,24], it was demonstrated that the pulsars are the fastest population in the galaxy ($< v > \sim 450$ km s^{-1}). Such speeds are far larger than can result generically from orbital motion due to birth in a binary (the "so-called" Blaauw effect). An extra "kick" is required, probably during the supernova explosion itself[25]. In the pulsar binaries, PSR J0045-7319 and PSR 1913+16, the spin axes and the orbital axes are misaligned, suggesting that the explosions that created the pulsars were not spherical[26,27]. In fact, for the former the orbital motion seems retrograde relative to the spin[28] and the explosion may have kicked the pulsar backwards. In addition, the orbital eccentricities of Be star/pulsar binaries are higher than one would expect from a spherical explosion, also implying an extra kick[29]. Furthermore, low–mass X–ray binaries (LMXB) are bound neutron star/low-mass star systems that would have been completely disrupted during the supernova explosion that left the neutron star, had that explosion been spherical[30]. In those few cases, a countervailing kick may have been required to keep the system bound. The kick had to act on a timescale shorter than the orbit period and the explosion orbit crossing time. Otherwise, the process would have been uselessly adiabatic. One is tempted to evoke as further proof the fact that pulsars seen around young (age $\leq 10^4$ years) supernova remnants are on average far from the remnant centers, but here ambiguities in the pulsar ages and distances and legitimate questions concerning the reality of many of the associations make this argument rather less convincing[31,32]. However, the ROSAT observations of the 3700 year–old supernova remnant Puppis A show an X–ray spot that has been interpreted as its neutron star[33]. This object has a large X–ray to optical flux ratio, but no pulsations are seen. If this interpretation is legitimate, then the inferred neutron star transverse speed is ~ 1000 km s^{-1}. Interestingly, the spot is opposite to the position of the fast, oxygen–rich knots, as one might expect in some models of neutron star recoil during the supernova explosion. Whatever the correct interpretation of the Puppis A data, it is clear that many neutron stars are given a hefty extra kick at birth (though the distribution of these kicks is broad) and that it is reasonable to implicate asymmetries in the supernova explosion itself.

5.1   Theories of Kicks

As stated, supernova theorists have determined that protoneutron star/supernova cores are grossly unstable to Rayleigh–Taylor–like instabilities[1,2,4]. During the post–bounce delay to explosion that might last 100 to 1000 milliseconds, these cores with 100– to 200–kilometer radii are strongly convective, boiling and churning at sonic ($\sim 3 \times 10^4$ km s^{-1}) speeds. Any slight asymmetry in collapse can amplify this jostling and result in vigorous kicks and torques[1,5,34] to the residue that can be either systematic or stochastic. Whatever the details, it would seem odd if the nascent neutron star were not left with a net recoil and spin, though whether pulsar speeds as high as 1500 km s^{-1} can be reached through this mechanism is unknown. Furthermore, asymmetries in the matter field may result in asymmetries in the emission of the neutrinos that carry away most of the binding energy of the neutron star. A net angular asymmetry in the neutrino radiation of only 1% would give the residue a recoil of $\sim$300 km s^{-1}. Not surprisingly, many theorists have focussed on producing such a net asymmetry in the neutrino field, either evoking anisotropic accretion, exotic neutrino flavor physics, or the influence of strong magnetic fields on neutrino cross sections and transport. The latter is particularly interesting, but generally requires magnetic fields of $\sim 10^{16}$ gauss[35], far larger than the canonical pulsar surface field of 10^{12} gauss. Perhaps, the pre–explosion convective motions themselves can generate via dynamo action the required fields. Perhaps, these fields are transient and subside to the observed fields after the agitation of the explosive phase. It would be hard to hide large fields of 10^{16} gauss in the inner core of an old neutron star, while still maintaining standard surface fields of 10^{12} gauss. In this context, it is interesting to note that surface fields as high as 10^{15} gauss are very indirectly being inferred for the so–called soft gamma repeaters[36], but these are a very small fraction of all neutron stars.

Whether the kick mechanism is hydrodynamic or due to neutrino momentum, one might expect that the more massive progenitors would give birth to speedier neutron stars. More massive progenitors generally have more massive cores. If the kick mechanism relies on the anisotropic ejection of matter[5], then for a given explosion energy and degree of anisotropy we might expect the core ejecta mass and, hence, the dipole component of the ejecta momentum to be larger ("$p \sim \sqrt{2ME}$"), resulting in a larger kick. The explosion energy may also be larger for the more massive progenitors, enhancing the effect (see §8). If the mechanism relies on anisotropic neutrino emission, the residues of more massive progenitors are likely to be more massive and have a greater binding energy ($E_B \propto M_{NS}^2$) to radiate. Hence, for a given degree of neutrino anisotropy, the impulse and kick ($\propto E_B/M_{NS}$) would be greater. In either case, despite the primitive nature of our current understanding of kick mechanisms, given the above arguements it is not unreasonable to speculate that the heaviest massive stars might yield the fastest neutron stars.

6 Neutrinos

6.1 Supernova Neutrino Burst Signature

Despite the ambiguities in supernova theory, there is a broad consensus on the basic features of the neutrino light curve from a supernova. However, it should be recalled that the luminosities and timescales for different massive star progenitors will be different. Generically, infall may last from 200 to 600 milliseconds during which time electron neutrinos will predominate. They will have roughly a capture spectrum that gradually hardens until shock breakout. The rise time of the associated luminosity depends upon the nuclear symmetry energy, but is approximately 5 milliseconds. The total energy radiated during this phase is roughly 10^{51} ergs. Bounce is almost immediately followed by the formation of the shock in the neutrino-opaque regions (at near 20 kilometers). The shock starts with a velocity near 50,000 km/s and so very quickly achieves the neutrinosphere (50-100 kilometers) and breaks out. Shock breakout is announced by a prodiguous burst of electron neutrinos produced by electron capture on free protons newly liberated by shock dissociation of the infalling nuclei. The electron neutrino luminosity may achieve 10^{54} ergs/second. The characteristic time of the breakout burst is 3-10 milliseconds and the total energy radiated in electron-type neutrinos during breakout is 3×10^{51} ergs. The magnitude of the latter will depend on the density structure of the collapsing Chandrasekhar core, and will be higher for the more massive progenitors. During this phase, perhaps 10 events in both SuperK and in ICARUS can be expected from a collapse at 10 kpc.

During breakout, the matter is heated to such a degree that anti-electron neutrinos and mu and tau neutrinos and anti-neutrinos (hereafter "mus") are thermally produced and radiated. The turnon timescale of this component is less than one millisecond, but the initial luminosity of the antis and the mus depends upon the degree of degeneracy of the electrons near the neutrinospheres and the magnitude of the production sources, still poorly known. It is thought that the initial anti-electron neutrino luminosity is within about one order of magnitude of its peak value (20-50 milliseconds after breakout). Even at such a level, at 10 kpc both SuperK and SNO will register 100's of anti-e events per second, in SuperK perhaps a kilohertz. After the abrupt rise, the anti-electron neutrino luminosity rises further to approximately meet the falling electron-type luminosity. After 20-50 milliseconds, the two decay together as the light curve transitions to the longer-term protoneutron star cooling and neutronization phase. Similarly, the mu neutrino luminosity per species achieves a value not more than 30% away from the electron neutrino luminosity.

The decay is gradual and there may be some quasi-periodic pulsation of the luminosities during this phase. However, the shock wave launched with such fanfare stalls into an accretion shock at 100-200 kilometers within 10-20 milliseconds of breakout. There is a delay to explosion, that may last between a few hundred milliseconds and a second, during which time perhaps $\geq 10^{53}$ ergs of neutrinos may be radiated. Explosion, when it comes, and there is some debate about the details of explosion, should be accompanied by a decrease by about a factor of two over about 20 milliseconds in all the neutrino luminosities. This may be detectable. After explosion, the luminosities decay on timescales of seconds to a minute. Indeed, after as long as a minute, the event rate at 10 kpc in SuperK may still be as high as one per second. After breakout, the spectra of all the neutrino species first harden on timescales of hundreds of milliseconds, then soften, particularly after explosion, as the luminosity inexorably decays. The rise and fall timescales, as well as the explosion time, are not known theoretically with sufficient precision.

Hence, the features detectors should key in on are: the infall rise, the breakout, the early anti-electron neutrino rise, the production of mus, the signature of explosion, the rise and fall of the average neutrino energies, and the late-time persistence. In addition, if a black hole forms during the high-luminosity phase, the prediction is that the signal will stop within less than a millisecond. Such a phenomenon will be detectable.

Given this generic neutrino light curve, can we use accurate timing of the features in the burst to triangulate on the supernova? This will depend upon the signal strength (and, hence, the distance). At the canonical distance of 10 kpc, with the forward-peaked SuperK electron neutrino events (100-150), one should be able to achieve 4-degree pointing without the aide of the network. If the initial anti-electron-type signal is indeed as abrupt as we believe and if it starts at a high luminosity, then initial count rates near one kilohertz in SuperK, SNO, and LVD/MACRO/ICARUS might enable the network to locate a supernova to within ~10 degrees at 10 kpc. The fact that the current detectors are all in the northern hemisphere is a problem, as is the possible fuzziness of the initial luminosity rise. However, there is general excitement that a network of coordinated neutrino telescopes might indeed be able to announce, with whatever angular precision, the advent of a galactic supernova and allow the astronomical community the early warning it has never before enjoyed.

6.2 Neutrino Transport and Some Useful Opacities and Rates

Though much of the recent excitement in supernova theory has concerned its multi–dimensional aspects, neutrino heating and transport are still central to the mechanism. The coupling between matter and radiation in the semi-transparent region between the stalled shock and the neutrinospheres determines the viability and characteristics of the explosion. Unfortunately, this is the most problematic regime. Diffusion algorithms and/or flux–limiters do not adequately reproduce the effects of variations in the Eddington factors and the spectrum as the neutrinos decouple. Hence, a multi–group full transport scheme is desirable.

The multi–group equations of neutrino transfer are the basic equations of radiative transfer for the various individual neutrino types, with the appropriate sources and sinks, and the full complement of scattering and absorption cross sections. The latter are suitably corrected for stimulated absorption. In addition, there is a composition equation for the electron fraction, importantly affected by electron neutrino, anti–neutrino, electron, and positron capture on nucleons, in or out of nuclei. These equations are solved simultaneously with the equations of hydrodynamics. Aside from the unique microphysics associated with neutrinos, the conserved electron lepton number (but oscillations?), and the multiplicity of neutrino types, neutrino transfer is similar to photon transfer. Here, I will not delve too deeply into the specifics of the equations, but do think it useful to summarize some of the dominant cross sections and processes. For a more complete review, I refer the reader to Bruenn[37].

Following Eastman and Pinto[38], a single transport equation for the specific intensity of a neutrino beam, physically equivalent to the Boltzmann equation (ignoring redshifts and acceleration terms) is,

$$\frac{1}{c}\frac{DI_\nu}{Dt} + \mu\frac{\partial I_\nu}{\partial r} + \frac{1-\mu^2}{r}(1-\beta Q\mu)\frac{\partial I_\nu}{\partial \mu} + \frac{\beta}{r}\left(1+Q\mu^2\right)\left(3 - \frac{\partial}{\partial\ln\varepsilon}\right)I_\nu$$

$$= \eta_\nu - \chi_\nu I_\nu + \frac{\kappa_s}{4\pi}\int \Phi(\mathbf{\Omega},\mathbf{\Omega}')\,I_\nu(\mathbf{\Omega}')\,d\mathbf{\Omega}' \,, \tag{1}$$

where $Q \equiv \partial\ln v/\partial\ln r - 1$ and v is the velocity. In eqn. (1), $\mu = \mathbf{\Omega}\cdot\mathbf{\Omega}' = \cos\theta$, ε is the neutrino energy, η_ν is the emissivity of the medium, and all other symbols have their standard meanings. The subscript ν indicates neutrino energy dependence and Φ is a phase function for neutrino scattering into the beam.

Other forms of this equation, good to $O(v/c)$, can be found in Mihalas[39,40]. $\chi_\nu(= \kappa_a + \kappa_s)$ is the total extinction coefficient and κ_a and κ_s contain contributions from all absorption and scattering processes, respectively:

$$\kappa_s = \sum_i n_i \, \sigma_i^s \quad \text{and} \quad \kappa_a = \sum_i n_i \, \sigma_i^a \ . \tag{2}$$

For neutrinos, the phase function, Φ, for a scattering process i is well approximated by

$$\Phi_i(\Omega, \Omega') = \Phi_i(\Omega \cdot \Omega') = (1 + \delta_i \, \Omega \cdot \Omega') = (1 + \delta_i \mu) \ . \tag{3}$$

Hence, we can write the differential cross section for a scattering process i in terms of the total scattering cross section:

$$\frac{d\sigma_i^s}{d\Omega} = \frac{\sigma_i^s}{4\pi}(1 + \delta_i \mu) \ . \tag{4}$$

Some important cross sections and rates are summarized below:

$\nu_e + n \rightarrow e^- + p$:

Here and below the convenient reference neutrino cross section is σ_o, given by

$$\sigma_o = \frac{4G^2 (m_e c^2)^2}{\pi(\hbar c)^4} \simeq 1.705 \times 10^{-44} \, cm^2 \ . \tag{5}$$

The total $\nu_e - n$ absorption cross section is given by

$$\sigma_{\nu_e n}^a = \sigma_o \left(\frac{1 + 3g_A^2}{4}\right) \left(\frac{\varepsilon_{\nu_e} + \Delta_{np}}{m_e c^2}\right)^2 \left[1 - \left(\frac{m_e c^2}{\varepsilon_{\nu_e} + \Delta_{np}}\right)^2\right]^{1/2} \ , \tag{6}$$

where $\Delta_{np} = m_n - m_p = 1.293318$ MeV for a collision in which the electron gets all of the kinetic energy, and $\varepsilon_{e^-} = \varepsilon_{\nu_e} + \Delta_{np}$. To calculate κ_a^*, $\sigma_{\nu_e n}^a$ must be multiplied by $1/(1 - f'_{\nu_e})$ and final–state blocking by the electrons and the protons must be included. $1/(1 - f'_{\nu_e})$ is the stimulated absorption correction, where f'_{ν_e} is given by

$$f'_{\nu_e} = [e^{(\varepsilon_{\nu_e} - (\mu_e - \hat{\mu}))\beta} + 1]^{-1} \ , \tag{7}$$

and $\hat{\mu} = \mu_n - \mu_p$.

$\bar{\nu}_e + p \rightarrow e^+ + n$:

The total $\bar{\nu}_e - p$ absorption cross section is given by

$$\sigma^a_{\bar{\nu}_e p} = \sigma_o \left(\frac{1 + 3g_A^2}{4} \right) \left(\frac{\varepsilon_{\bar{\nu}_e} - \Delta_{np}}{m_e c^2} \right)^2 \left[1 - \left(\frac{m_e c^2}{\varepsilon_{\bar{\nu}_e} - \Delta_{np}} \right)^2 \right]^{1/2} , \qquad (8)$$

where $\varepsilon_{e^+} = \varepsilon_{\bar{\nu}_e} - \Delta_{np}$. To calculate κ_a^*, $\sigma^a_{\bar{\nu}_e p}$ must also be corrected for stimulated absorption and blocking.

$\nu_i + p \rightarrow \nu_i + p$:

The total $\nu_i - p$ scattering cross section is

$$\sigma_p = \frac{\sigma_o}{4} \left(\frac{\varepsilon_\nu}{m_e c^2} \right)^2 \left(4 \sin^4 \theta_W - 2 \sin^2 \theta_W + \frac{(1 + 3g_A^2)}{4} \right) , \qquad (9)$$

which, in terms of $C'_V = 1/2 + 2\sin^2\theta_W$ and $C'_A = 1/2$, becomes

$$\sigma_p = \frac{\sigma_o}{4} \left(\frac{\varepsilon_\nu}{m_e c^2} \right)^2 \left[(C'_V - 1)^2 + 3g_A^2(C'_A - 1)^2 \right] . \qquad (10)$$

From eq. (4) we obtain the differential cross section

$$\frac{d\sigma_p}{d\Omega} = \frac{\sigma_p}{4\pi}(1 + \delta_p \mu) , \qquad (11)$$

where, from Schinder [41],

$$\delta_p = \frac{(C'_V - 1)^2 - g_A^2(C'_A - 1)^2}{(C'_V - 1)^2 + 3g_A^2(C'_A - 1)^2} . \qquad (12)$$

Note that δ_p, and δ_n below, are negative and that these processes are strongly backward–peaked.

The transport cross section, defined by

$$\sigma_i^{tr} = \int \frac{d\sigma_i}{d\Omega}(1 - \mu)\, d\Omega = \sigma_i \left(1 - \frac{1}{3}\delta_i \right) , \qquad (13)$$

is simply

$$\sigma_p^{tr} = \frac{\sigma_o}{6} \left(\frac{\varepsilon_\nu}{m_e c^2} \right)^2 \left[(C'_V - 1)^2 + 5g_A^2(C'_A - 1)^2 \right] . \qquad (14)$$

$\nu_i + n \rightarrow \nu_i + n$:

The total $\nu_i - n$ scattering cross section is

$$\sigma_n = \frac{\sigma_o}{4} \left(\frac{\varepsilon_\nu}{m_e c^2} \right)^2 \left(\frac{1 + 3g_A^2}{4} \right) . \tag{15}$$

From eq. (4), we obtain the differential cross section

$$\frac{d\sigma_n}{d\Omega} = \frac{\sigma_n}{4\pi}(1 + \delta_n \mu) , \tag{16}$$

where, from Schinder [41],

$$\delta_n = \frac{1 - g_A^2}{1 + 3g_A^2} . \tag{17}$$

The transport cross section is

$$\sigma_n^{tr} = \frac{\sigma_o}{4} \left(\frac{\varepsilon_\nu}{m_e c^2} \right)^2 \left(\frac{1 + 5g_A^2}{6} \right) . \tag{18}$$

$\nu_i + A \rightarrow \nu_i + A$:

The differential coherent $\nu_i - A$ scattering cross section may be written as

$$\frac{d\sigma_A}{d\Omega} = \frac{\sigma_o}{64\pi} \left(\frac{\varepsilon_\nu}{m_e c^2} \right)^2 A^2 \left\{ \mathcal{W} \mathcal{C}_{FF} + \mathcal{C}_{LOS} \right\}^2 \langle \mathcal{S}_{ion} \rangle (1 + \mu) , \tag{19}$$

where

$$\mathcal{W} = 1 - \frac{2Z}{A}(1 - 2\sin^2 \theta_W), \tag{20}$$

Z is the atomic number, A is the atomic weight, and $\langle \mathcal{S}_{ion} \rangle$ is the ion–ion correlation function, determined mostly by the Coulomb interaction between the nuclei during infall. Since the shock wave dissociates nuclei, $\nu_i - A$ scattering is important only during infall.

Leinson et al.[42] investigated the polarization correction, $\mathcal{C}_{LOS}$, and found that

$$\mathcal{C}_{LOS} = \frac{Z}{A} \left(\frac{1 + 4\sin^2 \theta_W}{1 + (kr_D)^2} \right) , \tag{21}$$

where

$$r_D = \sqrt{\frac{\pi \hbar^2 c}{4\alpha p_F E_F}} , \tag{22}$$

$k^2 = |\mathbf{p} - \mathbf{p}'|^2 = 2(\varepsilon_\nu/c)^2(1-\mu)$, $\alpha \simeq 137^{-1}$, and r_D is the Debye radius. Note that $r_D \sim 10\hbar/p_F$ in the ultra-relativistic limit ($p_F >> m_e c$). The $\mathcal{C}_{LOS}$ term is important only at low neutrino energies.

Following Tubbs and Schramm [43] and Burrows et al.[15], the form factor term, $\mathcal{C}_{FF}$, in eq. (19) is written as

$$\mathcal{C}_{FF} = e^{-y(1-\mu)/2} , \qquad (23)$$

where $y = 4b\varepsilon_\nu^2 = y \simeq \left(\frac{\varepsilon_\nu}{56\,\text{MeV}}\right)^2 \left(\frac{A}{100}\right)^{2/3}$, $b = \frac{1}{6}\langle r^2\rangle/(\hbar c)^2$, and r is the radius of the nucleus. $\mathcal{C}_{FF}$ differs from 1 for large A and ε_ν, when the de Broglie wavelength of the neutrino is smaller than the nuclear radius.

$e^+ + e^- \to \nu_i + \bar{\nu}_i$:

Ignoring phase space blocking of neutrinos in the final state and taking the relativistic limit ($m_e \to 0$), the total electron-positron annihilation rate for electron or mu- and tau-type neutrino production can be written in terms of the electron and positron distribution functions [44]:

$$Q_{\nu_e\bar{\nu}_e} = K_i \left(\frac{1}{m_e c^2}\right)^2 \left(\frac{1}{\hbar c}\right)^6 \int\int f_{e^-} f_{e^+}(\varepsilon_{e^-}^4 \varepsilon_{e^+}^3 + \varepsilon_{e^-}^3 \varepsilon_{e^+}^4)\, d\varepsilon_{e^-}\, d\varepsilon_{e^+} , \qquad (24)$$

where $K_i = (1/18\pi^4)c\sigma_o(C_V^2 + C_A^2)$. $C_V = 1/2 + 2\sin^2\theta_W$ for electron types, $C_V = -1/2 + 2\sin^2\theta_W$ for mu and tau types and $C_A^2 = (1/2)^2$. Rewriting eq. (24) in terms of the Fermi integral $F_n(\eta)$, we obtain:

$$Q_{\nu_e\bar{\nu}_e} = K_i (kT) \left(\frac{kT}{m_e c^2}\right)^2 \left(\frac{kT}{\hbar c}\right)^6 [F_4(\eta_e)F_3(-\eta_e) + F_4(-\eta_e)F_3(\eta_e)] , \qquad (25)$$

where $\eta_e \equiv \mu_e/kT$ and

$$F_n(\eta) \equiv \int_0^\infty \frac{x^n}{e^{x-\eta} + 1}\, dx . \qquad (26)$$

For $\nu_e\bar{\nu}_e$ production, eq. (24) can also be written as

$$Q_{\nu_e\bar{\nu}_e} \simeq 9.7615 \times 10^{24} \left[\frac{kT}{\text{MeV}}\right]^9 f(\eta_e)\, \text{ergs cm}^{-3}\text{s}^{-1} , \qquad (27)$$

where

$$f(\eta_e) = \frac{F_4(\eta_e)F_3(-\eta_e) + F_4(-\eta_e)F_3(\eta_e)}{2F_4(0)F_3(0)} . \qquad (28)$$

For $\nu_\mu\bar{\nu}_\mu$ and $\nu_\tau\bar{\nu}_\tau$ production combined,

$$Q_{\nu_{\mu,\tau}\bar{\nu}_{\mu,\tau}} \simeq 4.1724 \times 10^{24} \left[\frac{kT}{\text{MeV}}\right]^9 f(\eta_e)\, \text{ergs cm}^{-3}\text{s}^{-1} . \qquad (29)$$

$\nu_i + \bar{\nu}_i \rightarrow e^+ + e^-$:

In the limit of high temperatures and ignoring electron phase space blocking, the $\nu_i \bar{\nu}_i$ annihilation rate can be written [45]:

$$Q_{\nu_i \bar{\nu}_i} = 4K_i \pi^4 \left(\frac{1}{m_e c^2}\right) \left(\frac{4\pi}{c}\right)^2 \int \int \Phi' \, J_{\nu_i} J_{\bar{\nu}_i} (\varepsilon_{\nu_i} + \varepsilon_{\bar{\nu}_i}) \, d\varepsilon_{\nu_i} \, d\varepsilon_{\bar{\nu}_i} \, , \quad (30)$$

where J_ν is the zeroth moment of the radiation field, ε_ν is the neutrino energy, K_i is defined as before (i.e. $K_i = (1/18\pi^4)c\sigma_o(C_V^2 + C_A^2)$), and

$$\Phi'(\mathcal{F}_{\nu_i}, \mathcal{F}_{\bar{\nu}_i}, p_{\nu_i}, p_{\bar{\nu}_i}) = \frac{3}{4}\left[1 - 2\mathcal{F}_{\nu_i}\mathcal{F}_{\bar{\nu}_i} + p_{\nu_i}p_{\bar{\nu}_i} + \frac{1}{2}(1 - p_{\nu_i})(1 - p_{\bar{\nu}_i})\right] \, ,$$
$$(31)$$

where the flux factor $\mathcal{F}_\nu = \langle \mu_{\nu_i} \rangle = H_\nu / J_\nu$ and the Eddington factor $p_\nu = \langle \mu_{\nu_i}^2 \rangle = P_\nu / J_\nu$. Eq. (30) can be rewritten in terms of the invariant distribution functions f_ν:

$$Q_{\nu_i \bar{\nu}_i} = K_i \left(\frac{1}{m_e c^2}\right)^2 \left(\frac{1}{\hbar c}\right)^6 \int \int \Phi' \, f_{\nu_i} f_{\bar{\nu}_i} (\varepsilon_{\nu_i}^4 \varepsilon_{\bar{\nu}_i}^3 + \varepsilon_{\nu_i}^3 \varepsilon_{\bar{\nu}_i}^4) \, d\varepsilon_{\nu_i} \, d\varepsilon_{\bar{\nu}_i} \, . \quad (32)$$

Note that when the radiation field is isotropic ($\Phi' = 1$) and when $\eta_e = 0$ the total rate for $e^+ e^-$ annihilation given in eq. (24) equals that for $\nu_i \bar{\nu}_i$ annihilation given in eq. (32), as expected.

Despite what has been suggested in the past, the $\nu_e + \bar{\nu}_e \rightarrow e^+ + e$ and $\nu_\mu + \bar{\nu}_\mu \rightarrow e^+ + e^-$ energy deposition rates in the shocked region are no more than 0.01 and 0.001, respectively, times those of the dominant charged–current processes, $\nu_e + n \rightarrow e^- + p$ and $\bar{\nu}_e + p \rightarrow e^+ + n$. However, in the unshocked region ahead of the shock, depending upon the poorly–known ν–nucleus absorption rates, the ν–$\bar{\nu}$ annihilation rate can be competitive, though it is still irrelevant to the supernova.

It is thought that neutrino–electron scattering and inverse pair annihilation are the processes most responsible for the energy equilibration of the ν_μ's and their emergent spectra. However, recent calculations imply that the inverse of nucleon–nucleon bremsstrahlung (e.g., $n + n \rightarrow n + n + \nu\bar{\nu}$) is also important in equilibrating the ν_μ's [46]. This process has not heretofore been incorporated into supernova simulations. Our preliminary estimates suggest that inverse bremsstrahlung softens the emergent ν_μ spectrum, since the bremsstrahlung source spectrum is softer than that of pair annihilation. In addition, given the large ν_μ scattering albedo, one must properly distinguish absorption from scattering, in ways not possible with a flux–limiter. Since the relevant inelastic neutral–current processes are stiff functions of neutrino energy, these transport issues bear directly upon the viability of neutrino nucleosynthesis (*c.g.*, of ^{11}B and ^{19}F) [47,48].

Since neutrino–matter cross sections are higher for higher–energy neutrinos, the energy density spectrum is always harder than the flux spectrum. This hardness boosts the neutrino heating rates in the semi–transparent region. However, this effect is most pronounced in the cooling region below the gain region and tapers off as the shock is approached. Messer *et al.* [49], in particular, have highlighted this correction, but self–consistent calculations from collapse to explosion are needed, given the notorious feedbacks in the supernova problem. The same effect may be important in driving the protoneutron star wind (BHF) thought to be the site of the r–process [50,51]. Qian and Woosley [51] suggest that an extra heating source in the wind acceleration region may help establish the conditions for a successful r–process. Such a "source" may be a natural consequence of the proper handling of neutrino transport above the neutrinosphere. Indeed, full transport calculations of r–process winds and the supernova, even in 1–D, will be illuminating.

6.3 Many–Body Correlations in Neutrino–Matter Interactions

To focus exclusively on numerical and transport matters is frequently to lose sight of the important issues. After the ultimate algorithm is implemented, the results will depend on the initial progenitor models and the microphysics, in particular the neutrino cross sections. In this regard, the recent explorations into the effects of many–body correlations on neutrino–matter opacities at high densities are germane [52,53,54,55]. Though the final numbers have not yet been derived, indications are that we have been overestimating the neutral–current and the charged–current cross sections above 10^{14} gm cm^{-3} by factors of from two to five, depending upon density and the equation of state. The many–body corrections increase with density, decrease with temperature, and for neutral–current scattering are roughly independent of incident neutrino energy.

In the Ring approximation (RPA), using Fermi-Liquid Thoery (FLT) for the nuclear interaction, Burrows and Sawyer[52] derive for the neutral current scattering rate,

$$\frac{d^2\Gamma}{d\omega d\cos\theta} = (4\pi^2)^{-1} G_W^2 (E_1 - \omega)^2 [1 - f_\nu(E_1 - \omega)] \mathcal{I}_{NC} , \qquad (33)$$

where ω is the energy transfer to the medium and

$$\mathcal{I}_{NC} = (1 + \cos\theta) S_V(q, \omega) + (3 - \cos\theta)\alpha^2 S_A(q, \omega) . \qquad (34)$$

S_V and S_A are the Fermi and Gamow–Teller dynamic structure functions, α is the axial–vector coupling constant, and the other symbols have there standard meanings.

Using the approximations of Burrows and Sawyer[52], the Fermi and Gamow–Teller structure functions are given by

$$S_V(q,\omega) = 2\mathrm{Im}\Pi_n^{(0)}(1 - e^{-\beta\omega})^{-1}\mathcal{C}_V^{-1}$$
$$S_A(q,\omega) = 2\left[\mathrm{Im}\Pi_p^{(0)} + \mathrm{Im}\Pi_n^{(0)}\right](1 - e^{-\beta\omega})^{-1}\mathcal{C}_A^{-1},$$

$$(35)$$

where

$$\mathcal{C}_V = (1 - v_F\mathrm{Re}\Pi_n^{(0)})^2 + v_F^2(\mathrm{Im}\Pi_n^{(0)})^2$$
$$\mathcal{C}_A = \left[1 - v_{GT}(\mathrm{Re}\Pi_p^{(0)} + \mathrm{Re}\Pi_n^{(0)})\right]^2 + v_{GT}^2\left[\mathrm{Im}\Pi_p^{(0)} + \mathrm{Im}\Pi_n^{(0)}\right]^2$$

$$(36)$$

and

$$v_F = 1.76 \times 10^{-5} \text{ MeV}^{-2} \qquad \text{and} \qquad v_{GT} = 4.5 \times 10^{-5} \text{ MeV}^{-2}. \quad (37)$$

The polarization functions, $\Pi^{(0)}$, contain the full kinematics of the scattering and are easily derived:

$$\mathrm{Re}\Pi^{(0)}(q,\omega) = \frac{m^2}{2\pi^2 q\beta}\int_0^\infty \frac{ds}{s} log\left[\frac{1 + e^{-(s+Q_-)^2+\beta\mu}}{1 + e^{-(s-Q_-)^2+\beta\mu}}\right] + (\omega \to -\omega) \quad (38)$$

and

$$\mathrm{Im}\Pi^{(0)}(q,\omega) = \frac{m^2}{2\pi q\beta}log\left[\frac{1 + e^{-Q_+^2+\beta\mu}}{1 + e^{-Q_+^2+\beta\mu-\beta\omega}}\right], \quad (39)$$

where

$$Q_\pm = \left(\frac{m\beta}{2}\right)^{1/2}\left(\mp\frac{\omega}{q} + \frac{q}{2m}\right). \quad (40)$$

Importantly, the $\mathcal{C}_V$ and $\mathcal{C}_A$ act like dielectric constants. In fact, the spectrum of energy transfers in neutrino scattering is considerably broadened by the interactions in the medium. An identifiable component of this broadening comes from the absorption and emission of quanta of collective modes akin to the Gamow–Teller and Giant–Dipole resonances in nuclei (zero-sound; spin sound), with Čerenkov kinematics. This implies that all scattering processes may need to be handled with the full energy redistribution formalism and that ν-matter scattering at high densities can not be considered elastic. One consequence of this reevaluation, and a similar one for charged-current reactions [53] is that the late-time ($\geq$ 1000 milliseconds) neutrino luminosities may be as much as 50% larger for more than a second than heretofore estimated [52]. These luminosities reflect more the deep protoneutron star interiors than the early post-bounce luminosities of the outer mantle and the accretion phase. Since neutrinos drive the explosion, this may have a bearing on the specifics of the mechanism, but it is too soon to tell.

7   The Role of Progenitor Structure in Core-Collapse Supernovae

The density profiles of collapsing cores are functions of progenitor ZAMS mass [56,57]. The less massive progenitors (8 $M_\odot \lesssim M \lesssim$15-20(?)$M_\odot$) have compact cores with tenuous envelopes, while the more massive progenitors (20 (?) $M_\odot$ $\lesssim M$) have massive cores and dense envelopes. The density at 2 $M_\odot$ interior mass for the 38 $M_\odot$ progenitor is *six* orders of magnitude greater than that at the same point for the 11 $M_\odot$ progenitor. The structure of the core determines the mass accretion rate after bounce, as well as the binding energy of the inner mantle that an explosion must overcome to leave a neutron star rather than a black hole. A high mass accretion rate may smother, inhibit, or delay explosion. In a sense, a progenitor's inner density structure determines the outcome of its core's collapse. The cut between compact and extended cores may be at a different ZAMS mass, but a bifurcation into two classes seems to be an important ingredient of the collapse story. In the calculations of Burrows, Hayes, and Fryxell [1], while the 15 $M_\odot$ progenitor exploded, the 20 $M_\odot$ progenitor did not. However, the explosion in that calculation, as well as in that of Herant *et al.* [2], occurred too early to be consistent with nucleosynthetic constraints, ejecting ten [1] and one hundred [2] times as much neutron-rich material as the data allow. In and of itself, better transport will no doubt positively influence the duration of the delay to explosion [3] and the resulting nucleosynthetic yields, but more attention needs to be paid to progenitor stellar evolution. In this regard, Bazan and Arnett [58] are studying oxygen and silicon burning prior to collapse in 2-D. In the oxygen-burning zone, they are obtaining Mach #'s near 25% and anisotropies in the composition, density, and velocity that are far in excess of those inferred using the mixing-length prescription. These calculations suggest that the final word has not been uttered concerning supernova progenitor structures.

8 Speculations on Systematics

In addition to not being able to resolve current issues concerning progenitor structure, theory is also not yet adequate to determine the systematics with progenitor mass of the explosion energies, residue masses, ^{56}Ni yields, kicks, or, in fact, almost any parameter of a real supernova explosion. Despite this, there are hints, both observational and theoretical, some of which I would like to touch on here. The gravitational binding energy ($B.E.$) exterior to a given interior mass is an increasing function of progenitor mass, ranging at 1.5 $M_\odot$ interior mass from about 10^{50} ergs for a 10 $M_\odot$ progenitor to as much as 3×10^{51} ergs for a 40 $M_\odot$ progenitor [1,56]. This large range must affect the viability of explosion and its energy. It is not unreasonable to conclude, in a very crude way, that $B.E.$ sets the scale for the supernova explosion energy. When the "available" energy exceeds the "necessary" binding energy, both very poorly defined quantities at this stage, explosion is more "likely." However, how does the supernova, launched in the inner protoneutron star, know what binding energy it will be called upon to overcome when achieving larger radii? Since the post–bounce, pre-explosion accretion rate ($\dot{M}$) is a function of the star's inner density profile, as is the inner $B.E.$, and since a large $\dot{M}$ seems to inhibit explosion, it may be via $\dot{M}$ that $B.E.$, at least that of the inner star, is sensed. Furthermore, a neutrino–driven explosion requires a neutrino–absorbing mass and there is more mass available in the denser core of a more massive progenitor. One might think that binding energy and absorbing mass partially compensate or that a more massive progenitor can just wait longer to explode, until its binding energy problems are buried in the protoneutron star and $\dot{M}$ has subsided. The net effect in both cases may be similar explosion energies for different progenitors, though the residue mass could be systematically higher for the more massive stars. However, if these effects do not compensate, the fact that binding energy and absorbing mass are increasing functions of progenitor mass hints that the supernova explosion energy may also be an increasing function of mass. Since $B.E.$ varies so much along the progenitor continuum, the range in the explosion energy may not be small. Curiously, the amount of ^{56}Ni produced explosively also depends upon the mass between the residue and the radius at which the shock temperature goes below the explosive Si–burning temperature, a radius that depends upon explosion energy. Hence, the amount of ^{56}Ni produced may also increase with progenitor mass. Thermonuclear energy only partially compensates for the binding energy to be overcome, the former being about 10^{50} ergs for every 0.1 $M_\odot$ of ^{56}Ni produced.

Not all ^{56}Ni produced need be ejected. Fallback is possible and whether there is significant fallback must depend upon the binding energy profile. Personally, I think that there is not much fallback for the lighter progenitors, perhaps for masses below 15 $M_\odot$, but that there is significant fallback for the heaviest progenitors. The transition between the two classes may be abrupt. I base this surmise on the miniscule binding energies and tenuous envelopes of the lightest massive stars and on the theoretical prejudice that the r–process, or some fraction of it, originates in the protoneutron winds that follow the explosion for the lightest massive stars [59]. If there were significant fallback, these winds and their products would be smothered.

If there is significant fallback, the supernova may be in jeopardy and much of the ^{56}Ni produced will reimplode. There may be a narrow range of progenitor mass over which the supernova is still viable, while fallback is significant and both the mass of ^{56}Ni *ejected* and the supernova energy are decreasing. Above this mass range, a black hole may form. Hence, both low–mass and high–mass supernova progenitors may have low ^{56}Ni yields. Recently, two Type IIp supernovae have been detected, SN1994W [60] and SN1997D [61], which have very low ^{56}Ni yields ($\leq 0.0026\,M_\odot$ and $\leq 0.002\,M_\odot$, respectively), long–duration plateaus, and large inferred ejecta masses ($\geq 25M_\odot$). The estimated explosion energy for SN1997D is a slight 0.4×10^{51} ergs. (SN1987A's explosion energy was $1.5 \pm 0.5 \times 10^{51}$ ergs and its ^{56}Ni yield was 0.07 $M_\odot$.) These two supernovae may reside in the fallback gap and imply that the black hole cut–off is near 30 $M_\odot$.

In sum, supernova ^{56}Ni yields may vary by a factor of $\sim$100 and may peak at some intermediate progenitor mass, the supernova explosion energy may vary by a factor of $\sim$ 4 and also may peak at some intermediate progenitor mass, and the black hole hole cut–off mass may be near 30 $M_\odot$. Whether theoretical calculations will bear out these hinted–at systematics is unclear.

However, whatever the answers to the outstanding questions that still surround supernova theory, neutrino physics and transport will no doubt prove to be central elements. Even to those of us who have been involved with this topic for many years, the centrality of the weak interaction and of neutrinos to such a dramatic macroscopic phenomenon remains intriguing and a source of continuous fascination. Such has been the recent progress in both theory and observation that one can with no little confidence conclude that we are at the crossroads of a new understanding of both supernova neutrinos and supernova explosions, in all their diversity. With encouragement, I commend this subject to a new generation of avid practitioners.

Acknowledgments

I would like to acknowledge helpful and productive conversations with Phil Pinto, Ron Eastman, Steve Bruenn, Tony Mezzacappa, and Sanjay Reddy. In addition, thanks are extended to Todd Thompson for his aide with respect to the neutrino cross sections and to the NSF for its support through grant AST-96-17494.

References

1. A. Burrows, J. Hayes, and B.A. Fryxell, *Ap.J.* **450**, 830 (1995).
2. M. Herant, W. Benz, J. Hix, C. Fryer, and S.A. Colgate, *Ap.J.* **435**, 339 (1994).
3. A. Mezzacappa, *et al.*, *Ap.J.* **495**, 911 (1998).
4. H.-T. Janka and E. Müller, *A.&A.* **290**, 496 (1994).
5. A. Burrows and J. Hayes, *Phys. Rev. Lett.* **76**, 352 (1996).
6. A. Burrows, to be published in the proceedings of the 5'th CTIO/ESO/LCO Workshop entitled "SN1987A: Ten Years Later," eds. M.M. Phillips and N.B. Suntzeff, held in La Serena, Chile, February 24–28, (1997).
7. A. Burrows, in the proceedings of the 9'th Workshop on Nuclear Astrophysics, held at the Ringberg Castle, Germany, March 23–29, ed. E. Müller and W. Hillebrandt, p. 76 (1998).
8. D.S. Miller, J.R. Wilson, and R.W. Mayle, *Ap.J.* **415**, 278 (1993).
9. I. Lichtenstadt, A. Kholkhov, and J.C. Wheeler, *Ap.J.* , submitted (1998).
10. H. Bethe and J.R. Wilson, *Ap.J.* **295**, 14 (1985).
11. Mayle, R., Ph.D thesis. Univ. Calif., Berkeley (UCRL preprint no. 53713, 1985).
12. R.I. Epstein, *M.N.R.A.S.* **188**, 305 (1979).
13. R. Mayle and J.R. Wilson, *Ap.J.* **334**, 909 (1988).
14. S.W. Bruenn and T. Dineva, *Ap.J.* **458**, L71 (1996).
15. A. Burrows, T.J. Mazurek, and J.M. Lattimer, *Ap.J.* **251**, 325 (1981).
16. A. Burrows, *Ap.J.* **318**, L57 (1987).
17. W. Keil, H.-T. Janka, and E. Müller, *Ap.J.* **473**, 111 (1996).
18. S.S. Lawrence, *et al.*, *A.J.* **109**, 2635 (1995).
19. R.P. Kirshner, J.A. Morse, P.F. Winkler, and J.P. Blair, *Ap.J.* **342**, 260 (1989).
20. R. Strom, H.M. Johnston, F. Verbunt and B. Aschenbach, *Nature* **373**, 590 (1994).

21. C.S.J. Pun, R.P. Kirshner, P.M. Garnavich, and P.Challis, *B.A.A.S.* **191**, 9901 (1998).
22. D.H. Wooden, *et al.*, *Ap.J. Suppl.* **88**, 477 (1993).
23. P.A. Harrison, A.G. Lyne, and B. Anderson, *M.N.R.A.S.* **261**, 113 (1993).
24. A. Lyne and D.R. Lorimer, *Nature* **369**, 127 (1994).
25. C. Fryer, A. Burrows, and W. Benz, *Ap.J.* **496**, 333 (1998).
26. I. Wasserman, J. Cordes, and D. Chernoff, in preparation (1998).
27. V.M. Kaspi, *et al.*, *Nature* **381**, 584 (1996).
28. D. Lai, L. Bildsten, and V.M. Kaspi, *Ap.J.* **452**, 819 (1995).
29. E.P.J. van den Heuvel and S. Rappaport, in *I.A.U. Colloquium 92*, eds. A. Slettebak and T.D. Snow (Cambridge Univ. Press), pp. 291–308 (1987).
30. V. Kalogera, *P.A.S.P.* **109**, 1394 (1997).
31. P. Caraveo, *Ap.J.* **415**, L111 (1993).
32. D.A. Frail, W.M. Goss, and J.B.Z. Whiteoak, *Ap.J.* **437**, 781 (1994).
33. R. Petre, C.M. Becker, and P.F. Winkler, *Ap.J.* **465**, L43 (1996).
34. H. Spruit and E.S. Phinney, *Nature*, **393**, 139 (1998).
35. D. Lai and Y.-Z. Qian, *Ap.J.* **491**, 270 (1998).
36. M. Duncan and C. Thompson, *B.A.A.S.* **191**, 119.08 (1997).
37. S.W. Bruenn, *Ap.J. Suppl.* **58**, 771 (1985).
38. R. Eastman and P. Pinto, *Ap.J.* **412**, 731 (1993).
39. D. Mihalas, *Ap.J.* **238**, 1034 (1980).
40. D. Mihalas and B. Mihalas, *Foundations of Radiation Hydrodynamics*, New York, Oxford University Press (1984).
41. P.J. Schinder, *Ap.J. Suppl.* **74**, 249 (1990).
42. L.B. Leinson, V.N. Oraevsky, and V.B. Semikoz, *Phys. Lett.* B**209**, 80 (1988).
43. D.L. Tubbs and D.N. Schramm, *Ap.J.* **201**, 467 (1975).
44. D.A. Dicus, *Phys. Rev.* D**6**, 941 (1972).
45. H.-T. Janka, *A.&A.* **244**, 378 (1991).
46. Suzuki, H., in *Frontiers of Neutrino Astrophysics*, ed. Y. Suzuki and K. Nakamura (Tokyo: Universal Academy Press), p. 219 (1993).
47. W. Haxton, *Phys. Rev. Lett.* **60**, 1999 (1990).
48. S.E. Woosley, D. Hartmann, R. Hoffman, and W.C. Haxton, *Ap.J.* **356**, 272 (1990).
49. O.E.B. Messer, A. Mezzacappa, S.W. Bruenn, and M.W. Guidry, *Ap.J.* **507**, 353 (1998).
50. S.E. Woosley and R.D. Hoffman, *Ap.J.* **395**, 202 (1992).
51. Y.-Z. Qian and S.E. Woosley, *Ap.J.* **471**, 331 (1996).

52. A. Burrows and R.F. Sawyer, *Phys. Rev.* C58, 554 (1998).

53. A. Burrows and R. Sawyer, *Phys. Rev.* C, in press (1998).

54. S. Reddy, M. Prakash, and J.M. Lattimer, *Phys. Rev.* D58, no. 013009 (1998).

55. S. Yamada, in preparation (1998).

56. T.A. Weaver and S.E. Woosley, *Ap.J. Suppl.* 101, 181 (1995).

57. K. Nomoto and M. Hashimoto, *Phys. Repts.* 163, 13 (1988).

58. G. Bazan and D. Arnett, *Ap.J.* 433, L41 (1994).

59. G. Mathews, G. Bazan, and J. Cowan, *Ap.J.* 391, 719 (1992).

60. J. Sollerman, *et al.*, *Ap.J.* 493, 933 (1998).

61. M. Turatto, *et al.*, *Ap.J.* 498, L129 (1998).

GRAVITATIONAL WAVES

DANIEL SIGG

LIGO Hanford Observatory, P.O. Box 1970 S9-02,
Richland, WA 99352, USA
E-mail: sigg_d@ligo.mit.edu

A new generation of long baseline gravitational wave detectors is currently under construction (LIGO, VIRGO, GEO and TAMA). They incorporate high sensitive Michelson interferometers and have a design goal of measuring displacements of order 10^{-17} m r.m.s., integrated over a 100 Hz bandwidth centered at the minimum noise region. The purposes of these detectors is to observe gravitational waves from astrophysical sources at cosmological distances, and to open a new view to the universe by collecting information not accessible by conventional telescopes. These lectures present a description of the most promising candidate sources; and summarize the design characteristics of interferometric detectors—in particular, the Laser Interferometer Gravitational-wave Observatory (LIGO).

1 Introduction

According to general relativity theory gravity can be expressed as a space-time curvature[1,2]. One of the theory predictions is that a changing mass distribution can create ripples in space-time which propagate away from the source at the speed of light. These freely propagating ripples in space-time are called gravitational waves. Any attempts to directly detect gravitational waves have not been successful yet. However, their indirect influence has been measured in the binary neutron star system PSR1913+16[3,4,5,6].

This system consist of two neutron stars orbiting each other. One of the neutron stars is active and can be observed as a radio pulsar from earth. Since the observed radio pulses are Doppler shifted by the orbital velocity, the orbital period and its change over time can be determined precisely. If the system behaves according to general relativity theory, it will loose energy through the emission of gravitational waves. As a consequence the two neutron stars will decrease their separation and, thus, orbiting around each other at a higher frequency. From the observed orbital parameters one can first compute tho amount of emitted gravitational waves and then the inspiral rate. The calculated and the observed inspiral rates agree within experimental errors (better than 1%).

Gravitational waves are quite different from electro-magnetic waves. Most electro-magnetic waves originate from excited atoms and molecules, whereas observable gravitational waves are emitted by accelerated massive objects.

Also, electro-magnetic waves are easily scattered and absorbed by dust clouds between the object and the observer, whereas gravitational waves will pass through them almost unaffected. This gives rise to the expectation that the detection of gravitational waves will reveal a new and different view of the universe. In particular, it might lead to new insights in strong field gravity by observing black hole signatures, large scale nuclear matter (neutron stars) and the inner processes of supernova explosions. Of course, stepping into "uncharted territory" also carries the possibility to encounter the unexpected and to discover new kinds of astrophysical objects.

Table 1 shows an overview of the gravitational wave frequency bands, their most mature detection methods and their most likely sources.

Currently, a number of long baseline laser interferometers are under construction with the goal to be operational at the beginning of the new millennium. These interferometers incorporate high power stabilized laser sources, complicated optical configurations, suspended optical components and high performance seismic filters. They have arm lengths of up to 4 km and operate in a ultra high vacuum environment.

Table 1. Overview of frequency bands, detection methods and sources (see Ref.[7], and references therein). NS – neutron star and BH – black hole.

f(Hz)	λ	method	source
$\sim 10^{-16}$	$\sim 10^9$ lt.yrs.	anisotropy of μwave background	primordial
$\sim 10^{-9}$	~ 10 lt.yrs.	timing of milli second pulsars	primordial, cosmic strings
$\sim 10^{-4}$ to 10^{-1}	~ 0.01 AU to 10 AU	Doppler tracking of spacecraft, laser interferometer in space (LISA)	binary stars, supermassive black holes
~ 10 to 10^3	~ 300 km to 30,000	laser interferometers on earth (VIRGO, LIGO, GEO, TAMA)	inspirals: NS/NS, NS/BH, BH/BH
$\sim 10^3$	~ 300 km	Cryogenic resonant bar detectors	supernovæ spinning NS

Section 2 introduces gravitational waves and their general relativistic description; section 3 presents a summary of promising astrophysical sources which could be strong enough for a first direct detection. Section 4 describes laser interferometers and, in particular, the Laser Interferometer Gravitational-wave Observatory (LIGO) Project.

2 Waves in General Relativity

2.1 Weak field approximation

General Relativity predicts gravitational waves as freely propagating 'ripples' in space-time[1,2]. Far away from the source one can use the weak field approximation to express the curvature tensor $g_{\mu\nu}$ as a small perturbation $h_{\mu\nu}$ of the Minkowski metric $\eta_{\mu\nu}$ (see, for example, Ref.[8]):

$$g_{\mu\nu} = \eta_{\mu\nu} + h_{\mu\nu} \quad \text{with } |h_{\mu\nu}| \ll 1 \tag{1}$$

Using this ansatz to solve the Einstein field equations in vacuum yields a normal wave equation. Using the transverse-traceless gauge its general solutions can be written as

$$h_{\mu\nu} = h_+(t - z/c) + h_\times(t - z/c) \tag{2}$$

where z is the direction of propagation and h_+ and $h_\times$ are the two polarizations (pronounced 'plus' and 'cross'):

$$h_+(t - z/c) + h_\times(t - z/c) = \begin{pmatrix} 0 & 0 & 0 & 0 \\ 0 & h_+ & h_\times & 0 \\ 0 & -h_\times & h_+ & 0 \\ 0 & 0 & 0 & 0 \end{pmatrix} e^{(i\omega t - ikx)} \tag{3}$$

The above solution describes a quadrupole wave and has a particular simple physical interpretation (see Fig. 1): Let's assume two free masses are placed at positions x_1 and x_2 ($y = 0$) and a gravitational wave with + polarization is propagating along the z-axis, then the free masses will stay fixed at their coordinate positions, but the space in between—and therefore the distance between x_1 and x_2—will expand and shrink at the frequency of the gravitational wave. Similarly, along the y-axis the separation of two points will decrease and increase with opposite sign. The strength of a gravitational wave is then best expressed as a dimension-less quantity, the strain h which measures the relative length change $\Delta L/L$.

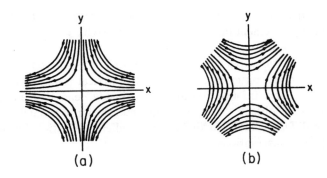

Figure 1. Direction of space deformation for a gravitational wave propagating along the z-axis, + polarization (a) and × polarization (b).

Table 2 shows a comparison between gravitational wave and electro-magnetic waves[9]. The combination of measuring the amplitude of a gravitational wave and having a large solid angle acceptance makes the event rate of gravi-

Table 2. Comparison between electro-magnetic and gravitational waves[9].

	electro-magnetic waves	gravitational waves
medium	space as medium	space-time itself
source	incoherent superposition of atoms and molecules	coherent motion of huge masses
resolution	imaging – λ small compared to source	$\lambda \geq$ scale of sources no spatial resolution
interaction	absorbed, scattered and dispersed by matter	very small interaction no shielding
frequency	10^7 Hz and up	10^4 Hz and down
detection	measure power (light) and amplitude (radio)	measure amplitude
acceptance	detectors are directional	detectors accept large solid angles

tational wave detectors scale with the third power of their sensitivity. In other words, every improvement of a factor of 2 in sensitivity will increase the event rate of astrophysical sources by a factor of 8.

Electro-magnetic waves which are visible to an observer on earth are usually produced in the outer layers of an astrophysical object, whereas gravitational waves carry information about the inside behaviour and the mass distribution of an object. Arguably, the information obtained by the two will be quite different; and it is difficult to predict gravitational sources from electro-magnetic observations.

2.2 Gravitational wave amplitudes

Before looking at possible detection techniques we (very) roughly estimate how large the observed effect of a gravitational wave form an astrophysical source could be. If we denote the quadrupole of the mass distribution of a source by Q, a dimensional argument—together with the assumption that gravitational radiation couples to the quadrupole moment only—yields:

$$h \sim \frac{G\ddot{Q}}{c^4 r} \sim \frac{G(E_{\text{kin}}^{\text{non}-\text{symm.}}/c^2)}{c^2 r} \tag{4}$$

with G the gravitational constant and $E_{\text{kin}}^{\text{non}-\text{symm.}}$ the non symmetric part of the kinetic energy. If one sets the non-symmetric kinetic energy equal to one solar mass

$$E_{\text{kin}}^{\text{non}-\text{symm.}}/c^2 \sim M_\odot \tag{5}$$

and if one assumes the source is located at inter-galactic or cosmological distance, respectively, one obtains a strain estimate of order

$$h \lesssim 10^{-21} \qquad \text{Virgo cluster} \tag{6}$$

$$h \lesssim 10^{-23} \qquad \text{Hubble distance} \tag{7}$$

By using a detector with a baseline of 10^4 m the relative length changes become of order:

$$\Delta L = hL \lesssim 10^{-19}\,\text{m to } 10^{-17}\,\text{m} \tag{8}$$

This is a rather optimistic estimate. Most sources will radiate significantly less energy in gravitational waves. We add that the observable effect is not small because the radiated energy is small—in contrary it is huge—but rather because space-time is a "stiff medium".

2.3 Gravitational wave frequencies

Similarly, one can estimate the upper bound for the frequencies of gravitational waves. A gravitational wave source can not be much smaller than its Schwarzshild radius $2GM/c^2$, and cannot emit strongly at periods shorter than the light travel time $4\pi GM/c^3$ around its circumference. This yields a maximum frequency of

$$f \leq \frac{c^3}{4\pi GM} \sim 10^4 \text{ Hz } \frac{M_\odot}{M} \qquad (9)$$

From the above equation one can see that the expected frequencies of emitted gravitational waves is the highest for massive compact objects, such as neutron stars or solar mass black holes.

2.4 Experimental evidence for gravitational waves

The only experimental evidence for gravitational waves comes from the timing of binary pulsar systems[6,10]. These systems consists of two neutron stars orbiting each other. To be observable one of them must be active and emit radio waves. Since pulsars emit radio waves mainly along their magnetic axis and since their rotation axis doesn't have to be aligned with the magnetic axis, earth-based radio antennæ can observe a periodic radio signal if the system is aligned so that the radio beacon passes over the earth. The frequency of this signal is determined by the rotation period of the pulsar and is typically of very high precision.

In a double neutron star system this periodic signal is modulated by the orbital frequency of the two neutron stars and can therefore be used to precisely determine the orbital period and phase. The first double pulsar system, PSR B1913+16, was discovered by Hulse and Taylor in 1974[3,4,5]. It is located in the Milky Way, its orbital period is $\sim$ 8 hours and the received radio signal repeats itself at a rate of $\sim$ 17/sec. The emission of gravitational waves brings the two neutron stars closer together, and thus increase the orbital frequency. Fig. 2 shows the advance of the orbital phase as function of time relative to a system which would have a constant orbital period. The loss of potential energy in this system is in agreement with the emission of gravitational waves predicted by general relativity theory[10,11]. As a consequence the two neutron stars will merge in about 300 million years.

Due to their tiny effect on space-time the direct observation of gravitational waves has not been successful until now. A list of the most mature methods, their applicable frequency band and the most likely sources in this band were already presented in Table 1.

598

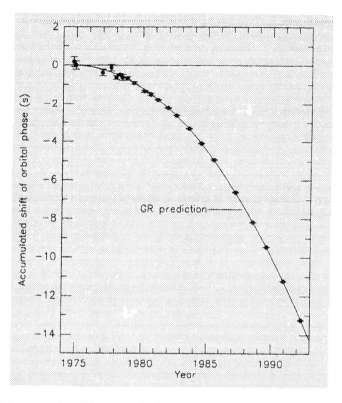

Figure 2. Advance of the orbital phase in the binary pulsar system PSR B1913+16. The plot is taken from Ref.[10].

3 Astrophysical sources

Only massive astrophysical object are good candidates for emitting gravitational waves which can be detected by an observer on earth. A more extensive overview of promising sources of gravitational waves can be found in Ref.[7]; we only give a brief summary here.

3.1 Coalescing compact binaries

Compact binaries are among the best candidates to be first seen by an earth-based gravitational-wave antenna. Compact binaries can consist of either two neutron stars, two black holes or one of each. Due to their small size

($\sim$ 20 km in case of a neutron star), they can orbit each other at close range and a high orbital frequency (up to $\sim$ 500 Hz). Being very close and rotating fast means that the second time derivative of the mass quadrupole moment is large and, hence, gravitational waves are emitted with a high efficiency. Indeed, the radiated energy is so large, that a double neutron star system which is 500 km or 100 km apart will loose all its potential energy within a couple of minutes or seconds, respectively. Since the emission of gravitational wave becomes more efficient at closer range, the waveform is a chirp signal (see Fig. 3 and Ref.[12]): increasing both in amplitude and frequency with time, until the two object are close enough to merge. To first order the the chirp signal can be described by the change of its frequency over time $\dot{f}$ and by its amplitude A:

$$\dot{f} \propto M_c^{5/3} f^{11/3} + \begin{pmatrix} \text{relativistic corrections} \\ M_1, M_2, S_1, S_2 \end{pmatrix} \qquad (10)$$

$$A \propto k_{\text{orbit}} M_c^{5/3} \frac{f^{2/3}}{r} \qquad (11)$$

with M_c the chirp mass

$$M_c = \frac{(M_1 M_2)^{3/5}}{(M_1 + M_2)^{1/5}}, \qquad (12)$$

f the orbital frequency, M_1, M_2, S_1 and S_2 the mass and spin of the two compact objects, respectively, k_{orbit} a constant accounting for the inclination of the source orbital plane and r the distance to the source. If enough binary systems are detected, one can average over orbital parameters and can use them as standard candles. (One can determine the distance from the second equation using the chirp mass from the first equation.)

Being able to determine the exact waveform of an inspiral event will also reveal additional information about the system itself (see, for example, Ref.[13]):

- harmonic content $\Rightarrow$ eccentricity of orbit

- even-odd modulation $\Rightarrow$ mass ratio of the two objects

- modulation of waveform $\Rightarrow$ spin-orbit coupling (mainly frame dragging in black-hole systems)

- higher-order corrections to waveform sweep$\Rightarrow$ individual mass and spin of constituents

600

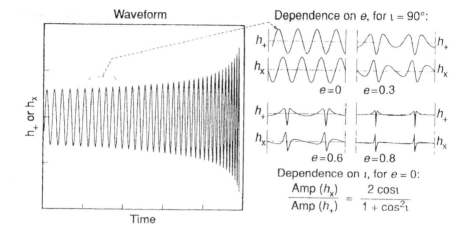

Figure 3. Chirp waveform from an inspiral event of a compact binary system. On the right hand side the dependency of the waveform on the orbital eccentricity e and the orbital inclination ι is demonstrated. The plot is taken from Ref.[14].

- end point (merger) $\Rightarrow$ large scale nuclear matter. If the nuclear state equation of a neutron star is soft the merger may happen earlier due to a hydrodynamic melting effect. On the other hand the gravitational field of the companion star may trigger the neutron start to fall into a black hole before the actual merger.

Calculating waveforms for coalescing compact binaries is straight forward, if the distance between the two objects is large, but for black hole mergers it is a formidable challenge. The coalescence of two black holes can be roughly divided into three phases:

- inspiral: The two black holes are well separated and the waveform of the emitted gravitational waveform is known,

- merger: The horizons of the two black holes merge together and the calculation of the exact waveform requires extensive simulations on a super computer, and

- linear pulsations: The two black holes have merged into a single black hole in an excited state. The excited state can be treated as a linear pulsation which decays by emitting gravitational waves.

3.2 Binary stars

Ordinary binary stars are one of the most reliably understood sources for periodic gravitational waves. Binary stars typically have orbital periods larger than an hour and, correspondingly, gravitational wave frequencies $\leq 10^{-3}$ Hz. This means that only space-based detectors will be able to detect them by integrating over long time periods (see section 4.4).

3.3 Rotating neutron stars

A rotating neutron star will emit gravitational waves if its mass distribution is non axis-symmetric along the rotation axis. A non axis-symmetric mass distribution could be due to extremely strong magnetic fields which deform the star, due to its past history which created the star in a deformed state, or due to accretion of matter from a companion star.

3.4 Neutron star instabilities

Only recently, it was recognized that gravitational radiation tends to drive the r-modes (hydrodynamic currents within the star's core) of all rotating stars unstable[15,16,17]. Gravitational radiation couples to these modes primarily through the current quadrupole, rather than the quadrupole of the mass distribution. These neutron stars can spin down to a fraction of their initial frequency within a relatively short period of time (~ 1 year).

3.5 Supernovæ

Supernovæ have all the attributes associated with a good gravitational wave source: they weigh several solar masses, they are compact and they experience large accelerations. However gravitational radiation only couples to a changing quadrupole moment and, hence, if a supernova collapse and the subsequent explosion have an axial symmetry, no gravitational waves are emitted.

There are several possible mechanisms which could overcome this deficit:

- Initial density and temperature fluctuations may trigger the collapse unevenly,

- High rotation speeds can lead to a bar instability,

- Hydrodynamic instabilities could introduce large convection streams which may effect the initial implosion

• A reminiscent neutron star may experience a strong boiling shortly after its formation.

It is unlikely that each and every supernova event will be exactly axis symmetric, but how large the asymmetries are and how often these asymmetries lead to detectable gravitational waves is very much uncertain at this time.

If a supernova is seen both in the electro-magnetic and the gravitational wave spectrum, one will also be able to compare the speed of light with the propagation speed of gravitational waves (general relativity theory predicts them to be the same).

3.6 Supermassive black holes

An other good sources of gravitational waves are supermassive black holes $(M > 10^5 M_\odot)$ eating surrounding objects. However, due to their size the frequency band of interest is lower than the one for the above sources. Typical frequencies are in the mHz regime and will not be accessible by earth-based observations due to limitations posed by seismic activities and gravity gradient noise (see next chapter). However, these sources are prime candidates for space-based antennæ.

3.7 Stochastic background

Density fluctuations in the early universe can lead to a stochastic background of gravitational waves (similar to the microwave background). Measuring the spectrum of the stochastic background would connect us to the Planck area and would be a good mean to discriminate different cosmological models (inflation, cosmic strings, QCD phase transitions). However, for most models the predicted amplitude of the stochastic background is well below the sensitivity of what is technologically achievable today or in the intermediate future.

4   Laser interferometers

The idea of detecting gravitational waves using a Michelson interferometer was discovered by several groups independently[18,19,20,21], and lead to the first prototype of an interferometric detector[22,23]. The idea took a significant step forward when R. Weiss[21] performed a study which identified all the important noise sources which limit the instrumental sensitivity (see next section).

There are two complementary approaches to detect gravitational waves with laser interferometers: space-based and earth-based. A space-based antenna is free from seismic excitations and can utilize long arm lengths of order 10^{10} m. It is best suited to detect gravitational waves in a frequency band between $\sim 10^{-4}$ Hz and $\sim 10^{-1}$ Hz. An earth-based antenna is limited by gravity gradient noise below a couple of Hz; in reality, seismic noise probably sets this limit even higher. Earth-based detectors have their best sensitivity in a frequency band between $\sim 10^1$ Hz and $\sim 10^3$ Hz.

4.1 Noise sources

Measuring length deviations smaller than the proton radius puts high requirements on the technology used to build these instruments. It also requires a good understanding of physical and technical noise sources which possibly limit the gravitational wave sensitivity. The design sensitivity of the Laser Interferometer gravitational wave Observatory (LIGO) Project is shown in Fig. 4. It shows that the sensitivity at low frequency, $f < 50$ Hz, is due to seismic noise, at intermediate frequencies, 50 Hz$< f < 150$ Hz, due to thermal noise and at high frequencies, $f > 150$ Hz, due to laser shot noise. The following paragraphs are listing noise sources influencing the strain measurement by directly affecting the laser light (limiting noise sources for initial earth-based interferometric detectors are shown in bold):

- **shot noise:** The fluctuations of the number of photons in the input beam causes fluctuations of the signal at the anti-symmetric port. For a power-recycled Michelson interferometer with Fabry-Perot arm cavities (see section 4.2) one obtains an equivalent shot noise limited strain sensitivity of

$$h_{\text{shot}}(f) \sim \frac{\sqrt{1 + (f/f_{\text{FPI}})^2}}{N_{\text{bounce}}} \frac{\lambda}{2\pi L} \sqrt{\frac{h\nu}{G_{\text{RC}} P_{\text{in}}}} \tag{13}$$

with f_{FPI} the cavity pole, N_{bounce} the average number of effective bounces in the arms, λ and ν the laser wavelength and frequency, respectively, L the arm length, G_{RC} the power-recycling gain and P_{in} the input laser power. Fig. 5 shows the sensitivity spectrum of the phase noise interferometer at MIT[24], demonstrating that it is technically possible to achieve shot noise limited sensitivity above a couple of 100 Hz.

- light amplitude and laser frequency noise: In a perfect Michelson interferometer common-mode noise sources such as the laser amplitude and frequency noise do not propagate to the anti-symmetric port. But in

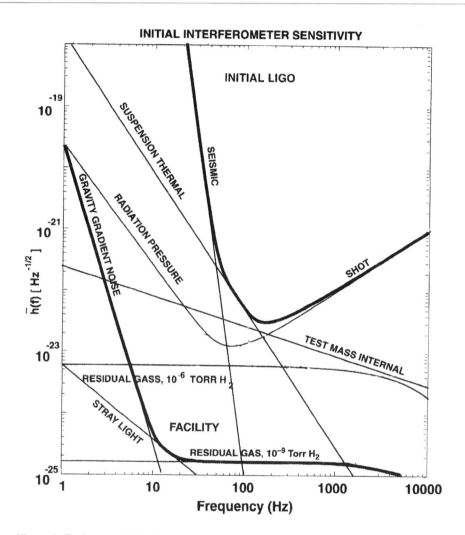

Figure 4. Design sensitivity for the Laser Interferometer Gravitational wave Observatory (LIGO). The plot shows that the initial strain sensitivity is limited by seismic, thermal and shot noise. These are technical noise sources which can be improved on in future designs. The plot also shows the gravity gradient, the scattered light and the residual gas noise which ultimately will limit the sensitivity of earth-based detectors.

reality, any small imbalance between the two Michelson arms will couple laser noise into the gravitational wave band. In a power-recycled Michelson interferometer with Fabry-Perot arm cavities, the amplitude noise mainly couples through differential deviation from resonance, whereas laser frequency noise couples through arm cavity differences in reflectivity and frequency response, and through differences in the path lengths of the Michelson. Even so these coupling coefficients are generally small, together with the required strain sensitivity, it still translates to very stringent requirements on the laser.

- oscillator phase and amplitude noise: A heterodyne detection scheme (see section 4.3) requires an oscillator to generate the rf modulation sidebands. Phase and amplitude noise of this oscillator can be coupled to the antisymmetric port through differential arm length deviations.

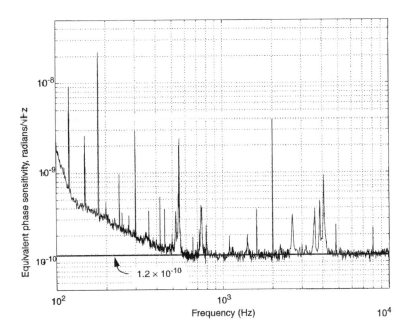

Figure 5. Spectral sensitivity of the MIT phase noise interferometer (see Ref.[24]). Above 500 Hz the spectrum is shot noise limited at a level close to the one needed for initial earth-based detectors. The additional features seen in the spectrum are due to 60 Hz powerline harmonics, wire resonances (500 Hz – 600 Hz), optical mount resonances (700 Hz – 800 Hz), calibration line (2 kHz) and resonances of the magnet standoffs ($\sim$ 4 kHz).

- scattered light: Light which scatters out of the beam path because of an imperfect optical surface and which then scatters from an outside surface back into the interferometer will produce a parasitic interference driven by the motion of the outside scattering surface. Extreme care is taken to isolate the optics of the interferometer from seismic excitations. It is important not to by-pass this isolation through parasitic interference from surfaces directly connected to the ground. Even if the motion is slow, it can be larger than a wavelength and, thus, cause an up-conversion of seismic noise to the gravitational wave band.

  Back scattering is the main reason the interferometer beams are contained in vacuum and not guided through fiber optics.

- beam jitter: Jitter of the input beam, both lateral and in angle, can couple to the anti-symmetric port through static angular misalignments of the interferometer.

- residual gas column density fluctuations: Density fluctuations in a gas induce fluctuations of the refractive index and lead to Rayleigh scattering.

Another set of noise sources cause displacement noise by introducing fluctuation forces which are moving the end points of the interferometer:

- **seismic noise:** The earth surface is in constant motion because of seismic and volcanic activities, because of ocean waves "hammering" on the shores, because of wind and because of the tidal forces of the moon. Seismic noise is most pronounced at low frequencies (0.1 Hz to 10 Hz) and falls off quickly at higher frequencies. Typical seismic noise levels are

$$x(f) \simeq 10^{-9} \text{ m}/\sqrt{\text{Hz}} \qquad \text{for } 1\,\text{Hz} < f \leq 10\,\text{Hz} \qquad (14)$$

$$x(f) \simeq \frac{10^{-7}}{f^2} \text{ m}/\sqrt{\text{Hz}} \qquad \text{for } f > 10\,\text{Hz} \qquad (15)$$

For initial earth-based interferometers roughly an attenuation of 9 orders of magnitude is required at frequencies around 100 Hz.

- **thermal noise in the suspension elements:** Thermally driven motions of the test masses (optical components) will limit the initial sensitivity of earth-based detectors in the intermediate frequency range around 100 Hz. The magnitude of these motions depends on $k_B T$, with k_B the Boltzmann constant and T the temperature. To investigate the effect of thermal noise one has to look at its spectral density. There is a deeper connection between the dissipation mechanism of a system and the power spectral density of the random displacements. Low loss systems typically have high Q resonances. Most of the random motion is concentrated in a small bandwidth around these resonances. By decreasing the dissipation of a system, one can increase the Q and at the same time reduce the spectral density of the random displacements away from resonance (for a more detailed description of thermal noise see for example[25]).

A simple way to obtain a low loss system is to suspend the test masses in form of a pendulum. The restoring force of a pendulum has two components: gravity and the elasticity of the suspension wire. For all practical purposes the "gravitational spring" is lossless, and only the elastic spring constant has a dissipative fraction. As long as the wire is reasonably fine, the elastic spring constant is much smaller than the gravitational spring constant. Typically, the pendulum frequency for a suspended test mass is around ~ 1 Hz. Above resonance the spectral density falls as $f^{5/2}$ (frictional damping).

The effect of thermal noise on the strain sensitivity of an interferometer is proportional to the (average) number of effective bounces of the laser beam. This is the main reason to favor a long baseline design with a low number of bounces over a shorter design with a higher number of bounces.

The sensitivity curve of the Caltech 40 m interferometer[26] is shown in Fig. 6. It clearly demonstrates the importance of thermal noise.

- **thermal noise driving mirror normal modes**: The equipartition theorem states that every eigenmode of a system is excited by thermal noise to a mean energy of $k_B T/2$. This is also true for the "drum" modes of the test masses. Typically, the frequencies of these eigenmodes is in the kHz regime.

- radiation pressure imbalance: The number of photons hitting either end test mass will fluctuate due to the photon count statistics. The recoil of these photons will then introduce a small force which pushes on the test masses. For a power-recycled Michelson interferometer with Fabry-Perot arm cavities one obtains a radiation pressure equivalent strain sensitivity of

$$h_{\gamma P}(f) \simeq \frac{2}{\pi^2} N_{\text{bounce}} \frac{\sqrt{G_{\text{RC}} P_{\text{in}} h\nu}}{LMcf^2} \qquad (16)$$

608

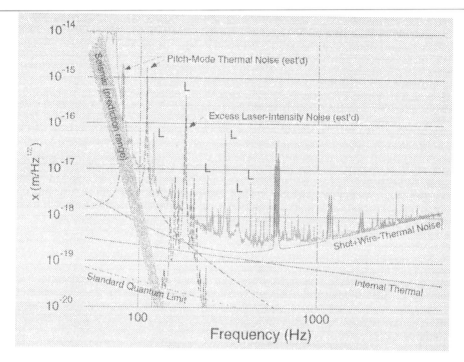

Figure 6. Spectral sensitivity of the Caltech 40 m interferometer (see Ref.[26]). It shows a displacement sensitivity comparable to the one of initial earth-based interferometers. The seismic noise prediction is an empirical one based on measurements of the ground noise and the transfer function of ground motion to interferometer displacement. The thermal noise prediction is a theoretical one based on measured frequencies and Q's for various modes and the assumption that the loss function is a constant for each mode. The shot noise curve is calculated theoretically and has been confirmed experimentally to within $\sim 20\%$. The broad peaks near 600, 1200 and 1800 Hz are sets of narrow violin-mode resonances of the test mass suspension wires excited by thermal noise which blend together in this relatively low resolution (approximately 1 Hz bandwidth) spectrum. The remaining peaks are largely instrumental artifacts. The most numerous are powerline frequency harmonics caused by electrical interference (marked "L") in the electronics used for this measurement. The peaks at 80 and 109 Hz are pendulum pitch-mode resonances.

with M the mass of the optical components. If one combines Eq. (13) and Eq. (16), one can see that without increasing the test mass there is a natural limit for a given frequency on how much one can increase the laser power to reduce the shot noise, before the radiation pressure noise becomes a problem.

- "radiometer" force: Photons absorbed in the mirror coating can transfer their energy to molecules which are bouncing off the mirror surface. The increased recoil of these molecules will apply an additional force to a test mass.

- gravity gradients: Any mass placed nearby an optical component will apply a force through gravity. Moving masses such as seismic waves compressing the earth and density fluctuations in the air are the main concerns, since they give rise to gravity gradients. For earth-based detectors this will set the ultimate limit in sensitivity at very low frequencies.

- electric field fluctuations: Varying external electric fields together with a (induced) surface charge can also affect a test mass.

- magnetic field fluctuations: Presently, most suspended test masses incorporate actuators for the active control system which consist of permanent magnets glued to the back of the test mass and a coil driver mounted to the suspension cage. External magnetic fields can then apply a force to a test mass, either, through an imbalance in the magnets or through field gradients.

- cosmic ray muons: Cosmic ray muons can be produced at high altitude when a high energy proton enters the earth atmosphere. Because the cross-section of muons is small they can reach the ground and in some rare cases stop in a test mass. The recoil of the muon then looks like a "random" force.

4.2 Interferometer configurations

Most modern designs implement improved versions of a simple Michelson interferometer (see Fig. 7). A simple Michelson interferometer has an antenna response function $A(\Omega)$ which is proportional to (see the appendix on how to derive an interferometer response function):

$$A(\Omega) \propto \text{sinc}\left(\frac{\Omega L}{c}\right) \qquad \text{Michelson} \qquad (17)$$

with $\text{sinc}\, x = \sin x / x$, Ω the angular frequency of the gravitational wave and L the length of each arm. Putting numbers into Eq. (17) shows that for frequency between 10 Hz and 1 kHz, the optimal antenna length is of order 10^5 m to 10^7 m. This is much larger than would be feasible for an earth-based detector. However, there is no reason that the arm of a Michelson

610

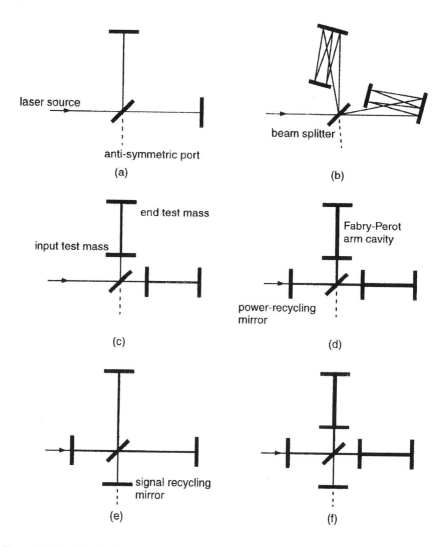

Figure 7. Possible interferometer configurations for gravitational wave detectors: simple Michelson interferometer (a), Michelson with delay lines (b), Michelson with Fabry-Perot arm cavities (c), power-recycled Michelson with Fabry-Perot arm cavities (d), dual recycled Michelson (e) and dual recycled Michelson with Fabry-Perot arm cavities (f). There are many more configurations; some of them are mentioned in the text.

interferometer cannot be folded[27]. Indeed, this configuration is known as a delay line and its antenna function is proportional to:

$$A(\Omega) \propto \text{sinc}\left(N\frac{\Omega L}{c}\right) \qquad \text{delay line} \qquad (18)$$

where N describes the number of folds (number of bounces). In practice, this configuration has a couple of disadvantages:

- If the number of folds is large, the mirror which is used to bounce the laser beams forth and back, has to be large as well. This is compounded by the fact that for long baseline interferometers the diffraction limited beam diameter is of order 10 mm to 100 mm.

- Light scattered by an imperfect mirror away from the nominal angle of reflection can interfere with the light from neighboring light passes and ruin the instrumental sensitivity.

A similar effect to folding the light pass N times can be achieved by inserting a Fabry-Perot cavity into each arm of the Michelson[28,29,30]. A Fabry-Perot cavity consists of a partially transmitting input mirror and a high reflective rear mirror. If the length of the Fabry-Perot is adjusted to a multiple of the laser wavelength the cavity becomes resonant. The light power inside the cavity builds up and simulates the effect of sending the light forth and back multiple times. However, in this case the number of bounces is not a fixed quantity, but rather an averaged effective value. Both the problem of the mirror size and the scattering is now much reduced, since the multiple light paths are now lying on top of each other. But, a Fabry-Perot cavity has to be hold on resonance during operations which requires an active control system. The antenna function of a Michelson interferometer with Fabry-Perot arm cavities can be written as[31,32]:

$$A(\Omega) \propto \text{sinc}\left(\frac{\Omega L}{c}\right) \text{FPI}\left(\frac{\Omega L}{\pi c}\right) \qquad \begin{array}{l} \text{Michelson with} \\ \text{Fabry-Perot arm cavities} \end{array} \qquad (19)$$

The Fabry-Perot response function (power build-up inside the cavity) depends on the input and rear mirror amplitude reflectivity coefficients, r_1 and r_2, respectively, and the input mirror amplitude transmission coefficient t_1 (see for example Ref.[33]).

$$\text{FPI}(x) = \left|\frac{t_1}{1 - r_1 r_2 e^{ix}}\right|^2 \qquad (20)$$

If the mirrors have low optical losses and if the rear mirror is a high reflector, most of the power incident to a Fabry-Perot arm cavity will be reflected back to the beam splitter. Ideally, the anti-symmetric port of the Michelson interferometer is set on a dark fringe to minimize shot noise. Then a differential length change induced by a gravitational wave will leave through the anti-symmetric port with the highest possible signal-to-noise ratio. This in turn means that most of the injected light will leave the interferometer through the symmetric port and be lost. But, by placing an other partially transmitting mirror at input one can form yet another cavity—the power recycling cavity—and recycle most of the otherwise lost light[34]. The interferometer response is then enhanced by the power recycling gain (additional power build-up in the power recycling cavity).

$$A(\Omega) \propto \text{sinc}\left(\frac{\Omega L}{c}\right) \text{FPI}(\frac{\Omega L}{\pi c}) G_{\text{RC}} \qquad \begin{array}{l} \text{Power-recycled} \\ \text{Michelson with} \\ \text{Fabry-Perot arm cavities} \end{array} \qquad (21)$$

By adding a partially transmitting mirror to the anti-symmetric output port the gravitational wave signal can be made resonant[35,36]. This makes it possible to shape the interferometer response, so that its sensitivity is improved in a narrow frequency band around the signal resonance. In general, this means that the sensitivity outside the resonant frequency band will be worse. This is not a problem at lower frequencies where the interferometer is usually limited by seismic noise. If both power and signal recycling are implemented the configuration is called dual recycled.

The above configurations are the most common ones currently implemented or designed, but there other possible layouts such as Sagnacs[37], configurations with an output mode cleaner, resonant recycling where the beam splitter is turned by 90° to directly couple the two arm cavities[34], and many more.

4.3 Detection schemes

Most interferometer configurations require an active control system to keep cavities locked to a resonance, or to keep the anti-symmetric port on a dark fringe, respectively. To be able to implement a feedback system one first needs an error signal which measures the microscopic longitudinal deviations. Neither of the above conditions would allow for simply monitoring the power levels, since moving away from resonance or away from a dark fringe will decrease or increase the power levels, respectively, without indicating the direction of the deviation. One could solve this problem by putting the "working point" of the feedback control system off resonance and towards mid fringe. But, this technique makes both power and signal recycling impossible.

All currently built and planned interferometers therefore implement a heterodyne detection scheme[21]. Historically, the first heterodyne detection schemes implemented a longitudinal dither of the cavities. This in turn modulates the power in the cavity and, if off-resonance, yields an error signal at the dither frequency. However, a laser source typically becomes shot noise limited at radio frequencies (rf) only, well above dither frequencies which are achievable in the lab.

A better scheme—the Pound-Drever-Hall reflection locking technique[28,29,30]—imposes phase modulate rf sidebands on the laser light itself. An off-resonance cavity then acts as an FM-to-AM converter yielding error signals at the rf frequency. The gravitational wave readout port usually implements a suppressed carrier scheme[39]. A differential length deviation will produce a signal at the carrier frequency leaking out the anti-symmetric port, which then beats against constant rf sidebands.

4.4 LISA: A space-based interferometer

The LISA (Laser Interferometer Space Antenna) Project[a,40] is a planned space mission, adopted by ESA and NASA, to deploy 3 satellites in solar orbit forming a large equilateral triangle with a base length of 5×10^6 km. The center of the triangle formation will be in the ecliptic plane 1 AU from the sun and 20 degrees behind the earth. The main objective of the LISA mission is to observe low frequency (10^{-4} Hz to 10^{-1} Hz) gravitational waves from galactic and extra-galactic binary systems, including gravitational waves generated in the vicinity of the very massive black holes found in the centers of many galaxies. The three LISA spacecrafts flying in formation will act as a giant Michelson interferometer, measuring the distortion of space caused by passing gravitational waves. Each spacecraft will contain two free-floating "proof masses". Lasers in each spacecraft will be used to measure changes in the optical path lengths with a precision of 20 pm. If approved, the project will start in the year 2005 with a planned launch in 2008.

A sensitivity plot is shown in Fig. 8. The primary goal of the LISA mission is to detect and study in detail gravitational wave signals from sources involving massive black holes. LISA will certainly observe distinguishable signals from thousands of binary systems containing compact stars, and be able to determine the number and distribution of such binaries in our galaxy.

---

[a] home pages: http://www.lisa.uni-hannover.de
http://lisa.jpl.nasa.gov

614

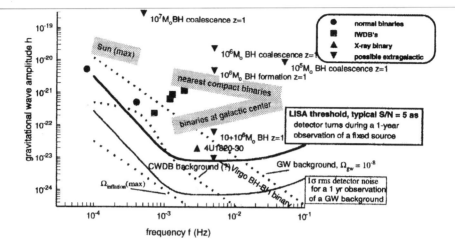

Figure 8. Sensitivity curve of the Laser Interferometer Space Antenna (LISA). Most of the LISA sources will be approximately monochromatic. The bold curve is the 1-year threshold curve, the amplitude that could be detected with confidence by a single (2-arm) LISA interferometer; it is drawn at the signal-to-noise (SNR) ratio of 5 for a fixed source that is observed by LISA for a full year.

The gravitational wave amplitude h is shown for different types of periodic and quasiperiodic sources. The expected signals from some known binaries are indicated. The nearest neutron star and white dwarf binaries at any frequency should lie in the band labeled "nearest compact binaries"; the band below that shows the amplitude expected from "typical" white dwarf binaries near the galactic center. Interacting white dwarf binaries (IWDB) are systems where a low-mass degenerate helium star fills its Roche lobe and transfers mass onto a more massive white dwarf. The shortest period stellar mass black hole binary in the Virgo cluster might be in the position shown.

The strongest sources in the diagram are binaries of massive black holes at cosmological distances, observed as they coalesce due to the orbital emission of gravitational waves. They have been placed in the diagram rather arbitrarily at their coalescence frequency and at an amplitude that correctly shows their SNR in relation to the heavy threshold curve, for a distance $z = 1$.

The 1-σ gravitational wave noise produced by a possible cosmological background left from the big bang is shown here at an energy density per decade of frequency today that is 10^{-8} of the total needed to close the universe. An upper limit to that generated by inflation is also shown. There may be more galactic close white dwarf binaries (CWDB) than LISA can resolve; a possible but uncertain effective noise level is shown. For comparison with these backgrounds, we have drawn the LISA rms noise level (faint lower curve).

The band labeled "SUN (max)" is where solar g−modes might produce strong near zone (Newtonian) gravitational perturbations observable by LISA. The plot is taken from Ref.[40].

Table 3. List of earth-based laser interferometers projects currently under construction worldwide.

project	length	site	configuration
GEO[b]	600 m	Hannover, Germany	dual recycled
LIGO[c]	4000 m 2000 m 4000 m	Hanford (WA), USA Livingston (LA), USA	power-rec. Fabry-Perot
TAMA[d]	300 m	Tokyo, Japan	dual recycled
VIRGO[e]	3000 m	Pisa, Italy	power-rec. Fabry-Perot

4.5 The LIGO Project

The Laser Interferometer Gravitational wave Observatory[14] is one of the new projects to build the next generation of gravitational wave detectors. A list of all projects currently under construction is shown in Table 3.

The essential attributes of the LIGO Project are:

- collaboration between the California Institute of Technology and the Massachusetts Institute of Technology,

- two widely separated sites under common management to make coincidence measurements,

- a vacuum system to accommodate interferometers with 4 km arm length,

- the capability to operate several interferometers at each site simultaneously,

- the ability to accommodate interferometers of two different arm lengths, 4 km and 2 km (at one site),

- a clear aperture for the laser beam of ~ 1 m,

- An ultimate vacuum of 10^{-9} torr hydrogen and 10^{-10} torr of other gases,

---

[b]home page: http://www.geo600.uni-hannover.de
[c]home page: http://www.ligo.caltech.edu
[d]home page: http://tamago.mtk.nao.ac.jp
[e]home page: http://www.pg.infn.it/virgo

- a physical environment monitor for detecting vetos caused by external disturbances and

- a facility lifetime of at least 20 years to do astrophysical research with gravitational waves.

All earth-based interferometric gravitational wave detectors share a similar design philosophy. The design of these detectors is driven by the goal to minimize the effect of noise on the instrumental sensitivity (see section 4.1). All designs use in-vacuum suspended optics build on top of a seismic isolation system for their main interferometer mirrors. Similarly, all designs use a highly stabilized laser source in conjunction with a mode cleaner to deliver a high quality laser beam to the interferometer. They all incorporate an optical configuration which requires an active control system for microscopically adjusting cavity and Michelson lengths, in order to counteract drifts and fluctuations introduced by seismic activities.

A brief description of the main detector components of LIGO is given below:

- laser source: The light source is a solid state diode-pumped Nd:YAG laser, consisting of a nonplaner master oscillator and a power amplifier. The nominal output power is 10 W single mode at a wavelength of 1064 nm. The laser is locked to a reference cavity to stabilize its frequency and is spatially filtered by a pre-mode cleaner. Pockels cells are used to impose phase modulated sidebands on the laser light before it is launched into the mode cleaner.

- input mode cleaner: The mode cleaner is a triangular cavity with the purpose to further filter and stabilize the laser beam.

- seismic isolation system: The seismic isolation system is a a vibration isolation stack, constructed of heavy steel plates separated by coil springs. This forms a coupled pendulum system, giving a damping factor proportional to f^{-2} above resonance for each stage.

- suspensions: All major optical components are suspended to form a pendulum using one single loop of 0.012 in. diameter steel music (piano) wire. The pendulum frequencies are typically below 1 Hz and the mass of a large mirrors is $\sim$ 10 kg. Four permanent magnets are glued to the back to control longitudinal and angular orientation of the test mass, and two magnets are glued to the side to control sideways motions. Corresponding coil drivers are mounted to the suspension cage, making it possible to adjust the force applied to the mirror by adjusting the electric current through the coil.

- optics: A diffraction limited laser beam which spot is of similar sizes at the input and end test mass must have a waist size of order ∼ 30 mm – ∼ 40 mm. This requires rather large optics. In case of LIGO the masses are circular cylinders fabricated from pieces of high-purity fused silica with bulk absorption of less than 5 ppm/cm. They are 25 cm in diameter and 10 cm thick. To minimize scatter and absorption losses it is crucial to have a very good surface figure and a very low loss (≤ 1 pmm), very high uniformity coating. Surface figures of $\lambda/1000$ rms over the the central 8 cm diameter have been achieved (after coating).

- sensing and control system: Multiple InGaAs photodetectors are used at anti-symmetric port and for the auxiliary extraction ports to sense 4 longitudinal degrees-of-freedom and 14 angular degrees-of-freedom. The signal is first down-converted into the baseband and then sampled by a digitizer. Most servo functions are implemented in software, and the signals are send to the suspension controllers through fiber optics.

The planned completion dates for the LIGO Hanford 2 km system are shown below:

beam tube	completed, currently baked
vacuum system	installed and baked
seismic isolation	installation started
laser source	installation started
mode cleaner	end of 1998
vertex Michelson	spring 1999
full interferometer	2000
engineering tests	2001
first data run	2002 and 2003

618

5 Conclusions

The direct observation of gravitational waves will allow to test general relativity theory by giving direct evidence of a time-dependent metric far away from the source and by independently probing strong field gravity[41,11]. It will also provide a new and different view of astrophysical processes hidden from electro-magnetic astronomy, such as the inner dynamics of supernova and neutron star cores, or such as the coalescence phase of neutron star and black hole mergers. Eventually, it may be possible to discriminate cosmological models by observing or setting a limit to the stochastic background.

The new generation of gravitational wave detectors, currently under construction, has the potential to open this new field of physics and may result in new and unexpected discoveries.

There are also certain risks associated with "stepping into a unknown territory": Are there enough strong astrophysical sources for gravitational waves? And, will the technology work at the required level? However, a direct detection of gravitational waves will almost certainly bring invaluable advance of our experimental knowledge of the universe.

Acknowledgments

I would like to thank my colleges of the LIGO collaboration for many stimulating discussions and for providing the opportunity of participating on a very interesting and challenging experiment. The LIGO Project is supported by the National Science Foundation grant PHY–9210038.

Appendix: Interferometer response function

This appendix demonstrates the techniques to calculate the antenna response function of an interferometric gravitational wave detector. We use a power recycled Michelson interferometer with Fabry-Perot arm cavities as an example.

The coordinate system is chosen to be aligned with the two arms of the interferometer, where the origin is positioned at the beam splitter and the z-axis points vertically upwards. Spherical coordinates are defined by

$$r = \begin{pmatrix} r\sin\theta\cos\phi \\ r\sin\theta\sin\phi \\ r\cos\theta \end{pmatrix} \quad \text{with} \begin{cases} 0 \leq \theta < 2\pi \\ 0 \leq \phi < \pi \end{cases} \tag{22}$$

We then define the rotation operator $O(\theta, \phi)$ which rotates the z-axis in the direction of r:

$$O(\theta, \phi) = O(\phi)O(\theta) \tag{23}$$

where $O(\phi) = \begin{pmatrix} \cos\phi & \sin\phi & 0 \\ \sin\phi & \cos\phi & 0 \\ 0 & 0 & 1 \end{pmatrix}$ and $O(\theta) = \begin{pmatrix} \cos\theta & 0 & \sin\theta \\ 0 & 1 & 0 \\ -\sin\theta & 0 & \cos\theta \end{pmatrix}$ \quad (24)

We write the phase of the light which is acquired in one round-trip in one of the interferometer arms as

$$\Phi_{rt}(t_0) = \int_{t_0}^{t_0+t(2L)} dt\, \omega \tag{25}$$

where L is the length of the arm, ω is the angular frequency of the light and t_0 the time the photon leaves the origin. We now change the integration over time into one over length by using

$$d\tau^2 = dx^\mu g_{\mu\nu} dx^\nu 0 \quad \text{with} \quad g_{\mu\nu} = \eta_{\mu\nu} + h_{\mu\nu} \tag{26}$$

where $\eta_{\mu\nu}$ is the Minkovski metric and $h_{\mu\nu}$ is the space-time ripple due to the gravitational wave. For a gravitational wave traveling along the z-axis, $h_{\mu\nu}$ becomes in the transverse-traceless gauge[8]

$$h_{\mu\nu} = \cos(\Omega t - kz) \begin{pmatrix} 0 & 0 & 0 & 0 \\ 0 & & & \\ 0 & & \hat{H}_{ik} & \\ 0 & & & \end{pmatrix} \quad \text{with } \hat{H}_{ik} = \begin{pmatrix} h_+ & h_\times & 0 \\ -h_\times & h_+ & 0 \\ 0 & 0 & 0 \end{pmatrix} \quad (27)$$

where Ω is the angular frequency of the gravitational wave, k is its wave vector, h_+ and $h_\times$ are the wave amplitudes for the "+" and the "×" polarization, respectively.

For arbitrary directions one has to rotate both z and $\hat{H}_{ik}$ in the direction of the wave vector k.

$$kz \rightarrow k(k_x x + k_y y + k_z z) \quad \text{with} \quad \begin{cases} k_x = \sin\theta\cos\phi \\ k_y = \sin\theta\sin\phi \\ k_z = \cos\theta \end{cases} \quad (28)$$

$$\hat{H}_{ik} \rightarrow H_{ik} = O(\theta,\phi)\hat{H}_{ik}O(\theta,\phi)^{-1} \equiv \begin{pmatrix} h_{xx} & h_{xy} & h_{xz} \\ h_{yx} & h_{yy} & h_{yz} \\ h_{zx} & h_{zy} & h_{zz} \end{pmatrix} \quad (29)$$

For an integration along the x-axis or the y-axis h_{xx} and h_{yy} are the only relevant matrix elements, respectively.

$$h_{xx} = -\cos\theta\sin 2\phi\, h_\times + (\cos^2\theta\cos^2\phi - \sin^2\phi)\, h_+ \quad (30)$$

$$h_{yy} = \cos\theta\sin 2\phi\, h_\times + (\cos^2\theta\sin^2\phi - \cos^2\phi)\, h_+ \quad (31)$$

Fig. 9 shows the angular dependence of $|\,h_{xx} - h_{yy}\,|$ for both polarization and the average. Using Eq. (25) we rewrite Eq. (26) as

$$\Phi_{rt}^x(t_0) = \frac{\Omega}{c} \int_0^L dx \left\{ \sqrt{1 + h_{xx}\cos(\Omega t_0 + k(1 - k_x)x)} + \right.$$

$$\left. \sqrt{1 + h_{xx}\cos(\Omega t_0 + k(2L - (1 + k_x))x)} \right\} \quad (32)$$

Similarly, $\Phi_{rt}^y(t_0)$ can be obtained by integrating along the y-axis. Since $h_{xx} \ll 1$ we can expand the square root of Eq. (32) in a Taylor series. Performing the integration, keeping only time-dependent terms, time-shift from departure to arrival, and changing to a complex notation where the absolute value denotes the amplitude and the argument denotes the phase shift, one gets:

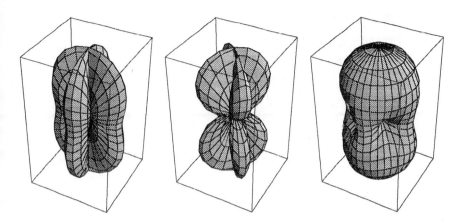

Figure 9. Antenna response function for an interferometric gravitational wave detector. The interferometer is placed at the center of the surrounding box with Michelson arms oriented along the horizontal axes. The distance from a point of the plot surface to the center of the box is a measure for the gravitational wave sensitivity in this direction. The plot to the left is for + polarization, the middle one for × polarization and the right one for unpolarized waves.

$$\Delta \Phi_{rt}^x = \frac{h_{xx} L \omega}{c} e^{i \Phi_\Omega} \frac{\sin \Phi_\Omega + i k_x \cos \Phi_\Omega - i k_x e^{i k_x \Phi_\Omega}}{\Phi_\Omega (1 - k_x^2)} \tag{33}$$

$$\simeq \frac{h_{xx} L \omega}{c} \operatorname{sinc} \Phi_\Omega \cos(\frac{k_x \Phi_\Omega}{\sqrt{12}}) e^{i(1 + k_x/2)\Phi_\Omega} \tag{34}$$

where $\Phi_\Omega = L\Omega/c$ and $\operatorname{sinc} x$ denotes $\sin x/x$. The approximation yields the exact solution for a gravitational wave traveling along the z-axis. From Eq. (34) one sees that the signal delay for photons arriving at the origin is $1 + k_x/2$ times half the round-trip time. The finite time a photon spends in a Michelson arm also leads to a small correction of the signal amplitude which would otherwise be determined by $h_{xx}L$ only. Fig. 10 shows the amplitude correction and time delay of the round trip phase of a gravitational wave as function of k_x relative to one of normal incident and strength h_{xx}. These effects are generally small and in most cases negligible.

To calculate the response of a cavity to a gravitational wave of a certain frequency Ω we write the electric field as a three-component vector denoting the carrier field, the upper audio sideband with frequency $+\Omega$ and the lower audio sideband with frequency $-\Omega$. The round-trip operator $X(\Omega)$ can be

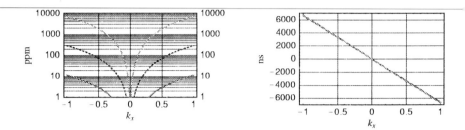

Figure 10. Amplitude correction and time delay for non normal incidence. For details see text.

expressed as[42]

$$X(\Omega) = \begin{pmatrix} 1 & 0 & 0 \\ -\frac{i}{2}\Delta\Phi_{rt}\,e^{-2i\Omega L/c} & 0 \\ -\frac{i}{2}\Delta\Phi_{rt}^* & 0 & e^{2i\Omega L/c} \end{pmatrix} \tag{35}$$

where we neglected the factor $e^{-2i\omega L/c+2i\eta}$ which is unity when the cavity is on resonance (with η the Gouy phase shift). The reflected field operator for a cavity then becomes

$$X_{\text{refl}}(\Omega) = \left(r_1 - (r_1^2 + t_1^2)\sqrt{1-\delta}\,X(\Omega)\right)\left(1 - r_1\sqrt{1-\delta}\,X(\Omega)\right)^{-1} \tag{36}$$

where r_1 and t_1 are the amplitude reflectivity and transmission coefficients of the input mirror and δ is the total round-trip loss (including the reflectivity of the rear mirror). Using a carrier only in the input field E_{in}, the reflected audio sidebands become

$$E_{\text{refl}}^{+\Omega} = \frac{i}{2}G_{\text{refl}}(\Omega)\,\Delta\Phi_{rt}\,e^{i\Omega t}E_{\text{in}} \quad\text{and}\quad E_{\text{refl}}^{-\Omega} = \frac{i}{2}G_{\text{refl}}^*(\Omega)\,\Delta\Phi_{rt}^*\,e^{-i\Omega t}E_{\text{in}} \tag{37}$$

$$G_{\text{refl}}(\Omega) = \frac{\sqrt{1-\delta}\,t_1^2}{(1-\sqrt{1-\delta}\,r_1))(1-\sqrt{1-\delta}\,r_1 e^{-2i\Omega L/c})} \tag{38}$$

$$\simeq \frac{\sqrt{1-\delta}\,t_1^2}{(1-\sqrt{1-\delta}\,r_1)^2}\frac{e^{i\Omega L/c}}{1+i\frac{\Omega}{\omega_{\text{cav}}}} \quad\text{for } \Omega \ll \frac{c}{2L} \tag{39}$$

and with the cavity pole at $\omega_{\text{cav}} = \dfrac{1 - r_1\sqrt{1-\delta}}{\sqrt{r_1\sqrt{1-\delta}}}$

$$\tag{40}$$

The audio sideband signal can be simplified to

$$E_{\text{refl}}^{+\Omega} + E_{\text{refl}}^{-\Omega} = \mid G(\Omega)\,\Delta\Phi_{rt}\mid \cos(\Omega t + \arg(G(\Omega)\,\Delta\Phi_{rt}))E_{\text{in}} \tag{41}$$

$$\equiv \mid g \mid \cos(\Omega t + \arg g) \tag{42}$$

The signal at the anti-symmetric port is then given by

$$E_{\text{anti}} = i\,t_{bs}r_{bs}\Big\{\sqrt{1-\delta_x}\,g_x\cos(\Omega t + \arg g_x)-$$

$$\sqrt{1-\delta_y}\,g_y\cos(\Omega t + \arg g_y)\Big\}E_{\text{RC}} \tag{43}$$

where r_{bs} and t_{bs} are the amplitude reflectivity and transmission coefficients for the beam splitter, respectively, δ_x and δ_y are the losses in the (short) inside Michelson arms for the incident carrier light and the reflected gravitational wave signal, g_x and g_y denote the signals from the in-line and the off-line arm cavities, respectively, and E_{RC} is the carrier field incident on the beam splitter. We now write the rf sideband signal at the anti-symmetric port as

$$E_{\text{dark}}^{\text{sb}} = 2i|E_{\text{sb}}|\sin\omega_m t \tag{44}$$

where $|E_{\text{sb}}|$ is the field strength of either rf sideband and ω_m is the angular modulation frequency. Down-converting the signal yields

$$V_{\text{dark}} = R\sin\omega_m t\,\epsilon_{\text{PD}}|E_{\text{dark}}^{\text{sb}} + E_{\text{dark}}|\,d^2 \tag{45}$$

$$\stackrel{\text{dc}}{=} \sqrt{32}R\,\epsilon_{\text{PD}}\sqrt{P_{\text{RC}}P_{\text{sb}}}\,t_{bs}r_{bs}\Big\{\sqrt{1-\delta_x}\,G_x(\Omega)\,\Delta\Phi_{rt}^x-$$

$$\sqrt{1-\delta_y}\,G_y(\Omega)\,\Delta\Phi_{rt}^y\Big\} \tag{46}$$

On the last line we returned to the complex notation where the absolute value denotes the signal amplitude and where the argument denotes the signal phase shift. R is the transimpedance gain of the mixer, filter, amplifier circuit chain, ϵ_{PD} is the efficiency of the photodetector, P_{RC} and P_{sb} are the carrier power on the beam splitter and the total sideband power at the anti-symmetric port, respectively.

References

1. A. Einstein, Preuss. Akad. Wiss. Berlin, Sitzungsberichte der physikalisch-mathematischen Klasse , 688 (1916).
2. A. Einstein, Preuss. Akad. Wiss. Berlin, Sitzungsberichte der physikalisch-mathematischen Klasse , 154 (1918).
3. R.A. Hulse and J.H. Taylor, Astrophys. J. **191**, L59–L61 (1974).
4. R.A. Hulse and J.H. Taylor, Astrophys. J. **195**, L51–L53 (1975).
5. R.A. Hulse and J.H. Taylor, Astrophys. J. **201**, L55–L59 (1975).
6. J.H Taylor, L.A. Fowler and P.M. McCulloch, Nature **277**, 437 (1979).
7. K.S. Thorne, *Gravitational Radiation*, in *300 Years of Gravitation*, eds. S.W. Hawking and W. Israel (Cambridge U. Press, Cambridge, UK, 1987); and references therein.
8. C.W. Misner, K.S. Thorne and J.A. Wheeler, *Gravitation* (W.H. Freeman and Copmpany, New York, 1973).
9. K.S. Thorne, in *Proceedings of the XXII SLAC Summer Institute on Particle Physics*, eds. J. Chan and L. DePorcel (SLAC Report 484, Standford, 1996), 41.
10. J.H Taylor, *Noble lecture: Binary Pulsars and Relativistic Gravity*, Rev. Mod. Phys. , 711–719 (1994).
11. T. Damour and G. Esposito-Farèse, Phys. Rev. D **58**, 042001 (1998).
12. B.F. Shutz, Nature **323**, 310 (1986).
13. C. Cutler and E.E. Flanagan, Phys.Rev. **D49**, 2658-2697 (1994).
14. A. Abramovici, W.E. Althouse, R.W.P. Drever, Y. Gürscl, S. Kawamura, F.J. Raab, D. Shoemaker, L. Sievers, R.E. Spero, K.S. Thorne, R.E. Vogt, R. Weiss, S.E. Whitcomb and M.E. Zucker, Science **256**, 325–333 (1992).
15. N. Anderson, gr-gc/9706075, to be published in Astrophys. J.
16. L. Lindblom, B.J. Owen and S.M. Morsink, Phys. Rev. Lett. **80**, 4843–4846 (1998).
17. N. Anderson, K.D. Kokkotas and B.F. Schutz, astro-ph/9805225.
18. F.A.E. Pirani, Acta Physica Polonica **15**, 389 (1956).
19. M.E. Gertsenshtein and V.I. Pustovoit, Soviet Phyics – JETP **14**, 433 (1962).
20. J. Weber, unpublished.
21. R. Weiss, Quarterly Progress Report of the Research Laboratory of Electronics of the Massachusetts Institute of Technology **105**, 54 (1972).
22. G.E. Moss, L.R. Miller and R.L. Forward, Appl. Opt. **10**, 2495 (1971).
23. R.L. Forward, Phys. Rev. D **17**, 379 (1978).
24. P. Fritschel, G. Gonzàlez, B. Lantz, P. Saha, M.E. Zucker, Phys. Rev. Lett. **80**, 3181–3184 (1998).

25. P.R. Saulson, *Fundamentals of Interferometric Gravitational Wave Detectors* (World Scientific, Singapore, 1994).
26. A. Abramovici, W. Althouse, J. Camp, J.A. Giaime, A. Gillespie, S. Kawamura, A. Kuhnert, T. Lyons, F.J. Raab, R.L. Savage Jr., D. Shoemaker, L. Sievers, R. Spero, R. Vogt, R. Weiss, S. Whitcomb and M. Zucker, Phys. Lett. A **218**, 157–163 (1996).
27. D. Herriot, H. Kogelnik and R. Kompfner, Appl. Opt. **3**, 523 (1964).
28. R.W.P. Drever, G.M. Ford, J. Hough, I. Kerr, A.J. Munley, J.R. Pugh, N.A. Robertson and H. Ward, *Proceedings of the Ninth International Conference on General Relativity and Gravitation*, ed. E. Schmutzer (VEB Deutscher Verlag der Wissenschaft, Berlin, Germany, 1980).
29. A. Schenzle, R. DeVoe and G. Brewer, Phys. Rev. A **25**, 2606–2621 (1982).
30. R.W.P. Drever, J.L Hall, F.V. Kowalski, J. Hough, G.M. Ford, A.J. Munley and H. Ward, Appl. Phys. B **31**, 97–105 (1983).
31. Y. Gürsel, P. Linsay, P. Saulson, P. Spero, R. Weiss and S. Whitcomb, unpublished Caltech/MIT manuscript (1983).
32. B.J. Meers, Phys. Lett. A **142**, 465–470 (1989).
33. A.E. Siegman, *Lasers* (University Science Books, Mill Valley, CA 94941, USA, 1986).
34. R.W.P. Drever, in *Gravitational Radiation*, eds. N. Deruelle and T. Piran (North-Holland, Amsterdam, Netherlands, 1983), 321.
35. B.J. Meers, Phys. Rev. D **38**, 2317–2326 (1988).
36. J. Mizuno. K.A. Strain, P.G. Nelson, J.M. Chen, R. Shilling, A. Rüdiger, W. Winkler and K. Danzmann, Phys. Lett. A **175**, 273–276 (1994).
37. K.-X. Sun, M.M. Fejer, E. Gustafson and R.L. Byer, Phys. Rev. Lett. **76**, 3053–3056 (1996).
38. P. Fritschel, D. Shoemaker and R. Weiss, Appl. Opt. **31**, 1412–1418 (1992).
39. L. Schnupp, Max Planck Institut für Quantenoptik, D–85748 Garching, Germany (personal communication, 1986).
40. LISA, *Pre-Phase A Report* (Max-Planck-Institut für Quantenoptik, 85748 Garching, Germany, 1995).
41. K.S. Thorne, *Probing Black Holes and Relativistic Stars with Gravitational Waves*, gr-qc-9706079 (1997).
42. J.-Y. Vinet, B.J. Meers, C.N. Man and A. Brillet, Phys. Rev. D **38**, 433–447 (1988).

The Beginning of Neutrino Astronomy

Alfred K. Mann
University of Pennsylvania
Department of Physics & Astronomy
209 S. 33rd Street
Philadelphia, PA 19104 USA
email: mann@dept.physics.upenn.edu

This is intended to be a semi-historical, semi-scientific paper on the origin of neutrino astronomy. In it we explore what has been learned about certain stars by using neutrinos to probe their interiors, and about the properties of neutrinos themselves when they traverse such dense matter and long distances.

Neutrino astronomy began with the observation of neutrinos from two extraterrestrial point sources: the Sun and Supernova 1987A. It used the property of the neutrino to probe the interiors of stellar bodies and provide information unattainable otherwise. But because the masses of those bodies are so large and their distances to Earth so great, other intrinsic neutrino properties are also manifested. This talk describes both the astronomical and elementary particle physics aspects of the phenomena and touches on cosmic ray neutrino physics as well.

Thirty years ago, the first attempt to detect neutrinos produced in the Sun began to yield positive results [1]. The chain of fusion reactions giving rise to solar neutrinos starts with the weak processes

$$p + p \to \, ^2H + e^+ + \nu_e \tag{1}$$

or

$$p + e^- + p \to \, ^2H + \nu_e. \tag{2}$$

The resultant deuterium reacts quickly through the reaction $p + \, ^2H \to \, ^3He + \gamma$, with two possibilities for ^3He: in Eq.1 it reacts with another ^3He to form ^4He and two protons or in Eq.2 it reacts about 15% of the time with ^4He present in the Sun through the process

$$^3He + \, ^4He \to \, ^7Be + \gamma. \tag{3}$$

Most of the ^7Be is converted to ^7Li by electron capture,

$$e^- + \, ^7Be \to \, ^7Li + \nu_e, \tag{4}$$

rapidly followed by

$$^7Li + p \rightarrow 2\ ^4He, \tag{5}$$

but about one ^7Be nucleus in a thousand undergoes the nuclear reaction

$$p +^7 Be \rightarrow^8 B + \gamma \tag{6}$$

These ^8B beta-decay into unstable $^8Be^*$:

$$^8\mathrm{B} \rightarrow\ ^8\mathrm{Be}^* + e^+ + \nu_e$$
$$\phantom{^8\mathrm{B} \rightarrow\ } \llcorner\!\!\rightarrow 2\,^4\mathrm{He}$$

$$\tag{7}$$

There are therefore three main sources of ν_e expected from the Sun:

1. By far the most copious source, called p-p neutrinos, is Eq. 1, which produces a continuous spectrum with an endpoint energy of 420 KeV.

2. Equation 4 produces a line source of ν_e with an energy of 862 KeV and a secondary line at 383 KeV. The integrated flux is thirteen times smaller than for Equation 1.

3. The ^8B decay (Equation 7) produces a continuous spectrum with an endpoint energy of approximately 15 MeV but a total flux 10^{-4} times that of p-p neutrinos.

These fluxes are derived from the standard model (SSM). The most important assumptions made in the model are as follows:

1. The interior of the Sun is in a state of hydrostatic equilibrium.

2. Energy is transported from the interior by radiative transfer except in the outer convective zone.

3. There is spherical symmetry and no significant rotation.

4. Chemical homogeneity exists at formation and is changed only locally as a result of nuclear reactions.

5. No unknown physics plays any significant role in the Sun.

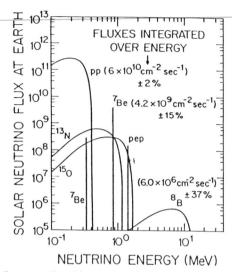

Figure 1: Solar neutrino fluxes predicted by the standard solar model. The energy thresholds for the gallium, chlorine, and Kamiokande-II neutrino detectors are 0.233, 0.814, and 9.3 MeV, respectively.

Detailed calculations based on that model[2] resulted in the neutrino fluxes shown in Figure 1. An independent calculation by another group[3] based on the same solar model used different nuclear and other parameters, particularly a different photon opacity as a function of temperature, density, and composition of the Sun's core, yielded precisely the same flux of p-p neutrinos as in[2], a flux of ^7Be neutrinos lower by 11% , and a ^8B neutrino flux lower by 35%. These differences may be taken as indicative of empirical uncertainties in the predictions of the *absolute* values of the solar neutrino fluxes.

The apparatus of[1], shown in Fig.2, was and still is located at a depth of 4850 ft in the Homestake Gold Mine in South Dakota, U.S.A. My description of Fig. 2 as the Homestake Neutrino Telescope has been criticized because traditionally a telescope provides an image of the object of interest, which the chlorine detector in Fig.2 did not do. Detection of solar neutrinos in that detector was done by producing the reaction

$$\nu_e + \,^{37}Cl \to e^- + \,^{37}Ar \tag{8}$$

where the 35-day lifetime of ^{37}Ar allowed it to be collected and counted long after the initial reaction happened. The use of ^{37}Cl for this purpose was first pointed out by Pontecorvo as early as 1948. Consequently, there was no

629

HOMESTAKE NEUTRINO TELESCOPE

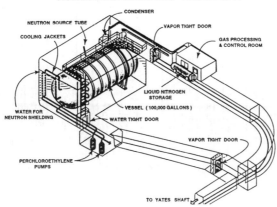

Figure 2: Schematic outline of the radiochemical solar neutrino telescope, one mile (1600 meters) deep underground in the Homestake Gold Mine in Lead, South Dakota, U.S.A. The tank holds 100,000 gallons of perchlorethylene (a dry cleaning fluid) which is both the target and detector for the solar neutrinos. The auxiliary equipment is for flushing helium gas through the perchlorethylene to remove the radioactive argon atoms produced by the solar neutrinos interacting with it, and for counting the individual argon atoms. This equipment began taking data in 1970, and continues to do so today. Courtesy Raymond Davis, Jr.

direct proof that the events occurring in the chlorine were caused by neutrinos from the Sun. Nevertheless, the Homestake Detector was the first to observe neutrinos that in all likelihood were from the Sun. The data accumulated by it are shown in Fig. 3 in which the average observed value for a 15-year period is compared with the value predicted in Reference[2]. The comparison, indicating a discrepancy of more than a factor of three between observed and calculated values, became known as the solar neutrino problem, a problem awaiting other experiments for its solution.

This was the situation when Masatoshi Koshiba and I attended a conference in Park City, Utah in January 1984 and met for the first time. Koshiba, a professor of physics in Tokyo University, had extensive experience in cosmic ray physics. A few years before the conference, he had completed construction of a massive water Cherenkov counter, named Kamiokande for the small town, Kamioka, near the mine in which the counter was located, to which was appended the letters n (for nucleon) and de for decay. Nucleon decay was an exciting prediction of grand unified theories of elementary forces (GUTs) which were current in the period 1970-1980s. The idea was to search for various decay modes of nucleons, e.g., $p \to e^+ + \pi^0$, that would be detected efficiently in a multikiloton water detector, even if the proton lifetime were as long as 10^{30} years, more or less the predicted value at the time. My own experience was

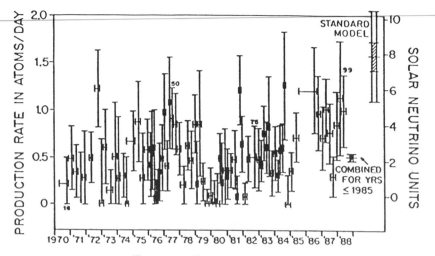

Figure 3: Production rate of ^{37}Ar in the ^{37}Cl detector as a function of time in years. The average rate in SNU for the period 1970-1985 is shown as 2.1 ± 0.3. Inclusion of the data for 1987-1988 raises the average to 2.3 ± 0.3. The prediction of the SSM from the calculation of Bahcall & Ulrich (1) is also shown, with the 3σ error quoted by them.

in elementary particle physics, particularly in neutrino scattering experiments of which the latest then was in a collaborative effort at Brookhaven National Laboratory in Long Island, NY, to measure the reaction $\nu_\mu + e \rightarrow \nu_\mu + e$. That elastic scattering reaction has the property shown in Fig. 4, namely that the square of the angle made by the recoiling electron relative to the incident neutrino is very small, varying inversely as the neutrino energy, E_ν.

Among other things, Koshiba and I discussed the possibility of observing the energetic (7–15 MeV) solar neutrinos from the ^8B decay (see Fig. 1) in his existing nucleon decay detector. Our idea was that knowledge of the angle and energy of the recoil electron from each elastic neutrino-electron scattering reaction in the detector—knowledge which could be obtained from measurement of the Cherenkov radiation for the electron—would specify the approximate direction of the incident neutrino, as indicated in Fig. 4. Furthermore, if the precise time of each event in the detector was also measured, the position of the Sun at that instant could be determined. Together, these would allow us to compare the angle made by the incident neutrino direction and the direction from the Sun to see if the neutrino was indeed from the Sun. This procedure would delineate the solar neutrino signal in the angular distribution of observed events, and in addition discriminate against background electrons from

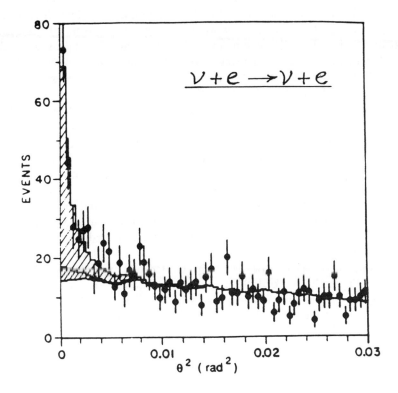

Figure 4: Differential distribution in θ_e^2 for the neutrino data. Data are points with error bars: y-independent term is light shaded; $(1-y)^2$ term is dark shaded and background is unshaded.

radioactive elements in the detector water whose Cherenkov radiation would show no angular correlation with the Sun.

Koshiba and I tended to think very much alike and found that we could talk easily and productively about that prospect. The enthusiasm it generated led us to go for a long walk in the late afternoon that induced an attack of laryngitis in Koshiba and prevented him from giving his scheduled lecture the following morning. By that time, we were friendly enough so that he felt able to ask me to give the talk for him using his material. The talk was on the future of Kamiokande and the promise of a ten times larger detector to be known as Super Kamiokande, and I felt at ease doing so. Before we left Park City, we arranged for several of us from the University of Pennsylvania to visit Tokyo and Kamioka to see the Kamiokande detector in action. It was the first of many such visits by Penn people, and by Tokyo people to Philadelphia.

We had convinced ourselves by calculation and observations in Kamiokande that ^8B solar neutrinos with energies in the region of 10 MeV could be measured if the detector was suitably modified. One modification entailed building a counter to enclose completely the entire area of the Kamiokande detector in order to record and veto the cosmic ray muons that penetrated the rock shielding and reached the detector at a rate of about 30,000 per day. The other modification was to provide new electronics for nanosecond timing of events and more sensitive measurement of the pulse amplitudes from the photomultiplier tubes in the detector. Both were big, moderately expensive tasks, and it was not until near the end of 1986 that the new detector—called Kamiokande II—was tested and ready to begin the solar neutrino observations we had planned two years earlier. The Kamioka Neutrino Telescope—Kamiokande II— is shown in Fig. 5.

The principal difficulty that we encountered in making Kamiokande II ready for data-taking was a large, time-varying background count rate which precluded solar neutrino measurements. That rate during a sample period in September 1986 is shown in Fig. 6. We had spent a number of months prior to that trying to reduce and steady the rate without success, and were becoming discouraged. Finally, almost at the end of 1986, we recognized how the radioactive background was getting into the detector and were able to reduce it to the much lower, steady level shown in Fig. 7. We were at last ready, we thought, to begin detecting solar neutrinos.

But that was to be delayed further. On February 23, 1987, to the astonishment of everyone in physics and astronomy, a supernova explosion took place in the Large Magellanic Cloud (LMC), the first supernova close enough to Earth to be seen by the naked eyes of observers in 384 years. Photographs of the star—Sanduleak 69^0202—are shown in Fig. 8 before and after it exploded.

KAMIOKA NEUTRINO TELESCOPE

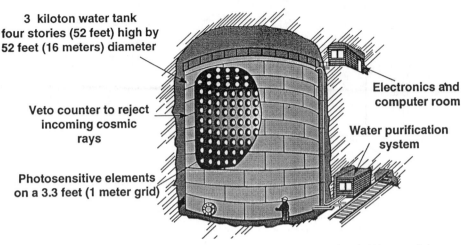

**3 kiloton water tank
four stories (52 feet) high by
52 feet (16 meters) diameter**

**Veto counter to reject
incoming cosmic
rays**

**Photosensitive elements
on a 3.3 feet (1 meter grid)**

**Electronics and
computer room**

**Water purification
system**

Figure 5: Schematic outline of the original neutrino observatory 3300 feet (1000 meters) deep in the Kamioka Mine in Japan. The tank holds 3000 tons of purified water viewed by 1000 photosensitive elements that detect light emitted by a rapidly moving electrically charged particle in the water. The counting room houses the sensitive electronics which processes signals from the photosensitive elements, and also the computers for preliminary analysis and storage of the data. The water purification system recirculates 125 tons of water each day to remove algae and radioactive matter. The railroad track is for the train to the mine entrance.

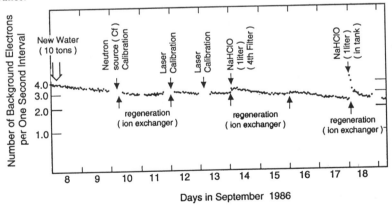

Figure 6: The high rate at which the Kamioka detector recorded electrons (entirely background events) during eleven days in September 1986, several months after the modified detector began operation. Each dot indicates the number of electrons recorded in a one second interval at a given time on a given day. Notice the several abrupt changes in rate and its slow but significant time dependence. The notations indicate maintenance procedures. Cf is the radioactive source Californium.

634

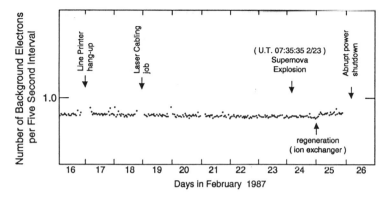

Figure 7: The rate corresponding to that in Fig. 6 measured in eleven days in February 1987 during which SN 1987A was observed. Steps taken before the end of 1986 stabilized the rate to the low constant value shown, less than one count per ten seconds. This made the observation of the Supernova Explosion possible on February 23rd. Note the abrupt, total power shutdown in the mine early on February 26th. U.T. stands for Universal or Greenwich time.

The neutrino signal that reached Kamiokande II from the supernova [4]—SN 1987A—the first ever to be observed from the collapsed core of a supernova— is shown in Fig. 9, and again in Fig. 10 with the confirming evidence from another massive water Cherenkov detector located in Lake Erie [5], near Cleveland, Ohio. The time order in which the neutrinos and the light from SN 1987A were observed is shown in Fig. 11. The earliest visual observation was recorded in photographic plates by McNaught in New South Wales, Australia, about three hours after the neutrinos were detected.

The process that gave rise to the staggeringly large neutrino emission began as the relatively young star, Sanduleak 69^0202, exhausted the supply of hydrogen in its core and was forced to burn helium as its energy generating fuel. This step required contraction of the core to increase its temperature, and further contraction and heating took place when the helium was used up and fusion of carbon became necessary to provide internal energy. The successive burning of higher mass fuels by the star continued until the core was almost entirely iron. The well-known dependence of nuclear binding energy on atomic mass makes clear that the binding energy per nucleon is a maximum at the element iron, and that consequently energy cannot be generated by the fusion of iron nuclei. The iron core of the star was without an energy generating source and unable therefore to withstand the force of gravity acting to collapse it. In a moment, the core changed its diameter from perhaps 10,000 km to

LARGE MAGELLANIC CLOUD

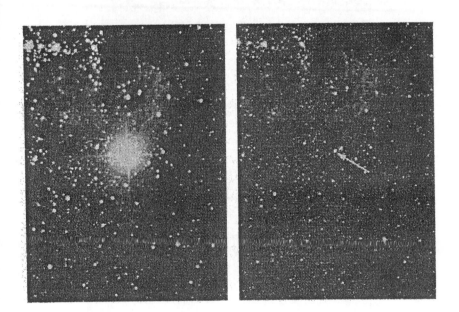

SN 1987A,
MARCH 1987

1984

Figure 8: Two photographs of the same portion of the Large Magellanic Cloud, a satellite galaxy of the Milky Way, by David Malin and Ray Sharples using the Anglo-Australian Telescope. The photograph on the right was taken in 1984. The arrow points towards the star Sanduleak. The photograph on the left was taken in March 1987, approximately one month after SN 1987A, which is evident at the position formerly occupied by Sanduleak.

636

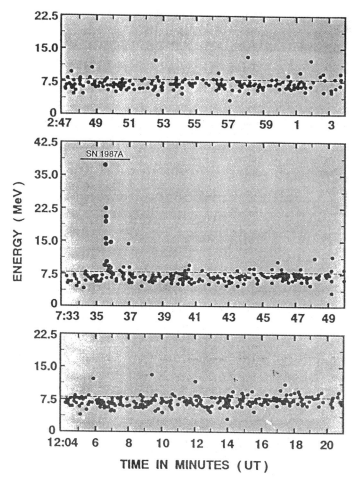

Figure 9: Individual events (electrons) observed in the Kamioka detector before, during, and after the occurrence of SN 1987A on February 23rd. Each dot indicates a single measured event whose energy can be read off the vertical axis and whose time of occurring is read off the horizontal axis. In the "before" and "after" plots, there is an occasional, single event with energy above the energy of almost all of the background events from radioactive elements in the water. The horizontal dashed lines are meant to guide the eye to the region containing most of the background events. The twelve events (positrons) produced by the neutrino burst from SN 1987A in the "during" plot are unmistakable.

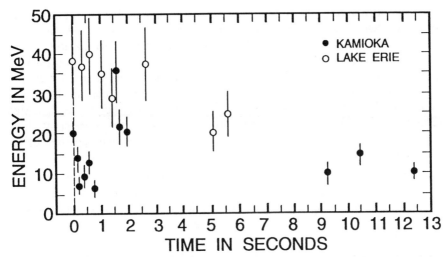

Figure 10: Plot showing the 12.4 second duration of the neutrino burst and energies of the produced events from both the Kamioka and Lake Erie detectors. This plot is similar to the "during" plot in Figure 9 but with a time scale in seconds rather than minutes. The earliest measured event in the Kamioka sample is adjusted to coincide in time with the earliest event in the Lake Erie sample. The vertical bars through the data points indicate the approximate errors in the energy measurements.

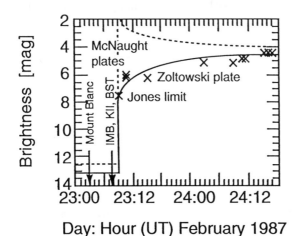

Day: Hour (UT) February 1987

Figure 11: Measured brightness of SN 1987A as a function of time on February 23rd and 24th. All visual observations fit well on the brightness curve that is initiated three hours after the Lake Erie (IMB), K II, and BST neutrino detection.

10 km, and was only restrained from becoming smaller by the repulsive force of the nucleon-nucleon potential at very short distance. The nuclei in the extremely hot collapsed core were disassociated into free neutrons and protons; the protons captured the electrons released in the collapse to form a neutron plus neutrino final state to join with the neutrons already there and give rise to a protoneutron star. The neutrinos from the electron capture by protons were not, however, the ones observed. Rather, in the hot ($T \simeq$ 3–5 MeV) protoneutron star, neutrino-antineutrino pairs of all flavors were created and, unlike the photons that produced them, were able to escape and thereby cool the star.

This model of the core collapse in a Type II supernova was the product of more than two decades of theorizing about the phenomenon [6]. It was thought that proof of the validity of the model would be observation of neutrinos carrying away the gravitational binding energy of the neutron star, and this is the way it turned out. Moreover, in addition to confirming the astrophysical model, the observation of neutrinos from a source 160,000 light years away made possible verification of several intrinsic properties of neutrinos. The three hour time interval between observation of the neutrinos and the light set an upper limit on the difference of the travel times of neutrinos and photons over that distance. This allowed a demonstration that the Weak Equivalence Principle is valid for both bosons and fermions [7]; and that the speeds of photons and neutrinos are the same within an uncertainty of 2 parts in 10^9. A number of estimates of neutrino mass were also made, the resultant upper limit being 23 eV at 95% confidence level [8].

After the excitement of SN 1987A had passed, Kamiokande II settled down to the business of observing solar neutrinos. This required patience and diligence because the recorded events were almost all due to radioactive background and each had to be analysed carefully to identify the few events for which the reconstructed direction of the incident neutrino was correlated with the direction of the Sun. Approximately two years of data-taking produced the plot shown in Fig. 12 which fully realized all of our hopes [9]. The angle θ_{Sun} in Fig. 12 was the angle between the observed direction of an electron in the detector and the radius vector from the Sun at the instant the electron was detected. In the absence of multiple scattering of the electron in the detector water, and measurement errors, the excess of events at $\cos \theta_{Sun} = 1 (\theta_{Sun} = 0^0)$ would be still more sharply peaked than it is, but even so the angular correlation clearly shows that the electrons in the peak were produced by neutrinos originating in the Sun. The isotropically distributed background in Fig. 12 was due to the residual radioactivity in the detector which we were unable to remove entirely.

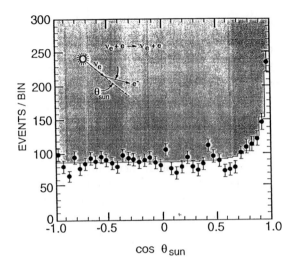

Figure 12: Plot showing the angular correlation of the solar neutrino events in the Kamioka detector with the direction from the Sun. The peak on the right hand side of the plot indicates solar neutrino events. The other data points are not correlated with the direction of the Sun.

The measured energy distribution of the electrons in the peak of the plot of Fig. 12 confirmed conclusively that the electrons were produced by neutrinos from ^8B decay in the Sun. Those energetic neutrinos are produced only by a process at the end of the nuclear reaction chain in the Sun—see Equations 6 and 7 and therefore were evidence that nuclear fusion reactions are the source of energy in the Sun and in all main sequence stars.

Furthermore, the observed rate of ^8B solar neutrino interactions in Kamiokande II was below the rate predicted by the solar models by about a factor of two, more or less in agreement with the result from the Homestake Detector when allowance was made for the lower energy detection threshold of the latter detector. The Homestake neutrino threshold at a value a little above 0.8 MeV allowed the detection of neutrino contributions from ^7Be, ^{13}N, ^{15}O, and pep, as well as ^8B, which were taken into account in a comparison of the Kamiokande II and Homestake results.

From one point of view, the agreement within a factor of two between the observations of the flux of neutrinos from ^8B and the predicted value represents a triumph of understanding that allows the absolute value of the flux to be calculated so accurately from a solar model involving such complexity. On the other hand, the discrepancy between the solar model and the data, and the difference between the two measurements themselves, were taken as suggesting some effect, either in the Sun or in the intrinsic nature of the neutrino, that was not understood. Later experiments using gallium [10] to observe neutrinos from the p-p reaction confirmed that suggestion, and experiments to resolve the solar neutrino anomaly remain at the forefront of neutrino physics and astronomy.

Still another set of measurements done in the Kamiokande II and Lake Erie detectors involved the so-called atmospheric neutrinos, birth partners of cosmic ray muons arising in the decay of mesons produced by the primary component of cosmic rays impinging on the Earth's atmosphere. Those neutrinos had much higher energy—1 to 1,000 GeV—than the supernova or solar neutrinos, and initially were not expected to be especially interesting. It turned out, however, that a departure of the measured number ratio of ν_μ to ν_e from the expected ratio was strong evidence of neutrino oscillations, most probably the conversion of ν_μ to ν_τ, and consequently of the mixing of those neutrino flavors and the non-zero mass of at least one of the neutrinos [11]. These results have been confirmed [12] and strengthened by data from SuperKamiokande, the larger, better resolution detector that is the successor to Kamiokande II.

The span of extra-terrestrial neutrino energies is likely to extend well beyond the energies of the atmospheric neutrinos. There is the prospect that Active Galactic Nuclei (AGN) and Gamma-Ray Bursters (GRBs) are sources of ultrahigh energy (UHE) neutrinos [13], that is, neutrinos above 10^6 GeV in energy, and that Topological Defects (TDs) in elementary particle theory may give rise to neutrinos of still higher energy. Detection of UHE neutrinos is a challenge for neutrino astronomy of the future that would open new areas in elementary particle physics as well as in astronomy. The promise of neutrino astronomy lies in the penetrating power of neutrinos, which in turn allows us to point them back to their origin, and possibly to understand the inner constitution of the stellar body from which they came. Coarse-grained, marginal resolution neutrino detectors of area greater than one square kilometer and kilometer thickness, that are deeply buried in ice or water for shielding against unwanted background, will be telescopes of choice for these explorations, just as finer-grained, more sophisticated telescopes will continue to explore the intrinsic properties of neutrinos. All of them will, of course, participate in the observation of neutrinos from the next nearby type II supernova. In short, the science of neutrino astronomy holds out the promise of our familiar universe to be better understood and new universes to be explored.

References

1. R. Davis, Jr., In Proc. of the Thirteenth International Conference on Neutrino Physics and Astrophysics, Neutrino '88. J. Schneps *et al.*, Eds. Singapore: World Scientific, p. 502.
2. J.N. Bahcall and R.K. Ulrich, *Rev. Mod. Phys* **60**, 297 (1988).
3. S. Turck-Chieże *et al. Astrophys. J.* **335**, 415 (1988).
4. K. Hirata *et al.* (Kamiokande II Collaboration), *Phys. Rev. Lett.* **58**, 1490 (1987).
5. R.M. Bionta *et al.*, *Phys. Rev. Lett.* **58**, 1494 (1987).
6. H.Y.Chiu, *Ann. Phys.* (NY) **26**, 364 (1964); S.A. Colgate and R.H. White, *Astrophys. J* **143**, 626 (1966); S.E. Woosley, James R. Wilson, and Ron Mayle, ibid. **302**, 19 (1986).
7. M.J. Longo, *Phys. Rev. Lett.* **60**, 173 (1988); L.M. Krauss and S. Tremaine. ibid. p. 176.
8. See, for example, T.J. Loredo and D.Q. Lamb, *Astrophys. J* (1989).
9. K.S. Hirata *et al.* (Kamiokande II Collaboration) *Phys. Rev. Lett.* **63**, 16 (1989).
10. P. Anselmann *et al.* (GALLEX Collaboration), *Phys. Lett.* **B342**, 440 (1995); ibid. **B388**, 364 (1996); V. Gavrin *et a.* (SAGE Collaboration),Proc. XVII Int'l Conf. on Neutrino Physics and Astrophysics, Helsinki, eds. K. Huitu *et al.* (World Scientific 1997), p. 14.
11. K.S. Hirata *et al.* (Kamiokande II Collaboration), *Phys. Lett.* **B 205**, 416 (1988).
12. R. Bionta *et al.*, *Phys. Rev.* **D38**, 768 (1988).
13. A.K. Mann, *Astropart. Phys.* **7**, 97 (1997).

University of Colorado
TASI - 98
June 1 - 26, 1998

STUDENT SEMINARS

1. Anja Werthenbach (Durham, U.K.), "Radiation Zeros," June 4, 1998

2. Michael Plumacher (DESY, Germany), "Baryon Asymmetry, neutrino mixing, and SO(10) Unification," June 4, 1998

3. Deidre Black (Syracuse , NY), "Evidence for a Kappa(900) Resonance in Pi-K Scattering," June 9, 1998

4. James Hormuzdiar (Yale, CT), "Signals of Disoriented Chiral Condensates and Pionic Breather States," June 9, 1998

5. Stefan Recksiegel (Karlsruhe, Germany), ""Flavor Changing Neutral Currents: The Decays B->s+gamma, and Lamda_b->Lamda+gamma," June 11,1998

6. Parvez Anandam (Eugene, OR), "Factorization Schemes of Parton Distribution Functions and Jet Cross Sections," June 11, 1998

7. Sven Bergmann (Weizmann, Israel), "Solar neutrinos in the Presence of FCNC," June 15, 1998

8. Sadek Mansour (Purdue, IN), "Supernova Neutrinos in the Presence of FCNC," June 15, 1998

9. Steen Hansen (Copenhagen, Denmark), "Massless /Massive Neutrinos, CMBR, Nucleosynthesis," June 17, 1998

10. Stuart Wick ((Vanderbilt , TN), "Relativistic Monopole Signatures in Antarctic Ice," June 17, 1998

11. Stephen Gibbons (Oklahoma State, OK), " A four-neutrino mixing scheme for observed neutrino data," June 22, 1998

12. Kirk Kaminsky (Edmonton, Canada), " Strings, M-Theory, and the Five-Dimensional Universe," June 22, 1998

13. Manoj Kaplinghat (Ohio State, OH), "Observational Constraints on Power Law Cosmologies," June 24, 1998

14. Jing Wang (Penn, PA), "Nu-Nu(bar) Transitions," June 24, 1998

TASI-98 PARTICIPANTS

Kevork Abazajian
University of California at San Diego
Department of Physics
MC 0350
La Jolla, CA   92093-0350
kev@physics.ucsd.edu

Parvez Anandam
University of Oregon
Institute of Theoretical Science
Eugene, OR   97403
anandam@darkwing.uoregon.edu

Alfredo Aranda
College of William and Mary
Physics Department
Williamsburg, VA   23187
fefo@physics.wm.edu

Sven Bergmann
Weizmann Institute of Science
Department of Particle Physics
Rehovot 76100
ISRAEL
ftsven@wicc.weizmann.ac.il

Deirdre Black
Syracuse University
Physics Department
Syracuse, NY   13244
black@physics.syr.edu

Mu-Chun Chen
University of Colorado
Department of Physics
Campus Box 390
Boulder, CO   80309
chun.chen@colorado.edu

Tyce DeYoung
University of Wisconsin
Department of Physics
1150 University Ave.
Madison, WI   53706
deyoung@alizarin.physics.wisc.edu

Stephen Gibbons
Oklahoma State University
Department of Physics
Stillwater, OK   74078-3072
Fabio@okstate.edu

Laith Haddad
University of Colorado
Department of Physics
Campus Box 390
Boulder, CO   80309
laith.haddad@colorado.edu

Steen Hansen
Juliane Maries Vej 30
DK-2100 Copenhagen
DENMARK
sthansen@tac.dk

643

Rellen Hardtke
University of Wisconsin
Department of Physics
1150 University Ave.
Madison, WI   53706-1390
rellen@lisa.physics.wisc.edu

James Hormuzdiar
Yale University
Physics Department
New Haven, CT   06520
jimh@minerva.cis.yale.edu

Sharada Iyer
The University of Arizona
Department of Physics
1118 E. 4th St.
Tucson, AZ   85721
iyers@soliton.physics.arizona.edu

Kirk Kaminsky
University of Alberta
Department of Physics
412A Physics Bldg.
Edmonton, AB T6G 2J1
CANADA
kaminsky@phys.ualberta.ca

Manoj Kaplinghat
Ohio State University
Smith Laboratory
174 W 18th Ave.
Columbus, OH 43210
manoj@hoyle.mps.ohio-state.edu

Hyun-Min Lee
Seoul National University
College of Natural Sciences
Department of Physics
Seoul 151-742
KOREA
minlee@phya.snu.ac.kr

Sadek Mansour
Purdue University
Department of Physics
1396 Physics Building
West Lafayette, IN   47907
mansour@physics.purdue.edu

Danny Marfatia
University of Wisconsin
Department of Physics
1150 University Ave.
Madison, WI   53706
marfatia@pheno.physics.wisc.edu

Irina Mocioiu
SUNY at Stony Brook
Physics Department
Stony Brook, NY   11794-3800
mocioiu@insti.physics.sunysb.edu

José Muñoz
CINVESTAV-IPN
Departamento de Fisica
Apartado Postal 14-740
MEXICO 07000 D.F.
herman@fis.cinvestav.mx

Chet Nieter
University of Colorado
Department of Physics
Campus Box 390
Boulder, CO 80309
nieter@pizero.colorado.edu

Eun Kyung Paik
Ewha Woman's University
Department of Physics
11-1 Daehyun-Dong Seodaemun-Gu
Seoul, 120-750
KOREA
ekpaik@physics.ewha.ac.kr

Lara Pasquali
University of Iowa
Physics & Astronomy Department
Van Allen Hall
Iowa City, IA 52242
pasquali@hepsun1.physics.uiowa.edu

Mitesh Patel
University of California at San Diego
Department of Physics
La Jolla, CA 92093-0350
mitesh@physics.ucsd.edu

Fernando Perez
University of Colorado
Department of Physics
Campus Box 390
Boulder, CO 80309
perez@longs.colorado.edu

Máximo Ave Pernas
Facultad de Fisica
Departamento Fisica de Particulas
15706 Santiago de Compostela
SPAIN
AVE@tpaxp1.usc.es

Michael Plümacher
DESY-T
Notkestr. 85
D-22603 Hamburg
GERMANY
pluemi@mail.desy.de

Norma Quiroz
CINVESTAV-IPN
Departamento de Fisica
AV. IPN 2508, Apdo. Postal 14-740
07000
MEXICO D.F.
nquiroz@fis.cinvestav.mx

Nurur Rahman
University of Kansas
Department of Physics
Lawrence, KS 66045
nurur@eagle.cc.ukans.edu

Stefan Recksiegel
Universität Karlsruhe
Institut für Theoretische Teilchenphysik
D-76128 Karlsruhe
GERMANY
Stefan.Recksiegel@physik.uni-karlsruhe.de

646

Kirk Schneider
Rutgers University
Department of Physics & Astronomy
136 Frelinghuysen Road
Piscataway, NJ 08854-8019
kpschnei@physics.rutgers.edu

Jing Wang
University of Pennsylvania
Department of Physics
Philadelphia, PA 19104
wangji@student.physics.upenn.edu

Anja Werthenbach
University of Durham
Department of Physics
South Road
DH1 3LE
UNITED KINGDOM
Anja.Werthenbach@durham.ac.uk

Stuart Wick
Vanderbilt University
Department of Physics & Astronomy
Box 1807-B
Nashville, TN 37235
wicksd@ctrvax.vanderbilt.edu

TASI-98 LECTURERS

Guido Altarelli
CERN
Theory Division
CH 1211 Geneva 23
SWITZERLAND
GUAL@mail.cern.ch

Sarbani Basu
Institute for Advanced Study
School of Natural Science
Princeton, NJ   08540
basu@sns.ias.edu

Tom Bowles
Los Alamos Nat'l Laboratory
Group P-23
MS H803
Los Alamos, NM   87545
tjb@lanl.gov

Adam Burrows
University of Arizona
Department of Astronomy
Tucson, AZ   85721
burrows@as.arizona.edu

Mirjam Cvetic
University of Pennsylvania
Department of Physics & Astronomy
Philadelphia, PA   19104-6396
cvetic@cvetic.hep.upenn.edu

Keith Dienes
CERN
Theory Division
1211 Geneva 23
SWITZERLAND
Keith.Dienes@cern.ch

Francis Halzen
University of Wisconsin
Physics Department
1150 University Ave.
Madison, WI   53706
halzen@pheno.physics.wisc.edu

Naoya Hata
Institute for Advanced Studies
School of Natural Science
Princeton, NJ   08540
hata@sns.ias.edu

Wick Haxton
University of Washington
Department of Physics
P.O. Box 351560
Seattle, WA   98195-1560
HAXTON@emmy.phys.washington.edu

Paul Langacker
University of Pennsylvania
Department of Physics & Astronomy
Philadelphia, PA   19104-6396
pgl@langacker.hep.upenn.edu

Chung-Pei Ma
University of Pennsylvania
Department of Physics & Astronomy
209 S. 33rd St.
Philadelphia, PA 19104-6396
cpma@strad.physics.upenn.edu

Alfred Mann
University of Pennsylvania
Department of Physics & Astronomy
209 S. 33rd St.
Philadelphia, PA 19104-6396
mann@dept.physics.upenn.edu

Marc Pinsonneault
Ohio State University
Department of Astronomy
174 W. 18th Ave.
Columbus, OH 43210
pinsono@astronomy.ohio-state.edu

Nir Polonsky
Rutgers University
Department of Physics & Astronomy
136 Frelinghuysen Road
Piscataway, NJ 08854-8019
nirp@physics.rutgers.edu

Daniel Sigg
LIGO Project
P.O. Box 1970, MS S9-02
Richland, WA 99352
sigg_d@ligo.mit.edu

Gary Steigman
Ohio State University
Physics Department
174 W. 18th Ave.
Columbus, OH 43210
steigman@mps.ohio-state.edu

Dieter Zeppenfeld
University of Wisconsin
Department of Physics
1150 University Ave.
Madison, WI 53706
dieter@pheno.physics.wisc.edu

TASI-98 LOCAL ORGANIZING COMMITTEE

Shanta deAlwis
University of Colorado
Department of Physics
Campus Box 390
Boudler, CO 80309
dealwis@gopika.colorado.edu

Sechul Oh
University of Colorado
Department of Physics
Campus Box 390
Boulder, CO 80309
soh@trishul.colorado.edu

Tom DeGrand
University of Colorado
Department of Physics
Campus Box 390
Boulder, CO 80309
degrand@aurinko.colorado.edu

Anna Hasenfratz
University of Colorado
Department of Physics
Campus Box 390
Boulder, CO 80309
anna@eotvos.colorado.edu

Tamas Kovacs
University of Colorado
Department of Physics
Campux Box 390
Boulder, CO 80309
kovacs@eotvos.colorado.edu

K.T. Mahanthappa
University of Colorado
Department of Physics
Campus Box 390
Boulder, CO 80309
ktm@verb.colorado.edu

TASI-98 DIRECTORS

Paul Langacker

K.T. Mahanthappa

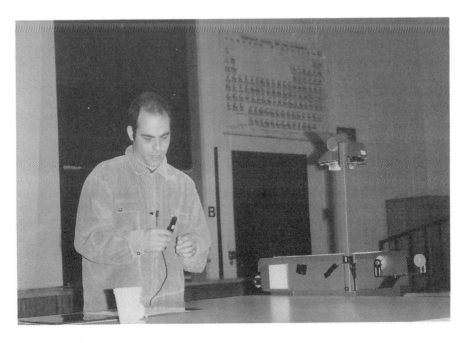